Numerical Methods for Engineers

EIGHTH EDITION

Steven C. Chapra

Berger Chair in Computing and Engineering
Tufts University

Raymond P. Canale

Professor Emeritus of Civil Engineering
University of Michigan

Mc
Graw
Hill

NUMERICAL METHODS FOR ENGINEERS, EIGHTH EDITION

Published by McGraw-Hill Education, 2 Penn Plaza, New York, NY 10121. Copyright © 2021 by McGraw-Hill Education. All rights reserved. Printed in the United States of America. Previous editions © 2015, 2010, and 2006. No part of this publication may be reproduced or distributed in any form or by any means, or stored in a database or retrieval system, without the prior written consent of McGraw-Hill Education, including, but not limited to, in any network or other electronic storage or transmission, or broadcast for distance learning.

Some ancillaries, including electronic and print components, may not be available to customers outside the United States.

This book is printed on acid-free paper.

2 3 4 5 6 7 8 9 LCR 24 23 22 21 20

ISBN 978-1-260-23207-3 (bound edition)
MHID 1-260-23207-7 (bound edition)
ISBN 978-1-260-48458-8 (loose-leaf edition)
MHID 1-260-48458-0 (loose-leaf edition)

Portfolio Manager: *Beth Bettcher*
Product Developer: *Heather Ervolino*
Senior Marketing Manager: *Shannon O'Donnell*
Content Project Managers: *Jeni McAtee, Samantha Donisi-Hamm*
Buyer: *Susan K. Culbertson*
Design: *Matt Diamond*
Content Licensing Specialist: *Abbey Jones*
Cover Image: © *iurii/Shutterstock*
Compositor: *Aptara®, Inc*

All credits appearing on page or at the end of the book are considered to be an extension of the copyright page.

Library of Congress Cataloging-in-Publication Data

Names: Chapra, Steven C., author. | Canale, Raymond P., author.
Title: Numerical methods for engineers / Steven C. Chapra, Berger Chair in
    Computing and Engineering, Tufts University, Raymond P. Canale Professor
    Emeritus of Civil Engineering, University of Michigan.
Description: Eighth edition. | New York, NY : McGraw-Hill Education, [2021] |
    Includes bibliographical references and index.
Identifiers: LCCN 2019017090| ISBN 9781260232073 (hard cover : acid-free
    paper) | ISBN 1260232077 (hard cover : acid-free paper)
Subjects: LCSH: Engineering mathematics—Data processing. | Numerical
    calculations—Data processing | Microcomputers—Programming.
Classification: LCC TA345 .C47 2021 | DDC 518.024/62—dc23 LC record
available at https://lccn.loc.gov/2019017090

The Internet addresses listed in the text were accurate at the time of publication. The inclusion of a website does not indicate an endorsement by the authors or McGraw-Hill Education, and McGraw-Hill Education does not guarantee the accuracy of the information presented at these sites.

mheducation.com/highered

To

Margaret and Gabriel Chapra

Helen and Chester Canale

ABOUT THE AUTHORS

Steve Chapra teaches in the Civil and Environmental Engineering Department at Tufts University, where he holds the Louis Berger Chair in Computing and Engineering. His other books include *Surface Water-Quality Modeling* and *Applied Numerical Methods with MATLAB*.

Dr. Chapra received engineering degrees from Manhattan College and the University of Michigan. Before joining the faculty at Tufts, he worked for the Environmental Protection Agency and the National Oceanic and Atmospheric Administration, and taught at Texas A&M University, the University of Colorado, and Imperial College London. His general research interests focus on surface water-quality modeling and advanced computer applications in environmental engineering.

He is a Fellow of the ASCE, and has received a number of awards for his scholarly contributions, including the Rudolph Hering Medal (ASCE), and the Meriam-Wiley Distinguished Author Award (American Society for Engineering Education). He has also been recognized as the outstanding teacher among the engineering faculties at Texas A&M University, the University of Colorado, and Tufts University.

Raymond P. Canale is an emeritus professor at the University of Michigan. During his over 20-year career at the university, he taught numerous courses in the areas of computers, numerical methods, and environmental engineering. He also directed extensive research programs in the area of mathematical and computer modeling of aquatic ecosystems. He has authored or coauthored several books and has published over 100 scientific papers and reports. He has also designed and developed personal computer software to facilitate engineering education and the solution of engineering problems. He has been given the Meriam-Wiley Distinguished Author Award by the American Society for Engineering Education for his books and software and several awards for his technical publications.

Professor Canale is now devoting his energies to applied problems, on which he works with engineering firms and industry and governmental agencies as a consultant and expert witness.

CONTENTS

PART EIGHT

PARTIAL DIFFERENTIAL EQUATIONS   858

PREFACE

It has been over thirty years since we published the first edition of this book. Over that period, our original contention that numerical methods and computers would figure more prominently in the engineering curriculum—particularly in the early parts—has been dramatically borne out. Many universities now offer freshman, sophomore, and junior courses in both introductory computing and numerical methods. In addition, many of our colleagues are integrating computer-oriented problems into other courses at all levels of the curriculum. Thus, this new edition is still founded on the basic premise that student engineers should be provided with a strong and early introduction to numerical methods. Consequently, we have endeavored to maintain many of the features that made previous editions accessible to both lower- and upper-level undergraduates. These include:

- **Problem Orientation.** Engineering students learn best when they are motivated by problems. This is particularly true for mathematics and computing. Consequently, we have approached numerical methods from a problem-solving perspective.
- **Student-Oriented Pedagogy.** We have developed several features to make this book as student-friendly as possible. These include the overall organization, the use of introductions and epilogues to consolidate major topics and the extensive use of worked examples and case studies from all areas of engineering. We have also endeavored to keep our explanations straightforward and oriented practically.
- **Computational Tools.** We empower our students by helping them utilize the standard "point-and-shoot" numerical problem-solving capabilities of packages like Excel, MATLAB, and Mathcad software. However, students are also shown how to develop simple, well-structured programs to extend the base capabilities of those environments. This knowledge carries over to standard programming languages such as Visual Basic, C/C++, Python, and modern versions of Fortran. We believe that the deemphasis of computer programming represents a "dumbing down" of the engineering curriculum. The bottom line is that if engineers are not content to be tool limited, they will have to write code. Only now they may be called "macros" or "scripts." This book is designed to empower them to do that.

Beyond these original principles, the eighth edition includes new material on cubic splines, Monte Carlo integration, and supplementary material on hyperbolic partial differential equations. It also has new and expanded problem sets. Many of the problems have been modified so that they yield different numerical solutions from previous editions. In addition, a variety of new problems have been included.

As always, our primary intent in writing this book is to provide students with a sound introduction to numerical methods. We believe that motivated students who enjoy numerical methods, computers, and mathematics will, in the end, make better engineers. If our book fosters an enthusiasm for these subjects, we will consider our efforts a success.

Acknowledgments. We would like to express our gratitude to our friends at McGraw-Hill. Special thanks to Heather Ervolino, who provided the positive and supportive atmosphere we needed to create this edition. Louis Poncz and Jane Hoover did a masterful job of copyediting and proofreading the manuscript as did Jeni McAtee as the book's production manager. Last but not least, our marketing manager, Shannon O'Donnell, has ensured that this new edition reaches the widest possible audience of educators and students. As in past editions, David Clough (University of Colorado), Mike Gustafson (Duke), and Jerry Stedinger (Cornell University) generously shared their insights. Useful suggestions were also made by Bill Philpot (Cornell University), Jim Guilkey (University of Utah), Dong-Il Seo (Chungnam National University, Korea), Niall Broekhuizen (NIWA, New Zealand), Marco Pilotti (University of Brescia, Italy), and Raymundo Cordero and Karim Muci (ITESM, Mexico). The text has also benefited from the reviews and suggestions of the following colleagues:

Kyle Arnold, University of Toronto
Betty Barr, University of Houston
Jalal Behzadi, Shahid Chamran University
Jordan Berg, Texas Tech University
Jacob Bishop, Utah State University
Estelle M. Eke, California State University, Sacramento
Mike Hovdesven, The George Washington University
Yazan A. Hussain, Jordan University of Science & Technology
Yogesh Jaluria, Rutgers University
S. Graham Kelly, The University of Akron
Subha Kumpaty, Milwaukee School of Engineering
Eckart Meiburg, University of California-Santa Barbara
Prashant Mhaskar, McMaster University
Luke Olson, University of Illinois at Urbana-Champaign
Richard Pates Jr., Old Dominion University
Joseph H. Pierluissi, University of Texas at El Paso
Juan Perán, Universidad Nacional de Educación a Distancia (UNED)
Ajaz Rashid, University of Bahrain
Scott A. Socolofsky, Texas A&M University

It should be stressed that although we received useful advice from the aforementioned individuals, we are responsible for any inaccuracies or mistakes you may detect in this edition. Please contact Steve Chapra via e-mail if you should detect any errors in this edition.

Finally, we would like to thank our family, friends, and students for their enduring patience and support. In particular, Cynthia Chapra, Danielle Husley, and Claire Canale are always there providing understanding, perspective, and love.

Steven C. Chapra
Medford, Massachusetts
steven.chapra@tufts.edu

Raymond P. Canale
Lake Leelanau, Michigan

ffordability & Outcomes = Academic Freedom!

You deserve choice, flexibility and control. You know what's best for your students and selecting the course materials that will help them succeed should be in your hands.

Thats why providing you with a wide range of options that lower costs and drive better outcomes is our highest priority.

 connect®

Students—study more efficiently, retain more and achieve better outcomes. Instructors—focus on what you love—teaching.

Laptop: McGraw-Hill Education

They'll thank you for it.

Study resources in Connect help your students be better prepared in less time. You can transform your class time from dull definitions to dynamic discussion. Hear from your peers about the benefits of Connect at **www.mheducation.com/highered/connect/smartbook**

Make it simple, make it affordable.

Connect makes it easy with seamless integration using any of the major Learning Management Systems—Blackboard®, Canvas, and D2L, among others—to let you organize your course in one convenient location. Give your students access to digital materials at a discount with our inclusive access program. Ask your McGraw-Hill representative for more information.

Learning for everyone.

McGraw-Hill works directly with Accessibility Services Departments and faculty to meet the learning needs of all students. Please contact your Accessibility Services office and ask them to email accessibility@mheducation.com, or visit **www.mheducation.com/about/accessibility.html** for more information.

A full array of affordable & effective solutions

- Digital Courseware
- Digital Print Bundles
- OER
- Inclusive Access
- Mobile Apps
- Discount Print Purchase
- Print/eBook Rentals

Learn more at: **www.mheducation.com/realvalue**

Rent It

Affordable print and digital rental options through our partnerships with leading textbook distributors including Amazon, Barnes & Noble, Chegg, Follett, and more.

Go Digital

A full and flexible range of affordable digital solutions ranging from Connect, ALEKS, inclusive access, mobile apps, OER and more.

Get Print

Students who purchase digital materials can get a loose-leaf print version at a significantly reduced rate to meet their individual preferences and budget.

Numerical Methods for Engineers

PART ONE

MODELING, COMPUTERS, AND ERROR ANALYSIS

PT1.1 MOTIVATION

Numerical methods are techniques by which mathematical problems are formulated so that they can be solved with arithmetic operations. Although there are many kinds of numerical methods, they have one common characteristic: they invariably involve large numbers of tedious arithmetic calculations. It is little wonder that with the development of fast, efficient digital computers, the role of numerical methods in engineering problem solving has increased dramatically in recent years.

PT1.1.1 Noncomputer Methods

Beyond providing increased computational firepower, the widespread availability of computers (especially personal computers) and their partnership with numerical methods has had a significant influence on the actual engineering problem-solving process. In the precomputer era there were generally three different ways in which engineers approached problem solving:

1. Solutions were derived for some problems using analytical, or exact, methods. These solutions were often useful and provided excellent insight into the behavior of some systems. However, analytical solutions can be derived for only a limited class of problems. These include those that can be approximated with linear models and those that have simple geometry and low dimensionality. Consequently, analytical solutions are of limited practical value because most real problems are nonlinear and involve complex shapes and processes.
2. Graphical solutions were used to characterize the behavior of systems. These graphical solutions usually took the form of plots or nomographs. Although graphical techniques can often be used to solve complex problems, the results are not very precise. Furthermore, graphical solutions (without the aid of computers) are extremely tedious and awkward to implement. Finally, graphical techniques are often limited to problems that can be described using three or fewer dimensions.
3. Calculators and slide rules were used to implement numerical methods manually. Although in theory such approaches should be perfectly adequate for solving complex

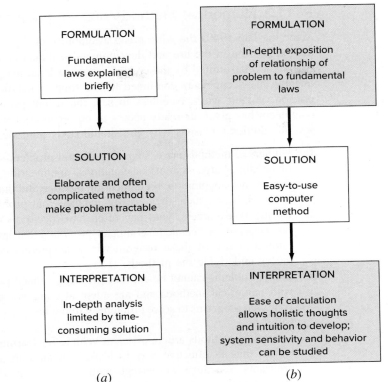

FIGURE PT1.1
The three phases of engineering problem solving in (a) the precomputer and (b) the computer era. The sizes of the boxes indicate the level of emphasis directed toward each phase. Computers facilitate the implementation of solution techniques and thus allow more emphasis to be placed on the creative aspects of problem formulation and interpretation of results.

problems, in actuality several difficulties are encountered. Manual calculations are slow and tedious. Furthermore, consistent results are elusive because of simple blunders that arise when numerous manual tasks are performed.

During the precomputer era, significant amounts of energy were expended on the solution technique itself, rather than on problem definition and interpretation (Fig. PT1.1a). This unfortunate situation existed because so much time and drudgery were required to obtain numerical answers using precomputer techniques.

Today, computers and numerical methods provide an alternative for such complicated calculations. Using computer power to obtain solutions directly, you can approach these calculations without recourse to simplifying assumptions or time-intensive techniques. Although analytical solutions are still extremely valuable both for problem solving and for providing insight, numerical methods represent alternatives that greatly enlarge your capabilities to confront and solve problems. As a result, more time is available for the use of your creative skills. Thus, more emphasis can be placed on problem formulation and solution interpretation and the incorporation of total system, or "holistic," awareness (Fig. PT1.1b).

PT1.1.2 Numerical Methods and Engineering Practice

Since the late 1940s the widespread availability of digital computers has led to a veritable explosion in the use and development of numerical methods. At first, this growth was somewhat limited by the cost of access to large mainframe computers, and, consequently, many engineers continued to use simple analytical approaches in a significant portion of their work. Needless to say, the recent evolution of inexpensive personal computers has given us ready access to powerful computational capabilities. There are several additional reasons why you should study numerical methods:

1. Numerical methods are extremely powerful problem-solving tools. They are capable of handling large systems of equations, nonlinearities, and complicated geometries that are not uncommon in engineering practice and that are often impossible to solve analytically. As such, they greatly enhance your problem-solving skills.
2. During your careers, you may often have occasion to use commercially available prepackaged, or "canned," computer programs that involve numerical methods. The intelligent use of these programs is often predicated on knowledge of the basic theory underlying the methods.
3. Many problems cannot be approached using canned programs. If you are conversant with numerical methods and are adept at computer programming, you can design your own programs to solve problems without having to buy or commission expensive software.
4. Numerical methods are an efficient vehicle for learning to use computers. It is well known that an effective way to learn programming is to actually write computer programs. Because numerical methods are for the most part designed for implementation on computers, they are ideal for this purpose. Further, they are especially well-suited to illustrate the power and the limitations of computers. When you successfully implement numerical methods on a computer and then apply them to solve otherwise intractable problems, you will be provided with a dramatic demonstration of how computers can serve your professional development. At the same time, you will also learn to acknowledge and control the errors of approximation that are part and parcel of large-scale numerical calculations.
5. Numerical methods provide a vehicle for you to reinforce your understanding of mathematics. Because one function of numerical methods is to reduce higher mathematics to basic arithmetic operations, they get at the "nuts and bolts" of some otherwise obscure topics. Enhanced understanding and insight can result from this alternative perspective.

PT1.2 MATHEMATICAL BACKGROUND

Every part in this book requires some mathematical background. Consequently, the introductory material for each part includes a section, such as the one you are reading, on mathematical background. Because Part One itself is devoted to background material on mathematics and computers, this section does not involve a review of a specific mathematical topic. Rather, we take this opportunity to introduce you to the types of mathematical subject areas covered in this book. As summarized in Fig. PT1.2, these are

FIGURE PT1.2
Summary of the numerical
methods covered in this book.

(a) *Part 2:* Roots of equations
   Solve $f(x) = 0$ for x.

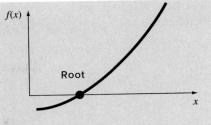

(b) *Part 3:* Linear algebraic equations
   Given the a's and the c's, solve

   $a_{11}x_1 + a_{12}x_2 = c_1$
   $a_{21}x_1 + a_{22}x_2 = c_2$
   for the x's.

(c) *Part 4:* Optimization
   Determine x that gives optimum $f(x)$.

(d) *Part 5:* Curve fitting

(e) *Part 6:* Integration
   $I = \int_a^b f(x)\, dx$
   Find the area under the curve.

(*f*) *Part 7:* Ordinary differential equations
Given

$$\frac{dy}{dt} \cong \frac{\Delta y}{\Delta t} = f(t, y)$$

solve for *y* as a function of *t*.

$$y_{i+1} = y_i + f(t_i, y_i)\,\Delta t$$

(*g*) *Part 8:* Partial differential equations
Given

$$\frac{\partial^2 u}{\partial x^2} + \frac{\partial^2 u}{\partial y^2} = f(x, y)$$

solve for *u* as a function of
x and *y*.

1. *Roots of Equations* (Fig. PT1.2a). These problems are concerned with the value of a variable or a parameter that satisfies a single nonlinear equation. These problems are especially valuable in engineering design contexts where it is often impossible to explicitly solve design equations for parameters.

2. *Systems of Linear Algebraic Equations* (Fig. PT1.2b). These problems are similar in spirit to roots of equations in the sense that they are concerned with values that satisfy equations. However, in contrast to satisfying a single equation, a set of values is sought that simultaneously satisfies a set of linear algebraic equations. Such equations arise in a variety of problem contexts and in all disciplines of engineering. In particular, they originate in the mathematical modeling of large systems of interconnected elements such as structures, electric circuits, and fluid networks. However, they are also encountered in other areas of numerical methods such as curve fitting and differential equations.

3. *Optimization* (Fig. PT1.2c). These problems involve determining a value or values of an independent variable that correspond to a "best," or optimal, value of a function. Thus, as in Fig. PT1.2c, optimization involves identifying maxima and minima. Such problems occur routinely in engineering design contexts. They also arise in a number of other numerical methods. We address both single- and multivariable unconstrained optimization. We also describe constrained optimization with particular emphasis on linear programming.

4. *Curve Fitting* (Fig. PT1.2d). You will often have occasion to fit curves to data points. The techniques developed for this purpose can be divided into two general categories: regression and interpolation. Regression is employed where there is a significant degree of error associated with the data. Experimental results are often of this kind. For these situations, the strategy is to derive a single curve that represents the general trend of the data without necessarily matching any individual points. In contrast,

interpolation is used where the objective is to determine intermediate values between relatively error-free data points. Such is usually the case for tabulated information. For these situations, the strategy is to fit a curve directly through the data points and use the curve to predict the intermediate values.

5. *Integration* (Fig. PT1.2e). As depicted, a physical interpretation of numerical integration is the determination of the area under a curve. Integration has many applications in engineering practice, ranging from the determination of the centroids of oddly shaped objects to the calculation of total quantities based on sets of discrete measurements. In addition, numerical integration formulas play an important role in the solution of differential equations.

6. *Ordinary Differential Equations* (Fig. PT1.2f). Ordinary differential equations are of great significance in engineering practice. This is because many physical laws are couched in terms of the rate of change of a quantity rather than the magnitude of the quantity itself. Examples range from population-forecasting models (rate of change of population) to the acceleration of a falling body (rate of change of velocity). Two types of problems are addressed: initial-value and boundary-value problems. In addition, the computation of eigenvalues is covered.

7. *Partial Differential Equations* (Fig. PT1.2g). Partial differential equations are used to characterize engineering systems where the behavior of a physical quantity is couched in terms of its rate of change with respect to two or more independent variables. Examples include the steady-state distribution of temperature on a heated plate (two spatial dimensions) or the time-variable temperature of a heated rod (time and one spatial dimension). Two fundamentally different approaches are employed to solve partial differential equations numerically. In the present text, we will emphasize finite-difference methods that approximate the solution in a pointwise fashion (Fig. PT1.2g). However, we will also present an introduction to finite-element methods, which use a piecewise approach.

PT1.3   ORIENTATION

Some orientation might be helpful before proceeding with our introduction to numerical methods. The following is intended as an overview of the material in Part One. In addition, some objectives have been included to focus your efforts when studying the material.

PT1.3.1 Scope and Preview

Figure PT1.3 is a schematic representation of the material in Part One. We have designed this diagram to provide you with a global overview of this part of the book. We believe that a sense of the "big picture" is critical to developing insight into numerical methods. When reading a text, it is often possible to become lost in technical details. Whenever you feel that you are losing the big picture, refer back to Fig. PT1.3 to reorient yourself. Every part of this book includes a similar figure.

Figure PT1.3 also serves as a brief preview of the material covered in Part One. *Chapter 1* is designed to orient you to numerical methods and to provide motivation by demonstrating how these techniques can be used in the engineering modeling process.

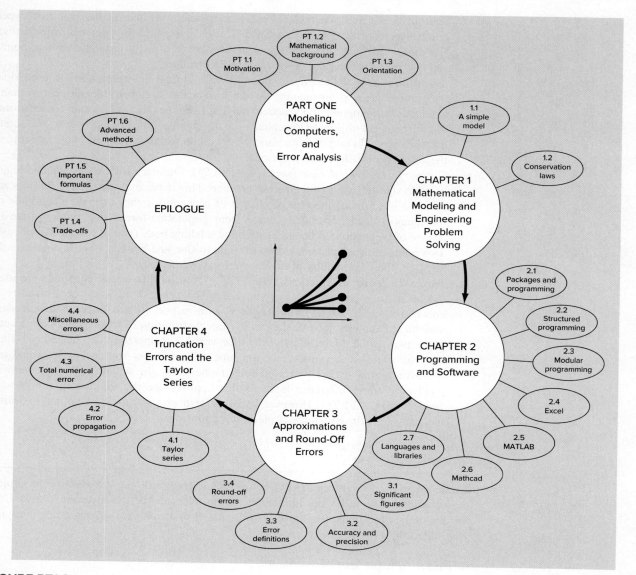

FIGURE PT1.3
Schematic of the organization of the material in Part One: Modeling, Computers, and Error Analysis.

TABLE PT1.1 Specific study objectives for Part One.

1. Recognize the difference between analytical and numerical solutions.
2. Understand how conservation laws are employed to develop mathematical models of physical systems.
3. Define top-down and modular design.
4. Delineate the rules that underlie structured programming.
5. Be capable of composing structured and modular programs in a high-level computer language.
6. Know how to translate structured flowcharts and pseudocode into code in a high-level language.
7. Start to familiarize yourself with any software packages that you will be using in conjunction with this text.
8. Recognize the distinction between truncation and round-off errors.
9. Understand the concepts of significant figures, accuracy, and precision.
10. Recognize the difference between true relative error ε_t, approximate relative error ε_a, and acceptable error ε_s, and understand how ε_a and ε_s are used to terminate an iterative computation.
11. Understand how numbers are represented in digital computers and how this representation induces round-off error. In particular, know the difference between single and extended precision.
12. Recognize how computer arithmetic can introduce and amplify round-off errors in calculations. In particular, appreciate the problem of subtractive cancellation.
13. Understand how the Taylor series and its remainder are employed to represent continuous functions.
14. Know the relationship between finite divided differences and derivatives.
15. Be able to analyze how errors are propagated through functional relationships.
16. Be familiar with the concepts of stability and condition.
17. Familiarize yourself with the trade-offs outlined in the Epilogue of Part One.

Chapter 2 is an introduction and review of computer-related aspects of numerical methods and suggests the level of computer skills you should acquire to efficiently apply succeeding information. *Chapters 3* and *4* deal with the important topic of error analysis, which must be understood for the effective use of numerical methods. In addition, an *epilogue* is included that introduces the trade-offs that have such great significance for the effective implementation of numerical methods.

PT1.3.2 Goals and Objectives

Study Objectives.   Upon completing Part One, you should be adequately prepared to embark on your studies of numerical methods. In general, you should have gained a fundamental understanding of the importance of computers and the role of approximations and errors in the implementation and development of numerical methods. In addition to these general goals, you should have mastered each of the specific study objectives listed in Table PT1.1.

Computer Objectives.   Upon completing Part One, you should have mastered sufficient computer skills to develop your own software for the numerical methods in this text. You should be able to develop well-structured and reliable computer programs on the basis of pseudocode, flowcharts, or other forms of algorithms. You should have developed the capability to document your programs so that they may be effectively employed by users. Finally, in addition to your own programs, you may be using software packages along with this book. Packages like Excel, Mathcad, or The MathWorks, Inc. MATLAB® program are examples of such software. You should become familiar with these packages, so that you will be comfortable using them to solve numerical problems later in the text.

Mathematical Modeling and Engineering Problem Solving

Knowledge and understanding are prerequisites for the effective implementation of any tool. No matter how impressive your tool chest, you will be hard-pressed to repair a car if you do not understand how it works.

This is particularly true when using computers to solve engineering problems. Although they have great potential utility, computers are practically useless without a fundamental understanding of how engineering systems work.

This understanding is initially gained by empirical means—that is, by observation and experiment. However, while such empirically derived information is essential, it is only half the story. Over years and years of observation and experiment, engineers and scientists have noticed that certain aspects of their empirical studies occur repeatedly. Such general behavior can then be expressed as fundamental laws that essentially embody the cumulative wisdom of past experience. Thus, most engineering problem solving employs the two-pronged approach of empiricism and theoretical analysis (Fig. 1.1).

It must be stressed that the two prongs are closely coupled. As new measurements are taken, the generalizations may be modified or new ones developed. Similarly, the generalizations can have a strong influence on the experiments and observations. In particular, generalizations can serve as organizing principles that can be employed to synthesize observations and experimental results into a coherent and comprehensive framework from which conclusions can be drawn. From an engineering problem-solving perspective, such a framework is most useful when it is expressed in the form of a mathematical model.

The primary objective of this chapter is to introduce you to mathematical modeling and its role in engineering problem solving. We will also illustrate how numerical methods figure in the process.

1.1 A SIMPLE MATHEMATICAL MODEL

A *mathematical model* can be broadly defined as a formulation or equation that expresses the essential features of a physical system or process in mathematical terms. In a very general sense, it can be represented as a functional relationship of the form

$$\begin{matrix} \text{Dependent} \\ \text{variable} \end{matrix} = f \left(\begin{matrix} \text{independent} \\ \text{variables} \end{matrix}, \text{parameters}, \begin{matrix} \text{forcing} \\ \text{functions} \end{matrix} \right) \qquad (1.1)$$

11

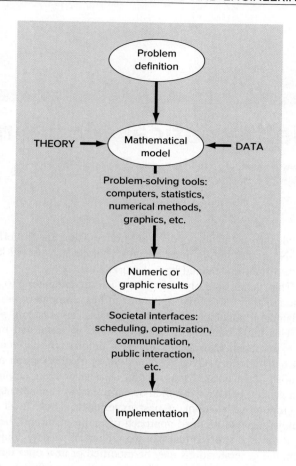

THEORY

FIGURE 1.1
The engineering problem-
solving process.

where the *dependent variable* is a characteristic that usually reflects the behavior or state of the system; the *independent variables* are usually dimensions, such as time and space, along which the system's behavior is being determined; the *parameters* are reflective of the system's properties or composition; and the *forcing functions* are external influences acting upon the system.

The actual mathematical expression of Eq. (1.1) can range from a simple algebraic relationship to large complicated sets of differential equations. For example, on the basis of his observations, Newton formulated his second law of motion, which states that the time rate of change of momentum of a body is equal to the resultant force acting on it. The mathematical expression, or model, of the second law is the well-known equation

$$F = ma \qquad (1.2)$$

where F = net force acting on the body (N, or kg m/s^2), m = mass of the object (kg), and a = its acceleration (m/s^2).

The second law can be recast in the format of Eq. (1.1) by merely dividing both sides by m to give

$$a = \frac{F}{m} \tag{1.3}$$

where a = the dependent variable reflecting the system's behavior, F = the forcing function, and m = a parameter representing a property of the system. Note that for this simple case there is no independent variable because we are not yet predicting how acceleration varies in time or space.

Equation (1.3) has several characteristics that are typical of mathematical models of the physical world:

1. It describes a natural process or system in mathematical terms.
2. It represents an idealization and simplification of reality. That is, the model ignores negligible details of the natural process and focuses on its essential manifestations. Thus, the second law does not include the effects of relativity that are of minimal importance when applied to objects and forces that interact on or about the earth's surface at velocities and on scales visible to humans.
3. Finally, it yields reproducible results and, consequently, can be used for predictive purposes. For example, if the force on an object and the mass of an object are known, Eq. (1.3) can be used to compute acceleration.

Because of its simple algebraic form, the solution of Eq. (1.2) can be obtained easily. However, other mathematical models of physical phenomena may be much more complex and either cannot be solved exactly or require more sophisticated mathematical techniques than simple algebra for their solution. To illustrate a more complex model of this kind, Newton's second law can be used to determine the terminal velocity of a free-falling body near the earth's surface. Our falling body will be a parachutist (Fig. 1.2). A model for this case can be derived by expressing the acceleration as the time rate of change of the velocity (dv/dt) and substituting it into Eq. (1.3) to yield

$$\frac{dv}{dt} = \frac{F}{m} \tag{1.4}$$

where v is velocity (m/s) and t is time (s). Thus, the mass multiplied by the rate of change of the velocity is equal to the net force acting on the body. If the net force is positive, the object will accelerate. If it is negative, the object will decelerate. If the net force is zero, the object's velocity will remain at a constant level.

Next, we will express the net force in terms of measurable variables and parameters. For a body falling within the vicinity of the earth (Fig. 1.2), the net force is composed of two opposing forces, the downward pull of gravity F_D and the upward force of air resistance F_U:

$$F = F_D + F_U \tag{1.5}$$

If the downward force is assigned a positive sign, the second law can be used to formulate the force due to gravity, as

$$F_D = mg \tag{1.6}$$

where g = the gravitational constant, or the acceleration due to gravity, which is approximately equal to 9.81 m/s^2.

FIGURE 1.2
Schematic diagram of the forces acting on a falling parachutist. F_D is the downward force due to gravity. F_U is the upward force due to air resistance.

Air resistance can be formulated in a variety of ways. A simple approach is to assume that it is linearly proportional to velocity[1] and acts in an upward direction, as in

$$F_U = -cv \tag{1.7}$$

where c = a proportionality constant called the *drag coefficient* (kg/s). Thus, the greater the fall velocity, the greater the upward force due to air resistance. The parameter c accounts for properties of the falling object, such as shape or surface roughness, that affect air resistance. For the present case, c might be a function of the type of jumpsuit or the orientation used by the parachutist during free-fall.

The net force is the difference between the downward and upward force. Therefore, Eqs. (1.4) through (1.7) can be combined to yield

$$\frac{dv}{dt} = \frac{mg - cv}{m} \tag{1.8}$$

or simplifying the right side,

$$\frac{dv}{dt} = g - \frac{c}{m}v \tag{1.9}$$

Equation (1.9) is a model that relates the acceleration of a falling object to the forces acting on it. It is a *differential equation* because it is written in terms of the differential rate of change (dv/dt) of the variable that we are interested in predicting. However, in contrast to the solution of Newton's second law in Eq. (1.3), the exact solution of Eq. (1.9) for the velocity of the falling parachutist cannot be obtained using simple algebraic manipulation. Rather, more advanced techniques, such as those of calculus, must be applied to obtain an exact or analytical solution. For example, if the parachutist is initially at rest ($v = 0$ at $t = 0$), calculus can be used to solve Eq. (1.9) for

$$v(t) = \frac{gm}{c}\left(1 - e^{-(c/m)t}\right) \tag{1.10}$$

Note that Eq. (1.10) is cast in the general form of Eq. (1.1), where $v(t)$ = the dependent variable, t = the independent variable, c and m = parameters, and g = the forcing function.

EXAMPLE 1.1 | **Analytical Solution to the Falling Parachutist Problem**

Problem Statement. A parachutist of mass 68.1 kg jumps out of a stationary hot air balloon. Use Eq. (1.10) to compute velocity prior to opening the chute. The drag coefficient is equal to 12.5 kg/s.

Solution. Inserting the parameters into Eq. (1.10) yields

$$v(t) = \frac{9.81(68.1)}{12.5}\left(1 - e^{-(12.5/68.1)t}\right) = 53.44\left(1 - e^{-0.18355t}\right)$$

which can be used to compute

[1] In fact, the relationship is actually nonlinear and might better be represented by a power relationship such as $F_U = -cv^2$. We will explore how such nonlinearities affect the model in problems at the end of this chapter.

t, s	v, m/s
0	0.00
2	16.42
4	27.80
6	35.68
8	41.14
10	44.92
12	47.54
∞	53.44

According to the model, the parachutist accelerates rapidly (Fig. 1.3). A velocity of 44.92 m/s is attained after 10 s. Note also that after a sufficiently long time, a constant velocity, called the *terminal velocity,* of 53.44 m/s is reached. This velocity is constant because, eventually, the force of gravity will be in balance with the air resistance. Thus, the net force is zero and acceleration has ceased.

Equation (1.10) is called an *analytical,* or *exact, solution* because it exactly satisfies the original differential equation. Unfortunately, there are many mathematical models that cannot be solved exactly. In many of these cases, the only alternative is to develop a numerical solution that approximates the exact solution.

As mentioned previously, *numerical methods* are those in which the mathematical problem is reformulated so that it can be solved by arithmetic operations. This can be

FIGURE 1.3
The analytical solution to the falling parachutist problem as computed in Example 1.1. Velocity increases with time and asymptotically approaches a terminal velocity.

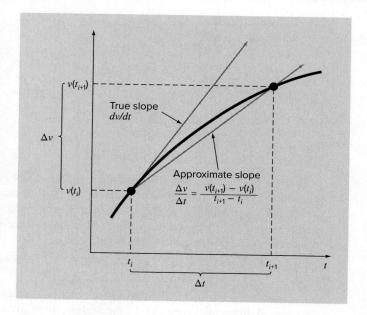

FIGURE 1.4

The use of a finite difference to approximate the first derivative of v with respect to t.

illustrated for Newton's second law by realizing that the time rate of change of velocity can be approximated by (Fig. 1.4):

$$\frac{dv}{dt} \cong \frac{\Delta v}{\Delta t} = \frac{v(t_{i+1}) - v(t_i)}{t_{i+1} - t_i} \tag{1.11}$$

where Δv and Δt = differences in velocity and time, respectively, computed over finite intervals, $v(t_i)$ = velocity at an initial time t_i, and $v(t_{i+1})$ = velocity at some later time t_{i+1}. Note that $dv/dt \cong \Delta v/\Delta t$ is approximate because Δt is finite. Remember from calculus that

$$\frac{dv}{dt} = \lim_{\Delta t \to 0} \frac{\Delta v}{\Delta t}$$

Equation (1.11) represents the reverse process.

Equation (1.11) is called a *finite divided difference* approximation of the derivative at time t_i. It can be substituted into Eq. (1.9) to give

$$\frac{v(t_{i+1}) - v(t_i)}{t_{i+1} - t_i} = g - \frac{c}{m} v(t_i)$$

This equation can then be rearranged to yield

$$v(t_{i+1}) = v(t_i) + \left[g - \frac{c}{m} v(t_i) \right] (t_{i+1} - t_i) \tag{1.12}$$

Notice that the term in brackets is the right-hand side of the differential equation itself [Eq. (1.9)]. That is, it provides a means to compute the rate of change or slope of v. Thus, the differential equation has been transformed into an equation that can be used to determine the velocity algebraically at t_{i+1} using the slope and previous values of v

and t. If you are given an initial value for velocity at some time t_i, you can easily compute velocity at a later time t_{i+1}. This new value of velocity at t_{i+1} can in turn be employed to extend the computation to velocity at t_{i+2} and so on. Thus, at any time along the way,

$$\text{New value} = \text{old value} + \text{slope} \times \text{step size}$$

Note that this approach is formally called *Euler's method*.

EXAMPLE 1.2 **Numerical Solution to the Falling Parachutist Problem**

Problem Statement.   Perform the same computation as in Example 1.1 but use Eq. (1.12) to compute the velocity. Employ a step size of 2 s for the calculation.

Solution.   At the start of the computation ($t_i = 0$), the velocity of the parachutist is zero. Using this information and the parameter values from Example 1.1, Eq. (1.12) can be used to compute velocity at $t_{i+1} = 2$ s:

$$v = 0 + \left[9.81 - \frac{12.5}{68.1}(0)\right]2 = 19.62 \text{ m/s}$$

For the next interval (from $t = 2$ to 4 s), the computation is repeated, with the result

$$v = 19.62 + \left[9.81 - \frac{12.5}{68.1}(19.62)\right]2 = 32.04 \text{ m/s}$$

The calculation is continued in a similar fashion to obtain additional values:

t, s	v, m/s
0	0.00
2	19.62
4	32.04
6	39.90
8	44.87
10	48.02
12	50.01
∞	53.44

The results are plotted in Fig. 1.5 along with the exact solution. It can be seen that the numerical method captures the essential features of the exact solution. However, because we have employed straight-line segments to approximate a continuously curving function, there is some discrepancy between the two results. One way to minimize such discrepancies is to use a smaller step size. For example, applying Eq. (1.12) at 1-s intervals results in a smaller error, as the straight-line segments track closer to the true solution. Using hand calculations, the effort associated with using smaller and smaller step sizes would make such numerical solutions impractical. However, with the aid of the computer, large numbers of calculations can be performed easily. Thus, you can accurately model the velocity of the falling parachutist without having to solve the differential equation exactly.

As in the previous example, a computational price must be paid for a more accurate numerical result. Each halving of the step size to attain more accuracy leads to a doubling

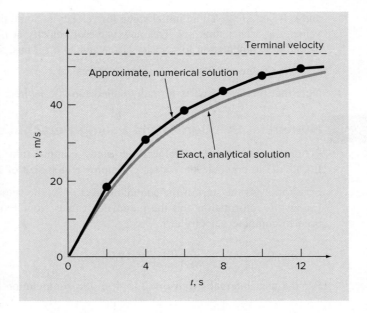

FIGURE 1.5
Comparison of the numerical and analytical solutions for the falling parachutist problem.

of the number of computations. Thus, we see that there is a trade-off between accuracy and computational effort. Such trade-offs figure prominently in numerical methods and constitute an important theme of this book. Consequently, we have devoted the Epilogue of Part One to an introduction to more of these trade-offs.

1.2 CONSERVATION LAWS AND ENGINEERING

Aside from Newton's second law, there are other major organizing principles in engineering. Among the most important of these are the conservation laws. Although they form the basis for a variety of complicated and powerful mathematical models, the great conservation laws of science and engineering are conceptually easy to understand. They all boil down to

$$\text{Change} = \text{increases} - \text{decreases} \tag{1.13}$$

This is precisely the format that we employed when using Newton's law to develop a force balance for the falling parachutist [Eq. (1.8)].

Although simple, Eq. (1.13) embodies one of the most fundamental ways in which conservation laws are used in engineering—that is, to predict changes with respect to time. We give Eq. (1.13) the special name *time-variable* (or *transient*) computation.

Aside from predicting changes, another way in which conservation laws are applied is for cases where change is nonexistent. If change is zero, Eq. (1.13) becomes

$$\text{Change} = 0 = \text{increases} - \text{decreases}$$

or

$$\text{Increases} = \text{decreases} \tag{1.14}$$

FIGURE 1.6

A flow balance for steady-state incompressible fluid flow at the junction of pipes.

Thus, if no change occurs, the increases and decreases must be in balance. This case, which is also given a special name—the *steady-state* computation—has many applications in engineering. For example, for steady-state incompressible fluid flow in pipes, the flow into a junction must be balanced by flow going out, as in

Flow in = flow out

For the junction in Fig. 1.6, the balance can be used to compute that the flow out of the fourth pipe must be 60.

For the falling parachutist, steady-state conditions would correspond to the case where the net force was zero, or [Eq. (1.8) with $dv/dt = 0$]

$$mg = cv \qquad (1.15)$$

Thus, at steady state, the downward and upward forces are in balance, and Eq. (1.15) can be solved for the terminal velocity

$$v = \frac{mg}{c}$$

Although Eqs. (1.13) and (1.14) might appear trivially simple, they embody the two fundamental ways that conservation laws are employed in engineering. As such, they will form an important part of our efforts in subsequent chapters to illustrate the connection between numerical methods and engineering. Our primary vehicles for making this connection are the engineering applications that appear at the end of each part of this book.

Table 1.1 summarizes some of the simple engineering models and associated conservation laws that will form the basis for many of these engineering applications. Most of the chemical engineering applications will focus on mass balances for reactors. The mass balance is derived from the conservation of mass. It specifies that the change of mass of a chemical in the reactor depends on the amount of mass flowing in minus the mass flowing out.

Both the civil and mechanical engineering applications will focus on models developed from the conservation of momentum. For civil engineering, force balances are utilized to analyze structures such as the simple truss in Table 1.1. The same principles are employed for the mechanical engineering applications to analyze the transient up-and-down motions, or vibrations, of an automobile.

TABLE 1.1 Devices and types of balances that are commonly used in the four major areas of engineering. For each case, the conservation law upon which the balance is based is specified.

Field	Device	Organizing Principle	Mathematical Expression
Chemical engineering	Reactors	Conservation of mass	Mass balance: Input → Output Over a unit of time period Δmass = inputs − outputs
Civil engineering	Structure	Conservation of momentum	Force balance: $+F_V$ $-F_H \longleftarrow \bullet \longrightarrow +F_H$ $-F_V$ At each node Σ horizontal forces $(F_H) = 0$ Σ vertical forces $(F_V) = 0$
Mechanical engineering	Machine	Conservation of momentum	Force balance: Upward force $x = 0$ Downward force $m \dfrac{d^2x}{dt^2}$ = downward force − upward force
Electrical engineering	Circuit	Conservation of charge	Current balance: For each node Σ current $(i) = 0$ $+i_1 \longrightarrow \bullet \longrightarrow -i_3$ $+i_2$
		Conservation of energy	Voltage balance: i_1R_1 i_2R_2 ξ i_3R_3 Around each loop Σ emf's − Σ voltage drops for resistors = 0 $\Sigma \xi - \Sigma iR = 0$

TABLE 1.2 Some practical issues that will be explored in the engineering applications at the end of each part of this book.

1. *Nonlinear versus linear.* Much of classical engineering depends on linearization to permit analytical solutions. Although this is often appropriate, expanded insight can often be gained if nonlinear problems are examined.
2. *Large versus small systems.* Without a computer, it is often not feasible to examine systems with over three interacting components. With computers and numerical methods, more realistic multicomponent systems can be examined.
3. *Nonideal versus ideal.* Idealized laws abound in engineering. Often there are nonidealized alternatives that are more realistic but more computationally demanding. Approximate numerical approaches can facilitate the application of these nonideal relationships.
4. *Sensitivity analysis.* Because they are so involved, many manual calculations require a great deal of time and effort for successful implementation. This sometimes discourages the analyst from implementing the multiple computations that are necessary to examine how a system responds under different conditions. Such sensitivity analyses are facilitated when numerical methods allow the computer to assume the computational burden.
5. *Design.* It is often a straightforward proposition to determine the performance of a system as a function of its parameters. It is usually more difficult to solve the inverse problem—that is, determining the parameters when the required performance is specified. Numerical methods and computers often permit this task to be implemented in an efficient manner.

Finally, the electrical engineering applications employ both current and energy balances to model electric circuits. The current balance, which results from the conservation of charge, is similar in spirit to the flow balance depicted in Fig. 1.6. Just as flow must balance at the junction of pipes, electric current must balance at the junction of electric wires. The energy balance specifies that the changes of voltage around any loop of the circuit must add up to zero. The engineering applications are designed to illustrate how numerical methods are actually employed in the engineering problem-solving process. As such, they will permit us to explore practical issues (Table 1.2) that arise in real-world applications. Making these connections between mathematical techniques such as numerical methods and engineering practice is a critical step in tapping their true potential. Careful examination of the engineering applications will help you to take this step.

PROBLEMS

1.1 Use calculus to solve Eq. (1.9) for the case where the initial velocity $v(0)$ is nonzero.

1.2 Repeat Example 1.2. Compute the velocity to $t = 10$ s, with a step size of **(a)** 1 and **(b)** 0.5 s. Can you make any statement regarding the errors of the calculation based on the results?

1.3 Rather than the linear relationship of Eq. (1.7), you might choose to model the upward force on the parachutist as a second-order relationship,

$$F_U = -c'v|v|$$

where c' = a bulk second-order drag coefficient (kg/m). Note that the second-order term could be represented as v^2 if the parachutist always fell in the downward direction. For the present case, we use the more general representation, $v|v|$, so that the proper sign is obtained for both the downward and the upward directions.

(a) Using calculus, obtain the closed-form solution for the case where the jumper is initially at rest ($v = 0$ at $t = 0$).

(b) Repeat the numerical calculation in Example 1.2 with the same initial condition and parameter values, but with second-order drag. Use a value of 0.225 kg/m for c_d'.

1.4 For the free-falling parachutist with linear drag, assume a first jumper is 70 kg and has a drag coefficient of 12 kg/s. If a second jumper has a drag coefficient of 15 kg/s and a mass of 75 kg, how long will it take him to reach the same velocity the first jumper reached in 10 s?

1.5 Compute the velocity of a parachutist using Euler's method for the case where $m = 80$ kg and $c = 10$ kg/s. Perform the calculation from $t = 0$ to 20 s with a step size of 1 s. Use an initial condition that the parachutist has an upward velocity of 20 m/s at $t = 0$. At $t = 10$ s, assume that the chute is instantaneously deployed so that the drag coefficient jumps to 50 kg/s.

1.6 The following information is available for a bank account:

Date	Deposits	Withdrawals	Interest	Balance
5/1				1512.33
	220.13	327.26		
6/1				
	216.80	378.61		
7/1				
	450.25	106.80		
8/1				
	127.31	350.61		
9/1				

Note that the money earns interest, which is computed as

Interest = iB_i

where i = the interest rate expressed as a fraction per month and $B_{i=}$ the initial balance at the beginning of the month.

(a) Use the conservation of cash to compute the balance on 6/1, 7/1, 8/1, and 9/1 if the interest rate is 1% per month (i = 0.01/month). Show each step in the computation.

(b) Write a differential equation for the cash balance in the form

$$\frac{dB}{dt} = f(D(t), W(t), i)$$

where t = time (months), $D(t)$ = deposits as a function of time ($/month), $W(t)$ = withdrawals as a function of time ($/month). For this case, assume that interest is compounded continuously; that is, interest = iB.

(c) Use Euler's method with a time step of 0.5 month to simulate the balance. Assume that the deposits and withdrawals are applied uniformly over the month.

(d) Develop a plot of balance versus time for **(a)** and **(c)**.

1.7 The amount of a uniformly distributed radioactive contaminant contained in a closed reactor is measured by its concentration c (becquerel/liter, or Bq/L). The contaminant decreases at a decay rate proportional to its concentration—that is,

decay rate = $-kc$

where k is a constant with units of day^{-1}. Therefore, according to Eq. (1.13), a mass balance for the reactor can be written as

$$\frac{dc}{dt} = -kc$$

$$\begin{pmatrix} \text{change} \\ \text{in mass} \end{pmatrix} = \begin{pmatrix} \text{decrease} \\ \text{by decay} \end{pmatrix}$$

(a) Use Euler's method to solve this equation from $t = 0$ to 1 d with $k = 0.2$ d^{-1}. Employ a step size of $\Delta t = 0.1$. The concentration at $t = 0$ is 10 Bq/L.

(b) Plot the solution on a semilog graph (i.e., ln c versus t) and determine the slope. Interpret your results.

1.8 A group of 35 students attend a class in a room that measures 11 m by 8 m by 3 m. Each student takes up about 0.075 m^3 and gives out about 80 W of heat (1 W = 1 J/s). Calculate the air temperature rise during the first 15 minutes of the class if the room is completely sealed and insulated. Assume the heat capacity, C_v, for air is 0.718 kJ/(kg K). Assume air is an ideal gas at 20°C and 101.325 kPa. Note that the heat absorbed by the air Q is related to the mass of the air m, the heat capacity, and the change in temperature by the following relationship:

$$Q = m \int_{T_1}^{T_2} C_v dT = mC_v (T_2 - T_1)$$

The mass of air can be obtained from the ideal gas law:

$$PV = \frac{m}{Mwt} RT$$

where P is the gas pressure, V is the volume of the gas, Mwt is the molecular weight of the gas (for air, 28.97 kg/kmol), and R is the ideal gas constant [8.314 kPa m^3/(kmol K)].

1.9 A storage tank contains a liquid at depth y, where $y = 0$ when the tank is half full (Fig. P1.9). Liquid is withdrawn at a constant flow rate Q to meet demands. The contents are resupplied at a sinusoidal rate $3Q \sin^2(t)$.

FIGURE P1.9

Equation (1.13) can be written for this system as

$$\frac{d(Ay)}{dt} = 3Q\sin^2(t) - Q$$

$$\begin{pmatrix} \text{change in} \\ \text{volume} \end{pmatrix} = (\text{inflow}) - (\text{outflow})$$

or, since the surface area A is constant,

$$\frac{dy}{dt} = 3\frac{Q}{A}\sin^2(t) - \frac{Q}{A}$$

Use Euler's method to solve for the depth y from $t = 0$ to 10 d with a step size of 0.5 d. The parameter values are $A = 1200$ m^2 and $Q = 500$ m^3/d. Assume that the initial condition is $y = 0$.

1.10 For the same storage tank described in Prob. 1.9, suppose that the outflow is not constant but rather depends on the depth. For this case, the differential equation for depth can be written as

$$\frac{dy}{dt} = 3\frac{Q}{A}\sin^2(t) - \frac{\alpha(1 + y)^{1.5}}{A}$$

Use Euler's method to solve for the depth y from $t = 0$ to 10 d with a step size of 0.5 d. The parameter values are $A = 1200$ m^2, $Q = 500$ m^3/d, and $\alpha = 150$. Assume that the initial condition is $y = 0$.

1.11 Apply the conservation of volume (see Prob. 1.9) to simulate the level of liquid in a conical storage tank (Fig. P1.11). The liquid flows in at a sinusoidal rate of $Q_{in} = 3 \sin^2(t)$ and flows out according to

$$Q_{out} = 3(y - y_{out})^{1.5} \qquad y > y_{out}$$
$$Q_{out} = 0 \qquad\qquad\qquad y \leq y_{out}$$

where flow has units of m^3/d and $y =$ the elevation of the water surface above the bottom of the tank (m). Use Euler's method to solve for the depth y from $t = 0$ to 8 d with a step size of 0.4 d. The parameter values are $r_{top} = 2.5$ m, $y_{top} = 4$ m, and $y_{out} = 1$ m. Assume that the level is initially below the outlet pipe with $y(0) = 0.75$ m.

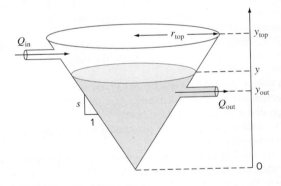

FIGURE P1.11

1.12 In our example of the free-falling parachutist, we assumed that the acceleration due to gravity was a constant value. Although this is a decent approximation when we are examining falling objects near the surface of the earth, the gravitational force decreases as we move above sea level. A more general representation based on *Newton's inverse square law* of gravitational attraction can be written as

$$g(x) = g(0)\frac{R^2}{(R + x)^2}$$

where $g(x) =$ gravitational acceleration at altitude x (in m) measured upward from the earth's surface (m/s^2), $g(0) =$ gravitational acceleration at the earth's surface ($\cong 9.81$ m/s^2), and $R =$ the earth's radius ($\cong 6.37 \times 10^6$ m).

(a) In a fashion similar to the derivation of Eq. (1.9), use a force balance to derive a differential equation for velocity as a function of time that utilizes this more complete representation of gravitation. However, for this derivation, assume that upward velocity is positive.

(b) For the case where drag is negligible, use the chain rule to express the differential equation as a function of altitude rather than time. Recall that the chain rule is

$$\frac{dv}{dt} = \frac{dv}{dx}\frac{dx}{dt}$$

(c) Use calculus to obtain the closed-form solution where $v = v_0$ at $x = 0$.

(d) Use Euler's method to obtain a numerical solution from $x = 0$ to 100,000 m using a step size of 10,000 m where the initial velocity is 1400 m/s upward. Compare your result with the analytical solution.

1.13 Suppose that a spherical droplet of liquid evaporates at a rate that is proportional to its surface area,

$$\frac{dV}{dt} = -kA$$

where $V =$ volume (mm^3), $t =$ time (min), $k =$ the evaporation rate (mm/min), and $A =$ surface area (mm^2). Use Euler's method to compute the volume of the droplet from $t = 0$ to 10 min using a step size of 0.25 min. Assume that $k = 0.1$ mm/min and that the droplet initially has a radius of 3 mm. Assess the validity of your results by determining the radius of your final computed volume and verifying that it is consistent with the evaporation rate.

1.14 *Newton's law of cooling* says that the temperature of a body changes at a rate proportional to the difference between its temperature and that of the surrounding medium (the ambient temperature),

$$\frac{dT}{dt} = -k(T - T_a)$$

where $T =$ the temperature of the body (°C), $t =$ time (min), $k =$ the proportionality constant (per minute), and $T_a =$ the ambient temperature (°C). Suppose that a cup of coffee originally has a temperature of 68°C. Use Euler's method to compute the temperature from $t = 0$ to 10 min using a step size of 1 min if $T_a = 21$°C and $k = 0.019$/min.

1.15 As depicted in Fig. P1.15, an *RLC circuit* consists of three elements: a resistor (R), an inductor (L), and a capacitor (C). The flow of current across each element induces a voltage drop.

Kirchhoff's second voltage law states that the algebraic sum of these voltage drops around a closed circuit is zero,

$$iR + L\frac{di}{dt} + \frac{q}{C} = 0$$

where i = current, R = resistance, L = inductance, t = time, q = charge, and C = capacitance. In addition, the current is related to charge as in

$$\frac{dq}{dt} = i$$

(a) If the initial values are $i(0) = 0$ and $q(0) = 0.5$ C, use Euler's method to solve this pair of differential equations from $t = 0$ to 0.1 s using a step size of $\Delta t = 0.01$ s. Employ the following parameters for your calculation: $R = 250\ \Omega$, $L = 5$ H, and $C = 10^{-4}$ F.

(b) Develop a plot of i and q versus t.

FIGURE P1.15

1.16 A fluid is pumped into the network shown in Fig. P1.16. If $Q_2 = 0.7$, $Q_3 = 0.5$, $Q_7 = 0.1$, and $Q_8 = 0.3$ m³/s, determine the other flows.

FIGURE P1.16

1.17 The velocity is equal to the rate of change of distance x (m),

$$\frac{dx}{dt} = v(t) \tag{P1.17}$$

(a) Substitute in Eq. (1.10) and develop an analytical solution for distance as a function of time. Assume that $x(0) = 0$.

(b) Use Euler's method to numerically integrate Eqs. (P1.17) and (1.9) in order to determine both the velocity and distance fallen as a function of time for the first 10 s of free-fall using the same parameters as in Example 1.2.

(c) Develop a plot of your numerical results together with the analytical solution.

1.18 You are working as a crime-scene investigator and must predict the temperature of a homicide victim over a 5-hr period. You know that the room where the victim was found was at 12°C when the body was discovered.

(a) Use *Newton's law of cooling* (Prob. 1.14) and Euler's method to compute the victim's body temperature for the 5-hr period using values of $k = 0.12$/hr and $\Delta t = 0.5$ hr. Assume that the victim's body temperature at the time of death was 37°C and that the room temperature was at a constant value of 12°C over the 5-hr period.

(b) Further investigation reveals that the room temperature had actually dropped linearly from 20 to 12°C over the 5-hr period. Repeat the same calculation as in (a) but incorporate this new information.

(c) Compare the results from (a) and (b) by plotting them on the same graph.

1.19 Suppose that a parachutist with linear drag ($m = 70$ kg, $c = 12.5$ kg/s) jumps from an airplane flying at an altitude of 200 m with a horizontal velocity of 180 m/s relative to the ground.

(a) Write a system of four differential equations for x, y, $v_x = dx/dt$, and $v_y = dy/dt$.

(b) If the initial horizontal position is defined as $x = 0$, use Euler's methods with $\Delta t = 1$ s to compute the jumper's position over the first 10 s.

(c) Develop plots of y versus t and y versus x. Use the plots to graphically estimate when and where the jumper would hit the ground if the chute failed to open.

(d) At what angle would the parachutist be traveling in the last whole second before impact?

1.20 As noted in Prob. 1.3, drag is more accurately represented as depending on the square of velocity. A more fundamental representation of the drag force, which assumes turbulent conditions (i.e., a high Reynolds number), can be formulated as

$$F_d = -\frac{1}{2}\rho A C_d v|v|$$

where F_d = the drag force (N), ρ = fluid density (kg/m³), A = the frontal area of the object on a plane perpendicular to the direction of motion (m²), v = velocity (m/s), and C_d = a dimensionless drag coefficient.

(a) Write the pair of differential equations for velocity and position (see Prob. 1.17) to describe the vertical motion of a sphere with diameter d (m) and a density of ρ_s (kg/m³). The differential equation for velocity should be written as a function of the sphere's diameter.

(b) Use Euler's method with a step size of $\Delta t = 2$ s to compute the position and velocity of a sphere over the first 14 s. Employ the following parameters in your calculation: $d = 125$ cm, $\rho = 1.3$ kg/m^3, $\rho_s = 2650$ kg/m^3, and $C_d = 0.475$. Assume that the sphere has the initial conditions: $y(0) = -100$ m and $v(0) = -45$ m/s.

(c) Develop a plot of your results (i.e., y and v versus t) and use it to graphically estimate when the sphere would hit the ground.

(d) Compute the value for the bulk second-order drag coefficient c_d' (kg/m). Note that, as described in Prob. 1.3, the bulk second-order drag coefficient is the term in the final differential equation for velocity that multiplies the term $v|v|$.

1.21 As depicted in Fig. P1.21, a spherical particle settling through a quiescent fluid is subject to three forces: the downward force of gravity (F_G), and the upward forces of buoyancy (F_B) and drag (F_D). Both the gravity and buoyancy forces can be computed with Newton's second law, with the latter equal to the weight of the displaced fluid. For laminar flow, the drag force can be computed with *Stokes's law*,

$$F_D = 3\pi\mu dv$$

where μ = the dynamic viscosity of the fluid (N s/m^2), d = the particle diameter (m), and v = the particle's settling velocity (m/s). Note that the mass of the particle can be expressed as the product of the particle's volume and density ρ_s (kg/m^3) and the mass of the displaced fluid can be computed as the product of the particle's volume and the fluid's density ρ (kg/m^3). The volume of a sphere is $\pi d^3/6$. In addition, laminar flow corresponds to the case where the dimensionless Reynolds number, Re, is less than 1, where Re $= \rho dv/\mu$.

(a) Use a force balance for the particle to develop the differential equation for dv/dt as a function of d, ρ, ρ_s, and μ.

(b) At steady-state, use this equation to solve for the particle's terminal velocity.

(c) Employ the result of **(b)** to compute the particle's terminal velocity in m/s for a spherical silt particle settling in water: $d = 8$ µm, $\rho = 1$ g/cm^3, $\rho_s = 2.7$ g/cm^3, and $\mu = 0.014$ g/(cm s).

(d) Check whether flow is laminar.

(e) Use Euler's method to compute the velocity from $t = 0$ to 2^{-15} s with $\Delta t = 2^{-18}$ s given the provided parameters along with the initial condition: $v(0) = 0$.

1.22 As described in Prob. 1.21, in addition to the downward force of gravity (weight) and drag, an object falling through a fluid is also subject to a buoyancy force that is proportional to the displaced volume. For example, for a sphere with diameter d (m), the sphere's volume is $V = \pi d^3/6$ and its projected area is $A = \pi d^2/4$. The buoyancy force can then be computed as $F_B = -\rho Vg$. We neglected buoyancy in our derivation of Eq. (1.9) because it is relatively small for an object like a parachutist moving through air. However, for a more dense fluid like water, it becomes more prominent.

(a) Derive a differential equation in the same fashion as Eq. (1.9), but include the buoyancy force and represent the drag force as described in Prob. 1.20.

(b) Rewrite the differential equation from **(a)** for the special case of a sphere.

(c) Use the equation developed in **(b)** to compute the terminal velocity (i.e., for the steady-state case). Use the following parameter values for a sphere falling through water: sphere diameter $= 1.1$ cm, sphere density $= 2650$ kg/m^3, water density $= 1000$ kg/m^3, and $C_d = 0.47$.

(d) Use Euler's method with a step size of $\Delta t = 0.03125$ s to numerically solve for the velocity from $t = 0$ to 0.25 s with an initial velocity of zero.

1.23 As depicted in Fig. P1.23, the downward deflection y (m) of a cantilever beam with a uniform load w (kg/m) can be computed as

$$y = \frac{w}{24EI}(x^4 - 4Lx^3 + 6L^2x^2)$$

where x = distance (m), E = the modulus of elasticity = 2×10^{11} Pa, I = moment of inertia = 3.3×10^{-4} m^4, $w = 12{,}000$ N/m, and L = length = 4 m. This equation can be differentiated to yield the slope of the downward deflection as a function of x:

$$\frac{dy}{dx} = \frac{w}{24EI}(4x^3 - 12Lx^2 + 12L^2x)$$

If $y = 0$ at $x = 0$, use this equation with Euler's method ($\Delta x = 0.125$ m) to compute the deflection from $x = 0$ to L. Develop a plot of your results along with the analytical solution computed with the first equation.

FIGURE P1.23
A cantilever beam.

1.24 Use *Archimedes' principle* to develop a steady-state force balance for a spherical ball of ice floating in seawater (Fig. P1.24). The force balance should be expressed as a third-order polynomial (cubic)

FIGURE P1.21

in terms of height of the cap above the water line (h), the seawater's density (ρ_f), the ball's density (ρ_s), and the ball's radius (r).

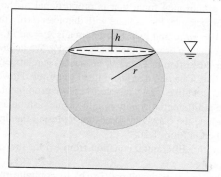

FIGURE P1.24

1.25 Beyond fluids, *Archimedes' principle* has proven useful in geology when applied to solids on the earth's crust. Figure P1.25 depicts one such case where a lighter conical granite mountain "floats on" a denser basalt layer at the earth's surface. Note that the part of the cone below the surface is formally referred to as a *frustum*. Develop a steady-state force balance for this case in terms of the following parameters: basalt's density (ρ_b), granite's density (ρ_g), the cone's bottom radius (r), and the height above (h_1) and below (h_2) the earth's surface.

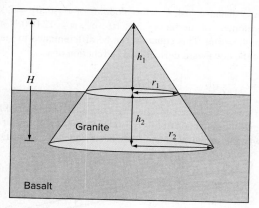

FIGURE P1.25

1.26 Figure P1.26 shows the forces exerted on a hot air balloon system.

FIGURE P1.26

Forces on a hot air balloon: F_B = buoyancy, F_G = weight of gas, F_P = weight of payload (including the balloon envelope), and F_D = drag. Note that the direction of the drag is downward when the balloon is rising.

Formulate the drag force as

$$F_D = \frac{1}{2} \rho_a v^2 A C_d$$

where ρ_a = air density (kg/m^3), v = velocity (m/s), A = projected frontal area (m^2), and C_d = the dimensionless drag coefficient ($\cong 0.47$ for a sphere). Note also that the total mass of the balloon consists of two components:

$$m = m_G + m_P$$

where m_G = the mass of the gas inside the expanded balloon (kg), and m_P = the mass of the payload (basket, passengers, and the unexpanded balloon = 265 kg). Assume that the ideal gas law holds ($P = \rho RT$), that the balloon is a perfect sphere with a diameter of

17.3 m, and that the heated air inside the envelope is at roughly the same pressure as the outside air.

Other necessary parameters are:

Normal atmospheric pressure, $P = 101,300$ Pa

The gas constant for dry air, $R = 287$ J/(kg K)

The air inside the balloon is heated to an average temperature, $T = 100°C$

The normal (ambient) air density, $\rho = 1.2$ kg/m^3

(a) Use a force balance to develop the differential equation for dv/dt as a function of the model's fundamental parameters.

(b) At steady-state, calculate the particle's terminal velocity.

(c) Use Euler's method and Excel to compute the velocity from $t = 0$ to 60 s with $\Delta t = 2$ s given the previous parameters along with the initial condition: $v(0) = 0$. Develop a plot of your results.

Programming and Software

In Chap. 1, we used a net force to develop a mathematical model to predict the fall velocity of a parachutist. This model took the form of a differential equation,

$$\frac{dv}{dt} = g - \frac{c}{m}v$$

We also learned that a solution to this equation could be obtained by a simple numerical approach called Euler's method,

$$v_{i+1} = v_i + \frac{dv_i}{dt}\Delta t$$

Given an initial condition, this equation can be implemented repeatedly to compute the velocity as a function of time. However, to obtain good accuracy, many small steps must be taken. This would be extremely laborious and time-consuming to implement by hand. However, with the aid of the computer, such calculations can be performed easily.

So our next task is to figure out how to do this. The present chapter will introduce you to how the computer is used as a tool to obtain such solutions.

2.1 PACKAGES AND PROGRAMMING

Today, there are two types of software users. On one hand, there are those who take what they are given. That is, they limit themselves to the capabilities found in the software's standard mode of operation. For example, it is a straightforward proposition to solve a system of linear equations or to generate a plot of x-y values with either Excel or MATLAB software. Because this usually involves a minimum of effort, most users tend to adopt this "vanilla" mode of operation. In addition, since the designers of these packages anticipate most typical user needs, many meaningful problems can be solved in this way.

But what happens when problems arise that are beyond the standard capability of the tool? Unfortunately, throwing up your hands and saying, "Sorry boss, no can do!" is not acceptable in most engineering circles. In such cases, you have two alternatives.

First, you can look for a different package and see if it is capable of solving the problem. That is one of the reasons we have chosen to cover both Excel and MATLAB in this book. As you will see, neither one is all encompassing and each has different strengths. By being conversant with both, you will greatly increase the range of problems you can address.

Second, you can grow and become a "power user" by learning to write Excel/VBA[1] macros or MATLAB M-files. And what are these? They are nothing more than computer programs that allow you to extend the capabilities of these tools. Because engineers should never be content to be tool limited, they will do whatever is necessary to solve their problems. A powerful way to do this is to learn to write programs in the Excel and MATLAB environments. Furthermore, the programming skills required for macros and M-files are the same as those needed to effectively develop programs in languages like Fortran 90 or C.

The major goal of the present chapter is to show you how this can be done. However, we do assume that you have been exposed to the rudiments of computer programming. Therefore, our emphasis here is on facets of programming that directly affect its use in engineering problem solving.

2.1.1 Computer Programs

Computer programs are merely a set of instructions that direct the computer to perform a certain task. Since many individuals write programs for a broad range of applications, most high-level computer languages, like Fortran 90 and C, have rich capabilities. Although some engineers might need to tap the full range of these capabilities, most merely require the ability to perform engineering-oriented numerical calculations.

Looked at from this perspective, we can narrow down the complexity to a few programming topics. These are

- Simple information representation (constants, variables, and type declarations).
- Advanced information representation (data structure, arrays, and records).
- Mathematical formulas (assignment, priority rules, and intrinsic functions).
- Input/output.
- Logical representation (sequence, selection, and repetition).
- Modular programming (functions and subroutines).

Because we assume that you have had some prior exposure to programming, we will not spend time on the first four of these areas. At best, we offer them as a checklist that covers what you will need to know to implement the programs that follow.

However, we will devote some time to the last two topics. We emphasize logical representation because it is the single area that most influences an algorithm's coherence and understandability. We include modular programming because it also contributes greatly to a program's organization. In addition, modules provide a means to archive useful algorithms in a convenient format for subsequent applications.

2.2 STRUCTURED PROGRAMMING

In the early days of computers, programmers usually did not pay much attention to whether their programs were clear and easy to understand. Today, it is recognized that there are many benefits to writing organized, well-structured code. Aside from the obvious benefit

---

[1]VBA is the acronym for Visual Basic for Applications.

of making software much easier to share, it also helps generate much more efficient program development. That is, well-structured algorithms are invariably easier to debug and test, resulting in programs that take a shorter time to develop, test, and update.

Computer scientists have systematically studied the factors and procedures needed to develop high-quality software of this kind. In essence, *structured programming* is a set of rules that prescribe good style habits for the programmer. Although structured programming is flexible enough to allow considerable creativity and personal expression, its rules impose enough constraints to render the resulting codes far superior to unstructured versions. In particular, the finished product is more elegant and easier to understand.

A key idea behind structured programming is that any numerical algorithm can be composed using the three fundamental control structures: sequence, selection, and repetition. By limiting ourselves to these structures, the resulting computer code will be clearer and easier to follow.

In the following paragraphs, we will describe each of these structures. To keep this description generic, we will employ flowcharts and pseudocode. A *flowchart* is a visual or graphical representation of an algorithm. The flowchart employs a series of blocks and arrows, each of which represents a particular operation or step in the algorithm (Fig. 2.1). The arrows represent the sequence in which the operations are implemented.

Not everyone involved with computer programming agrees that flowcharting is a productive endeavor. In fact, some experienced programmers do not advocate flowcharts. However, we feel that there are three good reasons for studying them. First, they are still used for expressing and communicating algorithms. Second, even if they are not employed routinely, there will be times when they will prove useful in planning, unraveling, or communicating the logic of your own or someone else's program. Finally, and most important for our purposes, they are excellent pedagogical tools. From a

FIGURE 2.1
Symbols used in flowcharts.

SYMBOL	NAME	FUNCTION
	Terminal	Represents the beginning or end of a program.
	Flowlines	Represent the flow of logic. The humps on the horizontal arrow indicate that it passes over and does not connect with the vertical flowlines.
	Process	Represents calculations or data manipulations.
	Input/output	Represents inputs or outputs of data and information.
	Decision	Represents a comparison, question, or decision that determines alternative paths to be followed.
	Junction	Represents the confluence of flowlines.
	Off-page connector	Represents a break that is continued on another page.
	Count-controlled loop	Used for loops that repeat a prespecified number of iterations.

teaching perspective, they are ideal vehicles for visualizing some of the fundamental control structures employed in computer programming.

An alternative approach to express an algorithm that bridges the gap between flowcharts and computer code is called *pseudocode*. This technique uses code-like statements in place of the graphical symbols of the flowchart. We have adopted some style conventions for the pseudocode in this book. Keywords such as IF, DO, INPUT, etc., are capitalized, whereas the conditions, processing steps, and tasks are in lowercase. Additionally, the processing steps are indented. Thus the keywords form a "sandwich" around the steps to visually define the extent of each control structure.

One advantage of pseudocode is that it is easier to develop a program with it than with a flowchart. The pseudocode is also easier to modify and share with others. However, because of their graphic form, flowcharts sometimes are better suited for visualizing complex algorithms. In the present text, we will use flowcharts for pedagogical purposes. Pseudocode will be our principal vehicle for communicating algorithms related to numerical methods.

2.2.1 Logical Representation

Sequence.   The sequence structure expresses the trivial idea that unless you direct it otherwise, the computer code is to be implemented one instruction at a time. As in Fig. 2.2, the structure can be expressed generically as a flowchart or as pseudocode.

Selection.   In contrast to the step-by-step sequence structure, selection provides a means to split the program's flow into branches based on the outcome of a logical condition. Figure 2.3 shows the two most fundamental ways for doing this.

The single-alternative decision, or *IF/THEN structure* (Fig. 2.3a), allows for a detour in the program flow if a logical condition is true. If it is false, nothing happens and the program moves directly to the next statement following the ENDIF. The double-alternative decision, or *IF/THEN/ELSE structure* (Fig. 2.3b), behaves in the same manner for a true condition. However, if the condition is false, the program implements the code between the ELSE and the ENDIF.

FIGURE 2.2
(a) Flowchart and
(b) pseudocode for the
sequence structure.

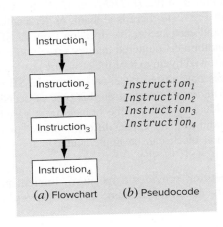

$Instruction_1$
$Instruction_2$
$Instruction_3$
$Instruction_4$

(a) Flowchart   (b) Pseudocode

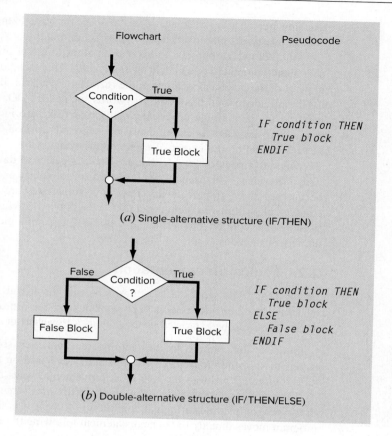

Flowchart   Pseudocode

```
IF condition THEN
    True block
ENDIF
```

(*a*) Single-alternative structure (IF/THEN)

```
IF condition THEN
    True block
ELSE
    False block
ENDIF
```

FIGURE 2.3

Flowchart and pseudocode for simple selection constructs. (a) Single-alternative selection (IF/THEN) and (b) double-alternative selection (IF/THEN/ELSE).

(*b*) Double-alternative structure (IF/THEN/ELSE)

Although the IF/THEN and the IF/THEN/ELSE constructs are sufficient to construct any numerical algorithm, two other variants are commonly used. Suppose that the ELSE clause of an IF/THEN/ELSE contains another IF/THEN. For such cases, the ELSE and the IF can be combined in the *IF/THEN/ELSEIF* structure shown in Fig. 2.4*a*.

Notice how in Fig. 2.4*a* there is a chain, or "cascade," of decisions. The first one is the IF statement, and each successive decision is an ELSEIF statement. Going down the chain, the first condition encountered that tests true will cause a branch to its corresponding code block followed by an exit of the structure. At the end of the chain of conditions, if all the conditions have tested false, an optional ELSE block can be included.

The *CASE structure* is a variant on this type of decision making (Fig. 2.4*b*). Rather than testing individual conditions, the branching is based on the value of a single *test expression*. Depending on its value, different blocks of code will be implemented. In addition, an optional block can be implemented if the expression takes on none of the prescribed values (CASE ELSE).

Repetition.   Repetition provides a means to implement instructions repeatedly. The resulting constructs, called *loops,* come in two "flavors" distinguished by how they are terminated.

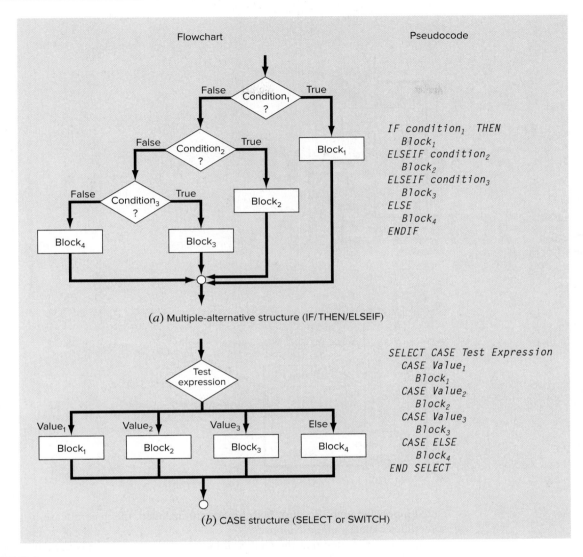

IF condition₁ THEN
    Block₁
ELSEIF condition₂
    Block₂
ELSEIF condition₃
    Block₃
ELSE
    Block₄
ENDIF

(a) Multiple-alternative structure (IF/THEN/ELSEIF)

SELECT CASE Test Expression
    CASE Value₁
        Block₁
    CASE Value₂
        Block₂
    CASE Value₃
        Block₃
    CASE ELSE
        Block₄
END SELECT

(b) CASE structure (SELECT or SWITCH)

FIGURE 2.4
Flowchart and pseudocode for supplementary selection or branching constructs. (a) Multiple-alternative selection (IF/THEN/ELSEIF) and (b) CASE construct.

The first and most fundamental type is called a *decision loop* because it terminates based on the result of a logical condition. Figure 2.5 shows the most generic type of decision loop, the *DOEXIT construct,* also called a *break loop.* This structure repeats until a logical condition is true.

It is not necessary to have two blocks in this structure. If the first block is not included, the structure is sometimes called a *pretest loop* because the logical test is performed before anything occurs. Alternatively, if the second block is omitted, it is

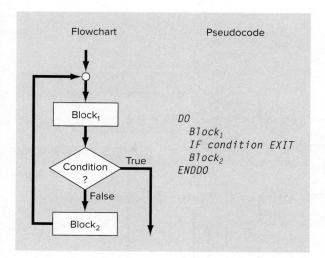

FIGURE 2.5
The DOEXIT or break loop.

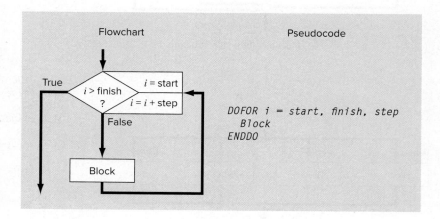

FIGURE 2.6
The count-controlled or DOFOR construct.

called a *posttest loop*. Because both blocks are included, the general case in Fig. 2.5 is sometimes called a *midtest loop*.

It should be noted that the DOEXIT loop was introduced in Fortran 90 in an effort to simplify decision loops. This control construct is a standard part of the Excel/VBA macro language but is not standard in C or MATLAB, which use the so-called WHILE structure. Because we believe that the DOEXIT is superior, we have adopted it as our decision loop structure throughout this book. In order to ensure that our algorithms are directly implemented in both MATLAB and Excel, we will show how the break loop can be simulated with the WHILE structure later in this chapter (see Sec. 2.5).

The break loop in Fig. 2.5 is called a *logical loop* because it terminates on a logical condition. In contrast, a *count-controlled* or *DOFOR loop* (Fig. 2.6) performs a specified number of repetitions, or iterations.

The count-controlled loop works as follows. The index (represented as *i* in Fig. 2.6) is a variable that is set at an initial value of *start*. The program then tests whether the *index* is less than or equal to the final value, *finish*. If so, it executes the body of the loop, and then cycles back to the DO statement. Every time the ENDDO statement is encountered, the *index* is automatically increased by the *step*. Thus the index acts as a counter. Then, when the *index* is greater than the final value (*finish*), the computer automatically exits the loop and transfers control to the line following the ENDDO statement. Note that for nearly all computer languages, including those of Excel and MATLAB, if the *step* is omitted, the computer assumes it is equal to 1.[2]

The numerical algorithms outlined in the following pages will be developed exclusively from the structures outlined in Figs. 2.2 through 2.6. The following example illustrates the basic approach by developing an algorithm to determine the roots of a quadratic equation.

EXAMPLE 2.1

Algorithm for Roots of a Quadratic

Problem Statement.   The roots of a quadratic equation,

$$ax^2 + bx + c = 0$$

can be determined with the quadratic formula,

$$\frac{x_1}{x_2} = \frac{-b \pm \sqrt{|b^2 - 4ac|}}{2a} \tag{E2.1.1}$$

Develop an algorithm that does the following:

Step 1: Prompts the user for the coefficients, a, b, and c.
Step 2: Implements the quadratic formula, guarding against all eventualities (for example, avoiding division by zero and allowing for complex roots).
Step 3: Displays the solution, that is, the values for x.
Step 4: Allows the user the option to return to step 1 and repeat the process.

Solution.   We will use a top-down approach to develop our algorithm. That is, we will successively refine the algorithm rather than trying to work out all the details the first time around.

To do this, let us assume for the present that the quadratic formula is foolproof regardless of the values of the coefficients (obviously not true, but good enough for now). A structured algorithm to implement the scheme is

```
DO
    INPUT a, b, c
    r1 = (-b + SQRT(b² - 4ac))/(2a)
    r2 = (-b - SQRT(b² - 4ac))/(2a)
    DISPLAY r1, r2
    DISPLAY 'Try again? Answer yes or no'
    INPUT response
    IF response = 'no' EXIT
ENDDO
```

---

[2]A negative step can be used. In such cases, the loop terminates when the index is less than the final value.

A DOEXIT construct is used to implement the quadratic formula repeatedly as long as the condition is false. The condition depends on the value of the character variable *response*. If *response* is equal to 'yes' the calculation is implemented. If not, that is, *response* = 'no' the loop terminates. Thus, the user controls termination by inputting a value for *response*.

Now although the above algorithm works for certain cases, it is not foolproof. Depending on the values of the coefficients, the algorithm might not work. Here is what can happen:

- If $a = 0$, an immediate problem arises because of division by zero. In fact, close inspection of Eq. (E2.1.1) indicates that two different cases can arise. That is,

  If $b \neq 0$, the equation reduces to a linear equation with one real root, $-c/b$.

  If $b = 0$, then no solution exists. That is, the problem is trivial.
- If $a \neq 0$, two possible cases occur depending on the value of the discriminant, $d = b^2 - 4ac$. That is,

  If $d \geq 0$, two real roots occur.

  If $d < 0$, two complex roots occur.

Notice how we have used indentation to highlight the decisional structure that underlies the mathematics. This structure then readily translates to a set of coupled IF/THEN/ELSE structures that can be inserted in place of the shaded statements in the previous code to give the final algorithm:

```
DO
  INPUT a, b, c
  r1 = 0: r2 = 0: i1 = 0: i2 = 0
  IF a = 0 THEN
    IF b ≠ 0 THEN
      r1 = -c/b
    ELSE
      DISPLAY "Trivial solution"
    ENDIF
  ELSE
    discr = b² - 4 * a * c
    IF discr ≥ 0 THEN
      r1 = (-b + Sqrt(discr))/(2 * a)
      r2 = (-b - Sqrt(discr))/(2 * a)
    ELSE
      r1 = -b/(2 * a)
      r2 = r1
      i1 = Sqrt(Abs(discr))/(2 * a)
      i2 = -i1
    ENDIF
  ENDIF
  DISPLAY r1, r2, i1, i2
  DISPLAY 'Try again? Answer yes or no'
  INPUT response
  IF response = 'no' EXIT
ENDDO
```

The approach in the foregoing example can be employed to develop an algorithm for the parachutist problem. Recall that, given an initial condition for time and velocity, the problem involved iteratively solving the formula

$$v_{i+1} = v_i + \frac{dv_i}{dt}\Delta t \tag{2.1}$$

Now also remember that if we desired to attain good accuracy, we would need to employ small steps. Therefore, we would probably want to apply the formula repeatedly from the initial time to the final time. Consequently, an algorithm to solve the problem would be based on a loop.

For example, suppose that we started the computation at $t = 0$ and wanted to predict the velocity at $t = 4$ s using a time step of $\Delta t = 0.5$ s. We would, therefore, need to apply Eq. (2.1) eight times, that is,

$$n = \frac{4}{0.5} = 8$$

where n = the number of iterations of the loop. Because this result is exact, that is, the ratio is an integer, we can use a count-controlled loop as the basis for the algorithm. Here is an example of the pseudocode:

```
g = 9.81
INPUT cd, m
INPUT ti, vi, tf, dt
t = ti
v = vi
n = (tf - ti) / dt
DOFOR i = 1 TO n
  dvdt = g - (cd / m) * v
  v = v + dvdt * dt
  t = t + dt
ENDDO
DISPLAY v
```

Although this scheme is simple to program, it is not foolproof. In particular, it will work only if the computation interval is evenly divisible by the time step.[3] In order to cover such cases, a decision loop can be substituted in place of the shaded area in the previous pseudocode. The final result is

```
g = 9.81
INPUT cd, m
INPUT ti, vi, tf, dt
t = ti
v = vi
```

---

[3]This problem is compounded by the fact that computers use base-2 number representation for their internal math. Consequently, some apparently evenly divisible numbers do not yield integers when the division is implemented on a computer. We will cover this in Chap. 3.

```
h = dt
DO
  IF t + dt > tf THEN
    h = tf - t
  ENDIF
  dvdt = g - (cd / m) * v
  v = v + dvdt * h
  t = t + h
  IF t ≥ tf EXIT
ENDDO
DISPLAY v
```

As soon as we enter the loop, we use an IF/THEN structure to test whether adding $t + dt$ will take us beyond the end of the interval. If it does not, which would usually be the case at first, we do nothing. If it does, we would need to shorten the interval by setting the variable step h to $tf - t$. By doing this, we guarantee that the next step falls exactly on tf. After we implement this final step, the loop will terminate because the condition $t \geq tf$ will test true.

Notice that before entering the loop, we assign the value of the time step, dt, to another variable, h. We create this dummy variable so that our routine does not change the given value of dt if and when we shorten the time step. We do this in anticipation that we might need to use the original value of dt somewhere else in the event that this code is integrated within a larger program.

It should be noted that the algorithm is still not foolproof. For example, the user could have mistakenly entered a step size greater than the calculation interval, for example, $tf - ti = 5$ and $dt = 20$. Thus, you might want to include error traps in your code to catch such errors and to then allow the user to correct the mistake.

2.3 MODULAR PROGRAMMING

Imagine how difficult it would be to study a textbook that had no chapters, sections, or paragraphs. Breaking complicated tasks or subjects into more manageable parts is one way to make them easier to handle. In the same spirit, computer programs can be divided into small subprograms, or modules, that can be developed and tested separately. This approach is called *modular programming*.

The most important attribute of modules is that they be as independent and self-contained as possible. In addition, they are typically designed to perform a specific, well-defined function and have one entry and one exit point. As such, they are usually short (generally 50 to 100 instructions in length) and highly focused.

In standard high-level languages such as Fortran 90 or C, the primary programming element used to represent each module is the procedure. A procedure is a series of computer instructions that together perform a given task. Two types of procedures are commonly employed: *functions* and *subroutines*. The former usually returns a single result, whereas the latter returns several.

In addition, it should be mentioned that much of the programming related to software packages like Excel and MATLAB involves the development of subprograms. Hence,

Excel macros and MATLAB functions are designed to receive some information, perform a calculation, and return results. Thus, modular thinking is also consistent with how programming is implemented in package environments.

Modular programming has a number of advantages. The use of small, self-contained units makes the underlying logic easier to devise and to understand for both the developer and the user. Development is facilitated because each module can be perfected in isolation. In fact, for large projects, different programmers can work on individual parts. Modular design also increases the ease with which a program can be debugged and tested because errors can be more easily isolated. Finally, program maintenance and modification are facilitated. This is primarily due to the fact that new modules can be developed to perform additional tasks and then easily incorporated into the already coherent and organized scheme.

While all these attributes are reason enough to use modules, the most important reason related to numerical engineering problem solving is that they allow you to maintain your own library of useful modules for later use in other programs. This will be the philosophy of this book: All the algorithms will be presented as modules.

This approach is illustrated in Fig. 2.7, which shows a function developed to implement Euler's method. Notice that this function application and the previous versions differ in how they handle input/output. In the former versions, input and output directly come from (via INPUT statements) and to (via DISPLAY statements) the user. In the function, the inputs are passed into the FUNCTION via its argument list

$$Function\ Euler(dt,\ ti,\ tf,\ yi)$$

and the output is returned via the assignment statement

$$y = Euler(dt,\ ti,\ tf,\ yi)$$

In addition, recognize how generic the routine has become. There are no references to the specifics of the parachutist problem. For example, rather than calling the dependent

FIGURE 2.7

Pseudocode for a function that solves a differential equation using Euler's method.

```
FUNCTION Euler(dt, ti, tf, yi)
t = ti
y = yi
h = dt
DO
   IF t + dt > tf THEN
      h = tf - t
   ENDIF
   dydt = dy(t, y)
   y = y + dydt * h
   t = t + h
   IF t ≥ tf EXIT
ENDDO
Euler = y
END Euler
```

variable v for velocity, the more generic label, y, is used within the function. Further, notice that the derivative is not computed within the function by an explicit equation. Rather, another function, dy, must be invoked to compute it. This acknowledges the fact that we might want to use this function for many different problems beyond solving for the parachutist's velocity.

2.4 EXCEL

Excel is the spreadsheet produced by Microsoft, Inc. Spreadsheets are a special type of mathematical software that allow the user to enter and perform calculations on rows and columns of data. As such, they are a computerized version of a large accounting worksheet on which large interconnected calculations can be implemented and displayed. Because the entire calculation is updated when any value on the sheet is changed, spreadsheets are ideal for "what if?" sorts of analysis.

Excel has some built-in numerical capabilities, including equation solving, curve fitting, and optimization. It also includes VBA as a macro language that can be used to implement numerical calculations. Finally, it has several visualization tools, such as graphs and three-dimensional surface plots, that serve as valuable adjuncts for numerical analysis. In the present section, we will show how these capabilities can be used to solve the parachutist problem.

To do this, let us first set up a simple spreadsheet. As shown below, the first step involves entering labels and numbers into the spreadsheet cells.

	A	B	C	D
1	Parachutist Problem			
2				
3	m	68.1	kg	
4	cd	12.5	kg/s	
5	dt	0.1	s	
6				
7	t	vnum (m/s)	vanal (m/s)	
8	0	0.000		
9	2			

Before we write a macro program to calculate the numerical value, we can make our subsequent work easier by attaching names to the parameter values. To do this, select cells A3:B5 (the easiest way to do this is by moving the mouse to A3, holding down the left mouse button and dragging down to B5). Next, go to the **Formulas** tab and in the **Defined Names** group, click **Create from Selection**. This will open the **Create Names from Selection** dialog box, where the **Left column** box should be automatically selected. Then click OK to create the names. To verify that this has worked properly, select cell B3 and check that the label "m" appears in the name box (located on the left side of the sheet just below the menu bars).

Move to cell C8 and enter the analytical solution (Eq. 1.9),

```
=9.81*m/cd*(1-exp(-cd/m*A8))
```

When this formula is entered, the value 0 should appear in cell C8. Then copy the formula down to cell C9 to give a value of 16.405 m/s.

All the above is typical of the standard use of Excel. For example, at this point you could change parameter values and see how the analytical solution changes.

Now, we will illustrate how VBA macros can be used to extend the standard capabilities. Figure 2.8 lists pseudocode alongside Excel/VBA code for all the control structures described in Sec. 2.2 (Figs. 2.3 through 2.6). Notice how, although the details differ, the structure of the pseudocode and the VBA code are identical.

We can now use some of the constructs from Fig. 2.8 to write a macro function to numerically compute velocity. Open VBA by selecting[4]

```
Tools Macro Visual Basic Editor
```

Once inside the *Visual Basic Editor* (VBE), select

```
Insert Module
```

and a new code window will open up. The following VBA function can be developed directly from the pseudocode in Fig. 2.7. Type it into the code window.

```
Option Explicit

Function Euler(dt, ti, tf, yi, m, cd)
Dim h As Double, t As Double, y As Double, dydt As Double
t = ti
y = yi
h = dt
Do
   If t + dt > tf Then
      h = tf - t
   End If
   dydt = dy(t, y, m, cd)
   y = y + dydt * h
   t = t + h
   If t >= tf Then Exit Do
Loop
Euler = y
End Function
```

Compare this macro with the pseudocode from Fig. 2.7 and recognize how similar they are. Also, see how we have expanded the function's argument list to include the necessary parameters for the parachutist velocity model. The resulting velocity, v, is then passed back to the spreadsheet via the function name.

---

[4]The hot key combination Alt-F11 is even quicker!

(a) Pseudocode	(b) Excel/VBA
IF/THEN:	
IF condition THEN	If b <> 0 Then
True block	r1 = -c / b
ENDIF	End If
IF/THEN/ELSE:	
IF condition THEN	If a < 0 Then
True block	b = Sqr(Abs(a))
ELSE	Else
False block	b = Sqr(a)
ENDIF	End If
IF/THEN/ELSEIF:	
IF condition$_1$ THEN	If class = 1 Then
Block$_1$	x = x + 8
ELSEIF condition$_2$	ElseIf class < 1 Then
Block$_2$	x = x - 8
ELSEIF condition$_3$	ElseIf class < 10 Then
Block$_3$	x = x - 32
ELSE	Else
Block$_4$	x = x - 64
ENDIF	End If
CASE:	
SELECT CASE Test Expression	Select Case a + b
CASE Value$_1$	Case Is < -50
Block$_1$	x = -5
CASE Value$_2$	Case Is < 0
Block$_2$	x = -5 - (a + b) / 10
CASE Value$_3$	Case Is < 50
Block$_3$	x = (a + b) / 10
CASE ELSE	Case Else
Block$_4$	x = 5
END SELECT	End Select
DOEXIT:	
DO	Do
Block$_1$	i = i + 1
IF condition EXIT	If i >= 10 Then Exit Do
Block$_2$	j = i*x
ENDDO	Loop
COUNT-CONTROLLED LOOP:	
DOFOR i = start, finish, step	For i = 1 To 10 Step 2
Block	x = x + i
ENDDO	Next i

FIGURE 2.8
The fundamental control structures in (a) pseudocode and (b) Excel/VBA.

Also notice how we have included another function to compute the derivative. This can be entered in the same module by typing it directly below the Euler function,

```
Function dy(t, v, m, cd)
Const g As Double = 9.81
dy = g - (cd / m) * v
End Function
```

The final step is to return to the spreadsheet and invoke the function by entering the following formula in cell B9

```
=Euler(dt,A8,A9,B8,m,cd)
```

The result of the numerical integration, 16.531, will appear in cell B9.

You should appreciate what has happened here. When you enter the function into the spreadsheet cell, the parameters are passed into the VBA program where the calculation is performed and the result is then passed back and displayed in the cell. In effect, the VBA macro language allows you to use Excel as your input/output mechanism. All sorts of benefits arise from this fact.

For example, now that you have set up the calculation, you can play with it. Suppose that the jumper was much heavier, say, $m = 100$ kg (about 220 lb). Enter 100 into cell B3 and the spreadsheet will update immediately to show a value of 17.438 in cell B9. Change the mass back to 68.1 kg and the previous result, 16.531, automatically reappears in cell B9.

Now let us take the process one step further by filling in some additional numbers for the time. Enter the numbers 4, 6, . . . , 16 in cells A10 through A16. Then copy the formulas from cells B9:C9 down to rows 10 through 16. Notice how the VBA program calculates the numerical result correctly for each new row. (To verify this, change dt to 2 and compare with the results previously computed by hand in Example 1.2.) An additional embellishment would be to develop an x-y plot of the results using the Excel Chart Wizard.

The final spreadsheet is shown below. We now have created a pretty nice problem-solving tool. You can perform sensitivity analyses by changing the values for each of the parameters. As each new value is entered, the computation and the graph will be automatically updated. It is this interactive nature that makes Excel so powerful. However, recognize that the ability to solve this problem hinges on being able to write the macro with VBA.

It is the combination of the Excel environment with the VBA programming language that truly opens up a world of possibilities for engineering problem solving. In the coming chapters, we will illustrate how this is accomplished.

2.5 MATLAB

MATLAB is the flagship software product of The MathWorks, Inc., which was cofounded by the numerical analysts Cleve Moler and John N. Little. As the name implies, MATLAB was originally developed as a *mat*rix *lab*oratory. To this day, the major element of MATLAB is still the matrix. Mathematical manipulations of matrices are very conveniently implemented in an easy-to-use, interactive environment. To these matrix manipulations, MATLAB has added a variety of numerical functions, symbolic computations, and visualization tools. As a consequence, the present version represents a fairly comprehensive technical computing environment.

MATLAB has a variety of functions and operators that allow convenient implementation of many of the numerical methods developed in this book. These will be described in detail in the individual chapters that follow. In addition, programs can be written as so-called *M-files* that can be used to implement numerical calculations. Let us explore how this is done.

First, you should recognize that normal MATLAB use is closely related to programming. For example, suppose that we wanted to determine the analytical solution to the parachutist problem. This could be done with the following series of MATLAB commands

```
>> g=9.81;
>> m=68.1;
>> cd=12.5;
>> tf=2;
>> v=g*m/cd*(1-exp(-cd/m*tf))
```

with the result being displayed as

```
v =
   16.4217
```

Thus, the sequence of commands is just like the sequence of instructions in a typical programming language.

Now what if you want to deviate from the sequential structure. Although there are some neat ways to inject some nonsequential capabilities in the standard command mode, the inclusion of decisions and loops is best done by creating a MATLAB document called an M-file. To do this, make the menu selection

```
File New Script
```

and a new window will open with a heading "MATLAB Editor/Debugger." In this window, you can type and edit MATLAB programs. Type the following code there:

```
g=9.81;
m=68.1;
cd=12.5;
tf=2;
v=g*m/cd*(1-exp(-cd/m*tf))
```

Notice how the commands are written in exactly the way they would be written in the front end of MATLAB. Save the program with the name: analpara. MATLAB will automatically attach the extension .m to denote it as an M-file: analpara.m.

To run the program, you must go back to the command mode. The most direct way to do this is to click on the "MATLAB Command Window" button on the task bar (which is usually at the bottom of the screen).

The program can now be run by typing the name of the M-file, analpara, which should look like

```
>> analpara
```

If you have done everything correctly, MATLAB should respond with the correct answer:

```
v =
    16.4217
```

Now one problem with the foregoing is that it is set up to compute one case only. You can make it more flexible by having the user input some of the variables. For example, suppose that you wanted to assess the impact of mass on the velocity at 2 s. The M-file could be rewritten as the following to accomplish this

```
g=9.81;
m=input('mass (kg): ');
cd=12.5;
tf=2;
v=g*m/cd*(1-exp(-cd/m*tf))
```

Save this as analpara2.m. If you typed analpara2 while being in command mode, the prompt would show

```
mass (kg):
```

The user could then enter a value like 100, and the result will be displayed as

```
v =
    17.3597
```

Now it should be pretty clear how we can program a numerical solution with an M-file. In order to do this, we must first understand how MATLAB handles logical and looping structures. Figure 2.9 lists pseudocode alongside MATLAB code for all the

(a) Pseudocode	(b) MATLAB
IF/THEN:	
IF condition THEN	`if b ~= 0`
True block	`  r1 = -c / b;`
ENDIF	`end`
IF/THEN/ELSE:	
IF condition THEN	`if a < 0`
True block	`  b = sqrt(abs(a));`
ELSE	`else`
False block	`  b = sqrt(a);`
ENDIF	`end`
IF/THEN/ELSEIF:	
IF condition$_1$ THEN	`if class == 1`
Block$_1$	`  x = x + 8;`
ELSEIF condition$_2$	`elseif class < 1`
Block$_2$	`  x = x - 8;`
ELSEIF condition$_3$	`elseif class < 10`
Block$_3$	`  x = x - 32;`
ELSE	`else`
Block$_4$	`  x = x - 64;`
ENDIF	`end`
CASE:	
SELECT CASE Test Expression	`switch a + b`
CASE Value$_1$	`  case 1`
Block$_1$	`    x = -25;`
CASE Value$_2$	`  case 2`
Block$_2$	`    x = -5 - (a + b) / 10;`
CASE Value$_3$	`  case 3`
Block$_3$	`    x = (a + b) / 10;`
CASE ELSE	`  otherwise`
Block$_4$	`    x = 5;`
END SELECT	`end`
DOEXIT:	
DO	`while (1)`
Block$_1$	`  i = i + 1;`
IF condition EXIT	`  if i >= 10, break, end`
Block$_2$	`  j = i*x;`
ENDDO	`end`
COUNT-CONTROLLED LOOP:	
DOFOR i = start, finish, step	`for i = 1:2:10`
Block	`  x = x + i;`
ENDDO	`end`

FIGURE 2.9
The fundamental control structures in (a) pseudocode and (b) the MATLAB programming language.

control structures from Sec. 2.2. Although the structures of the pseudocode and the MATLAB code are very similar, there are some slight differences that should be noted.

In particular, look at how we have represented the DOEXIT structure. In place of the DO, we use the statement while (1). Because MATLAB interprets the number 1 as corresponding to "true," this statement will repeat infinitely in the same manner as the DO statement. The loop is terminated with a *break* command. This command transfers control to the statement following the *end* statement that terminates the loop.

Also notice that the parameters of the count-controlled loop are ordered differently. For the pseudocode, the loop parameters are specified as *start, finish, step*. For MATLAB, the parameters are ordered as *start:step:finish*.

The following MATLAB M-file can now be developed directly from the pseudocode in Fig. 2.7. Type it into the MATLAB Editor/Debugger:

```
g=9.81;
m=input('mass (kg): ');
cd=12.5;
ti=0;
tf=2;
vi=0;
dt=0.1;
t = ti;
v = vi;
h = dt;
while (1)
   if t + dt > tf
     h = tf - t;
   end
   dvdt = g - (cd / m) * v;
   v = v + dvdt * h;
   t = t + h;
   if t >= tf, break, end
end
disp('velocity (m/s):')
disp(v)
```

Save this file as numpara.m and return to the command mode and run it by entering: numpara. The following output should result:

```
mass (kg): 100

velocity (m/s):
17.4559
```

As a final step in this development, let us take the above M-file and convert it into a proper function. This can be done in the following M-file based on the pseudocode from Fig. 2.7:

```
function yy = euler(dt,ti,tf,yi,m,cd)
t = ti;
y = yi;
h = dt;
```

```
while (1)
  if t + dt > tf
    h = tf - t;
  end
  dydt = dy(t, y, m, cd);
  y = y + dydt * h;
  t = t + h;
  if t >= tf, break, end
end
yy = y;
```

Save this file as euler.m and then create another M-file to compute the derivative,

```
function dydt = dy(t, v, m, cd)
g = 9.81;
dydt = g - (cd / m) * v;
```

Save this file as dy.m and return to the command mode. In order to invoke the function and see the result, you can type in the following commands:

```
>> m=68.1;
>> cd=12.5;
>> ti=0;
>> tf=2.;
>> vi=0;
>> dt=0.1;
>> euler(dt,ti,tf,vi,m,cd)
```

When the last command is entered, the answer will be displayed as

```
ans =
16.5478
```

It is the combination of the MATLAB environment with the M-file programming language that truly opens up a world of possibilities for engineering problem solving. In the coming chapters we will illustrate how this is accomplished.

2.6 MATHCAD

Mathcad attempts to bridge the gap between spreadsheets like Excel and notepads. It was originally developed by Allen Razdow of MIT who cofounded Mathsoft, Inc., which published the first commercial version in 1986. Today, Mathsoft is part of Parametric Technology Corporation (PTC) and Mathcad is in version 15.

Mathcad is essentially an interactive notepad that allows engineers and scientists to perform a number of common mathematical, data-handling, and graphical tasks. Information and equations are input to a "whiteboard" design environment that is similar in spirit to a page of paper. Unlike a programming tool or spreadsheet, Mathcad's interface accepts and displays natural mathematical notation using keystrokes or menu palette clicks—with no programming required. Because the worksheets contain live calculations, a single keystroke that changes an input or equation instantly returns an updated result.

Mathcad can perform tasks in either numeric or symbolic mode. In numeric mode, Mathcad functions and operators give numerical responses, whereas in symbolic mode results are given as general expressions or equations. Maple V, a comprehensive symbolic math package, is the basis of the symbolic mode and was incorporated into Mathcad in 1993.

Mathcad has a variety of functions and operators that allow convenient implementation of many of the numerical methods developed in this book. These will be described in detail in succeeding chapters. In the event that you are unfamiliar with Mathcad, Appendix C also provides a primer on using this powerful software.

2.7 OTHER LANGUAGES AND LIBRARIES

In Secs. 2.4 and 2.5, we showed how Excel and MATLAB function procedures for Euler's method could be developed from an algorithm expressed as pseudocode. You should recognize that similar functions can be written in high-level languages like Fortran 90 and C++. For example, a Fortran 90 function for Euler's method is

```
Function Euler(dt, ti, tf, yi, m, cd)

REAL dt, ti, tf, yi, m, cd
Real h, t, y, dydt

t = ti
y = yi
h = dt
Do
   If (t + dt > tf) Then
      h = tf - t
   End If
   dydt = dy(t, y, m, cd)
   y = y + dydt * h
   t = t + h
   If (t >= tf) Exit
End Do
Euler = y
End Function
```

For C, the result would look quite similar to the MATLAB function. The point is that once a well-structured algorithm is developed in pseudocode form, it can be readily implemented in a variety of programming environments.

In this book, our approach will be to provide you with well-structured procedures written as pseudocode. This collection of algorithms then constitutes a numerical library that can be accessed to perform specific numerical tasks in a range of software tools and programming languages.

Beyond your own programs, you should be aware that commercial programming libraries contain many useful numerical procedures. For example, the *Numerical Recipe* library includes a large range of algorithms written in Fortran and C.[5] These procedures are described in both book (for example, Press et al. 2007) and electronic form.

[5]Numerical Recipe procedures are also available in book and electronic format for Pascal, MS BASIC, and MATLAB. Information on all the Numerical Recipe products can be found at http://www.nr.com/.

PROBLEMS

2.1 Write pseudocode to implement the flowchart depicted in Fig. P2.1. Make sure that proper indentation is included to make the structure clear.

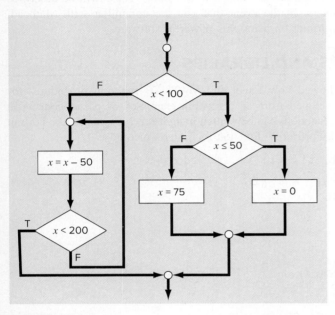

FIGURE P2.1

2.2 Rewrite the following pseudocode using proper indentation:

```
DO
i = i + 1
IF z > 50 EXIT
x = x + 5
IF x > 5 THEN
y = x
ELSE
y = 0
ENDIF
z = x + y
ENDDO
```

2.3 Develop, debug, and document a program to determine the roots of a quadratic equation, $ax^2 + bx + c$, in either a high-level language or a macro language of your choice. Use a subroutine procedure to compute the roots (either real or complex). Perform test runs for the cases **(a)** $a = 1$, $b = 6$, $c = 2$; **(b)** $a = 0$, $b = -4$, $c = 1.6$; **(c)** $a = 3$, $b = 2.5$, $c = 7$.

2.4 The sine function can be evaluated by the following infinite series:

$$\cos x = 1 - \frac{x^2}{2!} + \frac{x^4}{4!} - \frac{x^6}{6!} + \cdots$$

Write an algorithm to implement this formula so that it computes and prints out the values of cos x as each term in the series is added. In other words, compute and print in sequence the values for

$$\cos x = 1$$

$$\cos x = 1 - \frac{x^2}{2!}$$

$$\cos x = 1 - \frac{x^2}{2!} + \frac{x^4}{4!}$$

up to the order term n of your choosing. For each of the preceding, compute and display the percent relative error as

$$\% \text{ error} = \frac{\text{true} - \text{series approximation}}{\text{true}} \times 100\%$$

Write the algorithm as **(a)** a structured flowchart and **(b)** pseudocode.

2.5 Develop, debug, and document a program for Prob. 2.4 in either a high-level language or a macro language of your choice. Employ the library function for the cosine in your computer to determine the true value. Have the program print out the series approximation and the error at each step. As a test case, employ the program to compute sin(1.25) for up to and including the term $x^{10}/10!$. Interpret your results.

2.6 The following algorithm is designed to determine a grade for a course that consists of quizzes, homework, and a final exam:

Step 1: Input course number and name.
Step 2: Input weighting factors for quizzes (WQ), homework (WH), and the final exam (WF).
Step 3: Input quiz grades and determine an average quiz grade (AQ).
Step 4: Input homework grades and determine an average homework grade (AH).
Step 5: If this course has a final exam grade, continue to step 6. If not, go to step 9.
Step 6: Input final exam grade (FE).
Step 7: Determine average grade AG according to

$$AG = \frac{WQ \times AQ + WH \times AH + WF \times FE}{WQ + WH + WF} \times 100\%$$

Step 8: Go to step 10.
Step 9: Determine average grade AG according to

$$AG = \frac{WQ \times AQ + WH \times AH}{WQ + WH} \times 100\%$$

Step 10: Print out course number, name, and average grade.
Step 11: Terminate computation.
(a) Write well-structured pseudocode to implement this algorithm.
(b) Write, debug, and document a structured computer program based on this algorithm. Test it using the following data to calculate a grade without the final exam and a grade with the final exam: WQ = 35; WH = 30; WF = 35; quizzes = 98, 85, 90, 65, 99; homework = 95, 90, 87, 100, 92, 77; and final exam = 92.

2.7 The "divide and average" method, an old-time method for approximating the square root of any positive number a, can be formulated as

$$x = \frac{x + a/x}{2}$$

(a) Write well-structured pseudocode to implement this algorithm as depicted in Fig. P2.7. Use proper indentation so that the structure is clear.
(b) Develop, debug, and document a program to implement this equation in either a high-level language or a macro language of your choice. Structure your code according to Fig. P2.7.

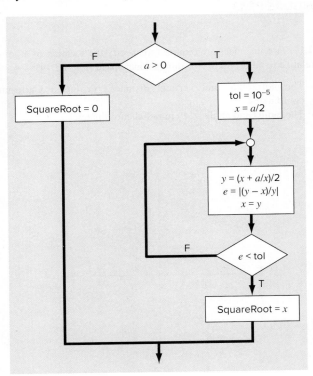

FIGURE P2.7

2.8 An amount of money P is invested in an account where interest is compounded at the end of the period. The future worth F yielded at an interest rate i after n periods may be determined from the following formula:

$$F = P(1 + i)^n$$

Write a program that will calculate the future worth of an investment for each year from 1 through n. The input to the function should include the initial investment P, the interest rate i (as a decimal), and the number of years n for which the future worth is to be calculated. The output should consist of a table with headings and columns for n and F. Run the program for $P = \$100,000$, $i = 0.06$, and $n = 5$ years.

2.9 Economic formulas are available to compute annual payments for loans. Suppose that you borrow an amount of money P and agree to repay it in n annual payments at an interest rate of i. The formula to compute the annual payment A is

$$A = P \frac{i(1 + i)^n}{(1 + i)^n - 1}$$

Write a program to compute A. Test it with $P = \$55,000$ and an interest rate of 5.6% ($i = 0.056$). Compute results for $n = 1, 2, 3, 4$, and 5 and display the results as a table with headings and columns for n and A.

2.10 The average daily temperature for an area can be approximated by the following function,

$$T = T_{mean} + (T_{peak} - T_{mean}) \cos(\omega(t - t_{peak}))$$

where T_{mean} = the average annual temperature, T_{peak} = the peak temperature, ω = the frequency of the annual variation (= $2\pi/365$), and t_{peak} = day of the peak temperature ($\cong$ 205 d). Develop a program that computes the average temperature between two days of the year for a particular city. Test it for **(a)** January–February ($t = 0$ to 59) in Miami, Florida ($T_{mean} = 22.1°C$; $T_{peak} = 28.3°C$), and **(b)** July–August ($t = 180$ to 242) in Boston, Massachusetts ($T_{mean} = 10.7°C$; $T_{peak} = 22.9°C$).

2.11 Develop, debug, and test a program in either a high-level language or a macro language of your choice to compute the velocity of the falling parachutist as outlined in Example 1.2. Design the program so that it allows the user to input values for the drag coefficient and mass. Test the program by duplicating the results from Example 1.2. Repeat the computation but employ step sizes of 1 and 0.5 s. Compare your results with the analytical solution obtained previously in Example 1.1. Does a smaller step size make the results better or worse? Explain your results.

2.12 The bubble sort is an inefficient, but easy-to-program, sorting technique. The idea behind the sort is to move down through an array comparing adjacent pairs and swapping the values if they are out of order. For this method to sort the array completely, it may need to pass through it many times. As the passes proceed for an ascending-order sort, the smaller elements in the array appear to rise toward the top like bubbles. Eventually, there will be a pass through the array where no swaps are required. Then, the array is sorted. After the first pass, the largest value in the array drops directly to the bottom. Consequently, the second pass only has to proceed to the second-to-last value, and so on. Develop a program to set up an array of 20 random numbers and sort them in ascending order with the bubble sort (Fig. P2.12).

decisional control structures (like IF/THEN, ELSEIF, ELSE, ENDIF). Design the function so that it returns the volume for all cases where the depth is less than $3R$. Return an error message ("Overtop") if you overtop the tank, that is, $d > 3R$. Test it with the following data:

R	1	1	1	1
d	0.5	1.2	3.0	3.1

FIGURE P2.13

2.14 Two distances are required to specify the location of a point relative to an origin in two-dimensional space (Fig. P2.14):

- The horizontal and vertical distances (x, y) in Cartesian coordinates
- The radius and angle (r, θ) in radial coordinates.

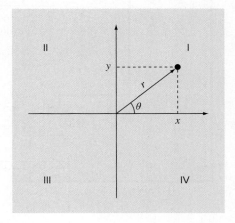

FIGURE P2.14

FIGURE P2.12

2.13 Figure P2.13 shows a cylindrical tank with a conical base. If the liquid level is quite low in the conical part, the volume is simply the conical volume of liquid. If the liquid level is mid-range in the cylindrical part, the total volume of liquid includes the filled conical part and the partially filled cylindrical part. Write a well-structured function procedure to compute the liquid's volume as a function of given values of R and d. Use

It is relatively straightforward to compute Cartesian coordinates (x, y) on the basis of polar coordinates (r, θ). The reverse process is not so simple. The radius can be computed by the following formula:

$$r = \sqrt{x^2 + y^2}$$

If the coordinates lie within the first and fourth coordinates (i.e., $x > 0$), then a simple formula can be used to compute θ

$$\theta = \tan^{-1}\left(\frac{y}{x}\right)$$

The difficulty arises for the other cases. The following table summarizes the possibilities:

x	y	θ
<0	>0	$\tan^{-1}(y/x) + \pi$
<0	<0	$\tan^{-1}(y/x) - \pi$
<0	=0	π
=0	>0	$\pi/2$
=0	<0	$-\pi/2$
=0	=0	0

(a) Write a well-structured flowchart for a subroutine procedure to calculate r and θ as a function of x and y. Express the final results for θ in degrees.
(b) Write a well-structured function procedure based on your flowchart. Test your program by using it to fill out the following table:

x	y	r	θ
1	0		
1	1		
0	1		
−1	1		
−1	0		
−1	−1		
0	−1		
1	−1		
0	0		

2.15 Develop a well-structured function procedure that is passed a numeric grade from 0 to 100 and returns a letter grade according to the following scheme:

Letter	Criteria
A	$90 \leq$ numeric grade ≤ 100
B	$80 \leq$ numeric grade < 90
C	$70 \leq$ numeric grade < 80
D	$60 \leq$ numeric grade < 70
F	numeric grade < 60

2.16 Develop well-structured function procedures to determine (a) the factorial; (b) the minimum value in a vector; and (c) the average of the values in a vector.
2.17 Develop well-structured programs to (a) determine the square root of the sum of the squares of the elements of a two-dimensional array (i.e., a matrix) and (b) normalize a matrix by dividing each row by the maximum absolute value in the row so that the maximum element in each row is 1.
2.18 *Piecewise functions* are sometimes useful when the relationship between a dependent and an independent variable cannot be adequately represented by a single equation. For example, the velocity of a rocket might be described by

$$v(t) = \begin{cases} 11t^2 - 5t & 0 \leq t \leq 10 \\ 1100 - 5t & 10 \leq t \leq 20 \\ 50t + 2(t - 20)^2 & 20 \leq t \leq 30 \\ 1520e^{-0.2(t-30)} & t > 30 \\ 0 & \text{otherwise} \end{cases}$$

Develop a well-structured function to compute v as a function of t. Then use this function to generate a table of v versus t for $t = -5$ to 50 at increments of 0.5.
2.19 Develop a well-structured function to determine the elapsed days in a year. The function should be passed three values: `mo` = the month (1–12), `da` = the day (1–31), and `leap` = (0 for non–leap year and 1 for leap year). Test it for January 1, 1999; February 29, 2000; March 1, 2001; June 21, 2002; and December 31, 2004. *Hint:* A nice way to do this combines the `for` and the `switch` structures.
2.20 Develop a well-structured function to determine the elapsed days in a year. The first line of the function should be set up as

```
function nd = days(mo, da, year)
```

where `mo` = the month (1–12), `da` = the day (1–31), and `year` = the year. Test it for January 1, 1999; February 29, 2000; March 1, 2001; June 21, 2002; and December 31, 2004.
2.21 Manning's equation can be used to compute the velocity of water in a rectangular open channel,

$$U = \frac{\sqrt{S}}{n}\left(\frac{BH}{B + 2H}\right)^{2/3}$$

where U = velocity (m/s), S = channel slope, n = roughness coefficient, B = width (m), and H = depth (m). The following data are available for five channels:

n	S	B	H
0.036	0.0001	10	2
0.020	0.0002	8	1
0.014	0.0012	19	1.7
0.030	0.0007	24	3
0.021	0.0004	15	2.6

Write a well-structured program that computes the velocity for each of these channels. Have the program display the input data along with the computed velocity in tabular form where velocity is the fifth column. Include headings on the table to label the columns.

2.22 A simply supported beam is loaded as shown in Fig. P2.22. Using singularity functions, the displacement along the beam can be expressed by the equation

$$u_y(x) = \frac{-5}{6}[\langle x - 0 \rangle^4 - \langle x - 5 \rangle^4] + \frac{15}{6}\langle x - 8 \rangle^3$$

$$+ 75\langle x - 7 \rangle^2 + \frac{57}{6}x^3 - 238.25x$$

By definition, the singularity function can be expressed as follows:

$$\langle x - a \rangle^n = \begin{cases} (x - a)^n & \text{when } x > a \\ 0 & \text{when } x \leq a \end{cases}$$

Develop a program that creates a plot of displacement versus distance along the beam x. Note that $x = 0$ at the left end of the beam.

FIGURE P2.22

2.23 The volume V of liquid in a hollow horizontal cylinder of radius r and length L is related to the depth of the liquid h by

$$V = \left[r^2 \cos^{-1}\left(\frac{r - h}{r} \right) - (r - h)\sqrt{2rh - h^2} \right] L$$

Develop a well-structured function to create a plot of volume versus depth. Test the program for $r = 2.5$ m and $L = 6$ m.

2.24 Develop a well-structured program to compute the velocity of a parachutist as a function of time using Euler's method. Test your program for the case where $m = 85$ kg and $c = 11$ kg/s. Perform the calculation from $t = 0$ to 20 s with a step size of 2 s. Use an initial condition that the parachutist has an upward velocity of 25 m/s at $t = 0$. At $t = 10$ s, assume that the parachute is instantaneously deployed so that the drag coefficient jumps to 55 kg/s.

2.25 The pseudocode in Fig. P2.25 computes the factorial. Express this algorithm as a well-structured function in the language of your choice. Test it by computing 0! and 6!. In addition, test the error trap by trying to evaluate −2!.

```
FUNCTION fac(n)
IF n ≥ 0 THEN
  x = 1
  DOFOR i = 1, n
    x = x · i
  END DO
  fac = x
ELSE
  display error message
  terminate
ENDIF
END fac
```

FIGURE P2.25

2.26 The height of a small rocket y can be calculated as a function of time after blastoff with the following piecewise function:

$$y = 38.1454t + 0.13743t^3 \qquad 0 \leq t < 15$$

$$y = 1036 + 130.909(t - 15) + 6.18425(t - 15)^2$$
$$- 0.428(t - 15)^3 \qquad 15 \leq t < 33$$

$$y = 2900 - 62.468(t - 33) - 16.9274(t - 33)^2$$
$$+ 0.41796(t - 33)^3 \qquad t \geq 33$$

Develop a well-structured pseudocode function to compute y as a function of t. Note that if the user enters a negative value of t or if the rocket has hit the ground ($y \le 0$), then a value of zero is returned for y. Also, the function should be invoked in the calling program as height(t). Write the algorithm (a) as pseudocode or (b) in a high-level language of your choice.

2.27 As depicted in Fig. P2.27, a water tank consists of a cylinder topped by the frustum of a cone. Develop a well-structured function in a high-level language or macro language of your choice to compute the volume given the water level h (m) above the tank's bottom. Design the function so that it returns a value of zero for negative h's and the value of the maximum filled volume for h's greater than the tank's maximum depth. Given the following parameters, $H_1 = 11$ m, $r_1 = 3.5$ m, $H_2 = 5$ m, and $r_2 = 6$ m, test your function by using it to compute the volumes and generate a graph of the volume as a function of level from $h = -1$ to 17 m.

2.28 Write a well-structured function procedure named Fnorm to calculate the Frobenius norm of an $m \times n$ matrix using nested count-controlled (for) loops:

$$\| A \|_f = \sqrt{\sum_{i=1}^{m} \sum_{j=1}^{n} a_{i,j}^2}$$

Test your function with

$$A = \begin{bmatrix} 5 & 7 & 9 \\ 1 & 8 & 4 \\ 7 & 6 & 2 \end{bmatrix}$$

2.29 The pressure and temperature of the atmosphere are constantly changing depending on a number of factors including

FIGURE P2.27

altitude, latitude/longitude, time of day, and season. To take all these variations into account when considering the design and performance of flight vehicles is impractical. Therefore, a *standard atmosphere* is frequently used to provide engineers and scientists with a common reference for their research and development. The *International Standard Atmosphere* is one such model of how conditions of the earth's atmosphere change over a wide range of altitudes, or elevations. The following table shows values of temperature and pressure at selected altitudes.

Layer index i	Layer name	Base geopotential altitude above MSL h (km)	Lapse rate (°C/km)	Base temperature T (°C)	Base pressure p (Pa)
1	Troposphere	0	−6.5	15	101325
2	Tropopause	11	0	−56.5	22632
3	Stratosphere	20	1	−56.5	5474.9
4	Stratosphere	32	2.8	−44.5	868.02
5	Stratopause	47	0	−2.5	110.91
6	Mesosphere	51	−2.8	−2.5	66.939
7	Mesosphere	71	−2.0	−58.5	3.9564
8	Mesopause	84.852	—	−86.28	0.3734

The temperature at each altitude can then be computed as

$$T(h) = T_i + \gamma_i(h - h_i) \qquad h_i < h \leq h_{i+1}$$

where $T(h)$ = temperature at altitude h (°C), T_i = the base temperature for layer i (°C), γ_i = lapse rate, or the rate at which atmospheric temperature decreases linearly with increase in altitude for layer i (°C/km), and h_i = base geopotential altitude above mean sea level (MSL) for layer i. The pressure at each altitude can then be computed as

$$p(h) = p_i + \frac{p_{i+1} - p_i}{h_{i+1} - h_i}(h - h_i)$$

where $p(h)$ = pressure at altitude h (Pa ≡ N/m²), p_i = the base pressure for layer i (Pa). The density, ρ (kg/m³), can then be calculated according to a molar form of the *ideal gas law:*

$$\rho = \frac{pM}{RT_a}$$

where M = molar mass (≅ 0.0289644 kg/mol), R = the universal gas constant (8.3144621 J/(mol K)), and T_a = absolute temperature (K) = $T + 273.15$.

Develop a function, StdAtmDens, to determine values of the density for a given altitude. If the user requests a value outside the range of altitudes, have the function display an error message and terminate the application. Test your function for altitudes of −200, 0, 11, 40, 84.852, and 100 km.

2.30 Develop a function to convert a vector of temperatures from Celsius to Fahrenheit and vice versa. Test it with the following data for the average monthly temperatures at Death Valley, CA, and at the South Pole.

2.31 As depicted in Fig. P2.31, the downward deflection, y (m), of a cantilever beam with a uniform load, w (kg/m), can be computed as

$$y = \frac{w}{24EI}(x^4 - 4Lx^3 + 6L^2x^2)$$

where x = distance (m), E = the modulus of elasticity = 2×10^{11} Pa, I = moment of inertia = 3.25×10^{-4} m⁴, w = 10,000 N/m, and L = length = 4 m. This equation can be differentiated to yield the slope of the downward deflection as a function of x,

$$\frac{dy}{dx} = \frac{w}{24EI}(4x^3 - 12Lx^2 + 12L^2x)$$

If $y = 0$ at $x = 0$, use this equation with Euler's method ($\Delta x = 0.125$ m) to compute the deflection from $x = 0$ to L. Develop a plot of your results along with the analytical solution computed with the first equation.

FIGURE P2.31

Day		15	45	75	105	135	165	195	225	255	285	315	345
Death Valley	°F	54	60	69	77	87	96	102	101	92	78	63	52
South Pole	°C	−27	−40	−53	−56	−57	−57	−59	−59	−59	−50	−38	−27

Approximations and Round-Off Errors

Because so many of the methods in this book are straightforward in description and application, it would be very tempting at this point for us to proceed directly to the main body of the text and teach you how to use these techniques. However, understanding the concept of error is so important to the effective use of numerical methods that we have chosen to devote the next two chapters to this topic.

The importance of error was introduced in our discussion of the falling parachutist in Chap. 1. Recall that we determined the velocity of a falling parachutist by both analytical and numerical methods. Although the numerical technique yielded estimates that were close to the exact analytical solution, there was a discrepancy, or *error,* because the numerical method involved an approximation. Actually, we were fortunate in that case because the availability of an analytical solution allowed us to compute the error exactly. For many applied engineering problems, we cannot obtain analytical solutions. Therefore, we cannot compute exactly the errors associated with our numerical methods. In these cases, we must settle for approximations or estimates of the errors.

Such errors are characteristic of most of the techniques described in this book. This statement might at first seem contrary to what one normally conceives of as sound engineering. Students and practicing engineers constantly strive to limit errors in their work. When taking examinations or doing homework problems, you are penalized, not rewarded, for your errors. In professional practice, errors can be costly and sometimes catastrophic. If a structure or device fails, lives can be lost.

Although perfection is a laudable goal, it is rarely, if ever, attained. For example, despite the fact that the model developed from Newton's second law is an excellent approximation, it would never in practice exactly predict the parachutist's fall. A variety of factors such as winds and slight variations in air resistance would result in deviations from the prediction. If these deviations are systematically high or low, then we might need to develop a new model. However, if they are randomly distributed and tightly grouped around the prediction, then the deviations might be considered negligible and the model deemed adequate. Numerical approximations also introduce similar discrepancies into the analysis. Again, the question is: How much error is present in our calculations and is it tolerable?

This chapter and Chap. 4 cover basic topics related to the identification, quantification, and minimization of these errors. In this chapter, general information concerned with the quantification of error is reviewed in the first sections. This is followed by a section on one of the two major forms of numerical error: round-off error. *Round-off error* is due to the fact that computers can represent only quantities with a finite number of digits. Then Chap. 4 deals with the other major form: truncation error. *Truncation error* is the discrepancy introduced by the fact that numerical methods may employ approximations to represent exact mathematical operations and quantities. Finally, we briefly discuss errors not directly connected with the numerical methods themselves. These include blunders, formulation or model errors, and data uncertainty.

3.1　SIGNIFICANT FIGURES

This book deals extensively with approximations connected with the manipulation of numbers. Consequently, before discussing the errors associated with numerical methods, it is useful to review basic concepts related to approximate representation of the numbers themselves.

Whenever we employ a number in a computation, we must have assurance that it can be used with confidence. For example, Fig. 3.1 depicts a speedometer and odometer from an automobile. Visual inspection of the speedometer indicates that the car is traveling between 48 and 49 km/h. Because the indicator is higher than the midpoint between the markers on the gauge, we can say with assurance that the car is traveling at approximately 49 km/h. We have confidence in this result because two or more reasonable individuals reading this gauge would arrive at the same conclusion. However, let us say that we insist that the speed be estimated to one decimal place. For this case,

FIGURE 3.1
An automobile speedometer and odometer illustrating the concept of a significant figure.

one person might say 48.8, whereas another might say 48.9 km/h. Therefore, because of the limits of this instrument, only the first two digits can be used with confidence. Estimates of the third digit (or higher) must be viewed as approximations. It would be ludicrous to claim, on the basis of this speedometer, that the automobile is traveling at 48.8642138 km/h. In contrast, the odometer provides up to six certain digits. From Fig. 3.1, we can conclude that the car has traveled slightly less than 87,324.5 km during its lifetime. In this case, the seventh digit (and higher) is uncertain.

The concept of a significant figure, or digit, has been developed to formally designate the reliability of a numerical value. The *significant digits* of a number are those that can be used with confidence. They correspond to the number of certain digits plus one estimated digit. For example, the speedometer and the odometer in Fig. 3.1 yield readings of three and seven significant figures, respectively. For the speedometer, the two certain digits are 48. It is conventional to set the estimated digit at one-half of the smallest scale division on the measurement device. Thus the speedometer reading would consist of the three significant figures: 48.5. In a similar fashion, the odometer would yield a seven-significant-figure reading of 87,324.45.

Although it is usually a straightforward procedure to ascertain the significant figures of a number, some cases can lead to confusion. For example, zeros are not always significant figures because they may be necessary just to locate a decimal point. The numbers 0.00001845, 0.0001845, and 0.001845 all have four significant figures. Similarly, when trailing zeros are used in large numbers, it is not clear how many, if any, of the zeros are significant. For example, at face value the number 45,300 may have three, four, or five significant digits, depending on whether the zeros are known with confidence. Such uncertainty can be resolved by using scientific notation, where 4.53×10^4, 4.530×10^4, 4.5300×10^4 designate that the number is known to three, four, and five significant figures, respectively.

The concept of significant figures has two important implications for our study of numerical methods:

1. As introduced in the falling parachutist problem, numerical methods yield approximate results. We must, therefore, develop criteria to specify how confident we are in our approximate result. One way to do this is in terms of significant figures. For example, we might decide that our approximation is acceptable if it is correct to four significant figures.
2. Although quantities such as π, e, or $\sqrt{7}$ represent specific quantities, they cannot be expressed exactly by a limited number of digits. For example,

   $\pi = 3.14159265358979323846 2643 \ldots$

   ad infinitum. Because computers retain only a finite number of significant figures, such numbers can never be represented exactly. The omission of the remaining significant figures is called round-off error.

Both round-off error and the use of significant figures to express our confidence in a numerical result will be explored in detail in subsequent sections. In addition, the concept of significant figures will have relevance to our definition of accuracy and precision in the next section.

3.2 ACCURACY AND PRECISION

The errors associated with both calculations and measurements can be characterized with regard to their accuracy and precision. *Accuracy* refers to how closely a computed or measured value agrees with the true value. *Precision* refers to how closely individual computed or measured values agree with each other.

These concepts can be illustrated graphically using an analogy from target practice. The bullet holes on each target in Fig. 3.2 can be thought of as the predictions of a numerical technique, whereas the bull's-eye represents the truth. *Inaccuracy* (also called *bias*) is defined as systematic deviation from the truth. Thus, although the shots in Fig. 3.2c are more tightly grouped than those in Fig. 3.2a, the two cases are equally biased because they are both centered on the upper left quadrant of the target. *Imprecision* (also called *uncertainty*), on the other hand, refers to the magnitude of the scatter. Therefore, although Fig. 3.2b and d are equally accurate (that is, centered on the bull's-eye), the latter is more precise because the shots are tightly grouped.

Numerical methods should be sufficiently accurate or unbiased to meet the requirements of a particular engineering problem. They also should be precise enough for adequate engineering design. In this book, we will use the collective term *error* to represent both

FIGURE 3.2

An example from marksmanship illustrating the concepts of accuracy and precision. (*a*) Inaccurate and imprecise; (*b*) accurate and imprecise; (*c*) inaccurate and precise; (*d*) accurate and precise.

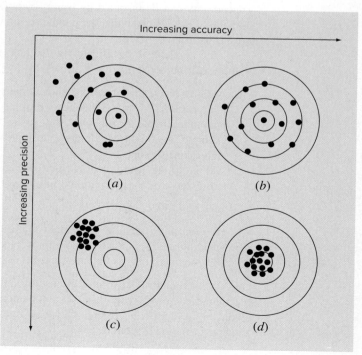

the inaccuracy and the imprecision of our predictions. With these concepts as background, we can now discuss the factors that contribute to the error of numerical computations.

3.3 ERROR DEFINITIONS

Numerical errors arise from the use of approximations to represent exact mathematical operations and quantities. These include *truncation errors,* which result when approximations are used to represent exact mathematical procedures, and *round-off errors,* which result when numbers having limited significant figures are used to represent exact numbers. For both types, the relationship between the exact, or true, result and the approximation can be formulated as

$$\text{True value} = \text{approximation} + \text{error} \tag{3.1}$$

By rearranging Eq. (3.1), we find that the numerical error is equal to the discrepancy between the true value and the approximation, as in

$$E_t = \text{true value} - \text{approximation} \tag{3.2}$$

where E_t is used to designate the exact value of the error. The subscript t is included to designate that this is the "true" error. This is in contrast to other cases, as described shortly, where an "approximate" estimate of the error must be employed.

A shortcoming of this definition is that it takes no account of the order of magnitude of the value under examination. For example, an error of a centimeter is much more significant if we are measuring a rivet rather than a bridge. One way to account for the magnitudes of the quantities being evaluated is to normalize the error to the true value, as in

$$\text{True fractional relative error} = \frac{\text{true error}}{\text{true value}}$$

where, as specified by Eq. (3.2), error = true value − approximation. The relative error can also be multiplied by 100 percent to express it as

$$\varepsilon_t = \frac{\text{true error}}{\text{true value}} 100\% \tag{3.3}$$

where ε_t designates the true percent relative error.

EXAMPLE 3.1 **Calculation of Errors**

Problem Statement. Suppose that you have the task of measuring the lengths of a bridge and a rivet and come up with 9999 and 9 cm, respectively. If the true values are 10,000 and 10 cm, respectively, compute (*a*) the true error and (*b*) the true percent relative error for each case.

Solution.

(a) The error for measuring the bridge is [Eq. (3.2)]

$$E_t = 10,000 - 9999 = 1 \text{ cm}$$

and for the rivet it is

$$E_t = 10 - 9 = 1 \text{ cm}$$

(b) The percent relative error for the bridge is [Eq. (3.3)]

$$\varepsilon_t = \frac{1}{10,000}100\% = 0.01\%$$

and for the rivet it is

$$\varepsilon_t = \frac{1}{10}100\% = 10\%$$

Thus, although both measurements have an error of 1 cm, the relative error for the rivet is much greater. We would conclude that we have done an adequate job of measuring the bridge, whereas our estimate for the rivet leaves something to be desired.

Notice that for Eqs. (3.2) and (3.3), E and ε are subscripted with a t to signify that the error is normalized to the true value. In Example 3.1, we were provided with this value. However, in actual situations such information is rarely available. For numerical methods, the true value will be known only when we deal with functions that can be solved analytically. Such will typically be the case when we investigate the theoretical behavior of a particular technique for simple systems. However, in real-world applications, we will obviously not know the true answer *a priori*. For these situations, an alternative is to normalize the error using the best available estimate of the true value, that is, to the approximation itself, as in

$$\varepsilon_a = \frac{\text{approximate error}}{\text{approximation}}100\% \tag{3.4}$$

where the subscript a signifies that the error is normalized to an approximate value. Note also that for real-world applications, Eq. (3.2) cannot be used to calculate the error term for Eq. (3.4). One of the challenges of numerical methods is to determine error estimates in the absence of knowledge regarding the true value. For example, certain numerical methods use an *iterative approach* to compute answers. In such an approach, a present approximation is made on the basis of a previous approximation. This process is performed repeatedly, or iteratively, to successively compute (we hope) better and better approximations. For such cases, the error is often estimated as the difference between previous and current approximations. Thus, percent relative error is determined according to

$$\varepsilon_a = \frac{\text{current approximation} - \text{previous approximation}}{\text{current approximation}}100\% \tag{3.5}$$

This and other approaches for expressing errors will be elaborated on in subsequent chapters.

The signs of the error values from Eqs. (3.2) through (3.5) may be either positive or negative. If the approximation is greater than the true value (or the previous approximation is greater than the current approximation), the error is negative; if the approximation is less than the true value, the error is positive. Also, for Eqs. (3.3) to

(3.5), the denominator may be less than zero, which can also lead to a negative error. Often, when performing computations, we may not be concerned with the sign of the error, but we are interested in whether the percent absolute value is lower than a pre-specified percent tolerance ε_s. Therefore, it is often useful to employ the absolute value of Eqs. (3.2) through (3.5). For such cases, the computation is repeated until

$$|\varepsilon_a| < \varepsilon_s \tag{3.6}$$

If this relationship holds, our result is assumed to be within the prespecified acceptable level ε_s. Note that for the remainder of this text, we will almost exclusively employ absolute values when we use relative errors.

It is also convenient to relate these errors to the number of significant figures in the approximation. It can be shown (Scarborough 1966) that if the following criterion is met, we can be assured that the result is correct to *at least n* significant figures.

$$\varepsilon_s = (0.5 \times 10^{2-n})\% \tag{3.7}$$

EXAMPLE 3.2

Error Estimates for Iterative Methods

Problem Statement. In mathematics, functions can often be represented by infinite series. For example, the exponential function can be computed using

$$e^x = 1 + x + \frac{x^2}{2} + \frac{x^3}{3!} + \cdots + \frac{x^n}{n!} \tag{E3.2.1}$$

Thus, as more terms are added in sequence, the approximation becomes a better and better estimate of the true value of e^x. Equation (E3.2.1) is called a *Maclaurin series expansion*.

Starting with the simplest version, $e^x = 1$, add terms one at a time to estimate $e^{0.5}$. After each new term is added, compute the true and approximate percent relative errors with Eqs. (3.3) and (3.5), respectively. Note that the true value is $e^{0.5} = 1.648721\ldots$. Add terms until the absolute value of the approximate error estimate ε_a falls below a prespecified error criterion ε_s conforming to three significant figures.

Solution. First, Eq. (3.7) can be employed to determine the error criterion that ensures a result is correct to at least three significant figures:

$$\varepsilon_s = (0.5 \times 10^{2-3})\% = 0.05\%$$

Thus, we will add terms to the series until ε_a falls below this level.

The first estimate is simply equal to Eq. (E3.2.1) with a single term. Thus, the first estimate is equal to 1. The second estimate is then generated by adding the second term, as in

$$e^x = 1 + x$$

or for $x = 0.5$,

$$e^{0.5} = 1 + 0.5 = 1.5$$

This represents a true percent relative error of [Eq. (3.3)]

$$\varepsilon_t = \frac{1.648721 - 1.5}{1.648721}100\% = 9.02\%$$

Equation (3.5) can be used to determine an approximate estimate of the error, as in

$$\varepsilon_a = \frac{1.5 - 1}{1.5} 100\% = 33.3\%$$

Because ε_a is not less than the required value of ε_s, we would continue the computation by adding another term, $x^2/2!$, and repeating the error calculation. The process is continued until $\varepsilon_a < \varepsilon_s$. The entire computation can be summarized as

Terms	Result	ε_t (%)	ε_a (%)
1	1	39.3	
2	1.5	9.02	33.3
3	1.625	1.44	7.69
4	1.645833333	0.175	1.27
5	1.648437500	0.0172	0.158
6	1.648697917	0.00142	0.0158

Thus, after six terms are included, the approximate error falls below $\varepsilon_s = 0.05\%$ and the computation is terminated. However, notice that, rather than three significant figures, the result is accurate to five! This is because, for this case, both Eqs. (3.5) and (3.7) are conservative. That is, they ensure that the result is at least as good as they specify. Although, as discussed in Chap. 6, this is not always the case for Eq. (3.5), it is true most of the time.

3.3.1 Computer Algorithm for Iterative Calculations

Many of the numerical methods described in the remainder of this text involve iterative calculations of the sort illustrated in Example 3.2. These all entail solving a mathematical problem by computing successive approximations to the solution starting from an initial guess.

The computer implementation of such iterative solutions involves loops. As we saw in Sec. 2.1.1, these come in two basic flavors: count-controlled and decision loops. Most iterative solutions use decision loops. Thus, rather than employing a prespecified number of iterations, the process typically is repeated until an approximate error estimate falls below a stopping criterion, as in Example 3.2.

A pseudocode for a generic iterative calculation is presented in Fig. 3.3. The function is passed a value (`val`) along with a stopping error criterion (`es`) and a maximum allowable number of iterations (`maxit`). The value is typically either (1) an initial value or (2) the value for which the iterative calculation is to be made.

The function first initializes three variables. These include (1) a variable `iter` that keeps track of the number of iterations, (2) a variable `sol` that holds the current estimate of the solution, and (3) a variable `ea` that holds the approximate percent relative error. Note that `ea` is initially set to a value of 100 to ensure that the loop executes at least once.

These initializations are followed by the decision loop that actually implements the iterative calculation. Prior to generating a new solution, `sol` is first assigned to `solold`. Then a new value of `sol` is computed and the iteration counter is incremented. If the new value of `sol` is nonzero, the percent relative error `ea` is determined. The stopping

```
FUNCTION IterMeth(val, es, maxit)
iter = 1
sol = val
ea = 100
DO
   solold = sol
   sol = ...
   iter = iter + 1
   IF sol ≠ 0 ea=abs((sol – solold)/sol)*100
   IF ea ≤ es OR iter ≥ maxit EXIT
END DO
IterMeth = sol
END IterMeth
```

FIGURE 3.3
Pseudocode for a generic iterative calculation.

criteria are then tested. If both are false, the loop repeats. If either is true, the loop terminates and the final solution is sent back to the function call. The following example illustrates how the generic algorithm can be applied to a specific iterative calculation.

EXAMPLE 3.3

Computer Implementation of an Iterative Calculation

Problem Statement.   Develop a computer program based on the pseudocode from Fig. 3.3 to implement the calculation from Example 3.2.

Solution.   A function to implement the Maclaurin series expansion for e^x can be based on the general scheme in Fig. 3.3. To do this, we first formulate the series expansion as a formula:

$$e^x \cong \sum_{i=0}^{n} \frac{x^n}{n!}$$

Figure 3.4 shows functions to implement this series written in VBA and MATLAB software. Similar codes could be developed in other languages such as C++ or Fortran 95. Notice that whereas MATLAB has a built-in `factorial` function, it is necessary to compute the factorial as part of the VBA implementation with a simple product accumulator `fac`.

When the programs are run, they generate an estimate for the exponential function. For the MATLAB version, the answer is returned along with the approximate error and the number of iterations. For example, e^1 can be evaluated as

```
>> format long
>> [val, ea, iter] = IterMeth(1,1e-6,100)

val =
    2.718281826198493
ea =
    9.216155641522974e-007
iter =
    12
```

We can see that after 12 iterations, we obtain a result of 2.7182818 with an approximate error estimate of $= 9.2162 \times 10^{-7}\%$. The result can be verified by using the built-in exp function to directly calculate the exact value and the true percent relative error,

```
>> trueval=exp(1)

trueval =
    2.718281828459046

>> et=abs((trueval-val)/trueval)*100
et =
    8.316108397236229e-008
```

As was the case with Example 3.2, we obtain the desirable outcome that the true error is less than the approximate error.

(a) Excel/VBA

```
Function IterMeth(x, es, maxit)
' initialization
iter = 1
sol = 1
ea = 100
fac = 1
' iterative calculation
Do
  solold = sol
  fac = fac * iter
  sol = sol + x ^ iter / fac
  iter = iter + 1
  If sol <> 0 Then
    ea = Abs((sol - solold) / sol) * 100
  End If
  If ea <= es Or iter >= maxit Then Exit Do
Loop
IterMeth = sol
End Function
```

(b) MATLAB

```
function [v,ea,iter] = IterMeth(x,es,maxit)
% initialization
iter = 1;
sol = 1;
ea = 100;

% iterative calculation
while (1)
  solold = sol;

  sol = sol + x ^ iter / factorial(iter);
  iter = iter + 1;
  if sol~=0
    ea=abs((sol - solold)/sol)*100;
  end
  if ea<=es || iter>=maxit,break,end
end
v = sol;
end
```

FIGURE 3.4
(a) Excel/VBA and (b) MATLAB functions based on the pseudocode from Fig. 3.3.

With the preceding definitions as background, we can now proceed to the two types of error connected directly with numerical methods: round-off errors and truncation errors.

3.4 ROUND-OFF ERRORS

As mentioned previously, round-off errors originate from the fact that computers retain only a fixed number of significant figures during a calculation. Numbers such as π, e, or $\sqrt{7}$ cannot be expressed by a fixed number of significant figures. Therefore, they cannot be represented exactly by the computer. In addition, because computers use a base-2 representation, they cannot precisely represent certain exact base-10 numbers. The discrepancy introduced by this omission of significant figures is called *round-off error.*

3.4.1 Computer Representation of Numbers

Numerical round-off errors are directly related to the manner in which numbers are stored in a computer. The fundamental unit whereby information is represented is called a *word.* This is an entity that consists of a string of *binary digits,* or *bits.* Numbers are typically stored in one or more words. To understand how this is accomplished, we must first review some material related to number systems.

Number Systems.   A *number system* is merely a convention for representing quantities. Because we have 10 fingers and 10 toes, the number system that we are most familiar with is the *decimal,* or *base*-10, number system. A base is the number used as the reference for constructing the system. The base-10 system uses the 10 digits—0, 1, 2, 3, 4, 5, 6, 7, 8, 9—to represent numbers. By themselves, these digits are satisfactory for counting from 0 to 9.

For larger quantities, combinations of these basic digits are used, with the position or *place value* specifying the magnitude. The right-most digit in a whole number represents a number from 0 to 9. The second digit from the right represents a multiple of 10. The third digit from the right represents a multiple of 100 and so on. For example, if we have the number 86,409, then we have eight groups of 10,000, six groups of 1000, four groups of 100, zero groups of 10, and nine more units, or

$$(8 \times 10^4) + (6 \times 10^3) + (4 \times 10^2) + (0 \times 10^1) + (9 \times 10^0) = 86,409$$

Figure 3.5*a* provides a visual representation of how a number is formulated in the base-10 system. This type of representation is called *positional notation.*

Because the decimal system is so familiar, it is not commonly realized that there are alternatives. For example, if human beings happened to have had eight fingers and eight toes, we would undoubtedly have developed an *octal,* or *base*-8, representation. In the same sense, our friend the computer is like a two-fingered animal who is limited to two states—either 0 or 1. This relates to the fact that the primary logic units of digital computers are on/off electronic components. Hence, numbers on the computer are represented with a *binary,* or *base*-2, system. Just as with the decimal system, quantities can be represented using positional notation. For example, the binary number 11 is equivalent to $(1 \times 2^1) + (1 \times 2^0) = 2 + 1 = 3$ in the decimal system. Figure 3.5*b* illustrates a more complicated example.

Integer Representation.   Now that we have reviewed how base-10 numbers can be represented in binary form, it is simple to conceive of how integers are represented on a computer. The most straightforward approach, called the *signed magnitude method,*

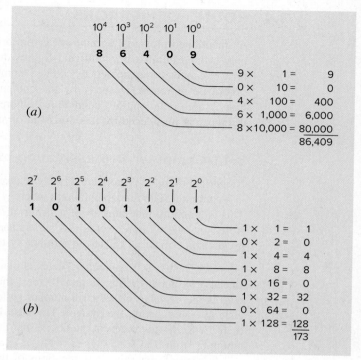

FIGURE 3.5
How the (a) decimal (base-10) and the (b) binary (base-2) systems work. In (b), the binary number 10101101 is equivalent to the decimal number 173.

FIGURE 3.6
The representation of the decimal integer −173 on a 16-bit computer using the signed magnitude method.

employs the first bit of a word to indicate the sign, with a 0 for positive and a 1 for negative. The remaining bits are used to store the number. For example, the integer value of −173 would be stored on a 16-bit computer as in Fig. 3.6.

EXAMPLE 3.4   **Range of Integers**

Problem Statement.   Determine the range of integers in base-10 that can be represented on a 16-bit computer.

Solution. Of the 16 bits, the first bit holds the sign. The remaining 15 bits can hold binary numbers from 0 to 111111111111111. The upper limit can be converted to a decimal integer, as in

$$(1 \times 2^{14}) + (1 \times 2^{13}) + \cdots + (1 \times 2^1) + (1 \times 2^0)$$

which equals 32,767 (note that this expression can be simply evaluated as $2^{15} - 1$). Thus, a 16-bit computer word can store decimal integers ranging from $-32,767$ to 32,767. In addition, because zero is already defined as 0000000000000000, it is redundant to use the number 1000000000000000 to define a "minus zero." Therefore, it is usually employed to represent an additional negative number: $-32,768$, and the range is from $-32,768$ to 32,767.

Note that the signed magnitude method described above is not used to represent integers on conventional computers. A preferred approach called the *2's complement* technique directly incorporates the sign into the number's magnitude rather than providing a separate bit to represent plus or minus. However, Example 3.4 still serves to illustrate how all digital computers are limited in their capability to represent integers. That is, numbers above or below the range cannot be represented. A more serious limitation is encountered in the storage and manipulation of fractional quantities as described next.

Floating-Point Representation. Fractional quantities are typically represented in computers using floating-point form. In this approach, the number is expressed as a fractional part, called a *mantissa* or *significand,* and an integer part, called an *exponent* or *characteristic,* as in

$$m \cdot b^e$$

where m = the mantissa, b = the base of the number system being used, and e = the exponent. For instance, the number 156.78 could be represented as 0.15678×10^3 in a floating-point base-10 system.

Figure 3.7 shows one way that a floating-point number could be stored in a word. The first bit is reserved for the sign, the next series of bits for the signed exponent, and the last bits for the mantissa.

FIGURE 3.7
The manner in which a floating-point number is stored in a word.

Note that the mantissa is usually *normalized* if it has leading zero digits. For example, suppose the quantity $1/34 = 0.029411765 \ldots$ was stored in a floating-point base-10 system that allowed only four decimal places to be stored. Thus, $1/34$ would be stored as

$$0.0294 \times 10^0$$

However, in the process of doing this, the inclusion of the useless zero to the right of the decimal forces us to drop the digit 1 in the fifth decimal place. The number can be normalized to remove the leading zero by multiplying the mantissa by 10 and lowering the exponent by 1 to give

$$0.2941 \times 10^{-1}$$

Thus, we retain an additional significant figure when the number is stored.

The consequence of normalization is that the absolute value of m is limited. That is,

$$\frac{1}{b} \leq m < 1 \tag{3.8}$$

where b = the base. For example, for a base-10 system, m would range between 0.1 and 1, and for a base-2 system, between 0.5 and 1.

Floating-point representation allows both fractions and very large numbers to be expressed on the computer. However, it has some disadvantages. For example, floating-point numbers take up more room and take longer to process than integer numbers. More significantly, however, their use introduces a source of error because the mantissa holds only a finite number of significant figures. Thus, a round-off error is introduced.

EXAMPLE 3.5

Hypothetical Set of Floating-Point Numbers

Problem Statement. Create a hypothetical floating-point number set for a machine that stores information using 7-bit words. Employ the first bit for the sign of the number, the next three for the sign and the magnitude of the exponent, and the last three for the magnitude of the mantissa (Fig. 3.8).

FIGURE 3.8

The smallest possible positive floating-point number from Example 3.5.

Solution. The smallest possible positive number is depicted in Fig. 3.8. The initial 0 indicates that the quantity is positive. The 1 in the second place designates that the exponent has a negative sign. The 1's in the third and fourth places give a maximum value to the exponent of

$$1 \times 2^1 + 1 \times 2^0 = 3$$

Therefore, the exponent will be −3. Finally, the mantissa is specified by the 100 in the last three places, which conforms to

$$1 \times 2^{-1} + 0 \times 2^{-2} + 0 \times 2^{-3} = 0.5$$

Although a smaller mantissa is possible (e.g., 000, 001, 010, 011), the value of 100 is used because of the limit imposed by normalization [Eq. (3.8)]. Thus, the smallest possible positive number for this system is $+0.5 \times 2^{-3}$, which is equal to 0.0625 in the base-10 system. The next highest numbers are developed by increasing the mantissa, as in

$$0111101 = (1 \times 2^{-1} + 0 \times 2^{-2} + 1 \times 2^{-3}) \times 2^{-3} = (0.078125)_{10}$$
$$0111110 = (1 \times 2^{-1} + 1 \times 2^{-2} + 0 \times 2^{-3}) \times 2^{-3} = (0.093750)_{10}$$
$$0111111 = (1 \times 2^{-1} + 1 \times 2^{-2} + 1 \times 2^{-3}) \times 2^{-3} = (0.109375)_{10}$$

Notice that the base-10 equivalents are spaced evenly with an interval of 0.015625.

At this point, to continue increasing, we must decrease the exponent to 10, which gives a value of

$$1 \times 2^1 + 0 \times 2^0 = 2$$

The mantissa is decreased back to its smallest value of 100. Therefore, the next number is

$$0110100 = (1 \times 2^{-1} + 0 \times 2^{-2} + 0 \times 2^{-3}) \times 2^{-2} = (0.125000)_{10}$$

This still represents a gap of $0.125000 - 0.109375 = 0.015625$. However, now when higher numbers are generated by increasing the mantissa, the gap is lengthened to 0.03125,

$$0110101 = (1 \times 2^{-1} + 0 \times 2^{-2} + 1 \times 2^{-3}) \times 2^{-2} = (0.156250)_{10}$$
$$0110110 = (1 \times 2^{-1} + 1 \times 2^{-2} + 0 \times 2^{-3}) \times 2^{-2} = (0.187500)_{10}$$
$$0110111 = (1 \times 2^{-1} + 1 \times 2^{-2} + 1 \times 2^{-3}) \times 2^{-2} = (0.218750)_{10}$$

This pattern is repeated as each larger quantity is formulated until a maximum number is reached,

$$0011111 = (1 \times 2^{-1} + 1 \times 2^{-2} + 1 \times 2^{-3}) \times 2^3 = (7)_{10}$$

The final number set is depicted graphically in Fig. 3.9.

Figure 3.9 manifests several aspects of floating-point representation that have significance regarding computer round-off errors:

1. *There Is a Limited Range of Quantities That May Be Represented.* Just as for the integer case, there are large positive and negative numbers that cannot be represented. Attempts to employ numbers outside the acceptable range will result in what is called

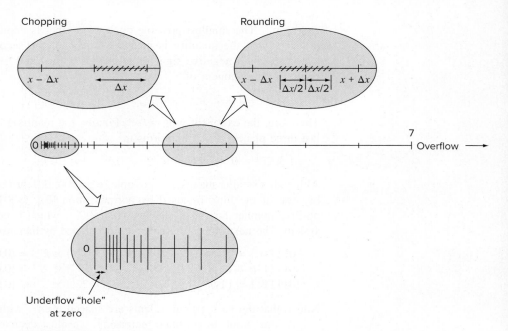

FIGURE 3.9
The hypothetical number system developed in Example 3.5. Each value is indicated by a tick mark. Only the positive numbers are shown. An identical set would also extend in the negative direction.

an *overflow error*. However, in addition to large quantities, the floating-point representation has the added limitation that very small numbers cannot be represented. This is illustrated by the *underflow* "hole" between zero and the first positive number in Fig. 3.9. It should be noted that this hole is enlarged because of the normalization constraint of Eq. (3.8).

2. *There Are Only a Finite Number of Quantities That Can Be Represented within the Range.* Thus, the degree of precision is limited. Obviously, irrational numbers cannot be represented exactly. Furthermore, rational numbers that do not exactly match one of the values in the set also cannot be represented precisely. The errors introduced by approximating both these cases are referred to as *quantizing errors*. The actual approximation is accomplished in either of two ways: chopping or rounding. For example, suppose that the value of $\pi = 3.14159265358\ldots$ is to be stored on a base-10 number system carrying seven significant figures. One method of approximation would be to merely omit, or "chop off," the eighth and higher terms, as in $\pi = 3.141592$, with the introduction of an associated error of [Eq. (3.2)]

$$E_t = 0.00000065\ldots$$

This technique of retaining only the significant terms was originally dubbed "truncation" in computer jargon. We prefer to call it *chopping* to distinguish it from the truncation errors discussed in Chap. 4. Note that for the base-2 number system in Fig. 3.9, chopping means that any quantity falling within an interval of length Δx

will be stored as the quantity at the lower end of the interval. Thus, the upper error bound for chopping is Δx. Additionally, a bias is introduced because all errors are positive. The shortcomings of chopping are attributable to the fact that the higher terms in the complete decimal representation have no impact on the shortened version. For instance, in our example of π, the first discarded digit is 6. Thus, the last retained digit should be rounded up to yield 3.141593. Such *rounding* reduces the error to

$$E_t = -0.00000035 \ldots$$

Consequently, rounding yields a lower absolute error than chopping. Note that for the base-2 number system in Fig. 3.9, rounding means that any quantity falling within an interval of length Δx will be represented as the nearest allowable number. Thus, the upper error bound for rounding is $\Delta x/2$. Additionally, no bias is introduced because some errors are positive and some are negative. Some computers employ rounding. However, this adds to the computational overhead, and, consequently, many machines use simple chopping. This approach is justified under the supposition that the number of significant figures is large enough that resulting round-off error is usually negligible.

3. *The Interval between Numbers, Δx, Increases as the Numbers Grow in Magnitude.* It is this characteristic, of course, that allows floating-point representation to preserve significant digits. However, it also means that quantizing errors will be proportional to the magnitude of the number being represented. For normalized floating-point numbers, this proportionality can be expressed, for cases where chopping is employed, as

$$\frac{|\Delta x|}{|x|} \le \mathscr{E} \tag{3.9}$$

and, for cases where rounding is employed, as

$$\frac{|\Delta x|}{|x|} \le \frac{\mathscr{E}}{2} \tag{3.10}$$

where $\mathscr{E}$ is referred to as the *machine epsilon*, which can be computed as

$$\mathscr{E} = b^{1-t} \tag{3.11}$$

where b is the number base and t is the number of significant digits in the mantissa. Notice that the inequalities in Eqs. (3.9) and (3.10) signify that these are error bounds. That is, they specify the worst cases.

EXAMPLE 3.6 **Machine Epsilon**

Problem Statement. Determine the machine epsilon and verify its effectiveness in characterizing the errors of the number system from Example 3.5. Assume that chopping is used.

Solution. The hypothetical floating-point system from Example 3.5 employed values of the base $b = 2$ and the number of mantissa bits $t = 3$. Therefore, the machine epsilon would be [Eq. (3.11)]

$$\mathscr{E} = 2^{1-3} = 0.25$$

FIGURE 3.10
The largest quantizing error will occur for those values falling just below the upper bound of the first of a series of equispaced intervals.

Consequently, the relative quantizing error should be bounded by 0.25 for chopping. The largest relative errors should occur for those quantities that fall just below the upper bound of the first interval between successive equispaced numbers (Fig. 3.10). Those numbers falling in the succeeding higher intervals would have the same value of Δx but a greater value of x and, hence, would have a lower relative error. An example of a maximum error would be a value falling just below the upper bound of the interval between $(0.125000)_{10}$ and $(0.156250)_{10}$. For this case, the error would be less than

$$\frac{0.03125}{0.125000} = 0.25$$

Thus, the error is as predicted by Eq. (3.9).

The magnitude dependence of quantizing errors has a number of practical applications in numerical methods. Most of these relate to the commonly employed operation of testing whether two numbers are equal. This occurs when testing convergence of quantities as well as in the stopping mechanism for iterative processes (recall Example 3.2). For these cases, it should be clear that, rather than test whether the two quantities are equal, it is advisable to test whether their difference is less than an acceptably small tolerance. Further, it should also be evident that normalized rather than absolute difference should be compared, particularly when dealing with numbers of large magnitude. In addition, the machine epsilon can be employed in formulating stopping or convergence criteria. This ensures that programs are portable—that is, they are not dependent on the computer on which they are implemented. Figure 3.11 lists pseudocode to automatically determine the machine epsilon of a binary computer.

Extended Precision. It should be noted at this point that, although round-off errors can be important in contexts such as testing convergence, the number of significant digits carried on most computers allows most engineering computations to be performed with more than acceptable precision. For example, the hypothetical number system in Fig. 3.9 is a gross exaggeration that was employed for illustrative purposes. Commercial computers use much larger words and, consequently, allow numbers to be expressed with more than adequate precision. For example, computers that use IEEE format allow 24 bits to be used for the mantissa, which translates into about seven significant base-10 digits of precision[1] with a range of about 10^{-38} to 10^{39}.

```
epsilon = 1
DO
  IF (epsilon + 1 ≤ 1)EXIT
  epsilon = epsilon/2
END DO
epsilon = 2 × epsilon
```

FIGURE 3.11
Pseudocode to determine machine epsilon for a binary computer.

[1]Note that only 23 bits are actually used to store the mantissa. However, because of normalization, the first bit of the mantissa is always 1 and is, therefore, not stored. Thus, this first bit together with the 23 stored bits gives the 24 total bits of precision for the mantissa.

With this acknowledged, there are still cases where round-off error becomes critical. For this reason most computers allow the specification of extended precision. The most common of these is double precision, in which the number of words used to store floating-point numbers is doubled. It provides about 15 to 16 decimal digits of precision and a range of approximately 10^{-308} to 10^{308}.

In many cases, the use of double-precision quantities can greatly mitigate the effect of round-off errors. However, a price is paid for such remedies in that they also require more memory and execution time. The difference in execution time for a small calculation might seem insignificant. However, as your programs become larger and more complicated, the added execution time could become considerable and have a negative impact on your effectiveness as a problem solver. Therefore, extended precision should not be used frivolously. Rather, it should be selectively employed where it will yield the maximum benefit at the least cost in terms of execution time. In the following sections, we will look closer at how round-off errors affect computations, and in so doing provide a foundation of understanding to guide your use of the double-precision capability.

Before proceeding, it should be noted that some of the commonly used software packages (for example, Excel, Mathcad) routinely use double precision to represent numerical quantities. Thus, the developers of these packages decided that mitigating round-off errors would take precedence over any loss of speed incurred by using extended precision. Other packages, like MATLAB software, allow you to use extended precision, if you desire.

3.4.2 Arithmetic Manipulations of Computer Numbers

Aside from the limitations of a computer's number system, the actual arithmetic manipulations involving these numbers can also result in round-off error. In the following section, we will first illustrate how common arithmetic operations affect round-off errors. Then we will investigate a number of particular manipulations that are especially prone to round-off errors.

Common Arithmetic Operations. Because of their familiarity, normalized base-10 numbers will be employed to illustrate the effect of round-off errors on simple addition, subtraction, multiplication, and division. Other number bases would behave in a similar fashion. To simplify the discussion, we will employ a hypothetical decimal computer with a 4-digit mantissa and a 1-digit exponent. In addition, chopping is used. Rounding would lead to similar though less dramatic errors.

When two floating-point numbers are added, the mantissa of the number with the smaller exponent is modified so that the exponents are the same. This has the effect of aligning the decimal points. For example, suppose we want to add $0.1557 \cdot 10^1 + 0.4381 \cdot 10^{-1}$. The decimal of the mantissa of the second number is shifted to the left a number of places equal to the difference of the exponents $[1 - (-1) = 2]$, as in

$$0.4381 \cdot 10^{-1} \rightarrow 0.004381 \cdot 10^1$$

Now the numbers can be added,

$$
\begin{array}{r}
0.1557 \quad\ \cdot 10^1 \\
\underline{0.004381 \cdot 10^1} \\
0.160081 \cdot 10^1
\end{array}
$$

and the result chopped to $0.1600 \cdot 10^1$. Notice how the last two digits of the second number that were shifted to the right have essentially been lost from the computation.

Subtraction is performed identically to addition except that the sign of the subtrahend is reversed. For example, suppose that we are subtracting 26.86 from 36.41. That is,

$$
\begin{array}{r}
0.3641 \cdot 10^2 \\
-0.2686 \cdot 10^2 \\
\hline
0.0955 \cdot 10^2
\end{array}
$$

For this case the result is not normalized, and so we must shift the decimal one place to the right to give $0.9550 \cdot 10^1 = 9.550$. Notice that the zero added to the end of the mantissa is not significant but is merely appended to fill the empty space created by the shift. Even more dramatic results would be obtained when the numbers are very close, as in

$$
\begin{array}{r}
0.7642 \cdot 10^3 \\
-0.7641 \cdot 10^3 \\
\hline
0.0001 \cdot 10^3
\end{array}
$$

which would be converted to $0.1000 \cdot 10^0 = 0.1000$. Thus, for this case, three nonsignificant zeros are appended. This introduces a substantial computational error because subsequent manipulations would act as if these zeros were significant. As we will see in a later section, the loss of significance during the subtraction of nearly equal numbers is among the greatest source of round-off error in numerical methods.

Multiplication and division are somewhat more straightforward than addition or subtraction. The exponents are added and the mantissas multiplied. Because multiplication of two n-digit mantissas will yield a $2n$-digit result, most computers hold intermediate results in a double-length register. For example,

$$0.1363 \cdot 10^3 \times 0.6423 \cdot 10^{-1} = 0.08754549 \cdot 10^2$$

If, as in this case, a leading zero is introduced, the result is normalized,

$$0.08754549 \cdot 10^2 \rightarrow 0.8754549 \cdot 10^1$$

and chopped to give

$$0.8754 \cdot 10^1$$

Division is performed in a similar manner, but the mantissas are divided and the exponents are subtracted. Then the results are normalized and chopped.

Large Computations. Certain methods require extremely large numbers of arithmetic manipulations to arrive at their final results. In addition, these computations are often interdependent. That is, the later calculations are dependent on the results of earlier ones. Consequently, even though an individual round-off error could be small, the cumulative effect over the course of a large computation can be significant.

EXAMPLE 3.7 Large Numbers of Interdependent Computations

Problem Statement. Investigate the effect of round-off error on large numbers of interdependent computations. Develop a program to sum a number 100,000 times. Sum the number 1 in single precision, and 0.00001 in single and double precision.

FIGURE 3.12

Fortran 90 program to sum a number 10^5 times. The case sums the number 1 in single precision and the number 10^{-5} in single and double precision.

```
PROGRAM fig0312
IMPLICIT none
INTEGER::i
REAL::sum1, sum2, x1, x2
DOUBLE PRECISION::sum3, x3
sum1=0.
sum2=0.
sum3=0.
x1=1.
x2=1.e-5
x3=1.d-5
DO i=1,100000
  sum1=sum1+x1
  sum2=sum2+x2
  sum3=sum3+x3
END DO
PRINT *, sum1
PRINT *, sum2
PRINT *, sum3
END
output:
100000.000000
      1.000990
  9.999999999980838E-001
```

Solution.   Figure 3.12 shows a Fortran 90 program that performs the summation. Whereas the single-precision summation of 1 yields the expected result, the single-precision summation of 0.00001 yields a large discrepancy. This error is reduced significantly when 0.00001 is summed in double precision.

Quantizing errors are the source of the discrepancies. Because the integer 1 can be represented exactly within the computer, it can be summed exactly. In contrast, 0.00001 cannot be represented exactly and is quantized by a value that is slightly different from its true value. Whereas this very slight discrepancy would be negligible for a small computation, it accumulates after repeated summations. The problem still occurs in double precision but is greatly mitigated because the quantizing error is much smaller.

Note that the type of error illustrated by the previous example is somewhat atypical in that all the errors in the repeated operation are of the same sign. In most cases the errors of a long computation alternate sign in a random fashion and, thus, often cancel out. However, there are also instances where such errors do not cancel but, in fact, lead to a spurious final result. The following sections are intended to provide insight into ways in which this may occur.

Adding a Large and a Small Number.   Suppose we add a small number, 0.0010, to a large number, 4000, using a hypothetical computer with the 4-digit mantissa and the 1-digit exponent. We modify the smaller number so that its exponent matches the larger,

$$
\begin{array}{r}
0.4000 \quad\ \cdot 10^4 \\
0.0000001 \cdot 10^4 \\
\hline
0.4000001 \cdot 10^4
\end{array}
$$

which is chopped to $0.4000 \cdot 10^4$. Thus, we might as well have not performed the addition!

This type of error can occur in the computation of an infinite series. The initial terms in such series are often relatively large in comparison with the later terms. Thus, after a few terms have been added, we are in the situation of adding a small quantity to a large quantity.

One way to mitigate this type of error is to sum the series in reverse order—that is, in ascending rather than descending order. In this way, each new term will be of comparable magnitude to the accumulated sum (see Prob. 3.6).

Subtractive Cancellation. This term refers to the round-off induced when subtracting two nearly equal floating-point numbers.

One common instance where this can occur involves finding the roots of a quadratic equation or parabola with the quadratic formula,

$$\begin{matrix} x_1 \\ x_2 \end{matrix} = \frac{-b \pm \sqrt{b^2 - 4ac}}{2a} \tag{3.12}$$

For cases where $b^2 \gg 4ac$, the difference in the numerator can be very small. In such cases, double precision can mitigate the problem. In addition, an alternative formulation can be used to minimize subtractive cancellation,

$$\begin{matrix} x_1 \\ x_2 \end{matrix} = \frac{-2c}{b \pm \sqrt{b^2 - 4ac}} \tag{3.13}$$

An illustration of the problem and the use of this alternative formula are provided in the following example.

| EXAMPLE 3.8 | **Subtractive Cancellation** |

Problem Statement. Compute the values of the roots of a quadratic equation with $a = 1$, $b = 3000.001$, and $c = 3$. Check the computed values versus the true roots of $x_1 = -0.001$ and $x_2 = -3000$.

Solution. Figure 3.13 shows an Excel/VBA program that computes the roots x_1 and x_2 on the basis of the quadratic formula [(Eq. (3.12)]. Note that both single- and double-precision versions are given. Whereas the results for x_2 are adequate, the percent relative error for x_1 is significant for the single-precision version, $\varepsilon_t = 2.4\%$. This level could be inadequate for many applied engineering problems. This result is particularly surprising because we are employing an analytical formula to obtain our solution!

The loss of significance occurs in the line of both programs where two relatively large numbers are subtracted. Similar problems do not occur when the same numbers are added.

On the basis of the above, we can draw the general conclusion that the quadratic formula will be susceptible to subtractive cancellation whenever $b^2 \gg 4ac$. One way to circumvent this problem is to use double precision. Another is to recast the quadratic formula in the format of Eq. (3.13). As in the program output, both options give a much smaller error because the subtractive cancellation is minimized or avoided.

```
Option Explicit

Sub fig0313()
Dim a As Single, b As Single
Dim c As Single, d As Single
Dim x1 As Single, x2 As Single
Dim x1r As Single
Dim aa As Double, bb As Double
Dim cc As Double, dd As Double
Dim x11 As Double, x22 As Double

'Single precision:
a = 1: b = 3000.001: c = 3
d = Sqr(b * b - 4 * a * c)
x1 = (-b + d) / (2 * a)
x2 = (-b - d) / (2 * a)

'Double precision:
aa = 1: bb = 3000.001: cc = 3
dd = Sqr(bb * bb - 4 * aa * cc)
x11 = (-bb + dd) / (2 * aa)
x22 = (-bb - dd) / (2 * aa)

'Modified formula for first root
'single precision:
x1r = -2 * c / (b + d)
```

```
'Display results
Sheets("sheet1").Select
Range("b2").Select
ActiveCell.Value = x1
ActiveCell.Offset(1, 0).Select
ActiveCell.Value = x2
ActiveCell.Offset(2, 0).Select
ActiveCell.Value = x11
ActiveCell.Offset(1, 0).Select
ActiveCell.Value = x22
ActiveCell.Offset(2, 0).Select
ActiveCell.Value = x1r
End Sub
```

OUTPUT:

	A	B	C
1	Single-precision results:		
2	x1	-0.000976563	
3	x2	-3000.00000000	
4	Double-precision results:		
5	x1	-0.00100000	
6	x2	-3000.00000000	
7	Modified formula for first root (single precision):		
8	x1	-0.00100000	

FIGURE 3.13
Excel/VBA program to determine the roots of a quadratic.

Note that, as in the foregoing example, there are times when subtractive cancellation can be circumvented by using a transformation. However, the only general remedy is to employ extended precision.

Smearing. Smearing occurs whenever the individual terms in a summation are larger than the summation itself. As in the following example, one case where this occurs is in series of mixed signs.

EXAMPLE 3.9 Evaluation of e^x Using Infinite Series

Problem Statement. The exponential function $y = e^x$ is given by the infinite series

$$y = 1 + x + \frac{x^2}{2} + \frac{x^3}{3!} + \cdots$$

Evaluate this function for $x = 10$ and $x = -10$, and be attentive to the problems of round-off error.

Solution. Figure 3.14a gives an Excel/VBA program that uses the infinite series to evaluate e^x. The variable i is the number of terms in the series, *term* is the value of the

(a) Program

```
Option Explicit

Sub fig0314()
Dim term As Single, test As Single
Dim sum As Single, x As Single
Dim i As Integer
i = 0: term = 1#: sum = 1#: test = 0#
Sheets("sheet1").Select
Range("b1").Select
x = ActiveCell.Value
Range("a3:c1003").ClearContents
Range("a3").Select
Do
  If sum = test Then Exit Do
  ActiveCell.Value = i
  ActiveCell.Offset(0, 1).Select
  ActiveCell.Value = term
  ActiveCell.Offset(0, 1).Select
  ActiveCell.Value = sum
  ActiveCell.Offset(1, -2).Select
  i = i + 1
  test = sum
  term = x ^ i / _
    Application.WorksheetFunction.Fact(i)
  sum = sum + term
Loop
ActiveCell.Offset(0, 1).Select
ActiveCell.Value = "Exact value = "
ActiveCell.Offset(0, 1).Select
ActiveCell.Value = Exp(x)
End Sub
```

(b) Evaluation of e^{10}

	A	B	C
1	x		10
2	i	term	sum
3	0	1.000000	1.000000
4	1	10.000000	11.000000
5	2	50.000000	61.000000
6	3	166.666672	227.666672
7	4	416.666656	644.333313
8	5	833.333313	1477.666626
30	27	9.183690E-02	22026.416016
31	28	3.279889E-02	22026.449219
32	29	1.130996E-02	22026.460938
33	30	3.769988E-03	22026.464844
34	31	1.216125E-03	22026.466797
35		Exact value =	22026.465795

(c) Evaluation of e^{-10}

	A	B	C
1	x		-10
2	i	term	sum
3	0	1.000000	1.000000
4	1	-10.000000	-9.000000
5	2	50.000000	41.000000
6	3	-166.666672	-125.666672
7	4	416.666656	291.000000
8	5	-833.333313	-542.333313
44	41	-2.989311E-09	1.103359E-04
45	42	7.117407E-10	1.103366E-04
46	43	-1.655211E-10	1.103365E-04
47	44	3.761843E-11	1.103365E-04
48	45	-8.359651E-12	1.103365E-04
49		Exact value =	4.539993E-05

FIGURE 3.14
(a) An Excel/VBA program to evaluate e^x using an infinite series. (b) Evaluation of e^x.
(c) Evaluation of e^{-x}.

current term added to the series, and *sum* is the accumulative value of the series. The variable *test* is the preceding accumulative value of the series prior to adding *term*. The series is terminated when the computer cannot detect the difference between *test* and *sum*.

Figure 3.14*b* shows the results of running the program for $x = 10$. Note that this case is completely satisfactory. The final result is achieved in 31 terms with the series identical to the library function value within seven significant figures.

Figure 3.14*c* shows similar results for $x = -10$. However, for this case, the results of the series calculation are not even the same sign as the true result. As a matter of fact, the negative results are open to serious question because e^x can never be less than zero. The problem here is caused by round-off error. Note that many of the terms that make up the sum are much larger than the final result of the sum. Furthermore, unlike the previous case, the individual terms vary in sign. Thus, in effect we are adding and subtracting large numbers (each with some small error) and placing great significance on the differences—that is, subtractive cancellation. Thus, we can see that the culprit behind this example of smearing is, in fact, subtractive cancellation. For such cases it is appropriate to seek some other computational strategy. For example, one might try to compute $y = e^{-10}$ as $y = (e^{-1})^{10}$. Other than such a reformulation, the only general recourse is extended precision.

Inner Products. As should be clear from the last sections, some infinite series are particularly prone to round-off error. Fortunately, the calculation of series is not one of the more common operations in numerical methods. A far more ubiquitous manipulation is the calculation of inner products, as in

$$\sum_{i=1}^{n} x_i y_i = x_1 y_1 + x_2 y_2 + \cdots + x_n y_n$$

This operation is very common, particularly in the solution of simultaneous linear algebraic equations. Such summations are prone to round-off error. Consequently, it is often desirable to compute such summations in extended precision.

Although the foregoing sections should provide rules of thumb to mitigate round-off error, they do not provide a direct means beyond trial and error to actually determine the effect of such errors on a computation. In Chap. 4, we will introduce the Taylor series, which will provide a mathematical approach for estimating these effects.

PROBLEMS

3.1 Convert the following base-2 numbers to base-10: **(a)** 1011001, **(b)** 110.0101, and **(c)** 0.01011.
3.2 Convert the following base-8 numbers to base-10: 61,565 and 2.71.
3.3 (a) Convert the following base-6 number to base-10: 1203.504.
(b) Hexadecimal, or base-16, numbers are based on the digits: 0 1 2 3 4 5 6 7 8 9 A B C D E F. Convert the following base-16 number to base-10: 2C.0B7.
3.4 Compose your own program based on Fig. 3.11 and use it to determine your computer's machine epsilon.
3.5 In a fashion similar to that in Fig. 3.11, write a short program to determine the smallest number, x_{min}, used on the computer you will be employing along with this book. Note that your computer will be unable to reliably distinguish between zero and a quantity that is smaller than this number.
3.6 The infinite series

$$f(n) = \sum_{i=1}^{n} \frac{1}{i^4}$$

converges on a value of $f(n) = \pi^4/90$ as n approaches infinity. Write a program in single precision to calculate $f(n)$ for $n = 10,000$ by computing the sum from $i = 1$ to 10,000. Then repeat the calculation but in reverse order—that is, from $i = 10,000$ to 1 using increments of -1. In each case, compute the true percent relative error. Explain the results.
3.7 Evaluate e^{-5} using two approaches,

$$e^{-x} = 1 - x + \frac{x^2}{2} - \frac{x^3}{3!} + \cdots$$

and

$$e^{-x} = \frac{1}{e^x} = \frac{1}{1 + x + \frac{x^2}{2} + \frac{x^3}{3!} + \cdots}$$

and compare with the true value of 6.737947×10^{-3}. Use 20 terms to evaluate each series and compute true and approximate relative errors as terms are added.

3.8 The derivative of $f(x) = 1/(1 - 3x^2)$ is given by

$$\frac{6x}{(1 - 3x^2)^2}$$

Do you expect to have difficulties evaluating this function at $x = 0.577$? Try it using 3- and 4-digit arithmetic with chopping.

3.9 (a) Evaluate the polynomial

$$y = x^3 - 7x^2 + 8x + 0.35$$

at $x = 1.37$. Use 3-digit arithmetic with chopping. Evaluate the percent relative error.
(b) Repeat **(a)** but express y as

$$y = ((x - 7)x + 8)x + 0.35$$

Evaluate the error and compare with part **(a)**.
3.10 Calculate the random access memory (RAM) in megabytes necessary to store a multidimensional array that is $20 \times 40 \times 120$. This array is double precision, and each value requires a 64-bit word. Recall that a 64-bit word = 8 bytes and 1 kilobyte = 2^{10} bytes. Assume that the index starts at 1.

3.11 Determine the number of terms necessary to approximate $\sin x$ to 8 significant figures using the Maclaurin series approximation,

$$\sin x = x - \frac{x^3}{3!} + \frac{x^5}{5!} - \frac{x^7}{7!} + \frac{x^9}{9!} - \cdots$$

Calculate the approximation using a value of $x = 0.3\pi$. Write a program to determine your result.
3.12 Use 5-digit arithmetic with chopping to determine the roots of the following equation with Eqs. (3.12) and (3.13):

$$x^2 - 5000.002x + 10$$

Compute percent relative errors for your results.
3.13 How can the machine epsilon be employed to formulate a stopping criterion ε_s for your programs? Provide an example.
3.14 The "divide and average" method, an old-time method for approximating the square root of any positive number a, can be formulated as

$$x = \frac{x + a/x}{2}$$

Write a well-structured function to implement this algorithm based on the algorithm outlined in Fig. 3.3. Test it for $a = 25$.

Truncation Errors and the Taylor Series

Truncation errors are those that result from using an approximation in place of an exact mathematical procedure. For example, in Chap. 1 we approximated the derivative of velocity of a falling parachutist by a finite-divided-difference equation of the form [Eq. (1.11)]

$$\frac{dv}{dt} \cong \frac{\Delta v}{\Delta t} = \frac{v(t_{i+1}) - v(t_i)}{t_{i+1} - t_i} \tag{4.1}$$

A truncation error was introduced into the numerical solution because the difference equation only approximates the true value of the derivative (recall Fig. 1.4). In order to gain insight into the properties of such errors, we now turn to a mathematical formulation that is used widely in numerical methods to express functions in an approximate fashion—the Taylor series.

4.1 THE TAYLOR SERIES

Taylor's theorem (Box 4.1) and its associated formula, the Taylor series, are of great value in the study of numerical methods. In essence, the *Taylor series* provides a means to predict a function value at one point in terms of the function value and its derivatives at another point. In particular, the theorem states that any smooth function can be approximated as a polynomial.

A useful way to gain insight into the Taylor series is to build it term by term. For example, the first term in the series is

$$f(x_{i+1}) \cong f(x_i) \tag{4.2}$$

This relationship, called the *zero-order approximation,* indicates that the value of f at the new point is the same as its value at the old point. This result makes intuitive sense because if x_i and x_{i+1} are close to each other, it is likely that the new value is probably similar to the old value.

Equation (4.2) provides a perfect estimate if the function being approximated is, in fact, a constant. However, if the function changes at all over the interval, additional terms

---

Box 4.1 Taylor's Theorem

Taylor's Theorem

If the function f and its first $n + 1$ derivatives are continuous on an interval containing a and x, then the value of the function at x is given by

$$f(x) = f(a) + f'(a)(x - a) + \frac{f''(a)}{2!}(x - a)^2$$

$$+ \frac{f^{(3)}(a)}{3!}(x - a)^3 + \cdots$$

$$+ \frac{f^{(n)}(a)}{n!}(x - a)^n + R_n \qquad (B4.1.1)$$

where the remainder R_n is defined as

$$R_n = \int_a^x \frac{(x - t)^n}{n!} f^{(n+1)}(t) dt \qquad (B4.1.2)$$

where $t = $ a dummy variable. Equation (B4.1.1) is called the *Taylor series* or *Taylor's formula.* If the remainder is omitted, the right side of Eq. (B4.1.1) is the Taylor polynomial approximation to $f(x)$. In essence, the theorem states that any smooth function can be approximated as a polynomial.

Equation (B4.1.2) is but one way, called the *integral form,* by which the remainder can be expressed. An alternative formulation can be derived on the basis of the integral mean-value theorem.

First Theorem of Mean for Integrals

If the function g is continuous and integrable on an interval containing a and x, then there exists a point ξ between a and x such that

$$\int_a^x g(t) \, dt = g(\xi)(x - a) \qquad (B4.1.3)$$

In other words, this theorem states that the integral can be represented by an average value for the function $g(\xi)$ times the interval length $x - a$. Because the average must occur between the minimum and maximum values for the interval, there is a point $x = \xi$ at which the function takes on the average value.

The first theorem is in fact a special case of a second mean-value theorem for integrals.

Second Theorem of Mean for Integrals

If the functions g and h are continuous and integrable on an interval containing a and x, and h does not change sign in the interval, then there exists a point ξ between a and x such that

$$\int_a^x g(t)h(t) dt = g(\xi) \int_a^x h(t) \, dt \qquad (B4.1.4)$$

Thus, Eq. (B4.1.3) is equivalent to Eq. (B4.1.4) with $h(t) = 1$.

The second theorem can be applied to Eq. (B4.1.2) with

$$g(t) = f^{(n+1)}(t) \qquad h(t) = \frac{(x - t)^n}{n!}$$

As t varies from a to x, $h(t)$ is continuous and does not change sign. Therefore, if $f^{(n+1)}(t)$ is continuous, then the integral mean-value theorem holds and

$$R_n = \frac{f^{(n+1)}(\xi)}{(n + 1)!}(x - a)^{n+1}$$

This equation is referred to as the *derivative,* or *Lagrange, form* of the remainder.

---

of the Taylor series are required to provide a better estimate. For example, the *first-order approximation* is developed by adding another term to yield

$$f(x_{i+1}) \cong f(x_i) + f'(x_i)(x_{i+1} - x_i) \qquad (4.3)$$

The additional first-order term consists of a slope $f'(x_i)$ multiplied by the difference between x_{i+1} and x_i. Thus, the expression is now in the form of a straight line and is capable of predicting an increase or decrease of the function between x_i and x_{i+1}.

Although Eq. (4.3) can predict a change, it is exact only for a straight-line, or *linear,* trend. Therefore, a *second-order* term is added to the series to capture some of the curvature that the function might exhibit:

$$f(x_{i+1}) \cong f(x_i) + f'(x_i)(x_{i+1} - x_i) + \frac{f''(x_i)}{2!} (x_{i+1} - x_i)^2 \qquad (4.4)$$

In a similar manner, additional terms can be included to develop the complete Taylor series expansion:

$$f(x_{i+1}) = f(x_i) + f'(x_i)(x_{i+1} - x_i) + \frac{f''(x_i)}{2!}(x_{i+1} - x_i)^2$$

$$+ \frac{f^{(3)}(x_i)}{3!}(x_{i+1} - x_i)^3 + \cdots + \frac{f^{(n)}(x_i)}{n!}(x_{i+1} - x_i)^n + R_n \qquad (4.5)$$

Note that because Eq. (4.5) is an infinite series, an equal sign replaces the approximate sign that was used in Eqs. (4.2) through (4.4). A remainder term is included to account for all terms from $n + 1$ to infinity:

$$R_n = \frac{f^{(n+1)}(\xi)}{(n+1)!}(x_{i+1} - x_i)^{n+1} \qquad (4.6)$$

where the subscript n connotes that this is the remainder for the nth-order approximation and ξ is a value of x that lies somewhere between x_i and x_{i+1}. The introduction of the ξ is so important that we will devote an entire section (Sec. 4.1.1) to its derivation. For the time being, it is sufficient to recognize that there is such a value that provides an exact determination of the error.

It is often convenient to simplify the Taylor series by defining a step size $h = x_{i+1} - x_i$ and expressing Eq. (4.5) as

$$f(x_{i+1}) = f(x_i) + f'(x_i)h + \frac{f''(x_i)}{2!}h^2 + \frac{f^{(3)}(x_i)}{3!}h^3 + \cdots + \frac{f^n(x_i)}{n!}h^n + R_n \qquad (4.7)$$

where the remainder term is now

$$R_n = \frac{f^{(n+1)}(\xi)}{(n+1)!}h^{n+1} \qquad (4.8)$$

EXAMPLE 4.1   **Taylor Series Approximation of a Polynomial**

Problem Statement.   Use zero- through fourth-order Taylor series expansions to approximate the function

$$f(x) = -0.1x^4 - 0.15x^3 - 0.5x^2 - 0.25x + 1.2$$

from $x_i = 0$ with $h = 1$. That is, predict the function's value at $x_{i+1} = 1$.

Solution.   Because we are dealing with a known function, we can compute values for $f(x)$ between 0 and 1. The results (Fig. 4.1) indicate that the function starts at $f(0) = 1.2$ and then curves downward to $f(1) = 0.2$. Thus, the true value that we are trying to predict is 0.2.

The Taylor series approximation with $n = 0$ is [Eq. (4.2)]

$$f(x_{i+1}) \cong 1.2$$

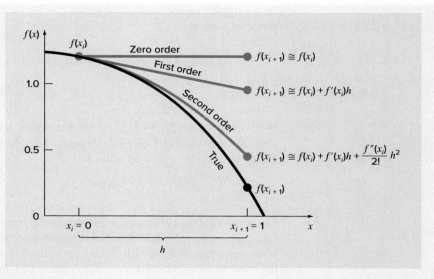

FIGURE 4.1
The approximation of $f(x) = -0.1x^4 - 0.15x^3 - 0.5x^2 - 0.25x + 1.2$ at $x = 1$ by zero-order, first-order, and second-order Taylor series expansions.

Thus, as in Fig. 4.1, the zero-order approximation is a constant. Using this formulation results in a truncation error [recall Eq. (3.2)] of

$$E_t = 0.2 - 1.2 = -1.0$$

at $x = 1$.

For $n = 1$, the first derivative must be determined and evaluated at $x = 0$:

$$f'(0) = -0.4(0.0)^3 - 0.45(0.0)^2 - 1.0(0.0) - 0.25 = -0.25$$

Therefore, the first-order approximation is [Eq. (4.3)]

$$f(x_{i+1}) \cong 1.2 - 0.25h$$

which can be used to compute $f(1) = 0.95$. Consequently, the approximation begins to capture the downward trajectory of the function in the form of a sloping straight line (Fig. 4.1). This results in a reduction of the truncation error to

$$E_t = 0.2 - 0.95 = -0.75$$

For $n = 2$, the second derivative is evaluated at $x = 0$:

$$f''(0) = -1.2(0.0)^2 - 0.9(0.0) - 1.0 = -1.0$$

Therefore, according to Eq. (4.4),

$$f(x_{i+1}) \cong 1.2 - 0.25h - 0.5h^2$$

and substituting $h = 1$, $f(1) = 0.45$. The inclusion of the second derivative now adds some downward curvature resulting in an improved estimate, as seen in Fig. 4.1. The truncation error is reduced further to $0.2 - 0.45 = -0.25$.

Additional terms would improve the approximation even more. In fact, the inclusion of the third and the fourth derivatives results in exactly the same equation we started with:

$$f(x) = 1.2 - 0.25h - 0.5h^2 - 0.15h^3 - 0.1h^4$$

where the remainder term is

$$R_4 = \frac{f^{(5)}(\xi)}{5!} h^5 = 0$$

because the fifth derivative of a fourth-order polynomial is zero. Consequently, the Taylor series expansion to the fourth derivative yields an exact estimate at $x_{i+1} = 1$:

$$f(1) = 1.2 - 0.25(1) - 0.5(1)^2 - 0.15(1)^3 - 0.1(1)^4 = 0.2$$

In general, the nth-order Taylor series expansion will be exact for an nth-order polynomial. For other differentiable and continuous functions, such as exponentials and sinusoids, a finite number of terms will not yield an exact estimate. Each additional term will contribute some improvement, however slight, to the approximation. This behavior will be demonstrated in Example 4.2. Only if an infinite number of terms are added will the series yield an exact result.

Although the above is true, the practical value of Taylor series expansions is that, in most cases, the inclusion of only a few terms will result in an approximation that is close enough to the true value for practical purposes. The assessment of how many terms are required to get "close enough" is based on the remainder term of the expansion. Recall that the remainder term is of the general form of Eq. (4.8). This relationship has two major drawbacks. First, ξ is not known exactly but merely lies somewhere between x_i and x_{i+1}. Second, to evaluate Eq. (4.8), we need to determine the $(n + 1)$th derivative of $f(x)$. To do this, we need to know $f(x)$. However, if we knew $f(x)$, there would be no need to perform the Taylor series expansion in the present context!

Despite this dilemma, Eq. (4.8) is still useful for gaining insight into truncation errors. This is because we *do* have control over the term h in the equation. In other words, we can choose how far away from x we want to evaluate $f(x)$, and we can control the number of terms we include in the expansion. Consequently, Eq. (4.8) is usually expressed as

$$R_n = O(h^{n+1})$$

where the nomenclature $O(h^{n+1})$ means that the truncation error is of the order of h^{n+1}. That is, the error is proportional to the step size h raised to the $(n + 1)$th power. Although this approximation implies nothing regarding the magnitude of the derivatives that multiply h^{n+1}, it is extremely useful in judging the comparative error of numerical methods based on Taylor series expansions. For example, if the error is $O(h)$, halving the step size will halve the error. On the other hand, if the error is $O(h^2)$, halving the step size will quarter the error.

In general, we can usually assume that the truncation error is decreased by the addition of terms to the Taylor series. In many cases, if h is sufficiently small, the first- and other lower-order terms usually account for a disproportionately high percent of the error. Thus, only a few terms are required to obtain an adequate estimate. This property is illustrated by the following example.

EXAMPLE 4.2 **Use of Taylor Series Expansion to Approximate a Function with an Infinite Number of Derivatives**

Problem Statement. Use Taylor series expansions with $n = 0$ to 6 to approximate $f(x) = \cos x$ at $x_{i+1} = \pi/3$ on the basis of the value of $f(x)$ and its derivatives at $x_i = \pi/4$. Note that this means that $h = \pi/3 - \pi/4 = \pi/12$.

Solution. As with Example 4.1, our knowledge of the true function means that we can determine the correct value, $f(\pi/3) = 0.5$.

The zero-order approximation is [Eq. (4.3)]

$$f\left(\frac{\pi}{3}\right) \cong \cos\left(\frac{\pi}{4}\right) = 0.707106781$$

which represents a percent relative error of

$$\varepsilon_t = \frac{0.5 - 0.707106781}{0.5}100\% = -41.4\%$$

For the first-order approximation, we add the first derivative term where $f'(x) = -\sin x$:

$$f\left(\frac{\pi}{3}\right) \cong \cos\left(\frac{\pi}{4}\right) - \sin\left(\frac{\pi}{4}\right)\left(\frac{\pi}{12}\right) = 0.521986659$$

which has $\varepsilon_t = -4.40\%$.

For the second-order approximation, we add the second derivative term where $f''(x) = -\cos x$:

$$f\left(\frac{\pi}{3}\right) \cong \cos\left(\frac{\pi}{4}\right) - \sin\left(\frac{\pi}{4}\right)\left(\frac{\pi}{12}\right) - \frac{\cos(\pi/4)}{2}\left(\frac{\pi}{12}\right)^2 = 0.497754491$$

with $\varepsilon_t = 0.449\%$. Thus, the inclusion of additional terms results in an improved estimate.

The process can be continued and the results listed, as in Table 4.1. Notice that the derivatives never go to zero, as was the case with the polynomial in Example 4.1. Therefore, each additional term results in some improvement in the estimate. However, also notice how most of the improvement comes with the initial terms. For this case, by the time we have added the third-order term, the error is reduced to 2.62×10^{-2} percent,

TABLE 4.1 Taylor series approximation of $f(x) = \cos x$ at $x_{i+1} = \pi/3$ using a base point of $\pi/4$. Values are shown for various orders (n) of approximation.

Order n	$f^{(n)}(x)$	$f(\pi/3)$	ε_t
0	$\cos x$	0.707106781	−41.4
1	$-\sin x$	0.521986659	−4.4
2	$-\cos x$	0.497754491	0.449
3	$\sin x$	0.499869147	2.62×10^{-2}
4	$\cos x$	0.500007551	-1.51×10^{-3}
5	$-\sin x$	0.500000304	-6.08×10^{-5}
6	$-\cos x$	0.499999988	2.44×10^{-6}

which means that we have attained 99.9738% of the true value. Consequently, although the addition of more terms will reduce the error further, the improvement becomes negligible.

4.1.1 The Remainder for the Taylor Series Expansion

Before demonstrating how the Taylor series is actually used to estimate numerical errors, we must explain why we included the argument ξ in Eq. (4.8). A mathematical derivation is presented in Box 4.1. We will now develop an alternative exposition based on a somewhat more visual interpretation. Then we can extend this specific case to the more general formulation.

Suppose that we truncated the Taylor series expansion [Eq. (4.7)] after the zero-order term to yield

$$f(x_{i+1}) \cong f(x_i)$$

A visual depiction of this zero-order prediction is shown in Fig. 4.2. The remainder, or error, of this prediction, which is also shown in the illustration, consists of the infinite series of terms that were truncated:

$$R_0 = f'(x_i)h + \frac{f''(x_i)}{2!}h^2 + \frac{f^{(3)}(x_i)}{3!}h^3 + \cdots$$

It is obviously inconvenient to deal with the remainder in this infinite series format. One simplification might be to truncate the remainder itself, as in

$$R_0 \cong f'(x_i)h \tag{4.9}$$

FIGURE 4.2
Graphical depiction of a zero-order Taylor series prediction and remainder.

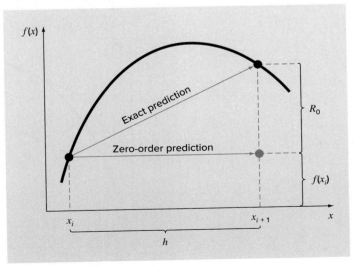

Although, as stated in the previous section, lower-order derivatives usually account for a greater share of the remainder than the higher-order terms, this result is still inexact because of the neglected second- and higher-order terms. This "inexactness" is implied by the approximate equality symbol ($\cong$) employed in Eq. (4.9).

An alternative simplification that transforms the approximation into an equivalence is based on a graphical insight. As in Fig. 4.3, the *derivative mean-value theorem* states that if a function $f(x)$ and its first derivative are continuous over an interval from x_i to x_{i+1}, then there exists at least one point on the function that has a slope, designated by $f'(\xi)$, that is parallel to the line joining $f(x_i)$ and $f(x_{i+1})$. The parameter ξ marks the x value where this slope occurs (Fig. 4.3). A physical illustration of this theorem is that, if you travel between two points with an average velocity, there will be at least one moment during the course of the trip when you will be moving at that average velocity.

By invoking this theorem, it is simple to realize that, as illustrated in Fig. 4.3, the slope $f'(\xi)$ is equal to the rise R_0 divided by the run h, or

$$f'(\xi) = \frac{R_0}{h}$$

which can be rearranged to give

$$R_0 = f'(\xi)h \tag{4.10}$$

Thus, we have derived the zero-order version of Eq. (4.8). The higher-order versions are merely a logical extension of the reasoning used to derive Eq. (4.10). The first-order version is

$$R_1 = \frac{f''(\xi)}{2!}h^2 \tag{4.11}$$

FIGURE 4.3
Graphical depiction of the derivative mean-value theorem.

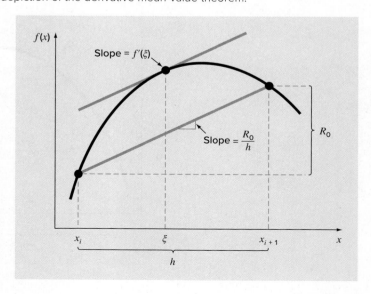

For this case, the value of ξ conforms to the x value corresponding to the second derivative that makes Eq. (4.11) exact. Similar higher-order versions can be developed from Eq. (4.8).

4.1.2 Using the Taylor Series to Estimate Truncation Errors

Although the Taylor series will be extremely useful in estimating truncation errors throughout this book, it may not be clear to you how the expansion can actually be applied to numerical methods. In fact, we have already done so in our example of the falling parachutist. Recall that the objective of both Examples 1.1 and 1.2 was to predict velocity as a function of time. That is, we were interested in determining $v(t)$. As specified by Eq. (4.5), $v(t)$ can be expanded in a Taylor series:

$$v(t_{i+1}) = v(t_i) + v'(t_i)(t_{i+1} - t_i) + \frac{v''(t_i)}{2!}(t_{i+1} - t_i)^2 + \cdots + R_n \tag{4.12}$$

Now let us truncate the series after the first derivative term:

$$v(t_{i+1}) = v(t_i) + v'(t_i)(t_{i+1} - t_i) + R_1 \tag{4.13}$$

Equation (4.13) can be solved for

$$v'(t_i) = \underbrace{\frac{v(t_{i+1}) - v(t_i)}{t_{i+1} - t_i}}_{\substack{\text{First-order} \\ \text{approximation}}} - \underbrace{\frac{R_1}{t_{i+1} - t_i}}_{\substack{\text{Truncation} \\ \text{error}}} \tag{4.14}$$

The first part of Eq. (4.14) is exactly the same relationship that was used to approximate the derivative in Example 1.2 [Eq. (1.11)]. However, because of the Taylor series approach, we have now obtained an estimate of the truncation error associated with this approximation of the derivative. Using Eqs. (4.6) and (4.14) yields

$$\frac{R_1}{t_{i+1} - t_i} = \frac{v''(\xi)}{2!}(t_{i+1} - t_i) \tag{4.15}$$

or

$$\frac{R_1}{t_{i+1} - t_i} = O(t_{i+1} - t_i) \tag{4.16}$$

Thus, the estimate of the derivative [Eq. (1.11) or the first part of Eq. (4.14)] has a truncation error of order $t_{i+1} - t_i$. In other words, the error of our derivative approximation should be proportional to the step size. Consequently, if we halve the step size, we would expect to halve the error of the derivative.

EXAMPLE 4.3

The Effect of Nonlinearity and Step Size on the Taylor Series Approximation

Problem Statement.   Figure 4.4 is a plot of the function

$$f(x) = x^m \tag{E4.3.1}$$

for $m = 1, 2, 3,$ and 4 over the range from $x = 1$ to 2. Notice that for $m = 1$ the function is linear, and as m increases, more curvature or nonlinearity is introduced into the function.

FIGURE 4.4
Plot of the function $f(x) = x^m$ for $m = 1, 2, 3,$ and 4. Notice that the function becomes more nonlinear as m increases.

Employ the first-order Taylor series to approximate this function for various values of the exponent m and the step size h.

Solution.   Equation (E4.3.1) can be approximated by a first-order Taylor series expansion, as in

$$f(x_{i+1}) = f(x_i) + mx_i^{m-1}h \qquad\qquad (E4.3.2)$$

which has a remainder

$$R_1 = \frac{f''(x_i)}{2!}h^2 + \frac{f^{(3)}(x_i)}{3!}h^3 + \frac{f^{(4)}(x_i)}{4!}h^4 + \cdots$$

First, we can examine how the approximation performs as m increases—that is, as the function becomes more nonlinear. For $m = 1$, the actual value of the function at $x = 2$ is 2.

The Taylor series yields

$$f(2) = 1 + 1(1) = 2$$

and

$$R_1 = 0$$

The remainder is zero because the second and higher derivatives of a linear function are zero. Thus, as expected, the first-order Taylor series expansion is perfect when the underlying function is linear.

For $m = 2$, the actual value is $f(2) = 2^2 = 4$. The first-order Taylor series approximation is

$$f(2) = 1 + 2(1) = 3$$

and

$$R_1 = \tfrac{2}{2}(1)^2 + 0 + 0 + \cdots = 1$$

Thus, because the function is a parabola, the straight-line approximation results in a discrepancy. Note that the remainder is determined exactly.

For $m = 3$, the actual value is $f(2) = 2^3 = 8$. The Taylor series approximation is

$$f(2) = 1 + 3(1)^2(1) = 4$$

and

$$R_1 = \tfrac{6}{2}(1)^2 + \tfrac{6}{6}(1)^3 + 0 + 0 + \cdots = 4$$

Again, there is a discrepancy that can be determined exactly from the Taylor series.

For $m = 4$, the actual value is $f(2) = 2^4 = 16$. The Taylor series approximation is

$$f(2) = 1 + 4(1)^3(1) = 5$$

and

$$R_1 = \tfrac{12}{2}(1)^2 + \tfrac{24}{6}(1)^3 + \tfrac{24}{24}(1)^4 + 0 + 0 + \cdots = 11$$

On the basis of these four cases, we observe that R_1 increases as the function becomes more nonlinear. Furthermore, R_1 accounts exactly for the discrepancy. This is because Eq. (E4.3.1) is a simple monomial with a finite number of derivatives. This permits a complete determination of the Taylor series remainder.

Next, we will examine Eq. (E4.3.2) for the case $m = 4$ and observe how R_1 changes as the step size h is varied. For $m = 4$, Eq. (E4.3.2) is

$$f(x + h) = f(x) + 4x_i^3 h$$

If $x = 1$, $f(1) = 1$ and this equation can be expressed as

$$f(1 + h) = 1 + 4h$$

with a remainder of

$$R_1 = 6h^2 + 4h^3 + h^4$$

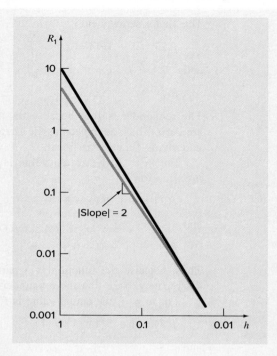

FIGURE 4.5
Log-log plot of the remainder R_1 of the first-order Taylor series approximation of the function $f(x) = x^4$ versus step size h. A line with a slope of 2 is also shown to indicate that as h decreases, the error becomes proportional to h^2.

TABLE 4.2   Comparison of the exact value of the function $f(x) = x^4$ with the first-order Taylor series approximation. Both the function and the approximation are evaluated at $x + h$, where $x = 1$.

h	Exact Value	First-Order Approximation	R_1
1	16	5	11
0.5	5.0625	3	2.0625
0.25	2.441406	2	0.441406
0.125	1.601807	1.5	0.101807
0.0625	1.274429	1.25	0.024429
0.03125	1.130982	1.125	0.005982
0.015625	1.063980	1.0625	0.001480

This leads to the conclusion that the discrepancy will decrease as h is reduced. Also, at sufficiently small values of h, the error should become proportional to h^2. That is, as h is halved, the error will be quartered. This behavior is confirmed by Table 4.2 and Fig. 4.5.

Thus, we conclude that the error of the first-order Taylor series approximation decreases as m approaches 1 and as h decreases. Intuitively, this means that the Taylor series becomes more accurate when the function we are approximating becomes more like a straight line over the interval of interest. This can be accomplished either by reducing the size of the interval or by "straightening" the function by reducing m. Obviously, the latter option is usually not available in the real world because the functions we analyze are typically dictated by the physical problem context. Consequently, we do not have control of their lack of linearity, and our only recourse is reducing the step size or including additional terms in the Taylor series expansion.

4.1.3 Numerical Differentiation

Equation (4.14) is given a formal label in numerical methods—it is called a *finite divided difference*. It can be represented generally as

$$f'(x_i) = \frac{f(x_{i+1}) - f(x_i)}{x_{i+1} - x_i} + O(x_{i+1} - x_i) \tag{4.17}$$

or

$$f'(x_i) = \frac{\Delta f_i}{h} + O(h) \tag{4.18}$$

where Δf_i is referred to as the *first forward difference* and h is called the step size, that is, the length of the interval over which the approximation is made. It is termed a "forward" difference because it utilizes data at i and $i + 1$ to estimate the derivative (Fig. 4.6a). The entire term $\Delta f/h$ is referred to as a *first finite divided difference*.

This forward divided difference is but one of many that can be developed from the Taylor series to approximate derivatives numerically. For example, backward and centered difference approximations of the first derivative can be developed in a fashion similar to the derivation of Eq. (4.14). The former utilizes values at x_{i-1} and x_i (Fig. 4.6b), whereas the latter uses values that are equally spaced around the point at which the derivative is estimated (Fig. 4.6c). More accurate approximations of the first derivative can be developed by including higher-order terms of the Taylor series. Finally, all the above versions can also be developed for second, third, and higher derivatives. The following sections provide brief summaries illustrating how some of these cases are derived.

Backward Difference Approximation of the First Derivative.   The Taylor series can be expanded backward to calculate a previous value on the basis of a present value, as in

$$f(x_{i-1}) = f(x_i) - f'(x_i)h + \frac{f''(x_i)}{2!}h^2 - \cdots \tag{4.19}$$

Truncating this equation after the first derivative and rearranging yields

$$f'(x_i) \cong \frac{f(x_i) - f(x_{i-1})}{h} = \frac{\nabla f_i}{h} \tag{4.20}$$

where the error is $O(h)$, and ∇f_i is referred to as the *first backward difference*. See Fig. 4.6b for a graphical representation.

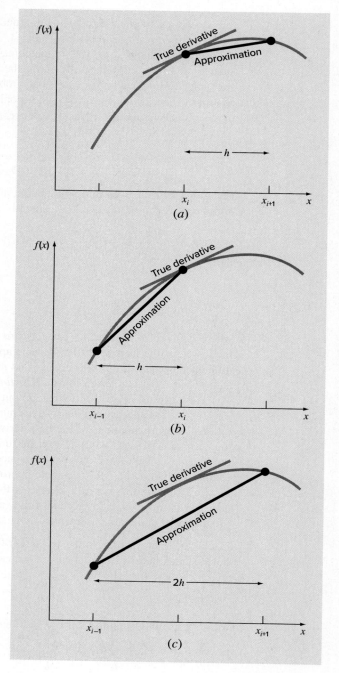

FIGURE 4.6
Graphical depiction of (*a*) forward, (*b*) backward, and (*c*) centered finite-divided-difference approximations of the first derivative.

Centered Difference Approximation of the First Derivative. A third way to approximate the first derivative is to subtract Eq. (4.19) from the forward Taylor series expansion:

$$f(x_{i+1}) = f(x_i) + f'(x_i)h + \frac{f''(x_i)}{2!}h^2 + \cdots \tag{4.21}$$

to yield

$$f(x_{i+1}) = f(x_{i-1}) + 2f'(x_i)h + \frac{2f^{(3)}(x_i)}{3!}h^3 + \cdots$$

which can be solved for

$$f'(x_i) = \frac{f(x_{i+1}) - f(x_{i-1})}{2h} - \frac{f^{(3)}(x_i)}{6}h^2 - \cdots$$

or

$$f'(x_i) = \frac{f(x_{i+1}) - f(x_{i-1})}{2h} - O(h^2) \tag{4.22}$$

Equation (4.22) is a *centered difference* representation of the first derivative. Notice that the truncation error is of the order of h^2 in contrast to the forward and backward approximations that were of the order of h. Consequently, the Taylor series analysis yields the practical information that the centered difference is a more accurate representation of the derivative (Fig. 4.6c). For example, if we halve the step size using a forward or backward difference, we would approximately halve the truncation error, whereas for the central difference, the error would be quartered.

EXAMPLE 4.4 Finite-Divided-Difference Approximations of Derivatives

Problem Statement. Use forward and backward difference approximations of $O(h)$ and a centered difference approximation of $O(h^2)$ to estimate the first derivative of

$$f(x) = -0.1x^4 - 0.15x^3 - 0.5x^2 - 0.25x + 1.25$$

at $x = 0.5$ using a step size of $h = 0.5$. Repeat the computation using $h = 0.25$. Note that the derivative can be calculated directly as

$$f'(x) = -0.4x^3 - 0.45x^2 - 1.0x - 0.25$$

and can be used to compute the true value as $f'(0.5) = -0.9125$.

Solution. For $h = 0.5$, the function can be employed to determine

$$
\begin{aligned}
x_{i-1} &= 0 & f(x_{i-1}) &= 1.2 \\
x_i &= 0.5 & f(x_i) &= 0.925 \\
x_{i+1} &= 1.0 & f(x_{i+1}) &= 0.2
\end{aligned}
$$

These values can be used to compute the forward divided difference [Eq. (4.17)],

$$f'(0.5) \cong \frac{0.2 - 0.925}{0.5} = -1.45 \qquad |\varepsilon_t| = 58.9\%$$

the backward divided difference [Eq. (4.20)],

$$f'(0.5) \cong \frac{0.925 - 1.2}{0.5} = -0.55 \qquad |\varepsilon_t| = 39.7\%$$

and the centered divided difference [Eq. (4.22)],

$$f'(0.5) \cong \frac{0.2 - 1.2}{1.0} = -1.0 \qquad |\varepsilon_t| = 9.6\%$$

For $h = 0.25$,

$$x_{i-1} = 0.25 \qquad f(x_{i-1}) = 1.10351563$$
$$x_i = 0.5 \qquad f(x_i) = 0.925$$
$$x_{i+1} = 0.75 \qquad f(x_{i+1}) = 0.63632813$$

which can be used to compute the forward divided difference,

$$f'(0.5) \cong \frac{0.63632813 - 0.925}{0.25} = -1.155 \qquad |\varepsilon_t| = 26.5\%$$

the backward divided difference,

$$f'(0.5) \cong \frac{0.925 - 1.10351563}{0.25} = -0.714 \qquad |\varepsilon_t| = 21.7\%$$

and the centered divided difference,

$$f'(0.5) \cong \frac{0.63632813 - 1.10351563}{0.5} = -0.934 \qquad |\varepsilon_t| = 2.4\%$$

For both step sizes, the centered difference approximation is more accurate than forward or backward differences. Also, as predicted by the Taylor series analysis, halving the step size approximately halves the error of the backward and forward differences and quarters the error of the centered difference.

Finite Difference Approximations of Higher Derivatives.   Besides first derivatives, the Taylor series expansion can be used to derive numerical estimates of higher derivatives. To do this, we write a forward Taylor series expansion for $f(x_{i+2})$ in terms of $f(x_i)$:

$$f(x_{i+2}) = f(x_i) + f'(x_i)(2h) + \frac{f''(x_i)}{2!}(2h)^2 + \cdots \tag{4.23}$$

Equation (4.21) can be multiplied by 2 and subtracted from Eq. (4.23) to give

$$f(x_{i+2}) - 2f(x_{i+1}) = -f(x_i) + f''(x_i)h^2 + \cdots$$

which can be solved for

$$f''(x_i) = \frac{f(x_{i+2}) - 2f(x_{i+1}) + f(x_i)}{h^2} + O(h) \tag{4.24}$$

This relationship is called the *second forward finite divided difference*. Similar manipulations can be employed to derive a backward version,

$$f''(x_i) = \frac{f(x_i) - 2f(x_{i-1}) + f(x_{i-2})}{h^2} + O(h)$$

and a centered version,

$$f''(x_i) = \frac{f(x_{i+1}) - 2f(x_i) + f(x_{i-1})}{h^2} + O(h^2)$$

As was the case with the first-derivative approximations, the centered difference is more accurate. Notice also that the centered version can be alternatively expressed as

$$f''(x_i) \cong \frac{\dfrac{f(x_{i+1}) - f(x_i)}{h} - \dfrac{f(x_i) - f(x_{i-1})}{h}}{h}$$

Thus, just as the second derivative is a derivative of a derivative, the second divided difference approximation is a difference of two first divided differences.

We will return to the topic of numerical differentiation in Chap. 23. We have introduced you to the topic at this point because it is a very good example of why the Taylor series is important in numerical methods. In addition, several of the formulas introduced in this section will be employed prior to Chap. 23.

4.2 ERROR PROPAGATION

The purpose of this section is to study how errors in numbers can propagate through mathematical functions. For example, if we multiply two numbers that have errors, we would like to estimate the error in the product.

4.2.1 Functions of a Single Variable

Suppose that we have a function $f(x)$ that is dependent on a single independent variable x. Assume that $\tilde{x}$ is an approximation of x. We, therefore, would like to assess the effect of the discrepancy between x and $\tilde{x}$ on the value of the function. That is, we would like to estimate

$$\Delta f(\tilde{x}) = |f(x) - f(\tilde{x})|$$

The problem with evaluating $\Delta f(\tilde{x})$ is that $f(x)$ is unknown because x is unknown. We can overcome this difficulty if $\tilde{x}$ is close to x and $f(\tilde{x})$ is continuous and differentiable. If these conditions hold, a Taylor series can be employed to compute $f(x)$ near $f(\tilde{x})$, as in

$$f(x) = f(\tilde{x}) + f'(\tilde{x})(x - \tilde{x}) + \frac{f''(\tilde{x})}{2}(x - \tilde{x})^2 + \cdots$$

Dropping the second- and higher-order terms and rearranging yields

$$f(x) - f(\tilde{x}) \cong f'(\tilde{x})(x - \tilde{x})$$

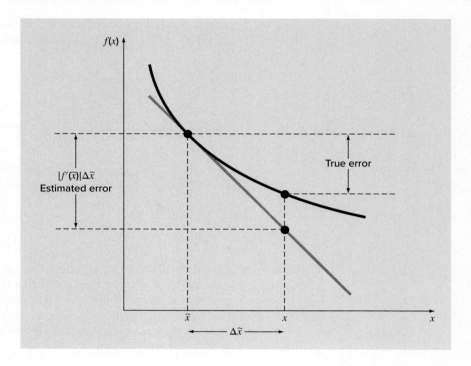

FIGURE 4.7
Graphical depiction of first-order error propagation.

or

$$\Delta f(\widetilde{x}) = |f'(\widetilde{x})|\Delta\widetilde{x} \tag{4.25}$$

where $\Delta f(\widetilde{x}) = |f(x) - f(\widetilde{x})|$ represents an estimate of the error of the function and $\Delta\widetilde{x} = |x - \widetilde{x}|$ represents an estimate of the error of x. Equation (4.25) provides the capability to approximate the error in $f(x)$ given the derivative of a function and an estimate of the error in the independent variable. Figure 4.7 is a graphical illustration of the operation.

EXAMPLE 4.5   **Error Propagation in a Function of a Single Variable**

Problem Statement.   Given a value of $\widetilde{x} = 2.5$ with an error of $\Delta\widetilde{x} = 0.01$, estimate the resulting error in the function $f(x) = x^3$.

Solution.   Using Eq. (4.25),

$$\Delta f(\widetilde{x}) \cong 3(2.5)^2(0.01) = 0.1875$$

Because $f(2.5) = 15.625$, we predict that

$$f(2.5) = 15.625 \pm 0.1875$$

or that the true value lies between 15.4375 and 15.8125. In fact, if x were actually 2.49, the function could be evaluated as 15.4382, and if x were 2.51, it would be 15.8132. For this case, the first-order error analysis provides a fairly close estimate of the true error.

4.2.2 Functions of More Than One Variable

The foregoing approach can be generalized to functions that are dependent on more than one independent variable. This is accomplished with a multivariable version of the Taylor series. For example, if we have a function of two independent variables u and v, the Taylor series can be written as

$$
f(u_{i+1}, v_{i+1}) = f(u_i, v_i) + \frac{\partial f}{\partial u}(u_{i+1} - u_i) + \frac{\partial f}{\partial v}(v_{i+1} - v_i)
$$

$$
+ \frac{1}{2!}\left[\frac{\partial^2 f}{\partial u^2}(u_{i+1} - u_i)^2 + 2\frac{\partial^2 f}{\partial u \partial v}(u_{i+1} - u_i)(v_{i+1} - v_i) \right.
$$

$$
\left. + \frac{\partial^2 f}{\partial v^2}(v_{i+1} - v_i)^2 \right] + \cdots \tag{4.26}
$$

where all partial derivatives are evaluated at the base point i. If all second-order and higher terms are dropped, Eq. (4.26) can be solved for

$$
\Delta f(\tilde{u}, \tilde{v}) = \left| \frac{\partial f}{\partial u} \right| \Delta \tilde{u} + \left| \frac{\partial f}{\partial v} \right| \Delta \tilde{v}
$$

where $\Delta \tilde{u}$ and $\Delta \tilde{v}$ = estimates of the errors in u and v, respectively.

For n independent variables $\tilde{x}_1, \tilde{x}_2, \ldots, \tilde{x}_n$ having errors $\Delta \tilde{x}_1, \Delta \tilde{x}_2, \ldots, \Delta \tilde{x}_n$, the following general relationship holds:

$$
\Delta f(\tilde{x}_1, \tilde{x}_2, \ldots, \tilde{x}_n) \cong \left| \frac{\partial f}{\partial x_1} \right| \Delta \tilde{x}_1 + \left| \frac{\partial f}{\partial x_2} \right| \Delta \tilde{x}_2 + \cdots + \left| \frac{\partial f}{\partial x_n} \right| \Delta \tilde{x}_n \tag{4.27}
$$

EXAMPLE 4.6   **Error Propagation in a Multivariable Function**

Problem Statement.   The deflection y of the top of a sailboat mast is

$$
y = \frac{FL^4}{8EI}
$$

where F = a uniform side loading (N/m), L = height (m), E = the modulus of elasticity (N/m^2), and I = the moment of inertia (m^4). Estimate the error in y given the following data:

$\tilde{F} = 750$ N/m	$\Delta \tilde{F} = 30$ N/m
$\tilde{L} = 9$ m	$\Delta \tilde{L} = 0.03$ m
$\tilde{E} = 7.5 \times 10^9$ N/m^2	$\Delta \tilde{E} = 5 \times 10^7$ N/m^2
$\tilde{I} = 0.0005$ m^4	$\Delta \tilde{I} = 0.000005$ m^4

Solution.   Employing Eq. (4.27) gives

$$
\Delta y(\tilde{F}, \tilde{L}, \tilde{E}, \tilde{I}) = \left| \frac{\partial y}{\partial F} \right| \Delta \tilde{F} + \left| \frac{\partial y}{\partial L} \right| \Delta \tilde{L} + \left| \frac{\partial y}{\partial E} \right| \Delta \tilde{E} + \left| \frac{\partial y}{\partial I} \right| \Delta \tilde{I}
$$

or

$$
\Delta y(\tilde{F}, \tilde{L}, \tilde{E}, \tilde{I}) \cong \frac{\tilde{L}^4}{8\tilde{E}\tilde{I}} \Delta \tilde{F} + \frac{\tilde{F}\tilde{L}^3}{2\tilde{E}\tilde{I}} \Delta \tilde{L} + \frac{\tilde{F}\tilde{L}^4}{8\tilde{E}^2\tilde{I}} \Delta \tilde{E} + \frac{\tilde{F}\tilde{L}^4}{8\tilde{E}\tilde{I}^2} \Delta \tilde{I}
$$

Substituting the appropriate values gives

$$\Delta y = 0.006561 + 0.002187 + 0.001094 + 0.00164 = 0.011482$$

Therefore, $y = 0.164025 \pm 0.011482$. In other words, y is between 0.152543 and 0.175507 m. The validity of these estimates can be verified by substituting the extreme values for the variables into the equation to generate an exact minimum of

$$y_{min} = \frac{720(8.97)^4}{8(7.55 \times 10^9)0.000505} = 0.152818$$

and an exact maximum of

$$y_{max} = \frac{780(9.03)^4}{8(7.45 \times 10^9)0.000495} = 0.175790$$

Thus, the first-order estimates are reasonably close to the exact values.

Equation (4.27) can be employed to define error propagation relationships for common mathematical operations. The results are summarized in Table 4.3. We will leave the derivation of these formulas as a homework exercise.

In addition, note that the first-order error analysis provides an upper bound on the impact of parameter uncertainties on computations. That is, it generates an estimate of the range (i.e., maximum minus minimum) of possible results. In the Introduction to Part Five, we will describe more advanced Monte Carlo methods that yield superior uncertainty estimates. Nevertheless, the first-order error analysis described here does provide a nice illustration of how the Taylor Series can be employed to quantify the sensitivity of computations to uncertainties.

4.2.3 Stability and Condition

The *condition* of a mathematical problem relates to its sensitivity to changes in its input values. We say that a computation is *numerically unstable* if the uncertainty of the input values is grossly magnified by the numerical method.

These ideas can be studied using a first-order Taylor series:

$$f(x) = f(\tilde{x}) + f'(\tilde{x})(x - \tilde{x})$$

This relationship can be employed to estimate the *relative error* of $f(x)$ as in

$$\frac{f(x) - f(\tilde{x})}{f(x)} \cong \frac{f'(\tilde{x})(x - \tilde{x})}{f(\tilde{x})}$$

The *relative error* of x is given by

$$\frac{x - \tilde{x}}{\tilde{x}}$$

TABLE 4.3 Estimated error bounds associated with common mathematical operations using inexact numbers $\tilde{u}$ and $\tilde{v}$.

Operation		Estimated Error
Addition	$\Delta(\tilde{u} + \tilde{v})$	$\Delta\tilde{u} + \Delta\tilde{v}$
Subtraction	$\Delta(\tilde{u} - \tilde{v})$	$\Delta\tilde{u} + \Delta\tilde{v}$
Multiplication	$\Delta(\tilde{u} \times \tilde{v})$	$\lvert\tilde{u}\rvert\Delta\tilde{v} + \lvert\tilde{v}\rvert\Delta\tilde{u}$
Division	$\Delta\left(\dfrac{\tilde{u}}{\tilde{v}}\right)$	$\dfrac{\lvert\tilde{u}\rvert\Delta\tilde{v} + \lvert\tilde{v}\rvert\Delta\tilde{u}}{\lvert\tilde{v}\rvert^2}$

A *condition number* can be defined as the ratio of these relative errors

$$\text{Condition number} = \frac{\tilde{x}f'(\tilde{x})}{f(\tilde{x})} \tag{4.28}$$

The condition number provides a measure of the extent to which an uncertainty in x is magnified by $f(x)$. A value of 1 tells us that the function's relative error is identical to the relative error in x. A value greater than 1 tells us that the relative error is amplified, whereas a value less than 1 tells us that it is attenuated. Functions with very large values are said to be *ill-conditioned*. Any combination of factors in Eq. (4.28) that increases the numerical value of the condition number will tend to magnify uncertainties in the computation of $f(x)$.

EXAMPLE 4.7

Condition Number

Problem Statement.   Compute and interpret the condition number for

$$f(x) = \tan x \qquad \text{for } \tilde{x} = \frac{\pi}{2} + 0.1\left(\frac{\pi}{2}\right)$$

$$f(x) = \tan x \qquad \text{for } \tilde{x} = \frac{\pi}{2} + 0.01\left(\frac{\pi}{2}\right)$$

Solution.   The condition number is computed as

$$\text{Condition number} = \frac{\tilde{x}(1/\cos^2 x)}{\tan \tilde{x}}$$

For $\tilde{x} = \pi/2 + 0.1(\pi/2)$,

$$\text{Condition number} = \frac{1.7279(40.86)}{-6.314} = -11.2$$

Thus, the function is ill-conditioned. For $\tilde{x} = \pi/2 + 0.01(\pi/2)$, the situation is even worse:

$$\text{Condition number} = \frac{1.5865(4053)}{-63.66} = -101$$

For this case, the major cause of ill conditioning appears to be the derivative. This makes sense because in the vicinity of $\pi/2$, the tangent approaches both positive and negative infinity.

4.3 TOTAL NUMERICAL ERROR

The *total numerical error* is the summation of the truncation and round-off errors. In general, the only way to minimize round-off errors is to increase the number of significant figures of the computer. Further, we have noted that round-off error will *increase* due to subtractive cancellation or due to an increase in the number of computations in an analysis. In contrast, Example 4.4 demonstrated that the truncation error can be reduced by decreasing the step size. Because a decrease in step size can lead to subtractive cancellation or to an increase in computations, the truncation errors are *decreased* as the round-off errors are *increased*. Therefore, we are faced by the following dilemma: The strategy for decreasing one component of the total error leads to an increase of the other component. In a computation, we could conceivably decrease the step size to minimize truncation error only to discover that in doing so, the round-off error begins to dominate the solution and the total error grows! Thus, our remedy becomes our problem (Fig. 4.8). One challenge that we face is to determine an appropriate step size for a particular computation. We would like to choose a large step size in order to decrease the amount of calculations and round-off errors without incurring the penalty of a large truncation error. If the total error is as shown in Fig. 4.8, the challenge is to identify the point of diminishing returns where round-off error begins to negate the benefits of step-size reduction.

In actual cases, however, such situations are relatively uncommon because most computers carry enough significant figures that round-off errors do not predominate. Nevertheless, they sometimes do occur and suggest a sort of "numerical uncertainty principle" that places an absolute limit on the accuracy that may be obtained using certain computerized numerical methods. We explore such a case in the following section.

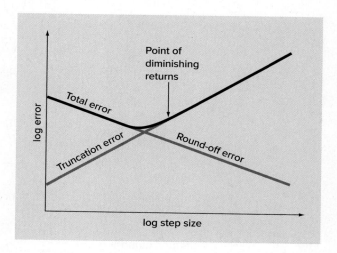

FIGURE 4.8
A graphical depiction of the trade-off between round-off and truncation error that sometimes comes into play in the course of a numerical method. The point of diminishing returns is shown, where round-off error begins to negate the benefits of step-size reduction.

4.3.1 Error Analysis of Numerical Differentiation

As described in the Sec. 4.1.3, a centered difference approximation of the first derivative can be written as (Eq. 4.22):

$$f'(x_i) = \frac{f(x_{i+1}) - f(x_{i-1})}{2h} - \frac{f^{(3)}(\xi)}{6}h^2 \qquad (4.29)$$

| True value | Finite-difference approximation | Truncation error |

Thus, if the two function values in the numerator of the finite-difference approximation have no round-off error, the only error is due to truncation.

However, because we are using digital computers, the function values do include round-off error as in

$$f(x_{i-1}) = \tilde{f}(x_{i-1}) + e_{i-1}$$
$$f(x_{i+1}) = \tilde{f}(x_{i+1}) + e_{i+1}$$

where the $\tilde{f}$'s are the rounded function values and the e's are the associated round-off errors. Substituting these values into Eq. (4.29) gives

$$f'(x_i) = \frac{\tilde{f}(x_{i+1}) - \tilde{f}(x_{i-1})}{2h} + \frac{e_{i+1} - e_{i-1}}{2h} - \frac{f^{(3)}(\xi)}{6}h^2$$

| True value | Finite-difference approximation | Round-off error | Truncation error |

We can see that the total error of the finite-difference approximation consists of a round-off error that increases with step size and a truncation error that decreases with step size.

Assuming that the absolute value of each component of the round-off error has an upper bound of ε, the maximum possible value of the difference $e_{i+1} - e_i$ will be 2ε. Further, assume that the third derivative has a maximum absolute value of M. An upper bound on the absolute value of the total error can therefore be represented as

$$\text{Total error} = \left| f'(x_i) - \frac{\tilde{f}(x_{i+1}) - \tilde{f}(x_{i-1})}{2h} \right| \leq \frac{\varepsilon}{h} + \frac{h^2 M}{6} \qquad (4.30)$$

An optimal step size can be determined by differentiating Eq. (4.30), setting the result equal to zero, and solving for

$$h_{\text{opt}} = \sqrt[3]{\frac{3\varepsilon}{M}} \qquad (4.31)$$

EXAMPLE 4.8

Round-off and Truncation Errors in Numerical Differentiation

Problem Statement. In Example 4.4, we used a centered difference approximation of $O(h^2)$ to estimate the first derivative of the following function at $x = 0.5$:

$$f(x) = -0.1x^4 - 0.15x^3 - 0.5x^2 - 0.25x + 1.2$$

Perform the same computation starting with $h = 1$. Then progressively divide the step size by a factor of 10 to demonstrate how round-off becomes dominant as the step size is reduced. Relate your results to Eq. (4.31). Recall that the true value of the derivative is -0.9125.

Solution.   We can develop a program to perform the computations and plot the results. For the present example, we have done this with a MATLAB software M-file. Notice that we pass both the function and its analytical derivative as arguments. In addition, the function generates a plot of the results.

```
function diffex(func,dfunc,x,n)
format long
dftrue=dfunc(x);
h=1;
H(1)=h;
D(1)=(func(x+h)-func(x-h))/(2*h);
E(1)=abs(dftrue-D(1));
for i=2:n
  h=h/10;
  H(i)=h;
  D(i)=(func(x+h)-func(x-h))/(2*h);
  E(i)=abs(dftrue-D(i));
end
L=[H' D' E']';
fprintf(' step size finite difference    true error\n');
fprintf('%14.10f %16.14f %16.13f\n',L);
loglog(H,E),xlabel('Step Size'),ylabel('Error')
title('Plot of Error Versus Step Size')
format short
```

The M-file can then be run using the following commands:

```
>> ff=@(x) -0.1*x^4-0.15*x^3-0.5*x^2-0.25*x+1.2;
>> df=@(x) -0.4*x^3-0.45*x^2-x-0.25;
>> diffex(ff,df,0.5,11)
```

When the function is run, the following numeric output is generated along with the plot (Fig. 4.9):

step size	finite difference	true error
1.0000000000	-1.26250000000000	0.3500000000000
0.1000000000	-0.91600000000000	0.0035000000000
0.0100000000	-0.91253500000000	0.0000350000000
0.0010000000	-0.91250035000001	0.0000003500000
0.0001000000	-0.91250000349985	0.0000000034998
0.0000100000	-0.91250000003318	0.0000000000332
0.0000010000	-0.91250000000542	0.0000000000054
0.0000001000	-0.91249999945031	0.0000000005497
0.0000000100	-0.91250000333609	0.0000000033361
0.0000000010	-0.91250001998944	0.0000000199894
0.0000000001	-0.91250007550059	0.0000000755006

The results are as expected. At first, round-off is minimal and the estimate is dominated by truncation error. Hence, as in Eq. (4.30), the total error drops by a factor of 100 each

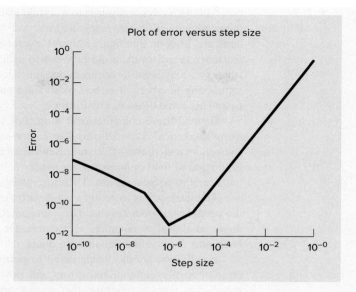

FIGURE 4.9
Plot of error versus step size.

time we divide the step by 10. However, starting at $h = 0.0001$, we see round-off error begin to creep in and erode the rate at which the error diminishes. A minimum error is reached at $h = 10^{-6}$. Beyond this point, the error increases as round-off dominates.

Because we are dealing with an easily differentiable function, we can also investigate whether these results are consistent with Eq. (4.31). First, we can estimate M by evaluating the function's third derivative as

$$M = |f^3(0.5)| = |-2.4(0.5) - 0.9| = 2.1$$

Because MATLAB has a precision of about 15 to 16 base-10 digits, a rough estimate of the upper bound on round-off would be about $\varepsilon = 0.5 \times 10^{-16}$. Substituting these values into Eq. (4.31) gives

$$h_{opt} = \sqrt[3]{\frac{3(0.5 \times 10^{-16})}{2.1}} = 4.3 \times 10^{-6}$$

which is on the same order as the result of 1×10^{-6} obtained with our computer program.

4.3.2 Control of Numerical Errors

For most practical cases, we do not know the exact error associated with numerical methods. The exception, of course, is when we have obtained the exact solution that makes our numerical approximations unnecessary. Therefore, for most engineering applications we must settle for some estimate of the error in our calculations.

There are no systematic and general approaches to evaluating numerical errors for all problems. In many cases, error estimates are based on the experience and judgment of the engineer.

Although error analysis is to a certain extent an art, there are several practical programming guidelines we can suggest. First and foremost, avoid subtracting two nearly equal numbers. Loss of significance almost always occurs when this is done. Sometimes you can rearrange or reformulate the problem to avoid subtractive cancellation. If this is not possible, you may want to use extended-precision arithmetic. Furthermore, when adding and subtracting numbers, it is best to sort the numbers and work with the smallest numbers first. This avoids loss of significance.

Beyond these computational hints, one can attempt to predict total numerical errors using theoretical formulations. The Taylor series is our primary tool for analysis of both truncation and round-off errors. Several examples have been presented in this chapter. Prediction of total numerical error is very complicated for even moderately sized problems and tends to be pessimistic. Therefore, it is usually attempted for only small-scale tasks.

The tendency is to push forward with the numerical computations and try to estimate the accuracy of your results. This can sometimes be done by seeing if the results satisfy some condition or equation as a check. Or it may be possible to substitute the results back into the original equation to check that it is actually satisfied.

Finally you should be prepared to perform numerical experiments to increase your awareness of computational errors and possible ill-conditioned problems. Such experiments may involve repeating the computations with a different step size or method and comparing the results. We may employ sensitivity analysis to see how our solution changes when we change model parameters or input values. We may want to try different numerical algorithms that have different theoretical foundations, are based on different computational strategies, or have different convergence properties and stability characteristics.

When the results of numerical computations are extremely critical and may involve loss of human life or have severe economic ramifications, it is appropriate to take special precautions. This may involve the use of two or more independent groups to solve the same problem so that their results can be compared.

The roles of errors will be a topic of concern and analysis in all sections of this book. We will leave these investigations to specific sections.

4.4 BLUNDERS, FORMULATION ERRORS, AND DATA UNCERTAINTY

Although the following sources of error are not directly connected with most of the numerical methods in this book, they can sometimes have great impact on the success of a modeling effort. Thus, they must always be kept in mind when applying numerical techniques in the context of real-world problems.

4.4.1 Blunders

We are all familiar with gross errors, or blunders. In the early years of computers, erroneous numerical results could sometimes be attributed to malfunctions of the computer itself. Today, this source of error is highly unlikely, and most blunders must be attributed to human imperfection.

Blunders can occur at any stage of the mathematical modeling process and can contribute to all the other components of error. They can be avoided only by sound knowledge of fundamental principles and by the care with which you approach and design your solution to a problem.

Blunders are usually disregarded in discussions of numerical methods. This is no doubt due to the fact that, try as we may, mistakes are to a certain extent unavoidable. However, we believe that there are a number of ways in which their occurrence can be minimized. In particular, the good programming habits that were outlined in Chap. 2 are extremely useful for mitigating programming blunders. In addition, there are usually simple ways to check whether a particular numerical method is working properly. Throughout this book, we discuss ways to check the results of numerical calculations.

4.4.2 Formulation Errors

Formulation, or *model, errors* relate to bias that can be ascribed to incomplete mathematical models. An example of a negligible formulation error is the fact that Newton's second law does not account for relativistic effects. This does not detract from the adequacy of the solution in Example 1.1 because these errors are minimal on the time and space scales associated with the falling parachutist problem.

However, suppose that air resistance is not linearly proportional to fall velocity, as in Eq. (1.7), but is a function of the square of velocity. If this were the case, both the analytical and numerical solutions obtained in the Chap. 1 would be erroneous because of formulation error. Further consideration of formulation error is included in some of the engineering applications in the remainder of the book. You should be cognizant of these problems and realize that, if you are working with a poorly conceived model, no numerical method will provide adequate results.

4.4.3 Data Uncertainty

Errors sometimes enter into an analysis because of uncertainty in the physical data upon which a model is based. For instance, suppose we wanted to test the falling parachutist model by having an individual make repeated jumps and then measuring his or her velocity after a specified time interval. Uncertainty would undoubtedly be associated with these measurements, since the parachutist would fall faster during some jumps than during others. These errors can exhibit both inaccuracy and imprecision. If our instruments consistently underestimate or overestimate the velocity, we are dealing with an inaccurate, or biased, device. On the other hand, if the measurements are randomly high and low, we are dealing with a question of precision.

Measurement errors can be quantified by summarizing the data with one or more well-chosen statistics that convey as much information as possible regarding specific characteristics of the data. These descriptive statistics are most often selected to represent (1) the location of the center of the distribution of the data and (2) the degree of spread of the data. As such, they provide a measure of the bias and imprecision, respectively. We will return to the topic of characterizing data uncertainty in Part Five.

Although you must be cognizant of blunders, formulation errors, and uncertain data, the numerical methods used for building models can be studied, for the most part, independently of these errors. Therefore, for most of this book, we will assume that we have not made gross errors, we have a sound model, and we are dealing with error-free measurements. Under these conditions, we can study numerical errors without complicating factors.

PROBLEMS

4.1 The following infinite series can be used to approximate e^x:

$$e^x = 1 + x + \frac{x^2}{2} + \frac{x^3}{3!} + \cdots + \frac{x^n}{n!}$$

(a) Prove that this Maclaurin series expansion is a special case of the Taylor series expansion [(Eq. (4.7)] with $x_i = 0$ and $h = x$.
(b) Use the Taylor series to estimate $f(x) = e^{-x}$ at $x_{i+1} = 1$ for $x_i = 0.2$. Employ the zero-, first-, second-, and third-order versions and compute $|\varepsilon_t|$ for each case.

4.2 The Maclaurin series expansion for $\cos x$ is

$$\cos x = 1 - \frac{x^2}{2} + \frac{x^4}{4!} - \frac{x^6}{6!} + \frac{x^8}{8!} - \cdots$$

Starting with the simplest version, $\cos x = 1$, add terms one at a time to estimate $\cos(\pi/6)$. After each new term is added, compute the true and approximate percent relative errors. Use your pocket calculator to determine the true value. Add terms until the absolute value of the approximate error estimate falls below an error criterion conforming to two significant figures.

4.3 Perform the same computation as in Prob. 4.2, but use the Maclaurin series expansion for $\sin x$ to estimate $\sin(\pi/6)$.

$$\sin x = x - \frac{x^3}{3!} + \frac{x^5}{5!} - \frac{x^7}{7!} + \cdots$$

4.4 The Maclaurin series expansion for the arctangent of x is defined for $|x| \leq 1$ as

$$\arctan x = \sum_{n=0}^{\infty} \frac{(-1)^n}{2n + 1} x^{2n+1}$$

(a) Write out the first four terms ($n = 0, \ldots, 3$).
(b) Starting with the simplest version, $\arctan x = x$, add terms one at a time to estimate $\arctan(\pi/4)$. After each new term is added, compute the true and approximate percent relative errors. Use your calculator to determine the true value. Add terms until the absolute value of the approximate error estimate falls below an error criterion conforming to two significant figures.

4.5 Use zero- through third-order Taylor series expansions to predict $f(2.5)$ for

$$f(x) = 25x^3 - 6x^2 + 7x - 88$$

using a base point at $x = 1$. Compute the true percent relative error ε_t for each approximation.

4.6 Use zero- through fourth-order Taylor series expansions to predict $f(2.5)$ for $f(x) = \ln x$ using a base point at $x = 1$. Compute the

true percent relative error ε_t for each approximation. Discuss the meaning of the results.

4.7 Use forward and backward difference approximations of $O(h)$ and a centered difference approximation of $O(h^2)$ to estimate the first derivative of the function examined in Prob. 4.5. Evaluate the derivative at $x = 2$ using a step size of $h = 0.2$. Compare your results with the true value of the derivative. Interpret your results on the basis of the remainder term of the Taylor series expansion.

4.8 Use a centered difference approximation of $O(h^2)$ to estimate the second derivative of the function examined in Prob. 4.5. Perform the evaluation at $x = 2$ using step sizes of $h = 0.25$ and 0.125. Compare your estimates with the true value of the second derivative. Interpret your results on the basis of the remainder term of the Taylor series expansion.

4.9 The Stefan-Boltzmann law can be employed to estimate the rate of radiation of energy H from a surface, as in

$$H = Ae\sigma T^4$$

where H is in watts, A = the surface area (m^2), e = the emissivity that characterizes the emitting properties of the surface (dimensionless), σ = a universal constant called the Stefan-Boltzmann constant ($= 5.67 \times 10^{-8}$ W m^{-2} K^{-4}), and T = absolute temperature (K). Determine the error of H for a steel plate with $A = 0.15$ m^2, $e = 0.90$, and $T = 650 \pm 25$. Compare your results with the exact error. Repeat the computation but with $T = 650 \pm 45$. Interpret your results.

4.10 Repeat Prob. 4.9 but for a copper sphere with radius $= 0.15 \pm 0.02$ m, $e = 0.90 \pm 0.04$, and $T = 550 \pm 17.5$.

4.11 Recall that the velocity of the falling parachutist can be computed by [Eq. (1.10)],

$$v(t) = \frac{gm}{c}\left(1 - e^{-(c/m)t}\right)$$

Use a first-order error analysis to estimate the error of v at $t = 6$ if $g = 9.81$ and $m = 50$ but $c = 12.5 \pm 1.5$.

4.12 Repeat Prob. 4.11 with $g = 9.81$, $t = 6$, $c = 12.5 \pm 1.5$, and $m = 50 \pm 2.5$.

4.13 Evaluate and interpret the condition numbers for

(a) $f(x) = \sqrt{|x - 1|} + 1$ for $x = 1.00001$
(b) $f(x) = e^{-x}$ for $x = 10$
(c) $f(x) = \sqrt{x^2 + 1} - x$ for $x = 300$
(d) $f(x) = \dfrac{e^{-x} - 1}{x}$ for $x = 0.001$
(e) $f(x) = \dfrac{\sin x}{1 + \cos x}$ for $x = 1.0001\pi$

4.14 Employing ideas from Sec. 4.2, derive the relationships from Table 4.3.

4.15 Prove that Eq. (4.4) is exact for all values of x if $f(x) = ax^2 + bx + c$.

4.16 Manning's formula for a rectangular channel can be written as

$$Q = \frac{1}{n} \frac{(BH)^{5/3}}{(B + 2H)^{2/3}} \sqrt{S}$$

where Q = flow (m³/s), n = a roughness coefficient, B = width (m), H = depth (m), and S = slope. You are applying this formula to a stream where you know that the width = 20 m and the depth = 0.3 m. Unfortunately, you know the roughness and the slope to only a ±10% precision. That is, you know that the roughness is about 0.03 with a range from 0.027 to 0.033 and the slope is 0.0003 with a range from 0.00027 to 0.00033. Use a first-order error analysis to determine the sensitivity of the flow prediction to each of these two factors. Which one should you attempt to measure with more precision?

4.17 If $|x| < 1$, it is known that

$$\frac{1}{1 - x} = 1 + x + x^2 + x^3 + \cdots$$

Repeat Prob. 4.1 for this series for $x = 0.1$.

4.18 A missile leaves the ground with an initial velocity v_0 forming an angle ϕ_0 with the vertical as shown in Fig. P4.18. The maximum desired altitude is αR, where R is the radius of the earth. The laws of mechanics can be used to show that

$$\sin \phi_0 = (1 + \alpha) \sqrt{1 - \frac{\alpha}{1 + \alpha} \left(\frac{v_e}{v_0}\right)^2}$$

where v_e = the escape velocity of the missile. It is desired to fire the missile and reach the design maximum altitude within an accuracy of ±2%. Determine the range of values for ϕ_0 if $v_e/v_0 = 2$ and $\alpha = 0.25$.

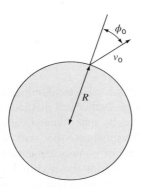

FIGURE P4.18

4.19 To calculate a planet's space coordinates, we have to solve the function

$$f(x) = x - 1 - 0.5 \sin x$$

Let the base point be $a = x_i = \pi/2$ on the interval $[0, \pi]$. Determine the highest-order Taylor series expansion resulting in a maximum error of 0.015 on the specified interval. The error is equal to the absolute value of the difference between the given function and the specific Taylor series expansion. (*Hint:* Solve graphically.)

4.20 Consider the function $f(x) = x^3 - 2x + 4$ on the interval $[-2, 2]$ with $h = 0.25$. Use the forward, backward, and centered finite difference approximations for the first and second derivatives so as to graphically illustrate which approximation is most accurate. Graph all three first derivative finite difference approximations along with the theoretical, and do the same for the second derivative as well.

4.21 Derive Eq. (4.31).

4.22 Repeat Example 4.8, but for $f(x) = \cos(x)$ at $x = \pi/6$.

4.23 Repeat Example 4.8, but for the forward divided difference (Eq. 4.17).

4.24 Develop a well-structured program to compute the Maclaurin series expansion for the sine function as described in Prob. 4.3. The function should have the following features:

- Iterate until the relative error falls below a stopping criterion (es) or exceeds a maximum number of iterations (maxit). Allow the user to specify values for these parameters.
- Include default values of es (= 0.000001) and maxit (= 100) in the event that they are not specified by the user.
- Return the estimate of sin x, the approximate relative error, the number of iterations, and the true relative error (which you can calculate based on the built-in sine function).

4.25 Under laminar flow conditions, the steady-state velocity of a spherical particle settling in a fluid can be computed with *Stokes law,*

$$v_s = \frac{g}{18} \left(\frac{\rho_s - \rho_f}{\mu}\right) d^2$$

where g = gravitational acceleration (= 981 cm s⁻²), ρ_s and ρ_f = densities of the particle and the fluid, respectively (g cm⁻³), μ = dynamic viscosity (g cm⁻¹ s⁻¹), and d = the sphere's diameter (cm). Suppose that you have two types of spherical particles: a phytoplankton cell ($d = 30$ μm, $\rho_s = 1.027$ g cm⁻³) and a silt particle ($d = 30$ μm, $\rho_s = 2.65$ g cm⁻³) settling in water ($\mu = 0.013$ g cm⁻¹ s⁻¹ and $\rho_f = 0.99973$ g cm⁻³). For each of these particles, (**a**) compute the settling velocity and (**b**) perform a first-order error analysis for the settling velocity. For (**b**) assume that the parameters ρ_s, ρ_f, μ, and d have uncertainties of ±2% around their mean values.

EPILOGUE: PART ONE

PT1.4   TRADE-OFFS

Numerical methods are scientific in the sense that they represent systematic techniques for solving mathematical problems. However, there is a certain degree of art, subjective judgment, and compromise associated with their effective use in engineering practice. For each problem, you may be confronted with several alternative numerical methods and many different types of computers. Thus, the elegance and efficiency of different approaches to problems is highly individualistic and correlated with your ability to choose wisely among options. Unfortunately, as with any intuitive process, the factors influencing this choice are difficult to communicate. Only by experience can these skills be fully comprehended and honed. However, because these skills play such a prominent role in the effective implementation of the methods, we have included this section as an introduction to some of the trade-offs that you must consider when selecting a numerical method and the tools for implementing the method. It is hoped that the discussion that follows will influence your orientation when approaching subsequent material. Also, it is hoped that you will refer back to this material when you are confronted with choices and trade-offs in the remainder of the book.

1. *Type of Mathematical Problem.* As delineated previously in Fig. PT1.2, several types of mathematical problems are discussed in this book:
   (a) Roots of equations.
   (b) Systems of simultaneous linear algebraic equations.
   (c) Optimization.
   (d) Curve fitting.
   (e) Numerical integration.
   (f) Ordinary differential equations.
   (g) Partial differential equations.

   You will probably be introduced to the applied aspects of numerical methods by confronting a problem in one of the above areas. Numerical methods will be required because the problem cannot be solved efficiently using analytical techniques. You should be cognizant of the fact that your professional activities will eventually involve problems in all the above areas. Thus, the study of numerical methods and the selection of automatic computation equipment should, at the minimum, consider these basic types of problems. More advanced problems may require capabilities of handling areas such as functional approximation, integral equations, etc. These areas typically demand greater computation power or advanced methods not covered in this text. Other references such as Carnahan, Luther, and Wilkes (1969); Hamming (1973); Ralston and Rabinowitz (1978); Burden and Faires (2005); and Moler (2004) should be consulted for problems beyond the scope of this book. In addition, at the end of each part of this text, we include a brief summary and references for advanced methods to provide you with avenues for pursuing further studies of numerical methods.

2. *Type, Availability, Precision, Cost, and Speed of Computer.* You may have the option of working with a variety of computation tools. These range from pocket calculators to large mainframe computers. Of course, any of the tools can be used to implement any numerical method (including simple paper and pencil). It is usually not a question of ultimate capability but rather of cost, convenience, speed, dependability, repeatability, and precision. Although each of the tools will continue to have utility, the recent rapid advances in the performance of personal computers have already had a major impact on the engineering profession. We expect this revolution will spread as technological improvements continue because personal computers offer an excellent compromise in convenience, cost, precision, speed, and storage capacity. Furthermore, they can be readily applied to most practical engineering problems.

3. *Program Development Cost versus Software Cost versus Run-Time Cost.* Once the types of mathematical problems to be solved have been identified and the computer system has been selected, it is appropriate to consider software and run-time costs. Software development may represent a substantial effort in many engineering projects and may therefore be a significant cost. In this regard, it is particularly important that you be very well acquainted with the theoretical and practical aspects of the relevant numerical methods. In addition, you should be familiar with professionally developed software. Low-cost software is widely available to implement numerical methods that may be readily adapted to a broad variety of problems.

4. *Characteristics of the Numerical Method.* When computer hardware and software costs are high, or if computer availability is limited (for example, on some timeshare systems), it pays to choose carefully the numerical method to suit the situation. On the other hand, if the problem is still at the exploratory stage and computer access and cost are not problems, it may be appropriate for you to select a numerical method that always works but may not be the most computationally efficient. The numerical methods available to solve any particular type of problem involve the types of trade-offs just discussed and others:

   (a) *Number of Initial Guesses or Starting Points.* Some of the numerical methods for finding roots of equations or solving differential equations require the user to specify initial guesses or starting points. Simple methods usually require one value, whereas complicated methods may require more than one value. The advantages of complicated methods that are computationally efficient may be offset by the requirement for multiple starting points. You must use your experience and judgment to assess the trade-offs for each particular problem.

   (b) *Rate of Convergence.* Certain numerical methods converge more rapidly than others. However, this rapid convergence may require more refined initial guesses and more complex programming than a method with slower convergence. Again, you must use your judgment in selecting a method. Faster is not always better.

   (c) *Stability.* Some numerical methods for finding roots of equations or solutions for systems of linear equations may diverge rather than converge on the correct answer for certain problems. Why would you tolerate this possibility when confronted with design or planning problems? The answer is that these methods may be highly efficient when they work. Thus, trade-offs again emerge. You must decide if your problem requirements justify the effort needed to apply a method that may not always converge.

(d) *Accuracy and Precision.* Some numerical methods are simply more accurate or precise than others. Good examples are the various equations available for numerical integration. Usually, the performance of low-accuracy methods can be improved by decreasing the step size or increasing the number of applications over a given interval. Is it better to use a low-accuracy method with small step sizes or a high-accuracy method with large step sizes? This question must be addressed on a case-by-case basis taking into consideration the additional factors such as cost and ease of programming. In addition, you must also be concerned with round-off errors when you are using multiple applications of low-accuracy methods and when the number of computations becomes large. Here the number of significant figures handled by the computer may be the deciding factor.

(e) *Breadth of Application.* Some numerical methods can be applied to only a limited class of problems or to problems that satisfy certain mathematical restrictions. Other methods are not affected by such limitations. You must evaluate whether it is worth your effort to develop programs that employ techniques that are appropriate for only a limited number of problems. The fact that such techniques may be widely used suggests that they have advantages that will often outweigh their disadvantages. Obviously, trade-offs are occurring.

(f) *Special Requirements.* Some numerical techniques attempt to increase accuracy and rate of convergence using additional or special information. An example would be the use of estimated or theoretical values of errors to improve accuracy. However, these improvements are generally not achieved without some inconvenience in terms of added computing costs or increased program complexity.

(g) *Programming Effort Required.* Efforts to improve rates of convergence, stability, and accuracy can be creative and ingenious. When improvements can be made without increasing the programming complexity, they may be considered elegant and will probably find immediate use in the engineering profession. However, if they require more complicated programs, you are again faced with a trade-off situation that may or may not favor the new method.

It is clear that the above discussion concerning a choice of numerical methods reduces to one of cost and accuracy. The costs are those involved with computer time and program development. Appropriate accuracy is a question of professional judgment and ethics.

5. *Mathematical Behavior of the Function, Equation, or Data.* In selecting a particular numerical method, type of computer, and type of software, you must consider the complexity of your functions, equations, or data. Simple equations and smooth data may be appropriately handled by simple numerical algorithms and inexpensive computers. The opposite is true for complicated equations and data exhibiting discontinuities.

6. *Ease of Application (User-Friendly?).* Some numerical methods are easy to apply; others are difficult. This may be a consideration when choosing one method over

another. This same idea applies to decisions regarding program development costs versus professionally developed software. It may take considerable effort to convert a difficult program to one that is user-friendly. Ways to do this were introduced in Chap. 2 and are elaborated throughout the book.

7. *Maintenance.* Programs for solving engineering problems require maintenance because during application, difficulties invariably occur. Maintenance may require changing the program code or expanding the documentation. Simple programs and numerical algorithms are simpler to maintain.

The chapters that follow involve the development of various types of numerical methods for various types of mathematical problems. Several alternative methods will be given in each chapter. These various methods (rather than a single method chosen by the authors) are presented because there is no single "best" method. There is no best method because there are many trade-offs that must be considered when applying the methods to practical problems. A table that highlights the trade-offs involved in each method will be found at the end of each part of the book. This table should assist you in selecting the appropriate numerical procedure for your particular problem context.

PT1.5 IMPORTANT RELATIONSHIPS AND FORMULAS

Table PT1.2 summarizes important information that was presented in Part One. The table can be consulted to quickly access important relationships and formulas. The epilogue of each part of the book will contain such a summary.

PT1.6 ADVANCED METHODS AND ADDITIONAL REFERENCES

The epilogue of each part of the book will also include a section designed to facilitate and encourage further studies of numerical methods. This section will reference other books on the subject as well as material related to more advanced methods.[1]

To extend the background provided in Part One, numerous manuals on computer programming are available. It would be difficult to reference all the excellent books and manuals pertaining to specific languages and computers. In addition, you probably already have material from your previous exposure to programming. However, if this is your first experience with computers, your instructor and fellow students should also be able to advise you regarding good reference books for the machines and languages available at your school.

As for error analysis, any good introductory calculus book will include supplementary material related to subjects such as the Taylor series expansion. Texts by Swokowski (1979), Thomas and Finney (1979), and Simmons (1985) provide very readable discussions of these subjects. In addition, Taylor (1982) presents a nice introduction to error analysis.

Finally, although we hope that our book serves you well, it is always good to consult other sources when trying to master a new subject. Burden and Faires (2005); Ralston

---

[1]Books are referenced only by author here; a complete bibliography will be found at the back of this text.

TABLE PT1.2 Summary of important information presented in Part One.

Error Definitions

True error

$$E_t = \text{true value} - \text{approximation}$$

True percent relative error

$$\varepsilon_t = \frac{\text{true value} - \text{approximation}}{\text{true value}} 100\%$$

Approximate percent relative error

$$\varepsilon_a = \frac{\text{present approximation} - \text{previous approximation}}{\text{present approximation}} 100\%$$

Stopping criterion

Terminate computation when

$$\varepsilon_a < \varepsilon_s$$

where ε_s is the desired percent relative error

Taylor Series

Taylor series expansion

$$f(x_{i+1}) = f(x_i) + f'(x_i)h + \frac{f''(x_i)}{2!}h^2$$
$$+ \frac{f'''(x_i)}{3!}h^3 + \cdots + \frac{f^{(n)}(x_i)}{n!}h^n + R_n$$

where

Remainder

$$R_n = \frac{f^{(n+1)}(\xi)}{(n+1)!}h^{n+1}$$

or

$$R_n = O(h^{n+1})$$

Numerical Differentiation

First forward finite divided difference

$$f'(x_i) = \frac{f(x_{i+1}) - f(x_i)}{h} + O(h)$$

(Other divided differences are summarized in Chaps. 4 and 23.)

Error Propagation

For n independent variables $x_1, x_2, \ldots, x_n$ having errors $\Delta \tilde{x}_1, \Delta \tilde{x}_2, \ldots, \Delta \tilde{x}_n$, the error in the function f can be estimated via

$$\Delta f = \left| \frac{\partial f}{\partial x_1} \right| \Delta \tilde{x}_1 + \left| \frac{\partial f}{\partial x_2} \right| \Delta \tilde{x}_2 + \cdots + \left| \frac{\partial f}{\partial x_n} \right| \Delta \tilde{x}_n$$

---

and Rabinowitz (1978); Hoffman (1992); and Carnahan, Luther, and Wilkes (1969) provide comprehensive discussions of most numerical methods. Other enjoyable books on the subject are Gerald and Wheatley (2004), and Cheney and Kincaid (2008). In addition, Press et al. (2007) include algorithms to implement a variety of methods, and Moler (2004) and Chapra (2007) are devoted to numerical methods with MATLAB software.

PART TWO

ROOTS OF EQUATIONS

PT2.1 MOTIVATION

Years ago, you learned to use the quadratic formula,

$$x = \frac{-b \pm \sqrt{b^2 - 4ac}}{2a} \tag{PT2.1}$$

to solve

$$f(x) = ax^2 + bx + c = 0 \tag{PT2.2}$$

The values calculated with Eq. (PT2.1) are called the "roots" of Eq. (PT2.2). They represent the values of x that make Eq. (PT2.2) equal to zero. Thus, we can define the root of an equation as the value of x that makes $f(x) = 0$. For this reason, roots are sometimes called the *zeros* of the equation.

Although the quadratic formula is handy for solving Eq. (PT2.2), there are many other functions for which the roots cannot be determined so easily. For these cases, the numerical methods described in Chaps. 5, 6, and 7 provide efficient means to obtain the answer.

PT2.1.1 Noncomputer Methods for Determining Roots

Before the advent of digital computers, there were several ways to solve for roots of algebraic and transcendental equations. For some cases, the roots could be obtained by direct methods, as was done with Eq. (PT2.1). Although there were equations like this that could be solved directly, there were many more that could not. For example, even an apparently simple function such as $f(x) = e^{-x} - x$ cannot be solved analytically. In such instances, the only alternative is an approximate solution technique.

One method to obtain an approximate solution is to plot the function and determine where it crosses the x axis. This point, which represents the x value for which $f(x) = 0$, is the root. Graphical techniques are discussed at the beginning of Chaps. 5 and 6.

Although graphical methods are useful for obtaining rough estimates of roots, they are limited because of their lack of precision. An alternative approach is to use trial and error. This "technique" consists of guessing a value of x and evaluating whether $f(x)$ is zero. If not (as is almost always the case), another guess is made, and $f(x)$ is again evaluated to determine whether the new value provides a better estimate of the root. The process is repeated until a guess is obtained that results in an $f(x)$ that is close to zero.

Such haphazard methods are obviously inefficient and inadequate for the requirements of engineering practice. The techniques described in Part Two represent alternatives that are also approximate but employ systematic strategies to home in on the true root. As elaborated on in the following pages, the combination of these systematic methods and computers makes the solution of most applied roots-of-equations problems a simple and efficient task.

PT2.1.2 Roots of Equations and Engineering Practice

Although they arise in other problem contexts, roots of equations frequently occur in the area of engineering design. Table PT2.1 lists several fundamental principles that are routinely used in design work. As introduced in Chap. 1, mathematical equations or models derived from these principles are employed to predict dependent variables as a function of independent variables, forcing functions, and parameters. Note that in each case, the dependent variables reflect the state or performance of the system, whereas the parameters represent its properties or composition.

TABLE PT2.1 Fundamental principles used in engineering design problems.

Fundamental Principle	Dependent Variable	Independent Variable	Parameters
Heat balance	Temperature	Time and position	Thermal properties of material and geometry of system
Mass balance	Concentration or quantity of mass	Time and position	Chemical behavior of material, mass transfer coefficients, and geometry of system
Force balance	Magnitude and direction of forces	Time and position	Strength of material, structural properties, and geometry of system
Energy balance	Changes in the kinetic- and potential-energy states of the system	Time and position	Thermal properties, mass of material, and system geometry
Newton's laws of motion	Acceleration, velocity, or location	Time and position	Mass of material, system geometry, and dissipative parameters such as friction or drag
Kirchhoff's laws	Currents and voltages in electric circuits	Time	Electrical properties of systems such as resistance, capacitance, and inductance

An example of such a model is the equation, derived from Newton's second law, used in Chap. 1 for the parachutist's velocity:

$$v = \frac{gm}{c}\left(1 - e^{-(c/m)t}\right)$$

(PT2.3)

where velocity v = the dependent variable, time t = the independent variable, the gravitational constant g = the forcing function, and the drag coefficient c and mass m = parameters. If the parameters are known, Eq. (PT2.3) can be used to predict the parachutist's velocity as a function of time. Such computations can be performed directly because v is expressed *explicitly* as a function of time. That is, it is isolated on one side of the equal sign.

However, suppose we had to determine the drag coefficient for a parachutist of a given mass to attain a prescribed velocity in a set time period. Although Eq. (PT2.3) provides a mathematical representation of the interrelationship among the model variables and parameters, it cannot be solved explicitly for the drag coefficient. Try it. There is no way to rearrange the equation so that c is isolated on one side of the equal sign. In such cases, c is said to be *implicit*.

This represents a real dilemma, because many engineering design problems involve specifying the properties or composition of a system (as represented by its parameters) to ensure that it performs in a desired manner (as represented by its variables). Thus, these problems often require the determination of implicit parameters.

The solution to the dilemma is provided by numerical methods for roots of equations. To solve the problem using numerical methods, it is conventional to reexpress Eq. (PT2.3). This is done by subtracting the dependent variable v from both sides of the equation to give

$$f(c) = \frac{gm}{c}\left(1 - e^{-(c/m)t}\right) - v$$

(PT2.4)

The value of c that makes $f(c) = 0$ is, therefore, the root of the equation. This value also represents the drag coefficient that solves the design problem.

Part Two of this book deals with a variety of numerical and graphical methods for determining roots of relationships such as Eq. (PT2.4). These techniques can be applied to engineering design problems that are based on the fundamental principles outlined in Table PT2.1 as well as to many other problems confronted routinely in engineering practice.

PT2.2 MATHEMATICAL BACKGROUND

For most of the subject areas in this book, there is usually some prerequisite mathematical background needed to successfully master the topic. For example, the concepts of error estimation and the Taylor series expansion discussed in Chaps. 3 and 4 have direct relevance to our discussion of roots of equations. Additionally, prior to this point we have mentioned "algebraic" and "transcendental" equations. It might be helpful to formally define these terms and discuss how they relate to the scope of this part of the book.

By definition, a function given by $y = f(x)$ is algebraic if it can be expressed in the form

$$f_n y^n + f_{n-1} y^{n-1} + \cdots + f_1 y + f_0 = 0$$

(PT2.5)

where f_i = an ith-order polynomial in x. *Polynomials* are a simple class of algebraic functions that are represented generally by

$$f_n(x) = a_0 + a_1 x + a_2 x^2 + \cdots + a_n x^n \qquad \text{(PT2.6)}$$

where n = the *order* of the polynomial and the a's = constants. Some specific examples are

$$f_2(x) = 1 - 2.37x + 7.5x^2 \qquad \text{(PT2.7)}$$

and

$$f_6(x) = 5x^2 - x^3 + 7x^6 \qquad \text{(PT2.8)}$$

A *transcendental* function is one that is nonalgebraic. These include trigonometric, exponential, logarithmic, and other, less familiar, functions. Examples are

$$f(x) = \ln x^2 - 1 \qquad \text{(PT2.9)}$$

and

$$f(x) = e^{-0.2x} \sin(3x - 0.5) \qquad \text{(PT2.10)}$$

Roots of equations may be either real or complex. Although there are cases where complex roots of nonpolynomials are of interest, such situations are less common than for polynomials. As a consequence, the standard methods for locating roots typically fall into two somewhat related but primarily distinct problem areas:

1. *The determination of the real roots of algebraic and transcendental equations.* These techniques are usually designed to determine the value of a single real root on the basis of foreknowledge of its approximate location.
2. *The determination of all real and complex roots of polynomials.* These methods are specifically designed for polynomials. They systematically determine all the roots of the polynomial rather than determining a single real root given an approximate location.

In this book we discuss both. Chapters 5 and 6 are devoted to the first category. Chapter 7 deals with polynomials.

PT2.3 ORIENTATION

Some orientation is helpful before proceeding to the numerical methods for determining roots of equations. The following is intended to give you an overview of the material in Part Two. In addition, some objectives have been included to help you focus your efforts when studying the material.

PT2.3.1 Scope and Preview

Figure PT2.1 is a schematic representation of the organization of Part Two. Examine this figure carefully, starting at the top and working clockwise.

After the present introduction, *Chap. 5* is devoted to *bracketing methods* for finding roots. These methods start with guesses that bracket, or contain, the root and then systematically reduce the width of the bracket. Two specific methods are covered: *bisection*

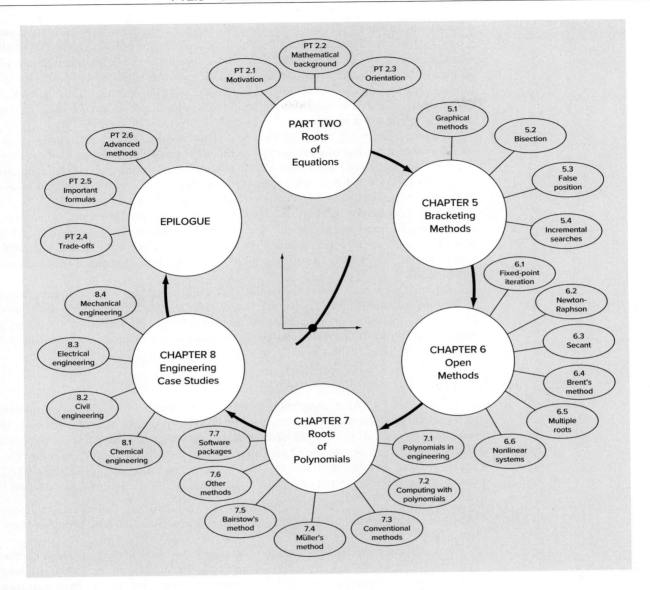

FIGURE PT2.1
Schematic of the organization of the material in Part Two: Roots of Equations.

and *false position*. Graphical methods are used to provide visual insight into the techniques. Error formulations are developed to help you determine how much computational effort is required to estimate the root to a prespecified level of precision.

   Chapter 6 covers *open methods*. These methods also involve systematic trial-and-error iterations but do not require that the initial guesses bracket the root. We will discover that these methods are usually more computationally efficient than bracketing methods but that they do not always work. *One-point iteration, Newton-Raphson,* and *secant methods* are described. Graphical methods are used to provide geometric insight into cases where the open methods do not work. Formulas are developed that provide an idea of how fast open methods home in on the root. An advanced approach, *Brent's method*, that combines the reliability of bracketing with the speed of open methods is described. In addition, an approach to extend the Newton-Raphson method to *systems of nonlinear equations* is explained.

Chapter 7 is devoted to finding the *roots of polynomials.* After background sections on polynomials, the use of conventional methods (in particular the open methods from Chap. 6) are discussed. Then two special methods for locating polynomial roots are described: *Müller's* and *Bairstow's methods.* The chapter ends with information related to finding roots with Excel, MATLAB software, and Mathcad.

Chapter 8 extends the above concepts to actual engineering problems. Engineering case studies are used to illustrate the strengths and weaknesses of each method and to provide insight into the application of the techniques in professional practice. The applications also highlight the trade-offs (as discussed in Part One) associated with the various methods.

An epilogue is included at the end of Part Two. It contains a detailed comparison of the methods discussed in Chaps. 5, 6, and 7. This comparison includes a description of trade-offs related to the proper use of each technique. This section also provides a summary of important formulas, along with references for some numerical methods that are beyond the scope of this text.

PT2.3.2 Goals and Objectives

Study Objectives.   After completing Part Two, you should have sufficient information to successfully approach a wide variety of engineering problems dealing with roots of equations. In general, you should have mastered the techniques, have learned to assess their reliability, and be capable of choosing the best method (or methods) for any particular problem. In addition to these general goals, the specific concepts in Table PT2.2 should be assimilated for a comprehensive understanding of the material in Part Two.

Computer Objectives.   The book provides you with software and simple computer algorithms to implement the techniques discussed in Part Two. All have utility as learning tools.

Pseudocodes for several methods are also supplied directly in the text. This information will allow you to expand your software library to include programs that are more efficient than the bisection method. For example, you may also want to have your own software for the false-position, Newton-Raphson, and secant techniques, which are often more efficient than the bisection method.

Finally, packages such as Excel, MATLAB, and Mathcad have powerful capabilities for locating roots. You can use this part of the book to become familiar with these capabilities.

TABLE PT2.2 Specific study objectives for Part Two.

1. Understand the graphical interpretation of a root.
2. Know the graphical interpretation of the false-position method and why it is usually superior to the bisection method.
3. Understand the difference between bracketing and open methods for root location.
4. Understand the concepts of convergence and divergence; use the two-curve graphical method to provide a visual manifestation of the concepts.
5. Know why bracketing methods always converge, whereas open methods may sometimes diverge.
6. Realize that convergence of open methods is more likely if the initial guess is close to the true root.
7. Understand the concepts of linear and quadratic convergence and their implications for the efficiencies of the fixed-point-iteration and Newton-Raphson methods.
8. Know the fundamental difference between the false-position and secant methods and how it relates to convergence.
9. Understand how Brent's method combines the reliability of bisection with the speed of open methods.
10. Understand the problems posed by multiple roots and the modifications available to mitigate them.
11. Know how to extend the single-equation Newton-Raphson approach to solve systems of nonlinear equations.

Bracketing Methods

This chapter on roots of equations deals with methods that exploit the fact that a function typically changes sign in the vicinity of a root. These techniques are called *bracketing methods* because two initial guesses for the root are required. As the name implies, these guesses must "bracket," or be on either side of, the root. The particular methods described herein employ different strategies to systematically reduce the width of the bracket and, hence, home in on the correct answer.

As a prelude to these techniques, we will briefly discuss graphical methods for depicting functions and their roots. Beyond their utility for providing rough guesses, graphical techniques are also useful for visualizing the properties of the functions and the behavior of the various numerical methods.

5.1 GRAPHICAL METHODS

A simple method for obtaining an estimate of the root of the equation $f(x) = 0$ is to make a plot of the function and observe where it crosses the x axis. This point, which represents the x value for which $f(x) = 0$, provides a rough approximation of the root.

EXAMPLE 5.1 The Graphical Approach

Problem Statement.   Use the graphical approach to determine the drag coefficient c needed for a parachutist of mass $m = 68.1$ kg to have a velocity of 40 m/s after free-falling for time $t = 10$ s. *Note:* The acceleration due to gravity is 9.81 m/s².

Solution.   This problem can be solved by determining the root of Eq. (PT2.4) using the parameters $t = 10$, $g = 9.81$, $v = 40$, and $m = 68.1$:

$$f(c) = \frac{9.81(68.1)}{c}(1 - e^{-(c/68.1)10}) - 40$$

or

$$f(c) = \frac{668.06}{c}(1 - e^{-0.146843c}) - 40 \tag{E5.1.1}$$

Various values of c can be substituted into this equation to compute the following points:

c	f(c)
4	34.190
8	17.712
12	6.114
16	−2.230
20	−8.368

These points are plotted in Fig. 5.1. The resulting curve crosses the c axis between 12 and 16. Visual inspection of the plot provides a rough estimate of the root of 14.75. The validity of the graphical estimate can be checked by substituting it into Eq. (E5.1.1) to yield

$$f(14.75) = \frac{668.06}{14.75}(1 - e^{-0.146843(14.75)}) - 40 = 0.100$$

which is close to zero. It can also be checked by substituting it into Eq. (PT2.3) along with the parameter values from this example to give

$$v = \frac{9.81(68.1)}{14.75}(1 - e^{-(14.75/68.1)10}) = 40.100$$

which is very close to the desired fall velocity of 40 m/s.

FIGURE 5.1
The graphical approach for determining the roots of an equation.

Graphical techniques are of limited practical value because they are not precise. However, graphical methods can be utilized to obtain rough estimates of roots. These estimates can be employed as starting guesses for numerical methods discussed in this and the next chapter.

Aside from providing rough estimates of the roots, graphical interpretations are important tools for understanding the properties of the functions and anticipating the pitfalls of the numerical methods. For example, Fig. 5.2 shows a number of ways in which roots can occur (or be absent) in an interval prescribed by a lower bound x_l and an upper bound x_u. Figure 5.2b depicts the case where a single root is bracketed by negative and positive values of $f(x)$. However, Fig. 5.2d, where $f(x_l)$ and $f(x_u)$ are also on opposite sides of the x axis, shows three roots occurring within the interval. In general, if $f(x_l)$ and $f(x_u)$ have opposite signs, there are an odd number of roots in the interval. As indicated by Fig. 5.2a and c, if $f(x_l)$ and $f(x_u)$ have the same sign, there are either no roots or an even number of roots between the values.

Although these generalizations are usually true, there are cases where they do not hold. For example, functions that are tangential to the x axis (Fig. 5.3a) and discontinuous functions (Fig. 5.3b) can violate these principles. An example of a function that is tangential to the axis is the cubic equation $f(x) = (x - 2)(x - 2)(x - 4)$. Notice that $x = 2$ makes two terms in this polynomial equal to zero. Mathematically, $x = 2$ is called a *multiple root*. At the end of Chap. 6, we will present techniques that are expressly designed to locate multiple roots.

The existence of cases of the type depicted in Fig. 5.3 makes it difficult to develop general computer algorithms guaranteed to locate all the roots in an interval. However, when used in conjunction with graphical approaches, the methods described in the

FIGURE 5.2

Illustration of a number of general ways that a root may occur in an interval prescribed by a lower bound x_l and an upper bound x_u. Parts (a) and (c) indicate that if both $f(x_l)$ and $f(x_u)$ have the same sign, either there will be no roots or there will be an even number of roots within the interval. Parts (b) and (d) indicate that if the function has different signs at the end points, there will be an odd number of roots in the interval.

FIGURE 5.3

Illustration of some exceptions to the general cases depicted in Fig. 5.2. (a) Multiple root that occurs when the function is tangential to the x axis. For this case, although the end points are of opposite signs, there are an even number of axis intersections for the interval. (b) Discontinuous function where end points of opposite sign bracket an even number of roots. Special strategies are required for determining the roots for these cases.

following sections are extremely useful for solving many roots-of-equations problems confronted routinely by engineers and applied mathematicians.

EXAMPLE 5.2 **Use of Computer Graphics to Locate Roots**

Problem Statement. Computer graphics can expedite and improve your efforts to locate roots of equations. The function

$$f(x) = \sin 10x + \cos 3x$$

has several roots over the range $x = 0$ to $x = 5$. Use computer graphics to gain insight into the behavior of this function.

Solution. Packages such as Excel and MATLAB software can be used to generate plots. Figure 5.4a is a plot of $f(x)$ from $x = 0$ to $x = 5$. This plot suggests the presence of several roots, including a possible double root at about $x = 4.2$ where $f(x)$ appears to be

FIGURE 5.4
The progressive enlargement of $f(x) = \sin 10x + \cos 3x$ by the computer. Such interactive graphics permit the analyst to determine that two distinct roots exist between $x = 4.2$ and $x = 4.3$.

(a)

(b)

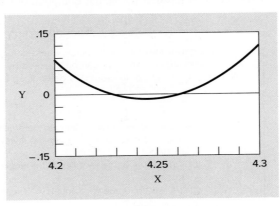

(c)

tangent to the x axis. A more detailed picture of the behavior of $f(x)$ is obtained by changing the plotting range from $x = 3$ to $x = 5$, as shown in Fig. 5.4b. Finally, in Fig. 5.4c, the vertical scale is narrowed further to $f(x) = -0.15$ to $f(x) = 0.15$ and the horizontal scale is narrowed to $x = 4.2$ to $x = 4.3$. This plot shows clearly that a double root does not exist in this region and that in fact there are two distinct roots at about $x = 4.23$ and $x = 4.26$.

Computer graphics will have great utility in your studies of numerical methods. This capability will also find many other applications in your other classes and professional activities as well.

5.2   THE BISECTION METHOD

When applying the graphical technique in Example 5.1, you have observed (Fig. 5.1) that $f(x)$ changed sign on opposite sides of the root. In general, if $f(x)$ is real and continuous in the interval from x_l to x_u and $f(x_l)$ and $f(x_u)$ have opposite signs, that is,

$$f(x_l)f(x_u) < 0 \tag{5.1}$$

then there is at least one real root between x_l and x_u.

Incremental search methods capitalize on this observation by locating an interval where the function changes sign. Then the location of the sign change (and consequently, the root) is identified more precisely by dividing the interval into a number of subintervals. Each of these subintervals is searched to locate the sign change. The process is repeated and the root estimate refined by dividing the subintervals into finer increments. We will return to the general topic of incremental searches in Sec. 5.4.

The *bisection method,* which is alternatively called binary chopping, interval halving, or Bolzano's method, is one type of incremental search method in which the interval is always divided in half. If a function changes sign over an interval, the function value at the midpoint is evaluated. The location of the root is then determined as lying at the midpoint of the subinterval within which the sign change occurs. The process is repeated to obtain refined estimates. A simple algorithm for the bisection calculation is listed in Fig. 5.5, and a graphical depiction of the method is provided in Fig. 5.6. The following example goes through the actual computations involved in the method.

FIGURE 5.5

Step 1: Choose lower x_l and upper x_u guesses for the root such that the function changes sign over the interval. This can be checked by ensuring that $f(x_l)f(x_u) < 0$.

Step 2: An estimate of the root x_r is determined by

$$x_r = \frac{x_l + x_u}{2}$$

Step 3: Make the following evaluations to determine in which subinterval the root lies:

(a) If $f(x_l)f(x_r) < 0$, the root lies in the lower subinterval. Therefore, set $x_u = x_r$ and return to step 2.

(b) If $f(x_l)f(x_r) > 0$, the root lies in the upper subinterval. Therefore, set $x_l = x_r$ and return to step 2.

(c) If $f(x_l)f(x_r) = 0$, the root equals x_r; terminate the computation.

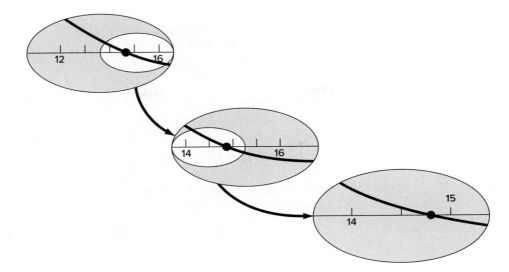

FIGURE 5.6
A graphical depiction of the
bisection method. This plot
conforms to the first three
iterations from Example 5.3.

EXAMPLE 5.3 **Bisection**

Problem Statement.   Use bisection to solve the same problem approached graphically in Example 5.1.

Solution.   The first step in bisection is to guess two values of the unknown (in the present problem, c) that give values for $f(c)$ with different signs. From Fig. 5.1, we can see that the function changes sign between values of 12 and 16. Therefore, the initial estimate of the root x_r lies at the midpoint of the interval

$$x_r = \frac{12 + 16}{2} = 14$$

This estimate represents a true percent relative error of $\varepsilon_t = 5.3\%$ (note that the true value of the root is 14.8011). Next we compute the product of the function value at the lower bound and at the midpoint:

$$f(12)f(14) = 6.114(1.611) = 9.850$$

which is greater than zero, and hence no sign change occurs between the lower bound and the midpoint. Consequently, the root must be located between 14 and 16. Therefore, we create a new interval by redefining the lower bound as 14 and determining a revised root estimate as

$$x_r = \frac{14 + 16}{2} = 15$$

which represents a true percent error of $\varepsilon_t = 1.3\%$. The process can be repeated to obtain refined estimates. For example,

$$f(14)f(15) = 1.611(-0.384) = -0.619$$

Therefore, the root is between 14 and 15. The upper bound is redefined as 15, and the root estimate for the third iteration is calculated as

$$x_r = \frac{14 + 15}{2} = 14.5$$

which represents a percent relative error of $\varepsilon_t = 2.0\%$. The method can be repeated until the result is accurate enough to satisfy your needs.

In the previous example, you may have noticed that the true error does not decrease with each iteration. However, the interval within which the root is located is halved with each step in the process. As discussed in the next section, the interval width provides an exact estimate of the upper bound of the error for the bisection method.

5.2.1 Termination Criteria and Error Estimates

We ended Example 5.3 with the statement that the method could be continued to obtain a refined estimate of the root. We must now develop an objective criterion for deciding when to terminate the method.

An initial suggestion might be to end the calculation when the true error falls below some prespecified level. For instance, in Example 5.3, the relative error dropped to 2.0 percent during the course of the computation. We might decide that we should terminate when the error drops below, say, 0.1 percent. This strategy is flawed because the error estimates in the example were based on knowledge of the true root of the function. This would not be the case in an actual situation because there would be no point in using the method if we already knew the root.

Therefore, we require an error estimate that is not contingent on foreknowledge of the root. As developed previously in Sec. 3.3, an approximate percent relative error ε_a can be calculated, as in [recall Eq. (3.5)]

$$\varepsilon_a = \left| \frac{x_r^{new} - x_r^{old}}{x_r^{new}} \right| 100\% \tag{5.2}$$

where x_r^{new} is the root for the present iteration and x_r^{old} is the root from the previous iteration. The absolute value is used because we are usually concerned with the magnitude of ε_a rather than with its sign. When ε_a becomes less than a prespecified stopping criterion ε_s, the computation is terminated.

EXAMPLE 5.4   **Error Estimates for Bisection**

Problem Statement.   Continue Example 5.3 until the approximate error falls below a stopping criterion of $\varepsilon_s = 0.5\%$. Use Eq. (5.2) to compute the errors.

Solution.   The results of the first two iterations for Example 5.3 were 14 and 15. Substituting these values into Eq. (5.2) yields

$$|\varepsilon_a| = \left| \frac{15 - 14}{15} \right| 100\% = 6.667\%$$

Recall that the true percent relative error for the root estimate of 15 was 1.3%. Therefore, ε_a is greater than ε_t. This behavior is manifested for the other iterations:

Iteration	x_l	x_u	x_r	ε_a (%)	ε_t (%)
1	12	16	14		5.413
2	14	16	15	6.667	1.344
3	14	15	14.5	3.448	2.035
4	14.5	15	14.75	1.695	0.345
5	14.75	15	14.875	0.840	0.499
6	14.75	14.875	14.8125	0.422	0.077

Thus, after six iterations ε_a finally falls below $\varepsilon_s = 0.5\%$, and the computation can be terminated.

These results are summarized in Fig. 5.7. The "ragged" nature of the true error is due to the fact that, for bisection, the true root can lie anywhere within the bracketing interval. The true and approximate errors are far apart when the interval happens to be centered on the true root. They are close when the true root falls at either end of the interval.

Although the approximate error does not provide an exact estimate of the true error, Fig. 5.7 suggests that ε_a captures the general downward trend of ε_t. In addition, the plot exhibits the extremely attractive characteristic that ε_a is always greater than ε_t. Thus,

FIGURE 5.7
Errors for the bisection method. True and estimated errors are plotted versus the number of iterations.

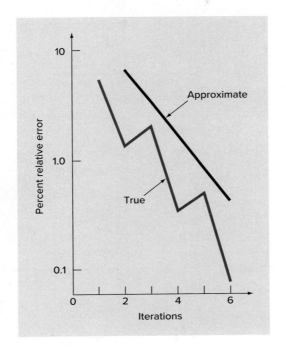

when ε_a falls below ε_s, the computation could be terminated with confidence that the root is known to be at least as accurate as the prespecified acceptable level.

Although it is always dangerous to draw general conclusions from a single example, it can be demonstrated that ε_a will always be greater than ε_t for the bisection method. This is because each time an approximate root is located using bisection as $x_r = (x_l + x_u)/2$, we know that the true root lies somewhere within an interval of $(x_u - x_l)/2 = \Delta x/2$. Therefore, the root must lie within $\pm\Delta x/2$ of our estimate (Fig. 5.8). For instance, when Example 5.3 was terminated, we could have made the definitive statement that

$$x_r = 14.5 \pm 0.5$$

Because $\Delta x/2 = x_r^{new} - x_r^{old}$ (Fig. 5.9), Eq. (5.2) provides an exact upper bound on the true error. For this bound to be exceeded, the true root would have to fall outside the bracketing interval, which, by definition, could never occur for the bisection method. As illustrated in a subsequent example (Example 5.6), other root-locating techniques do not always behave as nicely. Although bisection is generally slower than other methods,

FIGURE 5.8

Three ways in which the interval may bracket the root. In (*a*) the true value lies at the center of the interval, whereas in (*b*) and (*c*) the true value lies near the extreme. Notice that the discrepancy between the true value and the midpoint of the interval never exceeds half the interval length, or $\Delta x/2$.

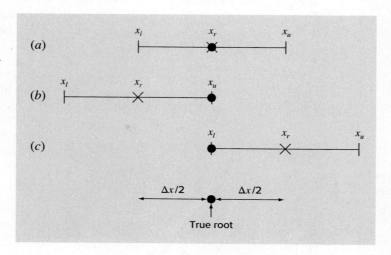

FIGURE 5.9

Graphical depiction of why the error estimate for bisection ($\Delta x/2$) is equivalent to the root estimate for the present iteration (x_r^{new}) minus the root estimate for the previous iteration (x_r^{old}).

the neatness of its error analysis is certainly a positive aspect that could make it attractive for certain engineering applications.

Before proceeding to the computer program for bisection, we should note that the relationships (Fig. 5.9)

$$x_r^{new} - x_r^{old} = \frac{x_u - x_l}{2}$$

and

$$x_r^{new} = \frac{x_l + x_u}{2}$$

can be substituted into Eq. (5.2) to develop an alternative formulation for the approximate percent relative error

$$\varepsilon_a = \left| \frac{x_u - x_l}{x_u + x_l} \right| 100\% \qquad (5.3)$$

This equation yields identical results to Eq. (5.2) for bisection. In addition, it allows us to calculate an error estimate on the basis of our initial guesses—that is, on our first iteration. For instance, on the first iteration of Example 5.2, an approximate error can be computed as

$$\varepsilon_a = \left| \frac{16 - 12}{16 + 12} \right| 100\% = 14.29\%$$

Another benefit of the bisection method is that the number of iterations required to attain an absolute error can be computed *a priori*—that is, before starting the iterations. This can be seen by recognizing that before starting the technique, the absolute error is

$$E_a^0 = x_u^0 - x_l^0 = \Delta x^0$$

where the superscript designates the iteration. Hence, before starting the method, we are at the "zero iteration." After the first iteration, the error becomes

$$E_a^1 = \frac{\Delta x^0}{2}$$

Because each succeeding iteration halves the error, a general formula relating the error and the number of iterations n is

$$E_a^n = \frac{\Delta x^0}{2^n} \qquad (5.4)$$

If $E_{a,d}$ is the desired error, this equation can be solved for

$$n = \frac{\log(\Delta x^0 / E_{a,d})}{\log 2} = \log_2 \left(\frac{\Delta x^0}{E_{a,d}} \right) \qquad (5.5)$$

Let us test the formula. For Example 5.4, the initial interval was $\Delta x_0 = 16 - 12 = 4$. After six iterations, the absolute error was

$$E_a = \frac{|14.875 - 14.75|}{2} = 0.0625$$

We can substitute these values into Eq. (5.5) to give

$$n = \frac{\log(4/0.0625)}{\log 2} = 6$$

Thus, if we knew beforehand that an error of less than 0.0625 was acceptable, the formula tells us that six iterations would yield the desired result.

Although we have emphasized the use of relative errors for obvious reasons, there will be cases where (usually through knowledge of the problem context) you will be able to specify an absolute error. For these cases, bisection along with Eq. (5.5) can provide a useful root-location algorithm. We will explore such applications in the end-of-chapter problems.

5.2.2 Bisection Algorithm

The algorithm in Fig. 5.5 can now be expanded to include the error check (Fig. 5.10). The algorithm employs user-defined functions to make root location and function evaluation more efficient. In addition, an upper limit is placed on the number of iterations. Finally, an error check is included to avoid division by zero during the error evaluation. Such would be the case when the bracketing interval is centered on zero. For this situation, Eq. (5.2) becomes infinite. If this occurs, the program skips over the error evaluation for that iteration.

The algorithm in Fig. 5.10 is not user-friendly; it is designed strictly to come up with the answer. In Prob. 5.21 at the end of this chapter, you will have the task of making it easier to use and understand.

FIGURE 5.10
Pseudocode for function to implement bisection.

```
FUNCTION Bisect(xl, xu, es, imax, xr, iter, ea)
  iter = 0
  DO
    xrold = xr
    xr = (xl + xu) / 2
    iter = iter + 1
    IF xr ≠ 0 THEN
      ea = ABS((xr - xrold) / xr) * 100
    END IF
    test = f(xl) * f(xr)
    IF test < 0 THEN
      xu = xr
    ELSE IF test > 0 THEN
      xl = xr
    ELSE
      ea = 0
    END IF
    IF ea < es OR iter ≥ imax EXIT
  END DO
  Bisect = xr
END Bisect
```

5.2.3 Minimizing Function Evaluations

The bisection algorithm in Fig. 5.10 is just fine if you are performing a single root evaluation for a function that is easy to evaluate. However, there are many instances in engineering when this is not the case. For example, suppose that you develop a computer program that must locate a root numerous times. In such cases you could call the algorithm from Fig. 5.10 thousands and even millions of times in the course of a single run.

Further, in its most general sense, a univariate function is merely an entity that returns a single value in return for a single value you send to it. Perceived in this sense, functions are not always simple formulas like the one-line equations solved in the preceding examples in this chapter. For example, a function might consist of many lines of code that could take a significant amount of execution time to evaluate. In some cases, the function might even represent an independent computer program.

Because of both these factors, it is imperative that numerical algorithms minimize function evaluations. In this light, the algorithm from Fig. 5.10 is deficient. In particular, notice that in making two function evaluations per iteration, it recalculates one of the functions that was determined on the previous iteration.

Figure 5.11 provides a modified algorithm that does not have this deficiency. We have highlighted the lines that differ from Fig. 5.10. In this case, only the new function value at

FIGURE 5.11

Pseudocode for bisection sub-program that minimizes function evaluations.

```
FUNCTION Bisect(xl, xu, es, imax, xr, iter, ea)
  iter = 0
  fl = f(xl)
  DO
    xrold = xr
    xr = (xl + xu) / 2
    fr = f(xr)
    iter = iter + 1
    IF xr ≠ 0 THEN
      ea = ABS((xr - xrold) / xr) * 100
    END IF
    test = fl * fr
    IF test < 0 THEN
      xu = xr
    ELSE IF test > 0 THEN
      xl = xr
      fl = fr
    ELSE
      ea = 0
    END IF
    IF ea < es OR iter ≥ imax EXIT
  END DO
  Bisect = xr
END Bisect
```

the root estimate is calculated. Previously calculated values are saved and merely reassigned as the bracket shrinks. Thus, $n + 1$ function evaluations are performed, rather than $2n$.

5.3   THE FALSE-POSITION METHOD

Although bisection is a perfectly valid technique for determining roots, its "brute-force" approach is relatively inefficient. False position is an alternative based on a graphical insight.

A shortcoming of the bisection method is that, in dividing the interval from x_l to x_u into equal halves, no account is taken of the magnitudes of $f(x_l)$ and $f(x_u)$. For example, if $f(x_l)$ is much closer to zero than $f(x_u)$, it is likely that the root is closer to x_l than to x_u (Fig. 5.12). An alternative method that exploits this graphical insight is to join $f(x_l)$ and $f(x_u)$ by a straight line. The intersection of this line with the x axis represents an improved estimate of the root. The fact that the replacement of the curve by a straight line gives a "false position" of the root is the origin of the name, *method of false position*, or in Latin, *regula falsi*. It is also called the *linear interpolation method*.

Using similar triangles (Fig. 5.12), the intersection of the straight line with the x axis can be estimated as

$$\frac{f(x_l)}{x_r - x_l} = \frac{f(x_u)}{x_r - x_u} \tag{5.6}$$

which can be solved for (see Box 5.1 for details).

$$x_r = x_u - \frac{f(x_u)(x_l - x_u)}{f(x_l) - f(x_u)} \tag{5.7}$$

FIGURE 5.12
A graphical depiction of the method of false position. Similar triangles used to derive the formula for the method are shaded.

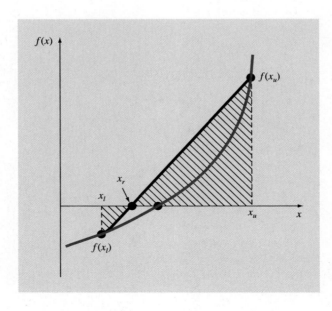

Box 5.1   Derivation of the Method of False Position

Cross-multiply Eq. (5.6) to yield

$$f(x_l)(x_r - x_u) = f(x_u)(x_r - x_l)$$

Collect terms and rearrange:

$$x_r[f(x_l) - f(x_u)] = x_u f(x_l) - x_l f(x_u)$$

Divide by $f(x_l) - f(x_u)$:

$$x_r = \frac{x_u f(x_l) - x_l f(x_u)}{f(x_l) - f(x_u)} \qquad (B5.1.1)$$

This is one form of the method of false position. Note that it allows the computation of the root x_r as a function of the lower and upper guesses x_l and x_u. It can be put in an alternative form by expanding it:

$$x_r = \frac{x_u f(x_l)}{f(x_l) - f(x_u)} - \frac{x_l f(x_u)}{f(x_l) - f(x_u)}$$

then adding and subtracting x_u on the right-hand side:

$$x_r = x_u + \frac{x_u f(x_l)}{f(x_l) - f(x_u)} - x_u - \frac{x_l f(x_u)}{f(x_l) - f(x_u)}$$

Collecting terms yields

$$x_r = x_u + \frac{x_u f(x_u)}{f(x_l) - f(x_u)} - \frac{x_l f(x_u)}{f(x_l) - f(x_u)}$$

or

$$x_r = x_u - \frac{f(x_u)(x_l - x_u)}{f(x_l) - f(x_u)}$$

which is the same as Eq. (5.7). We use this form because it involves one less function evaluation and one less multiplication than Eq. (B5.1.1). In addition, it is directly comparable with the secant method, which will be discussed in Chap. 6.

This is the *false-position formula.* The value of x_r computed with Eq. (5.7) then replaces whichever of the two initial guesses, x_l or x_u, yields a function value with the same sign as $f(x_r)$. In this way, the values of x_l and x_u always bracket the true root. The process is repeated until the root is estimated adequately. The algorithm is identical to the one for bisection (Fig. 5.5) with the exception that Eq. (5.7) is used for step 2. In addition, the same stopping criterion [Eq. (5.2)] is used to terminate the computation.

EXAMPLE 5.5   **False Position**

Problem Statement.   Use the false-position method to determine the root of the same equation investigated in Example 5.1 [Eq. (E5.1.1)].

Solution.   As in Example 5.3, initiate the computation with guesses of $x_l = 12$ and $x_u = 16$.

First iteration:

$$x_l = 12 \qquad f(x_l) = 6.1139$$
$$x_u = 16 \qquad f(x_u) = -2.2303$$
$$x_r = 16 - \frac{-2.2303(12 - 16)}{6.1139 - (-2.2303)} = 14.309$$

which has a true relative error of 0.88%.

Second iteration:

$$f(x_l)f(x_r) = -1.5376$$

Therefore, the root lies in the first subinterval, and x_r becomes the upper limit for the next iteration, $x_u = 14.9113$:

$$x_l = 12 \qquad\qquad f(x_l) = 6.1139$$
$$x_u = 14.9309 \qquad\qquad f(x_u) = -0.2515$$
$$x_r = 14.9309 - \frac{-0.2515(12 - 14.9309)}{6.1139 - (-0.2515)} = 14.8151$$

which has true and approximate relative errors of 0.09 and 0.78%. Additional iterations can be performed to refine the estimate of the roots.

A feeling for the relative efficiency of the bisection and false-position methods can be appreciated by referring to Fig. 5.13, where we have plotted the true percent relative errors for Examples 5.4 and 5.5. Note how the error for false position decreases much faster than for bisection because of the more efficient scheme for root location in the false-position method.

Recall in the bisection method that the interval between x_l and x_u grew smaller during the course of a computation. The interval, as defined by $\Delta x/2 = |x_u - x_l|/2$ for the first iteration, therefore provided a measure of the error for this approach. This is not the case

FIGURE 5.13

Comparison of the relative errors of the bisection and the false-position methods.

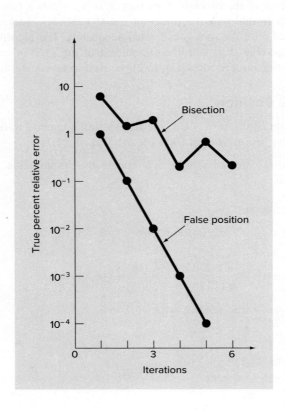

for the method of false position because one of the initial guesses may stay fixed throughout the computation as the other guess converges on the root. For instance, in Example 5.5 the lower guess x_l remained at 12 while x_u converged on the root. For such cases, the interval does not shrink but rather approaches a constant value.

Example 5.5 suggests that Eq. (5.2) represents a very conservative error criterion. In fact, Eq. (5.2) actually constitutes an approximation of the discrepancy of the previous iteration. This is because for a case such as Example 5.5, where the method is converging quickly (for example, the error is being reduced nearly an order of magnitude per iteration), the root for the present iteration x_r^{new} is a much better estimate of the true value than the result of the previous iteration x_r^{old}. Thus, the quantity in the numerator of Eq. (5.2) actually represents the discrepancy of the previous iteration. Consequently, we are assured that satisfaction of Eq. (5.2) ensures that the root will be known with greater accuracy than the prescribed tolerance. However, as described in the next section, there are cases where false position converges slowly. For these cases, Eq. (5.2) becomes unreliable, and an alternative stopping criterion must be developed.

5.3.1 Pitfalls of the False-Position Method

Although the false-position method would seem to always be the bracketing method of preference, there are cases where it performs poorly. In fact, as in the following example, there are certain cases where bisection yields superior results.

EXAMPLE 5.6

A Case Where Bisection Is Preferable to False Position

Problem Statement.   Use bisection and false position to locate the root of

$$f(x) = x^{10} - 1$$

between $x = 0$ and 1.3.

Solution.   Using bisection, the results can be summarized as follows:

Iteration	x_l	x_u	x_r	ε_a (%)	ε_t (%)
1	0	1.3	0.65	100.0	35
2	0.65	1.3	0.975	33.3	2.5
3	0.975	1.3	1.1375	14.3	13.8
4	0.975	1.1375	1.05625	7.7	5.6
5	0.975	1.05625	1.015625	4.0	1.6

Thus, after five iterations, the true error is reduced to less than 2%. For false position, a very different outcome is obtained:

Iteration	x_l	x_u	x_r	ε_a (%)	ε_t (%)
1	0	1.3	0.09430		90.6
2	0.09430	1.3	0.18176	48.1	81.8
3	0.18176	1.3	0.26287	30.9	73.7
4	0.26287	1.3	0.33811	22.3	66.2
5	0.33811	1.3	0.40788	17.1	59.2

FIGURE 5.14
Plot of $f(x) = x^{10} - 1$, illustrating slow convergence of the false-position method.

After five iterations, the true error has only been reduced to about 59 percent. In addition, note that $\varepsilon_a < \varepsilon_t$. Thus, the approximate error is misleading. Insight into these results can be gained by examining a plot of the function. As in Fig. 5.14, the curve violates the premise upon which false position was based—that is, if $f(x_l)$ is much closer to zero than $f(x_u)$, then the root is closer to x_l than to x_u (recall Fig. 5.12). Because of the shape of the present function, the opposite is true.

The forgoing example illustrates that blanket generalizations regarding root-location methods are usually not possible. Although a method such as false position is often superior to bisection, there are invariably cases that violate this general conclusion. Therefore, in addition to using Eq. (5.2), the results should always be checked by substituting the root estimate into the original equation and determining whether the result is close to zero. Such a check should be incorporated into all computer programs for root location.

The example also illustrates a major weakness of the false-position method: its one-sidedness. That is, as iterations are proceeding, one of the bracketing points will tend to stay fixed. This can lead to poor convergence, particularly for functions with significant curvature. The following section provides a remedy.

5.3.2 Modified False Position

One way to mitigate the "one-sided" nature of false position is to have the algorithm detect when one of the bounds is stuck. If this occurs, the function value at the stagnant bound can be divided in half. This is called the *modified false-position method.*

The algorithm in Fig. 5.15 implements this strategy. Notice how counters are used to determine when one of the bounds stays fixed for two iterations. If this occurs, the function value at this stagnant bound is halved.

The effectiveness of this algorithm can be demonstrated by applying it to Example 5.6. If a stopping criterion of 0.01% is used, the bisection and standard false-position

FIGURE 5.15

Pseudocode for the modified false-position method.

```
FUNCTION ModFalsePos(xl, xu, es, imax, xr, iter, ea)
  iter = 0
  fl = f(xl)
  fu = f(xu)
  DO
    xrold = xr
    xr = xu - fu * (xl - xu) / (fl - fu)
    fr = f(xr)
    iter = iter + 1
    IF xr <> 0 THEN
      ea = Abs((xr - xrold) / xr) * 100
    END IF
    test = fl * fr
    IF test < 0 THEN
      xu = xr
      fu = f(xu)
      iu = 0
      il = il + 1
      IF il ≥ 2 THEN fl = fl / 2
    ELSE IF test > 0 THEN
      xl = xr
      fl = f(xl)
      il = 0
      iu = iu + 1
      IF iu ≥ 2 THEN fu = fu / 2
    ELSE
      ea = 0
    END IF
    IF ea < es OR iter ≥ imax THEN EXIT
  END DO
  ModFalsePos = xr
END MODFALSEPOS
```

methods would converge in 14 and 39 iterations, respectively. In contrast, the modified false-position method would converge in 12 iterations. Thus, for this example, it is somewhat more efficient than bisection and is vastly superior to the unmodified false-position method.

5.4 INCREMENTAL SEARCHES AND DETERMINING INITIAL GUESSES

Besides checking an individual answer, you must determine whether all possible roots have been located. As mentioned previously, a plot of the function is usually very useful in guiding you in this task. Another option is to incorporate an incremental search at the beginning of the computer program. This consists of starting at one end of the region of interest and then making function evaluations at small increments across the region. When the function changes sign, it is assumed that a root falls within the increment. The x values at the beginning and the end of the increment can then serve as the initial guesses for one of the bracketing techniques described in this chapter.

A potential problem with an incremental search is the choice of the increment length. If the length is too small, the search can be very time-consuming. On the other hand, if the length is too great, there is a possibility that closely spaced roots might be missed (Fig. 5.16). The problem is compounded by the possible existence of multiple roots. A partial remedy for such cases is to compute the first derivative of the function $f'(x)$ at the beginning and the end of each interval. If the derivative changes sign, it suggests that a minimum or maximum may have occurred and that the interval should be examined more closely for the existence of a possible root.

Although such modifications or the employment of a very fine increment can alleviate the problem, it should be clear that brute-force methods such as incremental search are not foolproof. You would be wise to supplement such automatic techniques with any other information that provides insight into the location of the roots. Such information can be found in plotting and in understanding the physical problem from which the equation originated.

FIGURE 5.16
Cases where roots could be missed because the increment length of the search procedure is too large. Note that the last root on the right is multiple and would be missed regardless of increment length.

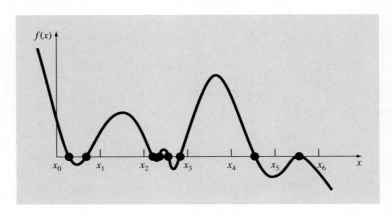

PROBLEMS

5.1 Determine the real roots of $f(x) = -0.6x^2 + 2.4x + 5.5$:
(a) Graphically.
(b) Using the quadratic formula.
(c) Using three iterations of the bisection method to determine the highest root. Employ initial guesses of $x_l = 5$ and $x_u = 10$. Compute the estimated error ε_a and the true error ε_t after each iteration.

5.2 Determine the real root of $f(x) = 4x^3 - 6x^2 + 7x - 2.3$:
(a) Graphically.
(b) Using bisection to locate the root. Employ initial guesses of $x_l = 0$ and $x_u = 1$ and iterate until the estimated error ε_a falls below a level of $\varepsilon_s = 10\%$.

5.3 Determine the real root of $f(x) = -26 + 85x - 91x^2 + 44x^3 - 8x^4 + x^5$:
(a) Graphically.
(b) Using bisection to determine the root to $\varepsilon_s = 10\%$. Employ initial guesses of $x_l = 0.5$ and $x_u = 1.0$.
(c) Perform the same computation as in **(b)** but use the false-position method and $\varepsilon_s = 0.2\%$.

5.4 **(a)** Determine the roots of $f(x) = -13 - 20x + 19x^2 - 3x^3$ graphically. In addition, determine the first root of the function with **(b)** bisection, and **(c)** false position. For **(b)** and **(c)**, use initial guesses of $x_l = -1$ and $x_u = 0$ and a stopping criterion of 1%.

5.5 Locate the first nontrivial root of $\sin x = x^3$ where x is in radians. Use a graphical technique and bisection with the initial interval from 0.5 to 1. Perform the computation until ε_a is less than $\varepsilon_s = 2\%$. Also perform an error check by substituting your final answer into the original equation.

5.6 Determine the positive real root of $\ln(x^4) = 0.7$ **(a)** graphically, **(b)** using three iterations of the bisection method, with initial guesses of $x_l = 0.5$ and $x_u = 2$, and **(c)** using three iterations of the false-position method, with the same initial guesses as in **(b)**.

5.7 Determine the real root of $f(x) = (0.8 - 0.3x)/x$:
(a) Analytically.
(b) Graphically.
(c) Using three iterations of the false-position method and initial guesses of 1 and 3. Compute the approximate error ε_a and the true error ε_t after each iteration. Is there a problem with the result?

5.8 Find the positive square root of 18 using the false-position method to within $\varepsilon_s = 0.5\%$. Employ initial guesses of $x_l = 4$ and $x_u = 5$.

5.9 Find the smallest positive root of the function (x is in radians) $x^2|\cos\sqrt{x}| = 5$ using the false-position method. To locate the region in which the root lies, first plot this function for values of x between 0 and 5. Perform the computation until ε_a falls below $\varepsilon_s = 1\%$. Check your final answer by substituting it into the original function.

5.10 Find the positive real root of $f(x) = x^4 - 8x^3 - 35x^2 + 450x - 1001$ using the false-position method. Use initial guesses of $x_l = 4.5$ and $x_u = 6$ and perform five iterations. Compute both the true and approximate errors based on the fact that the root is 5.60979. Use a plot to explain your results and perform the computation to within $\varepsilon_s = 1.0\%$.

5.11 Determine the real root of $x^{3.5} = 80$: **(a)** analytically and **(b)** with the false-position method to within $\varepsilon_s = 2.5\%$. Use initial guesses of 2.0 and 5.0.

5.12 Given

$$f(x) = -2x^6 - 1.6x^4 + 12x + 1$$

use bisection to determine the *maximum* of this function. Employ initial guesses of $x_l = 0$ and $x_u = 1$, and perform iterations until the approximate relative error falls below 5%.

5.13 The velocity v of a falling parachutist is given by

$$v = \frac{gm}{c}(1 - e^{-(c/m)t})$$

where $g = 9.81\,\text{m/s}^2$. For a parachutist with a drag coefficient $c = 15$ kg/s, compute the mass m so that the velocity is $v = 35$ m/s at $t = 9$ s. Use the false-position method to determine m to a level of $\varepsilon_s = 0.1\%$.

5.14 Use bisection to determine the drag coefficient needed so that an 80-kg parachutist has a velocity of 35 m/s after 3.9 s of free fall. Note: The acceleration due to gravity is 9.81 m/s². Start with initial guesses of $x_l = 3$ and $x_u = 5$ and iterate until the approximate relative error falls below 5%. Also perform an error check by substituting your final answer into the original equation.

5.15 As depicted in Fig. P5.15, the velocity of water, v (m/s), discharged from a cylindrical tank through a long pipe can be computed as

$$v = \sqrt{2gH}\tanh\left(\frac{\sqrt{2gH}}{2L}t\right)$$

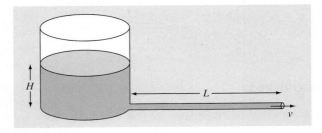

FIGURE P5.15

where $g = 9.81$ m/s^2, $H =$ initial head (m), $L =$ pipe length (m), and $t =$ elapsed time (s). Determine the head needed to achieve $v = 4$ m/s in 3 s for a 5-m-long pipe **(a)** graphically, **(b)** by bisection, and **(c)** with false position. Employ initial guesses of $x_l = 0$ and $x_u = 2$ m with a stopping criterion of $\varepsilon_s = 1\%$. Check your results.

5.16 Water is flowing in a trapezoidal channel at a rate of $Q = 20$ m^3/s. The critical depth y for such a channel must satisfy the equation

$$0 = 1 - \frac{Q^2}{gA_c^3}B$$

where $g = 9.81$ m/s^2, $A_c =$ the cross-sectional area (m^2), and $B =$ the width of the channel at the surface (m). For this case, the width and the cross-sectional area can be related to depth y by

$$B = 3 + y \quad \text{and} \quad A_c = 3y + \frac{y^2}{2}$$

Solve for the critical depth using **(a)** the graphical method, **(b)** bisection, and **(c)** false position. For **(b)** and **(c)**, use initial guesses of $x_l = 0.5$ and $x_u = 3$, and iterate until the approximate error falls below 1% or the number of iterations exceeds 10. Discuss your results.

5.17 You are designing a spherical tank (Fig. P5.17) to hold water for a small village in a developing country. The volume of liquid it can hold can be computed as

$$V = \pi h^2 \frac{[3R - h]}{3}$$

where $V =$ volume (m^3), $h =$ depth of water in tank (m), and $R =$ the tank radius (m).

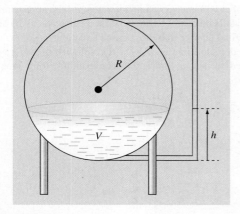

FIGURE P5.17

If $R = 4$ m, to what depth must the tank be filled so that it holds 40 m^3? Use three iterations of the false-position method to determine your answer. Determine the approximate relative error after each iteration. Employ initial guesses of 0 and R.

5.18 The saturation concentration of dissolved oxygen in freshwater can be calculated with the equation (APHA 1992)

$$\ln o_{sf} = -139.34411 + \frac{1.575701 \times 10^5}{T_a} - \frac{6.642308 \times 10^7}{T_a^2}$$
$$+ \frac{1.243800 \times 10^{10}}{T_a^3} - \frac{8.621949 \times 10^{11}}{T_a^4}$$

where $o_{sf} =$ the saturation concentration of dissolved oxygen in freshwater at 1 atm (mg/L) and $T_a =$ absolute temperature (K). Remember that $T_a = T + 273.15$, where $T =$ temperature (°C). According to this equation, saturation decreases with increasing temperature. For typical natural waters in temperate climates, the equation can be used to determine that oxygen concentration ranges from 14.621 mg/L at 0°C to 6.413 mg/L at 40°C. Given a value of oxygen concentration, this formula and the bisection method can be used to solve for temperature in °C.

(a) If the initial guesses are set as 0 and 40°C, how many bisection iterations would be required to determine temperature to an absolute error of 0.05°C?

(b) Develop and test a bisection program to determine T as a function of a given oxygen concentration to a prespecified absolute error as in **(a)**. Given initial guesses of 0 and 40°C, test your program for an absolute error $= 0.05$°C and the following cases: $o_{sf} = 8$, 10, and 13 mg/L. Check your results.

5.19 According to *Archimedes principle*, the *buoyancy* force is equal to the weight of fluid displaced by the submerged portion of an object. For the sphere depicted in Fig. P5.19, use bisection to determine the height h of the portion that is above water. Employ the following values for your computation: $r = 1.25$ m, $\rho_s =$ density of sphere $= 250$ kg/m^3, and $\rho_w =$ density of water $= 1000$ kg/m^3. Note that the volume of the above-water portion of the sphere can be computed with

$$V = \frac{\pi h^2}{3}(3r - h)$$

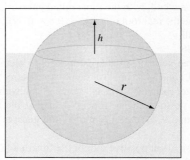

FIGURE P5.19

5.20 Perform the same computation as in Prob. 5.19, but for the *frustrum* of a cone, as depicted in Fig. P5.20. Employ the following values for your computation: $r_1 = 0.4$ m, $r_2 = 1$ m, $h = 1$ m, $\rho_f =$ frustrum density $= 250$ kg/m³, and $\rho_w =$ water density $= 1000$ kg/m³. Note that the volume of a frustrum is given by

$$V = \frac{\pi h}{3}(r_1^2 + r_2^2 + r_1 r_2)$$

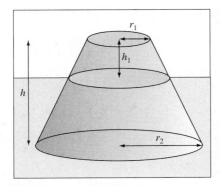

FIGURE P5.20

5.21 Integrate the algorithm outlined in Fig. 5.10 into a complete, user-friendly bisection subprogram. Among other things:

(a) Place documentation statements throughout the subprogram to identify what each section is intended to accomplish.

(b) Label the input and output.

(c) Add an answer check that substitutes the root estimate into the original function to verify whether the final result is close to zero.

(d) Test the subprogram by duplicating the computations from Examples 5.3 and 5.4.

5.22 Develop a subprogram for the bisection method that minimizes function evaluations based on the pseudocode from Fig. 5.11. Determine the number of function evaluations (n) per total iterations. Test the program by duplicating Example 5.6.

5.23 Develop a user-friendly program for the false-position method. The structure of your program should be similar to the bisection algorithm outlined in Fig. 5.10. Test the program by duplicating Example 5.5.

5.24 Develop a subprogram for the false-position method that minimizes function evaluations in a fashion similar to Fig. 5.11. Determine the number of function evaluations (n) per total iterations. Test the program by duplicating Example 5.6.

5.25 Develop a user-friendly subprogram for the modified false-position method based on Fig. 5.15. Test the program by determining the root of the function described in Example 5.6. Perform a number of runs until the true percent relative error falls below 0.01%. Plot the true and approximate percent relative errors versus number of iterations on semilog paper. Interpret your results.

5.26 Develop a function for bisection in a similar fashion to Fig. 5.10. However, rather than using the maximum iterations and Eq. (5.2), employ Eq. (5.5) as your stopping criterion. Make sure to round the result of Eq. (5.5) up to the next highest integer. Test your function by solving Example 5.3 using $E_{a,d} = 0.0001$.

Open Methods

For the bracketing methods in Chap. 5, the root is located within an interval prescribed by a lower and an upper bound. Repeated application of these methods always results in closer estimates of the true value of the root. Such methods are said to be *convergent* because they move closer to the truth as the computation progresses (Fig. 6.1*a*).

In contrast, the *open methods* described in this chapter are based on formulas that require only a single starting value of x or two starting values that do not

FIGURE 6.1
Graphical depiction of the fundamental difference between the (*a*) bracketing and (*b*) and (*c*) open methods for root location. In (*a*), which is the bisection method, the root is constrained within the interval prescribed by x_l and x_u. In contrast, for the open method depicted in (*b*) and (*c*), a formula is used to project from x_i to x_{i+1} in an iterative fashion. Thus, the method can either (*b*) diverge or (*c*) converge rapidly, depending on the value of the initial guess.

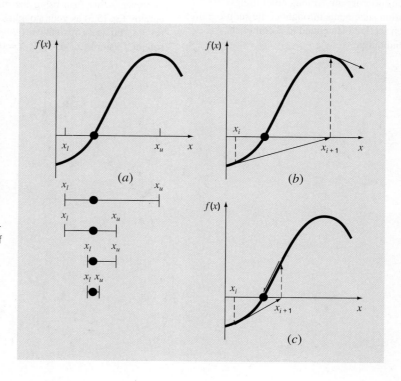

necessarily bracket the root. As such, they sometimes *diverge* or move away from the true root as the computation progresses (Fig. 6.1*b*). However, when the open methods converge (Fig. 6.1*c*), they usually do so much more quickly than the bracketing methods. We will begin our discussion of open techniques with a simple version that is useful for illustrating their general form and also for demonstrating the concept of convergence.

6.1 SIMPLE FIXED-POINT ITERATION

As mentioned above, open methods employ a formula to predict the root. Such a formula can be developed for simple *fixed-point iteration* (or, as it is also called, one-point iteration or successive substitution) by rearranging the function $f(x) = 0$ so that x is on the left-hand side of the equation:

$$x = g(x) \tag{6.1}$$

This transformation can be accomplished either by algebraic manipulation or by simply adding x to both sides of the original equation. For example,

$$x^2 - 2x + 3 = 0$$

can be simply manipulated to yield

$$x = \frac{x^2 + 3}{2}$$

whereas $\sin x = 0$ could be put into the form of Eq. (6.1) by adding x to both sides to yield

$$x = \sin x + x$$

The utility of Eq. (6.1) is that it provides a formula to predict a new value of x as a function of an old value of x. Thus, given an initial guess at the root x_i, Eq. (6.1) can be used to compute a new estimate x_{i+1} as expressed by the iterative formula

$$x_{i+1} = g(x_i) \tag{6.2}$$

As with other iterative formulas in this book, the approximate error for this equation can be determined using the error estimator [Eq. (3.5)]:

$$\varepsilon_a = \left| \frac{x_{i+1} - x_i}{x_{i+1}} \right| 100\%$$

EXAMPLE 6.1   **Simple Fixed-Point Iteration**

Problem Statement.   Use simple fixed-point iteration to locate the root of $f(x) = e^{-x} - x$.

Solution.   The function can be separated directly and expressed in the form of Eq. (6.2) as

$$x_{i+1} = e^{-x_i}$$

Starting with an initial guess of $x_0 = 0$, this iterative equation can be applied to compute

i	x_i	ε_a (%)	ε_t (%)
0	0		100.0
1	1.000000	100.0	76.3
2	0.367879	171.8	35.1
3	0.692201	46.9	22.1
4	0.500473	38.3	11.8
5	0.606244	17.4	6.89
6	0.545396	11.2	3.83
7	0.579612	5.90	2.20
8	0.560115	3.48	1.24
9	0.571143	1.93	0.705
10	0.564879	1.11	0.399

Thus, each iteration brings the estimate closer to the true value of the root: 0.56714329.

6.1.1 Convergence

Notice that the true percent relative error for each iteration of Example 6.1 is roughly proportional (by a factor of about 0.5 to 0.6) to the error from the previous iteration. This property, called *linear convergence,* is characteristic of fixed-point iteration.

Aside from the "rate" of convergence, we must comment at this point about the "possibility" of convergence. The concepts of convergence and divergence can be depicted graphically. Recall that in Sec. 5.1, we graphed a function to visualize its structure and behavior (Example 5.1). Such an approach is employed in Fig. 6.2a for the function $f(x) = e^{-x} - x$. An alternative graphical approach is to separate the equation into two component parts, as in

$$f_1(x) = f_2(x)$$

Then the two equations

$$y_1 = f_1(x) \tag{6.3}$$

and

$$y_2 = f_2(x) \tag{6.4}$$

can be plotted separately (Fig. 6.2b). The x values corresponding to the intersections of these functions represent the roots of $f(x) = 0$.

EXAMPLE 6.2 | **The Two-Curve Graphical Method**

Problem Statement.   Separate the equation $e^{-x} - x = 0$ into two parts and determine its root graphically.

Solution. Reformulate the equation as $y_1 = x$ and $y_2 = e^{-x}$. The following values can be computed:

x	y_1	y_2
0.0	0.0	1.000
0.2	0.2	0.819
0.4	0.4	0.670
0.6	0.6	0.549
0.8	0.8	0.449
1.0	1.0	0.368

These points are plotted in Fig. 6.2b. The intersection of the two curves indicates a root estimate of approximately $x = 0.57$, which corresponds to the point where the single curve in Fig. 6.2a crosses the x axis.

FIGURE 6.2

Two alternative graphical methods for determining the root of $f(x) = e^{-x} - x$. (a) Root at the point where the graph of the function crosses the x axis; (b) root at the intersection of the component functions.

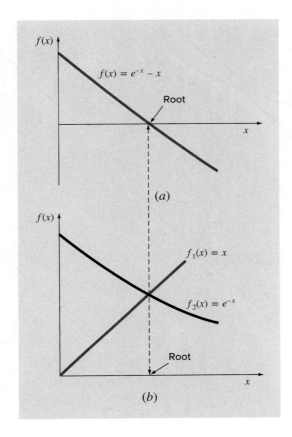

The two-curve method can now be used to illustrate the convergence and divergence of fixed-point iteration. First, Eq. (6.1) can be re-expressed as a pair of equations: $y_1 = x$ and $y_2 = g(x)$. These two equations can then be plotted separately. As was the case with Eqs. (6.3) and (6.4), the roots of $f(x) = 0$ correspond to the abscissa value at the intersection of the two curves. The function $y_1 = x$ and four different shapes for $y_2 = g(x)$ are plotted in Fig. 6.3.

For the first case (Fig. 6.3a), the initial guess of x_0 is used to determine the corresponding point on the y_2 curve, $[x_0, g(x_0)]$. The point (x_1, x_1) is located by moving left horizontally to the y_1 curve. These movements are equivalent to the first iteration in the fixed-point method:

$$x_1 = g(x_0)$$

FIGURE 6.3
Iteration cobwebs depicting convergence (a and b) and divergence (c and d) of simple fixed-point iteration. Graphs (a) and (c) are called monotone patterns, whereas (b) and (d) are called oscillating, or spiral, patterns. Note that convergence occurs when $|g'(x)| < 1$.

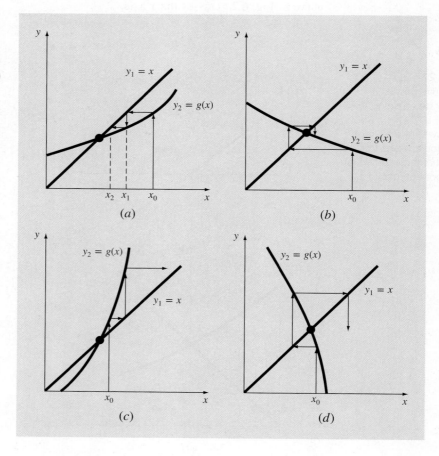

Box 6.1 Convergence of Fixed-Point Iteration

From studying Fig. 6.3, it should be clear that fixed-point itera-
tion converges if, in the region of interest, $|g'(x)| < 1$. In other
words, convergence occurs if the magnitude of the slope of $g(x)$
is less than the slope of the line $f(x) = x$. This observation can be
demonstrated theoretically. Recall that the iterative equation is

$$x_{i+1} = g(x_i)$$

Suppose that the true solution is

$$x_r = g(x_r)$$

Subtracting these equations yields

$$x_r - x_{i+1} = g(x_r) - g(x_i) \qquad (B6.1.1)$$

The *derivative mean-value theorem* (recall Sec. 4.1.1) states
that if a function $g(x)$ and its first derivative are continuous over
an interval $a \le x \le b$, then there exists at least one value of $x = \xi$ within the interval such that

$$g'(\xi) = \frac{g(b) - g(a)}{b - a} \qquad (B6.1.2)$$

The right-hand side of this equation is the slope of the line joining
$g(a)$ and $g(b)$. Thus, the mean-value theorem states that there is at
least one point between a and b that has a slope, designated by $g'(\xi)$,
which is parallel to the line joining $g(a)$ and $g(b)$ (recall Fig. 4.3).

Now, if we let $a = x_i$ and $b = x_r$, the right-hand side of Eq.
(B6.1.1) can be expressed as

$$g(x_r) - g(x_i) = (x_r - x_i)g'(\xi)$$

where ξ is somewhere between x_i and x_r. This result can then be
substituted into Eq. (B6.1.1) to yield

$$x_r - x_{i+1} = (x_r - x_i)g'(\xi) \qquad (B6.1.3)$$

If the true error for iteration i is defined as

$$E_{t,i} = x_r - x_i$$

then Eq. (B6.1.3) becomes

$$E_{t,i+1} = g'(\xi)E_{t,i}$$

Consequently, if $|g'(x)| < 1$, the errors decrease with each itera-
tion. For $|g'(x)| > 1$, the errors grow. Notice also that if the deriva-
tive is positive, the errors will be positive, and hence, the iterative
solution will be monotonic (Fig. 6.3a and c). If the derivative is
negative, the errors will oscillate (Fig. 6.3b and d).

An offshoot of the analysis is that it also demonstrates that
when the method converges, the error is roughly proportional to
and less than the error of the previous step. For this reason,
simple fixed-point iteration is said to be *linearly convergent*.

Thus, in both the equation and in the plot, a starting value of x_0 is used to obtain an
estimate of x_1. The next iteration consists of moving to $[x_1, g(x_1)]$ and then to (x_2, x_2).
This iteration is equivalent to the equation

$$x_2 = g(x_1)$$

The solution in Fig. 6.3a is *convergent* because the estimates of x move closer to the
root with each iteration. The same is true for Fig. 6.3b. However, this is not the case
for Fig. 6.3c and d, where the iterations diverge from the root. Notice that convergence
seems to occur only when the absolute value of the slope of $y_2 = g(x)$ is less than
the slope of $y_1 = x$, that is, when $|g'(x)| < 1$. Box 6.1 provides a theoretical derivation
of this result.

6.1.2 Algorithm for Fixed-Point Iteration

The computer algorithm for fixed-point iteration is extremely simple. It consists of a loop
to iteratively compute new estimates until the termination criterion has been met. Figure 6.4

```
FUNCTION Fixpt(x0, es, imax, iter, ea)
  xr = x0
  iter = 0
  DO
    xrold = xr
    xr = g(xrold)
    iter = iter + 1
    IF xr ≠ 0 THEN
```

$$ea = \left| \frac{xr - xrold}{xr} \right| \cdot 100$$

```
    END IF
    IF ea < es OR iter ≥ imax EXIT
  END DO
  Fixpt = xr
END Fixpt
```

FIGURE 6.4
Pseudocode for fixed-point iteration. Note that other open methods can be cast in this general format.

presents pseudocode for the algorithm. Other open methods can be programmed in a similar way, the major modification being to change the iterative formula that is used to compute the new root estimate.

6.2 THE NEWTON-RAPHSON METHOD

Perhaps the most widely used of all root-locating formulas is the Newton-Raphson equation (Fig. 6.5). If the initial guess at the root is x_i, a tangent can be extended from the point $[x_i, f(x_i)]$. The point where this tangent crosses the x axis usually represents an improved estimate of the root.

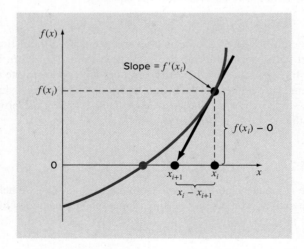

FIGURE 6.5
Graphical depiction of the Newton-Raphson method. A tangent to the function of x_i [that is, $f'(x_i)$] is extrapolated down to the x axis to provide an estimate of the root at x_{i+1}.

The Newton-Raphson method can be derived on the basis of this geometrical interpretation (an alternative method based on the Taylor series is described in Box 6.2). As in Fig. 6.5, the first derivative at x is equivalent to the slope:

$$f'(x_i) = \frac{f(x_i) - 0}{x_i - x_{i+1}} \tag{6.5}$$

which can be rearranged to yield

$$x_{i+1} = x_i - \frac{f(x_i)}{f'(x_i)} \tag{6.6}$$

which is called the *Newton-Raphson formula*.

EXAMPLE 6.3 Newton-Raphson Method

Problem Statement.   Use the Newton-Raphson method to estimate the root of $f(x) = e^{-x} - x$, employing an initial guess of $x_0 = 0$.

Solution.   The first derivative of the function can be evaluated as

$$f'(x) = -e^{-x} - 1$$

which can be substituted along with the original function into Eq. (6.6) to give

$$x_{i+1} = x_i - \frac{e^{-x_i} - x_i}{-e^{-x_i} - 1}$$

Starting with an initial guess of $x_0 = 0$, this iterative equation can be applied to compute

i	x_i	ε_t (%)	E_t
0	0.000000000	100	0.567143290
1	0.500000000	11.8	0.067143290
2	0.566311003	0.147	0.000832287
3	0.567143165	0.0000221	0.000000125
4	0.567143290	$< 10^{-8}$	0.000000000

Thus, the approach rapidly converges on the true root. Notice that the true percent relative error at each iteration decreases much faster than it does in simple fixed-point iteration (compare with Example 6.1).

6.2.1 Termination Criteria and Error Estimates

As with other root-location methods, Eq. (3.5) can be used as a termination criterion. In addition, however, the Taylor series derivation of the method (Box 6.2) provides theoretical insight regarding the rate of convergence as expressed by $E_{i+1} = O(E_i^2)$. Thus the error should be roughly proportional to the square of the previous error. In other words,

Box 6.2   Derivation and Error Analysis of the Newton-Raphson Method

Aside from the geometric derivation [Eqs. (6.5) and (6.6)], the Newton-Raphson method may also be developed from the Taylor series expansion. This alternative derivation is useful in that it also provides insight into the rate of convergence of the method.

Recall from Chap. 4 that the Taylor series expansion can be represented as

$$f(x_{i+1}) = f(x_i) + f'(x_i)(x_{i+1} - x_i)$$
$$+ \frac{f''(\xi)}{2!}(x_{i+1} - x_i)^2 \qquad \text{(B6.2.1)}$$

where ξ lies somewhere in the interval from x_i to x_{i+1}. An approximate version is obtainable by truncating the series after the first derivative term:

$$f(x_{i+1}) \cong f(x_i) + f'(x_i)(x_{i+1} - x_i)$$

At the intersection with the x axis, $f(x_{i+1})$ would be equal to zero, or

$$0 = f(x_i) + f'(x_i)(x_{i+1} - x_i) \qquad \text{(B6.2.2)}$$

which can be solved for

$$x_{i+1} = x_i - \frac{f(x_i)}{f'(x_i)}$$

which is identical to Eq. (6.6). Thus, we have derived the Newton-Raphson formula using a Taylor series.

Aside from the derivation, the Taylor series can also be used to estimate the error of the formula. This can be done by realizing that if the complete Taylor series were employed, an exact result would be obtained. For this situation $x_{i+1} = x_r$, where x

is the true value of the root. Substituting this value along with $f(x_r) = 0$ into Eq. (B6.2.1) yields

$$0 = f(x_i) + f'(x_i)(x_r - x_i) + \frac{f''(\xi)}{2!}(x_r - x_i)^2 \qquad \text{(B6.2.3)}$$

Equation (B6.2.2) can be subtracted from Eq. (B6.2.3) to give

$$0 = f'(x_i)(x_r - x_{i+1}) + \frac{f''(\xi)}{2!}(x_r - x_i)^2 \qquad \text{(B6.2.4)}$$

Now, realize that the error is equal to the discrepancy between x_{i+1} and the true value x_r, as in

$$E_{t,i+1} = x_r - x_{i+1}$$

and Eq. (B6.2.4) can be expressed as

$$0 = f'(x_i)E_{t,i+1} + \frac{f''(\xi)}{2!}E_{t,i}^2 \qquad \text{(B6.2.5)}$$

If we assume convergence, both x_i and ξ should eventually be approximated by the root x_r, and Eq. (B6.2.5) can be rearranged to yield

$$E_{t,i+1} \cong \frac{-f''(x_r)}{2f'(x_r)}E_{t,i}^2 \qquad \text{(B6.2.6)}$$

According to Eq. (B6.2.6), the error is roughly proportional to the square of the previous error. This means that the number of correct decimal places approximately doubles with each iteration. Such behavior is referred to as *quadratic convergence*. Example 6.4 manifests this property.

the number of significant figures of accuracy approximately doubles with each iteration. This behavior is examined in the following example.

EXAMPLE 6.4   ### Error Analysis of Newton-Raphson Method

Problem Statement.   As derived in Box 6.2, the Newton-Raphson method is quadratically convergent. That is, the error is roughly proportional to the square of the previous error, as in

$$E_{t,i+1} \cong \frac{-f''(x_r)}{2f'(x_r)}E_{t,i}^2 \qquad \text{(E6.4.1)}$$

Examine this formula and see if it applies to the results of Example 6.3.

Solution.   The first derivative of $f(x) = e^{-x} - x$ is

$$f'(x) = -e^{-x} - 1$$

which can be evaluated at $x_r = 0.56714329$ as $f'(0.56714329) = -1.56714329$. The second derivative is

$$f''(x) = e^{-x}$$

which can be evaluated as $f''(0.56714329) = 0.56714329$. These results can be substituted into Eq. (E6.4.1) to yield

$$E_{t,i+1} \cong -\frac{0.56714329}{2(-1.56714329)} E_{t,i}^2 = 0.18095 E_{t,i}^2$$

From Example 6.3, the initial error was $E_{t,0} = 0.56714329$, which can be substituted into the error equation to predict

$$E_{t,1} \cong 0.18095(0.56714329)^2 = 0.0582$$

which is close to the true error of 0.06714329. For the next iteration,

$$E_{t,2} \cong 0.18095(0.06714329)^2 = 0.0008158$$

which also compares favorably with the true error of 0.0008323. For the third iteration,

$$E_{t,3} \cong 0.18095(0.0008323)^2 = 0.000000125$$

which is the error obtained in Example 6.3. The error estimate improves in this manner because, as we come closer to the root, x and ξ are better approximated by x_r [recall our assumption in going from Eq. (B6.2.5) to Eq. (B6.2.6) in Box 6.2]. Finally,

$$E_{t,4} \cong 0.18095(0.000000125)^2 = 2.84 \times 10^{-15}$$

Thus, this example illustrates that the error of the Newton-Raphson method for this case is, in fact, roughly proportional (by a factor of 0.18095) to the square of the error of the previous iteration.

6.2.2 Pitfalls of the Newton-Raphson Method

Although the Newton-Raphson method is often very efficient, there are situations where it performs poorly. A special case—multiple roots—will be addressed later in this chapter. However, even when dealing with simple roots, difficulties can also arise, as in the following example.

EXAMPLE 6.5 | **Example of a Slowly Converging Function with Newton-Raphson**

Problem Statement.   Determine the positive root of $f(x) = x^{10} - 1$ using the Newton-Raphson method and an initial guess of $x = 0.5$.

Solution.   The Newton-Raphson formula for this case is

$$x_{i+1} = x_i - \frac{x_i^{10} - 1}{10 x_i^9}$$

which can be used to compute

Iteration	x
0	0.5
1	51.65
2	46.485
3	41.8365
4	37.65285
5	33.887565
.	
.	
.	
∞	1.0000000

Thus, after the first poor prediction, the technique is converging on the true root of 1, but at a very slow rate.

Aside from slow convergence due to the nature of the function, other difficulties can arise, as illustrated in Fig. 6.6. For example, Fig. 6.6a depicts the case where an inflection point [that is, $f''(x) = 0$] occurs in the vicinity of a root. Notice that iterations beginning at x_0 progressively diverge from the root. Figure 6.6b illustrates the tendency of the Newton-Raphson technique to oscillate around a local maximum or minimum. Such oscillations may persist, or as in Fig. 6.6b, a near-zero slope is reached, whereupon the solution is sent far from the area of interest. Figure 6.6c shows how an initial guess that is close to one root can jump to a location several roots away. This tendency to move away from the area of interest occurs because near-zero slopes are encountered. Obviously, a zero slope [$f'(x) = 0$] is truly a disaster because it results in division by zero in the Newton-Raphson formula [Eq. (6.6)]. Graphically (see Fig 6.6d), it means that the solution shoots off horizontally and never hits the x axis.

Thus, there is no general convergence criterion for Newton-Raphson. Its convergence depends on the nature of the function and on the accuracy of the initial guess. The only remedy is to have an initial guess that is "sufficiently" close to the root. And for some functions, no guess will work! Good guesses are usually predicated on knowledge of the physical problem setting or on devices such as graphs that provide insight into the behavior of the solution. The lack of a general convergence criterion also suggests that good computer software should be designed to recognize slow convergence or divergence. The next section addresses some of these issues.

6.2.3 Algorithm for Newton-Raphson

An algorithm for the Newton-Raphson method is readily obtained by substituting Eq. (6.6) for the predictive formula [Eq. (6.2)] in Fig. 6.4. Note, however, that the program must also be modified to compute the first derivative. This can be simply accomplished by the inclusion of a user-defined function.

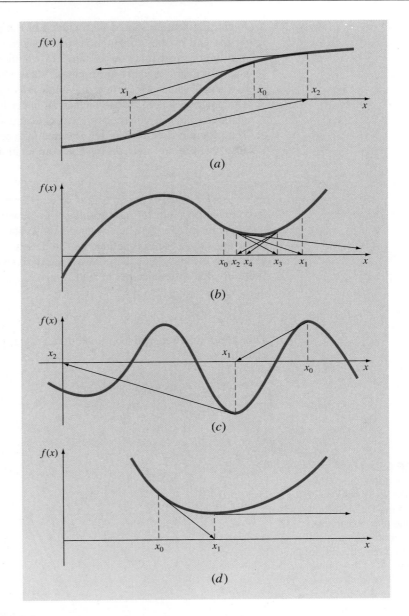

FIGURE 6.6
Four cases where the Newton-Raphson method exhibits poor convergence.

Additionally, in light of the foregoing discussion of potential problems of the Newton-Raphson method, the program would be improved by incorporating several additional features:

1. A plotting routine should be included in the program.
2. At the end of the computation, the final root estimate should always be substituted into the original function to compute whether the result is close to zero. This check partially guards against those cases where slow or oscillating convergence may lead to a small value of ε_a while the solution is still far from a root.
3. The program should always include an upper limit on the number of iterations to guard against oscillating, slowly convergent, or divergent solutions that could persist interminably.
4. The program should alert the user and take account of the possibility that $f'(x)$ might equal zero at any time during the computation.

6.3 THE SECANT METHOD

A potential problem in implementing the Newton-Raphson method is the evaluation of the derivative. Although this is not inconvenient for polynomials and many other functions, there are certain functions whose derivatives may be extremely difficult or inconvenient to evaluate. For these cases, the derivative can be approximated by a backward finite divided difference, as in (Fig. 6.7)

$$f'(x_i) \cong \frac{f(x_{i-1}) - f(x_i)}{x_{i-1} - x_i}$$

This approximation can be substituted into Eq. (6.6) to yield the following iterative equation:

$$x_{i+1} = x_i - \frac{f(x_i)(x_{i-1} - x_i)}{f(x_{i-1}) - f(x_i)} \tag{6.7}$$

FIGURE 6.7

Graphical depiction of the secant method. This technique is similar to the Newton-Raphson technique (Fig. 6.5) in the sense that an estimate of the root is predicted by extrapolating a tangent of the function to the x axis. However, the secant method uses a difference rather than a derivative to estimate the slope.

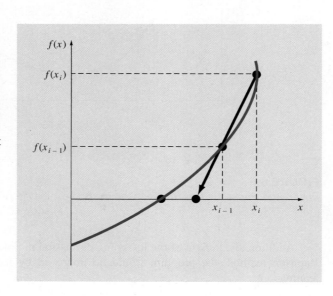

Equation (6.7) is the formula for the *secant method*. Notice that the approach requires two initial estimates of x. However, because $f(x)$ is not required to change signs between the estimates, it is not classified as a bracketing method.

EXAMPLE 6.6

The Secant Method

Problem Statement.   Use the secant method to estimate the root of $f(x) = e^{-x} - x$. Start with initial estimates of $x_{-1} = 0$ and $x_0 = 1.0$.

Solution.   Recall that the true root is 0.56714329. . . .

First iteration:

$$x_{-1} = 0 \qquad f(x_{-1}) = 1.00000$$
$$x_0 = 1 \qquad f(x_0) = -0.63212$$
$$x_1 = 1 - \frac{-0.63212(0 - 1)}{1 - (-0.63212)} = 0.61270 \qquad \varepsilon_t = 8.0\%$$

Second iteration:

$$x_0 = 1 \qquad f(x_0) = -0.63212$$
$$x_1 = 0.61270 \qquad f(x_1) = -0.07081$$

(Note that both estimates are now on the same side of the root.)

$$x_2 = 0.61270 - \frac{-0.07081(1 - 0.61270)}{-0.63212 - (-0.07081)} = 0.56384 \qquad \varepsilon_t = 0.58\%$$

Third iteration:

$$x_1 = 0.61270 \qquad f(x_1) = -0.07081$$
$$x_2 = 0.56384 \qquad f(x_2) = 0.00518$$
$$x_3 = 0.56384 - \frac{0.00518(0.61270 - 0.56384)}{-0.07081 - (-0.00518)} = 0.56717 \qquad \varepsilon_t = 0.0048\%$$

6.3.1 The Difference Between the Secant and False-Position Methods

Note the similarity between the secant method and the false-position method. For example, Eqs. (6.7) and (5.7) are identical on a term-by-term basis. Both use two initial estimates to compute an approximation of the slope of the function that is used to project to the x axis for a new estimate of the root. However, a critical difference between the methods is how one of the initial values is replaced by the new estimate. Recall that in the false-position method the latest estimate of the root replaces whichever of the original values yielded a function value with the same sign as $f(x_r)$. Consequently, the two estimates always bracket the root. Therefore, for all practical purposes, the method always converges because the root is kept within the bracket. In contrast, the secant method replaces the values in strict sequence, with the new value x_{i+1} replacing x_i and x_i replacing x_{i-1}. As a result, the two values can sometimes lie on the same side of the root. For certain cases, this can lead to divergence.

EXAMPLE 6.7 Comparison of Convergence of the Secant and False-Position Techniques

Problem Statement. Use the false-position and secant methods to estimate the root of $f(x) = \ln x$. Start the computation with values of $x_l = x_{-1} = 0.5$ and $x_u = x_0 = 5.0$.

Solution. For the false-position method, the use of Eq. (5.7) and the bracketing criterion for replacing estimates results in the following iterations:

Iteration	x_l	x_u	x_r
1	0.5	5.0	1.8546
2	0.5	1.8546	1.2163
3	0.5	1.2163	1.0585

As can be seen (Fig. 6.8*a* and *c*), the estimates are converging on the true root, which is equal to 1.

FIGURE 6.8
Comparison of the false-position and the secant methods. The first iterations (*a*) and (*b*) for both techniques are identical. However, for the second iterations (*c*) and (*d*), the points used differ. As a consequence, the secant method can diverge, as indicated in (*d*).

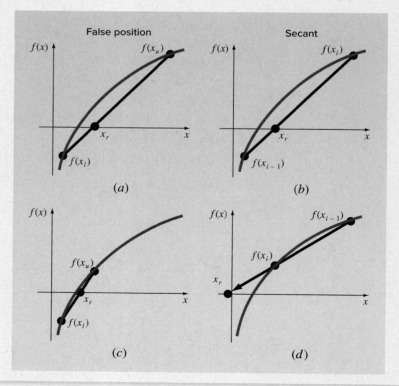

For the secant method, using Eq. (6.7) and the sequential criterion for replacing estimates results in

Iteration	x_{i-1}	x_i	x_{i+1}
1	0.5	5.0	1.8546
2	5.0	1.8546	−0.10438

As in Fig. 6.8d, the approach is divergent.

Although the secant method may be divergent, when it converges, it usually does so at a quicker rate than the false-position method. For instance, Fig. 6.9 demonstrates the superiority of the secant method in this regard. The inferiority of the false-position method is due to one end staying fixed to maintain the bracketing of the root. This property, which is an advantage in that it prevents divergence, is a shortcoming with regard to the rate of convergence; it makes the finite-difference estimate a less accurate approximation of the derivative.

FIGURE 6.9

Comparison of the true percent relative errors ε_t for the methods to determine the roots of $f(x) = e^{-x} - x$.

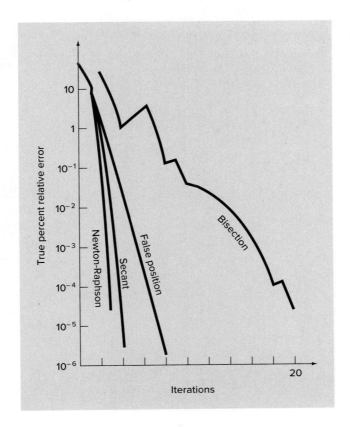

6.3.2 Algorithm for the Secant Method

As with the other open methods, an algorithm for the secant method is obtained simply by modifying Fig. 6.4 so that two initial guesses are inputted and by using Eq. (6.7) to calculate the root. In addition, the options suggested in Sec. 6.2.3 for the Newton-Raphson method can also be applied to good advantage for the secant program.

6.3.3 Modified Secant Method

Rather than using two arbitrary values to estimate the derivative, an alternative approach involves a fractional perturbation of the independent variable to estimate $f'(x)$,

$$f'(x_i) \cong \frac{f(x_i + \delta x_i) - f(x_i)}{\delta x_i}$$

where δ = a small perturbation fraction. This approximation can be substituted into Eq. (6.6) to yield the following iterative equation:

$$x_{i+1} = x_i - \frac{\delta x_i f(x_i)}{f(x_i + \delta x_i) - f(x_i)} \tag{6.8}$$

EXAMPLE 6.8

Modified Secant Method

Problem Statement. Use the modified secant method to estimate the root of $f(x) = e^{-x} - x$. Use a value of 0.01 for δ and start with $x_0 = 1.0$. Recall that the true root is 0.56714329. . . .

Solution.
First iteration:

$$x_0 = 1 \qquad\qquad f(x_0) = -0.63212$$
$$x_0 + \delta x_0 = 1.01 \qquad f(x_0 + \delta x_0) = -0.64578$$
$$x_1 = 1 - \frac{0.01(-0.63212)}{-0.64578 - (-0.63212)} = 0.537263 \qquad |\varepsilon_t| = 5.3\%$$

Second iteration:

$$x_0 = 0.537263 \qquad\qquad f(x_0) = 0.047083$$
$$x_0 + \delta x_0 = 0.542635 \qquad f(x_0 + \delta x_0) = 0.038579$$
$$x_1 = 0.537263 - \frac{0.005373(0.047083)}{0.038579 - 0.047083} = 0.56701 \qquad |\varepsilon_t| = 0.0236\%$$

Third iteration:

$$x_0 = 0.56701 \qquad\qquad f(x_0) = 0.000209$$
$$x_0 + \delta x_0 = 0.572680 \qquad f(x_0 + \delta x_0) = -0.00867$$
$$x_1 = 0.56701 - \frac{0.00567(0.000209)}{-0.00867 - 0.000209} = 0.567143 \qquad |\varepsilon_t| = 2.365 \times 10^{-5}\%$$

The choice of a proper value for δ is not automatic. If δ is too small, the method can be swamped by round-off error caused by subtractive cancellation in the denominator of Eq. (6.8). If it is too big, the technique can become inefficient and even divergent. However, if chosen correctly, it provides a nice alternative for cases where evaluating the derivative is difficult and developing two initial guesses is inconvenient.

6.4 BRENT'S METHOD

Wouldn't it be nice to have a hybrid approach that combined the reliability of bracketing with the speed of the open methods? *Brent's root-location method* is a clever algorithm that does just that by applying a speedy open method wherever possible, but reverting to a reliable bracketing method if necessary. The approach was developed by Richard Brent (1973) based on an earlier algorithm of Theodorus Dekker (1969).

The bracketing technique is the trusty bisection method (Sec. 5.2), and two different open methods are employed. The first is the secant method described in Sec. 6.3. As explained next, the second is inverse quadratic interpolation.

6.4.1 Inverse Quadratic Interpolation

Inverse quadratic interpolation is similar in spirit to the secant method. As in Fig. 6.10*a*, the secant method is based on computing a straight line that goes through two guesses. The intersection of this straight line with the x axis represents the new root estimate. For this reason, the secant method is sometimes referred to as a *linear interpolation method*.

Now suppose that we had three points. In that case, we could determine a quadratic function of x that goes through the three points (Fig. 6.10*b*). Just as with the linear secant method, the intersection of this parabola with the x axis would represent the new root estimate. And as illustrated in Fig. 6.10*b*, using a curve rather than a straight line often yields a better estimate.

Although this would seem to represent a great improvement, the approach has a fundamental flaw: It is possible that the parabola might not intersect the x axis! Such would be the case when the resulting parabola has complex roots. This is illustrated by the parabola, $y = f(x)$, in Fig. 6.11.

FIGURE 6.10
Comparison of (*a*) the secant method and (*b*) inverse quadratic interpolation. Note that the dark parabola passing through the three points in (*b*) is called "inverse" because it is written in *y* rather than in *x*.

(*a*)   (*b*)

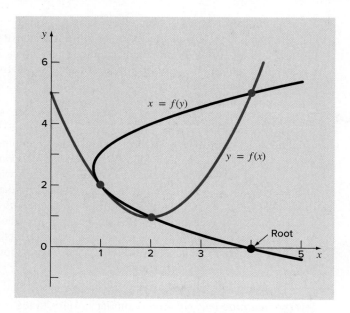

FIGURE 6.11

Two parabolas fit to three points. The parabola written as a function of x, $y = f(x)$, has complex roots and hence does not intersect the x axis. In contrast, if the variables are reversed, and the parabola developed as $x = f(y)$, the function does intersect the x axis.

The difficulty can be rectified by employing inverse quadratic interpolation. That is, rather than using a parabola in x, we can fit the points with a parabola in y. This amounts to reversing the axes and creating a "sideways" parabola [the curve $x = f(y)$, in Fig. 6.11].

If the three points are designated as (x_{i-2}, y_{i-2}), (x_{i-1}, y_{i-1}), and (x_i, y_i), a quadratic function of y that passes through the points can be generated as

$$g(y) = \frac{(y - y_{i-1})(y - y_i)}{(y_{i-2} - y_{i-1})(y_{i-2} - y_i)} x_{i-2} + \frac{(y - y_{i-2})(y - y_i)}{(y_{i-1} - y_{i-2})(y_{i-1} - y_i)} x_{i-1}$$
$$+ \frac{(y - y_{i-2})(y - y_{i-1})}{(y_i - y_{i-2})(y_i - y_{i-1})} x_i \qquad (6.9)$$

As we will learn in Sec. 18.2, this form is called a *Lagrange polynomial*. The root, x_{i+1}, corresponds to $y = 0$, which when substituted into Eq. (6.9) yields

$$x_{i+1} = \frac{y_{i-1} y_i}{(y_{i-2} - y_{i-1})(y_{i-2} - y_i)} x_{i-2} + \frac{y_{i-2} y_i}{(y_{i-1} - y_{i-2})(y_{i-1} - y_i)} x_{i-1}$$
$$+ \frac{y_{i-2} y_{i-1}}{(y_i - y_{i-2})(y_i - y_{i-1})} x_i \qquad (6.10)$$

As shown in Fig. 6.11, such a "sideways" parabola always intersects the x axis.

EXAMPLE 6.9   **Inverse Quadratic Interpolation**

Problem Statement.   Develop quadratic equations in both x and y for the data points depicted in Fig. 6.11: $(1, 2)$, $(2, 1)$, and $(4, 5)$. For the first, $y = f(x)$, employ the quadratic formula to illustrate that the roots are complex. For the latter, $x = g(y)$, use inverse quadratic interpolation (Eq. 6.10) to determine the root estimate.

Solution.   By reversing the x's and y's, Eq. (6.9) can be used to generate a quadratic in x as

$$f(x) = \frac{(x-2)(x-4)}{(1-2)(1-4)}2 + \frac{(x-1)(x-4)}{(2-1)(2-4)}1 + \frac{(x-1)(x-2)}{(4-1)(4-2)}5$$

or collecting terms

$$f(x) = x^2 - 4x + 5$$

This equation was used to generate the parabola, $y = f(x)$, in Fig. 6.11. The quadratic formula can be used to determine that the roots for this case are complex,

$$x = \frac{4 \pm \sqrt{(-4)^2 - 4(1)(5)}}{2} = 2 \pm i$$

Equation (6.9) can be used to generate the quadratic in y as

$$g(y) = \frac{(y-1)(y-5)}{(2-1)(2-5)}1 + \frac{(y-2)(y-5)}{(1-2)(1-5)}2 + \frac{(y-2)(y-1)}{(5-2)(5-1)}4$$

or collecting terms

$$g(y) = 0.5y^2 - 2.5y + 4$$

Finally, Eq. (6.10) can be used to determine the root as

$$x_{i+1} = \frac{-1(-5)}{(2-1)(2-5)}1 + \frac{-2(-5)}{(1-2)(1-5)}2 + \frac{-2(-1)}{(5-2)(5-1)}4 = 4$$

Before proceeding to Brent's algorithm, we need to mention one more case where inverse quadratic interpolation does not work. If the three y values are not distinct (that is, $y_{i-2} = y_{i-1}$ or $y_{i-1} = y_i$), an inverse quadratic function does not exist. So this is where the secant method comes into play. If we arrive at a situation where the y values are not distinct, we can always revert to the less efficient secant method to generate a root using two of the points. If $y_{i-2} = y_{i-1}$, we use the secant method with x_{i-1} and x_i. If $y_{i-1} = y_i$, we use x_{i-2} and x_{i-1}.

6.4.2 Brent's Method Algorithm

The general idea behind the *Brent's root-finding method* is whenever possible to use one of the quick open methods. In the event that these generate an unacceptable result (i.e., a root estimate that falls outside the bracket), the algorithm reverts to the more conservative bisection method. Although bisection may be slower, it generates an estimate guaranteed to fall within the bracket. This process is then repeated until the root is located to within an acceptable tolerance. As might be expected, bisection typically dominates at first but as the root is approached, the technique shifts to the faster open methods.

Figure 6.12 presents pseudocode for the algorithm based on a MATLAB software M-file developed by Cleve Moler (2004). It represents a stripped-down version of

```
FUNCTION fzerosimp(xl, xu)
eps = 2.22044604925031E-16
tol = 0.000001
a = xl: b = xu: fa = f(a): fb = f(b)
c = a: fc = fa: d = b - c: e = d
DO
  IF fb = 0 EXIT
  IF Sgn(fa) = Sgn(fb) THEN              (If necessary, rearrange points)
    a = c: fa = fc: d = b - c: e = d
  ENDIF
  IF |fa| < |fb| THEN
    c = b: b = a: a = c
    fc = fb: fb = fa: fa = fc
  ENDIF
  m = 0.5 * (a - b)        (Termination test and possible exit)
  tol = 2 * eps * max(|b|, 1)
  IF |m| ≤ tol OR fb = 0. THEN
    EXIT
  ENDIF
  (Choose open methods or bisection)
  IF |e| ≥ tol AND |fc| > |fb| THEN
   s = fb / fc
    IF a = c THEN                          (Secant method)
      p = 2 * m * s
      q = 1 - s
    ELSE                     (Inverse quadratic interpolation)
      q = fc / fa: r = fb / fa
      p = s * (2 * m * q * (q - r) - (b - c) * (r - 1))
      q = (q - 1) * (r - 1) * (s - 1)
    ENDIF
    IF p > 0 THEN q = -q ELSE p = -p
    IF 2 * p < 3 * m * q - |tol * q| AND p < |0.5 * e * q| THEN
      e = d: d = p / q
    ELSE
      d = m: e = m
    ENDIF
  ELSE                                      (Bisection)
   d = m: e = m
  ENDIF
  c = b: fc = fb
  IF |d| > tol THEN b = b + d ELSE b = b - Sgn(b - a) * tol
   fb = f(b)
ENDDO
fzerosimp = b
END fzerosimp
```

FIGURE 6.12

Pseudocode for Brent's root-finding algorithm based on a MATLAB M-file developed by Cleve Moler (2004).

the `fzero` function, which is the professional root-location function employed in MAT-LAB. For that reason, we call the simplified version: `fzerosimp`. Note that it requires another function, `f`, that holds the equation for which the root is being evaluated.

The `fzerosimp` function is passed two initial guesses that must bracket the root. After assigning values for machine epsilon and a tolerance, the three variables defining the search interval (`a`, `b`, `c`) are initialized, and `f` is evaluated at the endpoints.

A main loop is then implemented. If necessary, the three points are rearranged to satisfy the conditions required for the algorithm to work effectively. At this point, if the stopping criteria are met, the loop is terminated. Otherwise, a decision structure chooses among the three methods and checks whether the outcome is acceptable. A final section then evaluates `f` at the new point and the loop is repeated. Once the stopping criteria are met, the loop terminates and the final root estimate is returned.

Note that Sec. 7.7.2 presents an application of Brent's method where we illustrate how MATLAB's `fzero` function works. In addition, it is employed in Case Study 8.4 to determine the friction factor for airflow through a tube.

6.5   MULTIPLE ROOTS

A *multiple root* corresponds to a point where a function is tangent to the x axis. For example, a double root results from

$$f(x) = (x - 3)(x - 1)(x - 1) \tag{6.11}$$

or, multiplying terms, $f(x) = x^3 - 5x^2 + 7x - 3$. The equation has a *double root* because one value of x makes two terms in Eq. (6.11) equal to zero. Graphically, this corresponds to the curve touching the x axis tangentially at the double root. Examine Fig. 6.13a at $x = 1$. Notice that the function touches the axis but does not cross it at the root.

A *triple root* corresponds to the case where one x value makes three terms in an equation equal to zero, as in

$$f(x) = (x - 3)(x - 1)(x - 1)(x - 1)$$

or, multiplying terms, $f(x) = x^4 - 6x^3 + 12x^2 - 10x + 3$. Notice that the graphical depiction (Fig. 6.13b) again indicates that the function is tangent to the axis at the root, but that for this case the axis is crossed. In general, odd multiple roots cross the axis, whereas even ones do not. For example, the quadruple root in Fig. 6.13c does not cross the axis.

Multiple roots pose some difficulties for many of the numerical methods described in Part Two:

1. The fact that the function does not change sign at even multiple roots precludes the use of the reliable bracketing methods that were discussed in Chap. 5. Thus, of the methods covered in this book, you are limited to the open methods that may diverge.

2. Another possible problem is related to the fact that not only $f(x)$ but also $f'(x)$ goes to zero at the root. This poses problems for both the Newton-Raphson and secant methods, which both contain the derivative (or its estimate) in the denominator of

FIGURE 6.13
Examples of multiple roots where functions are tangential to the x axis. Notice that the function does not cross the axis on either side of even multiple roots (a) and (c), whereas it crosses the axis for odd cases (b).

their respective formulas. This could result in division by zero when the solution converges very close to the root. A simple way to circumvent these problems is based on the fact that it can be demonstrated theoretically (Ralston and Rabinowitz 1978) that $f(x)$ will always reach zero before $f'(x)$. Therefore, if a zero check for $f(x)$ is incorporated into the computer program, the computation can be terminated before $f'(x)$ reaches zero.

3. It can be demonstrated that the Newton-Raphson and secant methods are linearly, rather than quadratically, convergent for multiple roots (Ralston and Rabinowitz 1978). Modifications have been proposed to alleviate this problem. Ralston and Rabinowitz (1978) have indicated that a slight change in the formulation returns it to quadratic convergence, as in

$$x_{i+1} = x_i - m \frac{f(x_i)}{f'(x_i)} \tag{6.12}$$

where m is the multiplicity of the root (that is, $m = 2$ for a double root, $m = 3$ for a triple root, etc.). Of course, this may be an unsatisfactory alternative because it hinges on foreknowledge of the multiplicity of the root.

Another alternative, also suggested by Ralston and Rabinowitz (1978), is to define a new function $u(x)$ that is the ratio of the function to its derivative, as in

$$u(x) = \frac{f(x)}{f'(x)} \tag{6.13}$$

It can be shown that this function has roots at all the same locations as the original function. Therefore, Eq. (6.13) can be substituted into Eq. (6.6) to develop an alternative form of the Newton-Raphson formula:

$$x_{i+1} = x_i - \frac{u(x_i)}{u'(x_i)} \tag{6.14}$$

Equation (6.13) can be differentiated to give

$$u'(x) = \frac{f'(x)f'(x) - f(x)f''(x)}{[f'(x)]^2} \tag{6.15}$$

Equations (6.13) and (6.15) can be substituted into Eq. (6.14) and the result simplified to yield

$$x_{i+1} = x_i - \frac{f(x_i)f'(x_i)}{[f'(x_i)]^2 - f(x_i)f''(x_i)} \tag{6.16}$$

EXAMPLE 6.10   **Modified Newton-Raphson Method for Multiple Roots**

Problem Statement. Use both the standard and modified Newton-Raphson methods to evaluate the multiple root of Eq. (6.11), with an initial guess of $x_0 = 0$.

Solution.   The first derivative of Eq. (6.11) is $f'(x) = 3x^2 - 10x + 7$, and therefore, the standard Newton-Raphson formula for this problem is [Eq. (6.6)]

$$x_{i+1} = x_i - \frac{x_i^3 - 5x_i^2 + 7x_i - 3}{3x_i^2 - 10x_i + 7}$$

which can be solved iteratively for

i	x_i	ε_t (%)
0	0	100
1	0.4285714	57
2	0.6857143	31
3	0.8328654	17
4	0.9133290	8.7
5	0.9557833	4.4
6	0.9776551	2.2

As anticipated, the method is linearly convergent toward the true value of 1.0.

For the modified method, the second derivative is $f''(x) = 6x - 10$, and the iterative relationship is [Eq. (6.16)]

$$x_{i+1} = x_i - \frac{(x_i^3 - 5x_i^2 + 7x_i - 3)(3x_i^2 - 10x_i + 7)}{(3x_i^2 - 10x_i + 7)^2 - (x_i^3 - 5x_i^2 + 7x_i - 3)(6x_i - 10)}$$

which can be solved for

i	x_i	ε_t (%)
0	0	100
1	1.105263	11
2	1.003082	0.31
3	1.000002	0.00024

Thus, the modified formula is quadratically convergent.

We can also use both methods to search for the single root at $x = 3$. Using an initial guess of $x_0 = 4$ gives the following results:

i	Standard	ε_t (%)	Modified	ε_t (%)
0	4	33	4	33
1	3.4	13	2.636364	12
2	3.1	3.3	2.820225	6.0
3	3.008696	0.29	2.961728	1.3
4	3.000075	0.0025	2.998479	0.051
5	3.000000	2×10^{-7}	2.999998	7.7×10^{-5}

Thus, both methods converge quickly, with the standard method being somewhat more efficient.

The preceding example illustrates the trade-offs involved in opting for the modified Newton-Raphson method. Although it is preferable for multiple roots, it is somewhat less efficient and requires more computational effort than the standard method for simple roots.

It should be noted that a modified version of the secant method suited for multiple roots can also be developed by substituting Eq. (6.13) into Eq. (6.7). The resulting formula is (Ralston and Rabinowitz 1978)

$$x_{i+1} = x_i - \frac{u(x_i)(x_{i-1} - x_i)}{u(x_{i-1}) - u(x_i)}$$

6.6 SYSTEMS OF NONLINEAR EQUATIONS

To this point, we have focused on the determination of the roots of a single equation. A related problem is to locate the roots of a set of simultaneous equations,

$$f_1(x_1, x_2, \ldots, x_n) = 0$$
$$f_2(x_1, x_2, \ldots, x_n) = 0$$
$$\begin{matrix} \cdot & & \cdot \\ \cdot & & \cdot \\ \cdot & & \cdot \end{matrix} \qquad\qquad (6.17)$$
$$f_n(x_1, x_2, \ldots, x_n) = 0$$

The solution of this system consists of a set of x values that simultaneously result in all the equations equaling zero.

In Part Three, we will present methods for the case where the simultaneous equations are linear—that is, they can be expressed in the general form

$$f(x) = a_1 x_1 + a_2 x_2 + \cdots + a_n x_n - b = 0 \qquad\qquad (6.18)$$

where the b and the a's are constants. Algebraic and transcendental equations that do not fit this format are called *nonlinear equations*. For example,

$$x^2 + xy = 10$$

and

$$y + 3xy^2 = 57$$

are two simultaneous nonlinear equations with two unknowns, x and y. They can be expressed in the form of Eq. (6.17) as

$$u(x, y) = x^2 + xy - 10 = 0 \qquad\qquad (6.19a)$$
$$v(x, y) = y + 3xy^2 - 57 = 0 \qquad\qquad (6.19b)$$

Thus, the solution would be the values of x and y that make the functions $u(x, y)$ and $v(x, y)$ equal to zero. Most approaches for determining such solutions are extensions of the open methods for solving single equations. In this section, we will investigate two of these: fixed-point iteration and Newton-Raphson.

6.6.1 Fixed-Point Iteration

The fixed-point-iteration approach (Sec. 6.1) can be modified to solve two simultaneous, nonlinear equations. This approach will be illustrated in the following example.

EXAMPLE 6.11	Fixed-Point Iteration for a Nonlinear System

Problem Statement.   Use fixed-point iteration to determine the roots of Eq. (6.19). Note that a correct pair of roots is $x = 2$ and $y = 3$. Initiate the computation with guesses of $x = 1.5$ and $y = 3.5$.

Solution.   Equation (6.19*a*) can be solved for

$$x_{i+1} = \frac{10 - x_i^2}{y_i} \tag{E6.11.1}$$

and Eq. (6.19*b*) can be solved for

$$y_{i+1} = 57 - 3x_i y_i^2 \tag{E6.11.2}$$

Note that we will drop the subscripts for the remainder of the example.

On the basis of the initial guesses, Eq. (E6.11.1) can be used to determine a new value of x:

$$x = \frac{10 - (1.5)^2}{3.5} = 2.21429$$

This result and the initial value of $y = 3.5$ can be substituted into Eq. (E6.11.2) to determine a new value of y:

$$y = 57 - 3(2.21429)(3.5)^2 = -24.37516$$

Thus, the approach seems to be diverging. This behavior is even more pronounced on the second iteration:

$$x = \frac{10 - (2.21429)^2}{-24.37516} = -0.20910$$
$$y = 57 - 3(-0.20910)(-24.37516)^2 = 429.709$$

Obviously, the approach is deteriorating.

Now we will repeat the computation but with the original equations set up in a different format. For example, an alternative formulation of Eq. (6.19*a*) is

$$x = \sqrt{10 - xy}$$

and of Eq. (6.19*b*) is

$$y = \sqrt{\frac{57 - y}{3x}}$$

Now the results are more satisfactory:

$$x = \sqrt{10 - 1.5(3.5)} = 2.17945$$

$$y = \sqrt{\frac{57 - 3.5}{3(2.17945)}} = 2.86051$$

$$x = \sqrt{10 - 2.17945(2.86051)} = 1.94053$$

$$y = \sqrt{\frac{57 - 2.86051}{3(1.94053)}} = 3.04955$$

Thus, the approach is converging on the true values of $x = 2$ and $y = 3$.

The previous example illustrates the most serious shortcoming of simple fixed-point iteration—that is, convergence often depends on the manner in which the equations are formulated. Additionally, even in those instances where convergence is possible, divergence can occur if the initial guesses are insufficiently close to the true solution. Using reasoning similar to that in Box 6.1, it can be demonstrated that sufficient conditions for convergence for the two-equation case are

$$\left|\frac{\partial u}{\partial x}\right| + \left|\frac{\partial u}{\partial y}\right| < 1$$

and

$$\left|\frac{\partial v}{\partial x}\right| + \left|\frac{\partial v}{\partial y}\right| < 1$$

These criteria are so restrictive that fixed-point iteration has limited utility for solving nonlinear systems. However, as we will describe later in the book, it can be very useful for solving linear systems.

6.6.2 Newton-Raphson

Recall that the Newton-Raphson method was predicated on employing the derivative (that is, the slope) of a function to estimate its intercept with the axis of the independent variable—that is, the root (Fig. 6.5). This estimate was based on a first-order Taylor series expansion (recall Box 6.2),

$$f(x_{i+1}) = f(x_i) + (x_{i+1} - x_i)f'(x_i) \tag{6.20}$$

where x_i is the initial guess at the root and x_{i+1} is the point at which the slope intercepts the x axis. At this intercept, $f(x_{i+1})$ by definition equals zero and Eq. (6.20) can be rearranged to yield

$$x_{i+1} = x_i - \frac{f(x_i)}{f'(x_i)} \tag{6.21}$$

which is the single-equation form of the Newton-Raphson method.

The multiequation form is derived in an identical fashion. However, a multivariable Taylor series must be used to account for the fact that more than one independent variable contributes to the determination of the root. For the two-variable case, a first-order Taylor series can be written [recall Eq. (4.26)] for each nonlinear equation as

$$u_{i+1} = u_i + (x_{i+1} - x_i)\frac{\partial u_i}{\partial x} + (y_{i+1} - y_i)\frac{\partial u_i}{\partial y} \tag{6.22a}$$

and

$$v_{i+1} = v_i + (x_{i+1} - x_i)\frac{\partial v_i}{\partial x} + (y_{i+1} - y_i)\frac{\partial v_i}{\partial y} \tag{6.22b}$$

Just as for the single-equation version, the root estimate corresponds to the values of x and y where u_{i+1} and v_{i+1} equal zero. For this situation, Eq. (6.22) can be rearranged to give

$$\frac{\partial u_i}{\partial x}x_{i+1} + \frac{\partial u_i}{\partial y}y_{i+1} = -u_i + x_i\frac{\partial u_i}{\partial x} + y_i\frac{\partial u_i}{\partial y} \tag{6.23a}$$

$$\frac{\partial v_i}{\partial x}x_{i+1} + \frac{\partial v_i}{\partial y}y_{i+1} = -v_i + x_i\frac{\partial v_i}{\partial x} + y_i\frac{\partial v_i}{\partial y} \tag{6.23b}$$

Because all values subscripted with i's are known (they correspond to the latest guess or approximation), the only unknowns are x_{i+1} and y_{i+1}. Thus, Eq. (6.23) is a set of two linear equations with two unknowns [compare with Eq. (6.19)]. Consequently, algebraic manipulations (for example, Cramer's rule) can be employed to solve for

$$x_{i+1} = x_i - \frac{u_i\dfrac{\partial v_i}{\partial y} - v_i\dfrac{\partial u_i}{\partial y}}{\dfrac{\partial u_i}{\partial x}\dfrac{\partial v_i}{\partial y} - \dfrac{\partial u_i}{\partial y}\dfrac{\partial v_i}{\partial x}} \tag{6.24a}$$

$$y_{i+1} = y_i - \frac{v_i\dfrac{\partial u_i}{\partial x} - u_i\dfrac{\partial v_i}{\partial x}}{\dfrac{\partial u_i}{\partial x}\dfrac{\partial v_i}{\partial y} - \dfrac{\partial u_i}{\partial y}\dfrac{\partial v_i}{\partial x}} \tag{6.24b}$$

The denominator of each of these equations is formally referred to as the determinant of the *Jacobian* of the system.

Equation (6.24) is the two-equation version of the Newton-Raphson formula. As in the following example, it can be employed iteratively to home in on the roots of two simultaneous equations.

EXAMPLE 6.12 Newton-Raphson for a Nonlinear System

Problem Statement. Use the multiequation Newton-Raphson method to determine roots of Eq. (6.19). Note that a correct pair of roots is $x = 2$ and $y = 3$. Initiate the computation with guesses of $x = 1.5$ and $y = 3.5$.

Solution. First compute the partial derivatives and evaluate them at the initial guesses of x and y:

$$\frac{\partial u_0}{\partial x} = 2x + y = 2(1.5) + 3.5 = 6.5 \qquad \frac{\partial u_0}{\partial y} = x = 1.5$$

$$\frac{\partial v_0}{\partial x} = 3y^2 = 3(3.5)^2 = 36.75 \qquad \frac{\partial v_0}{\partial y} = 1 + 6xy = 1 + 6(1.5)(3.5) = 32.5$$

Thus, the determinant of the Jacobian for the first iteration is

$$6.5(32.5) - 1.5(36.75) = 156.125$$

The values of the functions can be evaluated at the initial guesses as

$$u_0 = (1.5)^2 + 1.5(3.5) - 10 = -2.5$$
$$v_0 = 3.5 + 3(1.5)(3.5)^2 - 57 = 1.625$$

These values can be substituted into Eq. (6.24) to give

$$x = 1.5 - \frac{-2.5(32.5) - 1.625(1.5)}{156.125} = 2.03603$$

$$y = 3.5 - \frac{1.625(6.5) - (-2.5)(36.75)}{156.125} = 2.84388$$

Thus, the results are converging to the true values of $x = 2$ and $y = 3$. The computation can be repeated until an acceptable accuracy is obtained.

Just as with fixed-point iteration, the Newton-Raphson approach will often diverge if the initial guesses are not sufficiently close to the true roots. Whereas graphical methods could be employed to derive good guesses for the single-equation case, no such simple procedure is available for the multiequation version. Although there are some advanced approaches for obtaining acceptable first estimates, often the initial guesses must be obtained on the basis of trial and error and knowledge of the physical system being modeled.

The two-equation Newton-Raphson approach can be generalized to solve n simultaneous equations. Because the most efficient way to do this involves matrix algebra and the solution of simultaneous linear equations, we will defer discussion of the general approach to Part Three.

PROBLEMS

6.1 Use simple fixed-point iteration to locate the root of

$$f(x) = 2\sin(\sqrt{x}) - x$$

Use an initial guess of $x_0 = 0.5$ and iterate until $\varepsilon_a \leq 0.01\%$. Verify that the process is linearly convergent as described in Box 6.1.

6.2 Determine the highest real root of

$$f(x) = 2.1x^3 - 11.6x^2 + 17.5x - 6$$

(a) Graphically.
(b) Fixed-point iteration method (three iterations, $x_0 = 3$). Note: Make certain that you develop a solution that converges on the root.
(c) Newton-Raphson method (three iterations, $x_0 = 3$).

(d) Secant method (three iterations, $x_{-1} = 3$, $x_0 = 4$).
(e) Modified secant method (three iterations, $x_0 = 3$, $\delta = 0.01$).
Compute the approximate percent relative errors for your solutions.
6.3 Use (a) fixed-point iteration and (b) the Newton-Raphson method to determine a root of $f(x) = -x^2 + 1.8x + 2.5$ using $x_0 = 5$. Perform the computation until ε_a is less than $\varepsilon_s = 0.01\%$. Also perform an error check of your final answer.
6.4 Determine the real roots of $f(x) = -1.5 + 6x - 3.9x^2 + 0.5x^3$:
(a) graphically and (b) using the Newton-Raphson method to within $\varepsilon_s = 0.01\%$.
6.5 Employ the Newton-Raphson method to determine a real root for $f(x) = -2 + 6x - 4x^2 + 0.5x^3$ using initial guesses of (a) 4.2 and

(b) 4.43. Discuss and use graphical and analytical methods to explain any peculiarities in your results.

6.6 Determine the lowest real root of $f(x) = -10 - 20x + 16x^2 - 2.5x^3$: **(a)** graphically and **(b)** using the secant method to a value of ε_s corresponding to three significant figures.

6.7 Locate the first positive root of

$$f(x) = \sin x + \cos(1 + x^2) - 1$$

where x is in radians. Use four iterations of the secant method with initial guesses of **(a)** $x_{-1} = 1.0$ and $x_0 = 3.0$; **(b)** $x_{-1} = 1.5$ and $x_0 = 2.5$, and **(c)** $x_{-1} = 1.5$ and $x_0 = 2.25$ to locate the root. **(d)** Use the graphical method to explain your results.

6.8 Determine the real root of $x^{3.6} = 75$, with the modified secant method to within $\varepsilon_s = 0.1\%$ using an initial guess of $x_0 = 3.5$ and $\delta = 0.01$.

6.9 Determine the highest real root of $f(x) = 0.95x^3 - 5.9x^2 + 10.9x - 6$:

(a) Graphically.
(b) Using the Newton-Raphson method (three iterations, $x_0 = 3.5$).
(c) Using the secant method (three iterations, $x_{-1} = 2.5$ and $x_0 = 3.5$).
(d) Using the modified secant method (three iterations, $x_0 = 3.5$, $\delta = 0.01$).

6.10 Determine the lowest positive root of $f(x) = 8 \sin(x)e^{-x} - 1$:
(a) Graphically.
(b) Using the Newton-Raphson method (three iterations, $x_0 = 0.3$).
(c) Using the secant method (five iterations, $x_{-1} = 0.5$ and $x_0 = 0.4$).
(d) Using the modified secant method (three iterations, $x_0 = 0.3$, $\delta = 0.01$).

6.11 Use the Newton-Raphson method to find the root of

$$f(x) = e^{-0.5x}(4 - x) - 2$$

Employ initial guesses of **(a)** 2, **(b)** 6, and **(c)** 8. Explain your results.

6.12 Given

$$f(x) = -2x^6 - 1.5x^4 + 10x + 2$$

use a root-location technique to determine the maximum of this function. Perform iterations until the approximate relative error falls below 5%. If you use a bracketing method, use initial guesses of $x_l = 0$ and $x_u = 1$. If you use the Newton-Raphson or the modified secant method, use an initial guess of $x_i = 1$. If you use the secant method, use initial guesses of $x_{-1} = 0$ and $x_0 = 1$. Assuming that convergence is not an issue, choose the technique that is best suited to this problem. Justify your choice.

6.13 You must determine the root of the following easily differentiable function,

$$e^{0.3x} = 6 - 6x$$

Pick the best numerical technique, justify your choice, and then use that technique to determine the root. Note that it is known that for positive initial guesses, all techniques except fixed-point iteration will eventually converge. Perform iterations until the approximate relative error falls below 2%. If you use a bracketing method, use initial guesses of $x_l = 0$ and $x_u = 2$. If you use the Newton-Raphson or the modified secant method, use an initial guess of $x_i = 0.8$. If you use the secant method, use initial guesses of $x_{i-1} = 0$ and $x_i = 2$.

6.14 Use **(a)** the Newton-Raphson method and **(b)** the modified secant method ($\delta = 0.05$) to determine a root of $f(x) = x^5 - 16.05x^4 + 88.75x^3 - 192.0375x^2 + 116.35x + 31.6875$ using an initial guess of $x_0 = 0.5825$ and $\varepsilon_s = 0.01\%$. Explain your results.

6.15 The "divide and average" method, an old-time method for approximating the square root of any positive number a, can be formulated as

$$x = \frac{x + a/x}{2}$$

Prove that this is equivalent to the Newton-Raphson algorithm.

6.16 **(a)** Apply the Newton-Raphson method to the function $f(x) = \tanh(x^2 - 9)$ to evaluate its known real root at $x = 3$. Use an initial guess of $x_0 = 3.2$ and take a minimum of four iterations. **(b)** Did the method exhibit convergence onto the real root? Sketch the plot with the results for each iteration shown.

6.17 The polynomial $f(x) = 0.0074x^4 - 0.284x^3 + 3.355x^2 - 12.183x + 5$ has a real root between 15 and 20. Apply the Newton-Raphson method to this function using an initial guess of $x_0 = 16.15$. Explain your results.

6.18 Use the secant method on the circle function $(x + 1)^2 + (y - 2)^2 = 16$ to find a positive real root. Set your initial guess to $x_0 = 3$ and $x_{-1} = 0.5$. Approach the solution from the first and fourth quadrants. When solving for $f(x)$ in the fourth quadrant, be sure to take the negative value of the square root. Why does your solution diverge?

6.19 You are designing a spherical tank (Fig. P6.19) to hold water for a small village in a developing country. The volume of liquid it can hold can be computed as

$$V = \pi h^2 \frac{[3R - h]}{3}$$

where $V =$ volume (m^3), $h =$ depth of water in tank (m), and $R =$ the tank radius (m). If $R = 4$ m, what depth must the tank be filled to so that it holds 50 m^3? Use three iterations of the Newton-Raphson method to determine your answer. Determine the approximate relative error after each iteration. Note that an initial guess of R will always converge.

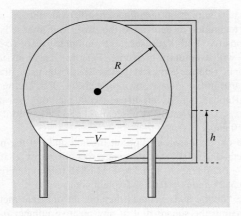

FIGURE P6.19

6.20 The Manning equation can be written for a rectangular open channel as

$$Q = \frac{\sqrt{S}(BH)^{5/3}}{n(B + 2H)^{2/3}}$$

where Q = flow (m³/s), S = slope (m/m), B = channel width (m), H = depth (m), and n = the Manning roughness coefficient. Develop a fixed-point iteration scheme to solve this equation for H given $Q = 10$, $S = 0.0001$, $B = 30$, and $n = 0.025$. Prove that your scheme converges for all initial guesses greater than or equal to zero.

6.21 The function $x^3 - 2x^2 - 4x + 8$ has a double root at $x = 2$. Use **(a)** the standard Newton-Raphson [Eq. (6.6)], **(b)** the modified Newton-Raphson [Eq. (6.12)], and **(c)** the modified Newton-Raphson [Eq. (6.16)] to solve for the root at $x = 2$. Compare and discuss the rate of convergence using an initial guess of $x_0 = 1.2$.

6.22 Determine the roots of the following simultaneous nonlinear equations using **(a)** fixed-point iteration and **(b)** the Newton-Raphson method:

$$y = -x^2 + x + 0.75$$

$$y + 5xy = x^2$$

Employ initial guesses of $x = y = 1.2$ and discuss the results.

6.23 Determine the roots of the simultaneous nonlinear equations

$$(x - 4)^2 + (y - 4)^2 = 5$$

$$x^2 + y^2 = 16$$

Use a graphical approach to obtain your initial guesses. Determine refined estimates with the two-equation Newton-Raphson method described in Sec. 6.6.2.

6.24 Repeat Prob. 6.23, except determine the positive root of

$$y = x^2 + 1$$

$$y = 2\cos x$$

6.25 A mass balance for a pollutant in a well-mixed lake can be written as

$$V\frac{dc}{dt} = W - Qc - kV\sqrt{c}$$

Given the parameter values $V = 2 \times 10^6$ m³, $Q = 2 \times 10^5$ m³/yr, $W = 2 \times 10^6$ g/yr, and $k = 0.33$ m$^{0.5}$/g$^{0.5}$/yr, use the modified secant method to solve for the steady-state concentration. Employ an initial guess of $c = 4$ g/m³ and $\delta = 0.5$. Perform three iterations and determine the percent relative error after the third iteration.

6.26 For Prob. 6.25, the root can be located with fixed-point iteration as

$$c = \left(\frac{W - Qc}{kV}\right)^2$$

or as

$$c = \frac{W - kV\sqrt{c}}{Q}$$

Only one will converge for initial guesses of $2 < c < 6$. Select the correct one and demonstrate why it will always work.

6.27 Develop a user-friendly program for the Newton-Raphson method based on Fig. 6.4 and Sec. 6.2.3. Test it by duplicating the computation from Example 6.3.

6.28 Develop a user-friendly program for the secant method based on Fig. 6.4 and Sec. 6.3.2. Test it by duplicating the computation from Example 6.6.

6.29 Develop a user-friendly program for the modified secant method based on Fig. 6.4 and Sec. 6.3.3. Test it by duplicating the computation from Example 6.8.

6.30 Develop a user-friendly program for Brent's root-location method based on Fig. 6.12. Test it by solving Prob. 6.6.

6.31 Develop a user-friendly program for the two-equation Newton-Raphson method based on Sec. 6.6.2. Test it by solving Example 6.12.

6.32 Use the program you developed in Prob. 6.31 to solve Probs. 6.22 and 6.23 to within a tolerance of $\varepsilon_s = 0.01\%$.

Roots of Polynomials

In this chapter, we will discuss methods used to find the roots of polynomial equations of the general form

$$f_n(x) = a_0 + a_1 x + a_2 x^2 + \cdots + a_n x^n \tag{7.1}$$

where n = the order of the polynomial and the a's are constant coefficients. Although the coefficients can be complex numbers, we will limit our discussion to cases where they are real. For such cases, the roots can be real and/or complex.

The roots of such polynomials follow these rules:

1. For an nth-order equation, there are n real or complex roots. It should be noted that these roots will not necessarily be distinct.
2. If n is odd, there is at least one real root.
3. If complex roots exist, they exist in conjugate pairs (that is, $\lambda + \mu i$ and $\lambda - \mu i$, where $i = \sqrt{-1}$).

Before describing the techniques for locating the roots of polynomials, we will provide some background. The first section offers some motivation for studying the techniques; the second deals with some fundamental computer manipulations involving polynomials.

7.1 POLYNOMIALS IN ENGINEERING AND SCIENCE

Polynomials have many applications in engineering and science. For example, they are used extensively in curve fitting. However, we believe that one of their most interesting and powerful applications is in characterizing dynamic systems and, in particular, linear systems. Examples include mechanical devices, structures, and electrical circuits. We will be exploring specific examples throughout the remainder of this text. In particular, they will be the focus of several of the engineering applications throughout the remainder of this text.

For the time being, we will keep the discussion simple and general by focusing on a simple second-order system defined by the following linear *ordinary differential equation* (or ODE):

$$a_2 \frac{d^2 y}{dt^2} + a_1 \frac{dy}{dt} + a_0 y = F(t) \tag{7.2}$$

where y and t are the dependent and independent variables, respectively, the a's are constant coefficients, and $F(t)$ is the forcing function.

In addition, it should be noted that Eq. (7.2) can be alternatively expressed as a pair of first-order ODEs by defining a new variable z,

$$z = \frac{dy}{dt} \tag{7.3}$$

Equation (7.3) can be substituted along with its derivative into Eq. (7.2) to remove the second-derivative term. This reduces the problem to solving

$$\frac{dz}{dt} = \frac{F(t) - a_1 z - a_0 y}{a_2} \tag{7.4}$$

$$\frac{dz}{dt} = z \tag{7.5}$$

In a similar fashion, an nth-order linear ODE can always be expressed as a system of n first-order ODEs.

Now let's look at the solution. The forcing function represents the effect of the external world on the system. The homogeneous solution, or *general solution*, of the equation deals with the case when the forcing function is set to zero,

$$a_2 \frac{d^2 y}{dt^2} + a_1 \frac{dy}{dt} + a_0 y = 0 \tag{7.6}$$

Thus, as the name implies, the general solution should tell us something very fundamental about the system being simulated—that is, how the system responds in the absence of external stimuli.

Now, the general solution to all unforced linear systems is of the form $y = e^{rt}$. If this function is differentiated and substituted into Eq. (7.6), the result is

$$a_2 r^2 e^{rt} + a_1 r e^{rt} + a_0 e^{rt} = 0$$

or canceling the exponential terms,

$$a_2 r^2 + a_1 r + a_0 = 0 \tag{7.7}$$

Notice that the result is a polynomial called the *characteristic equation*. The roots of this polynomial are the values of r that satisfy Eq. (7.7). These r's are referred to as the system's characteristic values, or *eigenvalues*.

So, here is the connection between roots of polynomials and engineering and science. The eigenvalue tells us something fundamental about the system we are modeling, and finding the eigenvalues involves finding the roots of polynomials. And, whereas finding the root of a second-order equation is easy with the quadratic formula, finding roots of higher-order systems (and hence, higher-order polynomials) is arduous analytically. Thus, the best general approach requires numerical methods of the type described in this chapter.

Before proceeding to these methods, let us take our analysis a bit farther by investigating what specific values of the eigenvalues might imply about the behavior of physical systems. First, let us evaluate the roots of Eq. (7.7) with the quadratic formula,

$$\begin{matrix} r_1 \\ r_2 \end{matrix} = \frac{-a_1 \pm \sqrt{a_1^2 - 4a_2a_0}}{a_0}$$

Thus, we get two roots. If the *discriminant* $(a_1^2 - 4a_2a_0)$ is positive, the roots are real and the general solution can be represented as

$$y = c_1 e^{r_1 t} + c_2 e^{r_2 t} \tag{7.8}$$

where the c's are constants that can be determined from the initial conditions. This is called the *overdamped case.*

If the discriminant is zero, a single real root results, and the general solution can be formulated as

$$y = (c_1 + c_2 t)e^{\lambda t} \tag{7.9}$$

This is called the *critically damped case.*

If the discriminant is negative, the roots will be complex conjugate numbers,

$$\begin{matrix} r_1 \\ r_2 \end{matrix} = \lambda \pm \mu i$$

and the general solution can be formulated as

$$y = c_1 e^{(\lambda + \mu i)t} + c_2 e^{(\lambda - \mu i)t}$$

The physical behavior of this solution can be elucidated by using Euler's formula,

$$e^{\mu i t} = \cos \mu t + i \sin \mu t$$

to reformulate the general solution as (see Boyce and DiPrima, 1992, for details of the derivation)

$$y = c_1 e^{\lambda t} \cos \mu t + c_2 e^{\lambda t} \sin \mu t \tag{7.10}$$

This is called the *underdamped case.*

Equations (7.8), (7.9), and (7.10) express the possible ways that linear systems respond dynamically. The exponential terms mean that the solutions are capable of decaying (negative real part) or growing (positive real part) exponentially with time (Fig. 7.1*a*). The sinusoidal terms (imaginary part) mean that the solutions can oscillate (Fig. 7.1*b*). If the eigenvalue has both real and imaginary parts, the exponential and sinusoidal shapes are combined (Fig. 7.1*c*). Because such knowledge is a key element in understanding, designing, and controlling the behavior of a physical system, characteristic polynomials are very important in engineering and many branches of science. We will explore the dynamics of several engineering systems in the applications covered in Chap. 8.

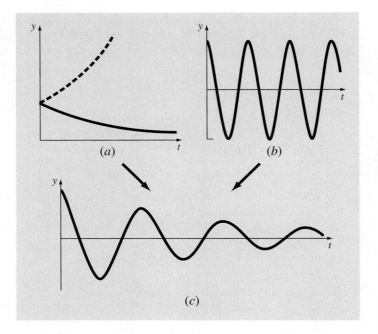

7.2 COMPUTING WITH POLYNOMIALS

Before describing root-location methods, we will discuss some fundamental computer
operations involving polynomials. These have utility in their own right as well as providing support for root finding.

7.2.1 Polynomial Evaluation and Differentiation

Although it is the most common format, Eq. (7.1) provides a poor means for determining the value of a polynomial for a particular value of *x*. For example, evaluating a
third-order polynomial as

$$f_3(x) = a_3 x^3 + a_2 x^2 + a_1 x + a_0 \qquad (7.11)$$

involves six multiplications and three additions. In general, for an *n*th-order polynomial,
this approach requires $n(n+1)/2$ multiplications and *n* additions.

In contrast, a nested format,

$$f_3(x) = ((a_3 x + a_2)x + a_1)x + a_0 \qquad (7.12)$$

involves three multiplications and three additions. For an *n*th-order polynomial, this approach requires *n* multiplications and *n* additions. Because the nested format minimizes
the number of operations, it also tends to minimize round-off errors. Note that, depending on your preference, the order of nesting can be reversed:

$$f_3(x) = a_0 + x(a_1 + x(a_2 + xa_3)) \qquad (7.13)$$

Succinct pseudocode to implement the nested form can be written simply as

```
DOFOR j = n, 0, -1
  p = p * x+a(j)
END DO
```

where p holds the value of the polynomial (defined by its coefficients, the a's) evaluated at x.

There are cases (such as in the Newton-Raphson method) where you might want to evaluate both the function and its derivative. This evaluation can also be neatly included by adding a single line to the preceding pseudocode,

```
DOFOR j = n, 0, -1
  df = df * x+p
  p = p * x+a(j)
END DO
```

where df holds the first derivative of the polynomial.

7.2.2 Polynomial Deflation

Suppose that you determine a single root of an nth-order polynomial. If you repeat your root-location procedure, you might find the same root. Therefore, it would be nice to remove the found root before proceeding. This removal process is referred to as *polynomial deflation*.

Before we show how this is done, some orientation might be useful. Polynomials are typically represented in the format of Eq. (7.1). For example, a fifth-order polynomial could be written as

$$f_5(x) = -120 - 46x + 79x^2 - 3x^3 - 7x^4 + x^5 \qquad (7.14)$$

Although this is a familiar format, it is not necessarily the best expression for understanding the polynomial's mathematical behavior. For example, this fifth-order polynomial might be expressed alternatively as

$$f_5(x) = (x + 1)(x - 4)(x - 5)(x + 3)(x - 2) \qquad (7.15)$$

This is called the *factored* form of the polynomial. If multiplication is completed and like terms collected, Eq. (7.14) would be obtained. However, the format of Eq. (7.15) has the advantage that it clearly indicates the function's roots. Thus, it is apparent that $x = -1, 4, 5, -3$, and 2 are all roots because each causes an individual term in Eq. (7.15) to become zero.

Now, suppose that we divide this fifth-order polynomial by any of its factors, for example, $x + 3$. For this case, the result would be a fourth-order polynomial,

$$f_4(x) = (x + 1)(x - 4)(x - 5)(x - 2) = -40 - 2x + 27x^2 - 10x^3 + x^4 \qquad (7.16)$$

with a remainder of zero.

In the distant past, you probably learned to divide polynomials using the approach called *synthetic division*. Several computer algorithms (based on both synthetic division and other methods) are available for performing the operation. One simple scheme is provided by the following pseudocode, which divides an nth-order polynomial by a monomial factor $x - t$:

```
r = a(n)
a(n) = 0
DOFOR i = n–1, 0, –1
    s = a(i)
    a(i) = r
    r = s + r * t
END DO
```

If the monomial is a root of the polynomial, the remainder r will be zero, and the coefficients of the quotient stored in a, at the end of the loop.

EXAMPLE 7.1 **Polynomial Deflation**

Problem Statement.   Divide the second-order polynomial

$$f(x) = (x - 4)(x + 6) = x^2 + 2x - 24$$

by the factor $x - 4$.

Solution.   Using the approach outlined in the above pseudocode, the parameters are $n = 2$, $a_0 = -24$, $a_1 = 2$, $a_2 = 1$, and $t = 4$. These can be used to compute

$$r = a_2 = 1$$
$$a_2 = 0$$

The loop is then iterated from $i = 2 - 1 = 1$ to 0. For $i = 1$,

$$s = a_1 = 2$$
$$a_1 = r = 1$$
$$r = s + rt = 2 + 1(4) = 6$$

For $i = 0$,

$$s = a_0 = -24$$
$$a_0 = r = 6$$
$$r = -24 + 6(4) = 0$$

Thus, the result is as expected—the quotient is $a_0 + a_1 x = 6 + x$, with a remainder of zero.

It is also possible to divide by polynomials of higher order. As we will see later in this chapter, the most common task involves dividing by a second-order polynomial or parabola. The subroutine in Fig. 7.2 addresses the more general problem of dividing an nth-order polynomial a by an mth-order polynomial d. The result is an $(n - m)$th-order polynomial q, with an $(m - 1)$th-order polynomial as the remainder.

Because each calculated root is known only approximately, it should be noted that deflation is sensitive to round-off errors. In some cases, round-off error can grow to the point that the results can become meaningless.

```
SUB poldiv(a, n, d, m, q, r)
  DOFOR j = 0, n
    r(j) = a(j)
    q(j) = 0
  END DO
  DOFOR k = n–m, 0, –1
    q(k+1) = r(m+k) / d(m)
    DOFOR j = m+k–1, k, –1
      r(j) = r(j)–q(k+1) * d(j–k)
    END DO
  END DO
  DOFOR j = m, n
    r(j) = 0
  END DO
  n = n–m
END SUB
```

FIGURE 7.2
Algorithm to divide a polynomial (defined by its coefficients a) by a lower-order polynomial d.

Some general strategies can be applied to minimize this problem. For example, round-off error is affected by the order in which the terms are evaluated. *Forward deflation* refers to the case where new polynomial coefficients are in order of descending powers of x (that is, from the highest-order to the zero-order term). For this case, it is preferable to divide by the roots of smallest absolute value first. Conversely, for *backward deflation* (that is, from the zero-order to the highest-order term), it is preferable to divide by the roots of largest absolute value first.

Another way to reduce round-off errors is to consider each successive root estimate obtained during deflation as a good first guess. Each estimate can then be used as a starting guess, and the root determined again with the original nondeflated polynomial. This is referred to as *root polishing*.

Finally, a problem arises when two deflated roots are inaccurate enough that they both converge on the same undeflated root. In that case, you might be erroneously led to believe that the polynomial has a multiple root (recall Sec. 6.5). One way to detect this problem is to compare each polished root with those that were located previously. Press et al. (2007) discuss this problem in more detail.

7.3 CONVENTIONAL METHODS

Now that we have covered some background material on polynomials, we can begin to describe methods to locate their roots. The obvious first step would be to investigate the viability of the bracketing and open approaches described in Chaps. 5 and 6.

The efficacy of these approaches depends on whether the problem being solved involves complex roots. If only real roots exist, any of the previously described methods could have utility. However, the problem of finding good initial guesses complicates both the bracketing and the open methods, whereas the open methods could be susceptible to divergence.

When complex roots are possible, the bracketing methods cannot be used because of the obvious problem that the criterion for defining a bracket (that is, sign change) does not translate to complex guesses.

Of the open methods, the conventional Newton-Raphson method would provide a viable approach. In particular, concise code including deflation can be developed. If a language that accommodates complex variables (like Fortran) is used, such an algorithm will locate both real and complex roots. However, as might be expected, it would be susceptible to convergence problems. For this reason, special methods have been developed to find the real and complex roots of polynomials. We describe two—the Müller and Bairstow methods—in the following sections. As you will see, both are related to the more conventional open approaches described in Chap. 6.

7.4 MÜLLER'S METHOD

Recall that the secant method obtains a root estimate by projecting a straight line to the x axis through two function values (Fig. 7.3a). Müller's method takes a similar approach, but projects a parabola through three points (Fig. 7.3b).

The method consists of deriving the coefficients of the parabola that goes through the three points. These coefficients can then be substituted into the quadratic formula to obtain the point where the parabola intercepts the x axis—that is, the root estimate. The approach is facilitated by writing the parabolic equation in a convenient form,

$$f_2(x) = a(x - x_2)^2 + b(x - x_2) + c \tag{7.17}$$

We want this parabola to intersect the three points $[x_0, f(x_0)]$, $[x_1, f(x_1)]$, and $[x_2, f(x_2)]$. The coefficients of Eq. (7.17) can be evaluated by substituting each of the three points to give

$$f(x_0) = a(x_0 - x_2)^2 + b(x_0 - x_2) + c \tag{7.18}$$
$$f(x_1) = a(x_1 - x_2)^2 + b(x_1 - x_2) + c \tag{7.19}$$
$$f(x_2) = a(x_2 - x_2)^2 + b(x_2 - x_2) + c \tag{7.20}$$

FIGURE 7.3
A comparison of two related approaches for locating roots: (a) the secant method and (b) Müller's method.

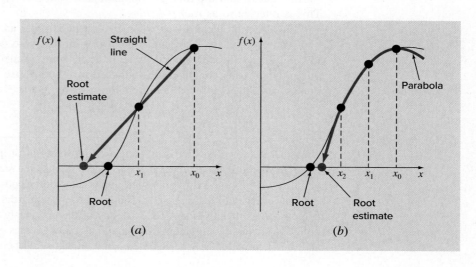

(a)   (b)

Note that we have dropped the subscript "2" from the function for conciseness. Because we have three equations, we can solve for the three unknown coefficients, a, b, and c. Because two of the terms in Eq. (7.20) are zero, it can be immediately solved for $c = f(x_2)$. Thus, the coefficient c is merely equal to the function value evaluated at the third guess, x_2. This result can then be substituted into Eqs. (7.18) and (7.19) to yield two equations with two unknowns:

$$f(x_0) - f(x_2) = a(x_0 - x_2)^2 + b(x_0 - x_2) \qquad (7.21)$$

$$f(x_1) - f(x_2) = a(x_1 - x_2)^2 + b(x_1 - x_2) \qquad (7.22)$$

Algebraic manipulation can then be used to solve for the remaining coefficients, a and b. One way to do this involves defining a number of differences,

$$h_0 = x_1 - x_0 \qquad\qquad h_1 = x_2 - x_1$$
$$\delta_0 = \frac{f(x_1) - f(x_0)}{x_1 - x_0} \qquad \delta_1 = \frac{f(x_2) - f(x_1)}{x_2 - x_1} \qquad (7.23)$$

These can be substituted into Eqs. (7.21) and (7.22) to give

$$(h_0 + h_1)b - (h_0 + h_1)^2 a = h_0\delta_0 + h_1\delta_1$$
$$h_1 \quad b - \quad h_1^2 \quad a = \quad h_1\delta_1$$

which can be solved for a and b. The results can be summarized as

$$a = \frac{\delta_1 - \delta_0}{h_1 + h_0} \qquad (7.24)$$

$$b = ah_1 + \delta_1 \qquad (7.25)$$

$$c = f(x_2) \qquad (7.26)$$

To find the root, we apply the quadratic formula to Eq. (7.17). However, because of potential round-off error, rather than using the conventional form, we use the alternative formulation [Eq. (3.13)] to yield

$$x_3 - x_2 = \frac{-2c}{b \pm \sqrt{b^2 - 4ac}} \qquad (7.27a)$$

or isolating the unknown x_3 on the left side of the equal sign,

$$x_3 = x_2 + \frac{-2c}{b \pm \sqrt{b^2 - 4ac}} \qquad (7.27b)$$

Note that the use of the quadratic formula means that both real and complex roots can be located. This is a major benefit of the method.

In addition, Eq. (7.27a) provides a neat means to determine the approximate error. Because the left side represents the difference between the present (x_3) and the previous (x_2) root estimate, the error can be calculated as

$$\varepsilon_a = \left| \frac{x_3 - x_2}{x_3} \right| 100\%$$

Now, a problem with Eq. (7.27a) is that it yields two roots, corresponding to the ± term in the denominator. In Müller's method, the sign is chosen to agree with the sign of b. This choice will result in the largest denominator, and hence, will give the root estimate that is closest to x_2.

Once x_3 is determined, the process is repeated. This brings up the issue of which point is discarded. Two general strategies are typically used:

1. If only real roots are being located, we choose the two original points that are nearest the new root estimate, x_3.
2. If both real and complex roots are being evaluated, a sequential approach is employed. That is, just as in the secant method, x_1, x_2, and x_3 take the place of x_0, x_1, and x_2.

EXAMPLE 7.2 **Müller's Method**

Problem Statement.   Use Müller's method with guesses of x_0, x_1, and $x_2 = 4.5$, 5.5, and 5, respectively, to determine a root of the equation

$$f(x) = x^3 - 13x - 12$$

Note that the roots of this equation are -3, -1, and 4.

Solution.   First, we evaluate the function at the guesses:

$$f(4.5) = 20.625 \qquad f(5.5) = 82.875 \qquad f(5) = 48$$

which can be used to calculate

$$h_0 = 5.5 - 4.5 = 1 \qquad\qquad h_1 = 5 - 5.5 = -0.5$$

$$\delta_0 = \frac{82.875 - 20.625}{5.5 - 4.5} = 62.25 \qquad \delta_1 = \frac{48 - 82.875}{5 - 5.5} = 69.75$$

These values in turn can be substituted into Eqs. (7.24) through (7.26) to compute

$$a = \frac{69.75 - 62.25}{-0.5 + 1} = 15 \qquad b = 15(-0.5) + 69.75 = 62.25 \qquad c = 48$$

The square root of the discriminant can be evaluated as

$$\sqrt{62.25^2 - 4(15)48} = 31.54461$$

Then, because $|62.25 + 31.54451| > |62.25 - 31.54451|$, a positive sign is employed in the denominator of Eq. (7.27b), and the new root estimate can be determined as

$$x_3 = 5 + \frac{-2(48)}{62.25 + 31.54451} = 3.976487$$

for which the approximate error is

$$\varepsilon_a = \left| \frac{-1.023513}{3.976487} \right| 100\% = 25.74\%$$

Because the error is large, new guesses are assigned; x_0 is replaced by x_1, x_1 is replaced by x_2, and x_2 is replaced by x_3. Therefore, for the new iteration,

$$x_0 = 5.5 \qquad x_1 = 5 \qquad x_2 = 3.976487$$

and the calculation is repeated. The results, tabulated below, show that the method converges rapidly on the root, $x_r = 4$:

i	x_r	ε_a (%)
0	5	
1	3.976487	25.74
2	4.00105	0.6139
3	4	0.0262
4	4	0.0000119

Pseudocode to implement Müller's method for real roots is presented in Fig. 7.4. Notice that this routine is set up to take a single initial nonzero guess that is then perturbed to develop the other two guesses. Of course, the algorithm can also be

FIGURE 7.4
Pseudocode for Müller's method.

```
SUB Muller(xr, h, eps, maxit)
x₂ = xr
x₁ = xr + h*xr
x₀ = xr − h*xr
DO
  iter = iter + 1
  h₀ = x₁ − x₀
  h₁ = x₂ − x₁
  d₀ = (f(x₁) − f(x₀)) / h₀
  d₁ = (f(x₂) − f(x₁)) / h₁
  a = (d₁ − d₀) / (h₁ + h₀)
  b = a*h₁ + d₁
  c = f(x₂)
  rad = SQRT(b*b − 4*a*c)
  IF |b+rad| > |b−rad| THEN
      den = b + rad
  ELSE
    den = b − rad
  END IF
  dxr = −2*c / den
  xr = x₂ + dxr
  PRINT iter, xr
  IF (|dxr| < eps*xr OR iter >= maxit) EXIT
  x₀ = x₁
  x₁ = x₂
  x₂ = xr
END DO
END Muller
```

programmed to accommodate three guesses. For languages like Fortran, the code will find complex roots if the proper variables are declared as complex.

7.5 BAIRSTOW'S METHOD

Bairstow's method is an iterative approach related loosely to both the Müller and Newton-Raphson methods. Before we launch into a mathematical description of the technique, recall the factored form of the polynomial,

$$f_5(x) = (x + 1)(x - 4)(x - 5)(x + 3)(x - 2) \tag{7.28}$$

If we divided by a factor that is not a root (for example, $x + 6$), the quotient would be a fourth-order polynomial. However, for this case, a remainder would result.

On the basis of the above, we can elaborate on an algorithm for determining a root of a polynomial: (1) guess a value for the root, $x = t$, (2) divide the polynomial by the factor $x - t$, and (3) determine whether there is a remainder. If not, the guess was perfect and the root is equal to t. If there is a remainder, the guess can be systematically adjusted and the procedure repeated until the remainder disappears and a root is located. After this is accomplished, the entire procedure can be repeated for the quotient to locate another root.

Bairstow's method is generally based on this approach. Consequently, it hinges on the mathematical process of dividing a polynomial by a factor. Recall from our discussion of polynomial deflation (Sec. 7.2.2) that synthetic division involves dividing a polynomial by a factor $x - t$. For example, the general polynomial [Eq. (7.1)]

$$f_n(x) = a_0 + a_1x + a_2x^2 + \cdots + a_nx^n \tag{7.29}$$

can be divided by the factor $x - t$ to yield a second polynomial that is one order lower,

$$f_{n-1}(x) = b_1 + b_2x + b_3x^2 + \cdots + b_nx^{n-1} \tag{7.30}$$

with a remainder $R = b_0$, where the coefficients can be calculated by the recurrence relationship

$$b_n = a_n$$
$$b_i = a_i + b_{i+1}t \qquad \text{for } i = n - 1 \text{ to } 0$$

Note that if t were a root of the original polynomial, the remainder b_0 would equal zero.

To permit the evaluation of complex roots, Bairstow's method divides the polynomial by a quadratic factor, $x^2 - rx - s$. If this is done to Eq. (7.29), the result is a new polynomial,

$$f_{n-2}(x) = b_2 + b_3x + \cdots + b_{n-1}x^{n-3} + b_nx^{n-2}$$

with the remainder

$$R = b_1(x - r) + b_0 \tag{7.31}$$

As with normal synthetic division, a simple recurrence relationship can be used to perform the division by the quadratic factor:

$$b_n = a_n \tag{7.32a}$$
$$b_{n-1} = a_{n-1} + rb_n \tag{7.32b}$$
$$b_i = a_i + rb_{i+1} + sb_{i+2} \qquad \text{for } i = n - 2 \text{ to } 0 \tag{7.32c}$$

The quadratic factor is introduced to allow the determination of complex roots. This relates to the fact that, if the coefficients of the original polynomial are real, the complex roots occur in conjugate pairs. If $x^2 - rx - s$ is an exact divisor of the polynomial, complex roots can be determined by the quadratic formula. Thus, the method reduces to determining the values of r and s that make the quadratic factor an exact divisor. In other words, we seek the values that make the remainder term equal to zero.

Inspection of Eq. (7.31) leads us to conclude that for the remainder to be zero, b_0 and b_1 must be zero. Because it is unlikely that our initial guesses at the values of r and s will lead to this result, we must determine a systematic way to modify our guesses so that b_0 and b_1 approach zero. To do this, Bairstow's method uses a strategy similar to the Newton-Raphson approach. Because both b_0 and b_1 are functions of both r and s, they can be expanded using a Taylor series, as in [recall Eq. (4.26)]

$$b_1(r + \Delta r, s + \Delta s) = b_1 + \frac{\partial b_1}{\partial r} \Delta r + \frac{\partial b_1}{\partial s} \Delta s$$

$$b_0(r + \Delta r, s + \Delta s) = b_0 + \frac{\partial b_0}{\partial r} \Delta r + \frac{\partial b_0}{\partial s} \Delta s$$

(7.33)

where the values on the right-hand side are all evaluated at r and s. Notice that second- and higher-order terms have been neglected. This represents an implicit assumption that Δr and Δs are small enough that the higher-order terms are negligible. Another way of expressing this assumption is to say that the initial guesses are adequately close to the values of r and s at the roots.

The changes, Δr and Δs, needed to improve our guesses can be estimated by setting Eq. (7.33) equal to zero to give

$$\frac{\partial b_1}{\partial r} \Delta r + \frac{\partial b_1}{\partial s} \Delta s = -b_1$$

(7.34)

$$\frac{\partial b_0}{\partial r} \Delta r + \frac{\partial b_0}{\partial s} \Delta s = -b_0$$

(7.35)

If the partial derivatives of the b's can be determined, these are a system of two equations that can be solved simultaneously for the two unknowns, Δr and Δs. Bairstow showed that the partial derivatives can be obtained by a synthetic division of the b's in a fashion similar to the way in which the b's themselves were derived:

$$c_n = b_n$$

(7.36a)

$$c_{n-1} = b_{n-1} + rc_n$$

(7.36b)

$$c_i = b_i + rc_{i+1} + sc_{i+2} \qquad \text{for } i = n - 2 \text{ to } 1$$

(7.36c)

where $\partial b_0/\partial r = c_1$, $\partial b_0/\partial s = \partial b_1/\partial r = c_2$, and $\partial b_1/\partial s = c_3$. Thus, the partial derivatives are obtained by synthetic division of the b's. Then the partial derivatives can be substituted into Eqs. (7.34) and (7.35) along with the b's to give

$$c_2 \Delta r + c_3 \Delta s = -b_1$$
$$c_1 \Delta r + c_2 \Delta s = -b_0$$

These equations can be solved for Δr and Δs, which can in turn be employed to improve the initial guesses of r and s. At each step, an approximate error in r and s can be estimated, as in

$$|\varepsilon_{a,r}| = \left| \frac{\Delta r}{r} \right| 100\% \tag{7.37}$$

and

$$|\varepsilon_{a,s}| = \left| \frac{\Delta s}{s} \right| 100\% \tag{7.38}$$

When both of these error estimates fall below a prespecified stopping criterion ε_s, the values of the roots can be determined by

$$x = \frac{r \pm \sqrt{r^2 + 4s}}{2} \tag{7.39}$$

At this point, three possibilities exist:

1. *The quotient is a third-order polynomial or greater.* For this case, Bairstow's method would be applied to the quotient to evaluate new values for r and s. The previous values of r and s can serve as the starting guesses for this application.
2. *The quotient is a quadratic.* For this case, the remaining two roots could be evaluated with Eq. (7.39).
3. *The quotient is a first-order polynomial.* For this case, the remaining single root can be evaluated simply as

$$x = -\frac{s}{r} \tag{7.40}$$

EXAMPLE 7.3

Bairstow's Method

Problem Statement.   Employ Bairstow's method to determine the roots of the polynomial

$$f_5(x) = x^5 - 3.5x^4 + 2.75x^3 + 2.125x^2 - 3.875x + 1.25$$

Use initial guesses of $r = s = -1$ and iterate to a level of $\varepsilon_s = 1\%$.

Solution.   Equations (7.32) and (7.36) can be applied to compute

$$b_5 = 1 \qquad b_4 = -4.5 \qquad b_3 = 6.25 \qquad b_2 = 0.375 \qquad b_1 = -10.5$$
$$b_0 = 11.375$$
$$c_5 = 1 \qquad c_4 = -5.5 \qquad c_3 = 10.75 \qquad c_2 = -4.875 \qquad c_1 = -16.375$$

Thus, the simultaneous equations to solve for Δr and Δs are

$$-4.875\Delta r + 10.75\Delta s = 10.5$$
$$-16.375\Delta r - 4.875\Delta s = -11.375$$

which can be solved for $\Delta r = 0.3558$ and $\Delta s = 1.1381$. Therefore, our original guesses can be corrected to

$$r = -1 + 0.3558 = -0.6442$$
$$s = -1 + 1.1381 = 0.1381$$

and the approximate errors can be evaluated by Eqs. (7.37) and (7.38),

$$|\varepsilon_{a,r}| = \left| \frac{0.3558}{-0.6442} \right| 100\% = 55.23\% \qquad |\varepsilon_{a,s}| = \left| \frac{1.1381}{0.1381} \right| 100\% = 824.1\%$$

Next, the computation is repeated using the revised values for r and s. Applying Eqs. (7.32) and (7.36) yields

$$b_5 = 1 \quad b_4 = -4.1442 \quad b_3 = 5.5578 \quad b_2 = -2.0276 \quad b_1 = -1.8013$$
$$b_0 = 2.1304$$
$$c_5 = 1 \quad c_4 = -4.7884 \quad c_3 = 8.7806 \quad c_2 = -8.3454 \quad c_1 = 4.7874$$

Therefore, we must solve

$$-8.3454\Delta r + 8.7806\Delta s = 1.8013$$
$$4.7874\Delta r - 8.3454\Delta s = -2.1304$$

for $\Delta r = 0.1331$ and $\Delta s = 0.3316$, which can be used to correct the root estimates:

$$r = -0.6442 + 0.1331 = -0.5111 \qquad |\varepsilon_{a,r}| = 26.0\%$$
$$s = 0.1381 + 0.3316 = 0.4697 \qquad |\varepsilon_{a,s}| = 70.6\%$$

The computation can be continued, with the result that after four iterations the method converges on values of $r = -0.5$ ($|\varepsilon_{a,r}| = 0.063\%$) and $s = 0.5$ ($|\varepsilon_{a,s}| = 0.040\%$). Equation (7.39) can then be employed to evaluate the roots as

$$x = \frac{-0.5 \pm \sqrt{(-0.5)^2 + 4(0.5)}}{2} = 0.5, -1.0$$

At this point, the quotient is the cubic equation

$$f(x) = x^3 - 4x^2 + 5.25x - 2.5$$

Bairstow's method can be applied to this polynomial using the results of the previous step, $r = -0.5$ and $s = 0.5$, as starting guesses. Five iterations yield estimates of $r = 2$ and $s = -1.249$, which can be used to compute

$$x = \frac{2 \pm \sqrt{2^2 + 4(-1.249)}}{2} = 1 \pm 0.499i$$

At this point, the quotient is a first-order polynomial that can be directly evaluated by Eq. (7.40) to determine the fifth root: 2.

Note that the heart of Bairstow's method is the evaluation of the b's and c's via Eqs. (7.32) and (7.36). One of the primary strengths of the method is the concise way in which these recurrence relationships can be programmed.

Figure 7.5 lists pseudocode to implement Bairstow's method. The heart of the algorithm consists of the loop to evaluate the b's and c's. Also notice that the code to solve the simultaneous equations checks to prevent division by zero. If this is the case, the values of r and s are perturbed slightly and the procedure is begun again. In addition, the algorithm places a user-defined upper limit on the number of iterations (*maxit*) and should be designed to avoid division by zero while calculating the error estimates.

(a) Bairstow Algorithm

```
SUB Bairstow (a,nn,es,rr,ss,maxit,re,im,ier)
DIMENSION b(nn), c(nn)
r = rr; s = ss; n = nn
ier = 0; ea1 = 1; ea2 = 1
DO
  IF n < 3 OR iter ≥ maxit EXIT
  iter = 0
  DO
    iter = iter + 1
    b(n) = a(n)
    b(n − 1) = a(n − 1) + r * b(n)
    c(n) = b(n)
    c(n − 1) = b(n − 1) + r * c(n)
    DO i = n − 2, 0, −1
      b(i) = a(i) + r * b(i + 1) + s * b(i + 2)
      c(i) = b(i) + r * c(i + 1) + s * c(i + 2)
    END DO
    det = c(2) * c(2) − c(3) * c(1)
    IF det ≠ 0 THEN
      dr = (−b(1) * c(2) + b(0) * c(3))/det
      ds = (−b(0) * c(2) + b(1) * c(1))/det
      r = r + dr
      s = s + ds
      IF r≠0 THEN ea1 = ABS(dr/r) * 100
      IF s≠0 THEN ea2 = ABS(ds/s) * 100
    ELSE
      r = r + 1
      s = s + 1
      iter = 0
    END IF
    IF ea1 ≤ es AND ea2 ≤ es OR iter ≥ maxit EXIT
  END DO
  CALL Quadroot(r,s,r1,i1,r2,i2)
  re(n) = r1
  im(n) = i1
  re(n − 1) = r2
  im(n − 1) = i2
  n = n − 2
  DO i = 0, n
    a(i) = b(i + 2)
  END DO
END DO
```

```
  IF iter < maxit THEN
    IF n = 2 THEN
      r = −a(1)/a(2)
      s = −a(0)/a(2)
      CALL Quadroot(r,s,r1,i1,r2,i2)
      re(n) = r1
      im(n) = i1
      re(n − 1) = r2
      im(n − 1) = i2
    ELSE
      re(n) = −a(0)/a(1)
      im(n) = 0
    END IF
  ELSE
    ier = 1
  END IF
END Bairstow
```

(b) Roots of Quadratic Algorithm

```
SUB Quadroot(r,s,r1,i1,r2,i2)
disc = r ^ 2 + 4 * s
IF disc > 0 THEN
  r1 = (r + SQRT(disc))/2
  r2 = (r − SQRT(disc))/2
  i1 = 0
  i2 = 0
ELSE
  r1 = r/2
  r2 = r1
  i1 = SQRT(ABS(disc))/2
  i2 = −i1
END IF
END Quadroot
```

FIGURE 7.5

(a) Algorithm for implementing Bairstow's method, along with (b) an algorithm to determine the roots of a quadratic.

Finally, the algorithm requires initial guesses for r and s (rr and ss in the code). If no prior knowledge of the roots exist, they can be set to zero in the calling program.

7.6 OTHER METHODS

Other methods are available to locate the roots of polynomials. The *Jenkins-Traub method* is commonly used in software libraries. It is fairly complicated, and a good starting point to understanding it is found in Ralston and Rabinowitz (1978).

Laguerre's method, which approximates both real and complex roots and has cubic convergence, is among the best approaches. A complete discussion can be found in Householder (1970). In addition, Press et al. (2007) present a nice algorithm to implement the method.

7.7 ROOT LOCATION WITH SOFTWARE PACKAGES

Software packages have great capabilities for locating roots. In this section, we will give you a taste of some of the more useful ones.

7.7.1 Excel

A spreadsheet like Excel can be used to locate a root by *trial and error.* For example, if you want to find a root of

$$f(x) = x - \cos x$$

first, you can enter a value for x in a cell. Then set up another cell for $f(x)$ that would obtain its value for x from the first cell. You can then vary the x cell until the $f(x)$ cell approaches zero. This process can be further enhanced by using Excel's plotting capabilities to obtain a good initial guess (Fig. 7.6).

FIGURE 7.6
A spreadsheet set up to determine the root of $f(x) = x - \cos x$ by trial and error. The plot is used to obtain a good initial guess.

Although Excel does facilitate a trial-and-error approach, it also has two standard tools that can be employed for root location: *Goal Seek* and *Solver*. Both these tools can be employed to systematically adjust the initial guesses. Goal Seek is expressly used to drive an equation to a value (in our case, zero) by varying a single parameter.

EXAMPLE 7.4 | **Using Excel's Goal Seek Tool to Locate a Single Root**

Problem Statement. Employ Goal Seek to determine the root of the transcendental function

$$f(x) = x - \cos x$$

Solution. As in Fig. 7.6, the key to solving a single equation with Excel is creating a cell to hold the value of the function in question and then making the value dependent on another cell. Once this is done, the selection Goal Seek is chosen from the What-If Analysis button on your Data ribbon. At this point a dialogue box will be displayed, asking you to set a cell to a value by changing another cell. For the example, suppose that as in Fig. 7.6 your guess is entered in cell A11 and your function result in cell B11. The Goal Seek dialogue box would be filled out as follows:

When the OK button is selected, a message box displays the results:

The cells on the spreadsheet would also be modified to the new values (as shown in Fig. 7.6).

The Solver tool is more sophisticated than Goal Seek in that (1) it can vary several cells simultaneously and (2) along with driving a target cell to a value, it can minimize and maximize its value. The next example illustrates how it can be used to solve a system of nonlinear equations.

EXAMPLE 7.5

Using Excel's Solver for a Nonlinear System

Problem Statement.   Recall that in Sec. 6.6 we obtained the solution of the following set of simultaneous equations,

$$u(x, y) = x^2 + xy - 10 = 0$$
$$v(x, y) = y + 3xy^2 - 57 = 0$$

Note that a correct pair of roots is $x = 2$ and $y = 3$. Use Solver to determine the roots using initial guesses of $x = 1$ and $y = 3.5$.

Solution.   As shown below, two cells (B1 and B2) can be created to hold the guesses for x and y. The function values themselves, $u(x, y)$ and $v(x, y)$ can then be entered into two other cells (B3 and B4). As can be seen, the initial guesses result in function values that are far from zero.

B6		⠇	✕	✓	*fx*	=B3^2+B4^2

	A	B	C	D
1	x	1		
2	y	3.5		
3	u(x,y)	-5.5		
4	v(x,y)	-16.75		
5				
6	Sum of Squares	310.8125		

Next, another cell can be created that contains a single value reflecting how close both functions are to zero. One way to do this is to sum the squares of the function values. This is done and the result entered in cell B6. If both functions are at zero, this function should also be at zero. Further, using the squared functions avoids the possibility that both functions could have the same nonzero value, but with opposite signs. For this case, the target cell (B6) would be zero, but the roots would be incorrect.

Once the spreadsheet is created, the selection Solver is chosen from the Data ribbon.[1] At this point a dialogue box will be displayed, querying you for pertinent information. The pertinent cells of the Solver dialogue box would be filled out as follows:

---

[1] Note that you may have to install Solver by choosing Office, Excel Options, Add-Ins. Select Excel Add-Ins from the Manage drop-down box at the bottom of the Excel options menu and click Go. Then, check the Solver box. The Solver then should be installed and a button to access it should appear on your Data ribbon.

When the OK button is selected, a dialogue box will open with a report on the success of the operation. For the present case, the Solver obtains the correct solution:

	A	B	C
1	x	1.999606	
2	y	3.000286	
3	u(x,y)	-0.00219	
4	v(x,y)	-8.5E-05	
5			
6	Sum of Squares	4.8E-06	

It should be noted that the Solver can fail. Its success depends on (1) the condition of the system of equations and/or (2) the quality of the initial guesses. Thus, the successful outcome of the previous example is not guaranteed. Despite this, we have found Solver useful enough to make it a feasible option for quickly obtaining roots in a wide range of engineering applications.

7.7.2 MATLAB

As summarized in Table 7.1, MATLAB software is capable of locating roots of single algebraic and transcendental equations. It is superb at manipulating and locating the roots of polynomials.

The `fzero` function is designed to locate one root of a single function. A simplified representation of its syntax is

$$\texttt{fzero(}f\texttt{,}x_0\texttt{,}options\texttt{)}$$

where f is the function you are analyzing, x_0 is the initial guess, and `options` are the optimization parameters (these are changed using the function `optimset`). If `options` are omitted, default values are employed. Note that one or two guesses can be employed. If two guesses are employed, they are assumed to bracket a root. The following example illustrates how `fzero` can be used.

TABLE 7.1 Common functions in MATLAB related to root location and polynomial manipulation.

Function	Description
fzero	Root of single function.
roots	Find polynomial roots.
poly	Construct polynomial with specified roots.
polyval	Evaluate polynomial.
polyvalm	Evaluate polynomial with matrix argument.
residue	Partial-fraction expansion (residues).
polyder	Differentiate polynomial.
conv	Multiply polynomials.
deconv	Divide polynomials.

EXAMPLE 7.6   **Using MATLAB for Root Location**

Problem Statement.   Use the MATLAB function `fzero` to find the roots of

$$f(x) = x^{10} - 1$$

within the interval $x_l = 0$ and $x_u = 4$. Obviously two roots occur at -1 and 1. Recall that in Example 5.6, we used the false-position and bisection methods with initial guesses of 0 and 1.3 to determine the positive root.

Solution.   Using the same initial conditions as in Example 5.6, we can use MATLAB to determine the positive root as in

```
>> x0= [0 1.3];
>> x=fzero(@(x) x^10-1,x0)

x =
    1
```

In a similar fashion, we can use initial guesses of −1.3 and 0 to determine the negative root,

```
>> x0= [−1.3 0];
>> x=fzero(@(x) x^10-1,x0)

x =
   −1
```

We can also employ a single guess. An interesting case would be to use an initial guess of 0,

```
>> x0=0;
>> x=fzero(@(x) x^10-1,x0)

x =
   −1
```

Thus, for this guess, the underlying algorithm happens to home in on the negative root.

The use of `optimset` can be illustrated by using it to display the actual iterations as the solution progresses:

```
>> x0=0;
>> option=optimset('DISP','ITER');
>> x=fzero(@(x) x^10-1,x0,option)
```

Func-count	x	f(x)	Procedure
1	0	−1	initial
2	−0.0282843	−1	search
3	0.0282843	−1	search
4	−0.04	−1	search
.			
.			
.			
21	0.64	−0.988471	search
22	−0.905097	−0.631065	search
23	0.905097	−0.631065	search
24	−1.28	10.8059	search

```
Looking for a zero in the interval [−1.28, 0.9051]
```

25	0.784528	−0.911674	interpolation
26	−0.247736	−0.999999	bisection
27	−0.763868	−0.932363	bisection

```
        28         -1.02193          0.242305       bisection
        29        -0.968701          -0.27239       interpolation
        30        -0.996873        -0.0308299       interpolation
        31        -0.999702       -0.00297526       interpolation
        32               -1       5.53132e-006       interpolation
        33               -1       -7.41965e-009      interpolation
        34               -1       -1.88738e-014      interpolation
        35               -1                 0        interpolation
 Zero found in the interval [-1.28, 0.9051].

    x =
        -1
```

These results illustrate the strategy used by `fzero` when it is provided with a single guess. First, it searches in the vicinity of the guess until it detects a sign change. Then it uses a combination of bisection and interpolation to home in on the root. The interpolation involves both the secant method and inverse quadratic interpolation (recall Sec. 6.4). It should be noted that the `fzero` algorithm has more to it than this basic description might imply. You can consult Press et al. (2007) for additional details.

EXAMPLE 7.7

Using MATLAB to Manipulate and Determine the Roots of Polynomials

Problem Statement.   Explore how MATLAB can be employed to manipulate and determine the roots of polynomials. Use the following equation from Example 7.3,

$$f_5(x) = x^5 - 3.5x^4 + 2.75x^3 + 2.125x^2 - 3.875x + 1.25 \qquad \text{(E7.7.1)}$$

which has three real roots: 0.5, −1.0, and 2, and one pair of complex roots: $1 \pm 0.5i$.

Solution.   Polynomials are entered into MATLAB by storing the coefficients as a vector. For example, at the MATLAB prompt (>>), typing and entering the follow line stores the coefficients in the vector a,

```
>> a= [1 -3.5 2.75 2.125 -3.875 1.25];
```

We can then proceed to manipulate the polynomial. For example, we can evaluate it at $x = 1$ by typing

```
>> polyval(a,1)
```

with the result $1(1)^5 - 3.5(1)^4 + 2.75(1)^3 + 2.125(1)^2 - 3.875(1) + 1.25 = -0.25$,

```
ans =
   -0.2500
```

We can evaluate the derivative $f'(x) = 5x^4 - 14x^3 + 8.25x^2 + 4.25x - 3.875$ by

```
>> polyder(a)
ans =
    5.0000   -14.0000    8.2500    4.2500   -3.8750
```

Next, let us create a quadratic polynomial that has roots corresponding to two of the original roots of Eq. (E7.7.1): 0.5 and -1. This quadratic is $(x - 0.5)(x + 1) = x^2 + 0.5x - 0.5$ and can be entered into MATLAB as the vector b,

```
>> b= [1  0.5 -0.5];
```

We can divide this polynomial into the original polynomial by

```
>> [d,e] =deconv(a,b)
```

with the result being a quotient (a third-order polynomial d) and a remainder (e),

```
d =
    1.0000   -4.0000    5.2500   -2.5000
e =
    0     0     0     0     0     0
```

Because the polynomial is a perfect divisor, the remainder polynomial has zero coefficients. Now, the roots of the quotient polynomial can be determined as

```
>> roots(d)
```

with the expected result that the remaining roots of the original polynomial (E7.7.1) are found,

```
ans =
    2.0000
    1.0000 + 0.5000i
    1.0000 - 0.5000i
```

We can now multiply d by b to come up with the original polynomial,

```
>> conv(d,b)
ans =
    1.0000   -3.5000    2.7500    2.1250   -3.8750    1.2500
```

Finally, we can determine all the roots of the original polynomial by

```
>> r=roots(a)
r =
   -1.0000
    2.0000
    1.0000 + 0.5000i
    1.0000 - 0.5000i
    0.5000
```

7.7.3 Mathcad

Mathcad has a numeric mode function called **root** that can be used to solve an equation of a single variable. The method requires that you supply a function $f(x)$ and either an initial guess or a bracket. When a single guess value is used, **root** uses the *secant* and

Müller methods. In the case where two guesses that bracket a root are supplied, it uses a combination of the *Ridder method* (a variation of false position) and *Brent's method.* It iterates until the magnitude of $f(x)$ at the proposed root is less than the predefined value of **TOL**. The Mathcad implementation has similar advantages and disadvantages to conventional root-location methods, such as issues concerning the quality of the initial guess and the rate of convergence.

Mathcad can find all the real or complex roots of polynomials with ***polyroots***. This numeric or symbolic mode function is based on the *Laguerre method.* This function does not require initial guesses, and all the roots are returned at the same time.

Mathcad contains a numeric mode function called ***Find*** that can be used to solve up to 50 simultaneous nonlinear algebraic equations. The **Find** function chooses an appropriate method from a group of available methods, depending on whether the problem is linear or nonlinear, and other attributes. Acceptable values for the solution may be unconstrained or constrained to fall within specified limits. If **Find** fails to locate a solution that satisfies the equations and constraints, it returns the error message "did not find solution." However, Mathcad also contains a similar function called **Minerr**. This function gives solution results that minimize the errors in the constraints even when exact solutions cannot be found. Thus, the problem of solving for the roots of nonlinear equations is closely related to both optimization and nonlinear least squares. These areas and **Minerr** are covered in detail in Parts Four and Five.

Figure 7.7 shows a typical Mathcad worksheet. The menus at the top provide quick access to common arithmetic operators and functions, various two- and three-dimensional plot types, and the environment to create subprograms. Equations, text, data, or graphs

FIGURE 7.7

Mathcad screen to find the root of a single equation.

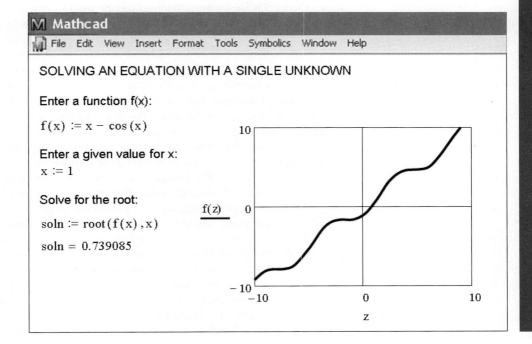

can be placed anywhere on the screen. You can use a variety of fonts, colors, and styles to construct worksheets with almost any design and format that pleases you. Consult the summary of the Mathcad User's manual in Appendix C or the full manual available from MathSoft. Note that in all our Mathcad examples, we have tried to fit the entire Mathcad session onto a single screen. You should realize that the graph would have to be placed below the commands to work properly.

Let's start with an example that solves for the root of $f(x) = x - \cos x$. The first step is to enter the function. This is done by typing f(x): which is automatically converted to f(x):= by Mathcad. The := is called the definition symbol. Next an initial guess is input in a similar manner using the definition symbol. Now, soln is defined as **root**(f(x), x), which invokes the secant method with a starting value of 1.0. Iteration is continued until $f(x)$ evaluated at the proposed root is less than **TOL**. The value of **TOL** is set from the Math/Options pull-down menu. Finally, the value of soln is displayed using a normal equal sign (=). The number of significant figures is set from the Format/Number pull-down menu. The text labels and equation definitions can be placed anywhere on the screen in a number of different fonts, styles, sizes, and colors. The graph can be placed anywhere on the worksheet by clicking to the desired location. This places a red crosshair at that location. Then use the Insert/Graph/X-Y Plot pull-down menu to place an empty plot on the worksheet with placeholders for the expressions to be graphed and for the ranges of the x and y axes. Simply type f(z) in the placeholder on the y axis and -10 and 10 for the z-axis range. Mathcad does all the rest to produce the graph shown in Fig. 7.7. Once the graph has been created, you can use the Format/Graph/X-Y Plot pull-down menu to vary the type of graph; change the color, type, and weight of the trace of the function; and add titles, labels and other features.

Figure 7.8 shows how Mathcad can be used to find the roots of a polynomial using the **polyroots** function. First, $p(x)$ and v are input using the := definition symbol. Note that v is a vector that contains the coefficients of the polynomial starting with the zero-order term and ending in this case with the third-order term. Next, r is defined (using :=) as polyroots(v), which invokes the Laguerre method. The roots contained in r are displayed as r^T using a normal equal sign (=). Next, a plot is constructed in a manner similar to the above, except that now two range variables, x and j, are used to define the range of the x axis and the location of the roots. The range variable for x is constructed by typing x and then ":" (which appears as :=) and then -4, and then "," and then -3.99, and then ";" (which is transformed into .. by Mathcad), and finally 4. This creates a vector of values of x ranging from -4 to 4 with an increment of 0.01 for the x axis with corresponding values for $p(x)$ on the y axis. The j range variable is used to create three values for r and $p(r)$ that are plotted as individual small circles. Note that again, in our effort to fit the entire Mathcad session onto a single screen, we have placed the graph above the commands. You should realize that the graph would have to be below the commands to work properly.

The last example shows the solution of a system of nonlinear equations using a Mathcad Solve Block (Fig. 7.9). The process begins with using the definition symbol to create initial guesses for x and y. The word **Given** then alerts Mathcad that what follows is a system of equations. Then comes the equations and inequalities (not used here). Note that for this application Mathcad requires the use of a symbolic equal sign typed as **[Ctrl]**= or < and > to separate the left and right sides of an equation. Now, the variable vec is defined as **Find**(x,y) and the value of vec is shown using an equal sign.

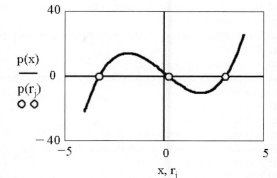

M Mathcad

File Edit View Insert Format Tools Symbolics Window Help

FINDING THE ROOT OF A POLYNOMIAL

Input a polynomial:

$p(x) := x^3 - 10 \cdot x + 2$

Input vector of coefficients,
beginning with the constant term:

$v := \begin{pmatrix} 2 \\ -10 \\ 0 \\ 1 \end{pmatrix}$

Determine the roots:

$r := polyroots(v)$

$r^T = (-3.257897 \quad 0.20081 \quad 3.057087)$

Create a plot:

$x := -4, -3.99..4$

$j := 0,1..2$

M Mathcad

File Edit View Insert Format Tools Symbolics Window Help

SOLVING A NONLINEAR SYSTEM OF EQUATIONS

Enter guesses for the n unknowns:

$x := 1$
$y := 3.5$

Given

Enter the n equations:

$x^2 + x \cdot y - 10 = 0$

$y + 3 \cdot x \cdot y - 57 = 0$

[Note: Use the symbolic equal sign [Ctrl=
           for equations within the Solve Block]

Solve the system:

$vec := Find(x, y)$

Solution:

$vec = \begin{pmatrix} 2 \\ 3 \end{pmatrix}$

PROBLEMS

7.1 Divide the polynomial $f(x) = x^4 - 5x^3 + 5x^2 + 5x - 6$ by the monomial factor $x - 2$. Is $x = 2$ a root?

7.2 Divide the polynomial $f(x) = x^5 - 6x^4 + x^3 - 7x^2 - 7x + 12$ by the monomial factor $x - 2$.

7.3 Use Müller's method to determine the positive real root of
(a) $f(x) = x^3 + x^2 - 4x - 4$
(b) $f(x) = x^3 - 0.5x^2 + 4x - 2$

7.4 Use Müller's method or MATLAB to determine the real and complex roots of
(a) $f(x) = x^3 - x^2 + 2x - 2$
(b) $f(x) = x^4 + 3x^2 + 5$
(c) $f(x) = x^4 - 2x^3 + 6x^2 - 8x + 8$

7.5 Use Bairstow's method to determine the roots of
(a) $f(x) = -2 + 6.2x - 4x^2 + 0.7x^3$
(b) $f(x) = 9 - 22x + 16x^2 - 3.6x^3$
(c) $f(x) = x^4 - 2x^3 + 21x^2 - 8x + 68$

7.6 Develop a program to implement Müller's method. Test it by duplicating Example 7.2.

7.7 Use the program developed in Prob. 7.6 to determine the real roots of Prob. 7.4**a**. Construct a graph (by hand or with a software package) to develop suitable starting guesses.

7.8 Develop a program to implement Bairstow's method. Test it by duplicating Example 7.3.

7.9 Use the program developed in Prob. 7.8 to determine the roots of the equations in Prob. 7.5.

7.10 Determine the real root of $x^{3.6} = 75$ with Excel, MATLAB, or Mathcad.

7.11 The velocity of a falling parachutist is given by

$$v = \frac{gm}{c}\left(1 - e^{-(c/m)t}\right)$$

where $g = 9.81$ m/s². For a parachutist with a drag coefficient $c = 16$ kg/s, compute the mass m so that the velocity is $v = 36$ m/s at $t = 7$ s. Use Excel, MATLAB, or Mathcad to determine m.

7.12 Perform the same calculation as in Prob. 7.11 but for the more accurate representation of drag as a function of the square of the velocity. For this case, the closed-form solution is

$$v = \sqrt{\frac{gm}{c'}}\tanh\left(\sqrt{\frac{gc'}{m}}\,t\right)$$

where c' = a second-order drag coefficient = 0.25 kg/m.

7.13 Determine the roots of the simultaneous nonlinear equations

$$y = -x^2 + x + 0.75$$
$$y + 5xy = x^2$$

Employ initial guesses of $x = y = 1.2$ and use the Solver tool from Excel or a software package of your choice.

7.14 Determine the roots of the simultaneous nonlinear equations

$$(x - 4)^2 + (y - 4)^4 = 5$$
$$x^2 + y^2 = 16$$

Use a graphical approach to obtain your initial guesses. Determine refined estimates with the Solver tool from Excel or a software package of your choice.

7.15 Perform the identical MATLAB operations used in Example 7.7 or use a software package of your choice to find all the roots of the polynomial

$$f(x) = (x - 6)(x + 2)(x - 1)(x + 4)(x - 8)$$

Note that the `poly` function can be used to convert the roots to a polynomial.

7.16 Use MATLAB or Mathcad to determine the roots for the equations in Prob. 7.5.

7.17 A two-dimensional circular cylinder is placed in a high-speed uniform flow. Vortices shed from the cylinder at a constant frequency, and pressure sensors on the rear surface of the cylinder detect this frequency by calculating how often the pressure oscillates. Given three data points, use Müller's method to find the time when the pressure was zero.

Time	0.60	0.62	0.64
Pressure	20	50	60

7.18 When trying to find the acidity of a solution of magnesium hydroxide in hydrochloric acid, we obtain the following equation

$$A(x) = x^3 + 3.5x^2 - 40$$

where x is the hydronium ion concentration. Find the hydronium ion concentration for a saturated solution (acidity equals zero) using two different methods in MATLAB (for example, graphically and the `roots` function).

7.19 Consider the following system with three unknowns a, u, and v:

$$u^2 - 2v^2 = a^2$$
$$u + v = 2$$
$$a^2 - 2a - u = 0$$

Solve for the real values of the unknowns using: **(a)** the Excel Solver and **(b)** a symbolic manipulator software package.

7.20 In control systems analysis, transfer functions are developed that mathematically relate the dynamics of a system's input to its output. A transfer function for a robotic-positioning system is given by

$$G(s) = \frac{C(s)}{N(s)} = \frac{s^3 + 12.5s^2 + 50.5s + 66}{s^4 + 19s^3 + 122s^2 + 296s + 192}$$

where $G(s)$ = system gain, $C(s)$ = system output, $N(s)$ = system input, and s = Laplace transform complex frequency. Use a numerical technique to find the roots of the numerator and denominator and factor these into the form

$$G(s) = \frac{(s + a_1)(s + a_2)(s + a_3)}{(s + b_1)(s + b_2)(s + b_3)(s + b_4)}$$

where a_i and b_i = the roots of the numerator and denominator, respectively.

7.21 Develop an M-file function for bisection similar to the algorithm in Fig. 5.10. Test the function by duplicating the computations from Examples 5.3 and 5.4.

7.22 Develop an M-file function for the false-position method. The structure of your function should be similar to the bisection algorithm outlined in Fig. 5.10. Test the program by duplicating Example 5.5.

7.23 Develop an M-file function for the Newton-Raphson method based on Fig. 6.4 and Sec. 6.2.3. Along with the initial guess, pass the function and its derivative as arguments. Test it by duplicating the computation from Example 6.3.

7.24 Develop an M-file function for the secant method based on Fig. 6.4 and Sec. 6.3.2. Along with the two initial guesses, pass the function as an argument. Test it by duplicating the computation from Example 6.6.

7.25 Develop an M-file function for the modified secant method based on Fig. 6.4 and Sec. 6.3.3. Along with the initial guess and the perturbation fraction, pass the function as an argument. Test it by duplicating the computation from Example 6.8.

Case Studies: Roots of Equations

The purpose of this chapter is to use the numerical procedures discussed in Chaps. 5, 6, and 7 to solve actual engineering problems. Numerical techniques are important for practical applications because engineers frequently encounter problems that cannot be approached using analytical techniques. For example, simple mathematical models that can be solved analytically may not be applicable when real problems are involved. Thus, more complicated models must be employed. For these cases, it is appropriate to implement a numerical solution on a computer. In other situations, engineering design problems may require solutions for implicit variables in complicated equations.

The following case studies are typical of those that are routinely encountered during upper-class courses and graduate studies. Furthermore, they are representative of problems you will address professionally. The problems are drawn from the four major disciplines of engineering: chemical, civil, electrical, and mechanical. These applications also serve to illustrate the trade-offs among the various numerical techniques.

The first application, taken from chemical engineering, provides an excellent example of how root-location methods allow you to use realistic formulas in engineering practice. In addition, it also demonstrates how the efficiency of the Newton-Raphson technique is used to advantage when a large number of root-location computations is required.

The following engineering design problems are taken from civil, electrical, and mechanical engineering. Section 8.2 uses bisection to determine changes in rainwater chemistry due to increases in atmospheric carbon dioxide. Section 8.3 shows how the roots of transcendental equations can be used in the design of an electrical circuit. Sections 8.2 and 8.3 also illustrate how graphical methods provide insight into the root-location process. Finally, Sec. 8.4 uses a variety of numerical methods to compute the friction factor for fluid flow in a pipe.

8.1 IDEAL AND NONIDEAL GAS LAWS (CHEMICAL/BIO ENGINEERING)

Background.   The *ideal gas law* is given by

$$pV = nRT \tag{8.1}$$

where p is the absolute pressure, V is the volume, n is the number of moles, R is the universal gas constant, and T is the absolute temperature. Although this equation is widely used by engineers and scientists, it is accurate over only a limited range of pressure and temperature. Furthermore, Eq. (8.1) is more appropriate for some gases than for others.

An alternative equation of state for gases is given by

$$\left(p + \frac{a}{v^2}\right)(v - b) = RT \tag{8.2}$$

and is known as the *van der Waals equation,* where $v = V/n$ is the molal volume and a and b are empirical constants that depend on the particular gas.

A chemical engineering design project requires that you accurately estimate the molal volume (v) of both carbon dioxide and oxygen for a number of different temperature and pressure combinations so that appropriate containment vessels can be selected. It is also of interest to examine how well each gas conforms to the ideal gas law by comparing the molal volumes as calculated by Eqs. (8.1) and (8.2). The following data are provided:

$R = 0.082054$ L atm/(mol K)

$\left.\begin{array}{l} a = 3.592 \\ b = 0.04267 \end{array}\right\}$ carbon dioxide

$\left.\begin{array}{l} a = 1.360 \\ b = 0.03183 \end{array}\right\}$ oxygen

The design pressures of interest are 1, 10, and 100 atm for temperature combinations of 300, 500, and 700 K.

Solution.   Molal volumes for both gases are calculated using the ideal gas law, with $n = 1$. For example, if $p = 1$ atm and $T = 300$ K,

$$v = \frac{V}{n} = \frac{RT}{p} = 0.082054 \ \frac{\text{L atm}}{\text{mol K}} \frac{300 \text{ K}}{1 \text{ atm}} = 24.6162 \text{ L/mol}$$

These calculations are repeated for all temperature and pressure combinations and presented in Table 8.1.

TABLE 8.1 Computations of molal volume.

Temperature, K	Pressure, atm	Molal Volume (Ideal Gas Law), L/mol	Molal Volume (van der Waals) Carbon Dioxide, L/mol	Molal Volume (van der Waals) Oxygen, L/mol
300	1	24.6162	24.5126	24.5928
	10	2.4616	2.3545	2.4384
	100	0.2462	0.0795	0.2264
500	1	41.0270	40.9821	41.0259
	10	4.1027	4.0578	4.1016
	100	0.4103	0.3663	0.4116
700	1	57.4378	57.4179	57.4460
	10	5.7438	5.7242	5.7521
	100	0.5744	0.5575	0.5842

The computation of molal volume from the van der Waals equation can be accomplished using any of the numerical methods for finding roots of equations discussed in Chaps. 5, 6, and 7, with

$$f(v) = \left(p + \frac{a}{v^2}\right)(v - b) - RT \tag{8.3}$$

In this case, the derivative of $f(v)$ is easy to determine and the Newton-Raphson method is convenient and efficient to implement. The derivative of $f(v)$ with respect to v is given by

$$f'(v) = p - \frac{a}{v^2} + \frac{2ab}{v^3} \tag{8.4}$$

The Newton-Raphson method is described by Eq. (6.6):

$$v_{i+1} = v_i - \frac{f(v_i)}{f'(v_i)}$$

which can be used to estimate the root. For example, using the initial guess of 24.6162, the molal volume of carbon dioxide at 300 K and 1 atm is computed as 24.5126 L/mol. This result was obtained after just two iterations and has an ε_a of less than 0.001%.

Similar computations for all combinations of pressure and temperature for both gases are presented in Table 8.1. It is seen that the results for the ideal gas law differ from those for the van der Waals equation for both gases, depending on specific values for p and T. Furthermore, because some of these results are significantly different, your design of the containment vessels would be quite different, depending on which equation of state was used.

In this case, a complicated equation of state was examined using the Newton-Raphson method. The results varied significantly from the ideal gas law for several cases. From a practical standpoint, the Newton-Raphson method was appropriate for this application because $f'(v)$ was easy to calculate. Thus, the rapid convergence properties of the Newton-Raphson method could be exploited.

In addition to demonstrating its power for a single computation, the present design problem also illustrates how the Newton-Raphson method is especially attractive when numerous computations are required. Because of the speed of digital computers, the efficiency of various numerical methods for determining most roots of equations is indistinguishable for a single computation. Even a 1-s difference between the crude bisection approach and the efficient Newton-Raphson method does not amount to a significant time loss when only one computation is performed. However, suppose that millions of root evaluations are required to solve a problem. In this case, the efficiency of the method could be a deciding factor in the choice of a technique.

For example, suppose that you are called upon to design an automatic computerized control system for a chemical production process. This system requires accurate estimates of molal volumes on an essentially continuous basis to properly manufacture the final product. Gauges are installed that provide instantaneous readings of pressure and temperature. Evaluations of v must be obtained for a variety of gases that are used in the process.

For such an application, bracketing methods such as bisection or false position would probably be too time-consuming. In addition, the two initial guesses that are required for these approaches may also interject a critical delay in the procedure. This shortcoming is relevant to the secant method, which also needs two initial estimates.

In contrast, the Newton-Raphson method requires only one guess for the root. The ideal gas law could be employed to obtain this guess at the initiation of the process. Then, assuming that the time frame is short enough so that pressure and temperature do not vary wildly between computations, the previous root solution would provide a good guess for the next application. Thus, the close guess that is often a prerequisite for convergence of the Newton-Raphson method would automatically be available. All the above considerations would greatly favor the Newton-Raphson technique for such problems.

8.2 GREENHOUSE GASES AND RAINWATER (CIVIL/ENVIRONMENTAL ENGINEERING)

Background.   Civil engineering is a broad field that includes such diverse areas as structural, geotechnical, transportation, water-resources, and environmental engineering. The last area has traditionally dealt with pollution control. However, in recent years, environmental engineers (as well as chemical engineers) have addressed broader problems such as climate change.

It is well documented that the atmospheric levels of several *greenhouse gases* have been increasing over the past 60 years. For example, Fig. 8.1 shows data for the partial pressure of carbon dioxide (CO_2) collected at Mauna Loa, Hawaii, from 1958 through 2003. The trend in the data can be nicely fit with a quadratic polynomial (in Part Five, we will learn how to determine such polynomials),

$$p_{CO_2} = 0.011825(t - 1980.5)^2 + 1.356975(t - 1980.5) + 339$$

where p_{CO_2} = the partial pressure of CO_2 in the atmosphere [ppm]. The data indicate that levels have increased during the period: from 315 to 376 ppm.

FIGURE 8.1

Average annual partial pressures of atmospheric carbon dioxide (ppm) measured at Mauna Loa, Hawaii.

Aside from global warming, greenhouse gases can also influence atmospheric chemistry. One question that we can address is how the carbon dioxide trend is affecting the pH of rainwater. Outside of urban and industrial areas, it is well documented that carbon dioxide is the primary determinant of the pH of the rain. pH is the measure of the activity of hydrogen ions and, therefore, of acidity. For dilute aqueous solutions, it can be computed as

$$pH = -\log_{10}[H^+] \tag{8.5}$$

where $[H^+]$ is the molar concentration of hydrogen ions.

The following five nonlinear system of equations govern the chemistry of rainwater,

$$K_1 = 10^6 \frac{[H^+][HCO_3^-]}{K_H p_{CO_2}} \tag{8.6}$$

$$K_2 = \frac{[H^+][CO_3^{2-}]}{[HCO_3^-]} \tag{8.7}$$

$$K_w = [H^+][OH^-] \tag{8.8}$$

$$c_T = \frac{K_H p_{CO_2}}{10^6} + [HCO_3^-] + [CO_3^{2-}] \tag{8.9}$$

$$0 = [HCO_3^-] + 2[CO_3^{2-}] + [OH^-] - [H^+] \tag{8.10}$$

where K_H = Henry's constant, and K_1, K_2, and K_w are equilibrium coefficients. The five unknowns in this system of five nonlinear equations are c_T = total inorganic carbon, $[HCO_3^-]$ = bicarbonate, $[CO_3^{2-}]$ = carbonate, $[H^+]$ = hydrogen ion, and $[OH^-]$ = hydroxyl ion. Notice how the partial pressure of CO_2 shows up in Eqs. (8.6) and (8.9).

Use these equations to compute the pH of rainwater given that $K_H = 10^{-1.46}$, $K_1 = 10^{-6.3}$, $K_2 = 10^{-10.3}$, and $K_w = 10^{-14}$. Compare the results for 1958 when the p_{CO_2} was 315 and for 2003 when it was 375 ppm. When selecting a numerical method for your computation, consider the following:

- You know with certainty that the pH of rain in pristine areas always falls between 2 and 12.
- You also know that your measurement devices can only measure pH to two places of decimal precision.

Solution. There are a variety of ways to solve this nonlinear system of five equations. One way is to eliminate unknowns by combining them to produce a single function that only depends on $[H^+]$. To do this, first solve Eqs. (8.6) and (8.7) for

$$[HCO_3^-] = \frac{K_1}{10^6[H^+]} K_H p_{CO_2} \tag{8.11}$$

$$[CO_3^{2-}] = \frac{K_2[HCO_3^-]}{[H^+]} \tag{8.12}$$

Substitute Eq. (8.11) into (8.12)

$$[CO_3^{2-}] = \frac{K_2 K_1}{10^6[H^+]^2} K_H p_{CO_2} \tag{8.13}$$

Equations (8.11) and (8.13) can be substituted along with Eq. (8.8) into Eq. (8.10) to give

$$0 = \frac{K_1}{10^6[H^+]}K_H p_{CO_2} + 2\frac{K_2 K_1}{10^6[H^+]^2}K_H p_{CO_2} + \frac{K_w}{[H^+]} - [H^+] \qquad (8.14)$$

Although it might not be apparent, this result is a third-order polynomial in $[H^+]$. Thus, its root can be used to compute the pH of the rainwater.

Now we must decide which numerical method to employ to obtain the solution. There are two reasons why bisection would be a good choice. First, the fact that the pH always falls within the range from 2 to 12 provides us with two good initial guesses. Second, because the pH can only be measured to two decimal places of precision, we will be satisfied with an absolute error of $E_{a,d} = 0.005$. Remember that given an initial bracket and the desired relative error, we can compute the number of iterations *a priori*. Using Eq. (5.5), the result is $n = \log_2(10)/0.005 = 10.9658$. Thus, eleven iterations of bisection will produce the desired precision.

If this is done, the result for 1958 will be a pH of 5.6279 with a relative error of 0.0868%. We can be confident that the rounded result of 5.63 is correct to two decimal places. This can be verified by performing another run with more iterations. For example, if we perform 35 iterations, a result of 5.6304 is obtained with an approximate relative error of $\varepsilon_a = 5.17 \times 10^{-9}$%. The same calculation can be repeated for the 2003 conditions to give pH = 5.59 with $\varepsilon_a = 0.0874$%.

Interestingly, these results indicate that the 19% rise in atmospheric CO_2 levels has produced only a 0.67% drop in pH. Although this is certainly true, remember that the pH represents a logarithmic scale as defined by Eq. (8.5). Consequently, a unit drop in pH represents a 10-fold increase in hydrogen ion. The concentration can be computed as $[H^+] = 10^{-pH}$ and the resulting percent change can be calculated as 9.1%. Therefore, the hydrogen ion concentration has increased about 9%.

There is quite a lot of controversy related to the true significance of the greenhouse gas trends. However, regardless of the ultimate implications, it is quite sobering to realize that something as large as our atmosphere has changed so much over a relatively short time period. This case study illustrates how numerical methods can be employed to analyze and interpret such trends. Over the coming years, engineers and scientists can hopefully use such tools to gain increased understanding and help rationalize the debate over the ramifications of these trends.

8.3 DESIGN OF AN ELECTRIC CIRCUIT (ELECTRICAL ENGINEERING)

Background. Electrical engineers often use Kirchhoff's laws to study the steady-state (not time-varying) behavior of electric circuits. Such steady-state behavior will be examined in Sec. 12.3. Another important problem involves circuits that are transient in nature, where sudden temporal changes take place. Such a situation occurs following the closing of the switch in Fig. 8.2. In this case, there will be a period of adjustment following the closing of the switch as a new steady state is reached. The length of this adjustment period is closely related to the storage properties of the capacitor and the inductor. Energy storage may oscillate between these two elements during a transient period. However, resistance in the circuit will dissipate the magnitude of the oscillations.

FIGURE 8.2
An electric circuit. When the switch is closed, the current will undergo a series of oscillations until a new steady state is reached.

The flow of current through the resistor causes a voltage drop (V_R) given by

$$V_R = iR$$

where i = the current and R = the resistance of the resistor. When R and i have units of ohms and amperes, respectively, V_R has units of volts.

Similarly, an inductor resists changes in current, such that the voltage drop V_L across it is

$$V_L = L\frac{di}{dt}$$

where L = the inductance. When L and i have units of henrys and amperes, respectively, V_L has units of volts and t has units of seconds.

The voltage drop across the capacitor (V_C) depends on the charge (q) on it:

$$V_C = \frac{q}{C} \tag{8.15}$$

where C = the capacitance. When the charge is expressed in units of coulombs, the unit of C is the farad.

Kirchhoff's second law states that the algebraic sum of voltage drops around a closed circuit is zero. After the switch is closed, we have

$$L\frac{di}{dt} + Ri + \frac{q}{C} = 0 \tag{8.16}$$

However, the current is related to the charge according to

$$i = \frac{dq}{dt} \tag{8.17}$$

Therefore,

$$L\frac{d^2q}{dt^2} + R\frac{dq}{dt} + \frac{1}{C}q = 0 \tag{8.18}$$

This is a second-order linear ordinary differential equation that can be solved using the methods of calculus. The solution is given by

$$q(t) = q_0 e^{-Rt/(2L)}\cos\left[\sqrt{\frac{1}{LC} - \left(\frac{R}{2L}\right)^2}\,t\right] \tag{8.19}$$

FIGURE 8.3
The charge on a capacitor as a function of time following the closing of the switch in Fig. 8.2.

where at $t = 0$, $q = q_0 = V_0 C$, and $V_0 = $ the voltage from the charging battery. Equation (8.19) describes the time variation of the charge on the capacitor. The solution $q(t)$ is plotted in Fig. 8.3.

A typical electrical engineering design problem might involve determining the proper resistor to dissipate energy at a specified rate, with known values for L and C. For this problem, assume the charge must be dissipated to 1% of its original value ($q/q_0 = 0.01$) in $t = 0.05$ s, with $L = 5$ H and $C = 10^{-4}$ F.

Solution. It is necessary to solve Eq. (8.19) for R, with known values of q, q_0, L, and C. However, a numerical approximation technique must be employed because R is an implicit variable in Eq. (8.19). The bisection method will be used for this purpose. The other methods discussed in Chaps. 5 and 6 are also appropriate, although the Newton-Raphson method might be deemed inconvenient because the derivative of Eq. (8.19) is a little cumbersome. Rearranging Eq. (8.19),

$$f(R) = e^{-Rt/(2L)} \cos\left[\sqrt{\frac{1}{LC} - \left(\frac{R}{2L}\right)^2}\, t\right] - \frac{q}{q_0}$$

or using the numerical values given,

$$f(R) = e^{-0.005R} \cos[\sqrt{2000 - 0.01R^2}\,(0.05)] - 0.01 \tag{8.20}$$

Examination of this equation suggests that a reasonable initial range for R is 0 to 400 Ω (because $2000 - 0.01R^2$ must be greater than zero). Figure 8.4, a plot of Eq. (8.20), confirms this. Twenty-one iterations of the bisection method give $R = 328.1515$ Ω, with an error of less than 0.0001%.

FIGURE 8.4
Plot of Eq. (8.20) used to obtain initial guesses for R that bracket the root.

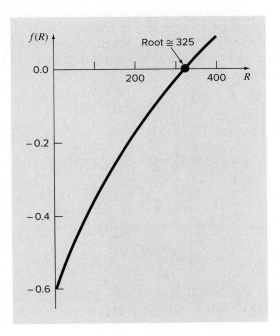

Thus, you can specify a resistor with this rating for the circuit shown in Fig. 8.2 and expect to achieve a dissipation performance that is consistent with the requirements of the problem. This design problem could not be solved efficiently without using the numerical methods in Chaps. 5 and 6.

8.4 PIPE FRICTION (MECHANICAL/AEROSPACE ENGINEERING)

Background.   Determining fluid flow through pipes and tubes has great relevance in many areas of engineering and science. In mechanical and aerospace engineering, typical applications include the flow of liquids and gases through cooling systems.

The resistance to flow in such conduits is parameterized by a dimensionless number called the *friction factor*. For turbulent flow, the *Colebrook equation* provides a means to calculate the friction factor,

$$0 = \frac{1}{\sqrt{f}} + 2.0 \log\left(\frac{\varepsilon}{3.7D} + \frac{2.51}{\text{Re}\sqrt{f}}\right) \tag{8.21}$$

where ε = the roughness (m), D = diameter (m), and Re = the *Reynolds number*,

$$\text{Re} = \frac{\rho V D}{\mu}$$

where ρ = the fluid's density (kg/m^3), V = its velocity (m/s), and μ = dynamic viscosity (N s/m^2). In addition to appearing in Eq. (8.21), the Reynolds number also serves as the criterion for whether flow is turbulent (Re > 4000).

In the present case study, we will illustrate how the numerical methods covered in this part of the book can be employed to determine f for air flow through a smooth, thin tube. For this case, the parameters are $\rho = 1.23$ kg/m^3, $\mu = 1.79 \times 10^{-5}$ N s/m^2, $D = 0.005$ m, $V = 40$ m/s, and $\varepsilon = 0.0015$ mm. Note that friction factors range from about 0.008 to 0.08. In addition, an explicit formulation called the *Swamee-Jain equation* provides an approximate estimate,

$$f = \frac{1.325}{\left[\ln\left(\dfrac{\varepsilon}{3.7D} + \dfrac{5.74}{\text{Re}^{0.9}}\right)\right]^2} \tag{8.22}$$

Solution.   The Reynolds number can be computed as

$$\text{Re} = \frac{\rho V D}{\mu} = \frac{1.23(40)0.005}{1.79 \times 10^{-5}} = 13{,}743$$

This value along with the other parameters can be substituted into Eq. (8.21) to give

$$g(f) = \frac{1}{\sqrt{f}} + 2.0 \log\left(\frac{0.0000015}{3.7(0.005)} + \frac{2.51}{13{,}743\sqrt{f}}\right)$$

Before determining the root, it is advisable to plot the function to estimate initial guesses and to anticipate possible difficulties. This can be done easily with tools such

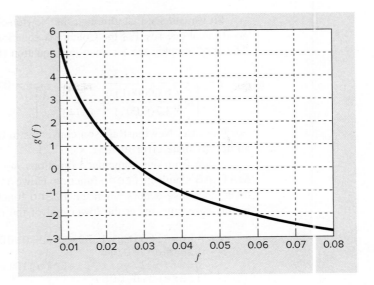

FIGURE 8.5

as MATLAB software, Excel, or Mathcad. For example, a plot of the function can be generated with the following MATLAB commands

```
>> rho=1.23;mu=1.79e-5;D=0.005;V=40;e=0.0015/1000;
>> Re=rho*V*D/mu;
>> g=@(f) 1/sqrt(f)+2*log10(e/(3.7*D)+2.51/(Re*sqrt(f)));
>> fplot(g,[0.008 0.08]),grid,xlabel('f'),ylabel('g(f)')
```

As you can see in Fig. 8.5, the root is located at about 0.03.

Because we are supplied initial guesses ($x_l = 0.008$ and $x_u = 0.08$), either of the bracketing methods from Chap. 5 could be used. For example, bisection gives a value of $f = 0.0289678$ with a percent relative error of error of 5.926×10^{-5} in 22 iterations. False position yields a result of similar precision in 26 iterations. Thus, although they produce the correct result, they are somewhat inefficient. This would not be important for a single application, but could become prohibitive if many evaluations were made.

We could try to attain improved performance by turning to an open method. Because Eq. (8.21) is relatively straightforward to differentiate, the Newton-Raphson method is a good candidate. For example, using an initial guess at the lower end of the range ($x_0 = 0.008$), Newton-Raphson converges quickly to 0.0289678 with an approximate error of $6.87 \times 10^{-6}\%$ in only six iterations. However, when the initial guess is set at the upper end of the range ($x_0 = 0.08$), the routine diverges!

As can be seen by inspecting Fig. 8.5, this occurs because the function's slope at the initial guess causes the first iteration to jump to a negative value. Further runs demonstrate that for this case, convergence only occurs when the initial guess is below about 0.066.

So we can see that although the Newton-Raphson method is very efficient, it requires good initial guesses. For the Colebrook equation, a good strategy might be to this and other "fzero" employ the Swamee-Jain equation (Eq. 8.22) to provide the initial guess, as in

$$f = \frac{1.325}{\left[\ln\left(\dfrac{0.0000015}{3.7(0.005)} + \dfrac{5.74}{13743^{0.9}}\right)\right]^2} = 0.029031$$

For this case, Newton-Raphson converges in only 3 iterations quickly to 0.0289678 with an approximate error of $8.51 \times 10^{-10}\%$.

Aside from our homemade functions, we can also use professional root finders like MATLAB's built-in fzero function. However, just as with the Newton-Raphson method, divergence also occurs when the fzero function is used with a single guess. However, in this case, guesses at the lower end of the range cause problems. For example,

```
>> rho=1.23;mu=1.79e-5;D=0.005;V=40;e=0.0015/1000;
>> Re=rho*V*D/mu
>> g=@(f) 1/sqrt(f)+2*log10(e/(3.7*D)+2.51/(Re*sqrt(f)));
>> fzero(g,0.008)

Exiting fzero: aborting search for an interval containing a
sign change because complex function value encountered
during search. (Function value at -0.0028 is -4.92028-
20.2423i.)
Check function or try again with a different starting value.
ans =
    NaN
```

If the iterations are displayed using optimset (recall Sec. 7.7.2), it is revealed that a negative value occurs during the search phase before a sign change is detected and the routine aborts. However, for single initial guesses above about 0.016, the routine works nicely. For example, for the guess of 0.08 that caused problems for Newton-Raphson, fzero does just fine,

```
>> fzero(g,0.08)

ans =
    0.02896781017144
```

As a final note, let's see whether convergence is possible for simple fixed-point iteration. The easiest and most straightforward version involves solving for the first f in Eq. (8.21),

$$f_{i+1} = \frac{0.25}{\left(\log\left(\dfrac{\varepsilon}{3.7D} + \dfrac{2.51}{\text{Re}\sqrt{f_i}}\right)\right)^2} \tag{8.23}$$

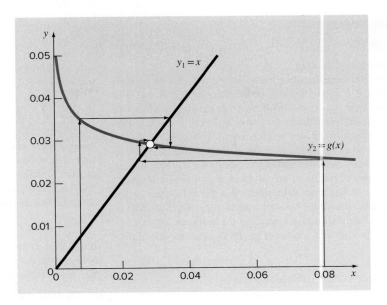

FIGURE 8.6

The cobweb display of this function indicates a surprising result (Fig. 8.6). Recall that fixed-point iteration converges when the y_2 curve has a relatively flat slope (i.e., $|g'(\xi)| < 1$). As indicated by Fig. 8.6, the fact that the y_2 curve is quite flat in the range from $f = 0.008$ to 0.08 means that not only does fixed-point iteration converge, but it converges fairly rapidly! In fact, for initial guesses anywhere between 0.008 and 0.08, fixed-point iteration yields predictions with percent relative errors less than 0.008% in six or fewer iterations. Thus, this simple approach that requires only one guess and no derivative estimates performs really well for this particular case.

The take-home message from this case study is that even great, professionally developed software like MATLAB is not always foolproof. Further, there is usually no single method that works best for all problems. Sophisticated users understand the strengths and weaknesses of the available numerical techniques. In addition, they understand enough of the underlying theory so that they can effectively deal with situations where a method breaks down.

PROBLEMS

Chemical/Bio Engineering

8.1 Perform the same computation as in Sec. 8.1, but for acetone ($a = 14.09$ and $b = 0.0994$) at a temperature of 400 K and p of 2.5 atm. Compare your results with that from the ideal gas law. Use any of the numerical methods discussed in Chaps. 5 and 6 to perform the computation. Justify your choice of technique.

8.2 In chemical engineering, plug flow reactors (that is, those in which fluid flows from one end to the other with minimal mixing along the longitudinal axis) are often used to convert reactants into products. It has been determined that the efficiency of the conversion can sometimes be improved by recycling a portion of the product stream so that it returns to the entrance for an additional pass through the reactor (Fig. P8.2). The recycle rate is defined as

$$R = \frac{\text{volume of fluid returned to entrance}}{\text{volume leaving the system}}$$

Suppose that we are processing a chemical A to generate a product B. For the case where A forms B according to an autocatalytic reaction (that is, in which one of the products acts as a catalyst or stimulus for the reaction), it can be shown that an optimal recycle rate must satisfy

$$\ln \frac{1 + R(1 - X_{Af})}{R(1 - X_{Af})} = \frac{R + 1}{R[1 + R(1 - X_{Af})]}$$

where X_{Af} = the fraction of reactant A that is converted to product B. The optimal recycle rate corresponds to the minimum-sized reactor needed to attain the desired level of conversion. Use a numerical method to determine the recycle rates needed to minimize reactor size for a fractional conversion of $X_{Af} = 0.96$.

8.3 In a chemical engineering process, water vapor (H_2O) is heated to sufficiently high temperatures that a significant portion of the water dissociates, or splits apart, to form oxygen (O_2) and hydrogen (H_2):

$$H_2O \rightleftharpoons H_2 + \tfrac{1}{2}O_2$$

FIGURE P8.2
Schematic representation of a plug flow reactor with recycle.

If it is assumed that this is the only reaction involved, the mole fraction x of H_2O that dissociates can be represented by

$$K = \frac{x}{1 - x} \sqrt{\frac{2p_t}{2 + x}} \tag{P8.3.1}$$

where K = the reaction equilibrium constant and p_t = the total pressure of the mixture. If $p_t = 3$ atm and $K = 0.05$, determine the value of x that satisfies Eq. (P8.3.1).

8.4 The following equation pertains to the concentration of a chemical in a completely mixed reactor:

$$c = c_{in}(1 - e^{-0.04t}) + c_0 e^{-0.04t}$$

If the initial concentration $c_0 = 4$ and the inflow concentration $c_{in} = 10$, compute the time required for c to be 93% of c_{in}.

8.5 A reversible chemical reaction

$$2A + B \rightleftharpoons C$$

can be characterized by the equilibrium relationship

$$K = \frac{c_c}{c_a^2 c_b}$$

where the nomenclature c_i represents the concentration of constituent i. Suppose that we define a variable x as representing the number of moles of C that are produced. Conservation of mass can be used to reformulate the equilibrium relationship as

$$K = \frac{(c_{c,0} + x)}{(c_{a,0} - 2x)^2 (c_{b,0} - x)}$$

where the subscript 0 designates the initial concentration of each constituent. If $K = 0.016$, $c_{a,0} = 42$, $c_{b,0} = 28$, and $c_{c,0} = 4$, determine the value of x. **(a)** Obtain the solution graphically. **(b)** On the basis of **(a)**, solve for the root with initial guesses of $x_l = 0$ and $x_u = 20$ to $\varepsilon_s = 0.5\%$. Choose either bisection or false position to obtain your solution. Justify your choice.

8.6 The following chemical reactions take place in a closed system

$$2A + B \rightleftharpoons C$$
$$A + D \rightleftharpoons C$$

At equilibrium, they can be characterized by

$$K_1 = \frac{c_c}{c_a^2 c_b}$$

$$K_2 = \frac{c_c}{c_a c_d}$$

where the nomenclature c_i represents the concentration of constituent i. If x_1 and x_2 are the numbers of moles of C that are produced due to the first and second reactions, respectively, use an approach similar to that of Prob. 8.5 to reformulate the equilibrium relationships in terms of the initial concentrations of the constituents. Then, use the Newton-Raphson method to solve the pair of simultaneous nonlinear equations for x_1 and x_2 if $K_1 = 4 \times 10^{-4}$, $K_2 = 3.7 \times 10^{-2}$, $c_{a,0} = 50$, $c_{b,0} = 20$, $c_{c,0} = 5$, and $c_{d,0} = 10$. Use a graphical approach to develop your initial guesses.

8.7 The *Redlich-Kwong equation* of state is given by

$$p = \frac{RT}{v - b} - \frac{a}{v(v + b)\sqrt{T}}$$

where $R =$ the universal gas constant [$= 0.518$ kJ/(kg K)], $T =$ absolute temperature (K), $p =$ absolute pressure (kPa), and $v =$ the volume of a kg of gas (m³/kg). The parameters a and b are calculated by

$$a = 0.427\frac{R^2 T_c^{2.5}}{p_c} \qquad b = 0.0866R\frac{T_c}{p_c}$$

where $p_c =$ critical pressure (kPa) and $T_c =$ critical temperature (K). As a chemical engineer, you are asked to determine the amount of methane fuel ($p_c = 4600$ kPa and $T_c = 191$ K) that can be held in a 3-m³ tank at a temperature of $-40°C$ with a pressure of 65,000 kPa. Use a root-locating method of your choice to calculate v and then determine the mass of methane contained in the tank.

8.8 The volume V of liquid in a hollow horizontal cylinder of radius r and length L is related to the depth of the liquid h by

$$V = \left[r^2 \cos^{-1}\left(\frac{r - h}{r}\right) - (r - h)\sqrt{2rh - h^2} \right] L$$

Determine h given $r = 2$ m, $L = 5$ m, and $V = 8$ m³. Note that if you are using a programming language or software tool that is not rich in trigonometric functions, the arccosine can be computed with

$$\cos^{-1} x = \frac{\pi}{2} - \tan^{-1}\left(\frac{x}{\sqrt{1 - x^2}}\right)$$

8.9 The volume V of liquid in a spherical tank of radius r is related to the depth h of the liquid by

$$V = \frac{\pi h^2 (3r - h)}{3}$$

Determine h given $r = 0.9$ m and $V = 0.6$ m³.
8.10 For the spherical tank in Prob. 8.9, it is possible to develop the following two fixed-point formulas:

$$h = \sqrt{\frac{h^3 + (3V/\pi)}{3r}}$$

and

$$h = \sqrt[3]{3\left(rh^2 - \frac{V}{\pi}\right)}$$

If $r = 1$ m and $V = 0.5$ m³, determine whether either of these is stable and the range of initial guesses for which they are stable.
8.11 The operation of a constant density plug flow reactor for the production of a substance via an enzymatic reaction is described by the equation below, where V is the volume of the reactor, F is the flow rate of reactant C, C_{in} and C_{out} are the concentrations of reactant entering and leaving the reactor, respectively, and K and k_{max} are constants. For a 500-L reactor, with an inlet concentration of $C_{in} = 0.5$ M, an inlet flow rate of 40 L/s, $k_{max} = 0.005$ s^{-1}, and $K = 0.1$ M, find the concentration of C at the outlet of the reactor.

$$\frac{V}{F} = -\int_{C_{in}}^{C_{out}} \frac{K}{k_{max}C} + \frac{1}{k_{max}} dC$$

8.12 The *Ergun equation*, shown below, is used to describe the flow of a fluid through a packed bed. ΔP is the pressure drop, ρ is the density of the fluid, G_o is the mass velocity (mass flow rate divided by cross-sectional area), D_p is the diameter of the particles within the bed, μ is the fluid viscosity, L is the length of the bed, and ε is the void fraction of the bed.

$$\frac{\Delta P \rho}{G_o^2} \frac{D_p}{L} \frac{\varepsilon^3}{1 - \varepsilon} = 150\frac{1 - \varepsilon}{(D_p G_o/\mu)} + 1.75$$

Given the parameter values listed below, find the void fraction ε of the bed.

$$\frac{D_p G_o}{\mu} = 1000$$

$$\frac{\Delta P \rho D_p}{G_o^2 L} = 10$$

8.13 The pressure drop in a section of pipe can be calculated as

$$\Delta p = f\frac{L\rho V^2}{2D}$$

where $\Delta p =$ the pressure drop (Pa), $f =$ the friction factor, $L =$ the length of pipe (m), $\rho =$ density (kg/m³) $V =$ velocity (m/s), and $D =$ diameter (m). For turbulent flow, the *Colebrook equation* provides a means to calculate the friction factor,

$$\frac{1}{\sqrt{f}} = -2.0 \log\left(\frac{\varepsilon}{3.7D} + \frac{2.51}{Re\sqrt{f}}\right)$$

where ε = the roughness (m) and Re = the *Reynolds number*,

$$Re = \frac{\rho V D}{\mu}$$

where μ = dynamic viscosity (N s/m^2).

(a) Determine Δp for a 0.2-m-long horizontal stretch of smooth drawn tubing given ρ = 1.23 kg/m^3, μ = 1.79 × 10^{-5} N s/m^2, D = 0.006 m, V = 45 m/s, and ε = 0.0014 mm. Use a numerical method to determine the friction factor. Note that for smooth pipes with Re < 10^5, a good initial guess can be obtained using the *Blasius formula*, f = 0.316/Re$^{0.25}$.

(b) Repeat the computation but for a rougher commercial steel pipe (ε = 0.045 mm).

8.14 Figure P8.14 shows the kinetic interactions governing the concentrations of a bacteria culture and their nutrition source (substrate) in a continuously stirred flow-through bioreactor.

The mass balances for the bacteria biomass, $X(gC/m^3)$, and the substrate concentration, $S(gC/m^3)$, can be written as

$$\frac{dX}{dt} = \left(Y k_{g,max} \frac{S}{K_s + S} - k_d - k_r - \frac{1}{\tau_w} \right) X$$

$$\frac{dS}{dt} = -k_{g,max} \frac{S}{K_s + S} X + k_d X + \frac{1}{\tau_w} (S_{in} - S)$$

where t = time (h), $k_{g,max}$ = maximum bacterial growth rate (d^{-1}), K_s = half-saturation constant (gC/m^3), k_d = death rate (d^{-1}), k_r = respiration rate (d^{-1}), Q = flow rate (m^3/h), V = reactor volume (m^3), Y = yield coefficient (gC$_{cell}$/gC$_{substrate}$), and S_{in} = inflow

substrate concentration (mgC/m^3). Use a root-location technique to determine the steady-state concentrations of the substrate, bacteria, and total organic carbon ($X + S$) for a residence time of τ_w = 20 h. Employ the following parameters for the simulation: $X(0)$ = 100 gC/m^3, $S(0)$ = 0, $k_{g,max}$ = 0.2/h, K_s = 150 gC/m^3, k_d = k_r = 0.01/h, Y = 0.5 gC$_{cell}$/gC$_{substrate}$, V = 0.01 m^3, and S_{in} = 1000 gC/m^3. Verify your results by substituting them into the steady-state versions of the mass balance equations.

8.15 *Enzymes* are proteins produced by living organisms ranging from bacteria to humans to catalyze or accelerate biochemical reactions. The molecules upon which enzymes may act are called substrates. For example, the enzyme amylase, found in saliva and pancreatic fluid, converts the substrate starch into simple sugars. The rate of the conversion when dealing with a single substrate can be formulated by the *Michaelis-Menten model*,

$$\frac{dS}{dt} = -\frac{V_{max} S}{K_m + S}$$

where S = substrate concentration (mM), t = time (min), V_{max} = the maximum conversion rate (mM/min), and K_m = the Michaelis constant, or *half-saturation constant* (mM). As the name implies, K_m equals the substrate concentration at which the rate is half the maximum rate (Fig. P8.15).

The above equation can be integrated to yield an implicit nonlinear equation for substrate concentration as a function of time.

$$K_m \ln\left(\frac{S}{S_0}\right) - (S_0 - S) = -V_{max} t$$

FIGURE P8.14

Continuously stirred flow-through bioreactor to grow a bacterial culture.

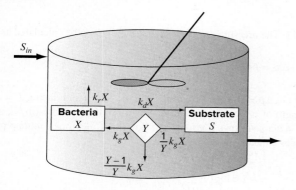

FIGURE P8.15

Michaelis-Menten curve for an enzyme reaction depicting the relation between the reaction rate ($-dS/dt$) and the substrate concentration (S).

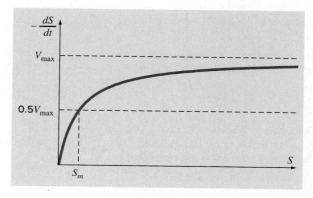

(a) Use a root-location method to determine the substrate concentration after $t = 8$ min given $S_0 = 2.5$ mM, $K_m = 1$ mM, and $V_{max} = 0.5$ mM/min.

(b) Using the same parameters as in (a), generate a plot of substrate concentration versus time for $t = 0$ to 10 min.

Civil and Environmental Engineering

8.16 In structural engineering, the *secant formula* defines the force per unit area, P/A, that causes a maximum stress σ_m in a column of given slenderness ratio L/k:

$$\frac{P}{A} = \frac{\sigma_m}{1 + (ec/k^2)\sec[0.5\sqrt{P/(EA)}(L/k)]}$$

where ec/k^2 = the eccentricity ratio and E = the modulus of elasticity. If for a steel beam, $E = 200{,}000$ MPa, $ec/k^2 = 0.4$, and $\sigma_m = 250$ MPa, compute P/A for $L/k = 50$. Recall that $\sec x = 1/\cos x$.

8.17 In environmental engineering (a specialty area in civil engineering), the following equation can be used to compute the oxygen level c (mg/L) in a river downstream from a sewage discharge:

$$c = 10 - 20(e^{-0.15x} - e^{-0.5x})$$

where x is the distance downstream in kilometers.

(a) Determine the distance downstream where the oxygen level first falls to a reading of 5 mg/L. (Hint: It is within 2 km of the discharge.) Determine your answer to a 1% error. Note that levels of oxygen below 5 mg/L are generally harmful to game-fish such as trout and salmon.

(b) Determine the distance downstream at which the oxygen is at a minimum. What is the concentration at that location?

8.18 The concentration of pollutant bacteria c in a lake decreases according to

$$c = 75e^{-1.5t} + 20e^{-0.075t}$$

Determine the time required for the bacteria concentration to be reduced to 9 using (a) the graphical method and (b) the Newton-Raphson method with an initial guess of $t = 6$ and a stopping criterion of 0.5%. Check your result.

8.19 A catenary cable is one that is hung between two points not in the same vertical line. As depicted in Fig. P8.19a, it is subject to no loads other than its own weight. Thus, its weight (N/m) acts as a uniform load per unit length along the cable. A free-body diagram of a section AB is depicted in Fig. P8.19b, where T_A and T_B are the tension forces at the ends. Based on horizontal and vertical force balances, the following differential equation model of the cable can be derived:

$$\frac{d^2y}{dx^2} = \frac{w}{T_A}\sqrt{1 + \left(\frac{dy}{dx}\right)^2}$$

FIGURE P8.19

(a) Forces acting on a section AB of a flexible hanging cable. The load is uniform along the cable (but not uniform per the horizontal distance x). (b) A free-body diagram of section AB.

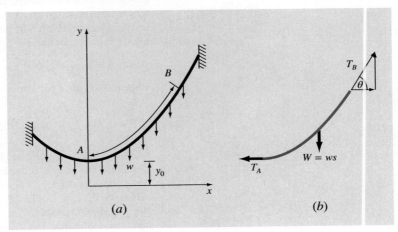

Calculus can be employed to solve this equation for the height y of the cable as a function of distance x,

$$y = \frac{T_A}{w} \cosh\left(\frac{w}{T_A}x\right) + y_0 - \frac{T_A}{w}$$

where the hyperbolic cosine can be computed by

$$\cosh x = \frac{1}{2}(e^x + e^{-x})$$

Use a numerical method to calculate a value for the parameter T_A given values for the parameters $w = 12$ and $y_0 = 6$, such that the cable has a height of $y = 15$ at $x = 50$.

8.20 Figure P8.20a shows a uniform beam subject to a linearly increasing distributed load. The equation for the resulting elastic curve is (see Fig. P8.20b)

$$y = \frac{w_0}{120EIL}(-x^5 + 2L^2x^3 - L^4x) \qquad \text{(P8.20)}$$

Use bisection to determine the point of maximum deflection (that is, the value of x where $dy/dx = 0$). Then substitute this value into Eq. (P8.20) to determine the value of the maximum deflection. Use the following parameter values in your computation: $L = 600$ cm, $E = 50,000$ kN/cm^2, $I = 30,000$ cm^4, and $w_0 = 2.5$ kN/cm.

8.21 The displacement of a structure is defined by the following equation for a damped oscillation:

$$y = 9e^{-kt}\cos \omega t$$

FIGURE P8.20

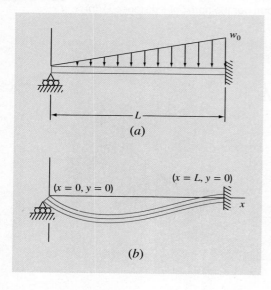

(a)

(b)

where $k = 0.7$ and $\omega = 4$.

(a) Use the graphical method to make an initial estimate of the time required for the displacement to decrease to 4.

(b) Use the Newton-Raphson method to determine the root to $\varepsilon_s = 0.01\%$.

(c) Use the secant method to determine the root to $\varepsilon_s = 0.01\%$.

8.22 The Manning equation can be written for a rectangular open channel as

$$Q = \frac{\sqrt{S}(BH)^{5/3}}{n(B + 2H)^{2/3}}$$

where Q = flow (m^3/s), S = slope (m/m), H = depth (m), and n = the Manning roughness coefficient. Develop a fixed-point iteration scheme to solve this equation for H given $Q = 5$, $S = 0.0002$, $B = 20$, and $n = 0.03$. Prove that your scheme converges for all initial guesses greater than or equal to zero.

8.23 In ocean engineering, the equation for a reflected standing wave in a harbor is given by $\lambda = 16$, $t = 12$, $v = 48$:

$$h = h_0\left[\sin\left(\frac{2\pi x}{\lambda}\right)\cos\left(\frac{2\pi tv}{\lambda}\right) + e^{-x}\right]$$

Solve for the lowest positive value of x if $h = 0.4h_0$.

8.24 You buy a $25,000 piece of equipment for nothing down and $5,500 per year for 6 years. What interest rate are you paying? The formula relating present worth P, annual payments A, number of years n, and interest rate i is

$$A = P\frac{i(1 + i)^n}{(1 + i)^n - 1}$$

8.25 Many fields of engineering require accurate population estimates. For example, transportation engineers might find it necessary to determine separately the population growth trends of a city and an adjacent suburb. The population of the urban area is declining with time according to

$$P_u(t) = P_{u,max}e^{-k_ut} + P_{u,min}$$

while the suburban population is growing, as in

$$P_s(t) = \frac{P_{s,max}}{1 + [P_{s,max}/P_0 - 1]e^{-k_st}}$$

where $P_{u,max}$, k_u, $P_{s,max}$, P_0, and k_s = empirically derived parameters. Determine the time and corresponding values of $P_u(t)$ and $P_s(t)$ when the suburbs are 20% larger than the city. The parameter values are $P_{u,max} = 75,000$, $k_u = 0.045$/yr, $P_{u,min} = 100,000$ people, $P_{s,max} = 300,000$ people, $P_0 = 10,000$ people, $k_s = 0.08$/yr. To obtain your solutions, use (a) graphical, (b) false-position, and (c) modified secant methods.

FIGURE P8.26

8.26 A simply supported beam is loaded as shown in Fig. P8.26. Using singularity functions, the shear along the beam can be expressed by the equation:

$$V(x) = 20[\langle x - 0 \rangle^1 - \langle x - 5 \rangle^1] - 15\langle x - 8 \rangle^0 - 57$$

By definition, the singularity function can be expressed as follows:

$$\langle x - a \rangle^n = \begin{cases} (x - a)^n & \text{when } x > a \\ 0 & \text{when } x \le a \end{cases}$$

Use a numerical method to find the point(s) where the shear equals zero.

8.27 For the simply supported beam from Prob. 8.26, the moment along the beam, $M(x)$, is given by:

$$M(x) = -10[\langle x - 0 \rangle^2 - \langle x - 5 \rangle^2] + 15\langle x - 8 \rangle^1$$
$$+ 150\langle x - 7 \rangle^0 + 57x$$

Use a numerical method to find the point(s) where the moment equals zero.

8.28 For the simply supported beam from Prob. 8.26, the slope along the beam is given by:

$$\frac{du_y}{dx}(x) = \frac{-10}{3}[\langle x - 0 \rangle^3 - \langle x - 5 \rangle^3] + \frac{15}{2}\langle x - 8 \rangle^2$$
$$+ 150\langle x - 7 \rangle^1 + \frac{57}{2}x^2 - 238.25$$

Use a numerical method to find the point(s) where the slope equals zero.

8.29 For the simply supported beam from Prob. 8.26, the displacement along the beam is given by:

$$u_y(x) = \frac{-5}{6}[\langle x - 0 \rangle^4 - \langle x - 5 \rangle^4] + \frac{15}{6}\langle x - 8 \rangle^3$$
$$+ 75\langle x - 7 \rangle^2 + \frac{57}{6}x^3 - 238.25x$$

(a) Find the point(s) where the displacement equals zero.

(b) How would you use a root-location technique to determine the location of the minimum displacement?

8.30 Although we did not mention it in Sec. 8.2, Eq. (8.10) is actually an expression of *electroneutrality*, that is, that positive and negative charges must balance. This can be seen more clearly by expressing the equation as

$$[H^+] = [HCO_3^-] + 2[CO_3^{2-}] + [OH^-]$$

In other words, the positive charges must equal the negative charges. Thus, when you compute the pH of a natural water body such as a lake, you must account for other ions that may be present. For the case where these ions originate from nonreactive salts, the net negative minus the positive charges due to these ions are lumped together in a quantity called *alkalinity*, and the equation is reformulated as

$$Alk + [H^+] = [HCO_3^-] + 2[CO_3^{2-}] + [OH^-] \tag{P8.30}$$

where Alk = alkalinity (eq/L). For example, the alkalinity of Lake Superior is approximately 0.4×10^{-3} eq/L. Perform the same calculations as in Sec. 8.2 to compute the pH of Lake Superior in 2008. Assume that just like the raindrops, the lake is in equilibrium with atmospheric CO_2, but account for the alkalinity as in Eq. (P8.30).

8.31 At sea level, the saturation vapor pressure of fresh water in air is a function of the air temperature, T_{air} (°C),

$$e_s = 4.596e^{17.27T_{air}/237.3+T_{air}}$$

If the vapor pressure exceeds this value, condensation will convert the vapor into liquid, producing phenomena such as fog, clouds, and dew. Of course, the water content of the air is often below saturation as can be calculated with

$$e_a = 4.596e^{17.27T_{dew}/237.3+T_{dew}}$$

where T_{dew} = the dew-point temperature, which is the temperature to which a parcel of air must be cooled at constant pressure for water vapor to condense into liquid water, hence the name "dew point." The ratio of these two formulas is called the relative humidity,

$$R_h = \frac{e_a}{e_s} \times 100\%$$

Develop a function to compute the dew-point temperature given the air temperature and the relative humidity. Test your function for $T_{air} = 25°C$ and $R_h = 50\%$.

8.32 The heat flux across a pond's surface, $J(W/m^2)$, which depends on a combination of five heat-transfer processes (Fig. P8.32), can be computed as:

$$J = J_{sn} + 0.97\left(0.6 + 0.031\sqrt{e_{air}}\right)\sigma(T_{air} + 273)^4 - 0.97\sigma(T_w + 273)^4$$

solar   atmospheric radiation   back radiation

$$- 0.47 f(U_w)(T_w - T_{air}) - f(U_w)(e_s - e_{air})$$

convection   evaporation

where J_{sn} = net solar shortwave radiation at the water surface (W/m^2), e_{air} = the air vapor pressure (mm Hg), σ = the Stefan-Boltzmann constant = 11.7×10^{-8} cal/(cm^2 d K^4), T_{air} = air temperature (°C), T_w = the pond's water temperature (°C), $f(U_w)$ = the dependence of the transfer on wind velocity over the water surface, U_w = the wind speed above the water surface (m/s), and e_s = the saturation vapor pressure at the water surface (mmHg). The saturation vapor pressure is computed as

$$e_{air} = 4.596 e^{17.27T_{dew}/237.3+T_{dew}} \tag{P8.32.1}$$

$$e_s = 4.596 e^{17.27T_w/237.3+T_w} \tag{P8.32.2}$$

$$f(U_w) = 19.0 + 0.95U_w^2 \tag{P8.32.3}$$

where T_{dew} = the dew-point temperature (°C). Determine the water temperature and evaporation flux at steady state $(J = 0)$ given the following values: $J_{sn} = 300$ cal/cm^2/d; $T_{air} = 25°C$; $T_{dew} = 16.7°C$; and $U_w = 3$ m/s.

Electrical Engineering

8.33 Perform the same computation as in Sec. 8.3, but determine the value of C required for the circuit to dissipate to 1% of its original value in $t = 0.05$ s, given $R = 280\Omega$ and $L = 7.5$ H. Use **(a)** a graphical approach, **(b)** bisection, and **(c)** root-location software such as Excel Solver, the MATLAB function `fzero`, or the Mathcad function root.

8.34 An oscillating current in an electric circuit is described by $i = 9e^{-t}\cos(2\pi t)$, where t is in seconds. Determine the lowest value of t such that $i = 3.5$.

8.35 The resistivity ρ of doped silicon is based on the charge q on an electron, the electron density n, and the electron mobility μ. The electron density is given in terms of the doping density N and the intrinsic carrier density n_i. The electron mobility is described by the temperature T, the reference temperature T_0, and the reference mobility μ_0. The equations required to compute the resistivity are

$$\rho = \frac{1}{qn\mu}$$

where

$$n = \frac{1}{2}(N + \sqrt{N^2 + 4n_i^2}) \quad \text{and} \quad \mu = \mu_0\left(\frac{T}{T_0}\right)^{-2.42}$$

Determine N, given $T_0 = 300$ K, $T = 1000$ K, $\mu_0 = 1350$ cm^2 (V s)$^{-1}$, $q_e = 1.602 \times 10^{-19}$ C, $n_i = 6.21 \times 10^9$ cm^{-3}, and a desired $\rho = 6.5 \times 10^6$ V s cm/C. Use **(a)** bisection and **(b)** the modified secant method.

8.36 A total charge Q is uniformly distributed around a ring-shaped conductor with radius a. A charge q is located at a distance x from the center of the ring (Fig. P8.36). The force exerted on the charge by the ring is given by

$$F = \frac{1}{4\pi e_0}\frac{qQx}{(x^2 + a^2)^{3/2}}$$

FIGURE P8.32

Components of heat exchange across the surface of a lake.

Radiation terms   Nonradiation terms

Air-water Interface

| Solar shortwave radiation | Atmospheric longwave radiation | Water longwave radiation | Conduction and convection | Evaporation and condensation |

Net absorbed radiation   Water-dependent terms

FIGURE P8.36

FIGURE P8.38
An embedded silicon chip insulated on all sides except its top face, where it exchanges heat with the surrounding air via radiation and convection.

where $e_0 = 8.854 \times 10^{-12}$ C^2/(N m^2). Find the distance x where the force is 1.2 N if q and Q are 2.25×10^{-5} C for a ring with a radius of 0.9 m.

8.37 Figure P8.37 shows a circuit with a resistor, an inductor, and a capacitor in parallel. Kirchhoff's rules can be used to express the impedance of the system as

$$\frac{1}{Z} = \sqrt{\frac{1}{R^2} + \left(\omega C - \frac{1}{\omega L}\right)^2}$$

where Z = impedance (Ω) and ω = the angular frequency. Find the ω that results in an impedance of 75 Ω using both bisection and false position with initial guesses of 1 and 1000 for the following parameters: $R = 225\ \Omega$, $C = 0.6 \times 10^{-6}$ F, and $L = 0.5$ H. Determine how many iterations of each technique are necessary to determine the answer to $\varepsilon_s = 0.1\%$. Use the graphical approach to explain any difficulties that arise.

surface temperature (K), and h = convective heat transfer coefficient (W/(m^2 K)). Use a steady-state heat balance to compute the chip's surface temperature given the following parameter values: $\varepsilon = 0.9$, $T_{a,\infty} = 293.15$ K, and $h = 50$ (W/(m^2 K)).

Mechanical and Aerospace Engineering
8.39 Beyond the *Colebrook equation*, other relationships, such as the *Fanning friction factor f*, are available to estimate friction in pipes. The Fanning friction factor is dependent on a number of parameters related to the size of the pipe and the fluid, which can all be represented by another dimensionless quantity, the *Reynolds number* Re. A formula that predicts f given Re is the *von Karman equation*,

$$\frac{1}{\sqrt{f}} = 4\log_{10}(\text{Re}\sqrt{f}) - 0.4$$

Typical values for the Reynolds number for turbulent flow are 10,000 to 500,000 and for the Fanning friction factor are 0.001 to 0.01. Develop a function that uses bisection to solve for f given a user-supplied value of Re between 2500 and 1,000,000. Design the function so that it ensures that the absolute error in the result is $E_{a,d} < 0.000005$.

8.40 Real mechanical systems may involve the deflection of nonlinear springs. In Fig. P8.40, a mass m is released a distance h above a nonlinear spring. The resistance force F of the spring is given by

$$F = -(k_1 d + k_2 d^{3/2})$$

FIGURE P8.37

8.38 A silicon chip measuring $5 \times 5 \times 1$ mm is embedded in a substrate (Fig. P8.38). At steady state, the chip generates 0.03 W of waste heat. Although the bottom and sides are insulated, the top surface is exposed to air flow and subject to both radiation and convective heat transfer. The radiation heat flux, J_{rad} (W/m^2), can be determined via the *Stefan-Boltzmann law*,

$$J_{rad} = \varepsilon\sigma(T_{a,\infty}^4 - T_{a,s}^4)$$

and the convective heat flux, J_{conv} (W/m^2), by

$$J_{conv} = h(T_{a,\infty} - T_{a,s})$$

where ε = emissivity, σ = Stefan Boltzmann constant [= 5.67 $\times$ 10^{-8} W/(m^2 K^4)], $T_{a,\infty}$ = ambient temperature (K), $T_{a,s}$ = the chip's

FIGURE P8.40

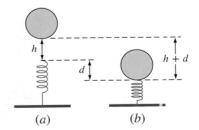

(a)   (b)

Conservation of energy can be used to show that

$$0 = \frac{2k_2 d^{5/2}}{5} + \frac{1}{2}k_1 d^2 - mgd - mgh$$

Solve for d given the following parameter values: $k_1 = 40{,}000$ g/s^2, $k_2 = 40$ g/(s^2 m$^{0.5}$), $m = 95$ g, $g = 9.81$ m/s^2, and $h = 0.43$ m.

8.41 Mechanical engineers, as well as most other engineers, use thermodynamics extensively in their work. The following polynomial can be used to relate the zero-pressure specific heat of dry air, c_p [kJ/(kg K)], to temperature (K):

$$c_p = 0.99403 + 1.671 \times 10^{-4}T + 9.7215 \times 10^{-8}T^2$$
$$-9.5838 \times 10^{-11}T^3 + 1.9520 \times 10^{-14}T^4$$

Determine the temperature that corresponds to a specific heat of 1.2 kJ/(kg K).

8.42 Aerospace engineers sometimes compute the trajectories of projectiles like rockets. A related problem deals with the trajectory of a thrown ball. The trajectory of a ball is defined by the (x, y) coordinates, as displayed in Fig. P8.42. The trajectory can be modeled as

$$y = (\tan \theta_0)x - \frac{g}{2v_0^2 \cos^2 \theta_0}x^2 + y_0$$

Find the appropriate initial angle θ_0 if the initial velocity $v_0 = 20$ m/s and the distance to the catcher x is 40 m. Note that the ball leaves the thrower's hand at an elevation of $y_0 = 1.8$ m and the catcher receives it at 1 m. Express the final result in degrees. Use a value of 9.81 m/s^2 for g and employ the graphical method to develop your initial guesses.

8.43 The general form for a three-dimensional stress field is given by

$$\begin{bmatrix} \sigma_{xx} & \sigma_{xy} & \sigma_{xz} \\ \sigma_{xy} & \sigma_{yy} & \sigma_{yz} \\ \sigma_{xz} & \sigma_{yz} & \sigma_{zz} \end{bmatrix}$$

FIGURE P8.42

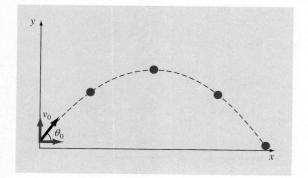

where the diagonal terms represent tensile or compressive stresses and the off-diagonal terms represent shear stresses. A stress field (in MPa) is given by

$$\begin{bmatrix} 10 & 14 & 25 \\ 14 & 7 & 15 \\ 25 & 15 & 16 \end{bmatrix}$$

To solve for the principal stresses, it is necessary to construct the following matrix (again in MPa):

$$\begin{bmatrix} 10 - \sigma & 14 & 25 \\ 14 & 7 - \sigma & 15 \\ 25 & 15 & 16 - \sigma \end{bmatrix}$$

σ_1, σ_2, and σ_3 can be found from the equation

$$\sigma^3 - I\sigma^2 + II\sigma - III = 0$$

where

$$I = \sigma_{xx} + \sigma_{yy} + \sigma_{zz}$$
$$II = \sigma_{xx}\sigma_{yy} + \sigma_{xx}\sigma_{zz} + \sigma_{yy}\sigma_{zz} - \sigma_{xy}^2 - \sigma_{xz}^2 - \sigma_{yz}^2$$
$$III = \sigma_{xx}\sigma_{yy}\sigma_{zz} - \sigma_{xx}\sigma_{yz}^2 - \sigma_{yy}\sigma_{xz}^2 - \sigma_{zz}\sigma_{xy}^2 + 2\sigma_{xy}\sigma_{xz}\sigma_{yz}$$

I, II, and III are known as the stress invariants. Find σ_1, σ_2, and σ_3 using a root-finding technique.

8.44 The upward velocity of a rocket can be computed with the following formula:

$$v = u \ln \frac{m_0}{m_0 - qt} - gt$$

where v = upward velocity, u = the velocity at which fuel is expelled relative to the rocket, m_0 = the initial mass of the rocket at time $t = 0$, q = the fuel consumption rate, and g = the downward acceleration of gravity (assumed constant = 9.81 m/s^2). If $u = 2200$ m/s, $m_0 = 160{,}000$ kg, and $q = 2680$ kg/s, compute the time at which $v = 1000$ m/s. (Hint: t is somewhere between 10 and 50 s.) Determine your result so that it is within 1% of the true value. Check your answer.

8.45 The phase angle ϕ between the forced vibration caused by the rough road and the motion of the car is given by

$$\tan \phi = \frac{2(c/c_c)(\omega/p)}{1 - (\omega/p)^2}$$

As a mechanical engineer, you would like to know if there are cases where $\phi = \omega/2 - 1$. Use the other parameters from the section to set up the equation as a root-location problem and solve for ω.

8.46 Two fluids at different temperatures enter a mixer and come out at the same temperature. The heat capacity of fluid A is given by:

$$c_p = 3.381 + 1.804 \times 10^{-2}T - 4.300 \times 10^{-6}T^2$$

and the heat capacity of fluid B is given by:

$$c_p = 8.592 + 1.290 \times 10^{-1}T - 4.078 \times 10^{-5}T^2$$

where c_p is in units of cal/mol K, and T is in units of K. Note that

$$\Delta H = \int_{T_1}^{T_2} c_p \, dT$$

A enters the mixer at 400°C. B enters the mixer at 600°C. There is twice as much A as there is B entering the mixer. At what temperature do the two fluids exit the mixer?

8.47 A compressor is operating at compression ratio R_c of 3.0 (the pressure of gas at the outlet is three times greater than the pressure of the gas at the inlet). The power requirement of the compressor H_p can be determined from the equation below. Assuming that the power requirement of the compressor is exactly equal to zRT_1/MW, find the polytropic efficiency n of the compressor. The parameter z is compressibility of the gas under operating conditions of the compressor, R is the gas constant, T_1 is the temperature of the gas at the compressor inlet, and MW is the molecular weight of the gas.

$$HP = \frac{zRT_1}{MW} \frac{n}{n-1} (R_c^{(n-1)/n} - 1)$$

8.48 In the thermos shown in Fig. P8.48, the innermost compartment is separated from the middle container by a vacuum. There is a final shell around the thermos. This final shell is separated from the middle layer by a thin layer of air. The outside of the final shell comes in contact with room air. Heat transfer from the inner compartment to the next layer q_1 is by radiation only (since the space is evacuated). Heat transfer between the middle layer and outside shell q_2 is by convection in a small space. Heat transfer from the outside shell to the air q_3 is by natural convection. The heat flux from each region of the thermos must be equal—that is, $q_1 = q_2 = q_3$. Find the temperatures T_1 and T_2 at steady state. T_0 is 500°C and $T_3 = 25$°C.

$$q_1 = 10^{-9}[(T_0 + 273)^4 - (T_1 + 273)^4]$$
$$q_2 = 4(T_1 - T_2)$$
$$q_3 = 1.3(T_2 - T_3)^{4/3}$$

8.49 Figure P8.49 shows three reservoirs connected by circular pipes. The pipes, which are made of asphalt-dipped cast iron ($\varepsilon = 0.0012$ m), have the following characteristics:

Pipe	1	2	3
Length, m	1800	500	1400
Diameter, m	0.4	0.25	0.2
Flow, m³/s	?	0.1	?

If the water surface elevations in Reservoirs A and C are 200 and 172.5 m, respectively, determine the elevation in Reservoir B and the flows in pipes 1 and 3. Note that the kinematic viscosity of water is 1×10^{-6} m²/s and use the *Colebrook equation* to determine the friction factor (recall Prob. 8.13).

FIGURE P8.48

FIGURE P8.49

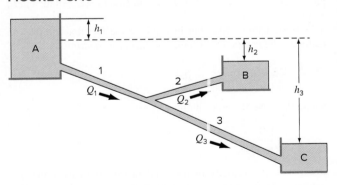

8.50 A fluid is pumped into the network of pipes shown in Fig. P8.50. At steady state, the following flow balances must hold,

$$Q_1 = Q_2 + Q_3$$
$$Q_3 = Q_4 + Q_5$$
$$Q_5 = Q_6 + Q_7$$

FIGURE P8.50

where Q_i = flow in pipe i (m³/s). In addition, the pressure drops around the three right-hand loops must equal zero. The pressure drop in each circular pipe length can be computed with

$$\Delta P = \frac{16}{\pi^2} \frac{fL\rho}{2D^5} Q^2$$

where ΔP = the pressure drop (Pa), f = the friction factor (dimensionless), L = the pipe length (m), ρ = the fluid density (kg/m³), and D = pipe diameter (m). Write a program (or develop an algorithm in a mathematics software package) that will allow you to compute the flow in every pipe length given that $Q_1 = 1$ m³/s and $\rho = 1.23$ kg/m³. All the pipes have $D = 500$ mm and $f = 0.005$. The pipe lengths are: $L_3 = L_5 = L_8 = L_9 = 2$ m; $L_2 = L_4 = L_6 = 4$ m; and $L_7 = 8$ m.

8.51 Repeat Prob. 8.50, but incorporate the fact that the friction factor can be computed with the *von Karman equation*,

$$\frac{1}{\sqrt{f}} = 4 \log_{10}(\text{Re} \sqrt{f}) - 0.4$$

where Re = the Reynolds number

$$\text{Re} = \frac{\rho V D}{\mu}$$

where V = the velocity of the fluid in the pipe (m/s) and μ = dynamic viscosity (N s/m²). Note that for a circular pipe $V = 4Q/\pi D^2$. Also, assume that the fluid has a viscosity of 1.79×10^{-5} N s/m².

8.52 The space shuttle, at liftoff from the launch pad, has four forces acting on it, which are shown on the free-body diagram (Fig. P8.52). The combined weight of the two solid rocket boosters and external fuel tank is $W_B = 1.663 \times 10^6$ lb. The weight of

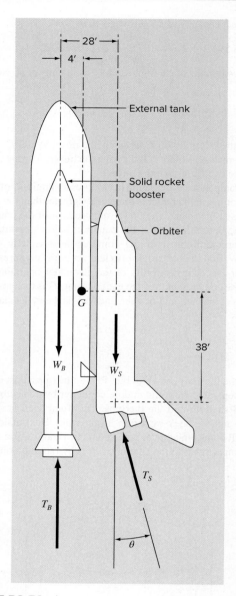

FIGURE P8.52

the orbiter with a full payload is $W_S = 0.23 \times 10^6$ lb. The combined thrust of the two solid rocket boosters is $T_B = 5.30 \times 10^6$ lb. The combined thrust of the three liquid fuel orbiter engines is $T_S = 1.125 \times 10^6$ lb.

At liftoff, the orbiter engine thrust is directed at angle θ to make the resultant moment acting on the entire craft assembly (external tank, solid rocket boosters, and orbiter) equal to zero. With the resultant moment equal to zero, the craft will not rotate about its mass center G at liftoff. With these forces, the craft will have a resultant force with components in both the vertical and horizontal direction. The vertical resultant force component is what allows the craft to lift off from the launch pad and fly vertically. The horizontal resultant force component causes the craft to fly horizontally. The resultant moment acting on the craft will be zero when θ is adjusted to the proper value. If this angle is not adjusted properly, and there is some resultant moment acting on the craft, the craft will tend to rotate about its mass center.

(a) Resolve the orbiter thrust T_S into horizontal and vertical components, and then sum moments about point G, the craft's mass center. Set the resulting moment equation equal to zero. This equation can now be solved for the value of θ required for liftoff.
(b) Derive an equation for the resultant moment acting on the craft in terms of the angle θ. Plot the resultant moment as a function of the angle θ over a range of -5 radians to $+5$ radians.
(c) Write a computer program to solve for the angle θ using Newton's method to find the root of the resultant moment equation. Make an initial first guess at the root of interest using the plot. Terminate your iterations when the value of θ has better than five significant figures.
(d) Repeat the program for the minimum payload weight of the orbiter: $W_S = 195{,}000$ lb.

8.53 Determining the velocity of particles settling through fluids is of great importance of many areas of engineering and science. Such calculations depend on the flow regime as represented by the dimensionless *Reynolds number*,

$$\text{Re} = \frac{\rho d v}{\mu} \tag{P8.53.1}$$

where ρ = the fluid's density (kg/m³), d = the particle diameter (m), v = the particle's settling velocity (m/s), and μ = the fluid's dynamic viscosity (N s/m²). Under laminar conditions (Re < 0.1), the settling velocity of a spherical particle can be computed with the following formula based on *Stokes law*,

$$v = \frac{g}{18}\left(\frac{\rho_s - \rho}{\mu}\right)d^2 \tag{P8.53.2}$$

where g = the gravitational constant (= 9.81 m/s²) and ρ_s = the particle's density (kg/m³). For turbulent conditions (i.e., higher Reynolds numbers), an alternative approach can be used based on the following formula:

$$v = \sqrt{\frac{4g(\rho_s - \rho)d}{3C_D\rho}} \tag{P8.53.3}$$

where C_D = the drag coefficient, which depends on the Reynolds number as in

$$C_D = \frac{24}{\text{Re}} + \frac{3}{\sqrt{\text{Re}}} + 0.34 \tag{P8.53.4}$$

(a) Combine Eqs. (P8.53.2), (P8.53.3), and (P8.53.4) to express the determination of v as a roots of equations problem. That is, express the combined formula in the format $f(v) = 0$.
(b) Use the modified secant method with $\delta = 10^{-3}$ and $\varepsilon_S = 0.05\%$ to determine v for a spherical iron particle settling in water, where $d = 200$ μm, $\rho = 1$ g/cm³, $\rho_s = 7.874$ g/cm³, and $\mu = 0.014$ g/(cm s). Employ Eq. (P8.53.2) to generate your initial guess.
(c) Based on the result of (b), compute the Reynolds number and the drag coefficient, and use the latter to confirm that the flow regime is not laminar.
(d) Develop a fixed-point iteration solution for the conditions outlined in (b).
(e) Use a graphical approach to illustrate that the formulation developed in (d) will converge for any positive guess.

8.54 The paths of two rockets in the x-y plane can be described by the following parametric equations

Rocket 1:	Rocket 2:
$x_1 = 200t$	$x_2 = 800 - 100(\cos\alpha)t$
$y_1 = 80t - 16t^2$	$y_2 = 50(\sin\alpha)t - 0.8t^2$

where t = time and α = the launch angle of the second rocket. As depicted in Fig. P8.54, the two paths intersect but the rockets will only collide if they simultaneously arrive at the same point. Determine the value of t and α so that this occurs.

8.55 *Bernoulli's equation* can be used to relate the pressure, velocity, and height of any two points in a steady-streamline, constant density flowing fluid. As depicted in Fig. P8.55, suppose we have

FIGURE P8.54

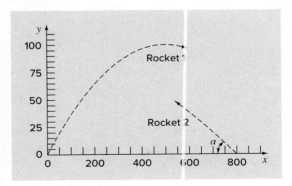

a liquid (density, $\rho = 1{,}264$ kg/m^3) pumped through a pipeline from points 1 to 2 with a pump adding 18 kW into the flow between the two points. Given $D_1 = 60$ cm, $p_1 = 310$ kPa, $D_2 = 30$ cm, $p_2 = 350$ kPa, $z_1 = 0$ and $z_2 = -100$ cm, determine Q assuming minimal friction losses. According to Bernoulli's equation

$$\left(\frac{p_1}{\rho g} + z_1 + \frac{V_1^2}{2g}\right) + h_p = \left(\frac{p_2}{\rho g} + z_2 + \frac{V_2^2}{2g}\right)$$

where g = the gravitational constant ($\cong 9.81$ m/s^2), and h_p = the energy head added to the flow by the pump (m). Note that the continuity equation, $Q = AV$, holds in each constant diameter pipe section, and the pump's power, P, is related to flow by

$$P = \rho g Q h_p$$

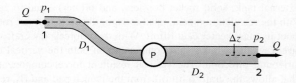

FIGURE P8.55

EPILOGUE: PART TWO

PT2.4 TRADE-OFFS

Table PT2.3 provides a summary of the trade-offs involved in solving for roots of algebraic and transcendental equations. Although graphical methods are time-consuming, they provide insight into the behavior of the function and are useful in identifying initial guesses and potential problems such as multiple roots. Therefore, if time permits, a quick sketch (or better yet, a computerized graph) can yield valuable information regarding the behavior of the function.

The numerical methods themselves are divided into two general categories: bracketing and open methods. The former requires two initial guesses that are on either side of a root. This "bracketing" is maintained as the solution proceeds, and thus, these techniques are always convergent. However, a price is paid for this property in that the rate of convergence is relatively slow.

TABLE PT2.3 Comparison of the characteristics of alternative methods for finding roots of algebraic and transcendental equations. The comparisons are based on general experience and do not account for the behavior of specific functions.

Method	Type	Guesses	Convergence	Stability	Programming	Comments
Direct	Analytical	—	—	—		
Graphical	Visual	—	—	—	—	Imprecise
Bisection	Bracketing	2	Slow	Always	Easy	
False-position	Bracketing	2	Slow/medium	Always	Easy	
Modified FP	Bracketing	2	Medium	Always	Easy	
Fixed-point iteration	Open	1	Slow	Possibly divergent	Easy	
Newton-Raphson	Open	1	Fast	Possibly divergent	Easy	Requires evaluation of $f'(x)$
Modified Newton-Raphson	Open	1	Fast (multiple), medium (single)	Possibly divergent	Easy	Requires evaluation of $f'(x)$ and $f''(x)$
Secant	Open	2	Medium/fast	Possibly divergent	Easy	Initial guesses do not have to bracket the root
Modified secant	Open	1	Medium/fast	Possibly divergent	Easy	
Brent	Hybrid	1 or 2	Medium	Always (for 2 guesses)	Moderate	Robust
Müller	Polynomials	2	Medium/fast	Possibly divergent	Moderate	
Bairstow	Polynomials	2	Fast	Possibly divergent	Moderate	

Open techniques differ from bracketing methods in that they use information at a single point (or two values that need not bracket the root to extrapolate to a new root estimate). This property is a double-edged sword. Although use of an open technique generally leads to quicker convergence, it also allows the possibility that the solution may diverge. In general, the convergence of open techniques is partially dependent on the quality of the initial guess and the nature of the function. The closer the guess is to the true root, the more likely the methods will converge.

Of the open techniques, the standard Newton-Raphson method is often used because of its property of quadratic convergence. However, its major shortcoming is that it requires that the derivative of the function be obtained analytically. For some functions this is impractical. In these cases, the secant method, which employs a finite-difference representation of the derivative, provides a viable alternative. Because of the finite-difference approximation, the rate of convergence of the secant method is initially slower than for the Newton-Raphson method. However, as the root estimate is refined, the difference approximation becomes a better representation of the true derivative, and convergence accelerates rapidly. The modified Newton-Raphson technique can be used to attain rapid convergence for multiple roots. However, this technique requires an analytical expression for both the first and second derivatives.

Of particular interest are hybrid methods that combine the reliability of bracketing with the speed of open methods. Brent's method does this by combining bisection with several open methods. All the methods are easy to moderate to program on computers and require minimal time to determine a single root. On this basis, you might conclude that simple methods such as bisection would be good enough for practical purposes. This would be true if you were exclusively interested in determining the root of an equation once. However, there are many cases in engineering where numerous root locations are required and where speed becomes important. For these cases, slow methods are very time-consuming and, hence, costly. On the other hand, the fast open methods may diverge, and the accompanying delays can also be costly. Some computer algorithms attempt to capitalize on the strong points of both classes of techniques by initially employing a bracketing method to approach the root, then switching to an open method to rapidly refine the estimate. Whether a single approach or a combination is used, the trade-offs between convergence and speed are at the heart of the choice of a root-location technique.

PT2.5 IMPORTANT RELATIONSHIPS AND FORMULAS

Table PT2.4 summarizes important information that was presented in Part Two. This table can be consulted to quickly access important relationships and formulas.

PT2.6 ADVANCED METHODS AND ADDITIONAL REFERENCES

The methods in this text have focused on determining a single real root of an algebraic or transcendental equation based on foreknowledge of its approximate location. In addition, we have also described methods expressly designed to determine both the real

TABLE PT2.4 Summary of important information presented in Part Two.

Method	Formulation	Graphical Interpretation	Errors and Stopping Criteria
		Bracketing methods:	
Bisection	$x_r = \dfrac{x_l + x_u}{2}$ If $f(x_l)f(x_r) < 0$, $x_u = x_r$ $f(x_l)f(x_r) > 0$, $x_l = x_r$		Stopping criterion: $\left\| \dfrac{x_r^{new} - x_r^{old}}{x_r^{new}} \right\| 100\% \leq \epsilon_s$
False position	$x_r = x_u - \dfrac{f(x_u)(x_l - x_u)}{f(x_l) - f(x_u)}$ If $f(x_l)f(x_r) < 0$, $x_u = x_r$ $f(x_l)f(x_r) > 0$, $x_l = x_r$		Stopping criterion: $\left\| \dfrac{x_r^{new} - x_r^{old}}{x_r^{new}} \right\| 100\% \leq \epsilon_s$
Newton-Raphson	$x_{i+1} = x_i - \dfrac{f(x_i)}{f'(x_i)}$		Stopping criterion: $\left\| \dfrac{x_{i+1} - x_i}{x_{i+1}} \right\| 100\% \leq \epsilon_s$ Error: $E_{i+1} = O(E_i^2)$
Secant	$x_{i+1} = x_i - \dfrac{f(x_i)(x_{i-1} - x_i)}{f(x_{i-1}) - f(x_i)}$		Stopping criterion: $\left\| \dfrac{x_{i+1} - x_i}{x_{i+1}} \right\| 100\% \leq \epsilon_s$

and complex roots of polynomials. Additional references on the subject are Ralston and Rabinowitz (1978) and Carnahan, Luther, and Wilkes (1969).

In addition to Müller's and Bairstow's methods, several techniques are available to determine all the roots of polynomials. In particular, the *quotient difference (QD) algorithm* (Henrici 1964; Gerald and Wheatley 2004) determines all roots without initial guesses. Ralston and Rabinowitz (1978) and Carnahan, Luther, and Wilkes (1969) contain discussions of this method as well as of other techniques for locating roots of polynomials. As discussed in the text, the *Jenkins-Traub* and *Laguerre's* methods are widely employed.

In summary, the foregoing is intended to provide you with avenues for deeper exploration of the subject. Additionally, all the above references provide descriptions of the basic techniques covered in Part Two. We urge you to consult these alternative sources to broaden your understanding of numerical methods for root location.[1]

---

[1]Books are referenced only by author here, a complete bibliography will be found at the back of this text.

PART THREE

LINEAR ALGEBRAIC EQUATIONS

PT3.1 MOTIVATION

In Part Two, we determined the value x that satisfied a single equation, $f(x) = 0$. Now, we deal with the case of determining the values $x_1, x_2, \ldots, x_n$ that simultaneously satisfy a set of equations

$$f_1(x_1, x_2, \ldots, x_n) = 0$$
$$f_2(x_1, x_2, \ldots, x_n) = 0$$
$$\begin{array}{cc} \cdot & \cdot \\ \cdot & \cdot \\ \cdot & \cdot \end{array}$$
$$f_n(x_1, x_2, \ldots, x_n) = 0$$

Such systems can be either linear or nonlinear. In Part Three, we deal with *linear algebraic equations* that are of the general form

$$a_{11}x_1 + a_{12}x_2 + \cdots + a_{1n}x_n = b_1$$
$$a_{21}x_1 + a_{22}x_2 + \cdots + a_{2n}x_n = b_2$$
$$\begin{array}{cc} \cdot & \cdot \\ \cdot & \cdot \\ \cdot & \cdot \end{array} \qquad \text{(PT3.1)}$$
$$a_{n1}x_1 + a_{n2}x_2 + \cdots + a_{nn}x_n = b_n$$

where the a's are constant coefficients, the b's are constants, and n is the number of equations. All other equations are nonlinear. Nonlinear systems were discussed in Chap. 6 and will be covered briefly again in Chap. 9.

PT3.1.1 Noncomputer Methods for Solving Systems of Equations

For small numbers of equations ($n \leq 3$), linear (and sometimes nonlinear) equations can be solved readily by simple techniques. Some of these methods will be reviewed at the beginning of Chap. 9. However, for four or more equations, solutions become arduous and computers must be utilized. Historically, the inability to solve all but the smallest sets of equations by hand has limited the scope of problems addressed in many engineering applications.

Before computers, techniques to solve linear algebraic equations were time-consuming and awkward. These approaches placed a constraint on creativity because the methods were often difficult to implement and understand. Consequently, the techniques were sometimes overemphasized at the expense of other aspects of the problem-solving process such as formulation and interpretation (recall Fig. PT1.1 and accompanying discussion).

The accessibility of computers makes it possible and practical for you to solve large sets of simultaneous linear algebraic equations. Thus, you can approach more complex and realistic examples and problems. Furthermore, you will have more time to test your creative skills because you will be able to place more emphasis on problem formulation and solution interpretation.

PT3.1.2 Linear Algebraic Equations and Engineering Practice

Many of the fundamental equations of engineering are based on conservation laws (recall Table 1.1). Some familiar quantities that conform to such laws are mass, energy, and momentum. In mathematical terms, these principles lead to balance or continuity equations that relate system *behavior* as represented by the *levels* or *response* of the quantity being modeled to the *properties* or *characteristics* of the system and the external *stimuli* or *forcing functions* acting on the system.

As an example, the principle of mass conservation can be used to formulate a model for a series of chemical reactors (Fig. PT3.1*a*). For this case, the quantity being modeled is the mass of the chemical in each reactor. The system properties are the reaction characteristics of the chemical and the reactors' sizes and flow rates. The forcing functions are the feed rates of the chemical into the system.

FIGURE PT3.1
Two types of systems that can be modeled using linear algebraic equations: (*a*) lumped variable system that involves coupled finite components and (*b*) distributed variable system that involves a continuum.

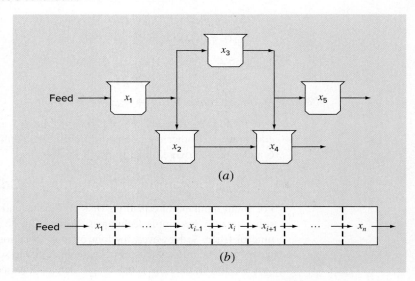

In Part Two, you saw how single-component systems result in a single equation that can be solved using root-location techniques. Multicomponent systems result in a coupled set of mathematical equations that must be solved simultaneously. The equations are coupled because the individual parts of the system are influenced by other parts. For example, in Fig. PT3.1a, reactor 4 receives chemical inputs from reactors 2 and 3. Consequently, its response is dependent on the quantity of chemical in these other reactors.

When these dependencies are expressed mathematically, the resulting equations are often of the linear algebraic form of Eq. (PT3.1). The x's are usually measures of the magnitudes of the responses of the individual components. Using Fig. PT3.1a as an example, x_1 might quantify the amount of mass in the first reactor, x_2 might quantify the amount in the second, and so forth. The a's typically represent the properties and characteristics that bear on the interactions between components. For instance, the a's for Fig. PT3.1a might be reflective of the flow rates of mass between the reactors. Finally, the b's usually represent the forcing functions acting on the system, such as the feed rate in Fig. PT3.1a. The applications in Chap. 12 provide other examples of such equations derived from engineering practice.

Multicomponent problems of the above types arise from both *lumped* (macro-) or *distributed* (micro-) variable mathematical models (Fig. PT3.1). Lumped variable problems involve coupled finite components. Examples include trusses (Sec. 12.2), reactors (Fig. PT3.1a and Sec. 12.1), and electric circuits (Sec. 12.3). These types of problems use models that provide little or no spatial detail.

Conversely, distributed variable problems attempt to describe spatial detail of systems on a continuous or semicontinuous basis. The distribution of chemicals along the length of an elongated, rectangular reactor (Fig. PT3.1b) is an example of a continuous variable model. Differential equations derived from the conservation laws specify the distribution of the dependent variable for such systems. These differential equations can be solved numerically by converting them to an equivalent system of simultaneous algebraic equations. The solution of such sets of equations represents a major engineering application area for the methods in the following chapters. These equations are coupled because the variables at one location are dependent on the variables in adjoining regions. For example, the concentration at the middle of the reactor is a function of the concentration in adjoining regions. Similar examples could be developed for the spatial distribution of temperature or momentum. We will address such problems when we discuss differential equations later in the book.

Aside from physical systems, simultaneous linear algebraic equations also arise in a variety of mathematical problem contexts. These result when mathematical functions are required to satisfy several conditions simultaneously. Each condition results in an equation that contains known coefficients and unknown variables. The techniques discussed in this part can be used to solve for the unknowns when the equations are linear and algebraic. Some widely used numerical techniques that employ simultaneous equations are regression analysis (Chap. 17) and spline interpolation (Chap. 18).

PT3.2   MATHEMATICAL BACKGROUND

All parts of this book require some mathematical background. For Part Three, matrix notation and algebra are useful because they provide a concise way to represent and manipulate linear algebraic equations. If you are already familiar with matrices, feel free

FIGURE PT3.2
A matrix.

to skip to Sec. PT3.3. For those who are unfamiliar or require a review, the following material provides a brief introduction to the subject.

PT3.2.1 Matrix Notation

A *matrix* consists of a rectangular array of elements represented by a single symbol. As depicted in Fig. PT3.2, [A] is the shorthand notation for the matrix and a_{ij} designates an individual *element* of the matrix.

A horizontal set of elements is called a *row* and a vertical set is called a *column*. The first subscript i always designates the number of the row in which the element lies. The second subscript j designates the column. For example, element a_{23} is in row 2 and column 3.

The matrix in Fig. PT3.2 has n rows and m columns and is said to have a dimension of n by m (or $n \times m$). It is referred to as an n by m matrix.

Matrices with row dimension $n = 1$, such as

$$[B] = [b_1 \quad b_2 \quad \cdots \quad b_m]$$

are called *row vectors*. Note that for simplicity, the first subscript of each element is dropped. Also, it should be mentioned that there are times when it is desirable to employ a special shorthand notation to distinguish a row matrix from other types of matrices. One way to accomplish this is to employ special open-topped brackets, as in $\lfloor B \rfloor$.

Matrices with column dimension $m = 1$, such as

$$[C] = \begin{bmatrix} c_1 \\ c_2 \\ \cdot \\ \cdot \\ \cdot \\ c_n \end{bmatrix}$$

are referred to as *column vectors*. For simplicity, the second subscript is dropped. As with the row vector, there are occasions when it is desirable to employ a special shorthand notation to distinguish a column matrix from other types of matrices. One way to accomplish this is to employ special brackets, as in $\{C\}$.

Matrices where $n = m$ are called *square matrices*. For example, a 4 by 4 matrix is

$$[A] = \begin{bmatrix} a_{11} & a_{12} & a_{13} & a_{14} \\ a_{21} & a_{22} & a_{23} & a_{24} \\ a_{31} & a_{32} & a_{33} & a_{34} \\ a_{41} & a_{42} & a_{43} & a_{44} \end{bmatrix}$$

The diagonal consisting of the elements a_{11}, a_{22}, a_{33}, and a_{44} is termed the *principal*, or *main, diagonal* of the matrix.

Square matrices are particularly important when solving sets of simultaneous linear equations. For such systems, the number of equations (corresponding to rows) and the number of unknowns (corresponding to columns) must be equal for a unique solution to be possible. Consequently, square matrices of coefficients are encountered when dealing with such systems. Some special types of square matrices are described in Box PT3.1.

Box PT3.1   Special Types of Square Matrices

There are a number of special forms of square matrices that are important and should be noted:

A *symmetric matrix* is one where $a_{ij} = a_{ji}$ for all i's and j's. For example,

$$[A] = \begin{bmatrix} 5 & 1 & 2 \\ 1 & 3 & 7 \\ 2 & 7 & 8 \end{bmatrix}$$

is a 3 by 3 symmetric matrix.

A *diagonal matrix* is a square matrix where all elements off the main diagonal are equal to zero, as in

$$[A] = \begin{bmatrix} a_{11} & & & \\ & a_{22} & & \\ & & a_{33} & \\ & & & a_{44} \end{bmatrix}$$

Note that where large blocks of elements are zero, they are left blank.

An *identity matrix* is a diagonal matrix where all elements on the main diagonal are equal to 1, as in

$$[I] = \begin{bmatrix} 1 & & & \\ & 1 & & \\ & & 1 & \\ & & & 1 \end{bmatrix}$$

The symbol $[I]$ is used to denote the identity matrix. The identity matrix has properties similar to unity.

An *upper triangular matrix* is one where all the elements below the main diagonal are zero, as in

$$[A] = \begin{bmatrix} a_{11} & a_{12} & a_{13} & a_{14} \\ & a_{22} & a_{23} & a_{24} \\ & & a_{33} & a_{34} \\ & & & a_{44} \end{bmatrix}$$

A *lower triangular matrix* is one where all elements above the main diagonal are zero, as in

$$[A] = \begin{bmatrix} a_{11} & & & \\ a_{21} & a_{22} & & \\ a_{31} & a_{32} & a_{33} & \\ a_{41} & a_{42} & a_{43} & a_{44} \end{bmatrix}$$

A *banded matrix* has all elements equal to zero, with the exception of a band centered on the main diagonal:

$$[A] = \begin{bmatrix} a_{12} & a_{12} & & \\ a_{21} & a_{22} & a_{23} & \\ & a_{32} & a_{33} & a_{34} \\ & & a_{43} & a_{44} \end{bmatrix}$$

The above matrix has a bandwidth of 3 and is given a special name—the *tridiagonal matrix*.

PT3.2.2 Matrix Operating Rules

Now that we have specified what we mean by a matrix, we can define some operating rules that govern its use. Two n by m matrices are equal if, and only if, every element in the first is equal to every element in the second, that is, $[A] = [B]$ if $a_{ij} = b_{ij}$ for all i and j.

Addition of two matrices, say, $[A]$ and $[B]$, is accomplished by adding corresponding terms in each matrix. The elements of the resulting matrix $[C]$ are computed as

$$c_{ij} = a_{ij} + b_{ij}$$

for $i = 1, 2, \ldots, n$ and $j = 1, 2, \ldots, m$. Similarly, the *subtraction* of two matrices, say, $[E]$ minus $[F]$, is obtained by subtracting corresponding terms, as in

$$d_{ij} = e_{ij} - f_{ij}$$

for $i = 1, 2, \ldots, n$ and $j = 1, 2, \ldots, m$. It follows directly from the above definitions that addition and subtraction can be performed only between matrices having the same dimensions.

Both addition and subtraction are *commutative:*

$$[A] + [B] = [B] + [A]$$

Addition and subtraction are also *associative,* that is,

$$([A] + [B]) + [C] = [A] + ([B] + [C])$$

The *multiplication* of a matrix $[A]$ by a scalar g is obtained by multiplying every element of $[A]$ by g, as in

$$[D] = g[A] = \begin{bmatrix} ga_{11} & ga_{12} & \cdots & ga_{1m} \\ ga_{21} & ga_{22} & \cdots & ga_{2m} \\ . & . & & . \\ . & . & & . \\ . & . & & . \\ ga_{n1} & ga_{n2} & \cdots & ga_{nm} \end{bmatrix}$$

The *product* of two matrices is represented as $[C] = [A][B]$, where the elements of $[C]$ are defined as (see Box PT3.2 for a simple way to conceptualize matrix multiplication)

$$c_{ij} = \sum_{k=1}^{n} a_{ik}b_{kj} \tag{PT3.2}$$

where n = the column dimension of $[A]$ and the row dimension of $[B]$. That is, the c_{ij} element is obtained by adding the product of individual elements from the ith row of the first matrix, in this case $[A]$, by the jth column of the second matrix $[B]$.

According to this definition, multiplication of two matrices can be performed only if the first matrix has as many columns as the number of rows in the second matrix. Thus, if $[A]$ is an n by m matrix, $[B]$ could be an m by l matrix. For this case, the resulting $[C]$ matrix would have the dimension of n by l. However, if $[B]$ were an l by m matrix, the multiplication could not be performed. Figure PT3.3 provides an easy way to check whether two matrices can be multiplied.

Box PT3.2   A Simple Method for Multiplying Two Matrices

Although Eq. (PT3.2) is well suited for implementation on a computer, it is not the simplest means for visualizing the mechanics of multiplying two matrices. What follows gives more tangible expression to the operation.

Suppose that we want to multiply $[X]$ by $[Y]$ to yield $[Z]$,

$$[Z] = [X][Y] = \begin{bmatrix} 3 & 1 \\ 8 & 6 \\ 0 & 4 \end{bmatrix} \begin{bmatrix} 5 & 9 \\ 7 & 2 \end{bmatrix}$$

A simple way to visualize the computation of $[Z]$ is to raise $[Y]$, as in

$$\begin{matrix} \Uparrow \\ \begin{bmatrix} 5 & 9 \\ 7 & 2 \end{bmatrix} \leftarrow [Y] \end{matrix}$$

$$[X] \rightarrow \begin{bmatrix} 3 & 1 \\ 8 & 6 \\ 0 & 4 \end{bmatrix} \begin{bmatrix} \ ? \ \end{bmatrix} \leftarrow [Z]$$

Now the answer $[Z]$ can be computed in the space vacated by $[Y]$. This format has utility because it aligns the appropriate rows and columns that are to be multiplied. For example, according to Eq. (PT3.2), the element z_{11} is obtained by multiplying the first row of $[X]$ by the first column of $[Y]$. This amounts to adding the product of x_{11} and y_{11} to the product of x_{12} and y_{21}, as in

$$\begin{bmatrix} 5 & 9 \\ 7 & 2 \end{bmatrix}$$
$$\downarrow$$
$$\begin{bmatrix} 3 & 1 \\ 8 & 6 \\ 0 & 4 \end{bmatrix} \rightarrow \begin{bmatrix} 3 \times 5 + 1 \times 7 = 22 \end{bmatrix}$$

Thus, z_{11} is equal to 22. Element z_{21} can be computed in a similar fashion, as in

$$\begin{bmatrix} 5 & 9 \\ 7 & 2 \end{bmatrix}$$
$$\downarrow$$
$$\begin{bmatrix} 3 & 1 \\ 8 & 6 \\ 0 & 4 \end{bmatrix} \rightarrow \begin{bmatrix} 22 \\ 8 \times 5 + 6 \times 7 = 82 \end{bmatrix}$$

The computation can be continued in this way, following the alignment of the rows and columns, to yield the result

$$[Z] = \begin{bmatrix} 22 & 29 \\ 82 & 84 \\ 28 & 8 \end{bmatrix}$$

Note how this simple method makes it clear why it is impossible to multiply two matrices if the number of columns of the first matrix does not equal the number of rows in the second matrix. Also, note how it demonstrates that the order of multiplication matters (that is, matrix multiplication is not commutative).

FIGURE PT3.3

$$[A]_{n \times m} \quad [B]_{m \times l} = [C]_{n \times l}$$

Interior dimensions are equal; multiplication is possible

Exterior dimensions define the dimensions of the result

If the dimensions of the matrices are suitable, matrix multiplication is *associative*,

$$([A][B])[C] = [A]([B][C])$$

and *distributive,*

$$[A]([B] + [C]) = [A][B] + [A][C]$$

or

$$([A] + [B])[C] = [A][C] + [B][C]$$

However, multiplication is not generally *commutative:*

$$[A][B] \neq [B][A]$$

That is, the order of multiplication is important.

Figure PT3.4 shows pseudocode to multiply an n by m matrix $[A]$ by an m by l matrix $[B]$, and store the result in an n by l matrix $[C]$. Notice that, instead of the inner product being directly accumulated in $[C]$, it is collected in a temporary variable, sum. This is done for two reasons. First, it is a bit more efficient, because the computer need determine the location of $c_{i,j}$ only $n \times l$ times rather than $n \times l \times m$ times. Second, the precision of the multiplication can be greatly improved by declaring sum as a double precision variable (recall the discussion of inner products in Sec. 3.4.2).

Although multiplication is possible, matrix division is not a defined operation. However, if a matrix $[A]$ is square and nonsingular, there is another matrix $[A]^{-1}$, called the *inverse* of $[A]$, for which

$$[A][A]^{-1} = [A]^{-1}[A] = [I] \tag{PT3.3}$$

Thus, the multiplication of a matrix by the inverse is analogous to division, in the sense that a number divided by itself is equal to 1. That is, multiplication of a matrix by its inverse leads to the identity matrix (recall Box PT3.1).

The inverse of a two-dimensional square matrix can be represented simply by

$$[A]^{-1} = \frac{1}{a_{11}a_{22} - a_{12}a_{21}} \begin{bmatrix} a_{22} & -a_{12} \\ -a_{21} & a_{11} \end{bmatrix} \tag{PT3.4}$$

FIGURE PT3.4

```
SUBROUTINE Mmult(a, b, c, m, n, l)
DOFOR i = 1, n
  DOFOR j = 1, l
    sum = 0.
    DOFOR k = 1, m
      sum = sum + a(i,k) · b(k,j)
    END DO
    c(i,j) = sum
  END DO
END DO
```

Similar formulas for higher-dimensional matrices are much more involved. Sections in Chaps. 10 and 11 will be devoted to techniques for using numerical methods and the computer to calculate the inverse for such systems.

Two other matrix manipulations that will have utility in our discussion are the transpose and the trace of a matrix. The transpose of a matrix involves transforming its rows into columns and its columns into rows. For example, for the 4×4 matrix,

$$[A] = \begin{bmatrix} a_{11} & a_{12} & a_{13} & a_{14} \\ a_{21} & a_{22} & a_{23} & a_{24} \\ a_{31} & a_{32} & a_{33} & a_{34} \\ a_{41} & a_{42} & a_{43} & a_{44} \end{bmatrix}$$

the transpose, designated $[A]^T$, is defined as

$$[A]^T = \begin{bmatrix} a_{11} & a_{21} & a_{31} & a_{41} \\ a_{12} & a_{22} & a_{32} & a_{42} \\ a_{13} & a_{23} & a_{33} & a_{43} \\ a_{14} & a_{24} & a_{34} & a_{44} \end{bmatrix}$$

In other words, the element a_{ij} of the transpose is equal to the element a_{ji} of the original matrix.

The *transpose* has a variety of functions in matrix algebra. One simple advantage is that it allows a column vector to be written as a row. For example, if

$$\{C\} = \begin{Bmatrix} c_1 \\ c_2 \\ c_3 \\ c_4 \end{Bmatrix}$$

then

$$\{C\}^T = \lfloor c_1 \quad c_2 \quad c_3 \quad c_4 \rfloor$$

where the superscript T designates the transpose. For example, this format can save space when writing a column vector in a manuscript. In addition, the transpose has numerous mathematical applications.

The *trace* of a matrix is the sum of the elements on its principal diagonal. It is designated as tr $[A]$ and is computed as

$$\text{tr }[A] = \sum_{i=1}^{n} a_{ii}$$

The trace will be used in our discussion of eigenvalues in Chap. 27.

The final matrix manipulation that will have utility in our discussion is *augmentation*. A matrix is augmented by the addition of a column (or columns) to the original matrix. For example, suppose we have a matrix of coefficients:

$$[A] = \begin{bmatrix} a_{11} & a_{12} & a_{13} \\ a_{21} & a_{22} & a_{23} \\ a_{31} & a_{32} & a_{33} \end{bmatrix}$$

We might wish to augment this matrix $[A]$ with an identity matrix (recall Box PT3.1) to yield a 3-by-6-dimensional matrix:

$$[A] = \begin{bmatrix} a_{11} & a_{12} & a_{13} & \vdots & 1 & 0 & 0 \\ a_{21} & a_{22} & a_{23} & \vdots & 0 & 1 & 0 \\ a_{31} & a_{32} & a_{33} & \vdots & 0 & 0 & 1 \end{bmatrix}$$

Such an expression has utility when we must perform a set of identical operations on two matrices. Thus, we can perform the operations on the single augmented matrix rather than on the two individual matrices.

PT3.2.3 Representing Linear Algebraic Equations in Matrix Form

It should be clear that matrices provide a concise notation for representing simultaneous linear equations. For example, Eq. (PT3.1) can be expressed as

$$[A]\{X\} = \{B\} \tag{PT3.5}$$

where $[A]$ is the n by n square matrix of coefficients,

$$[A] = \begin{bmatrix} a_{11} & a_{12} & \cdots & a_{1n} \\ a_{21} & a_{22} & \cdots & a_{2n} \\ \cdot & \cdot & & \cdot \\ \cdot & \cdot & & \cdot \\ \cdot & \cdot & & \cdot \\ a_{n1} & a_{n2} & \cdots & a_{nn} \end{bmatrix}$$

$\{B\}$ is the n by 1 column vector of constants,

$$\{B\}^T = \lfloor b_1 \quad b_2 \quad \cdots \quad b_n \rfloor$$

and $\{X\}$ is the n by 1 column vector of unknowns:

$$\{X\}^T = \lfloor x_1 \quad x_2 \quad \cdots \quad x_n \rfloor$$

Recall the definition of matrix multiplication [Eq. (PT3.2) or Box PT3.2] to convince yourself that Eqs. (PT3.1) and (PT3.5) are equivalent. Also, realize that Eq. (PT3.5) is a valid matrix multiplication because the number of columns, n, of the first matrix $[A]$ is equal to the number of rows, n, of the second matrix $\{X\}$.

This part of the book is devoted to solving Eq. (PT3.5) for $\{X\}$. A formal way to obtain a solution using matrix algebra is to multiply each side of the equation by the inverse of $[A]$ to yield

$$[A]^{-1}[A]\{X\} = [A]^{-1}\{B\}$$

Because $[A]^{-1}[A]$ equals the identity matrix, the equation becomes

$$\{X\} = [A]^{-1}\{B\} \tag{PT3.6}$$

Therefore, the equation has been solved for $\{X\}$. This is another example of how the inverse plays a role in matrix algebra that is similar to division. It should be noted that this is not a very efficient way to solve a system of equations. Thus, other approaches are employed in numerical algorithms. However, as discussed in Chap. 10, the matrix inverse itself has great value in the engineering analyses of such systems.

Finally, we will sometimes find it useful to augment $[A]$ with $\{B\}$. For example, if $n = 3$, this results in a 3-by-4-dimensional matrix:

$$[A] = \begin{bmatrix} a_{11} & a_{12} & a_{13} & \vdots & b_1 \\ a_{21} & a_{22} & a_{23} & \vdots & b_2 \\ a_{31} & a_{32} & a_{33} & \vdots & b_3 \end{bmatrix} \qquad \text{(PT3.7)}$$

Expressing the equations in this form is useful because several of the techniques for solving linear systems perform identical operations on a row of coefficients and the corresponding right-hand-side constant. As expressed in Eq. (PT3.7), we can perform the manipulation once on an individual row of the augmented matrix rather than separately on the coefficient matrix and the right-hand-side vector.

PT3.3   ORIENTATION

Before proceeding to the numerical methods, some further orientation might be helpful. The following is intended as an overview of the material discussed in Part Three. In addition, we have formulated some objectives to help focus your efforts when studying the material.

PT3.3.1 Scope and Preview

Figure PT3.5 provides an overview for Part Three. *Chapter 9* is devoted to the most fundamental technique for solving linear algebraic systems: *Gauss elimination*. Before launching into a detailed discussion of this technique, a preliminary section deals with simple methods for solving small systems. These approaches are presented to provide you with visual insight and because one of the methods—the elimination of unknowns—represents the basis for Gauss elimination.

After the preliminary material, "naive" Gauss elimination is discussed. We start with this "stripped-down" version because it allows the fundamental technique to be elaborated on without complicating details. Then, in subsequent sections, we discuss potential problems of the naive approach and present a number of modifications to minimize and circumvent these problems. The focus of this discussion will be the process of switching rows, or *partial pivoting*.

Chapter 10 begins by illustrating how Gauss elimination can be formulated as an *LU decomposition* solution. Such solution techniques are valuable for cases where many right-hand-side vectors need to be evaluated. It is shown how this attribute allows efficient calculation of the *matrix inverse,* which has tremendous utility in engineering practice. Finally, the chapter ends with a discussion of matrix condition. The *condition number* is introduced as a measure of the loss of significant digits of accuracy that can result when solving ill-conditioned matrices.

The beginning of *Chap. 11* focuses on special types of systems of equations that have broad engineering application. In particular, efficient techniques for solving *tridiagonal systems* are presented. Then, the remainder of the chapter focuses on an alternative to elimination methods called the *Gauss-Seidel method*. This technique is similar in spirit to

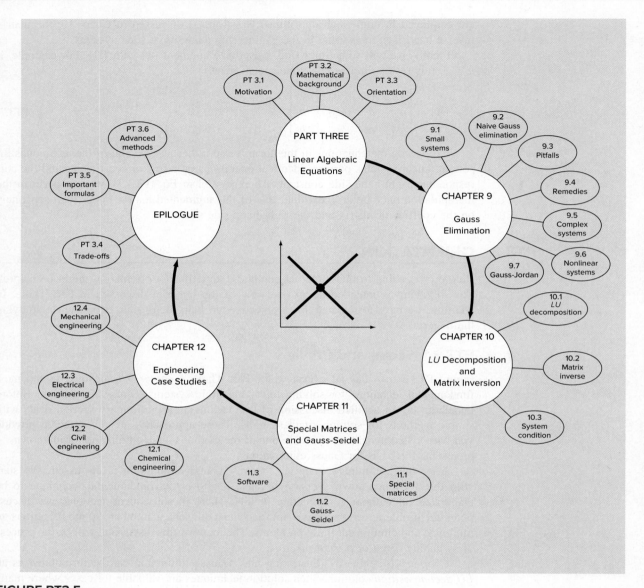

FIGURE PT3.5
Schematic of the organization of the material in Part Three: Linear Algebraic Equations.

the approximate methods for roots of equations that were discussed in Chap. 6. That is, the technique involves guessing a solution and then iterating to obtain a refined estimate. The chapter ends with information related to solving linear algebraic equations with software packages.

Chapter 12 demonstrates how the methods can actually be applied for problem solving. As with other parts of the book, applications are drawn from all fields of engineering.

Finally, an epilogue is included at the end of Part Three. This review includes discussion of trade-offs that are relevant to implementation of the methods in engineering practice. This section also summarizes the important formulas and advanced methods related to linear algebraic equations. As such, it can be used before exams or as a refresher after you have graduated and must return to linear algebraic equations as a professional.

PT3.3.2 Goals and Objectives

Study Objectives. After completing Part Three, you should be able to solve problems involving linear algebraic equations and appreciate the application of these equations in many fields of engineering. You should strive to master several techniques and assess their reliability. You should understand the trade-offs involved in selecting the "best" method (or methods) for any particular problem. In addition to these general objectives, the specific concepts listed in Table PT3.1 should be assimilated and mastered.

Computer Objectives. Your most fundamental computer objectives are to be able to solve a system of linear algebraic equations and to evaluate the matrix inverse. You will

TABLE PT3.1 Specific study objectives for Part Three.

1. Understand the graphical interpretation of ill-conditioned systems and how it relates to the determinant.
2. Be familiar with terminology: forward elimination, back substitution, pivot equation, and pivot coefficient.
3. Understand the problems of division by zero, round-off error, and ill-conditioning.
4. Know how to compute the determinant using Gauss elimination.
5. Understand the advantages of pivoting; realize the difference between partial and complete pivoting.
6. Know the fundamental difference between Gauss elimination and the Gauss-Jordan method and which is more efficient.
7. Recognize how Gauss elimination can be formulated as an *LU* decomposition.
8. Know how to incorporate pivoting and matrix inversion into an *LU* decomposition algorithm.
9. Know how to interpret the elements of the matrix inverse in evaluating stimulus-response computations in engineering.
10. Realize how to use the inverse and matrix norms to evaluate system condition.
11. Understand how banded and symmetric systems can be decomposed and solved efficiently.
12. Understand why the Gauss-Seidel method is particularly well suited for large, sparse systems of equations.
13. Know how to assess diagonal dominance of a system of equations and how diagonal dominance relates to whether the system can be solved with the Gauss-Seidel method.
14. Understand the rationale behind relaxation; know where underrelaxation and overrelaxation are appropriate.

want to have subprograms developed for *LU* decomposition of both full and tridiagonal matrices. You may also want to have your own software to implement the Gauss-Seidel method.

You should know how to use packages to solve linear algebraic equations and find the matrix inverse. You should become familiar with how the same evaluations can be implemented on popular software packages such as Excel, MATLAB software, and Mathcad.

CHAPTER
9

Gauss Elimination

This chapter deals with simultaneous linear algebraic equations that can be represented generally as

$$a_{11}x_1 + a_{12}x_2 + \cdots + a_{1n}x_n = b_1$$
$$a_{21}x_1 + a_{22}x_2 + \cdots + a_{2n}x_n = b_2$$

$$\vdots \qquad \vdots$$

$$a_{n1}x_1 + a_{n2}x_2 + \cdots + a_{nn}x_n = b_n$$

(9.1)

where the a's are constant coefficients and the b's are constants.

The technique described in this chapter is called *Gauss elimination* because it involves combining equations to eliminate unknowns. Although it is one of the earliest methods for solving simultaneous equations, it remains among the most important algorithms in use today and is the basis for linear equation solving on many popular software packages.

9.1 SOLVING SMALL NUMBERS OF EQUATIONS

Before proceeding to the computer methods, we will describe several methods that are appropriate for solving small ($n \le 3$) sets of simultaneous equations and that do not require a computer. These are the graphical method, Cramer's rule, and the elimination of unknowns.

9.1.1 The Graphical Method

A graphical solution is obtainable for two equations by plotting them on Cartesian coordinates with one axis corresponding to x_1 and the other to x_2. Because we are dealing with linear systems, each equation is a straight line. This can be easily illustrated for the general equations

$$a_{11}x_1 + a_{12}x_2 = b_1$$
$$a_{21}x_1 + a_{22}x_2 = b_2$$

Both equations can be solved for x_2:

$$x_2 = -\left(\frac{a_{11}}{a_{12}}\right)x_1 + \frac{b_1}{a_{12}}$$

$$x_2 = -\left(\frac{a_{21}}{a_{22}}\right)x_1 + \frac{b_2}{a_{22}}$$

Thus, the equations are now in the form of straight lines; that is, $x_2 = (\text{slope})x_1 + \text{intercept}$. These lines can be graphed on Cartesian coordinates with x_2 as the ordinate and x_1 as the abscissa. The values of x_1 and x_2 at the intersection of the lines represent the solution.

EXAMPLE 9.1

The Graphical Method for Two Equations

Problem Statement.   Use the graphical method to solve

$$3x_1 + 2x_2 = 18 \qquad\qquad\qquad (E9.1.1)$$
$$-x_1 + 2x_2 = 2 \qquad\qquad\qquad (E9.1.2)$$

Solution.   Let x_1 be the abscissa. Solve Eq. (E9.1.1) for x_2:

$$x_2 = -\frac{3}{2}x_1 + 9$$

which, when plotted on Fig. 9.1, is a straight line with an intercept of 9 and a slope of $-3/2$.

FIGURE 9.1
Graphical solution of a set of two simultaneous linear algebraic equations. The intersection of the lines represents the solution.

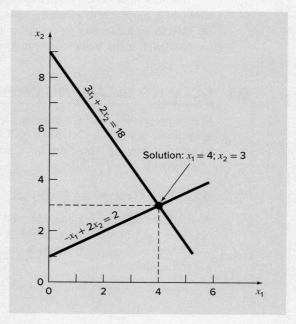

Equation (E9.1.2) can also be solved for x_2:

$$x_2 = \frac{1}{2}x_1 + 1$$

which is also plotted on Fig. 9.1. The solution is the intersection of the two lines at $x_1 = 4$ and $x_2 = 3$. This result can be checked by substituting these values into the original equations to yield

$$3(4) + 2(3) = 18$$
$$-(4) + 2(3) = 2$$

These results are equivalent to the right-hand sides of the original equations.

For three simultaneous equations, each equation would be represented by a plane in a three-dimensional coordinate system. The point where the three planes intersect would represent the solution. Beyond three equations, graphical methods break down and, consequently, have little practical value for solving simultaneous equations. However, they sometimes prove useful in visualizing properties of the solutions. For example, Fig. 9.2 depicts three cases that can pose problems when solving sets of linear equations. Figure 9.2a shows the case where the two equations represent parallel lines. For such situations, there is no solution because the lines never cross. Figure 9.2b depicts the case where the two lines are coincident. For such situations there is an infinite number of solutions. Both types of systems are said to be *singular*. In addition, systems that are very close to being singular (Fig. 9.2c) can also cause problems. These systems are said to be *ill-conditioned*. Graphically, this corresponds to the fact that it is difficult to identify the exact point at which the lines intersect. Ill-conditioned systems will also pose problems when they are encountered during the numerical solution of linear equations. This is because they will be extremely sensitive to round-off error (recall Sec. 4.2.3).

FIGURE 9.2
Graphical depiction of singular and ill-conditioned systems: (a) no solution, (b) infinite solutions, and (c) ill-conditioned system where the slopes are so close that the point of intersection is difficult to detect visually.

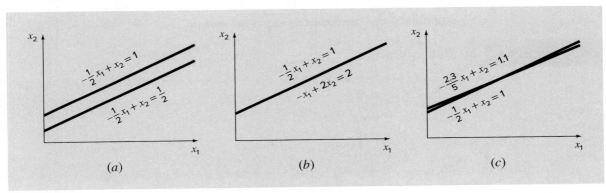

9.1.2 Determinants and Cramer's Rule

Cramer's rule is another solution technique that is best suited to small numbers of equations. Before describing this method, we will briefly introduce the concept of the determinant, which is used to implement Cramer's rule. In addition, the determinant has relevance to the evaluation of the ill-conditioning of a matrix.

Determinants.   The determinant can be illustrated for a set of three equations:

$$[A]\{X\} = \{B\}$$

where $[A]$ is the coefficient matrix:

$$[A] = \begin{bmatrix} a_{11} & a_{12} & a_{13} \\ a_{21} & a_{22} & a_{23} \\ a_{31} & a_{32} & a_{33} \end{bmatrix}$$

The determinant D of this system is formed from the coefficients of the equations, as in

$$D = \begin{vmatrix} a_{11} & a_{12} & a_{13} \\ a_{21} & a_{22} & a_{23} \\ a_{31} & a_{32} & a_{33} \end{vmatrix} \tag{9.2}$$

Although the determinant D and the coefficient matrix $[A]$ are composed of the same elements, they are completely different mathematical concepts. That is why they are distinguished visually by using brackets to enclose the matrix and straight lines to enclose the determinant. In contrast to a matrix, the determinant is a single number. For example, the value of the second-order determinant

$$D = \begin{vmatrix} a_{11} & a_{12} \\ a_{21} & a_{22} \end{vmatrix}$$

is calculated by

$$D = a_{11}a_{22} - a_{12}a_{21} \tag{9.3}$$

For the third-order case [Eq. (9.2)], a single numerical value for the determinant can be computed as

$$D = a_{11}\begin{vmatrix} a_{22} & a_{23} \\ a_{32} & a_{33} \end{vmatrix} - a_{12}\begin{vmatrix} a_{21} & a_{23} \\ a_{31} & a_{33} \end{vmatrix} + a_{13}\begin{vmatrix} a_{21} & a_{22} \\ a_{31} & a_{32} \end{vmatrix} \tag{9.4}$$

where the 2 by 2 determinants are called ***minors***.

EXAMPLE 9.2   **Determinants**

Problem Statement.   Compute values for the determinants of the systems represented in Figs. 9.1 and 9.2.

Solution.   For Fig. 9.1:

$$D = \begin{vmatrix} 3 & 2 \\ -1 & 2 \end{vmatrix} = 3(2) - 2(-1) = 8$$

For Fig. 9.2a:

$$D = \begin{vmatrix} -1/2 & 1 \\ -1/2 & 1 \end{vmatrix} = \frac{-1}{2}(1) - 1\left(\frac{-1}{2}\right) = 0$$

For Fig. 9.2b:

$$D = \begin{vmatrix} -1/2 & 1 \\ -1 & 2 \end{vmatrix} = \frac{-1}{2}(2) - 1(-1) = 0$$

For Fig. 9.2c:

$$D = \begin{vmatrix} -1/2 & 1 \\ -2.3/5 & 1 \end{vmatrix} = \frac{-1}{2}(1) - 1\left(\frac{-2.3}{5}\right) = -0.04$$

In the foregoing example, the singular systems had zero determinants. Additionally, the results suggest that the system that is almost singular (Fig. 9.2c) has a determinant that is close to zero. These ideas will be pursued further in our subsequent discussion of ill-conditioning (Sec. 9.3.3).

Cramer's Rule. This rule states that each unknown in a system of linear algebraic equations may be expressed as a fraction of two determinants with denominator D and with the numerator obtained from D by replacing the column of coefficients of the unknown in question by the constants $b_1, b_2, \ldots, b_n$. For example, x_1 would be computed as

$$x_1 = \frac{\begin{vmatrix} b_1 & a_{12} & a_{13} \\ b_2 & a_{22} & a_{23} \\ b_3 & a_{32} & a_{33} \end{vmatrix}}{D} \tag{9.5}$$

EXAMPLE 9.3 | **Cramer's Rule**

Problem Statement. Use Cramer's rule to solve

$$0.3x_1 + 0.52x_2 + x_3 = -0.01$$
$$0.5x_1 + x_2 + 1.9x_3 = 0.67$$
$$0.1x_1 + 0.3x_2 + 0.5x_3 = -0.44$$

Solution. The determinant D can be written as [Eq. (9.2)]

$$D = \begin{vmatrix} 0.3 & 0.52 & 1 \\ 0.5 & 1 & 1.9 \\ 0.1 & 0.3 & 0.5 \end{vmatrix}$$

The minors are [Eq. (9.3)]

$$A_1 = \begin{vmatrix} 1 & 1.9 \\ 0.3 & 0.5 \end{vmatrix} = 1(0.5) - 1.9(0.3) = -0.07$$

$$A_2 = \begin{vmatrix} 0.5 & 1.9 \\ 0.1 & 0.5 \end{vmatrix} = 0.5(0.5) - 1.9(0.1) = 0.06$$

$$A_3 = \begin{vmatrix} 0.5 & 1 \\ 0.1 & 0.3 \end{vmatrix} = 0.5(0.3) - 1(0.1) = 0.05$$

These can be used to evaluate the determinant, as in [Eq. (9.4)]

$$D = 0.3(-0.07) - 0.52(0.06) + 1(0.05) = -0.0022$$

Applying Eq. (9.5), the solution is

$$x_1 = \frac{\begin{vmatrix} -0.01 & 0.52 & 1 \\ 0.67 & 1 & 1.9 \\ -0.44 & 0.3 & 0.5 \end{vmatrix}}{-0.0022} = \frac{0.03278}{-0.0022} = -14.9$$

$$x_2 = \frac{\begin{vmatrix} 0.3 & -0.01 & 1 \\ 0.5 & 0.67 & 1.9 \\ 0.1 & -0.44 & 0.5 \end{vmatrix}}{-0.0022} = \frac{0.0649}{-0.0022} = -29.5$$

$$x_3 = \frac{\begin{vmatrix} 0.3 & 0.52 & -0.01 \\ 0.5 & 1 & 0.67 \\ 0.1 & 0.3 & -0.44 \end{vmatrix}}{-0.0022} = \frac{-0.04356}{-0.0022} = 19.8$$

For more than three equations, Cramer's rule becomes impractical because, as the number of equations increases, the determinants are time-consuming to evaluate by hand (or by computer). Consequently, more efficient alternatives are used. Some of these alternatives are based on the last noncomputer solution technique, covered in the next section—the elimination of unknowns.

9.1.3 The Elimination of Unknowns

The elimination of unknowns by combining equations is an algebraic approach that can be illustrated for a set of two equations:

$$a_{11}x_1 + a_{12}x_2 = b_1 \tag{9.6}$$

$$a_{21}x_1 + a_{22}x_2 = b_2 \tag{9.7}$$

The basic strategy is to multiply the equations by constants so that one of the unknowns will be eliminated when the two equations are combined. The result is a single equation that can be solved for the remaining unknown. This value can then be substituted into either of the original equations to compute the other variable.

For example, Eq. (9.6) might be multiplied by a_{21} and Eq. (9.7) by a_{11} to give

$$a_{11}a_{21}x_1 + a_{12}a_{21}x_2 = b_1 a_{21} \tag{9.8}$$

$$a_{21}a_{11}x_1 + a_{22}a_{11}x_2 = b_2 a_{11} \tag{9.9}$$

Subtracting Eq. (9.8) from Eq. (9.9) will, therefore, eliminate the x_1 term from the equations to yield

$$a_{22}a_{11}x_2 - a_{12}a_{21}x_2 = b_2a_{11} - b_1a_{21}$$

which can be solved for

$$x_2 = \frac{a_{11}b_2 - a_{21}b_1}{a_{11}a_{22} - a_{12}a_{21}} \tag{9.10}$$

Equation (9.10) can then be substituted into Eq. (9.6), which can be solved for

$$x_1 = \frac{a_{22}b_1 - a_{12}b_2}{a_{11}a_{22} - a_{12}a_{21}} \tag{9.11}$$

Notice that Eqs. (9.10) and (9.11) follow directly from Cramer's rule, which states

$$x_1 = \frac{\begin{vmatrix} b_1 & a_{12} \\ b_2 & a_{22} \end{vmatrix}}{\begin{vmatrix} a_{11} & a_{12} \\ a_{21} & a_{22} \end{vmatrix}} = \frac{b_1a_{22} - a_{12}b_2}{a_{11}a_{22} - a_{12}a_{21}}$$

$$x_2 = \frac{\begin{vmatrix} a_{11} & b_1 \\ a_{21} & b_2 \end{vmatrix}}{\begin{vmatrix} a_{11} & a_{12} \\ a_{21} & a_{22} \end{vmatrix}} = \frac{a_{11}b_2 - b_1a_{21}}{a_{11}a_{22} - a_{12}a_{21}}$$

EXAMPLE 9.4 **Elimination of Unknowns**

Problem Statement. Use the elimination of unknowns to solve (recall Example 9.1)

$$3x_1 + 2x_2 = 18$$
$$-x_1 + 2x_2 = 2$$

Solution. Using Eqs. (9.11) and (9.10),

$$x_1 = \frac{2(18) - 2(2)}{3(2) - 2(-1)} = 4$$

$$x_2 = \frac{3(2) - (-1)18}{3(2) - 2(-1)} = 3$$

which is consistent with our graphical solution (Fig. 9.1).

The elimination of unknowns can be extended to systems with more than two or three equations. However, the numerous calculations that are required for larger systems make the method extremely tedious to implement by hand. However, as described in the next section, the technique can be formalized and readily programmed for the computer.

9.2 NAIVE GAUSS ELIMINATION

In the previous section, the elimination of unknowns was used to solve a pair of simultaneous equations. The procedure consisted of two steps:

1. The equations were manipulated to eliminate one of the unknowns from the equations. The result of this *elimination* step was that we had one equation with one unknown.
2. Consequently, this equation could be solved directly and the result *back-substituted* into one of the original equations to solve for the remaining unknown.

This basic approach can be extended to large sets of equations by developing a systematic scheme or algorithm to eliminate unknowns and to back-substitute. *Gauss elimination* is the most basic of these schemes.

This section includes the systematic techniques for forward elimination and back substitution that comprise Gauss elimination. Although these techniques are ideally suited for implementation on computers, some modifications will be required to obtain a reliable algorithm. In particular, the computer program must avoid division by zero. The following method is called *"naive" Gauss elimination* because it does not avoid this problem. Subsequent sections will deal with the additional features required for an effective computer program.

The approach is designed to solve a general set of n equations:

$$a_{11}x_1 + a_{12}x_2 + a_{13}x_3 + \cdots + a_{1n}x_n = b_1 \qquad (9.12a)$$

$$a_{21}x_1 + a_{22}x_2 + a_{23}x_3 + \cdots + a_{2n}x_n = b_2 \qquad (9.12b)$$

$$\vdots \qquad \qquad \vdots$$

$$a_{n1}x_1 + a_{n2}x_2 + a_{n3}x_3 + \cdots + a_{nn}x_n = b_n \qquad (9.12c)$$

As was the case with the solution of two equations, the technique for n equations consists of two phases: elimination of unknowns and solution through back substitution.

Forward Elimination of Unknowns.   The first phase is designed to reduce the set of equations to an upper triangular system (Fig. 9.3). The initial step will be to eliminate the first unknown, x_1, from the second through the nth equations. To do this, multiply Eq. (9.12a) by a_{21}/a_{11} to give

$$a_{21}x_1 + \frac{a_{21}}{a_{11}}a_{12}x_2 + \cdots + \frac{a_{21}}{a_{11}}a_{1n}x_n = \frac{a_{21}}{a_{11}}b_1 \qquad (9.13)$$

Now, this equation can be subtracted from Eq. (9.12b) to give

$$\left(a_{22} - \frac{a_{21}}{a_{11}}a_{12}\right)x_2 + \cdots + \left(a_{2n} - \frac{a_{21}}{a_{11}}a_{1n}\right)x_n = b_2 - \frac{a_{21}}{a_{11}}b_1$$

or

$$a'_{22}x_2 + \cdots + a'_{2n}x_n = b'_2$$

where the prime indicates that the elements have been changed from their original values. The procedure is then repeated for the remaining equations. For instance, Eq. (9.12a) can be multiplied by a_{31}/a_{11} and the result subtracted from the third equation. Repeating

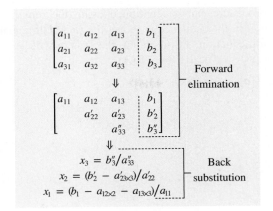

FIGURE 9.3
The two phases of Gauss elimination: forward elimination and back substitution. The primes indicate the number of times that the coefficients and constants have been modified.

the procedure for the remaining equations results in the following modified system:

$$a_{11}x_1 + a_{12}x_2 + a_{13}x_3 + \cdots + a_{1n}x_n = b_1 \tag{9.14a}$$

$$a'_{22}x_2 + a'_{23}x_3 + \cdots + a'_{2n}x_n = b'_2 \tag{9.14b}$$

$$a'_{32}x_2 + a'_{33}x_3 + \cdots + a'_{3n}x_n = b'_3 \tag{9.14c}$$

$$\vdots \qquad \qquad \vdots$$

$$a'_{n2}x_2 + a'_{n3}x_3 + \cdots + a'_{nn}x_n = b'_n \tag{9.14d}$$

For the foregoing steps, Eq. (9.12a) is called the *pivot equation* and a_{11} is called the *pivot coefficient* or *element*. Note that the process of multiplying the first row by a_{21}/a_{11} is equivalent to dividing it by a_{11} and multiplying it by a_{21}. Sometimes the division operation is referred to as normalization. We make this distinction because a zero pivot element can interfere with normalization by causing a division by zero. We will return to this important issue after we complete our description of naive Gauss elimination.

Now repeat the above to eliminate the second unknown from Eq. (9.14c) through (9.14d). To do this multiply Eq. (9.14b) by a'_{32}/a'_{22} and subtract the result from Eq. (9.14c). Perform a similar elimination for the remaining equations to yield

$$a_{11}x_1 + a_{12}x_2 + a_{13}x_3 + \cdots + a_{1n}x_n = b_1$$

$$a'_{22}x_2 + a'_{23}x_3 + \cdots + a'_{2n}x_n = b'_2$$

$$a''_{33}x_3 + \cdots + a''_{3n}x_n = b''_2$$

$$\vdots \qquad \qquad \vdots$$

$$a''_{n3}x_3 + \cdots + a''_{nn}x_n = b''_n$$

where the double prime indicates that the elements have been modified twice.

The procedure can be continued using the remaining pivot equations. The final manipulation in the sequence is to use the $(n - 1)$th equation to eliminate the x_{n-1} term from the nth equation. At this point, the system will have been transformed to an upper triangular system (recall Box PT3.1):

$$a_{11}x_1 + a_{12}x_2 + a_{13}x_3 + \cdots + a_{1n}x_n = b_1 \tag{9.15a}$$

$$a'_{22}x_2 + a'_{23}x_3 + \cdots + a'_{2n}x_n = b'_2 \tag{9.15b}$$

$$a''_{33}x_3 + \cdots + a''_{3n}x_n = b''_3 \tag{9.15c}$$

$$ \vdots \vdots$$

$$a_{nn}^{(n-1)}x_n = b_n^{(n-1)} \tag{9.15d}$$

Pseudocode to implement forward elimination is presented in Fig. 9.4a. Notice that three nested loops provide a concise representation of the process. The outer loop moves down the matrix from one pivot row to the next. The middle loop moves below the pivot row to each of the subsequent rows where elimination is to take place. Finally, the innermost loop progresses across the columns to eliminate or transform the elements of a particular row.

Back Substitution. Equation (9.15d) can now be solved for x_n:

$$x_n = \frac{b_n^{(n-1)}}{a_{nn}^{(n-1)}} \tag{9.16}$$

This result can be back-substituted into the $(n - 1)$th equation to solve for x_{n-1}. The procedure, which is repeated to evaluate the remaining x's, can be represented by the following formula:

$$x_i = \frac{b_i^{(i-1)} - \sum_{j=i+1}^{n} a_{ij}^{(i-1)}x_j}{a_{ii}^{(i-1)}} \qquad \text{for } i = n - 1, n - 2, \dots, 1 \tag{9.17}$$

FIGURE 9.4
Pseudocode to perform (a) forward elimination and (b) back substitution.

(a)
```
DOFOR k = 1, n - 1
    DOFOR i = k + 1, n
        factor = a_{i,k} / a_{k,k}
        DOFOR j = k + 1 to n
            a_{i,j} = a_{i,j} - factor · a_{k,j}
        END DO
        b_i = b_i - factor · b_k
    END DO
END DO
```

(b)
```
x_n = b_n / a_{n,n}
DOFOR i = n - 1, 1, -1
    sum = b_i
    DOFOR j = i + 1, n
        sum = sum - a_{i,j} · x_j
    END DO
    x_i = sum / a_{i,i}
END DO
```

Pseudocode to implement Eqs. (9.16) and (9.17) is presented in Fig. 9.4*b*. Notice the similarity between this pseudocode and that in Fig. PT3.4 for matrix multiplication. As with Fig. PT3.4, a temporary variable, *sum*, is used to accumulate the summation from Eq. (9.17). This results in a somewhat faster execution time than if the summation were accumulated in b_i. More importantly, it allows efficient improvement in precision if the variable *sum* is declared in double precision.

EXAMPLE 9.5   **Naive Gauss Elimination**

Problem Statement.   Use Gauss elimination to solve

$$3x_1 - 0.1x_2 - 0.2x_3 = 7.85 \tag{E9.5.1}$$
$$0.1x_1 + 7x_2 - 0.3x_3 = -19.3 \tag{E9.5.2}$$
$$0.3x_1 - 0.2x_2 + 10x_3 = 71.4 \tag{E9.5.3}$$

Carry six significant figures during the computation.

Solution.   The first part of the procedure is forward elimination. Multiply Eq. (E9.5.1) by 0.1/3 and subtract the result from Eq. (E9.5.2) to give

$$7.00333x_2 - 0.293333x_3 = -19.5617$$

Then multiply Eq. (E9.5.1) by 0.3/3 and subtract it from Eq. (E9.5.3) to eliminate x_1. After these operations, the set of equations is

$$3x_1 - \quad 0.1x_2 - \quad 0.2x_3 = 7.85 \tag{E9.5.4}$$
$$7.00333x_2 - 0.293333x_3 = -19.5617 \tag{E9.5.5}$$
$$- 0.190000x_2 + \quad 10.0200x_3 = 70.6150 \tag{E9.5.6}$$

To complete the forward elimination, x_2 must be removed from Eq. (E9.5.6). To accomplish this, multiply Eq. (E9.5.5) by $-0.190000/7.00333$ and subtract the result from Eq. (E9.5.6). This eliminates x_2 from the third equation and reduces the system to an upper triangular form, as in

$$3x_1 - \quad 0.1x_2 - \quad 0.2x_3 = 7.85 \tag{E9.5.7}$$
$$7.00333x_2 - 0.293333x_3 = -19.5617 \tag{E9.5.8}$$
$$10.0120x_3 = 70.0843 \tag{E9.5.9}$$

We can now solve these equations by back substitution. First, Eq. (E9.5.9) can be solved for

$$x_3 = \frac{70.0843}{10.0120} = 7.0000 \tag{E9.5.10}$$

This result can be back-substituted into Eq. (E9.5.8):

$$7.00333x_2 - 0.293333(7.0000) = -19.5617$$

which can be solved for

$$x_2 = \frac{-19.5617 + 0.293333(7.0000)}{7.00333} = -2.50000 \tag{E9.5.11}$$

Finally, Eqs. (E9.5.10) and (E9.5.11) can be substituted into Eq. (E9.5.4):

$$3x_1 - 0.1(-2.50000) - 0.2(7.0000) = 7.85$$

which can be solved for

$$x_1 = \frac{7.85 + 0.1(-2.50000) + 0.2(7.0000)}{3} = 3.00000$$

The results are identical to the exact solution of $x_1 = 3$, $x_2 = -2.5$, and $x_3 = 7$. This can be verified by substituting the results into the original equation set:

$$3(3) - 0.1(-2.5) - 0.2(7) = 7.85$$
$$0.1(3) + 7(-2.5) - 0.3(7) = -19.3$$
$$0.3(3) - 0.2(-2.5) + 10(7) = 71.4$$

9.2.1 Operation Counting

The execution time of Gauss elimination depends on the amount of *floating-point operations* (or *flops*) involved in the algorithm. On modern computers using math coprocessors, the time consumed to perform addition/subtraction and multiplication/division is about the same. Therefore, totaling up these operations provides insight into which parts of the algorithm are most time-consuming and how computation time increases as the system gets larger.

Before analyzing naive Gauss elimination, we will first define some quantities that facilitate operation counting:

$$\sum_{i=1}^{m} cf(i) = c \sum_{i=1}^{m} f(i) \qquad \sum_{i=1}^{m} f(i) + g(i) = \sum_{i=1}^{m} f(i) + \sum_{i=1}^{m} g(i) \qquad (9.18a,b)$$

$$\sum_{i=1}^{m} 1 = 1 + 1 + 1 + \cdots + 1 = m \qquad \sum_{i=k}^{m} 1 = m - k + 1 \qquad (9.18c,d)$$

$$\sum_{i=1}^{m} i = 1 + 2 + 3 + \cdots + m = \frac{m(m+1)}{2} = \frac{m^2}{2} + O(m) \qquad (9.18e)$$

$$\sum_{i=1}^{m} i^2 = 1^2 + 2^2 + 3^2 + \cdots + m^2 = \frac{m(m+1)(2m+1)}{6} = \frac{m^3}{3} + O(m^2) \qquad (9.18f)$$

where $O(m^n)$ means "terms of order m^n and lower."

Now let us examine the naive Gauss elimination algorithm (Fig. 9.4a) in detail. We will first count the flops in the elimination stage. On the first pass through the outer loop, $k = 1$. Therefore, the limits on the middle loop are from $i = 2$ to n. According to Eq. (9.18d), this means that the number of iterations of the middle loop will be

$$\sum_{i=2}^{n} 1 = n - 2 + 1 = n - 1 \qquad (9.19)$$

For every one of these iterations, there is one division to define the factor. The interior loop then performs a single multiplication and subtraction for each iteration from $j = 2$ to n. Finally, there is one additional multiplication and subtraction for the right-hand-side value.

Thus, for every iteration of the middle loop, the number of multiplications is

$$1 + [n - 2 + 1] + 1 = 1 + n \tag{9.20}$$

The total multiplications for the first pass through the outer loop is therefore obtained by multiplying Eq. (9.19) by (9.20) to give $[n - 1](1 + n)$. In like fashion, the number of subtractions is computed as $[n - 1](n)$.

Similar reasoning can be used to estimate the flops for the subsequent iterations of the outer loop. These can be summarized as

Outer Loop k	Middle Loop i	Addition/Subtraction Flops	Multiplication/Division Flops
1	2, n	$(n-1)(n)$	$(n-1)(n+1)$
2	3, n	$(n-2)(n-1)$	$(n-2)(n)$
.	.	.	
.	.	.	
.	.	.	
k	$k+1$, n	$(n-k)(n+1-k)$	$(n-k)(n+2-k)$
.	.		
.	.		
.	.		
$n-1$	n, n	$(1)(2)$	$(1)(3)$

Therefore, the total addition/subtraction flops for elimination can be computed as

$$\sum_{k=1}^{n-1}(n-k)(n+1-k) = \sum_{k=1}^{n-1}[n(n+1) - k(2n+1) + k^2]$$

or

$$n(n+1)\sum_{k=1}^{n-1}1 - (2n+1)\sum_{k=1}^{n-1}k + \sum_{k=1}^{n-1}k^2$$

Applying some of the relationships from Eq. (9.18) yields

$$[n^3 + O(n)] - [n^3 + O(n^2)] + \left[\frac{1}{3}n^3 + O(n^2)\right] = \frac{n^3}{3} + O(n) \tag{9.21}$$

A similar analysis for the multiplication/division flops yields

$$[n^3 + O(n^2)] - [n^3 + O(n)] + \left[\frac{1}{3}n^3 + O(n^2)\right] = \frac{n^3}{3} + O(n^2) \tag{9.22}$$

Summing these results gives

$$\frac{2n^3}{3} + O(n^2)$$

Thus, the total number of flops is equal to $2n^3/3$ plus an additional component proportional to terms of order n^2 and lower. The result is written in this way because as n gets large, the $O(n^2)$ and lower terms become negligible. We are therefore justified in concluding that for large n, the effort involved in forward elimination converges on $2n^3/3$.

Because only a single loop is used, back substitution is much simpler to evaluate. The number of addition/subtraction flops is equal to $n(n-1)/2$. Because of the extra

TABLE 9.1 Number of Flops for Gauss Elimination.

n	Elimination	Back Substitution	Total Flops	$2n^3/3$	Percent Due to Elimination
10	705	100	805	667	87.58%
100	671550	10000	681550	666667	98.53%
1000	6.67×10^8	1×10^6	6.68×10^8	6.67×10^8	99.85%

division prior to the loop, the number of multiplication/division flops is $n(n + 1)/2$. These can be added to arrive at a total of

$$n^2 + O(n)$$

Thus, the total effort in naive Gauss elimination can be represented as

$$\underbrace{\frac{2n^3}{3} + O(n^2)}_{\substack{\text{Forward} \\ \text{elimination}}} + \underbrace{n^2 + O(n)}_{\substack{\text{Backward} \\ \text{substitution}}} \xrightarrow{\text{as } n \text{ increases}} \frac{2n^3}{3} + O(n^2) \tag{9.23}$$

Two useful general conclusions can be drawn from this analysis:

1. As the system gets larger, the computation time increases greatly. As in Table 9.1, the amount of flops increases nearly three orders of magnitude for every order of magnitude increase in the dimension.
2. Most of the effort is incurred in the elimination step. Thus, efforts to make the method more efficient should probably focus on this step.

9.3 PITFALLS OF ELIMINATION METHODS

Whereas there are many systems of equations that can be solved with naive Gauss elimination, there are some pitfalls that must be explored before writing a general computer program to implement the method. Although the following material relates directly to naive Gauss elimination, the information is relevant for other elimination techniques as well.

9.3.1 Division by Zero

The primary reason that the foregoing technique is called "naive" is that during both the elimination and the back-substitution phases, it is possible that a division by zero can occur. For example, if we use naive Gauss elimination to solve

$$2x_2 + 3x_3 = 8$$
$$4x_1 + 6x_2 + 7x_3 = -3$$
$$2x_1 + x_2 + 6x_3 = 5$$

the normalization of the first row would involve division by $a_{11} = 0$. Problems also can arise when a coefficient is very close to zero. The technique of *pivoting* has been developed to partially avoid these problems. It will be described in Sec. 9.4.2.

9.3.2 Round-Off Errors

Even though the solution in Example 9.5 was close to the true answer, there was a slight discrepancy in the result for x_3 [Eq. (E9.5.10)]. This discrepancy, which amounted to a relative error of -0.00043%, was due to our use of six significant figures during the computation. If we had used more significant figures, the error in the results would be reduced further. If we had used fractions instead of decimals (and consequently avoided round-off altogether), the answers would have been exact. However, because computers carry only a limited number of significant figures (recall Sec. 3.4.1), round-off errors can occur and must be considered when evaluating the results.

The problem of round-off error can become particularly important when large numbers of equations are to be solved. This is due to the fact that every result is dependent on previous results. Consequently, an error in the early steps will tend to propagate—that is, it will cause errors in subsequent steps.

Specifying the system size where round-off error becomes significant is complicated by the fact that the type of computer and the properties of the equations are determining factors. A rough rule of thumb is that round-off error may be important when dealing with 100 or more equations. In any event, you should always substitute your answers back into the original equations to check whether a substantial error has occurred. However, as discussed below, the magnitudes of the coefficients themselves can influence whether such an error check ensures a reliable result.

9.3.3 Ill-Conditioned Systems

The adequacy of the solution depends on the condition of the system. In Sec. 9.1.1, a graphical depiction of system condition was developed. As discussed in Sec. 4.2.3, *well-conditioned systems* are those where a small change in one or more of the coefficients results in a similar small change in the solution. *Ill-conditioned systems* are those where small changes in coefficients result in large changes in the solution. An alternative interpretation of ill-conditioning is that a wide range of answers can approximately satisfy the equations. Because round-off errors can induce small changes in the coefficients, these artificial changes can lead to large solution errors for ill-conditioned systems, as illustrated in the following example.

EXAMPLE 9.6 | **Ill-Conditioned Systems**

Problem Statement.   Solve the following system:

$$x_1 + 2x_2 = 10 \tag{E9.6.1}$$
$$1.1x_1 + 2x_2 = 10.4 \tag{E9.6.2}$$

Then, solve it again, but with the coefficient of x_1 in the second equation modified slightly to 1.05.

Solution.   Using Eqs. (9.10) and (9.11), the solution is

$$x_1 = \frac{2(10) - 2(10.4)}{1(2) - 2(1.1)} = 4$$
$$x_2 = \frac{1(10.4) - 1.1(10)}{1(2) - 2(1.1)} = 3$$

However, with the slight change of the coefficient a_{21} from 1.1 to 1.05, the result is changed dramatically to

$$x_1 = \frac{2(10) - 2(10.4)}{1(2) - 2(1.05)} = 8$$

$$x_2 = \frac{1(10.4) - 1.1(10)}{1(2) - 2(1.05)} = 1$$

Notice that the primary reason for the discrepancy between the two results is that the denominator represents the difference of two almost equal numbers. As illustrated previously in Sec. 3.4.2, such differences are highly sensitive to slight variations in the numbers being manipulated.

At this point, you might suggest that substitution of the results into the original equations would alert you to the problem. Unfortunately, for ill-conditioned systems this is often not the case. Substitution of the erroneous values of $x_1 = 8$ and $x_2 = 1$ into Eqs. (E9.6.1) and (E9.6.2) yields

$$8 + 2(1) = 10 = 10$$

$$1.1(8) + 2(1) = 10.8 \cong 10.4$$

Therefore, although $x_1 = 8$ and $x_2 = 1$ is not the true solution to the original problem, the error check is close enough to possibly mislead you into believing that your solutions are adequate.

As was done previously in the section on graphical methods, a visual representative of ill-conditioning can be developed by plotting Eqs. (E9.6.1) and (E9.6.2) (recall Fig. 9.2). Because the slopes of the lines are almost equal, it is visually difficult to see exactly where they intersect. This visual difficulty is reflected quantitatively in the nebulous results of Example 9.6. We can mathematically characterize this situation by writing the two equations in general form:

$$a_{11}x_1 + a_{12}x_2 = b_1 \tag{9.24}$$

$$a_{21}x_1 + a_{22}x_2 = b_2 \tag{9.25}$$

Dividing Eq. (9.24) by a_{12} and Eq. (9.25) by a_{22} and rearranging yields alternative versions that are in the format of straight lines [$x_2 = (\text{slope})x_1 + \text{intercept}$]:

$$x_2 = -\frac{a_{11}}{a_{12}}x_1 + \frac{b_1}{a_{12}}$$

$$x_2 = -\frac{a_{21}}{a_{22}}x_1 + \frac{b_2}{a_{22}}$$

Consequently, if the slopes are nearly equal,

$$\frac{a_{11}}{a_{12}} \cong \frac{a_{21}}{a_{22}}$$

or, cross-multiplying,

$$a_{11}a_{22} \cong a_{12}a_{21}$$

which can be also expressed as

$$a_{11}a_{22} - a_{12}a_{21} \cong 0 \tag{9.26}$$

Now, recalling that $a_{11}a_{22} - a_{12}a_{21}$ is the determinant of a two-dimensional system [Eq. (9.3)], we arrive at the general conclusion that an ill-conditioned system is one with a determinant close to zero. In fact, if the determinant is exactly zero, the two slopes are identical, which connotes either no solution or an infinite number of solutions, as is the case for the singular systems depicted in Fig. 9.2a and b.

It is difficult to specify how close to zero the determinant must be to indicate ill-conditioning. This is complicated by the fact that the determinant can be changed by multiplying one or more of the equations by a scale factor without changing the solution. Consequently, the determinant is a relative value that is influenced by the magnitude of the coefficients.

EXAMPLE 9.7 | **Effect of Scale on the Determinant**

Problem Statement.   Evaluate the determinant of the following systems:

(a) From Example 9.1:

$$3x_1 + 2x_2 = 18 \tag{E9.7.1}$$
$$-x_1 + 2x_2 = 2 \tag{E9.7.2}$$

(b) From Example 9.6:

$$x_1 + 2x_2 = 10 \tag{E9.7.3}$$
$$1.1x_1 + 2x_2 = 10.4 \tag{E9.7.4}$$

(c) Repeat **(b)** but with the equations multiplied by 10.

Solution.

(a) The determinant of Eqs. (E9.7.1) and (E9.7.2), which are well-conditioned, is

$$D = 3(2) - 2(-1) = 8$$

(b) The determinant of Eqs. (E9.7.3) and (E9.7.4), which are ill-conditioned, is

$$D = 1(2) - 2(1.1) = -0.2$$

(c) The results of **(a)** and **(b)** seem to bear out the contention that ill-conditioned systems have near-zero determinants. However, suppose that the ill-conditioned system in **(b)** is multiplied by 10 to give

$$10x_1 + 20x_2 = 100$$
$$11x_1 + 20x_2 = 104$$

The multiplication of an equation by a constant has no effect on its solution. In addition, it is still ill-conditioned. This can be verified by the fact that multiplying by a constant has no effect on the graphical solution. However, the determinant is dramatically affected:

$$D = 10(20) - 20(11) = -20$$

Not only has it been raised two orders of magnitude, but it is now over twice as large as the determinant of the well-conditioned system in **(a)**.

As illustrated by the previous example, the magnitude of the coefficients interjects a scale effect that complicates the relationship between system condition and determinant size. One way to partially circumvent this difficulty is to scale the equations so that the maximum element in any row is equal to 1.

EXAMPLE 9.8   Scaling

Problem Statement.   Scale the systems of equations in Example 9.7 to a maximum value of 1 and recompute their determinants.

Solution.

(a) For the well-conditioned system, scaling results in

$$x_1 + 0.667x_2 = 6$$
$$-0.5x_1 + \quad x_2 = 1$$

for which the determinant is

$$D = 1(1) - 0.667(-0.5) = 1.333$$

(b) For the ill-conditioned system, scaling gives

$$0.5x_1 + x_2 = 5$$
$$0.55x_1 + x_2 = 5.2$$

for which the determinant is

$$D = 0.5(1) - 1(0.55) = -0.05$$

(c) For the last case, scaling changes the system to the same form as in **(b)** and the determinant is also −0.05. Thus, the scale effect is removed.

In a previous section (Sec. 9.1.2), we suggested that the determinant is difficult to compute for more than three simultaneous equations. Therefore, it might seem that it does not provide a practical means for evaluating system condition. However, as described in Box 9.1, there is a simple algorithm that results from Gauss elimination that can be used to evaluate the determinant.

Aside from the approach used in the previous example, there are a variety of other ways to evaluate system condition. For example, there are alternative methods for normalizing the elements (see Stark 1970). In addition, as described in the next chapter (Sec. 10.3), the matrix inverse and matrix norms can be employed to evaluate system condition. Finally, a simple (but time-consuming) test is to modify the coefficients slightly and

---

Box 9.1 Determinant Evaluation Using Gauss Elimination

In Sec. 9.1.2, we stated that determinant evaluation by expansion of minors was impractical for large sets of equations. Thus, we concluded that Cramer's rule would be applicable only to small systems. However, as mentioned in Sec. 9.3.3, the determinant has value in assessing system condition. It would, therefore, be useful to have a practical method for computing this quantity.

Fortunately, Gauss elimination provides a simple way to do this. The method is based on the fact that the determinant of a triangular matrix can be simply computed as the product of its diagonal elements:

$$D = a_{11}a_{22}a_{33} \cdots a_{nn} \qquad (B9.1.1)$$

The validity of this formulation can be illustrated for a 3 by 3 system:

$$D = \begin{vmatrix} a_{11} & a_{12} & a_{13} \\ 0 & a_{22} & a_{23} \\ 0 & 0 & a_{33} \end{vmatrix}$$

where the determinant can be evaluated as [recall Eq. (9.4)]

$$D = a_{11}\begin{vmatrix} a_{22} & a_{23} \\ 0 & a_{33} \end{vmatrix} - a_{12}\begin{vmatrix} 0 & a_{23} \\ 0 & a_{33} \end{vmatrix} + a_{13}\begin{vmatrix} 0 & a_{22} \\ 0 & 0 \end{vmatrix}$$

or, by evaluating the minors (that is, the 2 by 2 determinants),

$$D = a_{11}a_{22}a_{33} - a_{12}(0) + a_{13}(0) = a_{11}a_{12}a_{33}$$

Recall that the forward-elimination step of Gauss elimination results in an upper triangular system. Because the value of the determinant is not changed by the forward-elimination process, the determinant can be simply evaluated at the end of this step via

$$D = a_{11}a'_{22}a''_{33} \cdots a_{nn}^{(n-1)} \qquad (B9.1.2)$$

where the superscripts signify the number of times that the elements have been modified by the elimination process. Thus, we can capitalize on the effort that has already been expended in reducing the system to triangular form and, in the bargain, come up with a simple estimate of the determinant.

There is a slight modification to the above approach when the program employs partial pivoting (Sec. 9.4.2). For this case, the determinant changes sign every time a row is pivoted. One way to represent this is to modify Eq. (B9.1.2):

$$D = a_{11}a'_{22}a''_{33} \cdots a_{nn}^{(n-1)}(-1)^p \qquad (B9.1.3)$$

where p represents the number of times that rows are pivoted. This modification can be incorporated simply into a program; merely keep track of the number of pivots that take place during the course of the computation and then use Eq. (B9.1.3) to evaluate the determinant.

---

repeat the solution. If such modifications lead to drastically different results, the system is likely to be ill-conditioned.

As you might gather from the foregoing discussion, ill-conditioned systems are problematic. Fortunately, most linear algebraic equations derived from engineering problem settings are naturally well-conditioned. In addition, some of the techniques outlined in Sec. 9.4 help to alleviate the problem.

9.3.4 Singular Systems

In the previous section, we learned that one way in which a system of equations can be ill-conditioned is when two or more of the equations are nearly identical. Obviously, it is even worse when the equations are identical. In such cases, we would lose one degree of freedom, and would be dealing with the impossible case of $n - 1$ equations with n unknowns. Such cases might not be obvious to you, particularly when dealing with large equation sets. Consequently, it would be nice to have some way of automatically detecting singularity.

The answer to this problem is neatly offered by the fact that the determinant of a singular system is zero. This idea can, in turn, be connected to Gauss elimination by recognizing that after the elimination step, the determinant can be evaluated as the product of the diagonal elements (recall Box 9.1). Thus, a computer algorithm can test to discern whether a zero diagonal element is created during the elimination stage. If one is discovered, the calculation can be immediately terminated and a message displayed alerting the user. We will show the details of how this is done when we present a full algorithm for Gauss elimination later in this chapter.

9.4 TECHNIQUES FOR IMPROVING SOLUTIONS

The following techniques can be incorporated into the naive Gauss elimination algorithm to circumvent some of the pitfalls discussed in the previous section.

9.4.1 Use of More Significant Figures

The simplest remedy for ill-conditioning is to use more significant figures in the computation. If your application can be extended to handle larger word size, such a feature will greatly reduce the problem. However, a price must be paid in the form of the computational and memory overhead connected with using extended precision (recall Sec. 3.4.1).

9.4.2 Pivoting

As mentioned at the beginning of Sec. 9.3, obvious problems occur when a pivot element is zero because the normalization step leads to division by zero. Problems may also arise when the pivot element is close to, rather than exactly equal to, zero because if the magnitude of the pivot element is small compared to the other elements, then round-off errors can be introduced.

Therefore, before each row is normalized, it is advantageous to determine the largest available coefficient in the column below the pivot element. The rows can then be switched so that the largest element is the pivot element. This is called *partial pivoting*. If columns as well as rows are searched for the largest element and then switched, the procedure is called *complete pivoting*. Complete pivoting is rarely used because switching columns changes the order of the x's and, consequently, adds significant and usually unjustified complexity to the computer program. The following example illustrates the advantages of partial pivoting. Aside from avoiding division by zero, pivoting also minimizes round-off error. As such, it also serves as a partial remedy for ill-conditioning.

EXAMPLE 9.9   **Partial Pivoting**

Problem Statement.   Use Gauss elimination to solve

$$0.0003x_1 + 3.0000x_2 = 2.0001$$
$$1.0000x_1 + 1.0000x_2 = 1.0000$$

Note that in this form the first pivot element, $a_{11} = 0.0003$, is very close to zero. Then repeat the computation, but partial pivot by reversing the order of the equations. The exact solution is $x_1 = 1/3$ and $x_2 = 2/3$.

Solution.   Multiplying the first equation by $1/(0.0003)$ yields

$$x_1 + 10,000x_2 = 6667$$

which can be used to eliminate x_1 from the second equation:

$$-9999x_2 = -6666$$

which can be solved for

$$x_2 = \frac{2}{3}$$

This result can be substituted back into the first equation to evaluate x_1:

$$x_1 = \frac{2.0001 - 3(2/3)}{0.0003} \tag{E9.9.1}$$

However, due to subtractive cancellation, the result is very sensitive to the number of significant figures carried in the computation:

Significant Figures	x_2	x_1	Absolute Value of Percent Relative Error for x_1
3	0.667	−3.33	1099
4	0.6667	0.0000	100
5	0.66667	0.30000	10
6	0.666667	0.330000	1
7	0.6666667	0.3330000	0.1

Note how the solution for x_1 is highly dependent on the number of significant figures. This is because in Eq. (E9.9.1), we are subtracting two almost-equal numbers. On the other hand, if the equations are solved in reverse order, the row with the larger pivot element is normalized. The equations are

$$1.0000x_1 + 1.0000x_2 = 1.0000$$
$$0.0003x_1 + 3.0000x_2 = 2.0001$$

Elimination and substitution yield $x_2 = 2/3$. For different numbers of significant figures, x_1 can be computed from the first equation, as in

$$x_1 = \frac{1 - (2/3)}{1} \tag{E9.9.2}$$

This case is much less sensitive to the number of significant figures in the computation:

Significant Figures	x_2	x_1	Absolute Value of Percent Relative Error for x_1
3	0.667	0.333	0.1
4	0.6667	0.3333	0.01
5	0.66667	0.33333	0.001
6	0.666667	0.333333	0.0001
7	0.6666667	0.3333333	0.00001

Thus, a pivot strategy is much more satisfactory.

```
p = k
big = |a_k,k|
DOFOR ii = k+1, n
  dummy = |a_ii,k|
  IF (dummy > big)
    big = dummy
    p = ii
  END IF
END DO
IF (p ≠ k)
  DOFOR jj = k, n
    dummy = a_p,jj
    a_p,jj = a_k,jj
    a_k,jj = dummy
  END DO
  dummy = b_p
  b_p = b_k
  b_k = dummy
END IF
```

FIGURE 9.5
Pseudocode to implement
partial pivoting.

General-purpose computer programs must include a pivot strategy. Figure 9.5 provides a simple algorithm to implement such a strategy. Notice that the algorithm consists of two major loops. After storing the current pivot element and its row number as the variables, big and p, the first loop compares the pivot element with the elements below it to check whether any of these is larger than the pivot element. If so, the new largest element and its row number are stored in big and p. Then, the second loop switches the original pivot row with the one with the largest element so that the latter becomes the new pivot row. This pseudocode can be integrated into a program based on the other elements of Gauss elimination outlined in Fig. 9.4. The best way to do this is to employ a modular approach and write Fig. 9.5 as a subroutine (or procedure) that would be called directly after the beginning of the first loop in Fig. 9.4a.

Note that the second IF/THEN construct in Fig. 9.5 physically interchanges the rows. For large matrices, this can become quite time-consuming. Consequently, most codes do not actually exchange rows but rather keep track of the pivot rows by storing the appropriate subscripts in a vector. This vector then provides a basis for specifying the proper row ordering during the forward-elimination and back-substitution operations. Thus, the operations are said to be implemented *in place*.

9.4.3 Scaling

In Sec. 9.3.3, we proposed that scaling had value in standardizing the size of the determinant. Beyond this application, it has utility in minimizing round-off errors for those cases where some of the equations in a system have much larger coefficients than others. Such situations are frequently encountered in engineering practice when widely different units are used in the development of simultaneous equations. For instance, in electric-circuit problems, the unknown voltages can be expressed in units ranging from microvolts to kilovolts. Similar examples can arise in all fields of engineering. As long as each equation is consistent, the system will be technically correct and solvable. However, the use of widely differing units can lead to coefficients of widely differing magnitudes. This, in turn, can have an impact on round-off error as it affects pivoting, as illustrated by the following example.

EXAMPLE 9.10   **Effect of Scaling on Pivoting and Round-Off**

Problem Statement.

(a) Solve the following set of equations using Gauss elimination and a pivoting strategy:

$$2x_1 + 100,000x_2 = 100,000$$
$$x_1 + x_2 = 2$$

(b) Repeat the solution after scaling the equations so that the maximum coefficient in each row is 1.
(c) Finally, use the scaled coefficients to determine whether pivoting is necessary. However, actually solve the equations with the original coefficient values. For all cases, retain only three significant figures. Note that the correct answers are $x_1 = 1.00002$ and $x_2 = 0.99998$ or, for three significant figures, $x_1 = x_2 = 1.00$.

Solution.

(a) Without scaling, forward elimination is applied to give

$$2x_1 + 100{,}000x_2 = 100{,}000$$
$$-50{,}000x_2 = -50{,}000$$

which can be solved by back substitution for

$$x_2 = 1.00$$
$$x_1 = 0.00$$

Although x_2 is correct, x_1 is 100% in error because of round-off.

(b) Scaling transforms the original equations to

$$0.00002x_1 + x_2 = 1$$
$$x_1 + x_2 = 2$$

Therefore, the rows should be pivoted to put the greatest value on the diagonal.

$$x_1 + x_2 = 2$$
$$0.00002x_1 + x_2 = 1$$

Forward elimination yields

$$x_1 + x_2 = 2$$
$$x_2 = 1.00$$

which can be solved for

$$x_1 = x_2 = 1$$

Thus, scaling leads to the correct answer.

(c) The scaled coefficients indicate that pivoting is necessary. We therefore pivot but retain the original coefficients to give

$$x_1 + x_2 = 2$$
$$2x_1 + 100{,}000x_2 = 100{,}000$$

Forward elimination yields

$$x_1 + x_2 = 2$$
$$100{,}000x_2 = 100{,}000$$

which can be solved for the correct answer: $x_1 = x_2 = 1$. Thus, scaling was useful in determining whether pivoting was necessary, but the equations themselves did not require scaling to arrive at a correct result.

As in the previous example, scaling has utility in minimizing round-off. However, it should be noted that scaling itself also leads to round-off. For example, given the equation

$$2x_1 + 300,000x_2 = 1$$

and using three significant figures, scaling leads to

$$0.00000667x_1 + x_2 = 0.00000333$$

Thus, scaling introduces a round-off error to the first coefficient and the right-hand-side constant. For this reason, it is sometimes suggested that scaling should be employed only as in part (c) of the preceding example. That is, it is used to calculate scaled values for the coefficients solely as a criterion for pivoting, but the original coefficient values are retained for the actual elimination and substitution computations. This involves a trade-off if the determinant is being calculated as part of the program. That is, the resulting determinant will be unscaled. However, because many applications of Gauss elimination do not require determinant evaluation, it is the most common approach and will be used in the algorithm in the next section.

9.4.4 Computer Algorithm for Gauss Elimination

The algorithms from Figs. 9.4 and 9.5 can now be combined into a larger algorithm to implement the entire Gauss elimination algorithm. Figure 9.6 shows an algorithm for a general subroutine to implement Gauss elimination.

Note that the program includes modules for the three primary operations of the Gauss elimination algorithm: forward elimination, back substitution, and pivoting. In addition, there are several aspects of the code that differ and represent improvements over the pseudocodes from Figs. 9.4 and 9.5. These are

- The equations are not scaled, but scaled values of the elements are used to determine whether pivoting is to be implemented.
- The diagonal term is monitored during the pivoting phase to detect near-zero occurrences in order to flag singular systems. If it passes back a value of $er = -1$, a singular matrix has been detected and the computation should be terminated. A parameter tol is set by the user to a small number in order to detect near-zero occurrences.

EXAMPLE 9.11 **Solution of Linear Algebraic Equations Using the Computer**

Problem Statement.   A computer program to solve linear algebraic equations such as one based on Fig. 9.6 can be used to solve a problem associated with the falling parachutist example discussed in Chap. 1. Suppose that a team of three parachutists is connected by a weightless cord while free-falling at a velocity of 5 m/s (Fig. 9.7).

```
SUB Gauss(a, b, n, x, tol, er)
  DIMENSION s(n)
  er = 0
  DOFOR i = 1, n
    s_i = ABS(a_{i,1})
    DOFOR j = 2, n
      IF ABS(a_{i,j})>s_i THEN s_i = ABS(a_{i,j})
    END DO
  END DO
  CALL Eliminate(a, s, n, b, tol, er)
  IF er ≠ −1 THEN
      CALL Substitute(a, n, b, x)
  END IF
END Gauss


SUB Eliminate(a, s, n, b, tol, er)
  DOFOR k = 1, n − 1
    CALL Pivot (a, b, s, n, k)
    IF ABS (a_{k,k}/s_k) < tol THEN
      er = −1
      EXIT DO
    END IF
    DOFOR i = k + 1, n
      factor = a_{i,k}/a_{k,k}
      DOFOR j = k + 1, n
        a_{i,j} = a_{i,j} − factor*a_{k,j}
      END DO
      b_i = b_i − factor * b_k
    END DO
  END DO
  IF ABS(a_{n,n}/s_n) < tol THEN er = −1
END Eliminate
```

```
SUB Pivot(a, b, s, n, k)
  p = k
  big = ABS(a_{k,k}/s_k)
  DOFOR ii = k + 1, n
    dummy = ABS(a_{ii,k}/s_{ii})
    IF dummy > big THEN
      big = dummy
      p = ii
    END IF
  END DO
  IF p ≠ k THEN
      DOFOR jj = k, n
        dummy = a_{p,jj}
        a_{p,jj} = a_{k,jj}
        a_{k,jj} = dummy
      END DO
      dummy = b_p
      b_p = b_k
      b_k = dummy
      dummy = s_p
      s_p = s_k
      s_k = dummy
  END IF
END pivot


SUB Substitute(a, n, b, x)
  x_n = b_n/a_{n,n}
  DOFOR i = n − 1, 1, −1
    sum = 0
    DOFOR j = i + 1, n
      sum = sum + a_{i,j} * x_j
    END DO
    x_n = (b_n − sum) / a_{n,n}
  END DO
END Substitute
```

FIGURE 9.6
Pseudocode to implement Gauss elimination with partial pivoting.

FIGURE 9.7
Three parachutists free-falling while connected by weightless cords.

Calculate the tension in each section of cord and the acceleration of the team, given the following:

Parachutist	Mass, kg	Drag Coefficient, kg/s
1	70	10
2	60	14
3	40	17

Solution. Free-body diagrams for each of the parachutists are depicted in Fig. 9.8. Summing the forces in the vertical direction and using Newton's second law gives a set of three simultaneous linear equations:

$$
\begin{aligned}
m_1 g - T - c_1 v &= m_1 a \\
m_2 g + T - c_2 v - R &= m_2 a \\
m_3 g - c_3 v + R &= m_3 a
\end{aligned}
$$

These equations have three unknowns: a, T, and R. After substituting the known values, the equations can be expressed in matrix form as ($g = 9.81 \text{ m/s}^2$),

$$
\begin{bmatrix} 70 & 1 & 0 \\ 60 & -1 & 1 \\ 40 & 0 & -1 \end{bmatrix} \begin{Bmatrix} a \\ T \\ R \end{Bmatrix} = \begin{Bmatrix} 636.7 \\ 518.6 \\ 307.4 \end{Bmatrix}
$$

This system can be solved using your own software. The result is $a = 8.6041 \text{ m/s}^2$; $T = 34.4118 \text{ N}$; and $R = 36.7647 \text{ N}$.

FIGURE 9.8
Free-body diagrams for each of the three falling parachutists.

9.5   COMPLEX SYSTEMS

In some problems, it is possible to obtain a complex system of equations

$$[C]\{Z\} = \{W\} \tag{9.27}$$

where

$$[C] = [A] + i[B]$$
$$\{Z\} = \{X\} + i\{Y\}$$
$$\{W\} = \{U\} + i\{V\} \tag{9.28}$$

and $i = \sqrt{-1}$.

The most straightforward way to solve such a system is to employ one of the algorithms described in this part of the book, but replace all real operations with complex ones. Of course, this is only possible for those languages, such as Fortran, that allow complex variables.

For languages that do not permit the declaration of complex variables, it is possible to write a code to convert real to complex operations. However, this is not a trivial task. An alternative is to convert the complex system into an equivalent one dealing with real variables. This can be done by substituting Eq. (9.28) into Eq. (9.27) and equating real and complex parts of the resulting equation to yield

$$[A]\{X\} - [B]\{Y\} = \{U\} \tag{9.29}$$

and

$$[B]\{X\} + [A]\{Y\} = \{V\} \tag{9.30}$$

Thus, the system of n complex equations is converted to a set of $2n$ real ones. This means that storage and execution time will be increased significantly. Consequently, a trade-off exists regarding this option. If you evaluate complex systems infrequently, it is preferable to use Eqs. (9.29) and (9.30) because of their convenience. However, if you use them often and desire to employ a language that does not allow complex data types, it may be worth the up-front programming effort to write a customized equation solver that converts real to complex operations.

9.6   NONLINEAR SYSTEMS OF EQUATIONS

Recall that at the end of Chap. 6 we presented an approach for solving two nonlinear equations with two unknowns. This approach can be extended to the general case of solving n simultaneous nonlinear equations.

$$
\begin{aligned}
f_1(x_1, x_2, \ldots, x_n) &= 0 \\
f_2(x_1, x_2, \ldots, x_n) &= 0 \\
&\ \ \vdots \\
f_n(x_1, x_2, \ldots, x_n) &= 0
\end{aligned}
\tag{9.31}
$$

The solution of this system consists of the set of x values that simultaneously result in all the equations equaling zero.

As described in Sec. 6.6.2, one approach to solving such systems is based on a multiequation version of the Newton-Raphson method. Thus, a Taylor series expansion is written for each equation. For example, for the kth equation,

$$f_{k,i+1} = f_{k,i} + (x_{1,i+1} - x_{1,i})\frac{\partial f_{k,i}}{\partial x_1} + (x_{2,i+1} - x_{2,i})\frac{\partial f_{k,i}}{\partial x_2} + \cdots + (x_{n,i+1} - x_{n,i})\frac{\partial f_{k,i}}{\partial x_n}$$

(9.32)

where the first subscript, k, represents the equation or unknown and the second subscript denotes whether the value or function in question is at the present value (i) or at the next value ($i + 1$).

Equations of the form of (9.32) are written for each of the original nonlinear equations. Then, as was done in deriving Eq. (6.21) from (6.20), all $f_{k,i+1}$ terms are set to zero, as would be the case at the root, and Eq. (9.32) can be written as

$$-f_{k,i} + x_{1,i}\frac{\partial f_{k,i}}{\partial x_1} + x_{2,i}\frac{\partial f_{k,i}}{\partial x_2} + \cdots + x_{n,i}\frac{\partial f_{k,i}}{\partial x_n}$$

$$= x_{1,i+1}\frac{\partial f_{k,i}}{\partial x_1} + x_{2,i+1}\frac{\partial f_{k,i}}{\partial x_2} + \cdots + x_{n,i+1}\frac{\partial f_{k,i}}{\partial x_n}$$

(9.33)

Notice that the only unknowns in Eq. (9.33) are the $x_{k,i+1}$ terms on the right-hand side. All other quantities are located at the present value (i) and, thus, are known at any iteration. Consequently, the set of equations generally represented by Eq. (9.33) (that is, with $k = 1, 2, \ldots, n$) constitutes a set of linear simultaneous equations that can be solved by methods elaborated in this part of the book.

Matrix notation can be employed to express Eq. (9.33) concisely. The partial derivatives can be expressed as

$$[Z] = \begin{bmatrix} \dfrac{\partial f_{1,i}}{\partial x_1} & \dfrac{\partial f_{1,i}}{\partial x_2} & \cdots & \dfrac{\partial f_{1,i}}{\partial x_n} \\[2ex] \dfrac{\partial f_{2,i}}{\partial x_1} & \dfrac{\partial f_{2,i}}{\partial x_2} & \cdots & \dfrac{\partial f_{2,i}}{\partial x_n} \\[2ex] \cdot & \cdot & & \cdot \\ \cdot & \cdot & & \cdot \\ \cdot & \cdot & & \cdot \\[1ex] \dfrac{\partial f_{n,i}}{\partial x_1} & \dfrac{\partial f_{n,i}}{\partial x_2} & \cdots & \dfrac{\partial f_{n,i}}{\partial x_n} \end{bmatrix}$$

(9.34)

The initial and final values can be expressed in vector form as

$$\{X_i\}^T = \lfloor x_{1,i} \quad x_{2,i} \quad \cdots \quad x_{n,i} \rfloor$$

and

$$\{X_{i+1}\}^T = \lfloor x_{1,i+1} \quad x_{2,i+1} \quad \cdots \quad x_{n,i+1} \rfloor$$

$$\begin{bmatrix} a_{11} & a_{12} & a_{13} & \vdots & b_1 \\ a_{21} & a_{22} & a_{23} & \vdots & b_2 \\ a_{31} & a_{32} & a_{33} & \vdots & b_3 \end{bmatrix}$$

$$\downarrow$$

$$\begin{bmatrix} 1 & 0 & 0 & \vdots & b_1^{(n)} \\ 0 & 1 & 0 & \vdots & b_2^{(n)} \\ 0 & 0 & 1 & \vdots & b_3^{(n)} \end{bmatrix}$$

$$\downarrow$$

$$\begin{matrix} x_1 & & & = & b_1^{(n)} \\ & x_2 & & = & b_2^{(n)} \\ & & x_3 & = & b_3^{(n)} \end{matrix}$$

FIGURE 9.9
Graphical depiction of the Gauss-Jordan method. Compare with Fig. 9.3 to understand the differences between this technique and Gauss elimination. The superscript (n) means that the elements of the right-hand-side vector have been modified n times (for this case, $n = 3$).

Finally, the function values at i can be expressed as

$$\{F_i\}^T = \lfloor f_{1,i} \quad f_{2,i} \quad \cdots \quad f_{n,i} \rfloor$$

Using these relationships, Eq. (9.33) can be represented concisely as

$$[Z]\{X_{i+1}\} = -\{F_i\} + [Z]\{X_i\} \tag{9.35}$$

Equation (9.35) can be solved using a technique such as Gauss elimination. This process can be repeated iteratively to obtain refined estimates in a fashion similar to the two-equation case in Sec. 6.6.2.

It should be noted that there are two major shortcomings to the foregoing approach. First, Eq. (9.34) is often inconvenient to evaluate. Therefore, variations of the Newton-Raphson approach have been developed to circumvent this dilemma. As might be expected, most are based on using finite-difference approximations for the partial derivatives that comprise $[Z]$.

The second shortcoming of the multiequation Newton-Raphson method is that excellent initial guesses are usually required to ensure convergence. Because these are often difficult to obtain, alternative approaches that are slower than Newton-Raphson but which have better convergence behavior have been developed. One common approach is to reformulate the nonlinear system as a single function

$$F(x) = \sum_{i=1}^{n} [f_i(x_1, x_2, \ldots, x_n)]^2 \tag{9.36}$$

where $f_i(x_1, x_2, \ldots, x_n)$ is the ith member of the original system of Eq. (9.31). The values of x that minimize this function also represent the solution of the nonlinear system. As we will see in Chap. 17, this reformulation is used for a class of problems called *nonlinear regression*. As such, it can be approached with a number of optimization techniques such as the ones described later in this text (Part Four and specifically Chap. 14).

9.7   GAUSS-JORDAN

The Gauss-Jordan method is a variation of Gauss elimination. The major difference is that when an unknown is eliminated in the Gauss-Jordan method, it is eliminated from all other equations rather than just the subsequent ones. In addition, all rows are normalized by dividing them by their pivot elements. Thus, the elimination step results in an identity matrix rather than a triangular matrix (Fig. 9.9). Consequently, it is not necessary to employ back substitution to obtain the solution. The method is best illustrated by an example.

EXAMPLE 9.12   **Gauss-Jordan Method**

Problem Statement.   Use the Gauss-Jordan technique to solve the same system as in Example 9.5:

$$3x_1 - 0.1x_2 - 0.2x_3 = 7.85$$
$$0.1x_1 + 7x_2 - 0.3x_3 = -19.3$$
$$0.3x_1 - 0.2x_2 + 10x_3 = 71.4$$

Solution. First, express the coefficients and the constants on the right-hand side as an augmented matrix:

$$\begin{bmatrix} 3 & -0.1 & -0.2 & 7.85 \\ 0.1 & 7 & -0.3 & -19.3 \\ 0.3 & -0.2 & 10 & 71.4 \end{bmatrix}$$

Then normalize the first row by dividing it by the pivot element, 3, to yield

$$\begin{bmatrix} 1 & -0.0333333 & -0.066667 & 2.61667 \\ 0.1 & 7 & -0.3 & -19.3 \\ 0.3 & -0.2 & 10 & 71.4 \end{bmatrix}$$

The x_1 term can be eliminated from the second row by subtracting 0.1 times the first row from the second row. Similarly, subtracting 0.3 times the first row from the third row will eliminate the x_1 term from the third row:

$$\begin{bmatrix} 1 & -0.0333333 & -0.066667 & 2.61667 \\ 0 & 7.00333 & -0.293333 & -19.5617 \\ 0 & -0.190000 & 10.0200 & 70.6150 \end{bmatrix}$$

Next, normalize the second row by dividing it by 7.00333:

$$\begin{bmatrix} 1 & -0.0333333 & -0.066667 & 2.61667 \\ 0 & 1 & -0.0418848 & -2.79320 \\ 0 & -0.190000 & 10.0200 & 70.6150 \end{bmatrix}$$

Reduction of the x_2 terms from the first and third equations gives

$$\begin{bmatrix} 1 & 0 & -0.0680629 & 2.52356 \\ 0 & 1 & -0.0418848 & -2.79320 \\ 0 & 0 & 10.01200 & 70.0843 \end{bmatrix}$$

The third row is then normalized by dividing it by 10.0120:

$$\begin{bmatrix} 1 & 0 & -0.0680629 & 2.52356 \\ 0 & 1 & -0.0418848 & -2.79320 \\ 0 & 0 & 1 & 7.0000 \end{bmatrix}$$

Finally, the x_3 terms can be reduced from the first and the second equations to give

$$\begin{bmatrix} 1 & 0 & 0 & 3.0000 \\ 0 & 1 & 0 & -2.5000 \\ 0 & 0 & 1 & 7.0000 \end{bmatrix}$$

Thus, as depicted in Fig. 9.9, the coefficient matrix has been transformed to the identity matrix, and the solution is obtained in the right-hand-side vector. Notice that no back substitution was required to obtain the solution.

All the material in this chapter regarding the pitfalls and improvements in Gauss elimination also applies to the Gauss-Jordan method. For example, a similar pivoting strategy can be used to avoid division by zero and to reduce round-off error.

An algorithm to implement Gauss-Jordan for an augmented matrix (i.e., with the right-hand-side constants appended as an additional column to the coefficient matrix) without partial pivoting is displayed in Fig. 9.10. Although the Gauss-Jordan technique and Gauss elimination might appear almost identical, the former requires more work. Using a similar approach to Sec. 9.2.1, it can be determined that the number of flops involved in naive Gauss-Jordan is

$$n^3 + 3n^2 - 2n \xrightarrow{\text{as } n \text{ increases}} n^3 + O(n^2) \tag{9.37}$$

Thus, Gauss-Jordan involves approximately 50 percent more operations than Gauss elimination [compare with Eq. (9.23)]. Therefore, Gauss elimination is the simple elimination method of preference for obtaining solutions of linear algebraic equations. One of the primary reasons that we have introduced the Gauss-Jordan, however, is that it is still used in engineering as well as in some numerical algorithms.

FIGURE 9.10

Pseudocode to implement the Gauss-Jordan algorithm without partial pivoting.

```
SUB GaussJordan(aug, m, n, x)
DOFOR k = 1, m
  d = aug(k, k)
  DOFOR j = 1, n
    aug(k, j) = aug(k, j)/d
  END DO
  DOFOR i = 1, m
    IF i ≠ k THEN
      d = aug(i, k)
      DOFOR j = k, n
        aug(i, j) = aug(i, j) - d*aug(k, j)
      END DO
    END IF
  END DO
END DO
DOFOR k = 1, m
  x(k) = aug(k, n)
END DO
END GaussJordan
```

9.8 SUMMARY

In summary, we have devoted most of this chapter to Gauss elimination, the most fundamental method for solving simultaneous linear algebraic equations. Although it is one of the earliest techniques developed for this purpose, it is nevertheless an extremely effective algorithm for obtaining solutions for many engineering problems. Aside from this practical utility, focusing on this method also provided a context for our discussion of general issues such as round-off, scaling, and conditioning. In addition, we briefly presented material on the Gauss-Jordan method, as well as complex and nonlinear systems.

Answers obtained using Gauss elimination may be checked by substituting them into the original equations. However, this does not always represent a reliable check for ill-conditioned systems. Therefore, some measure of condition, such as the determinant of the scaled system, should be computed if round-off error is suspected. Using partial pivoting and more significant figures in the computation are two options for mitigating round-off error. In the next chapter, we will return to the topic of system condition when we discuss the matrix inverse.

PROBLEMS

9.1

(a) Write the following set of equations in matrix form:

$$40 = 5x_3 + 2x_1$$
$$10 - x_2 = x_3$$
$$4x_2 + 8x_1 = 21$$

(b) Multiply the matrix of coefficients by its transpose; i.e., $[A][A]^T$.

9.2 A number of matrices are defined as

$$[A] = \begin{bmatrix} 4 & 5 \\ 1 & 2 \\ 5 & 6 \end{bmatrix} \quad [B] = \begin{bmatrix} 4 & 3 & 7 \\ 1 & 2 & 6 \\ 2 & 0 & 4 \end{bmatrix}$$

$$\{C\} = \begin{Bmatrix} 3 \\ 5 \\ 1 \end{Bmatrix} \quad [D] = \begin{bmatrix} 9 & 4 & 3 & -6 \\ 2 & -1 & 6 & 5 \end{bmatrix}$$

$$[E] = \begin{bmatrix} 1 & 5 & 9 \\ 7 & 2 & 3 \\ 4 & 0 & 6 \end{bmatrix}$$

$$[F] = \begin{bmatrix} 2 & 0 & 1 \\ 1 & 7 & 3 \end{bmatrix} \quad \lfloor G \rfloor = \lfloor 7 \quad 5 \quad 4 \rfloor$$

Answer the following questions regarding these matrices:
(a) What are the dimensions of the matrices?
(b) Identify the square, column, and row matrices.
(c) What are the values of the elements: $a_{12}, b_{23}, d_{32}, e_{22}, f_{12}, g_{12}$?
(d) Perform the following operations:

(1) $[E] + [B]$
(2) $[A] \times [F]$
(3) $[B] - [E]$
(4) $8 \times [B]$
(5) $[A] \times [B]$
(6) $\{C\}^T$
(7) $[B] \times [A]$
(8) $[D]^T$
(9) $[A] \times \{C\}$
(10) $[I] \times [B]$
(11) $[E]^T[E]$
(12) $\{C\}^T\{C\}$

9.3 Three matrices are defined as

$$[A] = \begin{bmatrix} 1 & 5 \\ 3 & 10 \\ -4 & 3 \end{bmatrix} \quad [B] = \begin{bmatrix} 4 & 3 \\ 0.5 & 2 \end{bmatrix} \quad [C] = \begin{bmatrix} 2 & -2 \\ -3 & 5 \end{bmatrix}$$

(a) Perform all possible multiplications that can be computed between pairs of these matrices.
(b) Use the method in Box PT3.2 to justify why the remaining pairs cannot be multiplied.
(c) Use the results of **(a)** to illustrate why the order of multiplication is important.

9.4 Use the graphical method to solve

$$2x_1 - 6x_2 = -18$$
$$-x_1 + 8x_2 = 40$$

Check your results by substituting them back into the equations.

9.5 Given the system of equations

$$0.77x_1 + x_2 = 14.25$$
$$1.2x_1 + 1.7x_2 = 20$$

(a) Solve graphically and check your results by substituting them back into the equations.
(b) On the basis of the graphical solution, what do you expect regarding the condition of the system?
(c) Compute the determinant.
(d) Solve by the elimination of unknowns.

9.6 For the set of equations

$$2x_2 + 5x_3 = 1$$
$$2x_1 + x_2 + x_3 = 1$$
$$3x_1 + x_2 = 2$$

(a) Compute the determinant.
(b) Use Cramer's rule to solve for the x's.
(c) Substitute your results back into the original equations to check your results.

9.7 Given the equations

$$0.5x_1 - x_2 = -9.5$$
$$1.02x_1 - 2x_2 = -18.8$$

(a) Solve graphically.
(b) Compute the determinant.
(c) On the basis of **(a)** and **(b)**, what would you expect regarding the system's condition?
(d) Solve by the elimination of unknowns.
(e) Solve again, but with a_{11} modified slightly to 0.52. Interpret your results.

9.8 Given the equations

$$10x_1 + 2x_2 - x_3 = 27$$
$$-3x_1 - 6x_2 + 2x_3 = -61.5$$
$$x_1 + x_2 + 5x_3 = -21.5$$

(a) Solve by naive Gauss elimination. Show all steps of the computation.
(b) Substitute your results into the original equations to check your answers.

9.9 Use Gauss elimination to solve:

$$8x_1 + 2x_2 - 2x_3 = 8$$
$$10x_1 + 2x_2 + 4x_3 = 16$$
$$12x_1 + 2x_2 + 2x_3 = 16$$

Employ partial pivoting, and check your answers by substituting them into the original equations.

9.10 Given the system of equations

$$-3x_2 + 7x_3 = 2$$
$$x_1 + 2x_2 - x_3 = 3$$
$$5x_1 - 2x_2 = 2$$

(a) Compute the determinant.
(b) Use Cramer's rule to solve for the x's.
(c) Use Gauss elimination with partial pivoting to solve for the x's.
(d) Substitute your results back into the original equations to check your solution.

9.11 Given the equations

$$2x_1 - 6x_2 - x_3 = -38$$
$$-3x_1 - x_2 + 7x_3 = -34$$
$$-8x_1 + x_2 - 2x_3 = -20$$

(a) Solve by Gauss elimination with partial pivoting. Show all steps of the computation.
(b) Substitute your results into the original equations to check your answers.

9.12 Use Gauss-Jordan elimination to solve:

$$2x_1 + x_2 - x_3 = 2$$
$$5x_1 + 2x_2 + 2x_3 = 9$$
$$3x_1 + x_2 + x_3 = 5$$

Do not employ pivoting. Check your answers by substituting them into the original equations.

9.13 Solve:

$$x_1 + x_2 - x_3 = -3$$
$$6x_1 + 2x_2 + 2x_3 = 2$$
$$-3x_1 + 4x_2 + x_3 = 1$$

with **(a)** naive Gauss elimination, **(b)** Gauss elimination with partial pivoting, and **(c)** Gauss-Jordan without partial pivoting.

9.14 Perform the same computation as in Example 9.11, but use five parachutists with the following characteristics:

Parachutist	Mass, kg	Drag Coefficient, kg/s
1	60	10
2	75	12
3	60	15
4	75	16
5	90	10

The parachutists have a velocity of 9 m/s.

9.15 Solve

$$\begin{bmatrix} 3 + 2i & 4 \\ -i & 1 \end{bmatrix} \begin{Bmatrix} z_1 \\ z_2 \end{Bmatrix} = \begin{Bmatrix} 2 + i \\ 3 \end{Bmatrix}$$

9.16 Develop, debug, and test a program in either a high-level language or a macro language of your choice to multiply two matrices—that is, $[X] = [Y][Z]$, where $[Y]$ is m by n and $[Z]$ is n by p. Test the program using the matrices from Prob. 9.3.

9.17 Develop, debug, and test a program in either a high-level language or a macro language of your choice to generate the transpose of a matrix. Test it on the matrices from Prob. 9.3.

9.18 Develop, debug, and test a program in either a high-level language or a macro language of your choice to solve a system of equations with Gauss elimination with partial pivoting. Base the program on the pseudocode from Fig. 9.6. Test the program using the following system (which has an answer of $x_1 = x_2 = x_3 = 2$),

$$x_1 + 2x_2 - x_3 = 4$$
$$5x_1 + 2x_2 + 2x_3 = 18$$
$$-3x_1 + 5x_2 - x_3 = 2$$

9.19 Three masses are suspended vertically by a series of identical springs where mass 1 is at the top and mass 3 is at the bottom.

If $g = 9.81$ m/s^2, $m_1 = 2$ kg, $m_2 = 3$ kg, $m_3 = 2.5$ kg, and the k's $= 10$ kg/s^2, solve for the displacements x.

9.20 Develop, debug, and test a program in either a high-level language or a macro language of your choice to solve a system of n simultaneous nonlinear equations based on Sec. 9.6. Test the program by solving Prob. 7.13.

9.21 Recall from Sec. 8.2 that determining the chemistry of water exposed to atmospheric CO_2 can be determined by solving five nonlinear equations (Eqs. 8.6 through 8.10) for five unknowns: c_T, $[HCO_3^-]$, $[CO_3^{2-}]$, $[H^+]$, and $[OH^-]$. Employing the parameters from Sec. 8.2 and the program developed in Prob. 9.20, solve this system for conditions in 1958 when the partial pressure of CO_2 was 315 ppm. Use your results to compute the pH.

9.22 Develop, debug, and test a program in either a high-level language or a macro language of your choice to solve a system of equations with Gauss-Jordan elimination without partial pivoting. Base the program on the pseudocode from Fig. 9.10. Test the program using the same system as in Prob. 9.18. Compute the total number of flops in your algorithm to verify Eq. 9.37.

9.23 Suppose we want to add a number of vectors multiplied by scalars. For example,

$$\begin{Bmatrix} y_1 \\ y_2 \\ y_3 \end{Bmatrix} = 2 \begin{Bmatrix} 6 \\ -8 \\ -1 \end{Bmatrix} + 7 \begin{Bmatrix} -3 \\ 2 \\ 5 \end{Bmatrix} - 6 \begin{Bmatrix} 4 \\ 2 \\ 1 \end{Bmatrix} + 3 \begin{Bmatrix} -2 \\ -8 \\ 9 \end{Bmatrix}$$

or for the more general case where there can be m additions and the vectors can be any length, n,

$$\{y\} = c_1\{x\}_1 + c_2\{x\}_2 + \cdots + c_m\{x\}_m$$

Develop, debug, and test a program in either a high-level language or a macro language of your choice to perform this calculation using nested loops and subscripted variables. Test the program using the first equation above. Derive a simple formula to compute the number of flops as a function of m and n.

CHAPTER 10

LU Decomposition and Matrix Inversion

This chapter deals with a class of elimination methods called *LU* decomposition techniques, where the "*LU*" stands for "*Lower Upper*." The primary appeal of *LU* decomposition is that the time-consuming elimination step can be formulated so that it involves only operations on the matrix of coefficients, [*A*]. Thus, it is well suited for those situations where many right-hand-side vectors {*B*} must be evaluated for a single value of [*A*]. Although there are a variety of ways in which this is done, we will focus on showing how the Gauss elimination method can be implemented as an *LU* decomposition.

One motive for introducing *LU* decomposition is that it provides an efficient means to compute the matrix inverse. The inverse has a number of valuable applications in engineering practice. It also provides a means for evaluating system condition.

10.1 *LU* DECOMPOSITION

As described in Chap. 9, Gauss elimination is designed to solve systems of linear algebraic equations,

$$[A]\{X\} = \{B\} \tag{10.1}$$

Although it certainly represents a sound way to solve such systems, it becomes inefficient when solving equations with the same coefficients [*A*], but with different right-hand-side constants (the *b*'s).

Recall that Gauss elimination involves two steps: forward elimination and back substitution (Fig. 9.3). Of these, the forward-elimination step comprises the bulk of the computational effort (recall Table 9.1). This is particularly true for large systems of equations.

LU decomposition methods separate the time-consuming elimination of the matrix [*A*] from the manipulations of the right-hand side {*B*}. Thus, once [*A*] has been "decomposed," multiple right-hand-side vectors can be evaluated in an efficient manner.

Interestingly, Gauss elimination itself can be expressed as an *LU* decomposition. Before showing how this can be done, let us first provide a mathematical overview of the decomposition strategy.

10.1.1 Overview of *LU* Decomposition

Just as was the case with Gauss elimination, *LU* decomposition requires pivoting to avoid division by zero. However, to simplify the following description, we will defer the issue of pivoting until after the fundamental approach is elaborated. In addition, the following explanation is limited to a set of three simultaneous equations. The results can be directly extended to *n*-dimensional systems.

Equation (10.1) can be rearranged to give

$$[A]\{X\} - \{B\} = 0 \tag{10.2}$$

Suppose that Eq. (10.2) could be expressed as an upper triangular system:

$$\begin{bmatrix} u_{11} & u_{12} & u_{13} \\ 0 & u_{22} & u_{23} \\ 0 & 0 & u_{33} \end{bmatrix} \begin{Bmatrix} x_1 \\ x_2 \\ x_3 \end{Bmatrix} = \begin{Bmatrix} d_1 \\ d_2 \\ d_3 \end{Bmatrix} \tag{10.3}$$

Recognize that this is similar to the manipulation that occurs in the first step of Gauss elimination. That is, elimination is used to reduce the system to upper triangular form. Equation (10.3) can also be expressed in matrix notation and rearranged to give

$$[U]\{X\} - \{D\} = 0 \tag{10.4}$$

Now, assume that there is a lower diagonal matrix with 1's on the diagonal,

$$[L] = \begin{bmatrix} 1 & 0 & 0 \\ l_{21} & 1 & 0 \\ l_{31} & l_{32} & 1 \end{bmatrix} \tag{10.5}$$

that has the property that when Eq. (10.4) is premultiplied by it, Eq. (10.2) is the result. That is,

$$[L]\{[U]\{X\} - \{D\}\} = [A]\{X\} - \{B\} \tag{10.6}$$

If this equation holds, it follows from the rules for matrix multiplication that

$$[L][U] = [A] \tag{10.7}$$

and

$$[L]\{D\} = \{B\} \tag{10.8}$$

A two-step strategy (see Fig. 10.1) for obtaining solutions can be based on Eqs. (10.4), (10.7), and (10.8):

1. *LU decomposition step.* [A] is factored, or "decomposed," into lower [L] and upper [U] triangular matrices.
2. *Substitution step.* [L] and [U] are used to determine a solution {X} for the right-hand-side {B}. This step itself consists of two steps. First, Eq. (10.8) is used to generate an intermediate vector {D} by forward substitution. Then, the result is substituted into Eq. (10.4), which can be solved by back substitution for {X}.

Now, let us show how Gauss elimination can be implemented in this way.

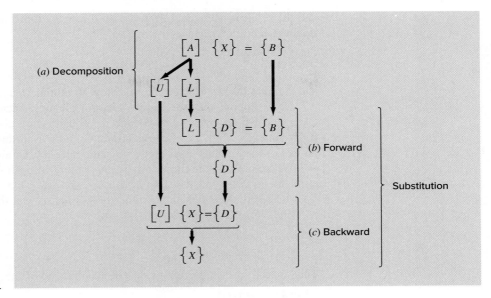

FIGURE 10.1
The steps in *LU* decomposition.

10.1.2 *LU* Decomposition Version of Gauss Elimination

Although it might appear at face value to be unrelated to *LU* decomposition, Gauss elimination can be used to decompose [A] into [L] and [U]. This can be easily seen for [U], which is a direct product of the forward elimination. Recall that the forward-elimination step is intended to reduce the original coefficient matrix [A] to the form

$$[U] = \begin{bmatrix} a_{11} & a_{12} & a_{13} \\ 0 & a'_{22} & a'_{23} \\ 0 & 0 & a''_{33} \end{bmatrix} \tag{10.9}$$

which is in the desired upper triangular format.

Though it might not be as apparent, the matrix [L] is also produced during the step. This can be readily illustrated for a three-equation system,

$$\begin{bmatrix} a_{11} & a_{12} & a_{13} \\ a_{21} & a_{22} & a_{23} \\ a_{31} & a_{32} & a_{33} \end{bmatrix} \begin{Bmatrix} x_1 \\ x_2 \\ x_3 \end{Bmatrix} = \begin{Bmatrix} b_1 \\ b_2 \\ b_3 \end{Bmatrix}$$

The first step in Gauss elimination is to multiply row 1 by the factor [recall Eq. (9.13)]

$$f_{21} = \frac{a_{21}}{a_{11}}$$

and subtract the result from the second row to eliminate a_{21}. Similarly, row 1 is multiplied by

$$f_{31} = \frac{a_{31}}{a_{11}}$$

and the result is subtracted from the third row to eliminate a_{31}. The final step is to multiply the modified second row by

$$f_{32} = \frac{a'_{32}}{a'_{22}}$$

and subtract the result from the third row to eliminate a'_{32}.

Now suppose that we merely perform all these manipulations on the matrix $[A]$. Clearly, if we do not want to change the equation, we also have to do the same to the right-hand side, $\{B\}$. But there is absolutely no reason that we have to perform the manipulations simultaneously. Thus, we could save the f's and manipulate $\{B\}$ later.

Where do we store the factors f_{21}, f_{31}, and f_{32}? Recall that the whole idea behind the elimination was to create zeros in a_{21}, a_{31}, and a_{32}. Thus, we can store f_{21} in a_{21}, f_{31} in a_{31}, and f_{32} in a_{32}. After elimination, the $[A]$ matrix can therefore be written as

$$\begin{bmatrix} a_{11} & a_{12} & a_{13} \\ f_{21} & a'_{22} & a'_{23} \\ f_{31} & f_{32} & a''_{33} \end{bmatrix} \tag{10.10}$$

This matrix, in fact, represents an efficient storage of the *LU* decomposition of $[A]$,

$$[A] \rightarrow [L][U] \tag{10.11}$$

where

$$[U] = \begin{bmatrix} a_{11} & a_{12} & a_{13} \\ 0 & a'_{22} & a'_{23} \\ 0 & 0 & a''_{33} \end{bmatrix}$$

and

$$[L] = \begin{bmatrix} 1 & 0 & 0 \\ f_{21} & 1 & 0 \\ f_{31} & f_{32} & 1 \end{bmatrix}$$

The following example confirms that $[A] = [L][U]$.

EXAMPLE 10.1 | *LU* **Decomposition with Gauss Elimination**

Problem Statement. Derive an *LU* decomposition based on the Gauss elimination performed in Example 9.5.

Solution. In Example 9.5, we solved the matrix

$$[A] = \begin{bmatrix} 3 & -0.1 & -0.2 \\ 0.1 & 7 & -0.3 \\ 0.3 & -0.2 & 10 \end{bmatrix}$$

After forward elimination, the following upper triangular matrix was obtained:

$$[U] = \begin{bmatrix} 3 & -0.1 & -0.2 \\ 0 & 7.00333 & -0.293333 \\ 0 & 0 & 10.0120 \end{bmatrix}$$

The factors employed to obtain the upper triangular matrix can be assembled into a lower triangular matrix. The elements a_{21} and a_{31} were eliminated by using the factors

$$f_{21} = \frac{0.1}{3} = 0.03333333 \qquad f_{31} = \frac{0.3}{3} = 0.1000000$$

and the element a'_{32} was eliminated by using the factor

$$f_{32} = \frac{-0.19}{7.00333} = -0.0271300$$

Thus, the lower triangular matrix is

$$[L] = \begin{bmatrix} 1 & 0 & 0 \\ 0.0333333 & 1 & 0 \\ 0.100000 & -0.0271300 & 1 \end{bmatrix}$$

Consequently, the *LU* decomposition is

$$[A] = [L][U] = \begin{bmatrix} 1 & 0 & 0 \\ 0.0333333 & 1 & 0 \\ 0.100000 & -0.0271300 & 1 \end{bmatrix} \begin{bmatrix} 3 & -0.1 & -0.2 \\ 0 & 7.00333 & -0.293333 \\ 0 & 0 & 10.0120 \end{bmatrix}$$

This result can be verified by performing the multiplication $[L][U]$ to give

$$[L][U] = \begin{bmatrix} 3 & -0.1 & -0.2 \\ 0.0999999 & 7 & -0.3 \\ 0.3 & -0.2 & 9.99996 \end{bmatrix}$$

where the minor discrepancies are due to round-off.

The following is pseudocode for a subroutine to implement the decomposition phase:

```
SUB Decompose(a, n)
  DOFOR k = 1, n - 1
    DOFOR i = k + 1, n
      factor = a_{i,k}/a_{k,k}
      a_{i,k} = factor
      DOFOR j = k + 1, n
        a_{i,j} = a_{i,j} - factor * a_{k,j}
      END DO
    END DO
  END DO
END Decompose
```

Notice that this algorithm is "naive" in the sense that pivoting is not included. This feature will be added later when we develop the full algorithm for *LU* decomposition.

After the matrix is decomposed, a solution can be generated for a particular right-hand-side vector $\{B\}$. This is done in two steps. First, a forward-substitution step is executed by solving Eq. (10.8) for $\{D\}$. It is important to recognize that this merely amounts to performing the elimination manipulations on $\{B\}$. Thus, at the end of this step, the right-hand side will be in the same state that it would have been had we performed forward manipulation on $[A]$ and $\{B\}$ simultaneously.

The forward-substitution step can be represented concisely as

$$d_i = b_i - \sum_{j=1}^{i-1} a_{ij} d_j \qquad \text{for } i = 2, 3, \dots, n \tag{10.12}$$

The second step then merely amounts to implementing back substitution, as in Eq. (10.4). Again, it is important to recognize that this is identical to the back-substitution phase of conventional Gauss elimination. Thus, in a fashion similar to Eqs. (9.16) and (9.17), the back-substitution step can be represented concisely as

$$x_n = d_n / a_{nn} \tag{10.13}$$

$$x_i = \frac{d_i - \displaystyle\sum_{j=i+1}^{n} a_{ij} x_j}{a_{ii}} \qquad \text{for } i = n-1, n-2, \dots, 1 \tag{10.14}$$

EXAMPLE 10.2 **The Substitution Steps**

Problem Statement.   Complete the problem initiated in Example 10.1 by generating the final solution with forward and back substitution.

Solution.   As stated above, the intent of forward substitution is to impose the elimination manipulations, which we had formerly applied to $[A]$, on the right-hand-side vector $\{B\}$. Recall that the system being solved in Example 9.5 was

$$\begin{bmatrix} 3 & -0.1 & -0.2 \\ 0.1 & 7 & -0.3 \\ 0.3 & -0.2 & 10 \end{bmatrix} \begin{Bmatrix} x_1 \\ x_2 \\ x_3 \end{Bmatrix} = \begin{Bmatrix} 7.85 \\ -19.3 \\ 71.4 \end{Bmatrix}$$

and that the forward-elimination phase of conventional Gauss elimination resulted in

$$\begin{bmatrix} 3 & -0.1 & -0.2 \\ 0 & 7.00333 & -0.293333 \\ 0 & 0 & 10.0120 \end{bmatrix} \begin{Bmatrix} x_1 \\ x_2 \\ x_3 \end{Bmatrix} = \begin{Bmatrix} 7.85 \\ -19.5617 \\ 70.0843 \end{Bmatrix} \tag{E10.2.1}$$

The forward-substitution phase is implemented by applying Eq. (10.7) to this case,

$$\begin{bmatrix} 1 & 0 & 0 \\ 0.0333333 & 1 & 0 \\ 0.100000 & -0.0271300 & 1 \end{bmatrix} \begin{Bmatrix} d_1 \\ d_2 \\ d_3 \end{Bmatrix} = \begin{Bmatrix} 7.85 \\ -19.3 \\ 71.4 \end{Bmatrix}$$

or, multiplying out the left-hand side,

$$d_1 = 7.85$$
$$0.0333333 d_1 + d_2 = -19.3$$
$$0.1 d_1 - 0.02713 d_2 + d_3 = 71.4$$

We can solve the first equation for d_1,

$$d_1 = 7.85$$

which can be substituted into the second equation to solve for

$$d_2 = -19.3 - 0.0333333(7.85) = -19.5617$$

Both d_1 and d_2 can be substituted into the third equation to give

$$d_3 = 71.4 - 0.1(7.85) + 0.02713(-19.5617) = 70.0843$$

Thus,

$$\{D\} = \left\{ \begin{array}{c} 7.85 \\ -19.5617 \\ 70.0843 \end{array} \right\}$$

which is identical to the right-hand side of Eq. (E10.2.1).

This result can then be substituted into Eq. (10.4), $[U]\{X\} = \{D\}$, to give

$$\begin{bmatrix} 3 & -0.1 & -0.2 \\ 0 & 7.00333 & -0.293333 \\ 0 & 0 & 10.0120 \end{bmatrix} \left\{ \begin{array}{c} x_1 \\ x_2 \\ x_3 \end{array} \right\} = \left\{ \begin{array}{c} 7.85 \\ -19.5617 \\ 70.0843 \end{array} \right\}$$

which can be solved by back substitution (see Example 9.5 for details) to obtain the final solution,

$$\{X\} = \left\{ \begin{array}{c} 3 \\ -2.5 \\ 7.00003 \end{array} \right\}$$

The following is pseudocode for a subroutine to implement both substitution phases:

```
SUB Substitute(a, n, b, x)
  'forward substitution
  DOFOR i = 2, n
    sum = b_i
    DOFOR j = 1, i - 1
      sum = sum - a_i,j * b_j
    END DO
    b_i = sum
  END DO
  'back substitution
  xn = bn/an,n
  DOFOR i = n - 1, 1, -1
    sum = 0
    DOFOR j = i + 1, n
      sum = sum + a_i,j * x_j
    END DO
    x_i = (b_i - sum)/a_i,i
  END DO
END Substitute
```

The *LU* decomposition algorithm requires the same total number of multiply/divide flops as for Gauss elimination. The only difference is that a little less effort is expended in the decomposition phase since the operations are not applied to the right-hand side. Thus, the number of multiply/divide flops involved in the decomposition phase can be calculated as

$$\frac{n^3}{3} - \frac{n}{3} \xrightarrow{\text{as } n \text{ increases}} \frac{n^3}{3} + O(n) \tag{10.15}$$

Conversely, the substitution phase takes a little more effort. Thus, the number of flops for forward and back substitution is n^2. The total effort is therefore identical to that involved in Gauss elimination

$$\frac{n^3}{3} - \frac{n}{3} + n^2 \xrightarrow{\text{as } n \text{ increases}} \frac{n^3}{3} + O(n^2) \tag{10.16}$$

10.1.3 *LU* Decomposition Algorithm

An algorithm to implement an *LU* decomposition version of Gauss elimination is listed in Fig. 10.2. Four features of this algorithm bear mention:

- The factors generated during the elimination phase are stored in the lower part of the matrix. This can be done because these are converted to zeros anyway and are unnecessary for the final solution. This storage saves space.
- This algorithm keeps track of pivoting by using an order vector *o*. This greatly speeds up the algorithm because only the order vector (as opposed to the whole row) is pivoted.
- The equations are not scaled, but scaled values of the elements are used to determine whether pivoting is to be implemented.
- The diagonal term is monitored during the pivoting phase to detect near-zero occurrences in order to flag singular systems. If the monitoring passes back a value of $er = -1$, a singular matrix has been detected and the computation should be terminated. A parameter *tol* is set by the user to a small number in order to detect near-zero occurrences.

10.1.4 Crout Decomposition

Notice that for the *LU* decomposition implementation of Gauss elimination, the [L] matrix has 1's on the diagonal. This approach is formally referred to as a *Doolittle decomposition*, or factorization. An alternative approach involves a [U] matrix with 1's on the diagonal. This is called *Crout decomposition*. Although there are some differences between the approaches (Atkinson 1978; Ralston and Rabinowitz 1978), their performance is comparable.

The Crout decomposition generates [U] and [L] by sweeping through the matrix by columns and rows, as depicted in Fig. 10.3. It can be implemented by the following concise series of formulas:

$$l_{i,1} = a_{i,1} \qquad \text{for } i = 1, 2, \dots, n \tag{10.17}$$

$$u_{1j} = \frac{a_{1j}}{l_{11}} \qquad \text{for } j = 2, 3, \dots, n \tag{10.18}$$

```
SUB Ludecomp(a, b, n, tol, x, er)
  DIM o_n, s_n
  er = 0
  CALL Decompose(a, n, tol, o, s, er)
  IF er <> -1 THEN
    CALL Substitute(a, o, n, b, x)
  END IF
END Ludecomp

SUB Decompose (a, n, tol, o, s, er)
  DOFOR i = 1, n
    o_i = i
    s_i = ABS(a_{i,1})
    DOFOR j = 2, n
      IF ABS(a_{i,j})>s_i THEN s_i = ABS(a_{i,j})
    END DO
  END DO
  DOFOR k = 1, n - 1
    CALL Pivot(a, o, s, n, k)
    IF ABS(a_{o(k),k}/s_{o(k)}) < tol THEN
      er = -1
      PRINT a_{o(k),k}/s_{o(k)}
      EXIT DO
    END IF
    DOFOR i = k + 1, n
      factor = a_{o(i),k}/a_{o(k),k}
      a_{o(i),k} = factor
      DOFOR j = k + 1, n
        a_{o(i),j} = a_{o(i),j} - factor * a_{o(k),j}
      END DO
    END DO
  END DO
  IF ABS(a_{o(k),k}/s_{o(k)}) < tol THEN
    er = -1
    PRINT a_{o(k),k}/s_{o(k)}
```

```
  END IF
END Decompose

SUB Pivot(a, o, s, n, k)
  p = k
  big = ABS(a_{o(k),k}/s_{o(k)})
  DOFOR ii = k + 1, n
    dummy = ABS(a_{o(ii),k}/s_{o(ii)})
    IF dummy > big THEN
      big = dummy
      p = ii
    END IF
  END DO
  dummy = o_p
  o_p = o_k
  o_k = dummy
END Pivot

SUB Substitute(a, o, n, b, x)
  DOFOR i = 2, n
    sum = b_{o(i)}
    DOFOR j = 1, i - 1
      sum = sum - a_{o(i),j} * b_{o(j)}
    END DO
    b_{o(i)} = sum
  END DO
  x_n = b_{o(n)}/a_{o(n),n}
  DOFOR i = n - 1, 1, -1
    sum = 0
    DOFOR j = i + 1, n
      sum = sum + a_{o(i),j} * x_j
    END DO
    x_i = (b_{o(i)} - sum)/a_{o(i),i}
  END DO
END Substitute
```

FIGURE 10.2
Pseudocode for an *LU* decomposition algorithm.

FIGURE 10.3
A schematic depicting the evaluations involved in Crout decomposition.

For $j = 2, 3, \ldots, n - 1$

$$l_{ij} = a_{ij} - \sum_{k=1}^{j-1} l_{ik}u_{kj} \quad \text{for } i = j, j + 1, \ldots, n \tag{10.19}$$

$$u_{jk} = \frac{a_{jk} - \sum_{i=1}^{j-1} l_{ji}u_{ik}}{l_{jj}} \quad \text{for } k = j + 1, j + 2, \ldots, n \tag{10.20}$$

```
DOFOR j = 2, n
  a_{1,j} = a_{1,j}/a_{1,1}
END DO
DOFOR j = 2, n − 1
  DOFOR i = j, n
    sum = 0
    DOFOR k = 1, j − 1
      sum = sum + a_{i,k} · a_{k,j}
    END DO
    a_{i,j} = a_{i,j} − sum
  END DO
  DOFOR k = j + 1, n
    sum = 0
    DOFOR i = 1, j − 1
      sum = sum + a_{j,i} · a_{i,k}
    END DO
    a_{j,k} = (a_{j,k} − sum)/a_{j,j}
  END DO
END DO
sum = 0
DOFOR k = 1, n − 1
  sum = sum + a_{n,k} · a_{k,n}
END DO
a_{n,n} = a_{n,n} − sum
```

FIGURE 10.4
Pseudocode for Crout's *LU* decomposition algorithm.

and

$$l_{nn} = a_{nn} - \sum_{k=1}^{n-1} l_{nk}u_{kn} \tag{10.21}$$

Aside from the fact that it consists of a few concise loops, the foregoing approach also has the benefit that storage space can be economized. There is no need to store the 1's on the diagonal of $[U]$ or the 0's for $[L]$ or $[U]$ because they are givens in the method. Consequently, the values of $[U]$ can be stored in the zero space of $[L]$. Further, close examination of the foregoing derivation makes it clear that after each element of $[A]$ is employed once, it is never used again. Therefore, as each element of $[L]$ and $[U]$ is computed, it can be substituted for the corresponding element (as designated by its subscripts) of $[A]$.

Pseudocode to accomplish this is presented in Fig. 10.4. Notice that Eq. (10.17) is not included in the pseudocode because the first column of $[L]$ is already stored in $[A]$. Otherwise, the algorithm directly follows from Eqs. (10.18) through (10.21).

10.2 THE MATRIX INVERSE

In our discussion of matrix operations (Sec. PT3.2.2), we introduced the notion that if a matrix $[A]$ is square, there is another matrix, $[A]^{-1}$, called the inverse of $[A]$, for which [Eq. (PT3.3)]

$$[A][A]^{-1} = [A]^{-1}[A] = [I]$$

Now we will focus on how the inverse can be computed numerically. Then we will explore how it can be used for engineering analysis.

10.2.1 Calculating the Inverse

The inverse can be computed in a column-by-column fashion by generating solutions with unit vectors as the right-hand-side constants. For example, if the right-hand-side constant has a 1 in the first position and zeros elsewhere,

$$\{b\} = \begin{Bmatrix} 1 \\ 0 \\ 0 \end{Bmatrix}$$

the resulting solution will be the first column of the matrix inverse. Similarly, if a unit vector with a 1 in the second row,

$$\{b\} = \begin{Bmatrix} 0 \\ 1 \\ 0 \end{Bmatrix}$$

is used, the result will be the second column of the matrix inverse.

The best way to implement such a calculation is with the LU decomposition algorithm described earlier in this chapter. Recall that one of the great strengths of LU decomposition is that it provides a very efficient means to evaluate multiple right-hand-side vectors. Thus, it is ideal for evaluating the multiple unit vectors needed to compute the inverse.

EXAMPLE 10.3   **Matrix Inversion**

Problem Statement.   Employ LU decomposition to determine the matrix inverse for the system from Example 10.2.

$$[A] = \begin{bmatrix} 3 & -0.1 & -0.2 \\ 0.1 & 7 & -0.3 \\ 0.3 & -0.2 & 10 \end{bmatrix}$$

Recall that the decomposition resulted in the following lower and upper triangular matrices:

$$[U] = \begin{bmatrix} 3 & -0.1 & -0.2 \\ 0 & 7.00333 & -0.293333 \\ 0 & 0 & 10.0120 \end{bmatrix} \quad [L] = \begin{bmatrix} 1 & 0 & 0 \\ 0.0333333 & 1 & 0 \\ 0.100000 & -0.0271300 & 1 \end{bmatrix}$$

Solution.   The first column of the matrix inverse can be determined by performing the forward-substitution procedure with a unit vector (with 1 in the first row) as the right-hand-side vector. Thus, Eq. (10.8), the lower-triangular system, can be set up as

$$\begin{bmatrix} 1 & 0 & 0 \\ 0.0333333 & 1 & 0 \\ 0.100000 & -0.0271300 & 1 \end{bmatrix} \begin{Bmatrix} d_1 \\ d_2 \\ d_3 \end{Bmatrix} = \begin{Bmatrix} 1 \\ 0 \\ 0 \end{Bmatrix}$$

and solved with forward substitution for $\{D\}^T = \lfloor 1 \quad -0.03333 \quad -0.1009 \rfloor$. This vector can then be used as the right-hand side of Eq. (10.3),

$$\begin{bmatrix} 3 & -0.1 & -0.2 \\ 0 & 7.00333 & -0.293333 \\ 0 & 0 & 10.0120 \end{bmatrix} \begin{Bmatrix} x_1 \\ x_2 \\ x_3 \end{Bmatrix} = \begin{Bmatrix} 1 \\ -0.03333 \\ -0.1009 \end{Bmatrix}$$

which can be solved by back substitution for $\{X\}^T = \lfloor 0.33249 \quad -0.00518 \quad -0.01008 \rfloor$, thus providing the first column of the matrix,

$$[A]^{-1} = \begin{bmatrix} 0.33249 & 0 & 0 \\ -0.00518 & 0 & 0 \\ -0.01008 & 0 & 0 \end{bmatrix}$$

To determine the second column, Eq. (10.8) is formulated as

$$\begin{bmatrix} 1 & 0 & 0 \\ 0.0333333 & 1 & 0 \\ 0.100000 & -0.0271300 & 1 \end{bmatrix} \begin{Bmatrix} d_1 \\ d_2 \\ d_3 \end{Bmatrix} = \begin{Bmatrix} 0 \\ 1 \\ 0 \end{Bmatrix}$$

This can be solved for $\{D\}$, and the results are used with Eq. (10.3) to determine $\{X\}^T = \lfloor 0.004944 \quad 0.142903 \quad 0.00271 \rfloor$, which is the second column of the matrix,

$$[A]^{-1} = \begin{bmatrix} 0.33249 & 0.004944 & 0 \\ -0.00518 & 0.142903 & 0 \\ -0.01008 & 0.00271 & 0 \end{bmatrix}$$

Finally, the forward- and back-substitution procedures can be implemented with $\{B\}^T = \lfloor 0 \quad 0 \quad 1 \rfloor$ to solve for $\{X\}^T = \lfloor 0.006798 \quad 0.004183 \quad 0.09988 \rfloor$, which is the final column of the matrix,

$$[A]^{-1} = \begin{bmatrix} 0.33249 & 0.004944 & 0.006798 \\ -0.00518 & 0.142903 & 0.004183 \\ -0.01008 & 0.00271 & 0.09988 \end{bmatrix}$$

The validity of this result can be checked by verifying that $[A][A]^{-1} = [I]$.

Pseudocode to generate the matrix inverse is shown in Fig. 10.5. Notice how the decomposition subroutine from Fig. 10.2 is called to perform the decomposition and then generates the inverse by repeatedly calling the substitution algorithm with unit vectors.

The effort required for this algorithm is simply computed as

$$\underbrace{\frac{n^3}{3} - \frac{n}{3}}_{\text{decomposition}} + \underbrace{n(n^2)}_{+\; n \times \text{ substitutions}} = \frac{4n^3}{3} - \frac{n}{4} \tag{10.22}$$

where from Sec. 10.1.2, the decomposition is defined by Eq. (10.15) and the effort involved with every right-hand-side evaluation involves n^2 multiply/divide flops.

```
CALL Decompose(a, n, tol, o, s, er)
IF er = 0 THEN
  DOFOR i = 1, n
    DOFOR j = 1, n
      IF i = j THEN
        b(j) = 1
      ELSE
        b(j) = 0
      END IF
    END DO
    CALL Substitute(a, o, n, b, x)
    DOFOR j = 1, n
      ai(j, i) = x(j)
    END DO
  END DO
  Output ai, if desired
ELSE
  PRINT "ill-conditioned system"
END IF
```

FIGURE 10.5
Driver program that uses some of the subprograms from Fig. 10.2 to generate a matrix inverse.

10.2.2 Stimulus-Response Computations

As discussed in Sec. PT3.1.2, many of the linear systems of equations confronted in engineering practice are derived from conservation laws. The mathematical expression of these laws is some form of balance equation to ensure that a particular property—mass, force, heat, momentum, or other—is conserved. For a force balance on a structure, the properties might be horizontal or vertical components of the forces acting on each node of the structure (see Sec. 12.2). For a mass balance, the properties might be the mass in each reactor of a chemical process (see Sec. 12.1). Other fields of engineering would yield similar examples.

A single balance equation can be written for each part of the system, resulting in a set of equations defining the behavior of the property for the entire system. These equations are interrelated, or coupled, in that each equation may include one or more of the variables from the other equations. For many cases, these systems are linear and, therefore, of the exact form dealt with in this chapter:

$$[A]\{X\} = \{B\} \tag{10.23}$$

Now, for balance equations, the terms of Eq. (10.23) have a definite physical interpretation. For example, the elements of $\{X\}$ are the levels of the property being balanced for each part of the system. In a force balance of a structure, they represent the horizontal and vertical forces in each member. For the mass balance, they are the mass of chemical in each reactor. In either case, they represent the system's state, or response, which we are trying to determine.

The right-hand-side vector $\{B\}$ contains those elements of the balance that are independent of behavior of the system—that is, they are constants. As such, they often represent the external forces or stimuli that drive the system.

Finally, the matrix of coefficients [*A*] usually contains the parameters that express how the parts of the system interact or are coupled. Consequently, Eq. (10.23) might be re-expressed as

[Interactions]{response} = {stimuli}

Thus, Eq. (10.23) can be seen as an expression of the fundamental mathematical model that we formulated previously as a single equation in Chap. 1 [recall Eq. (1.1)]. We can now see that Eq. (10.23) represents a version that is designed for coupled systems involving several dependent variables {*X*}.

As we know from this chapter and Chap. 9, there are a variety of ways to solve Eq. (10.23). However, using the matrix inverse yields a particularly interesting result. The formal solution can be expressed as

$$\{X\} = [A]^{-1}\{B\}$$

or (recalling our definition of matrix multiplication from Box PT3.2)

$$x_1 = a_{11}^{-1}b_1 + a_{12}^{-1}b_2 + a_{13}^{-1}b_3$$
$$x_2 = a_{21}^{-1}b_1 + a_{22}^{-1}b_2 + a_{23}^{-1}b_3$$
$$x_3 = a_{31}^{-1}b_1 + a_{32}^{-1}b_2 + a_{33}^{-1}b_3$$

Thus, we find that the inverted matrix itself, aside from providing a solution, has extremely useful properties. That is, each of its elements represents the response of a single part of the system to a unit stimulus of any other part of the system.

Notice that these formulations are linear and, therefore, superposition and proportionality hold. *Superposition* means that if a system is subject to several different stimuli (the *b*'s), the responses can be computed individually and the results summed to obtain a total response. *Proportionality* means that multiplying the stimuli by a quantity results in the response to those stimuli being multiplied by the same quantity. Thus, the coefficient a_{11}^{-1} is a proportionality constant that gives the value of x_1 due to a unit level of b_1. This result is independent of the effects of b_2 and b_3 on x_1, which are reflected in the coefficients a_{12}^{-1} and a_{13}^{-1}, respectively. Therefore, we can draw the general conclusion that the element a_{ij}^{-1} of the inverted matrix represents the value of x_i due to a unit quantity of b_j. Using the example of the structure, element a_{ij}^{-1} of the matrix inverse would represent the force in member *i* due to a unit external force at node *j*. Even for small systems, such behavior of individual stimulus-response interactions would not be intuitively obvious. As such, the matrix inverse provides a powerful technique for understanding the interrelationships of component parts of complicated systems. This power will be demonstrated in Secs. 12.1 and 12.2.

10.3 ERROR ANALYSIS AND SYSTEM CONDITION

Aside from its engineering applications, the matrix inverse also provides a means to discern whether systems are ill-conditioned. Three methods are available for this purpose:

1. Scale the matrix of coefficients [*A*] so that the largest element in each row is 1. Invert the scaled matrix and if there are elements of $[A]^{-1}$ that are several orders of magnitude greater than 1, it is likely that the system is ill-conditioned (see Box 10.1).

Box 10.1 Interpreting the Elements of the Matrix Inverse as a Measure of Ill-Conditioning

One method for assessing a system's condition is to scale $[A]$ so that the largest element in each row is 1 and then compute $[A]^{-1}$. If elements of $[A]^{-1}$ are several orders of magnitude greater than the elements of the original scaled matrix, it is likely that the system is ill-conditioned.

Insight into this approach can be gained by recalling that a way to check whether an approximate solution $\{X\}$ is acceptable is to substitute it into the original equations and see whether the original right-hand-side constants result. This is equivalent to

$$\{R\} = \{B\} - [A]\{\tilde{X}\} \tag{B10.1.1}$$

where $\{R\}$ is the residual between the right-hand-side constants and the values computed with the solution $\{\tilde{X}\}$. If $\{R\}$ is small, we might conclude that the $\{\tilde{X}\}$ values are adequate. However, suppose that $\{X\}$ is the exact solution that yields a zero residual, as in

$$\{0\} = \{B\} - [A]\{X\} \tag{B10.1.2}$$

Subtracting Eq. (B10.1.2) from (B10.1.1) yields

$$\{R\} = [A]\{\{X\} - \{\tilde{X}\}\}$$

Multiplying both sides of this equation by $[A]^{-1}$ gives

$$\{X\} - \{\tilde{X}\} = [A]^{-1}\{R\}$$

This result indicates why checking a solution by substitution can be misleading. For cases where elements of $[A]^{-1}$ are large, a small discrepancy in the right-hand-side residual $\{R\}$ could correspond to a large error $\{X\} - \{\tilde{X}\}$ in the calculated value of the unknowns. In other words, a small residual does not guarantee an accurate solution. However, we can conclude that if the largest element of $[A]^{-1}$ is on the order of magnitude of unity, the system can be considered to be well-conditioned. Conversely, if $[A]^{-1}$ includes elements much larger than unity, we conclude that the system is ill-conditioned.

2. Multiply the inverse by the original coefficient matrix and assess whether the result is close to the identity matrix. If not, this indicates ill-conditioning.
3. Invert the inverted matrix and assess whether the result is sufficiently close to the original coefficient matrix. If not, this also indicates that the system is ill-conditioned.

Although these methods can indicate ill-conditioning, it would be preferable to obtain a single number (such as the condition number from Sec. 4.2.3) that could serve as an indicator of the problem. Attempts to formulate such a matrix condition number are based on the mathematical concept of the norm.

10.3.1 Vector and Matrix Norms

A *norm* is a real-valued function that provides a measure of the size or "length" of multicomponent mathematical entities such as vectors and matrices (see Box 10.2).

A simple example is a vector in three-dimensional Euclidean space (Fig. 10.6) that can be represented as

$$\lfloor F \rfloor = \lfloor a \quad b \quad c \rfloor$$

where a, b, and c are the distances along the x, y, and z axes, respectively. The length of this vector—that is, the distance from the coordinate $(0, 0, 0)$ to (a, b, c)—can be simply computed as

$$\|F\|_e = \sqrt{a^2 + b^2 + c^2}$$

where the nomenclature $\|F\|_e$ indicates that this length is referred to as the Euclidean norm of $[F]$.

Box 10.2 Matrix Norms

As developed in this section, Euclidean norms can be employed to quantify the size of a vector,

$$\|X\|_e = \sqrt{\sum_{i=1}^{n} x_i^2}$$

or matrix,

$$\|A\|_e = \sqrt{\sum_{i=1}^{n} \sum_{j=1}^{n} a_{i,j}^2}$$

For vectors, there are alternatives called *p* norms that can be represented generally by

$$\|X\|_p = \left(\sum_{i=1}^{n} |x_i|^p \right)^{1/p}$$

We can also see that the Euclidean norm and the 2 norm, $\|X\|_2$, are identical for vectors.

Other important examples are

$$\|X\|_1 = \sum_{i=1}^{n} |x_i|$$

which represents the norm as the sum of the absolute values of the elements. Another is the *maximum-magnitude norm*, or *uniform-vector norm*,

$$\|X\|_\infty = \max_{1 \le i \le n} |x_i|$$

which defines the norm as the element with the largest absolute value.

Using a similar approach, norms can be developed for matrices. For example,

$$\|A\|_1 = \max_{1 \le j \le n} \sum_{i=1}^{n} |a_{ij}|$$

That is, a summation of the absolute values of the coefficients is performed for each column, and the largest of these summations is taken as the norm. This is called the *column-sum norm*.

A similar determination can be made for the rows, resulting in a *uniform-matrix norm*, or *row-sum norm*,

$$\|A\|_\infty = \max_{1 \le i \le n} \sum_{j=1}^{n} |a_{ij}|$$

It should be noted that, in contrast to the norms for vectors, the 2 norm and the Euclidean norm for a matrix are not the same. Whereas the Euclidean norm $\|A\|_e$ can be easily determined by Eq. (10.24), the matrix 2 norm $\|A\|_2$ is calculated as

$$\|A\|_2 = (\mu_{max})^{1/2}$$

where μ_{max} is the largest eigenvalue of $[A]^T [A]$. In Chap. 27, we will learn more about eigenvalues. For the time being, the important point is that the $\|A\|_2$, or *spectral, norm* is the minimum norm and, therefore, provides the tightest measure of size (Ortega 1972).

FIGURE 10.6

Graphical depiction of a vector $\lfloor F \rfloor = \lfloor a \quad b \quad c \rfloor$ in Euclidean space.

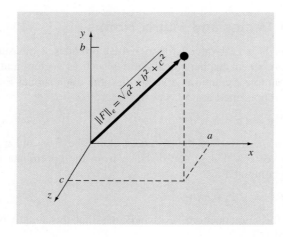

Similarly, for an n-dimensional vector $\lfloor X \rfloor = \lfloor x_1 \quad x_2 \quad \cdots \quad x_n \rfloor$, a Euclidean norm would be computed as

$$\|X\|_e = \sqrt{\sum_{i=1}^{n} x_i^2}$$

The concept can be extended to a matrix $[A]$, as in

$$\|A\|_e = \sqrt{\sum_{i=1}^{n} \sum_{j=1}^{n} a_{i,j}^2} \tag{10.24}$$

which is given a special name—the *Frobenius norm*. However, as with the other vector norms, it provides a single value to quantify the "size" of $[A]$.

It should be noted that there are alternatives to the Euclidean and Frobenius norms (see Box 10.2). For example, a *uniform-vector norm* is defined as

$$\|X\|_\infty = \max_{1 \leq i \leq n} |x_i|$$

That is, the element with the largest absolute value is taken as the measure of the vector's size. Similarly, a *uniform-matrix norm*, or *row-sum norm*, is defined as

$$\|A\|_\infty = \max_{1 \leq i \leq n} \sum_{j=1}^{n} |a_{ij}| \tag{10.25}$$

In this case, the sum of the absolute value of the elements is computed for each row, and the largest of these is taken as the norm.

Although there are theoretical benefits to using certain of the norms, the choice is sometimes influenced by practical considerations. For example, the uniform-row norm is widely used because of the ease with which it can be calculated and the fact that it usually provides an adequate measure of matrix size.

10.3.2 Matrix Condition Number

Now that we have introduced the concept of the norm, we can use it to define

$$\text{Cond } [A] = \|A\| \cdot \|A^{-1}\| \tag{10.26}$$

where Cond $[A]$ is called the *matrix condition number*. Note that for a matrix $[A]$, this number will be greater than or equal to 1. It can be shown (Ralston and Rabinowitz 1978; Gerald and Wheatley 2004) that

$$\frac{\|\Delta X\|}{\|X\|} \leq \text{Cond } [A] \frac{\|\Delta A\|}{\|A\|}$$

That is, the relative error of the norm of the computed solution can be as large as the relative error of the norm of the coefficients of $[A]$ multiplied by the condition number. For example, if the coefficients of $[A]$ are known to t-digit precision (that is, rounding errors are on the order of 10^{-t}) and Cond $[A] = 10^c$, the solution $[X]$ may be valid to only $t - c$ digits (rounding errors of $\sim 10^{c-t}$).

EXAMPLE 10.4   **Matrix Condition Evaluation**

Problem Statement.   The Hilbert matrix, which is notoriously ill-conditioned, can be represented generally as

$$\begin{bmatrix} 1 & 1/2 & 1/3 & \cdots & 1/n \\ 1/2 & 1/3 & 1/4 & \cdots & 1/(n+1) \\ \cdot & \cdot & \cdot & & \cdot \\ \cdot & \cdot & \cdot & & \cdot \\ \cdot & \cdot & \cdot & & \cdot \\ 1/n & 1/(n+1) & 1/(n+2) & \cdots & 1/(2n-1) \end{bmatrix}$$

Use the row-sum norm to estimate the matrix condition number for the 3×3 Hilbert matrix

$$[A] = \begin{bmatrix} 1 & 1/2 & 1/3 \\ 1/2 & 1/3 & 1/4 \\ 1/3 & 1/4 & 1/5 \end{bmatrix}$$

Solution.   First, the matrix can be normalized so that the maximum element in each row is 1,

$$[A] = \begin{bmatrix} 1 & 1/2 & 1/3 \\ 1 & 2/3 & 1/2 \\ 1 & 3/4 & 3/5 \end{bmatrix}$$

Summing each of the rows gives 1.833, 2.1667, and 2.35. Thus, the third row has the largest sum and the row-sum norm is

$$\|A\|_\infty = 1 + \frac{3}{4} + \frac{3}{5} = 2.35$$

The inverse of the scaled matrix can be computed as

$$[A]^{-1} = \begin{bmatrix} 9 & -18 & 10 \\ -36 & 96 & -60 \\ 30 & -90 & 60 \end{bmatrix}$$

Note that the elements of this matrix are larger than those of the original matrix. This is also reflected in its row-sum norm, which is computed as

$$\|A^{-1}\|_\infty = |-36| + |96| + |-60| = 192$$

Thus, the condition number can be calculated as

Cond $[A] = 2.35(192) = 451.2$

The fact that the condition number is considerably greater than unity suggests that the system is ill-conditioned. The extent of the ill-conditioning can be quantified by calculating $c = \log 451.2 = 2.65$. Computers using IEEE floating-point representation have approximately $t = \log 2^{-24} = 7.2$ significant base-10 digits (recall Sec. 3.4.1). Therefore, the solution could exhibit rounding errors of up to $10^{(2.65-7.2)} = 3 \times 10^{-5}$. Note that such estimates almost always overpredict the actual error. However, they are useful in alerting you to the possibility that round-off errors may be significant.

Practically speaking, the problem with implementing Eq. (10.26) is the computational price required to obtain $\|A^{-1}\|$. Rice (1983) outlines some possible strategies to mitigate this problem. Further, he suggests an alternative way to assess system condition: Run the same solution on two different compilers. Because the resulting codes will likely implement the arithmetic differently, the effect of ill-conditioning should be evident from such an experiment. Finally, it should be mentioned that software packages such as MATLAB software and Mathcad have the capability to conveniently compute matrix condition. We will cover these capabilities when we review such packages at the end of Chap. 11.

10.3.3 Iterative Refinement

In some cases, round-off errors can be reduced by the following procedure. Suppose that we are solving the following set of equations:

$$a_{11}x_1 + a_{12}x_2 + a_{13}x_3 = b_1$$
$$a_{21}x_1 + a_{22}x_2 + a_{23}x_3 = b_2 \tag{10.27}$$
$$a_{31}x_1 + a_{32}x_2 + a_{33}x_3 = b_3$$

For conciseness, we will limit the following discussion to this small (3×3) system. However, the approach is generally applicable to larger sets of linear equations.

Suppose an approximate solution vector is given by $\{\tilde{X}\}^T = \lfloor \tilde{x}_1 \; \tilde{x}_2 \; \tilde{x}_3 \rfloor$. This solution can be substituted into Eq. (10.27) to give

$$a_{11}\tilde{x}_1 + a_{12}\tilde{x}_2 + a_{13}\tilde{x}_3 = \tilde{b}_1$$
$$a_{21}\tilde{x}_1 + a_{22}\tilde{x}_2 + a_{23}\tilde{x}_3 = \tilde{b}_2 \tag{10.28}$$
$$a_{31}\tilde{x}_1 + a_{32}\tilde{x}_2 + a_{33}\tilde{x}_3 = \tilde{b}_3$$

Now, suppose that the exact solution $\{X\}$ is expressed as a function of the approximate solution and a vector of correction factors $\{\Delta X\}$, where

$$x_1 = \tilde{x}_1 + \Delta x_1$$
$$x_2 = \tilde{x}_2 + \Delta x_2 \tag{10.29}$$
$$x_3 = \tilde{x}_3 + \Delta x_3$$

If these results are substituted into Eq. (10.27), the following system results:

$$a_{11}(\tilde{x}_1 + \Delta x_1) + a_{12}(\tilde{x}_2 + \Delta x_2) + a_{13}(\tilde{x}_3 + \Delta x_3) = b_1$$
$$a_{21}(\tilde{x}_1 + \Delta x_1) + a_{22}(\tilde{x}_2 + \Delta x_2) + a_{23}(\tilde{x}_3 + \Delta x_3) = b_2 \tag{10.30}$$
$$a_{31}(\tilde{x}_1 + \Delta x_1) + a_{32}(\tilde{x}_2 + \Delta x_2) + a_{33}(\tilde{x}_3 + \Delta x_3) = b_3$$

Now, Eq. (10.28) can be subtracted from Eq. (10.30) to yield

$$a_{11}\Delta x_1 + a_{12}\Delta x_2 + a_{13}\Delta x_3 = b_1 - \tilde{b}_1 = E_1$$
$$a_{21}\Delta x_1 + a_{22}\Delta x_2 + a_{23}\Delta x_3 = b_2 - \tilde{b}_2 = E_2 \qquad (10.31)$$
$$a_{31}\Delta x_1 + a_{32}\Delta x_2 + a_{33}\Delta x_3 = b_3 - \tilde{b}_3 = E_3$$

This system itself is a set of simultaneous linear equations that can be solved to obtain the correction factors. The factors can then be applied to improve the solution, as specified by Eq. (10.29).

It is relatively straightforward to integrate an iterative refinement procedure into computer programs for elimination methods. It is especially effective for the *LU* decomposition approaches described earlier, which are designed to evaluate different right-hand-side vectors efficiently. Note that to be effective for correcting ill-conditioned systems, the *E*'s in Eq. (10.31) must be expressed in double precision.

PROBLEMS

10.1 Use the rules of matrix multiplication to prove that Eqs. (10.7) and (10.8) follow from Eq. (10.6).

10.2 (a) Use naive Gauss elimination to decompose the following system according to the description in Sec. 10.1.2.

$$7x_1 + 2x_2 - 3x_3 = -12$$
$$2x_1 + 5x_2 - 3x_3 = -20$$
$$x_1 - x_2 - 6x_3 = -26$$

Then, multiply the resulting [*L*] and [*U*] matrices to determine that [*A*] is produced. **(b)** Use *LU* decomposition to solve the system. Show all the steps in the computation. **(c)** Also solve the system for an alternative right-hand-side vector: $\{B\}^T = \lfloor 12 \ \ 18 \ \ -6 \rfloor$.

10.3
(a) Solve the following system of equations by *LU* decomposition without pivoting:

$$15x_1 + 7x_2 - 4x_3 = -51$$
$$4x_1 - 4x_2 + 9x_3 = 62$$
$$12x_1 - x_2 + 3x_3 = 8$$

(b) Determine the matrix inverse. Check your results by verifying that $[A][A]^{-1} = [I]$.

10.4 Solve the following system of equations using *LU* decomposition with partial pivoting:

$$2x_1 - 6x_2 - x_3 = -38$$
$$-3x_1 - x_2 + 7x_3 = -34$$
$$-8x_1 + x_2 - 2x_3 = -20$$

10.5 Determine the total flops as a function of the number of equations n for the **(a)** decomposition, **(b)** forward-substitution,

and **(c)** back-substitution phases of the *LU* decomposition version of Gauss elimination.

10.6 Use *LU* decomposition to determine the matrix inverse for the same system as in Prob. 10.2. Do not use a pivoting strategy, and check your results by verifying that $[A][A]^{-1} = [I]$.

10.7 Perform Crout decomposition on

$$2x_1 - 5x_2 + x_3 = 12$$
$$-x_1 + 3x_2 - x_3 = -8$$
$$3x_1 - 4x_2 + 2x_3 = 16$$

Then, multiply the resulting [*L*] and [*U*] matrices to determine that [*A*] is produced.

10.8 The following system of equations is designed to determine concentrations (the c's in g/m^3) in a series of coupled reactors as a function of the amount of mass input to each reactor (the right-hand sides are in g/day),

$$15c_1 - 3c_2 - c_3 = 3300$$
$$-3c_1 + 18c_2 - 6c_3 = 1200$$
$$-4c_1 - c_2 + 12c_3 = 2400$$

(a) Determine the matrix inverse.
(b) Use the inverse to determine the solution.
(c) Determine how much the rate of mass input to reactor 3 must be increased to induce a 10 g/m^3 rise in the concentration of reactor 1.
(d) How much will the concentration in reactor 3 be reduced if the rate of mass input to reactors 1 and 2 is reduced by 700 and 350 g/day, respectively?

10.9 Solve the following set of equations with *LU* decomposition:

$$3x_1 - 2x_2 + x_3 = -10$$
$$2x_1 + 6x_2 - 4x_3 = 44$$
$$-x_1 - 2x_2 + 5x_3 = -26$$

10.10 (a) Determine the *LU* decomposition without pivoting by hand for the following matrix, and check your results by validating that $[L][U] = [A]$.

$$\begin{bmatrix} 9 & 2 & -1 \\ 3 & 7 & 2 \\ 2 & 3 & 8 \end{bmatrix}$$

(b) Employ the result of **(a)** to compute the determinant.
(c) Repeat **(a)** and **(b)** using MATLAB.
10.11 Use the following *LU* decomposition to **(a)** compute the determinant and **(b)** solve $[A]\{x\} = \{b\}$ with $\{b\}^T = \lfloor -10\ 44\ -26 \rfloor$.

$$[A] = [L][U] = \begin{bmatrix} 1 & & \\ 0.6667 & 1 & \\ -0.3333 & -0.3636 & 1 \end{bmatrix} \begin{bmatrix} 3 & -2 & 1 \\ & 7.3333 & -4.6667 \\ & & 3.6364 \end{bmatrix}$$

10.12 Determine $\|A\|_e$, $\|A\|_1$, and $\|A\|_\infty$ for

$$[A] = \begin{bmatrix} -6 & -2 & 5 \\ 8 & 1.1 & -2.5 \\ -3 & -1 & 10.3 \end{bmatrix}$$

Scale the matrix by making the maximum element in each row equal to 1.
10.13 Determine the Frobenius and the row-sum norms for the systems in Probs. 10.3 and 10.4. Scale the matrices by making the maximum element in each row equal to 1.
10.14 A matrix $[A]$ is defined as

$$[A] = \begin{bmatrix} 0.125 & 0.25 & 0.5 & 1 \\ 0.015625 & 0.625 & 0.25 & 1 \\ 0.00463 & 0.02777 & 0.16667 & 1 \\ 0.001953 & 0.015625 & 0.125 & 1 \end{bmatrix}$$

Using the column-sum norm, compute the condition number and the number of suspect digits generated by this matrix.
10.15 (a) Determine the condition number for the following system using the row-sum norm. Do not normalize the system.

$$\begin{bmatrix} 1 & 4 & 9 & 16 & 25 \\ 4 & 9 & 16 & 25 & 36 \\ 9 & 16 & 25 & 36 & 49 \\ 16 & 25 & 36 & 49 & 64 \\ 25 & 36 & 49 & 64 & 81 \end{bmatrix}$$

How many digits of precision will be lost due to ill-conditioning? **(b)** Repeat part **(a)**, but scale the matrix by making the maximum element in each row equal to 1.
10.16 Determine the condition number based on the row-sum norm for the normalized 4×4 Hilbert matrix. How many significant digits of precision will be lost due to ill-conditioning?
10.17 Besides the Hilbert matrix, there are other matrices that are inherently ill-conditioned. One such case is the *Vandermonde matrix*, which has the following form:

$$\begin{bmatrix} x_1^2 & x_1 & 1 \\ x_2^2 & x_2 & 1 \\ x_3^2 & x_3 & 1 \end{bmatrix}$$

(a) Determine the condition number based on the row-sum norm for the case where $x_1 = 4$, $x_2 = 2$, and $x_3 = 7$.
(b) Use MATLAB or Mathcad software to compute the spectral and Frobenius condition numbers.
10.18 Develop a user-friendly program for *LU* decomposition based on the pseudocode from Fig. 10.2.
10.19 Develop a user-friendly program for *LU* decomposition, including the capability to evaluate the matrix inverse. Base the program on Figs. 10.2 and 10.5.
10.20 Use iterative refinement techniques to improve $x_1 = 2$, $x_2 = -3$, and $x_3 = 8$, which are approximate solutions of

$$2x_1 + 5x_2 + x_3 = -5$$
$$5x_1 + 2x_2 + x_3 = 12$$
$$x_1 + 2x_2 + x_3 = 3$$

10.21 Consider these vectors:

$$\vec{A} = 2\vec{i} - 3\vec{j} + a\vec{k}$$
$$\vec{B} = b\vec{i} + \vec{j} - 4\vec{k}$$
$$\vec{C} = 3\vec{i} + c\vec{j} + 2\vec{k}$$

Vector $\vec{A}$ is perpendicular to $\vec{B}$ as well as to $\vec{C}$. It is also known that $\vec{B} \cdot \vec{C} = 2$. Use any method presented in this chapter to solve for the three unknowns, a, b, and c.
10.22 Consider the following vectors:

$$\vec{A} = a\vec{i} + b\vec{j} + c\vec{k}$$
$$\vec{B} = -2\vec{i} + \vec{j} - 4\vec{k}$$
$$\vec{C} = \vec{i} + 3\vec{j} + 2\vec{k}$$

where $\vec{A}$ is an unknown vector. If

$$(\vec{A} \times \vec{B}) + (\vec{A} \times \vec{C}) = (5a + 6)\vec{i} + (3b - 2)\vec{j} + (-4c + 1)\vec{k}$$

use any method covered in this chapter to solve for the three unknowns, a, b, and c.

10.23 Let a function be defined on the interval $[0, 2]$ as follows:

$$f(x) = \begin{cases} ax + b, & 0 \le x \le 1 \\ cx + d, & 1 \le x \le 2 \end{cases}$$

Determine the constants a, b, c, and d so that the function f satisfies the following:

- $f(0) = f(2) = 1$.
- f is continuous on the entire interval.
- $a + b = 4$.

Derive and solve a system of linear algebraic equations with a matrix form identical to Eq. (10.1).

10.24

(a) Create a 3×3 Hilbert matrix. This will be matrix $[A]$. Multiply the matrix by the column vector $\{x\} = [1, 1, 1]^T$. The solution of $[A]\{x\}$ will be another column vector $\{b\}$. Using any numerical package and Gauss elimination, find the solution to $[A]\{x\} = \{b\}$ using the Hilbert matrix and the vector $\{b\}$ that you calculated. Compare the result to the known $\{x\}$ vector.

Use sufficient precision in displaying results to allow you to detect imprecision.

(b) Repeat part **(a)** using a 7×7 Hilbert matrix.

(c) Repeat part **(a)** using a 10×10 Hilbert matrix.

10.25 Polynomial interpolation consists of determining the unique $(n - 1)$th-order polynomial that fits n data points. Such polynomials have the general form,

$$f(x) = p_1 x^{n-1} + p_2 x^{n-2} + \cdots + p_{n-1} x + p_n \qquad \text{(P10.25)}$$

where the p's are constant coefficients. A straightforward way for computing the coefficients is to generate n linear algebraic equations that we can solve simultaneously for the coefficients. Suppose that we want to determine the coefficients of the fourth-order polynomial $f(x) = p_1 x^4 + p_2 x^3 + p_3 x^2 + p_4 x + p_5$ that passes through the following five points: (200, 0.746), (250, 0.675), (300, 0.616), (400, 0.525), and (500, 0.457). Each of these pairs can be substituted into Eq. (P10.25) to yield a system of five equations with five unknowns (the p's). Use this approach to solve for the coefficients. In addition, determine and interpret the condition number.

CHAPTER 11

Special Matrices and Gauss-Seidel

Certain matrices have a particular structure that can be exploited to develop efficient solution schemes. The first part of this chapter is devoted to two such systems: *banded* and *symmetric matrices*. Efficient elimination methods are described for both.

The second part of the chapter turns to an alternative to elimination methods, that is, approximate iterative methods. The focus is on the *Gauss-Seidel method,* which employs initial guesses and then iterates to obtain refined estimates of the solution. The Gauss-Seidel method is particularly well suited for large numbers of equations. In these cases, elimination methods can be subject to round-off errors. Because the error of the Gauss-Seidel method is controlled by the number of iterations, round-off error is not an issue of concern with this method. However, there are certain instances where the Gauss-Seidel technique will not converge on the correct answer. These and other trade-offs between elimination and iterative methods will be discussed in subsequent pages.

11.1 SPECIAL MATRICES

As mentioned in Box PT3.1, a *banded matrix* is a square matrix that has all elements equal to zero, with the exception of a band centered on the main diagonal. Banded systems are frequently encountered in engineering and scientific practice. For example, they typically occur in the solution of differential equations. In addition, other numerical methods such as cubic splines (Sec. 18.6) involve the solution of banded systems.

The dimensions of a banded system can be quantified by two parameters: the bandwidth BW and the half-bandwidth HBW (Fig. 11.1). These two values are related by BW = 2HBW + 1. In general, then, a banded system is one for which $a_{ij} = 0$ if $|i - j| >$ HBW.

Although Gauss elimination or conventional *LU* decomposition can be employed to solve banded systems, they are inefficient, because if pivoting is unnecessary none of the elements outside the band would change from their original values of zero. Thus, unnecessary space and time would be expended on the storage and manipulation of these useless zeros. If it is known beforehand that pivoting is unnecessary, very efficient algorithms can be developed that do not involve the zero elements outside the band. Because many problems involving banded systems do not require pivoting, these alternative algorithms, as described next, are the methods of choice.

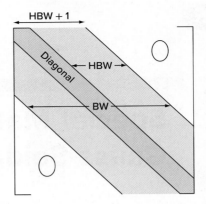

FIGURE 11.1
Parameters used to quantify the dimensions of a banded system. BW and HBW designate the bandwidth and the half-bandwidth, respectively.

(a) Decomposition

```
DOFOR k = 2, n
    e_k = e_k /f_{k-1}
    f_k = f_k - e_k · g_{k-1}
END DO
```

(b) Forward substitution

```
DOFOR k = 2, n
    r_k = r_k - e_k · r_{k-1}
END DO
```

(c) Back substitution

```
x_n = r_n /f_n
DOFOR k = n -1, 1, -1
    x_k = (r_k - g_k · x_{k+1})/f_k
END DO
```

FIGURE 11.2
Pseudocode to implement the Thomas algorithm, an *LU* decomposition method for tridiagonal systems.

11.1.1 Tridiagonal Systems

A tridiagonal system—that is, one with a bandwidth of 3—can be expressed generally as

$$\begin{bmatrix} f_1 & g_1 & & & & \\ e_2 & f_2 & g_2 & & & \\ & e_3 & f_3 & g_3 & & \\ & & \cdot & \cdot & \cdot & \\ & & & \cdot & \cdot & \cdot \\ & & & & e_{n-1} & f_{n-1} & g_{n-1} \\ & & & & & e_n & f_n \end{bmatrix} \begin{Bmatrix} x_1 \\ x_2 \\ x_3 \\ \cdot \\ \cdot \\ \cdot \\ x_{n-1} \\ x_n \end{Bmatrix} = \begin{Bmatrix} r_1 \\ r_2 \\ r_3 \\ \cdot \\ \cdot \\ \cdot \\ r_{n-1} \\ r_n \end{Bmatrix} \qquad (11.1)$$

Notice that we have changed our notation for the coefficients from a's and b's to e's, f's, g's, and r's. This was done to avoid storing large numbers of useless zeros in the square matrix of a's. This space-saving modification is advantageous because the resulting algorithm requires less computer memory.

Figure 11.2 shows pseudocode for an efficient method, called the *Thomas algorithm*, to solve Eq. (11.1). As with conventional *LU* decomposition, the algorithm consists of three steps: decomposition and forward and back substitution. Thus, all the advantages of *LU* decomposition, such as convenient evaluation of multiple right-hand-side vectors and the matrix inverse, can be accomplished by proper application of this algorithm.

EXAMPLE 11.1 | **Tridiagonal Solution with the Thomas Algorithm**

Problem Statement.   Solve the following tridiagonal system with the Thomas algorithm.

$$\begin{bmatrix} 2.04 & -1 & & \\ -1 & 2.04 & -1 & \\ & -1 & 2.04 & -1 \\ & & -1 & 2.04 \end{bmatrix} \begin{Bmatrix} T_1 \\ T_2 \\ T_3 \\ T_4 \end{Bmatrix} = \begin{Bmatrix} 40.8 \\ 0.8 \\ 0.8 \\ 200.8 \end{Bmatrix}$$

Solution.   First, the decomposition is implemented as

$$e_2 = -1/2.04 = -0.49$$
$$f_2 = 2.04 - (-0.49)(-1) = 1.550$$
$$e_3 = -1/1.550 = -0.645$$
$$f_3 = 2.04 - (-0.645)(-1) = 1.395$$
$$e_4 = -1/1.395 = -0.717$$
$$f_4 = 2.04 - (-0.717)(-1) = 1.323$$

Thus, the matrix has been transformed to

$$\begin{bmatrix} 2.04 & -1 & & \\ -0.49 & 1.550 & -1 & \\ & -0.645 & 1.395 & -1 \\ & & -0.717 & 1.323 \end{bmatrix}$$

and the *LU* decomposition is

$$[A] = [L][U] = \begin{bmatrix} 1 & & & \\ -0.49 & 1 & & \\ & -0.645 & 1 & \\ & & -0.717 & 1 \end{bmatrix} \begin{bmatrix} 2.04 & -1 & & \\ & 1.550 & -1 & \\ & & 1.395 & -1 \\ & & & 1.323 \end{bmatrix}$$

You can verify that this is correct by multiplying $[L][U]$ to yield $[A]$.

The forward substitution is implemented as

$$r_2 = 0.8 - (-0.49)40.8 = 20.8$$
$$r_3 = 0.8 - (-0.645)20.8 = 14.221$$
$$r_4 = 200.8 - (-0.717)14.221 = 210.996$$

Thus, the right-hand-side vector has been modified to

$$\begin{Bmatrix} 40.8 \\ 20.8 \\ 14.221 \\ 210.996 \end{Bmatrix}$$

which then can be used in conjunction with the $[U]$ matrix to perform back substitution and obtain the solution

$$T_4 = 210.996/1.323 = 159.480$$
$$T_3 = [14.221 - (-1)159.48]/1.395 = 124.538$$
$$T_2 = [20.800 - (-1)124.538]/1.550 = 93.778$$
$$T_1 = [40.800 - (-1)93.778]/2.040 = 65.970$$

11.1.2 Cholesky Decomposition

Recall from Box PT3.1 that a symmetric matrix is one where $a_{ij} = a_{ji}$ for all i and j. In other words, $[A] = [A]^T$. Such systems occur commonly in both mathematical and engineering problem contexts. They offer computational advantages because only half the storage is needed and, in most cases, only half the computation time is required for their solution.

One of the most popular approaches involves *Cholesky decomposition*. This algorithm is based on the fact that a symmetric matrix can be decomposed, as in

$$[A] = [L][L]^T \tag{11.2}$$

That is, the resulting triangular factors are the transposes of each other.

The terms of Eq. (11.2) can be multiplied out and set equal to each other. The result can be expressed simply by recurrence relations. For the kth row,

$$l_{ki} = \frac{a_{ki} - \sum_{j=1}^{i-1} l_{ij} l_{kj}}{l_{ii}} \qquad \text{for } i = 1, 2, \dots, k-1 \tag{11.3}$$

and

$$l_{kk} = \sqrt{a_{kk} - \sum_{j=1}^{k-1} l_{kj}^2} \tag{11.4}$$

EXAMPLE 11.2   **Cholesky Decomposition**

Problem Statement.   Apply Cholesky decomposition to the symmetric matrix

$$[A] = \begin{bmatrix} 6 & 15 & 55 \\ 15 & 55 & 225 \\ 55 & 225 & 979 \end{bmatrix}$$

Solution.   For the first row ($k = 1$), Eq. (11.3) is skipped and Eq. (11.4) is employed to compute

$$l_{11} = \sqrt{a_{11}} = \sqrt{6} = 2.4495$$

For the second row ($k = 2$), Eq. (11.3) gives

$$l_{21} = \frac{a_{21}}{l_{11}} = \frac{15}{2.4495} = 6.1237$$

and Eq. (11.4) yields

$$l_{22} = \sqrt{a_{22} - l_{21}^2} = \sqrt{55 - (6.1237)^2} = 4.1833$$

For the third row ($k = 3$), Eq. (11.3) gives ($i = 1$)

$$l_{31} = \frac{a_{31}}{l_{11}} = \frac{55}{2.4495} = 22.454$$

and ($i = 2$)

$$l_{32} = \frac{a_{32} - l_{21} l_{31}}{l_{22}} = \frac{225 - 6.1237(22.454)}{4.1833} = 20.917$$

```
DOFOR k = 1, n
  DOFOR i = 1, k - 1
    sum = 0.
    DOFOR j = 1, i - 1
      sum = sum + a_ij · a_kj
    END DO
    a_ki = (a_ki - sum)/a_ii
  END DO
  sum = 0.
  DOFOR j = 1, k - 1
    sum = sum + a²_kj
  END DO
  a_kk = √(a_kk - sum)
END DO
```

FIGURE 11.3

Pseudocode for the Cholesky decomposition algorithm.

and Eq. (11.4) yields

$$l_{33} = \sqrt{a_{33} - l_{31}^2 - l_{32}^2} = \sqrt{979 - (22.454)^2 - (20.917)^2} = 6.1101$$

Thus, the Cholesky decomposition yields

$$[L] = \begin{bmatrix} 2.4495 & & \\ 6.1237 & 4.1833 & \\ 22.454 & 20.917 & 6.1101 \end{bmatrix}$$

The validity of this decomposition can be verified by substituting it and its transpose into Eq. (11.2) to see if their product yields the original matrix [A]. This is left for an exercise.

Figure 11.3 presents pseudocode for implementing the Cholesky decomposition algorithm. It should be noted that the algorithm in Fig. 11.3 could result in an execution error if the evaluation of a_{kk} involves taking the square root of a negative number. However, for cases where the matrix is *positive definite*,[1] this will never occur. Because many symmetric matrices dealt with in engineering are, in fact, positive definite, the Cholesky algorithm has wide application. Another benefit of dealing with positive definite symmetric matrices is that pivoting is not required to avoid division by zero. Thus, we can implement the algorithm in Fig. 11.3 without the complication of pivoting.

11.2   GAUSS-SEIDEL

Iterative or approximate methods provide an alternative to the elimination methods described to this point. Such approaches are similar to the techniques we developed to obtain the roots of a single equation in Chap. 6. Those approaches consisted of guessing a value and then using a systematic method to obtain a refined estimate of the root. Because the present part of the book deals with a similar problem—obtaining the values that simultaneously satisfy a set of equations—we might suspect that such approximate methods could be useful in this context.

The *Gauss-Seidel method* is the most commonly used iterative method. Assume that we are given a set of *n* equations:

$$[A]\{X\} = \{B\}$$

Suppose that for conciseness we limit ourselves to a 3 × 3 set of equations. If the diagonal elements are all nonzero, the first equation can be solved for x_1, the second for x_2, and the third for x_3 to yield

$$x_1 = \frac{b_1 - a_{12}x_2 - a_{13}x_3}{a_{11}} \tag{11.5a}$$

[1]A *positive definite matrix* is one for which the product $\{X\}^T[A]\{X\}$ is greater than zero for all nonzero vectors $\{X\}$.

$$x_2 = \frac{b_2 - a_{21}x_1 - a_{23}x_3}{a_{22}} \qquad (11.5b)$$

$$x_3 = \frac{b_3 - a_{31}x_1 - a_{32}x_2}{a_{33}} \qquad (11.5c)$$

Now, we can start the solution process by choosing guesses for the x's. A simple way to obtain initial guesses is to assume that they are all zero. These zeros can be substituted into Eq. (11.5a), which can be used to calculate a new value for $x_1 = b_1/a_{11}$. Then, we substitute this new value of x_1 along with the previous guess of zero for x_3 into Eq. (11.5b) to compute a new value for x_2. The process is repeated for Eq. (11.5c) to calculate a new estimate for x_3. Then we return to the first equation and repeat the entire procedure until our solution converges closely enough to the true values. Convergence can be checked using the criterion [recall Eq. (3.5)]

$$|\varepsilon_{a,i}| = \left| \frac{x_i^j - x_i^{j-1}}{x_i^j} \right| 100\% < \varepsilon_s \qquad (11.6)$$

for all i, where j and $j - 1$ are the present and previous iterations.

EXAMPLE 11.3　　**Gauss-Seidel Method**

Problem Statement.　Use the Gauss-Seidel method to obtain the solution of the same system used in Example 10.2:

$$3x_1 - 0.1x_2 - 0.2x_3 = 7.85$$
$$0.1x_1 + 7x_2 - 0.3x_3 = -19.3$$
$$0.3x_1 - 0.2x_2 + 10x_3 = 71.4$$

Recall that the true solution is $x_1 = 3$, $x_2 = -2.5$, and $x_3 = 7$.

Solution.　First, solve each of the equations for its unknown on the diagonal.

$$x_1 = \frac{7.85 + 0.1x_2 + 0.2x_3}{3} \qquad (E11.3.1)$$

$$x_2 = \frac{-19.3 - 0.1x_1 + 0.3x_3}{7} \qquad (E11.3.2)$$

$$x_3 = \frac{71.4 - 0.3x_1 + 0.2x_2}{10} \qquad (E11.3.3)$$

By assuming that x_2 and x_3 are zero, Eq. (E11.3.1) can be used to compute

$$x_1 = \frac{7.85 + 0 + 0}{3} = 2.616667$$

This value, along with the assumed value of $x_3 = 0$, can be substituted into Eq. (E11.3.2) to calculate

$$x_2 = \frac{-19.3 - 0.1(2.616667) + 0}{7} = -2.794524$$

The first iteration is completed by substituting the calculated values for x_1 and x_2 into Eq. (E11.3.3) to yield

$$x_3 = \frac{71.4 - 0.3(2.616667) + 0.2(-2.794524)}{10} = 7.005610$$

For the second iteration, the same process is repeated to compute

$$x_1 = \frac{7.85 + 0.1(-2.794524) + 0.2(7.005610)}{3} = 2.990557 \qquad |\varepsilon_t| = 0.31\%$$

$$x_2 = \frac{-19.3 - 0.1(2.990557) + 0.3(7.005610)}{7} = -2.499625 \qquad |\varepsilon_t| = 0.015\%$$

$$x_3 = \frac{71.4 - 0.3(2.990557) + 0.2(-2.499625)}{10} = 7.000291 \qquad |\varepsilon_t| = 0.0042\%$$

The method is, therefore, converging on the true solution. Additional iterations could be applied to improve the answers. However, in an actual problem, we would not know the true answer a priori. Consequently, Eq. (11.6) provides a means to estimate the error. For example, for x_1,

$$|\varepsilon_{a,1}| = \left| \frac{2.990557 - 2.616667}{2.990557} \right| 100\% = 12.5\%$$

For x_2 and x_3, the error estimates are $|\varepsilon_{a,2}| = 11.8\%$ and $|\varepsilon_{a,3}| = 0.076\%$. Note that, as was the case when determining roots of a single equation, formulations such as Eq. (11.6) usually provide a conservative appraisal of convergence. Thus, when they are met, they ensure that the result is known to at least the tolerance specified by ε_s.

As each new x value is computed in the Gauss-Seidel method, it is immediately used in the next equation to determine another x value. Thus, if the solution is converging, the best available estimates will be employed. An alternative approach, called *Jacobi iteration,* utilizes a somewhat different tactic. Rather than using the latest available x's, this technique uses Eq. (11.5) to compute a set of new x's on the basis of a set of old x's. Thus, as new values are generated, they are not immediately used but rather are retained for the next iteration.

The difference between the Gauss-Seidel method and Jacobi iteration is depicted in Fig. 11.4. Although there are certain cases where the Jacobi method is useful, Gauss-Seidel's utilization of the best available estimates usually makes it the method of preference.

11.2.1 Convergence Criterion for the Gauss-Seidel Method

Note that the Gauss-Seidel method is similar in spirit to the technique of simple fixed-point iteration that was used in Sec. 6.1 to solve for the roots of a single equation. Recall that simple fixed-point iteration had two fundamental problems: (1) it was sometimes nonconvergent and (2) when it converged, it often did so very slowly. The Gauss-Seidel method can also exhibit these shortcomings.

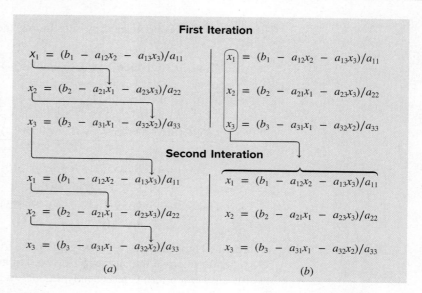

FIGURE 11.4
Graphical depiction of the difference between (a) the Gauss-Seidel and (b) the Jacobi iterative methods for solving simultaneous linear algebraic equations.

Convergence criteria can be developed by recalling from Sec. 6.6.1 that sufficient conditions for convergence of two nonlinear equations, $u(x, y)$ and $v(x, y)$, are

$$\left| \frac{\partial u}{\partial x} \right| + \left| \frac{\partial u}{\partial y} \right| < 1 \tag{11.7a}$$

and

$$\left| \frac{\partial v}{\partial x} \right| + \left| \frac{\partial v}{\partial y} \right| < 1 \tag{11.7b}$$

These criteria also apply to linear equations of the sort we are solving with the Gauss-Seidel method. For example, in the case of two simultaneous equations, the Gauss-Seidel algorithm [Eq. (11.5)] can be expressed as

$$u(x_1, x_2) = \frac{b_1}{a_{11}} - \frac{a_{12}}{a_{11}} x_2 \tag{11.8a}$$

and

$$v(x_1, x_2) = \frac{b_2}{a_{22}} - \frac{a_{21}}{a_{22}} x_1 \tag{11.8b}$$

The partial derivatives of these equations can be evaluated with respect to each of the unknowns as

$$\frac{\partial u}{\partial x_1} = 0 \qquad \frac{\partial u}{\partial x_2} = -\frac{a_{12}}{a_{11}}$$

and

$$\frac{\partial v}{\partial x_1} = -\frac{a_{21}}{a_{22}} \qquad \frac{\partial v}{\partial x_2} = 0$$

which can be substituted into Eq. (11.7) to give

$$\left|\frac{a_{12}}{a_{11}}\right| < 1 \tag{11.9a}$$

and

$$\left|\frac{a_{21}}{a_{22}}\right| < 1 \tag{11.9b}$$

In other words, the absolute values of the slopes of Eq. (11.8) must be less than unity to ensure convergence. This is displayed graphically in Fig. 11.5. Equation (11.9) can also be reformulated as

$$|a_{11}| > |a_{12}|$$

and

$$|a_{22}| > |a_{21}|$$

That is, the diagonal element must be greater than the off-diagonal element for each row.

The extension of the above to n equations is straightforward and can be expressed as

$$|a_{ii}| > \sum_{\substack{j=1 \\ j \neq i}}^{n} |a_{ij}| \tag{11.10}$$

FIGURE 11.5
Iteration cobwebs illustrating (*a*) convergence and (*b*) divergence of the Gauss-Seidel method. Notice that the same functions are plotted in both cases (*u*: $11x_1 + 13x_2 = 286$; *v*: $11x_1 - 9x_2 = 99$). Thus, the order in which the equations are implemented (as depicted by the direction of the first arrow from the origin) dictates whether the computation converges.

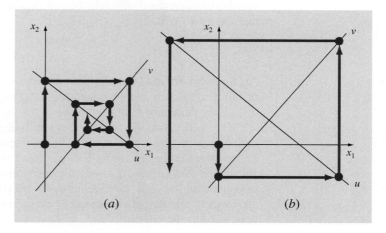

That is, the diagonal coefficient in each of the equations must be larger than the sum of the absolute values of the other coefficients in the equation. This criterion is sufficient but not necessary for convergence. That is, although the method may sometimes work if Eq. (11.10) does not hold, convergence is guaranteed if the condition is satisfied. Systems where Eq. (11.10) holds are called *diagonally dominant*. Fortunately, many engineering problems of practical importance fulfill this requirement.

11.2.2 Improvement of Convergence Using Relaxation

Relaxation represents a slight modification of the Gauss-Seidel method and is designed to enhance convergence. After each new value of x is computed using Eq. (11.5), that value is modified by a weighted average of the results of the previous and the present iterations:

$$x_i^{new} = \lambda x_i^{new} + (1 - \lambda)x_i^{old} \tag{11.11}$$

where λ is a weighting factor that is assigned a value between 0 and 2.

If $\lambda = 1$, $(1 - \lambda)$ is equal to 0 and the result is unmodified. However, if λ is set at a value between 0 and 1, the result is a weighted average of the present and the previous results. This type of modification is called *underrelaxation*. It is typically employed to make a nonconvergent system converge or to hasten convergence by dampening out oscillations.

For values of λ from 1 to 2, extra weight is placed on the present value. In this instance, there is an implicit assumption that the new value is moving in the correct direction toward the true solution but at too slow a rate. Thus, the added weight of λ is intended to improve the estimate by pushing it closer to the truth. Hence, this type of modification, which is called *overrelaxation,* is designed to accelerate the convergence of an already convergent system. The approach is also called *successive* or *simultaneous overrelaxation,* or *SOR.*

The choice of a proper value for λ is highly problem-specific and is often determined empirically. For a single solution of a set of equations, it is often unnecessary. However, if the system under study is to be solved repeatedly, the efficiency introduced by a wise choice of λ can be extremely important. Good examples are the very large systems of partial differential equations that often arise when modeling continuous variations of variables (recall the distributed system depicted in Fig. PT3.1*b*). We will return to this topic in Part Eight.

11.2.3 Algorithm for Gauss-Seidel

An algorithm for the Gauss-Seidel method, with relaxation, is depicted in Fig. 11.6. Note that this algorithm is not guaranteed to converge if the equations are not input in a diagonally dominant form.

The pseudocode has two features that bear mentioning. First, there is an initial set of nested loops to divide each equation by its diagonal element. This reduces the total number of operations in the algorithm. Second, notice that the error check is designated by a variable called `sentinel`. If any of the equations has an approximate error greater than the stopping criterion (`es`), then the iterations are allowed to continue. The use of `sentinel` allows us to circumvent unnecessary calculations of error estimates once one of the equations exceeds the criterion.

```
SUBROUTINE Gseid(a,b,n,x,imax,es,lambda)
  DOFOR i = 1,n
    dummy = aᵢ,ᵢ
    DOFOR j = 1,n
      aᵢ,ⱼ = aᵢ,ⱼ/dummy
    END DO
    bᵢ = bᵢ/dummy
  END DO
  DOFOR i = 1, n
    sum = bᵢ
    DOFOR j = 1, n
      IF i ≠ j THEN sum = sum − aᵢ,ⱼ*xⱼ
    END DO
    xᵢ=sum
  END DO
  iter=1
  DO
    sentinel = 1
    DOFOR i = 1,n
      old = xᵢ
      sum = bᵢ
      DOFOR j = 1,n
        IF i ≠ j THEN sum = sum − aᵢ,ⱼ*xⱼ
      END DO
      xᵢ = lambda*sum +(1.−lambda)*old
      IF sentinel = 1 AND xᵢ ≠ 0. THEN
        ea = ABS((xᵢ − old)/xᵢ)*100.
        IF ea > es THEN sentinel = 0
      END IF
    END DO
    iter = iter + 1
    IF sentinel = 1 OR (iter ≥ imax) EXIT
  END DO
END Gseid
```

FIGURE 11.6
Pseudocode for Gauss-Seidel
with relaxation.

11.2.4 Problem Contexts for the Gauss-Seidel Method

Aside from circumventing the round-off dilemma, the Gauss-Seidel technique has a number of other advantages that make it particularly attractive in the context of certain engineering problems. For example, when the matrix in question is very large and very sparse (that is, most of the elements are zero), elimination methods waste large amounts of computer memory by storing zeros.

At the beginning of this chapter, we saw how this shortcoming could be circumvented if the coefficient matrix is banded. For nonbanded systems, there is usually no simple way to avoid large memory requirements when using elimination methods.

Because all computers have a finite amount of memory, this inefficiency can place a constraint on the size of systems for which elimination methods are practical.

Although a general algorithm such as the one in Fig. 11.6 is prone to the same constraint, the structure of the Gauss-Seidel equations [Eq. (11.5)] permits concise programs to be developed for specific systems. Because only nonzero coefficients need be included in Eq. (11.5), large savings of computer memory are possible. Although this entails more up-front investment in software development, the long-term advantages are substantial when dealing with large systems for which many simulations are to be performed. Both lumped- and distributed-variable systems can result in large, sparse matrices for which the Gauss-Seidel method has utility.

11.3 LINEAR ALGEBRAIC EQUATIONS WITH SOFTWARE PACKAGES

Software packages have great capabilities for solving systems of linear algebraic equations. Before describing these tools, we should mention that the approaches described in Chap. 7 for solving nonlinear systems can be applied to linear systems. However, in this section, we will focus on the approaches that are expressly designed for linear equations.

11.3.1 Excel

There are two ways to solve linear algebraic equations with Excel: (1) using the Solver tool or (2) using matrix inversion and multiplication functions.

Recall that one way to determine the solution of linear algebraic equations is

$$\{X\} = [A]^{-1}\{B\} \tag{11.12}$$

Excel has built-in functions for both matrix inversion and multiplication that can be used to implement this formula.

EXAMPLE 11.4   **Using Excel to Solve Linear Systems**

Problem Statement.   Recall that in Chap. 10 we introduced the Hilbert matrix. The following system is based on the Hilbert matrix. Note that it is scaled, as was done previously in Example 10.3, so that the maximum coefficient in each row is unity.

$$\begin{bmatrix} 1 & 1/2 & 1/3 \\ 1 & 2/3 & 1/2 \\ 1 & 3/4 & 3/5 \end{bmatrix} \begin{Bmatrix} x_1 \\ x_2 \\ x_3 \end{Bmatrix} = \begin{Bmatrix} 1.833333 \\ 2.166667 \\ 2.35 \end{Bmatrix}$$

The solution to this system is $\{X\}^T = \lfloor 1 \ 1 \ 1 \rfloor$. Use Excel to obtain this solution.

Solution.   The spreadsheet to solve this problem is displayed in Fig. 11.7. First, the matrix $[A]$ and the right-hand-side constants $\{B\}$ are entered into the spreadsheet cells. Then, a set of cells of the proper dimensions (in our example 3×3) is highlighted by either clicking and dragging the mouse or by using the arrow keys while depressing the shift key. As in Fig. 11.7, we highlight the range: B5. .D7.

	A	B	C	D	E	F
1		1	0.5	0.33333333		1.83333333333333
2	[A] =	1	0.66666667	0.5	{B} =	2.16666666666667
3		1	0.75	0.6		2.35000000000000
4						
5		9	-18	10		0.99999999999992
6	[A]-1 =	-36	96	-60	{X} =	1.00000000000043
7		30	-90	60		0.99999999999960

=MINVERSE(B1:D3)   =MMULT(B5:D7,F1:F3)

FIGURE 11.7

Next, a formula invoking the matrix inverse function is entered,

```
=minverse(B1..D3)
```

Note that the argument is the range holding the elements of [A]. The *Ctrl* and *Shift* keys are held down while the *Enter* key is depressed. The resulting inverse of [A] will be calculated by Excel and displayed in the range B5. .D7 as shown in Fig. 11.7.

A similar approach is used to multiply the inverse by the right-hand-side vector. For this case, the range F5. .F7 is highlighted and the following formula is entered

```
=mmult(B5..D7,F1..F3)
```

where the first range is the first matrix to be multiplied, $[A]^{-1}$, and the second range is the second matrix to be multiplied, {B}. When the *Ctrl-Shift-Enter* combination is used again, the solution {X} will be calculated by Excel and displayed in the range F5. .F7, as shown in Fig. 11.7. As can be seen, the correct answer results.

Notice that we deliberately reformatted the results in Example 11.4 to show 15 digits. We did this because Excel uses double-precision to store numerical values. Thus, we see that round-off error occurs in the last two digits. This implies a condition number on the order of 100, which agrees with the result of 451.2 originally calculated in Example 10.3. Excel does not have the capability to calculate a condition number. In most cases, particularly because it employs double-precision numbers, this does not represent a problem. However, for cases where you suspect that the system is ill-conditioned, determination of the condition number is useful. MATLAB and Mathcad software are capable of computing this quantity.

11.3.2 MATLAB

As the name implies, MATLAB (short for MATrix LABoratory) was designed to facilitate matrix manipulations. Thus, as might be expected, its capabilities in this area are

TABLE 11.1 MATLAB functions to implement matrix analysis and numerical linear algebra.

Matrix Analysis		Linear Equations	
Function	Description	Function	Description
cond	Matrix condition number	\ and /	Linear equation solution; use "help slash"
norm	Matrix or vector norm	chol	Cholesky factorization
rcond	LINPACK reciprocal condition estimator	lu	Factors from Gauss elimination
rank	Number of linearly independent rows or columns	inv	Matrix inverse
det	Determinant	qr	Orthogonal-triangular decomposition
trace	Sum of diagonal elements	qrdelete	Delete a column from the QR factorization
null	Null space	qrinsert	Insert a column in the QR factorization
orth	Orthogonalization	nnls	Nonnegative least squares
rref	Reduced row echelon form	pinv	Pseudoinverse
		lscov	Least squares in the presence of known covariance

excellent. Some of the key MATLAB functions related to matrix operations are listed in Table 11.1. The following example illustrates a few of these capabilities.

EXAMPLE 11.5

Using MATLAB to Manipulate Linear Algebraic Equations

Problem Statement. Explore how MATLAB can be employed to solve and analyze linear algebraic equations. Use the same system as in Example 11.4.

Solution. First, we can enter the $[A]$ matrix and the $\{B\}$ vector,

```
>> A = [ 1   1/2   1/3 ; 1   2/3   1/2 ; 1   3/4   3/5 ]

A =
    1.0000    0.5000    0.3333
    1.0000    0.6667    0.5000
    1.0000    0.7500    0.6000

>> B = [1+1/2+1/3;1+2/3+2/4;1+3/4+3/5]

B =
    1.8333
    2.1667
    2.3500
```

Next, we can determine the condition number for $[A]$, as in

```
>> cond(A)

ans =
  366.3503
```

This result is based on the spectral, or $\|A\|_2$, norm discussed in Box 10.2. Note that it is of the same order of magnitude as the condition number 451.2 that was obtained based on the row-sum norm in Example 10.3. Both results imply that between two and three digits of precision could be lost.

Now we can solve the system of equations in two different ways. The most direct and efficient way is to employ backslash division, or "left division":

```
>> X=A\B

X =
    1.0000
    1.0000
    1.0000
```

For cases such as ours, MATLAB uses Gauss elimination to solve such systems.

As an alternative, we can implement Eq. (PT3.6) directly, as in

```
>> X=inv(A)*B

X =
    1.0000
    1.0000
    1.0000
```

This approach actually determines the matrix inverse first and then performs the matrix multiplication. Hence, it is more time-consuming than using the backslash approach.

11.3.3 Mathcad

Mathcad contains many special functions that manipulate vectors and matrices. These include common operations such as the dot product, matrix transpose, matrix addition, and matrix multiplication. In addition, it allows calculation of the matrix inverse, determinant, trace, various types of norms, and condition numbers based on different norms. It also has several functions that decompose matrices.

Systems of linear equations can be solved in two ways by Mathcad. First, it is possible to use matrix inversion and subsequent multiplication by the right-hand-side vector as discussed in Chap. 10. In addition, Mathcad has a special function called **lsolve(A,b)** that is specifically designed to solve linear equations. You can use other built-in functions to evaluate the condition of A to determine if A is nearly singular and thus possibly subject to round-off errors.

As an example, let's use **lsolve** to solve a system of linear equations. As shown in Fig. 11.8, the first step is to enter the coefficients of the A matrix using the definition symbol and the Insert/Matrix pull-down menu. This gives a box that allows you to

FIGURE 11.8
Mathcad screen showing how to solve a system of linear algebraic equations.

specify the dimensions of the matrix. For our case, we will select a dimension of 4×4, and Mathcad places a blank 4-by-4-size matrix on the screen. Now, simply click in each cell location and enter values. Repeat similar operations to create the right-hand-side b vector. Now the vector x is defined as **lsolve(A,b)** and the value of x is displayed with the equal sign.

We can also solve the same system using the matrix inverse. The inverse can be simply computed by merely raising A to the exponent -1. The result is shown on the right side of Fig. 11.8. The solution is then generated as the product of the inverse times b.

Next, let's use Mathcad to find the inverse and the condition number of the Hilbert matrix. As in Fig. 11.9, the scaled matrix can be entered using the definition symbol and the Insert/Matrix pull-down menu. The inverse can again be computed by simply raising H to the exponent -1. The result is shown in Fig. 11.9. We can then use some other Mathcad functions to determine condition numbers by using the definition symbol to define variables c1, c2, ce, and ci as the condition number based on the column-sum (**cond1**), spectral (**cond2**), the Euclidean (**conde**), and the row-sum (**condi**) norms, respectively. The resulting values are shown at the bottom of Fig. 11.9. As expected, the spectral norm provides the result with the smallest magnitude.

CALCULATING THE INVERSE AND CONDITION NUMBERS

Enter a scaled Hilbert matrix:

$$H := \begin{pmatrix} 1 & 0.5 & 0.3333333 \\ 1 & 0.6666667 & 0.5 \\ 1 & 0.75 & 0.6 \end{pmatrix}$$

Calculate matrix inverse:

$$I := H^{-1}$$

$$I := \begin{pmatrix} 9 & -18 & 10 \\ -36 & 96 & -60 \\ 30 & -90 & 60 \end{pmatrix}$$

Calculate condition numbers:

$c1 := cond1(H)$	$c2 := cond2(H)$	$ce := conde(H)$	$ci := condi(H)$
$c1 = 612$	$c2 = 366.35$	$ce = 368.087$	$ci = 451.2$

FIGURE 11.9
Mathcad screen showing how to determine the matrix inverse and condition numbers of a scaled 3×3 Hilbert matrix.

PROBLEMS

11.1 Perform the same calculations as in (a) Example 11.1 and (b) Example 11.3 but for the tridiagonal system

$$\begin{bmatrix} 0.8 & -0.4 & \\ -0.4 & 0.8 & -0.4 \\ & -0.4 & 0.8 \end{bmatrix} \begin{Bmatrix} x_1 \\ x_2 \\ x_3 \end{Bmatrix} = \begin{Bmatrix} 41 \\ 25 \\ 105 \end{Bmatrix}$$

11.2 Determine the matrix inverse for Example 11.1 based on the *LU* decomposition and unit vectors.

11.3 The following tridiagonal system must be solved as part of a larger algorithm (Crank-Nicolson) for solving partial differential equations:

$$\begin{bmatrix} 2.01475 & -0.020875 & & \\ -0.020875 & 2.01475 & -0.020875 & \\ & -0.020875 & 2.01475 & -0.020875 \\ & & -0.020875 & 2.01475 \end{bmatrix}$$

$$\times \begin{Bmatrix} T_1 \\ T_2 \\ T_3 \\ T_4 \end{Bmatrix} = \begin{Bmatrix} 4.175 \\ 0 \\ 0 \\ 2.0875 \end{Bmatrix}$$

Use the Thomas algorithm to obtain a solution.

11.4 Confirm the validity of the Cholesky decomposition of Example 11.2 by substituting the results into Eq. (11.2) to see if the product of $[L]$ and $[L]^T$ yields $[A]$.

11.5 Perform the same calculations as in Example 11.2, but for the symmetric system

$$\begin{bmatrix} 6 & 15 & 55 \\ 15 & 55 & 225 \\ 55 & 225 & 979 \end{bmatrix} \begin{Bmatrix} a_0 \\ a_1 \\ a_2 \end{Bmatrix} = \begin{Bmatrix} 152.6 \\ 585.6 \\ 2488.8 \end{Bmatrix}$$

In addition to performing the Cholesky decomposition, employ it to solve for the a's.

11.6 Perform a Cholesky decomposition and then use the result to solve the following symmetric system by hand:

$$\begin{bmatrix} 8 & 20 & 15 \\ 20 & 80 & 50 \\ 15 & 50 & 60 \end{bmatrix} \begin{Bmatrix} x_1 \\ x_2 \\ x_3 \end{Bmatrix} = \begin{Bmatrix} 100 \\ 250 \\ 100 \end{Bmatrix}$$

11.7 Compute the Cholesky decomposition of

$$[A] = \begin{bmatrix} 100 & 0 & 0 \\ 0 & 25 & 0 \\ 0 & 0 & 16 \end{bmatrix}$$

Do your results make sense in terms of Eqs. (11.3) and (11.4)?

11.8 Use the Gauss-Seidel method to solve the tridiagonal system from Prob. 11.1 ($\varepsilon_s = 5\%$). Use overrelaxation with $\lambda = 1.2$.

11.9 Recall from Prob. 10.8, that the following system of equations is designed to determine concentrations (the c's in g/m^3) in a series of coupled reactors as a function of amount of mass input to each reactor (the right-hand sides are in g/day):

$$15c_1 - 3c_2 - c_3 = 3300$$
$$-3c_1 + 18c_2 - 6c_3 = 1200$$
$$-4c_1 - c_2 + 12c_3 = 2400$$

Solve this problem with the Gauss-Seidel method to $\varepsilon_s = 5\%$.

11.10 Repeat Prob. 11.9, but use Jacobi iteration.

11.11 Use the Gauss-Seidel method to solve the following system until the percent relative error falls below $\varepsilon_s = 5\%$:

$$10x_1 + 2x_2 - x_3 = 22$$
$$-3x_1 - 6x_2 + 2x_3 = -14$$
$$x_1 + x_2 + 5x_3 = 14$$

11.12 Use the Gauss-Seidel method (a) without relaxation and (b) with relaxation ($\lambda = 0.95$) to solve the following system to a tolerance of $\varepsilon_s = 5\%$. If necessary, rearrange the equations to achieve convergence.

$$-3x_1 + x_2 + 15x_3 = 44$$
$$6x_1 - 2x_2 + x_3 = 5$$
$$5x_1 + 10x_2 + x_3 = 28$$

11.13 Use the Gauss-Seidel method (a) without relaxation and (b) with relaxation ($\lambda = 1.2$) to solve the following system to a tolerance of $\varepsilon_s = 5\%$. If necessary, rearrange the equations to achieve convergence.

$$2x_1 - 6x_2 - x_3 = -38$$
$$-3x_1 - x_2 + 7x_3 = -34$$
$$-8x_1 + x_2 - 2x_3 = -20$$

11.14 Redraw Fig. 11.5 for the case where the slopes of the equations are 1 and -1. What is the result of applying Gauss-Seidel to such a system?

11.15 Of the following three sets of linear equations, identify the set(s) that you could not solve using an iterative method such as Gauss-Seidel. Use the number of iterations that are necessary to show that your solution does not converge. Clearly state your convergence criteria (how you know it is not converging).

Set One	Set Two	Set Three
$9x + 3y + z = 13$	$x + y + 6z = 8$	$-3x + 4y + 5z = 6$
$-6x + 8z = 8$	$x + 5y - z = 5$	$-2x + 2y - 4z = -3$
$2x + 5y - z = 6$	$4x + 2y - 2z = 4$	$2y - z = 1$

11.16 Use the software package of your choice to obtain a solution, calculate the inverse, and determine the condition number (without scaling) based on the row-sum norm for

(a)

$$\begin{bmatrix} 1 & 4 & 9 \\ 4 & 9 & 16 \\ 9 & 16 & 25 \end{bmatrix} \begin{Bmatrix} x_1 \\ x_2 \\ x_3 \end{Bmatrix} = \begin{Bmatrix} 14 \\ 29 \\ 50 \end{Bmatrix}$$

(b)

$$\begin{bmatrix} 1 & 4 & 9 & 16 \\ 4 & 9 & 16 & 25 \\ 9 & 16 & 25 & 36 \\ 16 & 25 & 36 & 49 \end{bmatrix} \begin{Bmatrix} x_1 \\ x_2 \\ x_3 \\ x_4 \end{Bmatrix} = \begin{Bmatrix} 30 \\ 54 \\ 86 \\ 126 \end{Bmatrix}$$

In both cases, the answers for all the x's should be 1.

11.17 Given the pair of nonlinear simultaneous equations:

$$f(x, y) = 4 - y - 2x^2$$
$$g(x, y) = 8 - y^2 - 4x$$

(a) Use the Excel Solver to determine the two pairs of values of x and y that satisfy these equations.

(b) Using a range of initial guesses ($x = -6$ to 6 and $y = -6$ to 6), determine which initial guesses yield each of the solutions.

11.18 An electronics company produces transistors, resistors, and computer chips. Each transistor requires four units of copper, one unit of zinc, and two units of glass. Each resistor requires three,

three, and one units of the three materials, respectively, and each computer chip requires two, one, and three units of these materials, respectively. Putting this information into table form, we get:

Component	Copper	Zinc	Glass
Transistors	4	1	2
Resistors	3	3	1
Computer chips	2	1	3

Supplies of these materials vary from week to week, so the company needs to determine a different production run each week. For example, one week the total amounts of materials available are 960 units of copper, 510 units of zinc, and 610 units of glass. Set up the system of equations modeling the production run, and use Excel, MATLAB, or Mathcad to solve for the number of transistors, resistors, and computer chips to be manufactured this week.

11.19 Use MATLAB or Mathcad software to determine the spectral condition number for a 10-dimensional Hilbert matrix. How many digits of precision are expected to be lost due to ill-conditioning? Determine the solution for this system for the case where each element of the right-hand-side vector $\{b\}$ consists of the summation of the coefficients in its row. In other words, solve for the case where all the unknowns should be exactly 1. Compare the resulting errors with those expected based on the condition number.

11.20 Repeat Prob. 11.19, but for the case of a six-dimensional Vandermonde matrix (see Prob. 10.17) where $x_1 = 4$, $x_2 = 2$, $x_3 = 7$, $x_4 = 10$, $x_5 = 3$, and $x_6 = 5$.

11.21 Given a square matrix $[A]$, write a single-line MATLAB command that will create a new matrix $[Aug]$ that consists of the original matrix $[A]$ augmented by an identity matrix $[I]$.

11.22 Write the following set of equations in matrix form:

$$55 = 5x_3 - 7x_2$$

$$4x_2 + 6x_3 + 30 = 0$$

$$x_1 - 7x_3 = 40 - 2x_2 + 5x_1$$

Use Excel, MATLAB, or Mathcad to solve for the unknowns. In addition, compute the transpose and the inverse of the coefficient matrix.

11.23 In Sec. 9.2.1, we determined the number of operations required for Gauss elimination without partial pivoting. Make a similar determination for the Thomas algorithm (Fig. 11.2). Develop a plot of operations versus n (from 2 to 20) for both techniques.

11.24 Develop a user-friendly program in either a high-level or a macro language of your choice to obtain a solution for a tridiagonal system with the Thomas algorithm (Fig. 11.2). Test your program by duplicating the results of Example 11.1.

11.25 Develop a user-friendly program in either a high-level or a macro language of your choice for Cholesky decomposition based

on Fig. 11.3. Test your program by duplicating the results of Example 11.2.

11.26 Develop a user-friendly program in either a high-level or a macro language of your choice for the Gauss-Seidel method based on Fig. 11.6. Test your program by duplicating the results of Example 11.3.

11.27 As described in Sec. PT3.1.2, linear algebraic equations can arise in the solution of differential equations. For example, the following differential equation results from a steady-state mass balance for a chemical in a one-dimensional canal:

$$0 = D\frac{d^2c}{dx^2} - U\frac{dc}{dx} - kc$$

where c = concentration, x = distance, D = diffusion coefficient, U = fluid velocity, and k = a first-order decay rate. Convert this differential equation to an equivalent system of simultaneous algebraic equations. Given $D = 2.5$, $U = 0.75$, $k = 0.15$, $c(0) = 75$ and $c(10) = 25$, solve these equations from $x = 0$ to 10 with $\Delta x = 2$, and develop a plot of concentration versus distance.

11.28 A *pentadiagonal system* with a bandwidth of five can be expressed generally as

$$\begin{bmatrix} f_1 & g_1 & h_1 & & & & \\ e_2 & f_2 & g_2 & h_2 & & & \\ d_3 & e_3 & f_3 & g_3 & h_3 & & \\ & & \cdot & \cdot & \cdot & & \\ & & & \cdot & \cdot & \cdot & \\ & & & & \cdot & \cdot & \cdot \\ & & & d_{n-1} & e_{n-1} & f_{n-1} & g_{n-1} \\ & & & & d_n & e_n & f_n \end{bmatrix} \begin{Bmatrix} x_1 \\ x_2 \\ x_3 \\ \cdot \\ \cdot \\ \cdot \\ x_{n-1} \\ x_n \end{Bmatrix} = \begin{Bmatrix} r_1 \\ r_2 \\ r_3 \\ \cdot \\ \cdot \\ \cdot \\ r_{n-1} \\ r_n \end{Bmatrix}$$

Develop a program to efficiently solve such systems without pivoting in a similar fashion to the algorithm used for tridiagonal matrices in Sec. 11.1.1. Test your program for the following case:

$$\begin{bmatrix} 8 & -2 & -1 & 0 & 0 \\ -2 & 9 & -4 & -1 & 0 \\ -1 & -3 & 7 & -1 & -2 \\ 0 & -4 & -2 & 12 & -5 \\ 0 & 0 & -7 & -3 & 15 \end{bmatrix} \begin{Bmatrix} x_1 \\ x_2 \\ x_3 \\ x_4 \\ x_5 \end{Bmatrix} = \begin{Bmatrix} 5 \\ 2 \\ 0 \\ 1 \\ 5 \end{Bmatrix}$$

11.29 Besides solving simultaneous equations, linear algebra has many other applications in engineering and science. An example from computer graphics involves rotating an object in Euclidean space. The following *rotation matrix* can be employed to rotate a group of points counter clockwise through an angle θ about the origin of a Cartesian coordinate system:

$$R = \begin{bmatrix} \cos\theta & -\sin\theta \\ \sin\theta & \cos\theta \end{bmatrix}$$

To do this, each point's position must be represented by a column vector v, containing the coordinates of the point. For example, here are vectors for the x and y coordinates of the rectangle in Fig. P11.29

```
x = [1 4 4 1]; y = [1 1 4 4];
```

The rotated vector is then generated with matrix multiplication: $[R]\{v\}$. Develop a MATLAB function to perform this operation and display the initial and the rotated points as filled shapes on the same graph. Here is a script to test your function:

```
clc;clf;format compact
x = [1 4 4 1]; y = [1 1 4 4];
[xt, yt] = Rotate2D(45, x, y);
```

and here is a skeleton of the function:

FIGURE P11.29

```
function [xr, yr] = Rotate2D(thetad, x, y)
% two-dimensional rotation 2D rotate Cartesian
% [xr, yr] = rot2d(thetad, x, y)
% Rotation of a two-dimensional object the Cartesian coordinates
% of which are contained in the vectors x and y.
% input:
% thetad = angle of rotation (degrees)
% x = vector containing objects x coordinates
% y = vector containing objects y coordinates
% output:
% xr = vector containing objects rotated x coordinates
% yr = vector containing objects rotated y coordinates

% convert angle to radians and set up rotation matrix
    .
    .
    .
% close shape
    .
    .
    .
% plot original object
hold on, grid on
    .
    .
    .
% rotate shape
    .
    .
    .
% plot rotated object
    .
    .
    .
hold off
```

Case Studies: Linear Algebraic Equations

The purpose of this chapter is to use the numerical procedures discussed in Chaps. 9, 10, and 11 to solve systems of linear algebraic equations for some engineering case studies. These systematic numerical techniques have practical significance because engineers frequently encounter problems involving systems of equations that are too large to solve by hand. The numerical algorithms in these applications are particularly convenient to implement on personal computers.

Section 12.1 shows how a mass balance can be employed to model a system of reactors. *Section 12.2* places special emphasis on the use of the matrix inverse to determine the complex cause-effect interactions between forces in the members of a truss. *Section 12.3* presents an example of the use of Kirchhoff's laws to compute the currents and voltages in a resistor circuit. Finally, *Sec. 12.4* provides an illustration of how linear equations are employed to determine the steady-state configuration of a mass-spring system.

12.1 STEADY-STATE ANALYSIS OF A SYSTEM OF REACTORS (CHEMICAL/BIO ENGINEERING)

Background.   One of the most important organizing principles in chemical engineering is the *conservation of mass* (recall Table 1.1). In quantitative terms, the principle is expressed as a mass balance that accounts for all mass sources and sinks that pass in and out of a volume (Fig. 12.1). Over a finite period of time, this can be expressed as

$$\text{Accumulation} = \text{inputs} - \text{outputs} \qquad (12.1)$$

The mass balance represents a bookkeeping exercise for the particular substance being modeled. For the period of the computation, if the inputs are greater than the outputs, the mass of the substance within the volume increases. If the outputs are greater than the inputs, the mass decreases. If inputs are equal to the outputs, accumulation is zero and mass remains constant. For this stable condition, or steady state, Eq. (12.1) can be expressed as

$$\text{Inputs} = \text{outputs} \qquad (12.2)$$

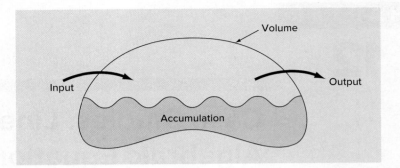

FIGURE 12.1
A schematic representation of mass balance.

Employ the conservation of mass to determine the steady-state concentrations of a system of coupled reactors.

Solution.   The mass balance can be used for engineering problem solving by expressing the inputs and outputs in terms of measurable variables and parameters. For example, if we were performing a mass balance for a conservative substance (that is, one that does not increase or decrease due to chemical transformations) in a reactor (Fig. 12.2), we would have to quantify the rates at which mass flows into the reactor through the two inflow pipes and out of the reactor through the outflow pipe. This can be done by taking the product of the flow rate Q (in cubic meters per minute) and the concentration c (in milligrams per cubic meter) for each pipe. For example, for pipe 1 in Fig. 12.2, $Q_1 = 2$ m^3/min and $c_1 = 25$ mg/m^3; therefore, the rate at which mass flows into the reactor through pipe 1 is $Q_1c_1 = (2$ m^3/min$)(25$ mg/m$^3) = 50$ mg/min. Thus, 50 mg of chemical flows into the reactor through this pipe each minute. Similarly, for pipe 2 the mass inflow rate can be calculated as $Q_2c_2 = (1.5$ m^3/min$)(10$ mg/m$^3) = 15$ mg/min.

Notice that the concentration out of the reactor through pipe 3 is not specified by Fig. 12.2. This is because we already have sufficient information to calculate it on the basis of the conservation of mass. Because the reactor is at steady state, Eq. (12.2) holds and the inputs should be in balance with the outputs, as in

$$Q_1c_1 + Q_2c_2 = Q_3c_3$$

Substituting the given values into this equation yields

$$50 + 15 = 3.5c_3$$

which can be solved for $c_3 = 18.6$ mg/m^3. Thus, we have determined the concentration in the third pipe. However, the computation yields an additional bonus. Because the reactor is well mixed (as represented by the propeller in Fig. 12.2), the concentration will be uniform, or homogeneous, throughout the tank. Therefore the concentration in pipe 3 should be identical to the concentration throughout the reactor. Consequently, the mass balance has allowed us to compute both the concentration in the reactor and in the outflow pipe. Such information is of great utility to chemical and petroleum engineers who must design reactors to yield mixtures of a specified concentration.

FIGURE 12.2

A steady-state, completely mixed reactor with two inflow pipes and one outflow pipe. The flows Q are in cubic meters per minute, and the concentrations c are in milligrams per cubic meter.

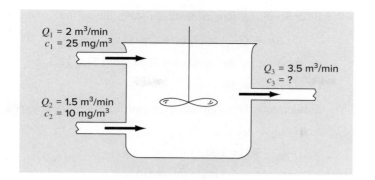

FIGURE 12.3

Five reactors linked by pipes.

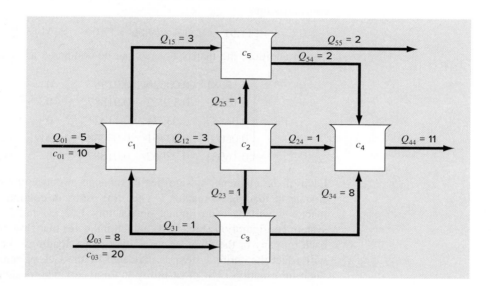

Because simple algebra was used to determine the concentration for the single reactor in Fig. 12.2, it might not be obvious how computers figure in mass-balance calculations. Figure 12.3 shows a problem setting where computers are not only useful but are a practical necessity. Because there are five interconnected, or coupled, reactors, five simultaneous mass-balance equations are needed to characterize the system. For reactor 1, the rate of mass flow in is

$$5(10) + Q_{31}c_3$$

and the rate of mass flow out is

$$Q_{12}c_1 + Q_{15}c_1$$

Because the system is at steady-state, the inflows and outflows must be equal:

$$5(10) + Q_{31}c_3 = Q_{12}c_1 + Q_{15}c_1$$

or, substituting the values for flow from Fig. 12.3,

$$6c_1 - c_3 = 50$$

Similar equations can be developed for the other reactors:

$$-3c_1 + 3c_2 = 0$$
$$-c_2 + 9c_3 = 160$$
$$-c_2 - 8c_3 + 11c_4 - 2c_5 = 0$$
$$-3c_1 - c_2 + 4c_5 = 0$$

A numerical method can be used to solve these five equations for the five unknown concentrations:

$$\{C\}^T = \lfloor 11.51 \quad 11.51 \quad 19.06 \quad 17.00 \quad 11.51 \rfloor$$

In addition, the matrix inverse can be computed as

$$[A]^{-1} = \begin{bmatrix} 0.16981 & 0.00629 & 0.01887 & 0 & 0 \\ 0.16981 & 0.33962 & 0.01887 & 0 & 0 \\ 0.01887 & 0.03774 & 0.11321 & 0 & 0 \\ 0.06003 & 0.07461 & 0.08748 & 0.09091 & 0.04545 \\ 0.16981 & 0.08962 & 0.01887 & 0 & 0.25000 \end{bmatrix}$$

Each of the elements a_{ij} signifies the change in concentration of reactor i due to a unit change in loading to reactor j. Thus, the zeros in column 4 indicate that a loading to reactor 4 will have no impact on reactors 1, 2, 3, and 5. This is consistent with the system configuration (Fig. 12.3), which indicates that flow out of reactor 4 does not feed back into any of the other reactors. In contrast, loadings to any of the first three reactors will affect the entire system as indicated by the lack of zeros in the first three columns. Such information is of great utility to engineers who design and manage such systems.

12.2 ANALYSIS OF A STATICALLY DETERMINATE TRUSS (CIVIL/ENVIRONMENTAL ENGINEERING)

Background.   An important problem in structural engineering is that of finding the forces and reactions associated with a statically determinate truss. Figure 12.4 shows an example of such a truss.

The forces (F) represent either tension or compression on the members of the truss. External reactions (H_2, V_2, and V_3) are forces that characterize how the truss interacts with the supporting surface. The hinge at node 2 can transmit both horizontal and vertical forces to the surface, whereas the roller at node 3 transmits only vertical forces. It is observed that the effect of the external loading of 1000 lb is distributed among the various members of the truss.

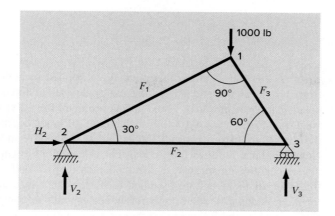

FIGURE 12.4
Forces on a statically determinate truss.

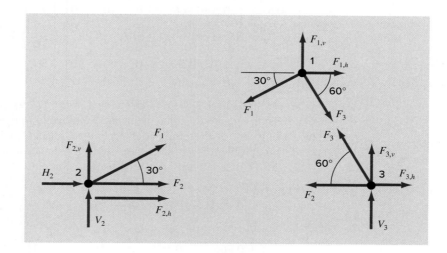

FIGURE 12.5
Free-body force diagrams for the nodes of a statically determinate truss.

Solution.   This type of structure can be described as a system of coupled linear algebraic equations. Free-body force diagrams are shown for all three nodes in Fig. 12.5. The sum of the forces in both horizontal and vertical directions must be zero at each node, because the system is at rest. Therefore, for node 1,

$$\Sigma F_H = 0 = -F_1 \cos 30° + F_3 \cos 60° + F_{1,h} \tag{12.3}$$

$$\Sigma F_V = 0 = -F_1 \sin 30° - F_3 \sin 60° + F_{1,v} \tag{12.4}$$

for node 2,

$$\Sigma F_H = 0 = F_2 + F_1 \cos 30° + F_{2,h} + H_2 \tag{12.5}$$

$$\Sigma F_V = 0 = F_1 \sin 30° + F_{2,v} + V_2 \tag{12.6}$$

for node 3,

$$\Sigma F_H = 0 = -F_2 - F_3 \cos 60° + F_{3,h} \tag{12.7}$$
$$\Sigma F_V = 0 = F_3 \sin 60° + F_{3,v} + V_3 \tag{12.8}$$

where $F_{i,h}$ is the external horizontal force applied to node i (where a positive force is from left to right) and $F_{i,v}$ is the external vertical force applied to node i (where a positive force is upward). Thus, in this problem, the 1000-lb downward force on node 1 corresponds to $F_{1,v} = -1000$. For this case all other $F_{i,v}$'s and $F_{i,h}$'s are zero. Note that the directions of the internal forces and reactions are unknown. Proper application of Newton's laws requires only consistent assumptions regarding direction. Solutions are negative if the directions are assumed incorrectly. Also note that in this problem, the forces in all members are assumed to be tension forces acting to pull adjoining nodes together. A negative solution therefore corresponds to compression. This problem can be represented as the following system of six equations and six unknowns:

$$\begin{bmatrix} 0.866 & 0 & -0.5 & 0 & 0 & 0 \\ 0.5 & 0 & 0.866 & 0 & 0 & 0 \\ -0.866 & -1 & 0 & -1 & 0 & 0 \\ -0.5 & 0 & 0 & 0 & -1 & 0 \\ 0 & 1 & 0.5 & 0 & 0 & 0 \\ 0 & 0 & -0.866 & 0 & 0 & -1 \end{bmatrix} \begin{Bmatrix} F_1 \\ F_2 \\ F_3 \\ H_2 \\ V_2 \\ V_3 \end{Bmatrix} = \begin{Bmatrix} 0 \\ -1000 \\ 0 \\ 0 \\ 0 \\ 0 \end{Bmatrix} \tag{12.9}$$

Notice that, as the system is formulated in Eq. (12.9), partial pivoting is required to avoid division by zero diagonal elements. Employing a pivot strategy, the system can be solved using any of the elimination techniques discussed in Chap. 9 or 10. However, because this problem is an ideal case study for demonstrating the utility of the matrix inverse, the LU decomposition can be used to compute

$$F_1 = -500 \qquad F_2 = 433 \qquad F_3 = -866$$
$$H_2 = 0 \qquad V_2 = 250 \qquad V_3 = 750$$

and the matrix inverse is

$$[A]^{-1} = \begin{bmatrix} 0.866 & 0.5 & 0 & 0 & 0 & 0 \\ 0.25 & -0.433 & 0 & 0 & 1 & 0 \\ -0.5 & 0.866 & 0 & 0 & 0 & 0 \\ -1 & 0 & -1 & 0 & -1 & 0 \\ -0.433 & -0.25 & 0 & -1 & 0 & 0 \\ 0.433 & -0.75 & 0 & 0 & 0 & -1 \end{bmatrix}$$

Now, realize that the right-hand-side vector represents the externally applied horizontal and vertical forces on each node, as in

$$\{F\}^T = \lfloor F_{1,h} \quad F_{1,v} \quad F_{2,h} \quad F_{2,v} \quad F_{3,h} \quad F_{3,v} \rfloor \tag{12.10}$$

Because the external forces have no effect on the LU decomposition, the method need not be implemented over and over again to study the effect of different external forces on the truss. Rather, all that we have to do is perform the forward- and backward-substitution steps for each right-hand-side vector to efficiently obtain alternative solutions. For example,

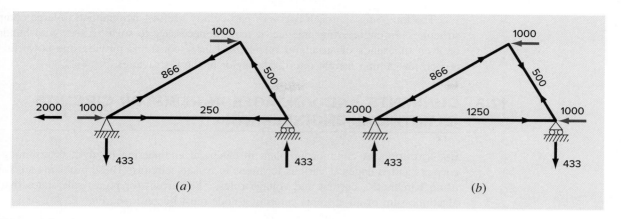

FIGURE 12.6
Two test cases showing (*a*) wind from the left and (*b*) wind from the right.

we might want to study the effect of horizontal forces induced by a wind blowing from left to right. If the wind force can be idealized as two point forces of 1000 lb on nodes 1 and 2 (Fig. 12.6*a*), the right-hand-side vector is

$$\{F\}^T = \lfloor -1000 \quad 0 \quad 1000 \quad 0 \quad 0 \quad 0 \rfloor$$

which can be used to compute

$$F_1 = 866 \qquad F_2 = 250 \qquad F_3 = -500$$
$$H_2 = -2000 \qquad V_2 = -433 \qquad V_3 = 433$$

For a wind from the right (Fig. 12.6*b*), $F_{1,h} = -1000$, $F_{3,h} = -1000$, and all other external forces are zero, with the result that

$$F_1 = -866 \qquad F_2 = -1250 \qquad F_3 = 500$$
$$H_2 = 2000 \qquad V_2 = 433 \qquad V_3 = -433$$

The results indicate that the winds have markedly different effects on the structure. Both cases are depicted in Fig. 12.6.

The individual elements of the inverted matrix also have direct utility in elucidating stimulus-response interactions for the structure. Each element represents the change of one of the unknown variables to a unit change of one of the external stimuli. For example, element a_{32}^{-1} indicates that the third unknown (F_3) will change 0.866 due to a unit change of the second external stimulus ($F_{1,v}$). Thus, if the vertical load at the first node were increased by 1, F_3 would increase by 0.866. The fact that elements are 0 indicates that certain unknowns are unaffected by some of the external stimuli. For instance $a_{13}^{-1} = 0$ means that F_1 is unaffected by changes in $F_{2,h}$. This ability to isolate interactions has a number of engineering applications, including the identification of those components that are most sensitive to external stimuli and, as a consequence, most prone to failure. In addition, it can be used to determine components that may be unnecessary (see Prob. 12.20).

The foregoing approach becomes particularly useful when applied to large complex structures. In engineering practice, it may be necessary to solve trusses with hundreds or even thousands of structural members. Linear equations provide one powerful approach for gaining insight into the behavior of these structures.

12.3 CURRENTS AND VOLTAGES IN RESISTOR CIRCUITS (ELECTRICAL ENGINEERING)

Background.   A common problem in electrical engineering involves determining the currents and voltages at various locations in resistor circuits. These problems are solved using Kirchhoff's current and voltage rules. The current (or point) rule states that the algebraic sum of all currents entering a node must be zero (see Fig. 12.7a), or

$$\Sigma i = 0 \qquad (12.11)$$

where all current entering the node is considered positive in sign. The current rule is an application of the principle of conservation of charge (recall Table 1.1).

The voltage (or loop) rule specifies that the algebraic sum of the potential differences (that is, voltage changes) in any loop must equal zero. For a resistor circuit, this is expressed as

$$\Sigma \xi - \Sigma iR = 0 \qquad (12.12)$$

where ξ is the emf (electromotive force) of the voltage sources and R is the resistance of any resistors in the loop. Note that the second term derives from Ohm's law (Fig. 12.7b), which states that the voltage drop across an ideal resistor is equal to the product of the current and the resistance. Kirchhoff's voltage rule is an expression of the conservation of energy.

Solution.   Application of these rules results in systems of simultaneous linear algebraic equations because the various loops within a circuit are coupled. For example, consider the circuit shown in Fig. 12.8. The currents associated with this circuit are unknown both in magnitude and direction. This presents no great difficulty because one simply assumes a direction for each current. If the resultant solution from Kirchhoff's laws is negative, then the assumed direction was incorrect. For example, Fig. 12.9 shows some assumed currents.

FIGURE 12.7
Schematic representations of (a) Kirchhoff's current rule and (b) Ohm's law.

(a)

(b)

FIGURE 12.8
A resistor circuit to be solved using simultaneous linear algebraic equations.

FIGURE 12.9
Assumed currents.

Given these assumptions, Kirchhoff's current rule is applied at each node to yield

$$i_{12} + i_{52} + i_{32} = 0$$
$$i_{65} - i_{52} - i_{54} = 0$$
$$i_{43} - i_{32} = 0$$
$$i_{54} - i_{43} = 0$$

Application of the voltage rule to each of the two loops gives

$$-i_{54}R_{54} - i_{43}R_{43} - i_{32}R_{32} + i_{52}R_{52} = 0$$
$$-i_{65}R_{65} - i_{52}R_{52} - i_{12}R_{12} - 200 = 0$$

or, substituting the resistances from Fig. 12.8 and bringing constants to the right-hand side,

$$-15i_{54} - 5i_{43} - 10i_{32} + 10i_{52} = 0$$
$$-20i_{65} - 10i_{52} + 5i_{12} = 200$$

Therefore, the problem amounts to solving the following set of six equations with six unknown currents:

$$
\begin{bmatrix}
1 & 1 & 1 & 0 & 0 & 0 \\
0 & -1 & 0 & 1 & -1 & 0 \\
0 & 0 & -1 & 0 & 0 & 1 \\
0 & 0 & 0 & 0 & 1 & -1 \\
0 & 10 & -10 & 0 & -15 & -5 \\
5 & -10 & 0 & -20 & 0 & 0
\end{bmatrix}
\begin{Bmatrix}
i_{12} \\
i_{52} \\
i_{32} \\
i_{65} \\
i_{54} \\
i_{43}
\end{Bmatrix}
=
\begin{Bmatrix}
0 \\
0 \\
0 \\
0 \\
0 \\
200
\end{Bmatrix}
$$

Although impractical to solve by hand, this system is easily handled using an elimination method. Proceeding in this manner, the solution is

$$i_{12} = 6.1538 \qquad i_{52} = -4.6154 \qquad i_{32} = -1.5385$$
$$i_{65} = -6.1538 \qquad i_{54} = -1.5385 \qquad i_{43} = -1.5385$$

Thus, with proper interpretation of the signs of the result, the circuit currents and voltages are as shown in Fig. 12.10. The advantages of using numerical algorithms and computers for problems of this type should be evident.

FIGURE 12.10
The solution for currents and voltages obtained using an elimination method.

12.4 SPRING-MASS SYSTEMS (MECHANICAL/AEROSPACE ENGINEERING)

Background.   Idealized spring-mass systems play an important role in mechanical and other engineering problems. Figure 12.11 shows such a system. After they are released, the masses are pulled downward by the force of gravity. Notice that the resulting displacement of each spring in Fig. 12.11b is measured along local coordinates referenced to its initial position in Fig. 12.11a.

As introduced in Chap. 1, Newton's second law can be employed in conjunction with force balances to develop a mathematical model of the system. For each mass, the second law can be expressed as

$$m\frac{d^2x}{dt^2} = F_D - F_U \tag{12.13}$$

To simplify the analysis, we will assume that all the springs are identical and follow Hooke's law. A free-body diagram for the first mass is depicted in Fig. 12.12a. The upward force is merely a direct expression of Hooke's law:

$$F_U = kx_1 \tag{12.14}$$

The downward component consists of the two spring forces along with the action of gravity on the mass,

$$F_D = k(x_2 - x_1) + k(x_2 - x_1) = m_1g \tag{12.15}$$

Note how the force component of the two springs is proportional to the displacement of the second mass, x_2, corrected for the displacement of the first mass, x_1.

Equations (12.14) and (12.15) can be substituted into Eq. (12.13) to give

$$m_1\frac{d^2x_1}{dt^2} = 2k(x_2 - x_1) + m_1g - kx_1 \tag{12.16}$$

Thus, we have derived a second-order ordinary differential equation to describe the displacement of the first mass with respect to time. However, notice that the solution cannot be obtained because the model includes a second dependent variable, x_2. Consequently, free-body diagrams must be developed for the second and the third masses (Fig. 12.12b

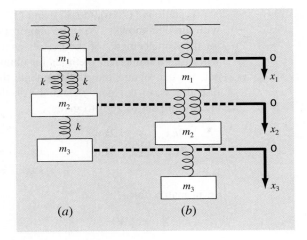

FIGURE 12.11
A system composed of three masses suspended vertically by a series of springs. (*a*) The system before release, that is, prior to extension or compression of the springs. (*b*) The system after release. Note that the positions of the masses are referenced to local coordinates with origins at their position before release.

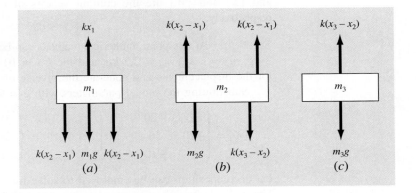

FIGURE 12.12
Free-body diagrams for the three masses from Fig. 12.11.

and *c*) that can be employed to derive

$$m_2 \frac{d^2x_2}{dt^2} = k(x_3 - x_2) + m_2g - 2k(x_2 - x_1) \tag{12.17}$$

and

$$m_3 \frac{d^2x_3}{dt^2} = m_3g - k(x_3 - x_2) \tag{12.18}$$

Equations (12.16), (12.17), and (12.18) form a system of three differential equations with three unknowns. With the appropriate initial conditions, they could be used to solve for the displacements of the masses as a function of time (that is, their oscillations). We will discuss numerical methods for obtaining such solutions in Part Seven. For the present, we can obtain the displacements that occur when the system eventually comes to rest, that is, to the steady state. To do this, the derivatives in Eqs. (12.16), (12.17), and (12.18) are set to zero to give

$$
\begin{array}{rcrcrcl}
3kx_1 & - & 2kx_2 & & & = & m_1 g \\
-2kx_1 & + & 3kx_2 & - & kx_3 & = & m_2 g \\
& & -kx_2 & + & kx_3 & = & m_3 g
\end{array}
$$

or, in matrix form,

$$[K]\{X\} = \{W\}$$

where $[K]$, called the *stiffness matrix*, is

$$
[K] = \begin{bmatrix}
3k & -2k & \\
-2k & 3k & -k \\
& -k & k
\end{bmatrix}
$$

and $\{X\}$ and $\{W\}$ are the column vectors of the unknowns x and the weights mg, respectively.

Solution.   At this point, numerical methods can be employed to obtain a solution. If $m_1 = 2$ kg, $m_2 = 3$ kg, $m_3 = 2.5$ kg, and the k's $= 10$ kg/s^2, use *LU* decomposition to solve for the displacements and generate the inverse of $[K]$.

Substituting the model parameters with $g = 9.81$ gives

$$
[K] = \begin{bmatrix}
30 & -20 & \\
-20 & 30 & -10 \\
& -10 & 10
\end{bmatrix}
\qquad
\{W\} = \begin{Bmatrix}
19.62 \\
29.43 \\
24.525
\end{Bmatrix}
$$

LU decomposition can be employed to solve for $x_1 = 7.36$, $x_2 = 10.06$, and $x_3 = 12.51$. These displacements were used to construct Fig. 12.11b. The inverse of the stiffness matrix is computed as

$$
[K]^{-1} = \begin{bmatrix}
0.1 & 0.1 & 0.1 \\
0.1 & 0.15 & 0.15 \\
0.1 & 0.15 & 0.25
\end{bmatrix}
$$

Each element of this matrix k_{ji}^{-1} tells us the displacement of mass i due to a unit force imposed on mass j. Thus, the values of 0.1 in column 1 tell us that a downward unit load to the first mass will displace all of the masses 0.1 m downward. The other elements can be interpreted in a similar fashion. Therefore, the inverse of the stiffness matrix provides a fundamental summary of how the system's components respond to externally applied forces.

PROBLEMS

Chemical/Bio Engineering

12.1 Perform the same computation as in Sec. 12.1, but change c_{01} to 20 and c_{03} to 6. Also change the following flows: $Q_{01} = 6$, $Q_{12} = 4$, $Q_{24} = 2$, and $Q_{44} = 12$.

12.2 If the input to reactor 3 in Sec. 12.1 is decreased 25 percent, use the matrix inverse to compute the percent change in the concentration of reactors 2 and 4.

12.3 Because the system shown in Fig. 12.3 is at steady state, what can be said regarding the four flows: Q_{01}, Q_{03}, Q_{44}, and Q_{55}?

12.4 Recompute the concentrations for the five reactors shown in Fig. 12.3, if the flows are changed to

$$Q_{01} = 5 \quad Q_{31} = 2 \quad Q_{25} = 3 \quad Q_{23} = 1$$
$$Q_{15} = 3 \quad Q_{55} = 4 \quad Q_{54} = 2 \quad Q_{34} = 9$$
$$Q_{12} = 4 \quad Q_{03} = 10 \quad Q_{24} = 0 \quad Q_{44} = 11$$

12.5 Solve the same system as specified in Prob. 12.4, but set $Q_{15} = 5$, $Q_{25} = 1$, $Q_{23} = 2$, $Q_{24} = 1$, $Q_{01} = 6$, $Q_{03} = 7$, $Q_{44} = 9$, and $Q_{55} = 4$. Use conservation of flow to recompute the values for the other flows. Employ the matrix inverse to determine the change in concentration for the fifth reactor if the inflow concentrations to the first and third reactors are doubled.

12.6 Figure P12.6 shows three reactors linked by pipes. As indicated, the rate of transfer of chemicals through each pipe is equal to a flow rate (Q, with units of cubic meters per second) multiplied by the concentration of the reactor from which the flow originates (c, with units of milligrams per cubic meter). If the system is at steady state, the transfer into each reactor will balance the transfer out. Develop mass-balance equations for the reactors, and solve the three simultaneous linear algebraic equations for their concentrations.

12.7 Employing the same basic approach as in Sec. 12.1, determine the concentration of chloride in each of the Great Lakes using the information shown in Fig. P12.7. Compute the matrix inverse, and use it to determine the percent reduction of the concentration in Lake Ontario due to a 50% reduction of the loadings to Lake Superior and Lake Michigan.

12.8 The Lower Colorado River consists of a series of four reservoirs as shown in Fig. P12.8. Mass balances can be written for each reservoir and the following set of simultaneous linear algebraic equations results:

$$\begin{bmatrix} 13.442 & 0 & 0 & 0 \\ -13.442 & 12.252 & 0 & 0 \\ 0 & -12.252 & 12.377 & 0 \\ 0 & 0 & -12.377 & 11.797 \end{bmatrix} \begin{Bmatrix} c_1 \\ c_2 \\ c_3 \\ c_4 \end{Bmatrix} = \begin{Bmatrix} 750.5 \\ 300 \\ 102 \\ 30 \end{Bmatrix}$$

where the right-hand-side vector consists of the loadings of chloride to each of the four lakes and c_1, c_2, c_3, and c_4 are the resulting chloride concentrations for Lakes Powell, Mead, Mohave, and Havasu, respectively.

(a) Use the matrix inverse to solve for the concentration in each of the four lakes.

(b) How much must the loading to Lake Powell be reduced in order for the chloride concentration of Lake Havasu to be 75?

(c) Using the column-sum norm, compute the condition number and determine how many suspect digits would be generated by solving this system.

12.9 A stage extraction process is depicted in Fig. P12.9. In such systems, a stream containing a weight fraction Y_{in} of a chemical enters from the left at a mass flow rate of F_1. Simultaneously, a solvent carrying a weight fraction X_{in} of the same chemical enters from the right at a flow rate of F_2. Thus, for stage i, a mass balance can be represented as

$$F_1 Y_{i-1} + F_2 X_{i+1} = F_1 Y_i + F_2 X_i \qquad \text{(P12.9.1)}$$

FIGURE P12.6

Three reactors linked by pipes. The rate of mass transfer through each pipe is equal to the product of flow Q and concentration c of the reactor from which the flow originates.

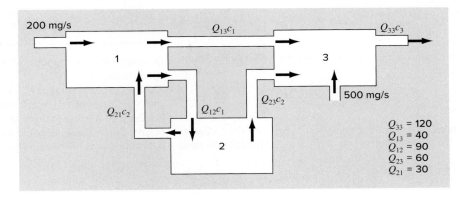

$Q_{33} = 120$
$Q_{13} = 40$
$Q_{12} = 90$
$Q_{23} = 60$
$Q_{21} = 30$

FIGURE P12.7

A chloride mass balance for the Great Lakes. Numbered arrows are direct inputs.

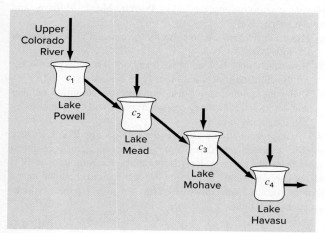

FIGURE P12.8

The Lower Colorado River.

At each stage, an equilibrium is assumed to be established between Y_i and X_i as in

$$K = \frac{X_i}{Y_i} \tag{P12.9.2}$$

where K is called a distribution coefficient. Equation (P12.9.2) can be solved for X_i and the result substituted into Eq. (P12.9.1) to yield

$$Y_{i-1} - \left(1 + \frac{F_2}{F_1}K\right)Y_i + \left(\frac{F_2}{F_1}K\right)Y_{i+1} = 0 \tag{P12.9.3}$$

If $F_1 = 400$ kg/h, $Y_{in} = 0.1$, $F_2 = 800$ kg/h, $X_{in} = 0$, and $K = 5$, determine the values of Y_{out} and X_{out} if a five-stage reactor is used. Note that Eq. (P12.9.3) must be modified to account for the inflow weight fractions when applied to the first and last stages.

12.10 An irreversible, first-order reaction takes place in four well-mixed reactors (Fig. P12.10),

$$A \xrightarrow{k} B$$

Thus, the rate at which A is transformed to B can be represented as

$$R_{ab} = kVc$$

The reactors have different volumes, and because they are operated at different temperatures, each has a different reaction rate:

Reactor	V, L	k, h^{-1}
1	25	0.075
2	75	0.15
3	100	0.4
4	25	0.1

Determine the concentration of A and B in each of the reactors at steady state.

12.11 A peristaltic pump delivers a unit flow (Q_1) of a highly viscous fluid. The network is depicted in Fig. P12.11. Every pipe section has the same length and diameter. The mass and mechanical energy balance can be simplified to obtain the flows in every pipe. Solve the following system of equations to obtain the flow in every pipe.

$$Q_3 + 2Q_4 - 2Q_2 = 0$$
$$Q_5 + 2Q_6 - 2Q_4 = 0$$

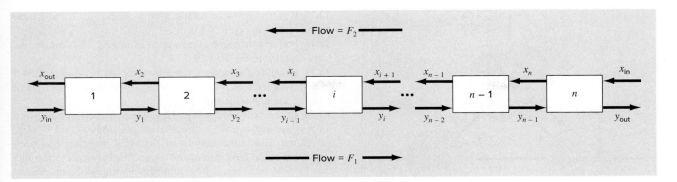

FIGURE P12.9

A stage extraction process.

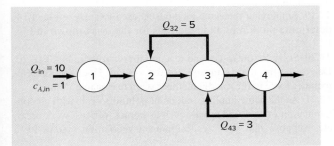

FIGURE P12.10

FIGURE P12.11

$$3Q_7 - 2Q_6 = 0$$
$$Q_1 = Q_2 + Q_3$$
$$Q_3 = Q_4 + Q_5$$
$$Q_5 = Q_6 + Q_7$$

12.12 Figure P12.12 depicts a chemical exchange process consisting of a series of reactors in which a gas flowing from left to right is passed over a liquid flowing from right to left. The transfer of a

FIGURE P12.12

chemical from the gas into the liquid occurs at a rate that is proportional to the difference between the gas and liquid concentrations in each reactor. At steady state, a mass balance for the first reactor can be written for the gas as

$$Q_G c_{G0} - Q_G c_{G1} + D(c_{L1} - c_{G1}) = 0$$

and for the liquid as

$$Q_L c_{L2} - Q_L c_{L1} + D(c_{G1} - c_{L1}) = 0$$

where Q_G and Q_L are the gas and liquid flow rates, respectively, and D = the gas-liquid exchange rate. Similar balances can be written for the other reactors. Solve for the concentrations given the following values: $Q_G = 2$, $Q_L = 1$, $D = 0.8$, $c_{G0} = 100$, $c_{L6} = 20$.

12.13 Figure P12.13 shows a system of three interconnected reactors. Note that aside from flow between the tanks, mass is lost via first-order decay reactions within each tank. Mass balances can be written for each segment, and the following set of simultaneous linear algebraic equations results:

$$\begin{bmatrix} 127 & -22 & 0 \\ -5 & 52 & -7 \\ -117 & 0 & 137 \end{bmatrix} \begin{Bmatrix} c_1 \\ c_2 \\ c_3 \end{Bmatrix} = \begin{Bmatrix} 1000 \\ 2000 \\ 0 \end{Bmatrix}$$

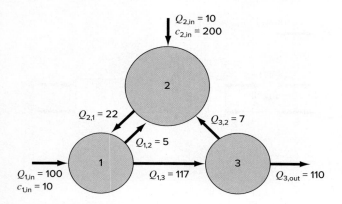

FIGURE P12.13

The *LU* factorization for this system is

$$[L][U] = \begin{bmatrix} 1 & 0 & 0 \\ -0.03937 & 1 & 0 \\ -0.92126 & -0.39637 & 1 \end{bmatrix} \begin{bmatrix} 127 & -22 & 0 \\ 0 & 51.13386 & -7 \\ 0 & 0 & 134.2254 \end{bmatrix}$$

(a) Here are the first two columns of the matrix inverse for this system:

$$[A]^{-1} = \begin{bmatrix} 0.008173 & 0.003458 & \rule{1.5cm}{0.4pt} \\ 0.001725 & 0.019961 & \rule{1.5cm}{0.4pt} \\ 0.006980 & 0.002953 & \rule{1.5cm}{0.4pt} \end{bmatrix}$$

Use the *LU* factorization to compute the third column of the matrix inverse.
(b) If the inflow concentration to the second tank is set to zero, compute the reduction in concentration in tank 1.

Civil/Environmental Engineering

12.14 A civil engineer involved in construction requires 4800, 5800, and 5700 m³ of sand, fine gravel, and coarse gravel, respectively, for a building project. There are three pits from which these materials can be obtained. The composition of these pits is

	Sand %	Fine Gravel %	Coarse Gravel %
Pit 1	52	30	18
Pit 2	20	50	30
Pit 3	25	20	55

How many cubic meters must be hauled from each pit in order to meet the engineer's needs?

12.15 Perform the same computation as in Sec. 12.2, but for the truss depicted in Fig. P12.15.
12.16 Perform the same computation as in Sec. 12.2, but for the truss depicted in Fig. P12.16.
12.17 Calculate the forces and reactions for the truss in Fig. 12.4 if a downward force of 3000 kg and a horizontal force to the right of 1500 kg are applied at node 1.
12.18 In the example for Fig. 12.4, where a 1000-lb downward force is applied at node 1, the external reactions V_2 and V_3 were calculated. But if the lengths of the truss members had been given, we could have calculated V_2 and V_3 by utilizing the fact that $V_2 + V_3$ must equal 1000 and by summing moments around node 2. However, because we do know V_2 and V_3, we can work backward to solve for the lengths of the truss members. Note that because there are three unknown lengths and only two equations, we can solve for only the relationship between lengths. Solve for this relationship.
12.19 Employing the same methods as used to analyze Fig. 12.4, determine the forces and reactions for the truss shown in Fig. P12.19.
12.20 Solve for the forces and reaction for the truss in Fig. P12.20. Determine the matrix inverse for the system. Does the vertical-member force in the middle member seem reasonable? Why?
12.21 As the name implies, indoor air pollution deals with air contamination in enclosed spaces such as homes, offices, work areas, etc. Suppose that you are designing a ventilation system for a

FIGURE P12.15

FIGURE P12.16

FIGURE P12.19

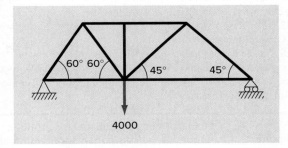

FIGURE P12.20

restaurant as shown in Fig. P12.21. The restaurant serving area consists of two square rooms and one elongated room. Room 1 and room 3 have sources of carbon monoxide from smokers and a faulty grill, respectively. Steady-state mass balances can be written

for each room. For example, for the smoking section (room 1), the balance can be written as

$$0 = W_{\text{smoker}} + Q_a c_a - Q_a c_1 + E_{13}(c_3 - c_1)$$
$$\text{(load)} + \text{(inflow)} - \text{(outflow)} + \text{(mixing)}$$

or substituting the parameters:

$$225c_1 - 25c_3 = 3400$$

Similar balances can be written for the other rooms.

(a) Solve for the steady-state concentration of carbon monoxide in each room.

(b) Determine what percent of the carbon monoxide in the kids' section is due to (i) the smokers, (ii) the grill, and (iii) the air in the intake vents.

(c) If the smoker and grill loads are increased to 4000 and 6000 mg/hr, respectively, use the matrix inverse to determine the increase in the concentration in the kids' section.

(d) How does the concentration in the kids' section change if a screen is constructed so that the mixing between areas 2 and 4 is decreased to 5 m³/hr?

12.22 An upward force of 20 kN is applied at the top of a tripod as depicted in Fig. P12.22. Determine the forces in the legs of the tripod.

12.23 A truss is loaded as shown in Fig. P12.23. Using the following set of equations, solve for the 10 unknowns: AB, BC, AD, BD, CD, DE, CE, A_x, A_y, and E_y.

$A_x + AD = 0$	$-24 - CD - (4/5)CE = 0$
$A_y + AB = 0$	$-AD + DE - (3/5)BD = 0$
$84 + BC + (3/5)BD = 0$	$CD + (4/5)BD = 0$
$-AB - (4/5)BD = 0$	$-DE - (3/5)CE = 0$
$-BC + (3/5)CE = 0$	$E_y + (4/5)CE = 0$

FIGURE P12.21
Overhead view of rooms in a restaurant. The one-way arrows represent volumetric airflows, whereas the two-way arrows represent diffusive mixing. The smoker and grill loads add carbon monoxide mass to the system but negligible airflow.

FIGURE P12.22

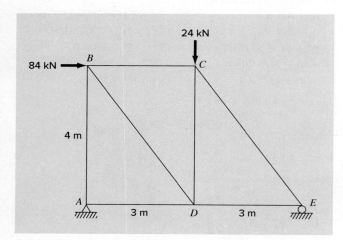

FIGURE P12.23

12.24 The following differential equation results from a force balance for a beam with a uniform loading:

$$0 = EI\frac{d^2y}{dx^2} - \frac{wLx}{2} + \frac{wx^2}{2}$$

where x = distance along the beam (m), y = deflection (m), L = length (m), E = modulus of elasticity (N/m^2), I = moment of inertia (m^4), and w = uniform load (N/m).

(a) Convert the given differential equation to an equivalent system of simultaneous algebraic equations using a centered-difference approximation for the second derivative.

(b) Solve these equations from $x = 0$ to L to generate the resulting distances and deflections. Test your solution for the following parameters: $L = 3$ m, $\Delta x = 0.2$ m, $E = 250 \times 10^9$ N/m^2, $I = 3 \times 10^{-4}$ m^4, $w = 22{,}500$ N/m, $y(0) = 0$, and $y(3) = 0$.

Electrical Engineering

12.25 Perform the same computation as in Sec. 12.3, but for the circuit depicted in Fig. P12.25.

12.26 Perform the same computation as in Sec. 12.3, but for the circuit depicted in Fig. P12.26.

12.27 Solve the circuit in Fig. P12.27 for the currents in each wire. Use Gauss elimination with pivoting.

12.28 An electrical engineer supervises the production of three types of electrical components. Three kinds of material—metal, plastic, and rubber—are required for their production. The amounts needed to produce each component are

Component	Metal, g/component	Plastic, g/component	Rubber, g/component
1	15	0.25	1.0
2	17	0.33	1.2
3	19	0.42	1.6

FIGURE P12.25

FIGURE P12.26

FIGURE P12.27

FIGURE P12.29

If totals of 3.89, 0.095, and 0.282 kg of metal, plastic, and rubber, respectively, are available each day, how many components can be produced per day?

12.29 Determine the currents for the circuit in Fig. P12.29.

12.30 Determine the currents for the circuit in Fig. P12.30.

12.31 The following system of equations was generated by applying the current law to the circuit in Fig. P12.31:

$$55I_1 - 25I_4 = -200$$
$$-37I_3 - 4I_4 = -250$$
$$-25I_1 - 4I_3 + 29I_4 = 100$$

Solve for I_1, I_3, and I_4.

12.32 The following system of equations was generated by applying the current law to the circuit in Fig. P12.32:

$$60I_1 - 40I_2 = 200$$
$$-40I_1 + 150I_2 - 100I_3 = 0$$
$$-100I_2 + 130I_3 = 230$$

Solve for I_1, I_2, and I_3.

12.33 Determine the currents for the circuit in Fig. P12.33 given $R_1 = 10 \, \Omega$, $R_2 = 5 \, \Omega$, $R_3 = 8 \, \Omega$, $R_4 = 15 \, \Omega$, $R_5 = 2 \, \Omega$, and $V = 100$ V.

FIGURE P12.30

FIGURE P12.31

FIGURE P12.32

FIGURE P12.33

Mechanical/Aerospace Engineering

12.34 Perform the same computation as in Sec. 12.4, but add a third spring between masses 1 and 2 and triple k for all springs.

12.35 Perform the same computation as in Sec. 12.4, but change the masses from 2, 3, and 2.5 kg to 10, 3.5, and 2 kg, respectively.

12.36 Idealized spring-mass systems have numerous applications throughout engineering. Figure P12.36 shows an arrangement of four springs in series being depressed with a force of 2000 kg. At equilibrium, force-balance equations can be developed defining the interrelationships between the springs,

$$k_2(x_2 - x_1) = k_1 x_1$$
$$k_3(x_3 - x_2) = k_2(x_2 - x_1)$$
$$k_4(x_4 - x_3) = k_3(x_3 - x_2)$$
$$F = k_4(x_4 - x_3)$$

FIGURE P12.36

where the k's are spring constants. If k_1 through k_4 are 150, 50, 75, and 225 N/m, respectively, compute the x's.

12.37 Three blocks are connected by a weightless cord and rest on an inclined plane (Fig. P12.37a). Employing a procedure similar to the one used in the analysis of the falling parachutists in Example 9.11 yields the following set of simultaneous equations (free-body diagrams are shown in Fig. P12.37b):

$$100a + T \qquad = 519.72$$
$$50a - T + R = 216.55$$
$$25a \qquad - R = 108.28$$

Solve for acceleration a and the tensions T and R in the two ropes.

12.38 Perform a computation similar to that called for in Prob. 12.37, but for the system shown in Fig. P12.38.

12.39 Perform the same computation as in Prob. 12.37, but for the system depicted in Fig. P12.39 (angles are 45°).

12.40 Consider the system of three masses and four springs in Fig. P12.40. Determining the equations of motion from $\Sigma F_x = ma$, for each mass using its free-body diagram, results in the following differential equations:

$$\ddot{x}_1 + \left(\frac{k_1 + k_2}{m_1}\right)x_1 - \left(\frac{k_2}{m_1}\right)x_2 = 0$$

$$\ddot{x}_2 - \left(\frac{k_2}{m_2}\right)x_1 + \left(\frac{k_2 + k_3}{m_2}\right)x_2 - \left(\frac{k_3}{m_2}\right)x_3 = 0$$

$$\ddot{x}_3 - \left(\frac{k_3}{m_3}\right)x_2 + \left(\frac{k_3 + k_4}{m_3}\right)x_3 = 0$$

where $k_1 = k_4 = 10$ N/m, $k_2 = k_3 = 30$ N/m, and $m_1 = m_2 = m_3 = 2$ kg. Write the three equations in matrix form:

$$0 = [\text{Acceleration vector}] + [k/m \text{ matrix}][\text{displacement vector } x]$$

At a specific time when $x_1 = 0.05$ m, $x_2 = 0.04$ m, and $x_3 = 0.03$ m, this is a tridiagonal matrix. Solve for the acceleration of each mass.

12.41 Linear algebraic equations can arise in the solution of differential equations. For example, the following differential equation derives from a heat balance for a long, thin rod (Fig. P12.41):

$$\frac{d^2 T}{dx^2} + h'(T_a - T) = 0 \qquad \text{(P12.41.1)}$$

where T = temperature (°C), x = distance along the rod (m), h' = a heat transfer coefficient between the rod and the ambient air (m^{-2}), and T_a = the temperature of the surrounding air (°C). This equation can be transformed into a set of linear algebraic equations by using a finite-divided-difference approximation for the second derivative (recall Section 4.1.3),

$$\frac{d^2 T}{dx^2} = \frac{T_{i+1} - 2T_i + T_{i-1}}{\Delta x^2}$$

FIGURE P12.37

FIGURE P12.38

FIGURE P12.39

FIGURE P12.40

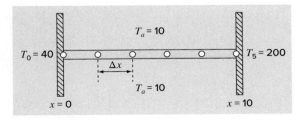

FIGURE P12.41

A noninsulated uniform rod positioned between two walls at constant but different temperatures. The finite-difference representation employs four interior nodes.

where T_i designates the temperature at node i. This approximation can be substituted into Eq. (P12.41.1) to give

$$-T_{i-1} + (2 + h'\Delta x^2)T_i - T_{i+1} = h'\Delta x^2 T_a$$

This equation can be written for each of the interior nodes of the rod, resulting in a tridiagonal system of equations. The first and last nodes at the rod's ends are fixed by boundary conditions.

(a) Develop an analytical solution for Eq. (P12.41.1) for a 10-m rod with $T_a = 20$, $T(x = 0) = 40$, $T(x = 10) = 200$, and $h' = 0.02$.

(b) Develop a numerical solution for the same parameter values employed in part (a) using a finite-difference solution with four interior nodes as shown in Fig. P12.41 ($\Delta x = 2$ m).

12.42 The steady-state distribution of temperature on a heated plate can be modeled by the *Laplace equation*,

$$0 = \frac{\partial^2 T}{\partial x^2} + \frac{\partial^2 T}{\partial y^2}$$

If the plate is represented by a series of nodes (Fig. P12.42), centered finite-divided differences can be substituted for the second derivatives, which results in a system of linear algebraic equations. Use the Gauss-Seidel method to solve for the temperatures of the nodes in Fig. P12.42.

12.43 A rod on a ball-and-socket joint is attached to cables A and B, as in Fig. P12.43.

(a) If a 50-N force is exerted on the massless rod at C, what is the tensile force at cables A and B?

(b) Solve for the reactant forces at the base of the rod. Call the base point P.

FIGURE P12.42

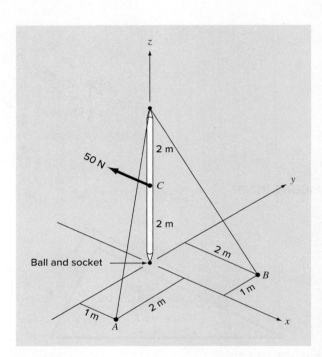

FIGURE P12.43

EPILOGUE: PART THREE

PT3.4 TRADE-OFFS

Table PT3.2 provides a summary of the trade-offs involved in solving simultaneous linear algebraic equations. Two methods—graphical and Cramer's rule—are limited to small (≤ 3) numbers of equations and thus have little utility for practical problem solving. However, these techniques are useful didactic tools for understanding the behavior of linear systems in general.

The numerical methods themselves are divided into two general categories: exact and approximate methods. As the name implies, the former are intended to yield exact answers. However, because they are affected by round-off errors, they sometimes yield imprecise results. The magnitude of the round-off error varies from system to system and is dependent on a number of factors. These include the system's dimensions, its condition, and whether the matrix of coefficients is sparse or full. In addition, computer precision will affect round-off error.

It is recommended that a pivoting strategy be employed in any computer program implementing exact elimination methods. The inclusion of such a strategy minimizes round-off error and avoids problems such as division by zero. All other things being equal, *LU* decomposition–based algorithms are the methods of choice because of their efficiency and flexibility.

TABLE PT3.2 Comparison of the characteristics of alternative methods for finding solutions of simultaneous linear algebraic equations.

Method	Stability	Precision	Breadth of Application	Programming Effort	Comments
Graphical	—	Poor	Limited	—	May take more time than the numerical method but can be useful for visualization
Cramer's rule	—	Affected by round-off error	Limited	—	Excessive computational effort required for more than three equations
Gauss elimination (with partial pivoting)	—	Affected by round-off error	General	Moderate	
LU decomposition	—	Affected by round-off error	General	Moderate	Preferred elimination method; allows computation of matrix inverse
Gauss-Seidel	May not converge if system is not diagonally dominant	Excellent	Appropriate only for diagonally dominant systems	Easy	

Although elimination methods have great utility, their use of the entire matrix of coefficients can be somewhat limiting when dealing with very large, sparse systems. This is due to the fact that large portions of computer memory would be devoted to storage of meaningless zeros. For banded systems, techniques are available to implement elimination methods without having to store the entire coefficient matrix.

The approximate technique described in this book is called the Gauss-Seidel method. It differs from the exact techniques in that it employs an iterative scheme to obtain progressively closer estimates of the solution. Thus, the effect of round-off is a moot point with the Gauss-Seidel method because the iterations can be continued as long as is necessary to obtain the desired precision. In addition, versions of the Gauss-Seidel method can be developed to efficiently utilize computer storage requirements for sparse systems. Consequently, the Gauss-Seidel technique has utility for large systems of equations where storage requirements would pose significant problems for the exact techniques.

The disadvantage of the Gauss-Seidel method is that it does not always converge or sometimes converges slowly on the true solution. It is strictly reliable only for those systems that are diagonally dominant. However, relaxation methods are available that sometimes offset these disadvantages. In addition, because many sets of linear algebraic equations originating from physical systems exhibit diagonal dominance, the Gauss-Seidel method has great utility for engineering problem solving.

In summary, a variety of factors will bear on your choice of a technique for a particular problem involving linear algebraic equations. However, as outlined above, the size and sparseness of the system are particularly important factors in determining your choice.

PT3.5 IMPORTANT RELATIONSHIPS AND FORMULAS

Every part of this book includes a section that summarizes important formulas. Although Part Three does not really deal with single formulas, we have used Table PT3.3 to summarize the algorithms that were covered. The table provides an overview that should be helpful for review and in elucidating the major differences between the methods.

PT3.6 ADVANCED METHODS AND ADDITIONAL REFERENCES

General references on the solution of simultaneous linear equations can be found in Fadeev and Fadeeva (1963), Stewart (1973), Varga (1962), and Young (1971). Ralston and Rabinowitz (1978) provide a general summary.

Many advanced techniques are available to increase the savings in time and/or space when solving linear algebraic equations. Most of these focus on exploiting properties of the equations such as symmetry and bandedness. In particular, algorithms are available to operate on sparse matrices to convert them to a minimum banded format. Jacobs (1977) and Tewarson (1973) include information on this area. Once they are in a minimum banded format, there are a variety of efficient solution strategies that are employed such as the active column storage approach of Bathe and Wilson (1976).

Aside from $n \times n$ sets of equations, there are other systems where the number of equations, m, and number of unknowns, n, are not equal. Systems where $m < n$ are called

TABLE PT3.3 Summary of important information presented in Part Three.

Method	Procedure		Potential Problems and Remedies

Gauss elimination

$$\begin{bmatrix} a_{11} & a_{12} & a_{13} & | & c_1 \\ a_{21} & a_{22} & a_{23} & | & c_2 \\ a_{31} & a_{32} & a_{33} & | & c_3 \end{bmatrix} \Rightarrow \begin{bmatrix} a_{11} & a_{12} & a_{13} & | & c_1 \\ & a'_{22} & a'_{23} & | & c'_2 \\ & & a''_{33} & | & c''_3 \end{bmatrix}$$

$$\Rightarrow \begin{array}{l} x_3 = c''_3 / a''_{33} \\ x_2 = (c'_2 - a'_{23}x_3)/a'_{22} \\ x_1 = (c_1 - a_{12}x_1 - a_{13}x_3)/a_{11} \end{array}$$

Problems:
Ill conditioning
Round-off
Division by zero
Remedies:
Higher precision
Partial pivoting

LU decomposition

Decomposition   Back Substitution

$$\begin{bmatrix} a_{11} & a_{12} & a_{13} \\ a_{21} & a_{22} & a_{23} \\ a_{31} & a_{32} & a_{33} \end{bmatrix} \Rightarrow \begin{bmatrix} 1 & 0 & 0 \\ l_{21} & 1 & 0 \\ l_{31} & l_{32} & 1 \end{bmatrix} \begin{Bmatrix} d_1 \\ d_2 \\ d_3 \end{Bmatrix} = \begin{Bmatrix} c_1 \\ c_2 \\ c_3 \end{Bmatrix} \Rightarrow \begin{bmatrix} u_{11} & u_{12} & u_{13} \\ 0 & u_{22} & u_{23} \\ 0 & 0 & u_{33} \end{bmatrix} \begin{Bmatrix} x_1 \\ x_2 \\ x_3 \end{Bmatrix} = \begin{Bmatrix} d_1 \\ d_2 \\ d_3 \end{Bmatrix} = \begin{Bmatrix} x_1 \\ x_2 \\ x_3 \end{Bmatrix}$$

Forward Substitution

Problems:
Ill conditioning
Round-off
Division by zero
Remedies:
Higher precision
Partial pivoting

Gauss-Seidel method

$$x_1^j = (c_1 - a_{12}x_2^{j-1} - a_{13}x_3^{j-1})/a_{11}$$
$$x_2^j = (c_2 - a_{21}x_1^j - a_{23}x_3^{j-1})/a_{22}$$
$$x_3^j = (c_3 - a_{31}x_1^j - a_{32}x_2^j)/a_{33}$$

continue iteratively until $\left| \dfrac{x_i^j - x_i^{j-1}}{x_i^j} \right| 100\% < \epsilon_s$ for all $x_i's$

Problems:
Divergent or converges slowly
Remedies:
Diagonal dominance
Relaxation

underdetermined. In such cases, there can be either no solution or more than one. Systems where $m > n$ are called *overdetermined.* For such situations, there is in general no exact solution. However, it is often possible to develop a compromise solution that attempts to determine answers that come "closest" to satisfying all the equations simultaneously. A common approach is to solve the equation in a "least-squares" sense (Lawson and Hanson 1974; Wilkinson and Reinsch 1971). Alternatively, one can use linear programming methods where the equations are solved in an "optimal" sense by minimizing some objective function (Dantzig 1963; Luenberger 1984; Rabinowitz 1968). We describe this approach in detail in Chap. 15.

PART FOUR

OPTIMIZATION

PT4.1 MOTIVATION

Root location (Part Two) and optimization are related in the sense that both involve guessing and searching for a point on a function. The fundamental difference between the two types of problems is illustrated in Fig. PT4.1. Root location involves searching for zeros of a function or functions. In contrast, *optimization* involves searching for either the minimum or the maximum.

The optimum is the point where the curve is flat. In mathematical terms, this corresponds to the x value where the derivative $f'(x)$ is equal to zero. Additionally, the second derivative, $f''(x)$, indicates whether the optimum is a minimum or a maximum: If $f''(x) < 0$, the point is a maximum; if $f''(x) > 0$, the point is a minimum.

Now, understanding the relationship between roots and optima would suggest a possible strategy for finding the latter. That is, you can differentiate the function and locate the root (that is, the zero) of the new function. In fact, some optimization methods seek to find an optima by solving the root problem: $f'(x) = 0$. It should be noted that such searches are often complicated because $f'(x)$ is not always available analytically. Thus, one must sometimes use finite-difference approximations to estimate the derivative.

FIGURE PT4.1

A function of a single variable illustrating the difference between roots and optima.

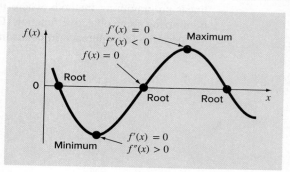

Beyond viewing optimization as a roots problem, it should be noted that the task of locating optima is aided by some extra mathematical structure that is not part of simple root finding. This tends to make optimization a more tractable task, particularly for multidimensional cases.

PT4.1.1 Noncomputer Methods and History

As mentioned above, differential calculus methods are still used to determine optimum solutions. All engineering and science students recall working maxima-minima problems by determining first derivatives of functions in their calculus courses. Bernoulli, Euler, Lagrange, and others laid the foundations of the calculus of variations, which deals with the minimization of functions. The Lagrange multiplier method was developed to optimize constrained problems, that is, optimization problems where the variables are bounded in some way.

The first major advances in numerical approaches occurred only with the development of digital computers after World War II. Koopmans in the United Kingdom and Kantorovich in the former Soviet Union independently worked on the general problem of least-cost distribution of supplies and products. In 1947, Koopman's student Dantzig invented the *simplex procedure* for solving linear programming problems. This approach paved the way for other methods of constrained optimization to be developed by a number of investigators, notably Charnes and his coworkers. Approaches for unconstrained optimization also developed rapidly following the widespread availability of computers.

PT4.1.2 Optimization and Engineering Practice

Most of the mathematical models we have dealt with to this point have been *descriptive* models. That is, they have been derived to simulate the behavior of an engineering device or system. In contrast, optimization typically deals with finding the "best result," or optimum solution, of a problem. Thus, in the context of modeling, they are often termed *prescriptive* models since they can be used to prescribe a course of action or the best design.

Engineers must continuously design devices and products that perform tasks in an efficient fashion. In doing so, they are constrained by the limitations of the physical world. Further, they must keep costs down. Thus, they are always confronting optimization problems that balance performance and limitations. Some common instances are listed in Table PT4.1. The following example has been developed to help you get a feel for the way in which such problems might be formulated.

EXAMPLE PT4.1   **Optimization of Parachute Cost**

Problem Statement.   Throughout the book, we have used the falling parachutist to illustrate the basic problem areas of numerical methods. You may have noticed that none of these examples concentrate on what happens after the chute opens. In this example, we will examine a case where the chute has opened and we are interested in predicting impact velocity at the ground.

You are an engineer working for an agency planning to airlift supplies to refugees in a war zone. The supplies will be dropped at low altitude (500 m) so that the drop is

TABLE PT4.1 Some common examples of optimization problems in engineering.

- Design aircraft for minimum weight and maximum strength.
- Optimal trajectories of space vehicles.
- Design civil engineering structures for minimum cost.
- Design water-resource projects like dams to mitigate flood damage while yielding maximum hydropower.
- Predict structural behavior by minimizing potential energy.
- Material-cutting strategy for minimum cost.
- Design pump and heat transfer equipment for maximum efficiency.
- Maximize power output of electrical networks and machinery while minimizing heat generation.
- Shortest route of salesperson visiting various cities during one sales trip.
- Optimal planning and scheduling.
- Statistical analysis and models with minimum error.
- Optimal pipeline networks.
- Inventory control.
- Maintenance planning to minimize cost.
- Minimize waiting and idling times.
- Design waste treatment systems to meet water-quality standards at least cost.

not detected and the supplies fall as close as possible to the refugee camp. The chutes open immediately upon leaving the plane. To reduce damage, the vertical velocity on impact must be below a critical value of $v_c = 20$ m/s.

The parachute used for the drop is depicted in Fig. PT4.2. The cross-sectional area of the chute is that of a half sphere,

$$A = 2\pi r^2 \tag{PT4.1}$$

The length of each of the 16 cords connecting the chute to the mass is related to the chute radius by

$$\ell = \sqrt{2r} \tag{PT4.2}$$

FIGURE PT4.2
A deployed parachute.

You know that the drag force for the chute is a linear function of its cross-sectional area described by the following formula

$$c = k_c A \qquad\qquad (PT4.3)$$

where c = drag coefficient (kg/s) and k_c = a proportionality constant parameterizing the effect of area on drag [kg/(s m^2)].

Also, you can divide the payload into as many parcels as you like. That is, the mass of each individual parcel can be calculated as

$$m = \frac{M_t}{n}$$

where m = mass of an individual parcel (kg), M_t = total load being dropped (kg), and n = total number of parcels.

Finally, the cost of each chute is related to chute size in a nonlinear fashion,

$$\text{Cost per chute} = c_0 + c_1\ell + c_2 A^2 \qquad\qquad (PT4.4)$$

where c_0, c_1, and c_2 = cost coefficients. The constant term, c_0, is the base price for the chutes. The nonlinear relationship between cost and area exists because larger chutes are much more difficult to construct than small chutes.

Determine the size (r) and number of chutes (n) that result in minimum cost while at the same time meeting the requirement of having a sufficiently small impact velocity.

Solution. The objective here is to determine the number and size of parachutes to minimize their cost. The problem is constrained because the parcels must have an impact velocity less than a critical value.

The cost can be computed by multiplying the cost of the individual parachute [Eq. (PT4.4)] by the number of parachutes (n). Thus, the function you wish to minimize, which is formally called the *objective function,* is written as

$$\text{Minimize } C = n(c_0 + c_1\ell + c_2 A^2) \qquad\qquad (PT4.5)$$

where C = cost ($) and A and ℓ are calculated by Eqs. (PT4.1) and (PT4.2), respectively.

Next, we must specify the *constraints*. For this problem there are two constraints. First, the impact velocity must be equal to or less than the critical velocity,

$$v \le v_c \qquad\qquad (PT4.6)$$

Second, the number of parcels must be an integer and greater than or equal to 1,

$$n \ge 1 \qquad\qquad (PT4.7)$$

where n is an integer.

At this point, the optimization problem has been formulated. As can be seen, it is a nonlinear constrained problem.

Although the problem has been broadly formulated, one more issue must be addressed: How do we determine the impact velocity v? Recall from Chap. 1 that the velocity of a falling object can be computed with

$$v = \frac{gm}{c}(1 - e^{-(c/m)t}) \qquad\qquad (1.10)$$

FIGURE PT4.3
The height z and velocity v of a deployed parachute as it falls to earth ($z = 0$).

where v = velocity (m/s), g = acceleration due to gravity (m/s^2), m = mass (kg), and t = time (s).

Although Eq. (1.10) provides a relationship between v and t, we need to know how long the mass falls. Therefore, we need a relationship between the drop distance z and the time of fall t. The drop distance can be calculated from the velocity in Eq. (1.10) by integration

$$z = \int_0^t \frac{gm}{c} \left(1 - e^{-(c/m)t}\right) dt \qquad (PT4.8)$$

This integral can be evaluated to yield

$$z = z_0 - \frac{gm}{c}t + \frac{gm^2}{c^2}\left(1 - e^{-(c/m)t}\right) \qquad (PT4.9)$$

where z_0 = initial height (m). This function, as plotted in Fig. PT4.3, provides a way to predict z given knowledge of t.

However, we do not need z as a function of t to solve this problem. Rather, we need to compute the time required for the parcel to fall the distance z_0. Thus, we recognize that we must reformulate Eq. (PT4.9) as a root-finding problem. That is, we must solve for the time at which z goes to zero,

$$f(t) = 0 = z_0 - \frac{gm}{c}t + \frac{gm^2}{c^2}\left(1 - e^{-(c/m)t}\right) \qquad (PT4.10)$$

Once the time to impact is computed, we can substitute it into Eq. (1.10) to solve for the impact velocity.

The final specification of the problem, therefore, would be

$$\text{Minimize } C = n(c_0 + c_1\ell + c_2 A^2) \qquad (PT4.11)$$

subject to

$$v \leq v_c \qquad\qquad\qquad\qquad\qquad\qquad\qquad\qquad\qquad\text{(PT4.12)}$$

$$n \geq 1 \qquad\qquad\qquad\qquad\qquad\qquad\qquad\qquad\qquad\text{(PT4.13)}$$

where

$$A = 2\pi r^2 \qquad\qquad\qquad\qquad\qquad\qquad\qquad\qquad\text{(PT4.14)}$$

$$\ell = \sqrt{2}r \qquad\qquad\qquad\qquad\qquad\qquad\qquad\qquad\text{(PT4.15)}$$

$$c = k_c A \qquad\qquad\qquad\qquad\qquad\qquad\qquad\qquad\text{(PT4.16)}$$

$$m = \frac{M_t}{n} \qquad\qquad\qquad\qquad\qquad\qquad\qquad\qquad\text{(PT4.17)}$$

$$t = \text{root}\left[z_0 - \frac{gm}{c}t + \frac{gm^2}{c^2}\left(1 - e^{-(c/m)t}\right) \right] \qquad\qquad\text{(PT4.18)}$$

$$v = \frac{gm}{c}\left(1 - e^{-(c/m)t}\right) \qquad\qquad\qquad\qquad\qquad\text{(PT4.19)}$$

We will solve this problem in Example 15.4 in Chap. 15. For the time being, recognize that it has most of the fundamental elements of other optimization problems you will routinely confront in engineering practice. These are

- The problem will involve an *objective function* that embodies your goal.
- There will be a number of *design variables*. These variables can be real numbers or they can be integers. In our example, these are r (real) and n (integer).
- The problem will include *constraints* that reflect the limitations you are working under.

We should make one more point before proceeding. Although the objective function and constraints may superficially appear to be simple equations [e.g., Eq. (PT4.11)], they may in fact be the "tip of the iceberg." That is, they may be underlain by complex dependencies and models. For instance, as in our example, they may involve other numerical methods [Eq. (PT4.18)]. This means that the functional relationships you will be using could actually represent large and complicated calculations. Thus, techniques that can find the optimal solution, while minimizing function evaluations, can be extremely valuable.

PT4.2 MATHEMATICAL BACKGROUND

There are a number of mathematical concepts and operations that underlie optimization. Because we believe that they will be more relevant to you in context, we will defer discussion of specific mathematical prerequisites until they are needed. For example, we will discuss the important concepts of the gradient and Hessians at the beginning of Chap. 14 on multivariate unconstrained optimization. In the meantime, we will limit ourselves here to the more general topic of how optimization problems are classified.

An *optimization*, or *mathematical programming*, problem generally can be stated as:

Find x, which minimizes or maximizes $f(x)$

subject to

$$d_i(\boldsymbol{x}) \leq a_i \qquad i = 1, 2, \ldots, m \qquad\qquad\qquad\qquad \text{(PT4.20)}$$

$$e_i(\boldsymbol{x}) = b_i \qquad i = 1, 2, \ldots, p \qquad\qquad\qquad\qquad \text{(PT4.21)}$$

where $\boldsymbol{x}$ is an *n-dimensional design vector*, $f(\boldsymbol{x})$ is the *objective function*, $d_i(\boldsymbol{x})$ are *inequality constraints*, $e_i(\boldsymbol{x})$ are *equality constraints*, and a_i and b_i are constants.

Optimization problems can be classified on the basis of the form of $f(\boldsymbol{x})$:

- If $f(\boldsymbol{x})$ and the constraints are linear, we have *linear programming*.
- If $f(\boldsymbol{x})$ is quadratic and the constraints are linear, we have *quadratic programming*.
- If $f(\boldsymbol{x})$ is not linear or quadratic and/or the constraints are nonlinear, we have *nonlinear programming*.

Further, when Eqs. (PT4.20) and (PT4.21) are included, we have a *constrained optimization* problem; otherwise, it is an *unconstrained optimization* problem.

Note that for constrained problems, the degrees of freedom are given by $n - p - m$. Generally, to obtain a solution, $p + m$ must be $\leq n$. If $p + m > n$, the problem is said to be *overconstrained*.

Another way in which optimization problems are classified is by dimensionality. This is most commonly done by dividing them into one-dimensional and multidimensional problems. As the name implies, *one-dimensional problems* involve functions that depend on a single dependent variable. As in Fig. PT4.4a, the search then consists of climbing or descending one-dimensional peaks and valleys. *Multidimensional problems* involve functions that depend on two or more dependent variables. In the same spirit, a two-dimensional optimization can again be visualized as searching out peaks and valleys

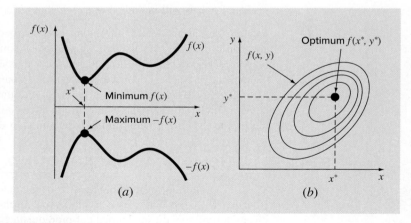

FIGURE PT4.4

(*a*) One-dimensional optimization. This figure also illustrates how minimization of $f(x)$ is equivalent to maximization of $-f(x)$. (*b*) Two-dimensional optimization. Note that this figure can be taken to represent either a maximization (contours increase in elevation up to the maximum like a mountain) or a minimization (contours decrease in elevation down to the minimum like a valley).

(Fig. PT4.4*b*). However, just as in real hiking, we are not constrained to walk in a single direction, instead the *topography* is examined to efficiently reach the goal.

Finally, the process of finding a maximum versus finding a minimum is essentially identical because the same value, x^*, both minimizes $f(x)$ and maximizes $-f(x)$. This equivalence is illustrated graphically for a one-dimensional function in Fig. PT4.4*a*.

PT4.3   ORIENTATION

Some orientation is helpful before proceeding to the numerical methods for optimization. The following is intended to provide an overview of the material in Part Four. In addition, some objectives have been included to help you focus your efforts when studying the material.

PT4.3.1   Scope and Preview

Figure PT4.5 is a schematic representation of the organization of Part Four. Examine this figure carefully, starting at the top and working clockwise.

After the present introduction, *Chap. 13* is devoted to *one-dimensional unconstrained optimization.* Methods are presented to find the minimum or maximum of a function of a single variable. Three methods are covered: *golden-section search, parabolic interpolation,* and *Newton's method.* An advanced hybrid approach, *Brent's method*, that combines the reliability of the golden-section search with the speed of parabolic interpolation is also described.

Chapter 14 covers two general types of methods to solve *multidimensional unconstrained optimization* problems. *Direct methods* such as *random searches, univariate searches,* and *pattern searches* do not require the evaluation of the function's derivatives. On the other hand, *gradient methods* use first and sometimes second derivatives to find the optimum. The chapter introduces the *gradient* and the *Hessian,* which are multidimensional representations of the first and second derivatives. The method of *steepest ascent/descent* is then covered in some detail. This is followed by descriptions of some advanced methods: *conjugate gradient, Newton's method, Marquardt's method,* and *quasi-Newton methods.*

Chapter 15 is devoted to *constrained optimization. Linear programming* is described in detail using both a graphical representation and the *simplex method.* The detailed analysis of *nonlinear constrained optimization* is beyond this book's scope, but we provide an overview of the major approaches. In addition, we illustrate how solutions of such problems (along with the problems covered in Chaps. 13 and 14) can be obtained with software packages such as Excel, MATLAB, and Mathcad.

Chapter 16 extends the above concepts to actual engineering problems. Engineering applications are used to illustrate how optimization problems are formulated and provide insight into the application of the solution techniques in professional practice.

An epilogue is included at the end of Part Four. It contains an overview of the methods discussed in Chaps. 13, 14, and 15. This overview includes a description of trade-offs related to the proper use of each technique. It also provides references for some numerical methods that are beyond the scope of this text.

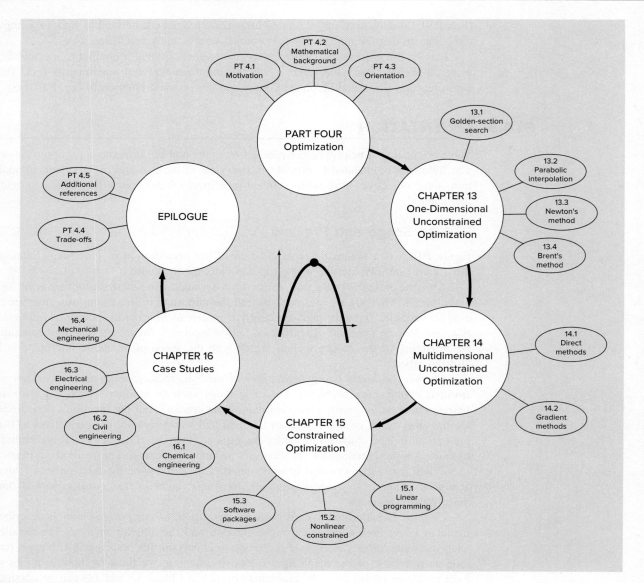

FIGURE PT4.5
Schematic of the organization of the material in Part Four: Optimization.

PT4.3.2 Goals and Objectives

Study Objectives. After completing Part Four, you should have sufficient information to successfully approach a wide variety of engineering problems dealing with optimization. In general, you should have mastered the techniques, have learned to assess their reliability, and be capable of analyzing alternative methods for any particular problem. In

TABLE PT4.2 Specific study objectives for Part Four.

1. Understand why and where optimization occurs in engineering problem solving.
2. Understand the major elements of the general optimization problem: objective function, decision variables, and constraints.
3. Be able to distinguish between linear and nonlinear optimization, and between constrained and unconstrained problems.
4. Be able to define the golden ratio and understand how it makes one-dimensional optimization efficient.
5. Locate the optimum of a single-variable function with the golden-section search, parabolic interpolation, and Newton's method. Also, recognize the trade-offs among these approaches, with particular attention to initial guesses and convergence.
6. Understand how Brent's optimization method combines the reliability of the golden-section search with the speed of parabolic interpolation.
7. Be capable of writing a program and solving for the optimum of a multivariable function using random searching.
8. Understand the ideas behind pattern searches, conjugate directions, and Powell's method.
9. Be able to define and evaluate the gradient and Hessian of a multivariable function both analytically and numerically.
10. Compute by hand the optimum of a two-variable function using the method of steepest ascent/descent.
11. Understand the basic ideas behind the conjugate gradient, Newton's, Marquardt's, and quasi-Newton methods. In particular, understand the trade-offs among the approaches and recognize how each improves on the steepest ascent/descent.
12. Be capable of recognizing and setting up a linear programming problem to represent applicable engineering problems.
13. Be able to solve a two-dimensional linear programming problem with both the graphical and simplex methods.
14. Understand the four possible outcomes of a linear programming problem.
15. Be able to set up and solve nonlinear constrained optimization problems using a software package.

addition to these general goals, the specific concepts in Table PT4.2 should be assimilated for a comprehensive understanding of the material in Part Four.

Computer Objectives. You should be able to write a subprogram to implement a simple one-dimensional (like golden-section search or parabolic interpolation) and multidimensional (like the random-search method) search. In addition, software packages such as Excel, MATLAB, or Mathcad have varying capabilities for optimization. You can use this part of the book to become familiar with these capabilities.

One-Dimensional Unconstrained Optimization

This chapter will describe techniques to find the minimum or maximum of a function of a single variable, $f(x)$. A useful image in this regard is the one-dimensional, "roller coaster"–like function depicted in Fig. 13.1. Recall from Part Two that root location was complicated by the fact that several roots can occur for a single function. Similarly, both local and global optima can occur in optimization. Such cases are called *multimodal*. In almost all instances, we will be interested in finding the absolute highest or lowest value of a function. Thus, we must take care that we do not mistake a local result for the global optimum.

Distinguishing a global from a local extremum can be a very difficult problem for the general case. There are three usual ways to approach this problem. First, insight into the behavior of low-dimensional functions can sometimes be obtained graphically. Second, it may be possible to find optima based on widely varying and perhaps randomly generated starting guesses, and then select the largest of these as global. Finally, we might perturb the starting point associated with a local optimum and see if the routine returns a better point or always returns to the same point. Although all these approaches can have utility, the fact is that in some problems (usually the large ones), there may be no practical way to ensure that you have located a global optimum. However, although you should always be sensitive to the issue, it is fortunate that there are numerous engineering problems where you can locate the global optimum in an unambiguous fashion.

FIGURE 13.1
A function that asymptotically approaches zero at plus and minus ∞ and has two maximum and two minimum points in the vicinity of the origin. The two points to the right are local optima, whereas the two to the left are global.

Just as in root location, optimization in one dimension can be divided into bracketing and open methods. As described in the next section, the golden-section search is an example of a bracketing method that depends on initial guesses that bracket a single optimum. This is followed by an alternative approach, parabolic interpolation, which often converges faster than the golden-section search, but sometimes diverges.

Another method described in this chapter is an open method based on the idea from calculus that the minimum or maximum can be found by solving $f'(x) = 0$. This reduces the optimization problem to finding the root of $f'(x)$ using techniques of the sort described in Part Two. We will demonstrate one version of this approach—Newton's method.

Finally, an advanced hybrid approach, *Brent's method*, is described. This approach combines the reliability of the golden-section search with the speed of parabolic interpolation.

13.1 GOLDEN-SECTION SEARCH

In solving for the root of a single nonlinear equation, the goal was to find the value of the variable x that yields a *zero* of the function $f(x)$. *Single-variable optimization* has the goal of finding the value of x that yields an *extremum,* either a maximum or minimum of $f(x)$.

The golden-section search is a simple, general-purpose, single-variable search technique. It is similar in spirit to the bisection approach for locating roots in Chap. 5. Recall that bisection hinged on defining an interval, specified by a lower guess (x_l) and an upper guess (x_u), that bracketed a single root. The presence of a root between these bounds was verified by determining that $f(x_l)$ and $f(x_u)$ had different signs. The root was then estimated as the midpoint of this interval,

$$x_r = \frac{x_l + x_u}{2}$$

The final step in a bisection iteration involved determining a new smaller bracket. This was done by replacing whichever of the bounds x_l or x_u had a function value with the same sign as $f(x_r)$. One advantage of this approach was that the new value x_r replaced one of the old bounds.

Now we can develop a similar approach for locating the optimum of a one-dimensional function. For simplicity, we will focus on the problem of finding a maximum. When we discuss the computer algorithm, we will describe the minor modifications needed to simulate a minimum.

As with bisection, we can start by defining an interval that contains a single answer. That is, the interval should contain a single maximum, and hence is called *unimodal.* We can adopt the same nomenclature as for bisection, where x_l and x_u defined the lower and upper bounds, respectively, of such an interval. However, in contrast to bisection, we need a new strategy for finding a maximum within the interval. Rather than using only two function values (which are sufficient to detect a sign change, and hence a zero), we would need three function values to detect whether a maximum occurred. Thus, an additional point within the interval has to be chosen. Next, we have to pick a fourth point. Then the test for the maximum could be applied to discern whether the maximum occurred within the first three or the last three points.

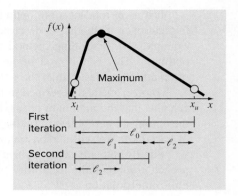

FIGURE 13.2
The initial step of the golden-section search algorithm involves choosing two interior points according to the golden ratio.

The key to making this approach efficient is the wise choice of the intermediate points. As in bisection, the goal is to minimize function evaluations by replacing old values with new values. This goal can be achieved by specifying that the following two conditions hold (Fig. 13.2):

$$\ell_0 = \ell_1 + \ell_2 \tag{13.1}$$

$$\frac{\ell_1}{\ell_0} = \frac{\ell_2}{\ell_1} \tag{13.2}$$

The first condition specifies that the sum of the two sublengths ℓ_1 and ℓ_2 must equal the original interval length. The second says that the ratios of the lengths must be equal. Equation (13.1) can be substituted into Eq. (13.2),

$$\frac{\ell_1}{\ell_1 + \ell_2} = \frac{\ell_2}{\ell_1} \tag{13.3}$$

If the reciprocal is taken and $R = \ell_2/\ell_1$, we arrive at

$$1 + R = \frac{1}{R} \tag{13.4}$$

or

$$R^2 + R - 1 = 0 \tag{13.5}$$

which can be solved for the positive root

$$R = \frac{-1 + \sqrt{1 - 4(-1)}}{2} = \frac{\sqrt{5} - 1}{2} = 0.61803\ldots \tag{13.6}$$

This value, which has been known since antiquity, is called the *golden ratio* (see Box 13.1). Because it allows optima to be found efficiently, it is the key element of the

---

Box 13.1 The Golden Ratio and Fibonacci Numbers

In many cultures, certain numbers are ascribed qualities. For example, we in the West are all familiar with "Lucky 7" and "Friday the 13th." Ancient Greeks called the following number the "golden ratio:"

$$\frac{\sqrt{5}-1}{2} = 0.61803\dots$$

This ratio was employed for a number of purposes, including the development of the rectangle in Fig. 13.3. These proportions were considered aesthetically pleasing by the Greeks. Among other things, many of their temples followed this shape.

The golden ratio is related to an important mathematical series known as the *Fibonacci numbers,* which are

0, 1, 1, 2, 3, 5, 8, 13, 21, 34, …

Thus, each number after the first two represents the sum of the preceding two. This sequence pops up in many diverse areas of science and engineering. In the context of the present discussion, an interesting property of the Fibonacci sequence relates to the ratio of consecutive numbers in the sequence; that is, $0/1 = 0$, $1/1 = 1$, $1/2 = 0.5$, $2/3 = 0.667$, $3/5 = 0.6$, $5/8 = 0.625$, $8/13 = 0.615$, and so on. As one proceeds, the ratio of consecutive numbers approaches the golden ratio!

FIGURE 13.3
The Parthenon in Athens, Greece, was constructed in the 5th century B.C. Its front dimensions can be fit almost exactly within a golden rectangle.

golden-section method we have been developing conceptually. Now let us derive an algorithm to implement this approach on the computer.

As mentioned above and as depicted in Fig. 13.4, the method starts with two initial guesses, x_l and x_u, that bracket one local extremum of $f(x)$. Next, two interior points x_1 and x_2 are chosen according to the golden ratio,

$$d = \frac{\sqrt{5}-1}{2}(x_u - x_l)$$

$$x_1 = x_l + d$$

$$x_2 = x_u - d$$

The function is evaluated at these two interior points. Two results can occur:

1. If, as is the case in Fig. 13.4, $f(x_1) > f(x_2)$, then the domain of x to the left of x_2, from x_l to x_2, can be eliminated because it does not contain the maximum. For this case, x_2 becomes the new x_l for the next round.
2. If $f(x_2) > f(x_1)$, then the domain of x to the right of x_1, from x_1 to x_u would be eliminated. In this case, x_1 becomes the new x_u for the next round.

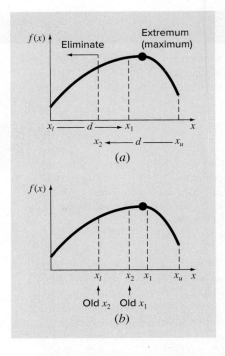

FIGURE 13.4
(*a*) The initial step of the golden-section search algorithm involves choosing two interior points according to the golden ratio. (*b*) The second step involves defining a new interval that includes the optimum.

Now, here is the real benefit from the use of the golden ratio. Because the original x_1 and x_2 were chosen using the golden ratio, we do not have to recalculate all the function values for the next iteration. For example, for the case illustrated in Fig. 13.4, the old x_1 becomes the new x_2. This means that we already have the value for the new $f(x_2)$, since it is the same as the function value at the old x_1.

To complete the algorithm, we now only need to determine the new x_1. This is done with the same proportionality as before,

$$x_1 = x_l + \frac{\sqrt{5} - 1}{2}(x_u - x_l)$$

A similar approach would be used for the alternate case where the optimum fell in the left subinterval.

As the iterations are repeated, the interval containing the extremum is reduced rapidly. In fact, each round the interval is reduced by a factor of the golden ratio (about 61.8%). That means that after 10 rounds, the interval is shrunk to about 0.618^{10} or 0.008 or 0.8% of its initial length. After 20 rounds, it is about 0.0066%. This is not quite as good as the reduction achieved with bisection, but this is a harder problem.

| EXAMPLE 13.1 | **Golden-Section Search** |

Problem Statement.   Use the golden-section search to find the maximum of

$$f(x) = 2 \sin x - \frac{x^2}{10}$$

within the interval $x_l = 0$ and $x_u = 4$.

Solution.   First, the golden ratio is used to create the two interior points

$$d = \frac{\sqrt{5} - 1}{2}(4 - 0) = 2.472$$

$$x_1 = 0 + 2.472 = 2.472$$

$$x_2 = 4 - 2.472 = 1.528$$

The function can be evaluated at the interior points

$$f(x_2) = f(1.528) = 2 \sin(1.528) - \frac{1.528^2}{10} = 1.765$$

$$f(x_1) = f(2.472) = 0.63$$

Because $f(x_2) > f(x_1)$, the maximum is in the interval defined by x_l, x_2, and x_1. Thus, for the new interval, the lower bound remains $x_l = 0$, and x_1 becomes the upper bound; that is, $x_u = 2.472$. In addition, the former x_2 value becomes the new x_1; that is, $x_1 = 1.528$. Further, we do not have to recalculate $f(x_1)$ because it was determined on the previous iteration as $f(1.528) = 1.765$.

All that remains is to compute the new values of d and x_2,

$$d = \frac{\sqrt{5} - 1}{2}(2.472 - 0) = 1.528$$

$$x_2 = 2.4721 - 1.528 = 0.944$$

The function evaluation at x_2 is $f(0.994) = 1.531$. Since this value is less than the function value at x_1, the maximum is in the interval prescribed by x_2, x_1, and x_u.

The process can be repeated, with the results tabulated below:

i	x_l	$f(x_l)$	x_2	$f(x_2)$	x_1	$f(x_1)$	x_u	$f(x_u)$	d
1	0	0	1.5279	1.7647	2.4721	0.6300	4.0000	−3.1136	2.4721
2	0	0	0.9443	1.5310	1.5279	1.7647	2.4721	0.6300	1.5279
3	0.9443	1.5310	1.5279	1.7647	1.8885	1.5432	2.4721	0.6300	0.9443
4	0.9443	1.5310	1.3050	1.7595	1.5279	1.7647	1.8885	1.5432	0.5836
5	1.3050	1.7595	1.5279	1.7647	1.6656	1.7136	1.8885	1.5432	0.3607
6	1.3050	1.7595	1.4427	1.7755	1.5279	1.7647	1.6656	1.7136	0.2229
7	1.3050	1.7595	1.3901	1.7742	1.4427	1.7755	1.5279	1.7647	0.1378
8	1.3901	1.7742	1.4427	1.7755	1.4752	1.7732	1.5279	1.7647	0.0851

Note that the current maximum is highlighted for every iteration. After the eighth iteration, the maximum occurs at $x = 1.4427$ with a function value of 1.7755. Thus, the result is converging on the true value of 1.7757 at $x = 1.4276$.

Recall that for bisection (Sec. 5.2.1), an exact upper bound for the error can be calculated at each iteration. Using similar reasoning, an upper bound for golden-section search can be derived as follows: Once an iteration is complete, the optimum will either fall in one of two intervals. If x_2 is the optimum function value, it will be in the lower interval, (x_l, x_2, x_1). If x_1 is the optimum function value, it will be in the upper interval, (x_2, x_1, x_u). Because the interior points are symmetrical, either case can be used to define the error.

Looking at the upper interval, if the true value were at the far left, the maximum distance from the estimate would be

$$\Delta x_a = x_1 - x_2$$
$$= x_l + R(x_u - x_l) - x_u + R(x_u - x_l)$$
$$= (x_l - x_u) + 2R(x_u - x_l)$$
$$= (2R - 1)(x_u - x_l)$$

or $0.236(x_u - x_l)$.

If the true value were at the far right, the maximum distance from the estimate would be

$$\Delta x_b = x_u - x_1$$
$$= x_u - x_l - R(x_u - x_l)$$
$$= (1 - R)(x_u - x_l)$$

or $0.382(x_u - x_l)$. Therefore, this case would represent the maximum error. This result can then be normalized to the optimal value for that iteration, x_{opt}, to yield

$$\varepsilon_a = (1 - R) \left| \frac{x_u - x_l}{x_{opt}} \right| 100\%$$

This estimate provides a basis for terminating the iterations.

Pseudocode for the golden-section-search algorithm for maximization is presented in Fig. 13.5a. The minor modifications to convert the algorithm to minimization are listed in Fig. 13.5b. In both versions the x value for the optimum is returned as the function value ($Gold$). In addition, the value of $f(x)$ at the optimum is returned as the variable (fx).

You may be wondering why we have stressed the reduced function evaluations of the golden-section search. Of course, for solving a single optimization, the speed savings would be negligible. However, there are two important contexts where minimizing the number of function evaluations can be important. These are

1. *Many evaluations.* There are cases where the golden-section-search algorithm may be a part of a much larger calculation. In such cases, it may be called many times. Therefore, keeping function evaluations to a minimum could pay great dividends for such cases.

FIGURE 13.5
Algorithm for the golden-section search.

```
FUNCTION Gold(xlow, xhigh, maxit, es, fx)
R = (5^0.5 − 1)/2
xℓ = xlow; xu = xhigh
iter = 1
d = R * (xu − xℓ)
x1 = xℓ + d; x2 = xu − d
f1 = f(x1)
f2 = f(x2)
IF f1 > f2 THEN                                  IF f1 < f2 THEN
  xopt = x1
  fx = f1
ELSE
  xopt = x2
  fx = f2
END IF
DO
  d = R*d; xint = xu − xℓ
  IF f1 > f2 THEN                                IF f1 < f2 THEN
    xℓ = x2
    x2 = x1
    x1 = xℓ+d
    f2 = f1
    f1 = f(x1)
  ELSE
    xu = x1
    x1 = x2
    x2 = xu−d
    f1 = f2
    f2 = f(x2)
  END IF
  iter = iter+1
  IF f1 > f2 THEN                                IF f1 < f2 THEN
    xopt = x1
    fx = f1
  ELSE
    xopt = x2
    fx = f2
  END IF
  IF xopt ≠ 0. THEN
    ea = (1.−R) *ABS(xint/xopt) *100.
  END IF
  IF ea ≤ es OR iter ≥ maxit EXIT
END DO
Gold = xopt
END Gold
```

  (*a*) **Maximization**   (*b*) **Minimization**

2. *Time-consuming evaluation.* For pedagogical reasons, we use simple functions in most of our examples. You should understand that a function can be very complex and time-consuming to evaluate. For example, in a later part of this book, we will describe how optimization can be used to estimate the parameters of a model consisting of a system of differential equations. For such cases, the "function" involves time-consuming model integration. Any method that minimizes such evaluations would be advantageous.

13.2 PARABOLIC INTERPOLATION

Parabolic interpolation takes advantage of the fact that a second-order polynomial often provides a good approximation to the shape of $f(x)$ near an optimum (Fig. 13.6).

Just as there is only one straight line connecting two points, there is only one quadratic polynomial, or parabola, connecting three points. Thus, if we have three points that jointly bracket an optimum, we can fit a parabola to the points. Then we can differentiate the quadratic polynomial, set the result equal to zero, and solve for an estimate of the optimal x. It can be shown through some algebraic manipulations that the result is

$$x_3 = \frac{f(x_0)(x_1^2 - x_2^2) + f(x_1)(x_2^2 - x_0^2) + f(x_2)(x_0^2 - x_1^2)}{2f(x_0)(x_1 - x_2) + 2f(x_1)(x_2 - x_0) + 2f(x_2)(x_0 - x_1)} \tag{13.7}$$

where x_0, x_1, and x_2 are the initial guesses, and x_3 is the value of x that corresponds to the maximum value of the parabolic fit to the guesses. After generating the new point, there are two strategies for selecting the points for the next iteration. The simplest approach, which is similar to the secant method, is to merely assign the new points sequentially. That is, for the new iteration, $z_0 = z_1$, $z_1 = z_2$, and $z_2 = z_3$. Alternatively, as illustrated in the following example, a bracketing approach, similar to bisection or the golden-section search, can be employed.

FIGURE 13.6
Graphical description of parabolic interpolation.

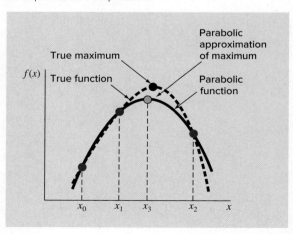

EXAMPLE 13.2   **Parabolic Interpolation**

Problem Statement.   Use parabolic interpolation to approximate the maximum of

$$f(x) = 2 \sin x - \frac{x^2}{10}$$

with initial guesses of $x_0 = 0$, $x_1 = 1$, and $x_2 = 4$.

Solution.   The function values at the three guesses can be evaluated,

$$x_0 = 0 \qquad f(x_0) = 0$$
$$x_1 = 1 \qquad f(x_1) = 1.5829$$
$$x_2 = 4 \qquad f(x_2) = -3.1136$$

and substituted into Eq. (13.7) to give

$$x_3 = \frac{0(1^2 - 4^2) + 1.5829(4^2 - 0^2) + (-3.1136)(0^2 - 1^2)}{2(0)(1 - 4) + 2(1.5829)(4 - 0) + 2(-3.1136)(0 - 1)} = 1.5055$$

which has a function value of $f(1.5055) = 1.7691$.

Next, a strategy similar to the golden-section search can be employed to determine which point should be discarded. Because the function value for the new point is higher than for the intermediate point (x_1) and the new x value is to the right of the intermediate point, the lower guess (x_0) is discarded. Therefore, for the next iteration,

$$x_0 = 1 \qquad\qquad f(x_0) = 1.5829$$
$$x_1 = 1.5055 \qquad f(x_1) = 1.7691$$
$$x_2 = 4 \qquad\qquad f(x_2) = -3.1136$$

which can be substituted into Eq. (13.7) to give

$$x_3 = \frac{1.5829(1.5055^2 - 4^2) + 1.7691(4^2 - 1^2) + (-3.1136)(1^2 - 1.5055^2)}{2(1.5829)(1.5055 - 4) + 2(1.7691)(4 - 1) + 2(-3.1136)(1 - 1.5055)}$$
$$= 1.4903$$

which has a function value of $f(1.4903) = 1.7714$.

The process can be repeated, with the results tabulated below:

i	x_0	$f(x_0)$	x_1	$f(x_1)$	x_2	$f(x_2)$	x_3	$f(x_3)$
1	0.0000	0.0000	1.0000	1.5829	4.0000	-3.1136	1.5055	1.7691
2	1.0000	1.5829	1.5055	1.7691	4.0000	-3.1136	1.4903	1.7714
3	1.0000	1.5829	1.4903	1.7714	1.5055	1.7691	1.4256	1.7757
4	1.0000	1.5829	1.4256	1.7757	1.4903	1.7714	1.4266	1.7757
5	1.4256	1.7757	1.4266	1.7757	1.4903	1.7714	1.4275	1.7757

Thus, within five iterations, the result is converging rapidly on the true value of 1.7757 at $x = 1.4276$.

We should mention that just like the false-position method, parabolic interpolation can get hung up with just one end of the interval converging. Thus, convergence can be slow. For example, notice that in our example, 1.0000 was an endpoint for most of the iterations.

This method, as well as others using third-order polynomials, can be formulated into algorithms that contain convergence tests, careful selection strategies for the points to retain on each iteration, and attempts to minimize round-off error accumulation.

13.3 NEWTON'S METHOD

Recall that the Newton-Raphson method of Chap. 6 is an open method that finds the root x of a function such that $f(x) = 0$. The method is summarized as

$$x_{i+1} = x_i - \frac{f(x_i)}{f'(x_i)}$$

A similar open approach can be used to find an optimum of $f(x)$ by defining a new function, $g(x) = f'(x)$. Thus, because the same optimal value x^* satisfies both,

$$f'(x^*) = g(x^*) = 0$$

we can use the following,

$$x_{i+1} = x_i - \frac{f'(x_i)}{f''(x_i)} \tag{13.8}$$

as a technique to find the minimum or maximum of $f(x)$. It should be noted that this equation can also be derived by writing a second-order Taylor series for $f(x)$ and setting the derivative of the series equal to zero. Newton's method is an open method similar to Newton-Raphson because it does not require initial guesses that bracket the optimum. In addition, it also shares the disadvantage that it may be divergent. Finally, it is usually a good idea to check that the second derivative has the correct sign to confirm that the technique is converging on the result you desire.

EXAMPLE 13.3 **Newton's Method**

Problem Statement. Use Newton's method to find the maximum of

$$f(x) = 2 \sin x - \frac{x^2}{10}$$

with an initial guess of $x_0 = 2.5$.

Solution. The first and second derivatives of the function can be evaluated as

$$f'(x) = 2 \cos x - \frac{x}{5}$$

$$f''(x) = -2 \sin x - \frac{1}{5}$$

which can be substituted into Eq. (13.8) to give

$$x_{i+1} = x_i - \frac{2 \cos x_i - x_i/5}{-2 \sin x_i - 1/5}$$

Substituting the initial guess yields

$$x_1 = 2.5 - \frac{2 \cos 2.5 - 2.5/5}{-2 \sin 2.5 - 1/5} = 0.99508$$

which has a function value of 1.57859. The second iteration gives

$$x_2 = 0.995 - \frac{2 \cos 0.995 - 0.995/5}{-2 \sin 0.995 - 1/5} = 1.46901$$

which has a function value of 1.77385.

The process can be repeated, with the results tabulated below:

i	x	f(x)	f'(x)	f''(x)
0	2.5	0.57194	−2.10229	−1.39694
1	0.99508	1.57859	0.88985	−1.87761
2	1.46901	1.77385	−0.09058	−2.18965
3	1.42764	1.77573	−0.00020	−2.17954
4	1.42755	1.77573	0.00000	−2.17952

Thus, within four iterations, the result converges rapidly on the true value.

Although Newton's method works well in some cases, it is impractical for cases where the derivatives cannot be conveniently evaluated. For these cases, other approaches that do not involve derivative evaluation are available. For example, a secant-like version of Newton's method can be developed by using finite-difference approximations for the derivative evaluations.

A bigger reservation regarding the approach is that it may diverge based on the nature of the function and the quality of the initial guess. Thus, it is usually employed only when we are close to the optimum. As described next, hybrid techniques that use *bracketing approaches* far from the optimum and *open methods* near the optimum attempt to exploit the strong points of both approaches.

13.4 BRENT'S METHOD

Recall that in Sec. 6.4, we described Brent's method for root location. This hybrid method combined several root-finding methods into a single algorithm that balanced reliability with efficiency.

Brent also developed a similar approach for one-dimensional minimization. It combines the slow, dependable golden-section search with the faster, but possibly unreliable, parabolic interpolation. It first attempts parabolic interpolation and keeps applying it as long as acceptable results are obtained. If not, it uses the golden-section search to get matters in hand.

Figure 13.7 presents pseudocode for the algorithm based on a MATLAB software M-file developed by Cleve Moler (2004). It represents a stripped-down version of the

FIGURE 13.7
Pseudocode for Brent's
minimum-finding algorithm
based on a MATLAB M-file
developed by Cleve
Moler (2004).

```
Function fminsimp(x1, xu)
tol = 0.000001; phi = (1 + √5)/2;; rho = 2 − phi
u = x1 + rho*(xu − x1); v = u; w = u; x = u
fu = f(u); fv = fu; fw = fu; fx = fu
xm = 0.5*(x1 + xu); d = 0; e = 0
DO
  IF |x − xm| ≤ tol EXIT
  para = |e| > tol
  IF para THEN                       (Try parabolic fit)
    r = (x − w)*(fx − fv); q = (x − v)*(fx − fw)
    p = (x − v)*q − (x − w)*r; s = 2*(q − r)
    IF s > 0 THEN p = −p
    s = |s|
    ' Is the parabola acceptable?
    para = |p| < |0.5*s*e| AND p > s*(x1 − x) AND p < s*(xu − x)
    IF para THEN
      e = d; d = p/s               (Parabolic interpolation step)
    ENDIF
  ENDIF
  IF NOT para THEN
    IF x ≥ xm THEN                   (Golden-section search step)
      e = x1 − x
    ELSE
      e = xu − x
    ENDIF
    d = rho*e
  ENDIF
  u = x + d; fu = f(u)
  IF fu ≤ fx THEN                    (Update x1, xu, x, v, w, xm)
    IF u ≥ x THEN
      x1 = x
    ELSE
      xu = x
    ENDIF
    v = w; fv = fw; w = x; fw = fx; x = u; fx = fu
  ELSE
    IF u < x THEN
      x1 = u
    ELSE
      xu = u
    ENDIF
    IF fu ≤ fw OR w = x THEN
      v = w; fv = fw; w = u; fw = fu
    ELSEIF fu ≤ fv OR v = x OR v = w THEN
      v = u; fv = fu
    ENDIF
  ENDIF
  xm = 0.5*(x1 + xu)
ENDDO
fminsimp = fu
END fminsimp
```

fminbnd function, which is the professional minimization function employed in MATLAB. For that reason, we call the simplified version fminsimp. Note that it requires another function f that holds the equation for which the minimum is being evaluated.

This concludes our treatment of methods to solve the optima of functions of a single variable. Some engineering examples are presented in Chap. 16. In addition, the techniques described here are an important element of some procedures to optimize multi-variable functions, as discussed in Chap. 14.

PROBLEMS

13.1 Given the formula

$$f(x) = -x^2 + 9x - 11$$

(a) Determine the maximum and the corresponding value of x for this function analytically (i.e., using differentiation).
(b) Verify that Eq. (13.7) yields the same results based on initial guesses of $x_0 = 0$, $x_1 = 2$, and $x_2 = 6$.

13.2 Given

$$f(x) = -1.6x^6 - 3x^4 + 10x$$

(a) Plot the function.
(b) Use analytical methods to prove that the function is concave for all values of x.
(c) Differentiate the function and then use a root-location method to solve for the maximum $f(x)$ and the corresponding value of x.

13.3 Solve for the value of x that maximizes $f(x)$ in Prob. 13.2 using the golden-section search. Employ initial guesses of $x_l = 0$ and $x_u = 1$ and perform three iterations.

13.4 Repeat Prob. 13.3, except use parabolic interpolation in the same fashion as Example 13.2. Employ initial guesses of $x_0 = 0$, $x_1 = 1$, and $x_2 = 2$ and perform three iterations.

13.5 Repeat Prob. 13.3 but use Newton's method. Employ an initial guess of $x_0 = 1$ and perform three iterations.

13.6 Employ the following methods to find the maximum of

$$f(x) = 4x - 1.8x^2 + 1.2x^3 - 0.3x^4$$

(a) Golden-section search ($x_l = -2$, $x_u = 4$, $\varepsilon_s = 1\%$).
(b) Parabolic interpolation ($x_0 = 1.75$, $x_1 = 2$, $x_2 = 2.5$, iterations = 4). Select new points sequentially as in the secant method.
(c) Newton's method ($x_0 = 3$, $\varepsilon_s = 1\%$).

13.7 Consider the following function:

$$f(x) = -x^4 - 2x^3 - 9x^2 - 6x$$

Use analytical and graphical methods to show that the function has a maximum for some value of x in the range $-2 \le x \le 1$.

13.8 Employ the following methods to find the maximum of the function from Prob. 13.7:
(a) Golden-section search ($x_l = -2$, $x_u = 1$, $\varepsilon_s = 1\%$).
(b) Parabolic interpolation ($x_0 = -2$, $x_1 = -1$, $x_2 = 1$, iterations = 4). Select new points sequentially as in the secant method.
(c) Newton's method ($x_0 = -1$, $\varepsilon_s = 1\%$).

13.9 Consider the following function:

$$f(x) = 2x + \frac{3}{x}$$

Perform 10 iterations of parabolic interpolation to locate the minimum. Select new points in the same fashion as in Example 13.2. Comment on the convergence of your results. ($x_0 = 0.1$, $x_1 = 0.5$, $x_2 = 5$)

13.10 Consider the following function:

$$f(x) = 3 + 6x + 5x^2 + 3x^3 + 4x^4$$

Locate the minimum by finding the root of the derivative of this function. Use bisection with initial guesses of $x_l = -2$ and $x_u = 1$.

13.11 Determine the minimum of the function from Prob. 13.10 with the following methods:
(a) Newton's method ($x_0 = -1$, $\varepsilon_s = 1\%$).
(b) Newton's method, but using a finite difference approximation for the derivative estimates.

$$f'(x) = \frac{f(x_i + \delta x_i) - f(x_i - \delta x_i)}{2\delta x_i}$$

$$f''(x) = \frac{f(x_i + \delta x_i) - 2f(x_i) - f(x_i - \delta x_i)}{(\delta x_i)^2}$$

where δ = a perturbation fraction (= 0.01). Use an initial guess of $x_0 = -1$ and iterate to $\varepsilon_s = 1\%$.

13.12 Develop a program using a programming or macro language to implement the golden-section search algorithm. Design the program so that it is expressly designed to locate a maximum. The subroutine should have the following features:

- Iterate until the relative error falls below a stopping criterion or a maximum number of iterations has been reached.
- Return both the optimal x and $f(x)$.
- Minimize the number of function evaluations.

Test your program with the same problem as Example 13.1.

13.13 Develop a program as described in Prob. 13.12, but make it perform minimization or maximization depending on the user's preference.

13.14 Develop a program using a programming or macro language to implement the parabolic interpolation algorithm. Design the program so that it is expressly designed to locate a maximum and selects new points as in Example 13.2. The subroutine should have the following features:

- Base it on two initial guesses, and have the program generate the third initial value at the midpoint of the interval.
- Check whether the guesses bracket a maximum. If not, the subroutine should not implement the algorithm, but should return an error message.
- Iterate until the relative error falls below a stopping criterion or a maximum number of iterations has been reached.
- Return both the optimal x and $f(x)$.
- Minimize the number of function evaluations.

Test your program with the same problem as Example 13.2.

13.15 Develop a program using a programming or macro language to implement Newton's method. The subroutine should have the following features:

- Iterate until the relative error falls below a stopping criterion or a maximum number of iterations has been reached.
- Return both the optimal x and $f(x)$.

Test your program with the same problem as Example 13.3.

13.16 Pressure measurements are taken at certain points behind an airfoil over time. These data best fit the curve $y = 5 \cos x - 1.2 \sin x$ from $x = 0$ to 6 s. Use four iterations of the golden-section search to find the minimum pressure. Set $x_l = 2$ and $x_u = 4$.

13.17 The trajectory of a ball can be computed with

$$y = (\tan \theta_0)x - \frac{g}{2v_0^2 \cos^2 \theta_0}x^2 + y_0$$

where y = the height (m), θ_0 = the initial angle (radians), v_0 = the initial velocity (m/s), g = the gravitational constant = 9.81 m/s^2, and y_0 = the initial height (m). Use the golden-section search to determine the maximum height given $y_0 = 2$ m, $v_0 = 24$ m/s, and $\theta_0 = 52°$. Iterate until the approximate error falls below $\varepsilon_s = 1\%$ using initial guesses of $x_l = 0$ and $x_u = 60$ m.

13.18 The deflection of a uniform beam subject to a linearly increasing distributed load can be computed as

$$y = \frac{w_0}{120EIL}(-x^5 + 2L^2x^3 - L^4x)$$

Given that $L = 500$ cm, $E = 50,000$ kN/cm^2, $I = 30,000$ cm^4, and $w_0 = 2.5$ kN/cm, determine the point of maximum deflection **(a)** graphically and **(b)** using the golden-section search until the approximate error falls below $\varepsilon_s = 5\%$ with initial guesses of $x_l = 0$ and $x_u = L$.

13.19 An object with a mass of 120 kg is projected upward from the surface of the earth at a velocity of 55 m/s. If the object is subject to linear drag ($c = 16$ kg/s), use the golden-section search to determine the maximum height the object attains and the time it takes to reach the maximum. Hint: Recall Sec. PT4.1.2.

13.20 The normal distribution is a bell-shaped curve defined by

$$y = e^{-x^2}$$

Use the golden-section search to determine the location of the inflection point of this curve for positive x.

13.21 An object can be projected upward at a specified velocity. If it is subject to linear drag, its altitude as a function of time can be computed as

$$z = z_0 + \frac{m}{c}\left(v_0 + \frac{mg}{c}\right)(1 - e^{-(c/m)t}) - \frac{mg}{c}t$$

where z = altitude (m) above the earth's surface (defined as $z = 0$), z_0 = the initial altitude (m), m = mass (kg), c = a linear drag coefficient (kg/s), v_0 = initial velocity (m/s), and t = time (s). Note that for this formulation, positive velocity is considered to be in the upward direction. Given the following parameter values: $g = 9.81$ m/s^2, $z_0 = 100$ m, $v_0 = 60$ m/s, $m = 90$ kg, and $c = 14$ kg/s, the equation can be used to calculate the object's altitude. Determine the time and altitude of the peak elevation **(a)** graphically, **(b)** analytically, and **(c)** with the golden-section search until the approximate error falls below $\varepsilon_s = 1\%$ with initial guesses of $t_l = 0$ and $t_u = 10$ s.

13.22 Use the golden-section search to determine the length of the shortest ladder that reaches from the ground over the fence to touch the building's wall (Fig. P13.22). Test it for the case where $h = d = 4$ m.

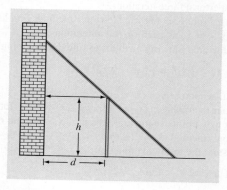

FIGURE P13.22
A ladder leaning against a fence and just touching a wall.

Multidimensional Unconstrained Optimization

This chapter describes techniques to find the minimum or maximum of a function of several variables. Recall from Chap. 13 that our visual image of a one-dimensional search was like a roller coaster. For two-dimensional cases, the image becomes that of mountains and valleys (Fig. 14.1). For higher-dimensional problems, convenient images are not possible.

We have chosen to limit this chapter to the two-dimensional case. We have adopted this approach because the essential features of multidimensional searches are often best communicated visually.

Techniques for multidimensional unconstrained optimization can be classified in a number of ways. For purposes of the present discussion, we will divide them depending on whether they require derivative evaluation. The approaches that do not require derivative evaluation are called *nongradient,* or *direct, methods*. Those that require derivatives are called *gradient,* or *descent* (or *ascent*), *methods*.

FIGURE 14.1

The most tangible way to visualize two-dimensional searches is in the context of ascending a mountain (maximization) or descending into a valley (minimization). (*a*) A 2-D topographic map that corresponds to the 3-D mountain in (*b*).

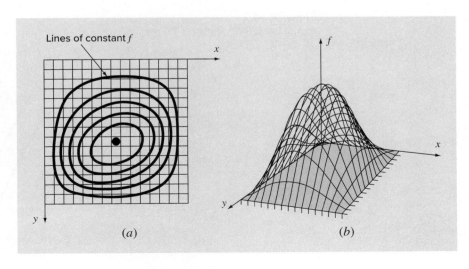

Lines of constant *f*

(*a*)   (*b*)

14.1 DIRECT METHODS

Direct methods vary from simple brute force approaches to more elegant techniques that attempt to exploit the nature of the function. We will start our discussion with a brute force approach.

14.1.1 Random Search

A simple example of a brute force approach is the *random search method*. As the name implies, this method repeatedly evaluates the function at randomly selected values of the independent variables. If a sufficient number of samples are conducted, the optimum will eventually be located.

EXAMPLE 14.1 | **Random Search Method**

Problem Statement.   Use a random number generator to locate the maximum of

$$f(x, y) = y - x - 2x^2 - 2xy - y^2 \qquad \text{(E14.1.1)}$$

in the domain bounded by $x = -2$ to 2 and $y = 1$ to 3. The domain is depicted in Fig. 14.2. Notice that a single maximum of 1.25 occurs at $x = -1$ and $y = 1.5$.

Solution.   Random number generators typically generate values between 0 and 1. If we designate such a number as r, the following formula can be used to generate x values randomly within a range from x_l to x_u:

$$x = x_l + (x_u - x_l)r$$

For the present application, $x_l = -2$ and $x_u = 2$, and the formula is

$$x = -2 + (2 - (-2))r = -2 + 4r$$

This can be tested by substituting 0 and 1 to yield -2 and 2, respectively.

FIGURE 14.2
Plot of Eq. (E14.1.1) showing the maximum at $x = -1$ and $y = 1.5$.

Similarly for y, a formula for the present example could be developed as

$$y = y_l + (y_u - y_l)r = 1 + (3 - 1)r = 1 + 2r$$

The following Excel/VBA macrocode uses the VBA random number function Rnd, to generate (x, y) pairs. These are then substituted into Eq. (E14.1.1). The maximum value from among these random trials is stored in the variable maxf, and the corresponding x and y values in maxx and maxy, respectively.

```
maxf = -1E9
For j = 1 To n
   x = -2 + 4 * Rnd
   y = 1 + 2 * Rnd
   fn = y - x - 2 * x ^ 2 - 2 * x * y - y ^ 2
   If fn > maxf Then
      maxf = fn
      maxx = x
      maxy = y
   End If
Next j
```

A number of iterations yields

Iterations	x	y	$f(x, y)$
1000	−0.9886	1.4282	1.2462
2000	−1.0040	1.4724	1.2490
3000	−1.0040	1.4724	1.2490
4000	−1.0040	1.4724	1.2490
5000	−1.0040	1.4724	1.2490
6000	−0.9837	1.4936	1.2496
7000	−0.9960	1.5079	1.2498
8000	−0.9960	1.5079	1.2498
9000	−0.9960	1.5079	1.2498
10000	−0.9978	1.5039	1.2500

The results indicate that the technique homes in on the true maximum.

This simple brute force approach works even for discontinuous and nondifferentiable functions. Furthermore, it always finds the global optimum rather than a local optimum. Its major shortcoming is that as the number of independent variables grows, the implementation effort required can become onerous. In addition, it is not efficient because it takes no account of the behavior of the underlying function. The remainder of the approaches described in this chapter do take function behavior into account as well as the results of previous trials to improve the speed of convergence. Thus, although the random search method can certainly prove useful in specific problem contexts, the following methods have more general utility and almost always lead to more efficient convergence.

It should be noted that more sophisticated search techniques are available. These are heuristic approaches that were developed to handle either nonlinear and/or discontinuous problems that classical optimization cannot usually handle well, if at all. Simulated annealing, tabu search, artificial neural networks, and genetic algorithms are a few. The most widely applied is the *genetic algorithm,* with a number of commercial packages available. Holland (1975) pioneered the genetic algorithm approach, and Davis (1991) and Goldberg (1989) provide good overviews of the theory and application of the method.

14.1.2 Univariate and Pattern Searches

It is very appealing to have an efficient optimization approach that does not require evaluation of derivatives. The random search method described above does not require derivative evaluation, but it is not very efficient. This section describes an approach, the univariate search method, that is more efficient and still does not require derivative evaluation.

The basic strategy underlying the *univariate search method* is to change one variable at a time to improve the approximation while the other variables are held constant. Since only one variable is changed, the problem reduces to a sequence of one-dimensional searches that can be solved using a variety of methods (including those described in Chap. 13).

Let us perform a univariate search graphically, as shown in Fig. 14.3. Start at point 1, and move along the x axis with y constant to the maximum at point 2. You can see that point 2 is a maximum by noticing that the trajectory along the x axis just touches a contour line at the point. Next, move along the y axis with x constant to point 3. Continue this process generating points 4, 5, 6, etc.

FIGURE 14.3
A graphical depiction of how a univariate search is conducted.

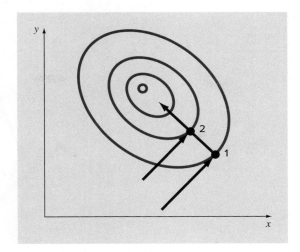

FIGURE 14.4
Conjugate directions.

Although we are gradually moving toward the maximum, the search becomes less efficient as we move along the narrow ridge toward the maximum. However, also note that lines joining alternate points such as 1-3, 3-5 or 2-4, 4-6 point in the general direction of the maximum. These trajectories present an opportunity to shoot directly along the ridge toward the maximum. Such trajectories are called *pattern directions*.

Formal algorithms are available that capitalize on the idea of pattern directions to find optimum values efficiently. The best known of these algorithms is called *Powell's method*. It is based on the observation (see Fig. 14.4) that if points 1 and 2 are obtained by one-dimensional searches in the same direction but from different starting points, then the line formed by 1 and 2 will be directed toward the maximum. Such lines are called *conjugate directions*.

In fact, it can be proved that if $f(x, y)$ is a quadratic function, sequential searches along conjugate directions will converge exactly in a finite number of steps regardless of the starting point. Since a general nonlinear function can often be reasonably approximated by a quadratic function, methods based on conjugate directions are usually quite efficient and are in fact quadratically convergent as they approach the optimum.

Let us graphically implement a simplified version of Powell's method to find the maximum of

$$f(x, y) = c - (x - a)^2 - (y - b)^2$$

where a, b, and c are positive constants. This equation results in circular contours in the xy plane, as shown in Fig. 14.5.

Initiate the search at point 0 with starting directions h_1 and h_2. Note that h_1 and h_2 are not necessarily conjugate directions. From zero, move along h_1 until a maximum is located

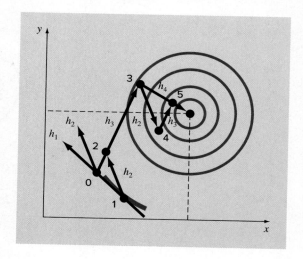

FIGURE 14.5
Powell's method.

at point 1. Then search from point 1 along direction h_2 to find point 2. Next, form a new search direction h_3 through points 0 and 2. Search along this direction until the maximum at point 3 is located. Then search from point 3 in the h_2 direction until the maximum at point 4 is located. From point 4 arrive at point 5 by again searching along h_3. Now, observe that both points 5 and 3 have been located by searching in the h_3 direction from two different points. Powell has shown that h_4 (formed by points 3 and 5) and h_3 are conjugate directions. Thus, searching from point 5 along h_4 brings us directly to the maximum.

Powell's method can be refined to make it more efficient, but the formal algorithms are beyond the scope of this text. However, it is an efficient method that is quadratically convergent without requiring derivative evaluation.

14.2 GRADIENT METHODS

As the name implies, *gradient methods* explicitly use derivative information to generate efficient algorithms to locate optima. Before describing specific approaches, we must first review some key mathematical concepts and operations.

14.2.1 Gradients and Hessians

Recall from calculus that the first derivative of a one-dimensional function provides a slope or tangent to the function being differentiated. From the standpoint of optimization, this is useful information. For example, if the slope is positive, it tells us that increasing the independent variable will lead to a higher value of the function we are exploring.

From calculus, also recall that the first derivative may tell us when we have reached an optimal value since this is the point that the derivative goes to zero. Further, the sign of the second derivative can tell us whether we have reached a minimum (positive second derivative) or a maximum (negative second derivative).

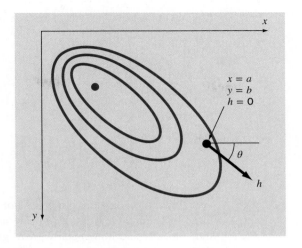

FIGURE 14.6
The directional gradient is defined along an axis h that forms an angle θ with the x axis.

These ideas were useful to us in the one-dimensional search algorithms we explored in Chap. 13. However, to fully understand multidimensional searches, we must first understand how the first and second derivatives are expressed in a multidimensional context.

The Gradient.   Suppose we have a two-dimensional function $f(x, y)$. An example might be your elevation on a mountain as a function of your position. Suppose that you are at a specific location on the mountain (a, b) and you want to know the slope in an arbitrary direction. One way to define the direction is along a new axis h that forms an angle θ with the x axis (Fig. 14.6). The elevation along this new axis can be thought of as a new function $g(h)$. If you define your position as being the origin of this axis (that is, $h = 0$), the slope in this direction would be designated as $g'(0)$. This slope, which is called the *directional derivative,* can be calculated from the partial derivatives along the x and y axis by

$$g'(0) = \frac{\partial f}{\partial x} \cos \theta + \frac{\partial f}{\partial y} \sin \theta \tag{14.1}$$

where the partial derivatives are evaluated at $x = a$ and $y = b$.

Assuming that your goal is to gain the most elevation with the next step, the next logical question would be: What direction is the steepest ascent? The answer to this question is provided very neatly by what is referred to mathematically as the *gradient,* which is defined as

$$\nabla f = \frac{\partial f}{\partial x} \mathbf{i} + \frac{\partial f}{\partial y} \mathbf{j} \tag{14.2}$$

This vector is also referred to as "del f." It represents the directional derivative of $f(x, y)$ at point $x = a$ and $y = b$.

Vector notation provides a concise means to generalize the gradient to n dimensions, as

$$\nabla f(\mathbf{x}) = \begin{Bmatrix} \dfrac{\partial f}{\partial x_1}(\mathbf{x}) \\[2mm] \dfrac{\partial f}{\partial x_2}(\mathbf{x}) \\[2mm] \cdot \\ \cdot \\ \cdot \\ \dfrac{\partial f}{\partial x_n}(\mathbf{x}) \end{Bmatrix}$$

How do we use the gradient? For the mountain-climbing problem, if we are interested in gaining elevation as quickly as possible, the gradient tells us what direction to move locally and how much we will gain by taking it. Note, however, that this strategy does not necessarily take us on a direct path to the summit! We will discuss these ideas in more depth later in this chapter.

EXAMPLE 14.2	**Using the Gradient to Evaluate the Path of Steepest Ascent**

Problem Statement.   Employ the gradient to evaluate the steepest ascent direction for the function

$$f(x, y) = xy^2$$

at the point (2, 2). Assume that positive x is pointed east and positive y is pointed north.

Solution.   First, our elevation can be determined as

$$f(2, 2) = 2(2)^2 = 8$$

Next, the partial derivatives can be evaluated,

$$\frac{\partial f}{\partial x} = y^2 = 2^2 = 4$$

$$\frac{\partial f}{\partial y} = 2xy = 2(2)(2) = 8$$

These values can be used to determine the gradient as

$$\nabla f = 4\mathbf{i} + 8\mathbf{j}$$

This vector can be sketched on a topographical map of the function, as in Fig. 14.7. This immediately tells us that the direction we must take is

$$\theta = \tan^{-1}\left(\frac{8}{4}\right) = 1.107 \text{ radians } (= 63.4°)$$

relative to the x axis. The slope in this direction, which is the magnitude of ∇f, can be calculated as

$$\sqrt{4^2 + 8^2} = 8.944$$

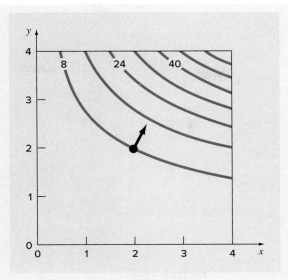

FIGURE 14.7
The arrow indicates the direction of steepest ascent calculated with the gradient.

Thus, during our first step, we will initially gain 8.944 units of elevation rise for a unit distance advanced along this steepest path. Observe that Eq. (14.1) yields the same result,

$$g'(0) = 4\cos(1.107) + 8\sin(1.107) = 8.944$$

Note that for any other direction, say $\theta = 1.107/2 = 0.5235$, $g'(0) = 4\cos(0.5235) + 8\sin(0.5235) = 7.608$, which is smaller.

As we move forward, both the direction and magnitude of the steepest path will change. These changes can be quantified at each step using the gradient, and your climbing direction modified accordingly.

A final insight can be gained by inspecting Fig. 14.7. As indicated, the direction of steepest ascent is perpendicular, or *orthogonal,* to the elevation contour at the coordinate (2, 2). This is a general characteristic of the gradient.

Aside from defining a steepest path, the first derivative can also be used to discern whether an optimum has been reached. As is the case for a one-dimensional function, if the partial derivatives with respect to both x and y are zero, a two-dimensional optimum has been reached.

The Hessian.   For one-dimensional problems, both the first and second derivatives provide valuable information for searching out optima. The first derivative (1) provides a steepest trajectory of the function and (2) tells us that we have reached an optimum. Once at an optimum, the second derivative tells us whether we are a maximum [negative $f''(x)$] or a minimum [positive $f''(x)$]. In the previous paragraphs, we illustrated how the

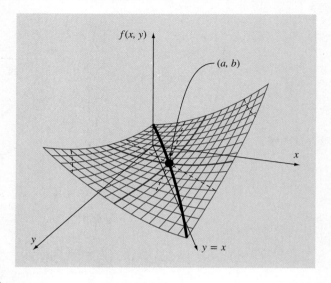

FIGURE 14.8

A saddle point ($x = a$ and $y = b$). Notice that when the curve is viewed along the x and y directions, the function appears to go through a minimum (positive second derivative), whereas when viewed along an axis $x = y$, it is concave downward (negative second derivative).

gradient provides best local trajectories for multidimensional problems. Now, we will examine how the second derivative is used in such contexts.

You might expect that if the partial second derivatives with respect to both x and y are both negative, then you have reached a maximum. Figure 14.8 shows a function where this is not true. The point (a, b) of this graph appears to be a minimum when observed along either the x dimension or the y dimension. In both instances, the second partial derivatives are positive. However, if the function is observed along the line $y = x$, it can be seen that a maximum occurs at the same point. This shape is called a *saddle*, and clearly, neither a maximum nor a minimum occurs at the point.

Whether a maximum or a minimum occurs involves not only the partials with respect to x and y but also the second partials with respect to x and y. Assuming that the partial derivatives are continuous at and near the point being evaluated, the following quantity can be computed:

$$|H| = \frac{\partial^2 f}{\partial x^2} \frac{\partial^2 f}{\partial y^2} - \left(\frac{\partial^2 f}{\partial x \partial y}\right)^2 \tag{14.3}$$

Three cases can occur

- If $|H| > 0$ and $\partial^2 f/\partial x^2 > 0$, then $f(x, y)$ has a local minimum.
- If $|H| > 0$ and $\partial^2 f/\partial x^2 < 0$, then $f(x, y)$ has a local maximum.
- If $|H| < 0$, then $f(x, y)$ has a saddle point.

The quantity $|H|$ is equal to the determinant of a matrix made up of the second derivatives,[1]

$$H = \begin{bmatrix} \dfrac{\partial^2 f}{\partial x^2} & \dfrac{\partial^2 f}{\partial x \partial y} \\[2ex] \dfrac{\partial^2 f}{\partial y \partial x} & \dfrac{\partial^2 f}{\partial y^2} \end{bmatrix} \tag{14.4}$$

This matrix is formally referred to as the *Hessian* of f.

Besides providing a way to discern whether a multidimensional function has reached an optimum, the Hessian has other uses in optimization (for example, for the multidimensional form of Newton's method). In particular, it allows searches to include second-order curvature to attain superior results.

Finite-Difference Approximations. It should be mentioned that, for cases where they are difficult or inconvenient to compute analytically, both the gradient and the determinant of the Hessian can be evaluated numerically. In most cases, the approach introduced in Sec. 6.3.3 for the modified secant method is employed. That is, the independent variables can be perturbed slightly to generate the required partial derivatives. For example, if a centered-difference approach is adopted, they can be computed as

$$\frac{\partial f}{\partial x} = \frac{f(x + \delta x, y) - f(x - \delta x, y)}{2\delta x} \tag{14.5}$$

$$\frac{\partial f}{\partial y} = \frac{f(x, y + \delta y) - f(x, y - \delta y)}{2\delta y} \tag{14.6}$$

$$\frac{\partial^2 f}{\partial x^2} = \frac{f(x + \delta x, y) - 2f(x, y) + f(x - \delta x, y)}{\delta x^2} \tag{14.7}$$

$$\frac{\partial^2 f}{\partial y^2} = \frac{f(x, y + \delta y) - 2f(x, y) + f(x, y - \delta y)}{\delta y^2} \tag{14.8}$$

$$\frac{\partial^2 f}{\partial x \partial y} =$$

$$\frac{f(x + \delta x, y + \delta y) - f(x + \delta x, y - \delta y) - f(x - \delta x, y + \delta y) + f(x - \delta x, y - \delta y)}{4\delta x \delta y}$$

$$\tag{14.9}$$

where δ is some small fractional value.

Note that the methods employed in commercial software packages also use forward differences. In addition, they are usually more complicated than the approximations listed in Eqs. (14.5) through (14.9). Dennis and Schnabel (1996) provide more detail on such approaches.

Regardless of how the approximation is implemented, the important point is that you may have the option of evaluating the gradient and/or the Hessian analytically. This can sometimes be an arduous task, but the performance of the algorithm may benefit

---

[1]Note that $\partial^2 f/(\partial x \partial y) = \partial^2 f/(\partial y \partial x)$.

enough to make your effort worthwhile. The closed-form derivatives will be exact, but more importantly, you will reduce the number of function evaluations. This latter point can have a critical impact on the execution time.

On the other hand, you will often exercise the option of having the quantities computed internally using numerical approaches. In many cases, the performance will be quite adequate and you will be saved the difficulty of numerous partial differentiations. Such would be the case on the optimizers used in certain spreadsheets and mathematical software packages (for example, Excel). In such cases, you may not even be given the option of entering an analytically derived gradient and Hessian. However, for small to moderately sized problems, this is usually not a major shortcoming.

14.2.2 Steepest Ascent Method

An obvious strategy for climbing a hill would be to determine the maximum slope at your starting position and then start walking in that direction. But clearly, another problem arises almost immediately. Unless you were really lucky and started on a ridge that pointed directly to the summit, as soon as you moved, your path would diverge from the steepest ascent direction.

Recognizing this fact, you might adopt the following strategy. You could walk a short distance along the gradient direction. Then you could stop, reevaluate the gradient and walk another short distance. By repeating the process you would eventually get to the top of the hill.

Although this strategy sounds superficially sound, it is not very practical. In particular, the continuous reevaluation of the gradient can be computationally demanding. A preferred approach involves moving in a fixed path along the initial gradient until $f(x, y)$ stops increasing, that is, becomes level along your direction of travel. This stopping point becomes the starting point where ∇f is reevaluated and a new direction followed. The process is repeated until the summit is reached. This approach is called the *steepest ascent method*.[2] It is the most straightforward of the gradient search techniques. The basic idea behind the approach is depicted in Fig. 14.9.

We start at an initial point (x_0, y_0) labeled "0" in the figure. At this point, we determine the direction of steepest ascent, that is, the gradient. We then search along the direction of the gradient, h_0, until we find a maximum, which is labeled "1" in the figure. The process is then repeated.

Thus, the problem boils down to two parts: (1) determining the "best" direction to search and (2) determining the "best value" along that search direction. As we will see, the effectiveness of the various algorithms described in the coming pages depends on how clever we are at both parts.

For the time being, the steepest ascent method uses the gradient approach as its choice for the "best" direction. We have already shown how the gradient is evaluated in Example 14.2. Now, before examining how the algorithm goes about locating the maximum along the steepest direction, we must pause to explore how to transform a function of x and y into a function of h along the gradient direction.

---

[2]Because of our emphasis on maximization here, we use the terminology *steepest ascent.* The same approach can also be used for minimization, in which case the terminology *steepest descent* is used.

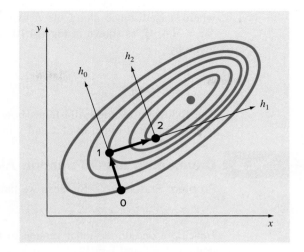

FIGURE 14.9
A graphical depiction of the method of steepest ascent.

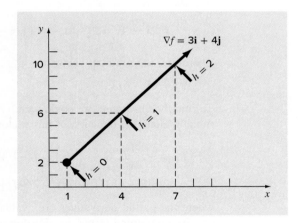

FIGURE 14.10
The relationship between an arbitrary direction h and x and y coordinates.

Starting at x_0, y_0 the coordinates of any point in the gradient direction can be expressed as

$$x = x_0 + \frac{\partial f}{\partial x} h \tag{14.10}$$

$$y = y_0 + \frac{\partial f}{\partial y} h \tag{14.11}$$

where h is distance along the h axis. For example, suppose $x_0 = 1$ and $y_0 = 2$ and $\nabla f = 3\mathbf{i} + 4\mathbf{j}$, as shown in Fig. 14.10. The coordinates of any point along the h axis are given by

$$x = 1 + 3h \tag{14.12}$$
$$y = 2 + 4h \tag{14.13}$$

The following example illustrates how we can use these transformations to convert a two-dimensional function of x and y into a one-dimensional function in h.

EXAMPLE 14.3

Developing a 1-D Function Along the Gradient Direction

Problem Statement.   Suppose we have the following two-dimensional function:

$$f(x, y) = 2xy + 2x - x^2 - 2y^2$$

Develop a one-dimensional version of this equation along the gradient direction at the point $x = -1$ and $y = 1$.

Solution.   The partial derivatives can be evaluated at $(-1, 1)$,

$$\frac{\partial f}{\partial x} = 2y + 2 - 2x = 2(1) + 2 - 2(-1) = 6$$

$$\frac{\partial f}{\partial y} = 2x - 4y = 2(-1) - 4(1) = -6$$

Therefore, the gradient vector is

$$\nabla f = 6\mathbf{i} - 6\mathbf{j}$$

To find the maximum, we could search along the gradient direction, that is, along an h axis running along the direction of this vector. The function can be expressed along this axis as

$$f\left(x_0 + \frac{\partial f}{\partial x}h, y_0 + \frac{\partial f}{\partial y}h\right) = f(-1 + 6h, 1 - 6h)$$
$$= 2(-1 + 6h)(1 - 6h) + 2(-1 + 6h) - (-1 + 6h)^2 - 2(1 - 6h)^2$$

where the partial derivatives are evaluated at $x = -1$ and $y = 1$.

By combining terms, we develop a one-dimensional function $g(h)$ that maps $f(x, y)$ along the h axis,

$$g(h) = -180h^2 + 72h - 7$$

Now that we have developed a function along the path of steepest ascent, we can explore how to answer the second question. That is, how far along this path do we travel? One approach might be to move along this path until we find the maximum of this function. We will call the location of this maximum h^*. This is the value of the step that maximizes g (and hence, f) in the gradient direction. This problem is equivalent to finding the maximum of a function of a single variable h. This can be done using different one-dimensional search techniques like the ones we discussed in Chap. 13. Thus, we convert from finding the optimum of a two-dimensional function to performing a one-dimensional search along the gradient direction.

This method is called *steepest ascent* when an arbitrary step size h is used. If a value of a single step $h*$ is found that brings us directly to the maximum along the gradient direction, the method is called the *optimal steepest ascent*.

EXAMPLE 14.4 **Optimal Steepest Ascent**

Problem Statement.   Maximize the following function:

$$f(x, y) = 2xy + 2x - x^2 - 2y^2$$

using initial guesses $x = -1$ and $y = 1$.

Solution.   Because this function is so simple, we can first generate an analytical solution. To do this, the partial derivatives can be evaluated as

$$\frac{\partial f}{\partial x} = 2y + 2 - 2x = 0$$

$$\frac{\partial f}{\partial y} = 2x - 4y = 0$$

This pair of equations can be solved for the optimum, $x = 2$ and $y = 1$. The second partial derivatives can also be determined and evaluated at the optimum,

$$\frac{\partial^2 f}{\partial x^2} = -2$$

$$\frac{\partial^2 f}{\partial y^2} = -4$$

$$\frac{\partial^2 f}{\partial x \partial y} = \frac{\partial^2 f}{\partial y \partial x} = 2$$

and the determinant of the Hessian is computed [Eq. (14.3)],

$$|H| = -2(-4) - 2^2 = 4$$

Therefore, because $|H| > 0$ and $\partial^2 f/\partial x^2 < 0$, the function value $f(2, 1)$ is a maximum.

Now let us implement optimal steepest ascent. Recall that, at the end of Example 14.3, we had already implemented the initial steps of the problem by generating

$$g(h) = -180h^2 + 72h - 7$$

Now, because this is a simple parabola, we can directly locate the maximum (that is, $h = h*$) by solving the problem

$$g'(h*) = 0$$

$$-360h* + 72 = 0$$

$$h* = 0.2$$

This means that if we travel along the h axis, $g(h)$ reaches a minimum value when $h = h* = 0.2$. This result can be placed back into Eqs. (14.10) and (14.11) to solve for the

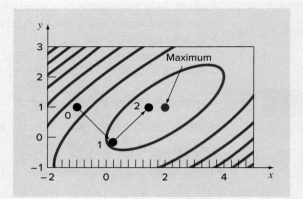

FIGURE 14.11
The method of optimal steepest ascent.

(x, y) coordinates corresponding to this point,

$$x = -1 + 6(0.2) = 0.2$$
$$y = 1 - 6(0.2) = -0.2$$

This step is depicted in Fig. 14.11 as the move from point 0 to 1.

The second step is implemented by merely repeating the procedure. First, the partial derivatives can be evaluated at the new starting point $(0.2, -0.2)$ to give

$$\frac{\partial f}{\partial x} = 2(-0.2) + 2 - 2(0.2) = 1.2$$

$$\frac{\partial f}{\partial y} = 2(0.2) - 4(-0.2) = 1.2$$

Therefore, the gradient vector is

$$\nabla f = 1.2\mathbf{i} + 1.2\mathbf{j}$$

This means that the steepest direction is now pointed up and to the right at a 45° angle with the x axis (see Fig. 14.11). The coordinates along this new h axis can now be expressed as

$$x = 0.2 + 1.2h$$
$$y = -0.2 + 1.2h$$

Substituting these values into the function yields

$$f(0.2 + 1.2h, -0.2 + 1.2h) = g(h) = -1.44h^2 + 2.88h + 0.2$$

The step h^* to take us to the maximum along the search direction can then be directly computed as

$$g'(h^*) = -2.88h^* + 2.88 = 0$$
$$h^* = 1$$

This result can be placed back into Eqs. (14.10) and (14.11) to solve for the (x, y) coordinates corresponding to this new point,

$$x = 0.2 + 1.2(1) = 1.4$$
$$y = -0.2 + 1.2(1) = 1$$

As depicted in Fig. 14.11, we move to the new coordinates, labeled point 2 in the plot, and in so doing move closer to the maximum. The approach can be repeated with the final result converging on the analytical solution, $x = 2$ and $y = 1$.

It can be shown that the method of steepest ascent is linearly convergent. Further, it tends to move very slowly along long, narrow ridges. This is because the new gradient at each maximum point will be perpendicular to the original direction. Thus, the technique takes many small steps criss-crossing the direct route to the summit. Hence, although it is reliable, there are other approaches that converge much more rapidly, particularly in the vicinity of an optimum. The remainder of the section is devoted to such methods.

14.2.3 Advanced Gradient Approaches

Conjugate Gradient Method (Fletcher-Reeves).   In Sec. 14.1.2, we have seen how conjugate directions in Powell's method greatly improved the efficiency of a univariate search. In a similar manner, we can also improve the linearly convergent steepest ascent using conjugate gradients. In fact, an optimization method that makes use of conjugate gradients to define search directions can be shown to be quadratically convergent. Use of conjugate gradients also ensures that the method will optimize a quadratic function exactly in a finite number of steps regardless of the starting point. Since most well-behaved functions can be approximated reasonably well by a quadratic in the vicinity of an optimum, quadratically convergent approaches are often very efficient near an optimum.

We have seen how, starting with two arbitrary search directions, Powell's method produced new conjugate search directions. This method is quadratically convergent and does not require gradient information. On the other hand, if evaluation of derivatives is practical, we can devise algorithms that combine the ideas of steepest descent and conjugate directions to achieve robust initial performance and rapid convergence as the technique gravitates toward the optimum. The *Fletcher-Reeves conjugate gradient algorithm* modifies the steepest ascent method by imposing the condition that successive gradient search directions be mutually conjugate. The proof and algorithm are beyond the scope of the text but are described by Rao (1996).

Newton's Method.   Newton's method for a single variable (recall Sec. 13.3) can be extended to multivariate cases. Write a second-order Taylor series for $f(\mathbf{x})$ near $\mathbf{x} = \mathbf{x}_i$,

$$f(\mathbf{x}) = f(\mathbf{x}_i) + \nabla f^T(\mathbf{x}_i)(\mathbf{x} - \mathbf{x}_i) + \frac{1}{2}(\mathbf{x} - \mathbf{x}_i)^T H_i(\mathbf{x} - \mathbf{x}_i)$$

where H_i is the Hessian matrix. At the minimum,

$$\frac{\partial f(\mathbf{x})}{\partial \mathbf{x}_j} = 0 \qquad \text{for } j = 1, 2, \dots, n$$

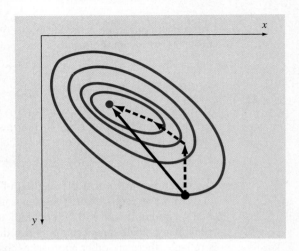

FIGURE 14.12
When the starting point is close to the optimal point, following the gradient can be inefficient. Newton's method attempts to search along a direct path to the optimum (solid line).

Thus,

$$\nabla f = \nabla f(\mathbf{x}_i) + H_i(\mathbf{x} - \mathbf{x}_i) = 0$$

If H is nonsingular,

$$\mathbf{x}_{i+1} = \mathbf{x}_i - H_i^{-1}\nabla f \qquad (14.14)$$

which can be shown to converge quadratically near the optimum. This method again performs better than the steepest ascent method (see Fig. 14.12). However, note that the method requires both the computation of second derivatives and matrix inversion at each iteration. Thus, the method is not very useful in practice for functions with large numbers of variables. Furthermore, Newton's method may not converge if the starting point is not close to the optimum.

Marquardt's Method.   We know that the method of steepest ascent increases the function value even if the starting point is far from an optimum. On the other hand, we have just described Newton's method, which converges rapidly near the maximum. *Marquardt's method* uses the steepest ascent method when **x** is far from **x***, and Newton's method when **x** closes in on an optimum. This is accomplished by modifying the diagonal of the Hessian in Eq. (14.14),

$$\tilde{H}_i = H_i + \alpha_i I$$

where α_i is a positive constant and I is the identity matrix. At the start of the procedure, α_i is assumed to be large and

$$\tilde{H}_i^{-1} \approx \frac{1}{\alpha_i} I$$

which reduces Eq. (14.14) to the steepest ascent method. As the iterations proceed, α_i approaches zero and the method becomes Newton's method.

Thus, Marquardt's method offers the best of both worlds: It plods along reliably from poor initial starting values yet accelerates rapidly when it approaches the optimum. Unfortunately, the method still requires Hessian evaluation and matrix inversion at each step. It should be noted that the Marquardt method is primarily used for nonlinear least-squares problems.

Quasi-Newton Methods. *Quasi-Newton,* or *variable metric, methods* seek to estimate the direct path to the optimum in a manner similar to Newton's method. However, notice that the Hessian matrix in Eq. (14.14) is composed of the second derivatives of f that vary from step to step. Quasi-Newton methods attempt to avoid these difficulties by approximating H with another matrix A using only first partial derivatives of f. The approach involves starting with an initial approximation of H^{-1} and updating and improving it with each iteration. The methods are called quasi-Newton because we do not use the true Hessian, rather an approximation. Thus, we have two approximations at work simultaneously: (1) the original Taylor-series approximation and (2) the Hessian approximation.

There are two primary methods of this type: the *Davidon-Fletcher-Powell* (DFP) and the *Broyden-Fletcher-Goldfarb-Shanno* (BFGS) algorithms. They are similar except for details concerning how they handle round-off error and convergence issues. BFGS is generally recognized as being superior in most cases. Rao (1996) provides details and formal statements of both the DFP and the BFGS algorithms.

PROBLEMS

14.1 Find the directional derivative of

$$f(x, y) = 2x^2 + y^2$$

at $x = 2$ and $y = 2$ in the direction of $h = 3i + 2j$.

14.2 Repeat Example 14.2 for the following function at the point (0.8, 1.2).

$$f(x, y) = 2.1xy + 1.6y - 1.3x^2 - 1.75y^2 + 4$$

14.3 Given

$$f(x, y) = 2.25xy + 1.75y - 1.5x^2 - 2y^2$$

construct and solve a system of linear algebraic equations that maximizes $f(x)$. Note that this is done by setting the partial derivatives of f with respect to both x and y to zero.

14.4

(a) Start with an initial guess of $x = 1$ and $y = 1$ and apply two applications of the steepest ascent method to $f(x, y)$ from Prob. 14.3.

(b) Construct a plot from the results of (a) showing the path of the search.

14.5 Find the gradient vector and Hessian matrix for each of the following functions:

(a) $f(x, y) = 2xy^2 + 3e^{xy}$

(b) $f(x, y, z) = x^2 + y^2 + 2z^2$

(c) $f(x, y) = \ln(x^2 + 2xy + 3y^2)$

14.6 Find the minimum value of

$$f(x, y) = (x - 3)^2 + (y - 2)^2$$

starting at $x = 1$ and $y = 1$, using the steepest descent method with a stopping criterion of $\varepsilon_s = 1\%$. Explain your results.

14.7 Perform one iteration of the steepest ascent method to locate the maximum of

$$f(x, y) = 3.5x + 2y + x^2 - x^4 + 2xy - y^2$$

using initial guesses $x = 0$ and $y = 0$. Employ bisection to find the optimal step size in the gradient search direction.

14.8 Perform one iteration of the optimal gradient steepest descent method to locate the minimum of

$$f(x, y) = -9x + x^2 + 11y + 4y^2 - 2xy$$

using initial guesses $x = 0$ and $y = 0$.

14.9 Develop a program using a programming or macro language to implement the random search method. Design the subprogram so that it is expressly designed to locate a maximum. Test the program with $f(x, y)$ from Prob. 14.7. Use a range of -2 to 2 for both x and y.

14.10 The grid search is another brute force approach to optimization. The two-dimensional version is depicted in Fig. P14.10. The x and y dimensions are divided into increments to create a grid. The function is then evaluated at each node of the grid. The denser the grid, the more likely it would be to locate the optimum.

Develop a program using a programming or macro language to implement the grid search method. Design the program so that it is expressly designed to locate a maximum. Test it with the same problem as in Example 14.1.

14.11 Develop a one-dimensional equation in the pressure gradient direction at the point (4, 2). The pressure function is

$$f(x, y) = 6x^2y - 9y^2 - 8x^2$$

14.12 A temperature function is

$$f(x, y) = 2x^3y^2 - 6xy - x^2 + 3y$$

Develop a one-dimensional function in the temperature gradient direction at the point (1, 1).

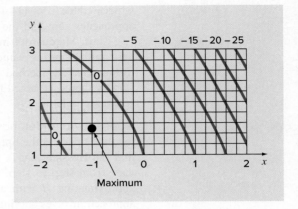

FIGURE P14.10
The grid search.

Constrained Optimization

This chapter deals with optimization problems where constraints come into play. We first discuss problems where both the objective function and the constraints are linear. For such cases, special methods are available that exploit the linearity of the underlying functions. Called linear programming methods, the resulting algorithms solve very large problems with thousands of variables and constraints with great efficiency. They are used in a wide range of problems in engineering and management.

Then we will turn briefly to the more general problem of nonlinear constrained optimization. Finally, we provide an overview of how software packages can be employed for optimization.

15.1 LINEAR PROGRAMMING

Linear programming (LP) is an optimization approach that deals with meeting a desired objective such as maximizing profit or minimizing cost in the presence of constraints such as limited resources. The term *linear* connotes that the mathematical functions representing both the objective and the constraints are linear. The term *programming* does not mean "computer programming," but rather connotes "scheduling" or "setting an agenda" (Revelle et al. 1997).

15.1.1 Standard Form

The basic linear programming problem consists of two major parts: the objective function and a set of constraints. For a maximization problem, the objective function is generally expressed as

$$\text{Maximize } Z = c_1x_1 + c_2x_2 + \cdots + c_nx_n \tag{15.1}$$

where c_j = payoff of each unit of the jth activity that is undertaken and x_j = magnitude of the jth activity. Thus, the value of the objective function, Z, is the total payoff due to the total number of activities, n.

The constraints can be represented generally as

$$a_{i1}x_1 + a_{i2}x_2 + \cdots + a_{in}x_n \leq b_i \tag{15.2}$$

where a_{ij} = amount of the ith resource that is consumed for each unit of the jth activity and b_i = amount of the ith resource that is available. That is, the resources are limited.

The second general type of constraint specifies that all activities must have a positive value,

$$x_i \geq 0 \tag{15.3}$$

In the present context, this expresses the realistic notion that, for some problems, negative activity is physically impossible (for example, we cannot produce negative goods).

Together, the objective function and the constraints specify the linear programming problem. They say that we are trying to maximize the payoff for a number of activities under the constraint that these activities utilize finite amounts of resources. Before showing how this result can be obtained, we will first develop an example.

| **EXAMPLE 15.1** | ## Setting Up the LP Problem |

Problem Statement. The following problem is developed from the area of chemical or petroleum engineering. However, it is relevant to all areas of engineering that deal with producing products with limited resources.

Suppose that a gas-processing plant receives a fixed amount of raw gas each week. The raw gas is processed into two grades of heating gas, regular and premium quality. These grades of gas are in high demand (that is, they are guaranteed to sell) and yield different profits to the company. However, their production involves both time and on-site storage constraints. For example, only one of the grades can be produced at a time, and the facility is open for only 80 hr/week. Further, there is limited on-site storage for each of the products. All these factors are listed below (note that a metric ton, or *tonne,* is equal to 1000 kg):

	Product		
Resource	**Regular**	**Premium**	**Resource Availability**
Raw gas	7 m³/tonne	11 m³/tonne	77 m³/week
Production time	10 hr/tonne	8 hr/tonne	80 hr/week
Storage	9 tonnes	6 tonnes	
Profit	$150/tonne	$175/tonne	

Develop a linear programming formulation to maximize the profits for this operation.

Solution. The engineer operating this plant must decide how much of each gas to produce to maximize profits. If the amounts of regular and premium produced weekly are designated as x_1 and x_2, respectively, the total weekly profit can be calculated as

$$\text{Total profit} = 150x_1 + 175x_2$$

or written as a linear programming objective function,

$$\text{Maximize } Z = 150x_1 + 175x_2$$

The constraints can be developed in a similar fashion. For example, the total raw gas used can be computed as

Total gas used $= 7x_1 + 11x_2$

This total cannot exceed the available supply of 77 m^3/week, so the constraint can be represented as

$7x_1 + 11x_2 \leq 77$

The remaining constraints can be developed in a similar fashion, with the resulting total LP formulation given by

Maximize $Z = 150x_1 + 175x_2$ (maximize profit)

subject to

$7x_1 + 11x_2 \leq 77$	(material constraint)
$10x_1 + 8x_2 \leq 80$	(time constraint)
$x_1 \leq 9$	("regular" storage constraint)
$x_2 \leq 6$	("premium" storage constraint)
$x_1, x_2 \geq 0$	(positivity constraints)

Note that the above set of equations constitute the total LP formulation. The parenthetical explanations at the right have been appended to clarify the meaning of each equation.

15.1.2 Graphical Solution

Because they are limited to two or three dimensions, graphical solutions have limited practical utility. However, they are very useful for demonstrating some basic concepts that underlie the general algebraic techniques used to solve higher-dimensional problems with the computer.

For a two-dimensional problem, such as the one in Example 15.1, the solution space is defined as a plane with x_1 measured along the abscissa and x_2 along the ordinate. Because they are linear, the constraints can be plotted on this plane as straight lines. If the LP problem was formulated properly (that is, it has a solution), these constraint lines will delineate a region, called the *feasible solution space,* encompassing all possible combinations of x_1 and x_2 that obey the constraints and hence represent feasible solutions. The objective function for a particular value of Z can then be plotted as another straight line and superimposed on this space. The value of Z can then be adjusted until it is at the maximum value while still touching the feasible space. This value of Z represents the optimal solution. The corresponding values of x_1 and x_2, where Z touches the feasible solution space, represent the optimal values for the activities. The following example should help clarify the approach.

EXAMPLE 15.2   **Graphical Solution**

Problem Statement. Develop a graphical solution for the gas-processing problem previously derived in Example 15.1:

Maximize $Z = 150x_1 + 175x_2$

subject to

$$7x_1 + 11x_2 \leq 77 \tag{1}$$
$$10x_1 + 8x_2 \leq 80 \tag{2}$$
$$x_1 \leq 9 \tag{3}$$
$$x_2 \leq 6 \tag{4}$$
$$x_1 \geq 0 \tag{5}$$
$$x_2 \geq 0 \tag{6}$$

We have numbered the constraints to identify them in the following graphical solution.

Solution.   First, the constraints can be plotted on the solution space. For example, the first constraint can be reformulated as a line by replacing the inequality by an equal sign and solving for x_2:

$$x_2 = -\frac{7}{11}x_1 + 7$$

Thus, as in Fig. 15.1a, the possible values of x_1 and x_2 that obey this constraint fall below this line (the direction designated in the plot by the small arrow). The other constraints can be evaluated similarly, as superimposed on Fig. 15.1a. Notice how they encompass a region where they are all met. This is the feasible solution space (the area $ABCDE$ in the plot).

Aside from defining the feasible space, Fig. 15.1a also provides additional insight. In particular, we can see that constraint 3 (storage of regular gas) is "redundant." That is, the feasible solution space is unaffected if it were deleted.

FIGURE 15.1
Graphical solution of a linear programming problem. (a) The constraints define a feasible solution space. (b) The objective function can be increased until it reaches the highest value that obeys all constraints. Graphically, the function moves up and to the right until it touches the feasible space at a single optimal point.

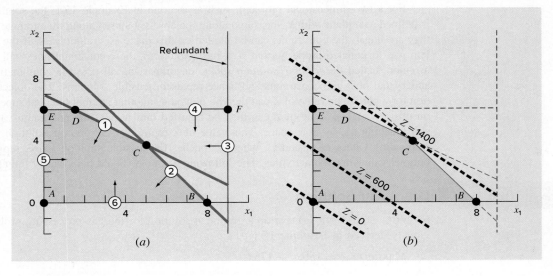

Next, the objective function can be added to the plot. To do this, a value of Z must be chosen. For example, for $Z = 0$, the objective function becomes

$$0 = 150x_1 + 175x_2$$

or, solving for x_2, we derive the line

$$x_2 = -\frac{150}{175}x_1$$

As displayed in Fig. 15.1*b*, this represents a dashed line intersecting the origin. Now, since we are interested in maximizing Z, we can increase it to say, 600, and the objective function is

$$x_2 = \frac{600}{175} - \frac{150}{175}x_1$$

Thus, increasing the value of the objective function moves the line away from the origin. Because the line still falls within the solution space, our result is still feasible. For the same reason, however, there is still room for improvement. Hence, Z can keep increasing until a further increase will take the value of the function beyond the feasible region. As shown in Fig. 15.1*b*, the maximum value of Z corresponds to approximately 1400. At this point, x_1 and x_2 are equal to approximately 4.9 and 3.9, respectively. Thus, the graphical solution tells us that if we produce these quantities of regular and premium, we will reap a maximum profit of about $1400.

Aside from determining optimal values, the graphical approach provides further insights into the problem. This can be appreciated by substituting the answers back into the constraint equations,

$$7(4.9) + 11(3.9) \cong 77$$
$$10(4.9) + 8(3.9) \cong 80$$
$$4.9 \leq 9$$
$$3.9 \leq 6$$

Consequently, as is also clear from the plot, producing the optimal amount of each product brings us right to the point where we just meet the resource (1) and time constraints (2). Such constraints are said to be *binding*. Further, as is also evident graphically, neither of the storage constraints [(3) and (4)] acts as a limitation. Such constraints are called *nonbinding*. This leads to the practical conclusion that, for this case, we can increase profits by either increasing our resource supply (the raw gas) or increasing our production time. Further, it indicates that increasing storage would have no impact on profit.

The result obtained in the previous example is one of four possible outcomes that can be generally obtained in a linear programming problem. These are

1. *Unique solution.* As in the example, the maximum objective function intersects a single point.
2. *Alternate solutions.* Suppose that the objective function in the example had coefficients so that it was precisely parallel to one of the constraints. In our example problem,

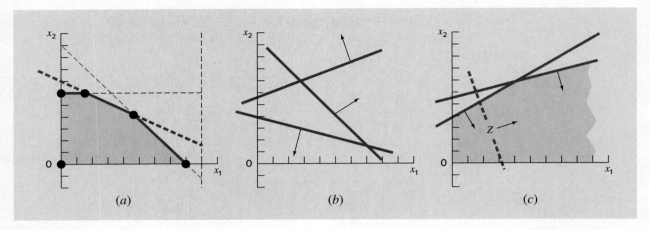

FIGURE 15.2
Aside from a single optimal solution (for example, Fig. 15.1b), there are three other possible outcomes of a linear programming problem: (a) alternative optima, (b) no feasible solution, and (c) an unbounded result.

one way in which this would occur would be if the profits were changed to $140/ tonne and $220/tonne. Then, rather than a single point, the problem would have an infinite number of optima corresponding to a line segment (Fig. 15.2a).

3. *No feasible solution.* As in Fig. 15.2b, it is possible that the problem is set up so that there is no feasible solution. This can be due to dealing with an unsolvable problem or due to errors in setting up the problem. The latter can result if the problem is over-constrained to the point that no solution can satisfy all the constraints.

4. *Unbounded problems.* As in Fig. 15.2c, this usually means that the problem is under-constrained and therefore open-ended. As with the no-feasible-solution case, it can often arise from errors committed during problem specification.

Now let us suppose that our problem involves a unique solution. The graphical approach might suggest an enumerative strategy for hunting down the maximum. From Fig. 15.1, it should be clear that the optimum always occurs at one of the corner points where two constraints meet. Such a point is known formally as an *extreme point*. Thus, out of the infinite number of possibilities in the decision space, focusing on extreme points clearly narrows down the possible options.

Further, we can recognize that not every extreme point is feasible, that is, satisfying all constraints. For example, notice that point F in Fig. 15.1a is an extreme point but is not feasible. Limiting ourselves to *feasible extreme points* narrows the field down still further.

Finally, once all feasible extreme points are identified, the one yielding the best value of the objective function represents the optimum solution. Finding this optimal solution could be done by exhaustively (and inefficiently) evaluating the value of the objective function at every feasible extreme point. The following section discusses the simplex method, which offers a preferable strategy that charts a selective course through a sequence of feasible extreme points to arrive at the optimum in an extremely efficient manner.

15.1.3 The Simplex Method

The simplex method is predicated on the assumption that the optimal solution will be an extreme point. Thus, the approach must be able to discern whether during problem solution an extreme point occurs. To do this, the constraint equations are reformulated as equalities by introducing what are called slack variables.

Slack Variables.   As the name implies, a *slack variable* measures how much of a constrained resource is available, that is, how much "slack" of the resource is available. For example, recall the resource constraint used in Examples 15.1 and 15.2,

$$7x_1 + 11x_2 \leq 77$$

We can define a slack variable S_1 as the amount of raw gas that is not used for a particular production level (x_1, x_2). If this quantity is added to the left side of the constraint, it makes the relationship exact,

$$7x_1 + 11x_2 + S_1 = 77$$

Now recognize what the slack variable tells us. If it is positive, it means that we have some "slack" for this constraint. That is, we have some surplus resource that is not being fully utilized. If it is negative, it tells us that we have exceeded the constraint. Finally, if it is zero, we exactly meet the constraint. That is, we have used up all the allowable resource. Since this is exactly the condition where constraint lines intersect, the slack variable provides a means to detect extreme points.

A different slack variable is developed for each constraint equation, resulting in what is called the *fully augmented version,*

$$\text{Maximize } Z = 150x_1 + 175x_2$$

subject to

$$7x_1 + 11x_2 + S_1 \qquad\qquad = 77 \qquad\qquad (15.4a)$$
$$10x_1 + 8x_2 \qquad + S_2 \qquad\quad = 80 \qquad\qquad (15.4b)$$
$$x_1 \qquad\qquad + S_3 \quad = 9 \qquad\qquad (15.4c)$$
$$x_2 \qquad\qquad + S_4 = 6 \qquad\qquad (15.4d)$$
$$x_1, \ x_2, \ S_1, \ S_2, \ S_3, \ S_4 \geq 0$$

Notice how we have set up the four equality equations so that the unknowns are aligned in columns. We did this to underscore that we are now dealing with a system of linear algebraic equations (recall Part Three). In the following section, we will show how these equations can be used to determine extreme points algebraically.

Algebraic Solution.   In contrast to Part Three, where we had n equations with n unknowns, our example system [Eqs. (15.4)] is *underspecified* or *underdetermined,* that is, it has more unknowns than equations. In general terms, there are n *structural variables* (that is, the original unknowns); m *surplus*, or *slack*, *variables* (one per constraint); and $n + m$ total variables (structural plus surplus). For the gas production problem, we have 2 structural variables, 4 slack variables, and 6 total variables. Thus, the problem involves solving 4 equations with 6 unknowns.

The difference between the number of unknowns and the number of equations (equal to 2 for our problem) is directly related to how we can distinguish a feasible extreme point. Specifically, every feasible extreme point has 2 variables out of 6 equal to zero. For example, the five corner points of the area *ABCDE* have the following zero values:

Extreme Point	Zero Variables
A	x_1, x_2
B	x_2, S_2
C	S_1, S_2
D	S_1, S_4
E	x_1, S_4

This observation leads to the conclusion that the extreme points can be determined from the standard form by setting two of the variables equal to zero. In our example, this reduces the problem to a solvable form of 4 equations with 4 unknowns. For example, for point *E*, setting $x_1 = S_4 = 0$ reduces the standard form to

$$11x_2 + S_1 \qquad\qquad\qquad\quad = 77$$
$$8x_2 \qquad\quad + S_2 \qquad\qquad = 80$$
$$\qquad\qquad\qquad + S_3 = 9$$
$$x_2 \qquad\qquad\qquad\qquad = 6$$

which can be solved for $x_2 = 6$, $S_1 = 11$, $S_2 = 32$, and $S_3 = 9$. Together with $x_1 = S_4 = 0$, these values define point *E*.

To generalize, a basic solution for *m* linear equations with *n* unknowns is developed by setting $n - m$ variables to zero, and solving the *m* equations for the *m* remaining unknowns. The zero variables are formally referred to as *nonbasic variables,* whereas the remaining *m* variables are called *basic variables.* If all the basic variables are nonnegative, the result is called a *basic feasible solution.* The optimum will be one of these.

Now a direct approach to determining the optimal solution would be to calculate all the basic solutions, determine which were feasible, and among those, which had the highest value of Z. There are two reasons why this is not a wise approach.

First, for even moderately sized problems, the approach can involve solving a great number of equations. For *m* equations with *n* unknowns, this approach requires solving

$$C_m^n = \frac{n!}{m!(n-m)!}$$

simultaneous equations. For example, if there are 10 equations ($m = 10$) with 16 unknowns ($n = 16$), you would have 8008 [$= 16!/(10! \ 6!)$] 10×10 systems of equations to solve!

Second, a significant portion of the solutions may be infeasible. For example, in the present problem, out of $C_4^6 = 15$ extreme points, only 5 are feasible. Clearly, if we could avoid solving all these unnecessary systems, a more efficient algorithm would be developed. Such an approach is described next.

Simplex Method Implementation. The simplex method avoids the inefficiencies outlined in the previous section. It does this by starting with a basic feasible solution. Then it moves through a sequence of other basic feasible solutions that successively improve the value of the objective function. Eventually, the optimal value is reached and the method is terminated.

We will illustrate the approach using the gas-processing problem from Examples 15.1 and 15.2. The first step is to start at a basic feasible solution (that is, at an extreme corner point of the feasible space). For cases like ours, an obvious starting point would be point A; that is, $x_1 = x_2 = 0$. The original 4 equations with 6 unknowns become

$$
\begin{aligned}
S_1 \qquad\qquad\qquad &= 77 \\
S_2 \qquad\qquad &= 80 \\
S_3 \qquad &= 9 \\
S_4 &= 6
\end{aligned}
$$

Thus, the starting values for the basic variables are given automatically as being equal to the right-hand sides of the constraints.

Before proceeding to the next step, the beginning information can now be summarized in a convenient tabular format called a *tableau.* As shown below, the tableau provides a concise summary of the key information constituting the linear programming problem.

Basic	Z	x_1	x_2	S_1	S_2	S_3	S_4	Solution	Intercept
Z	1	−150	−175	0	0	0	0	0	
S_1	0	7	11	1	0	0	0	77	11
S_2	0	10	8	0	1	0	0	80	8
S_3	0	1	0	0	0	1	0	9	9
S_4	0	0	1	0	0	0	1	6	∞

Notice that for the purposes of the tableau, the objective function is expressed as

$$ Z - 150x_1 - 175x_2 - 0S_1 - 0S_2 - 0S_3 - 0S_4 = 0 \tag{15.5} $$

The next step involves moving to a new basic feasible solution that leads to an improvement of the objective function. This is accomplished by increasing a current nonbasic variable (at this point, x_1 or x_2) above zero so that Z increases. Recall that, for the present example, extreme points must have two zero values. Therefore, one of the current basic variables (S_1, S_2, S_3, or S_4) must also be set to zero.

To summarize this important step: One of the current nonbasic variables must be made basic (nonzero). This variable is called the *entering variable.* In the process, one of the current basic variables is made nonbasic (zero). This variable is called the *leaving variable.*

Now, let us develop a mathematical approach for choosing the entering and leaving variables. Because of the convention by which the objective function is written [(Eq. (15.5)], the entering variable can be any variable in the objective function having a negative coefficient (because this will make Z bigger). The variable with the largest negative value is conventionally chosen because it usually leads to the largest increase

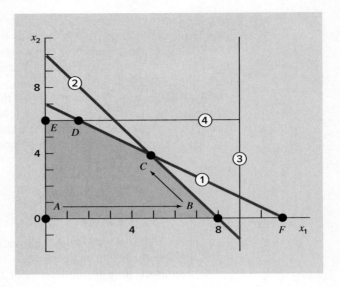

FIGURE 15.3
Graphical depiction of how the simplex method successively moves through feasible basic solutions to arrive at the optimum in an efficient manner.

in Z. For our case, x_2 would be the entering variable since its coefficient, -175, is more negative than the coefficient of x_1, -150.

At this point the graphical solution can be consulted for insight. As in Fig. 15.3, we start at the initial point A. Based on its coefficient, x_2 should be chosen to enter. However, to keep the present example brief, we choose x_1 since we can see from the graph that this will bring us to the maximum quicker.

Next, we must choose the leaving variable from among the current basic variables— S_1, S_2, S_3, or S_4. Graphically, we can see that there are two possibilities. Moving to point B will drive S_2 to zero, whereas moving to point F will drive S_1 to zero. However, the graph also makes it clear that F is not possible because it lies outside the feasible solution space. Thus, we decide to move from A to B.

How is the same result detected mathematically? One way is to calculate the values at which the constraint lines intersect the axis or line corresponding to the entering variable (in our case, the x_1 axis). We can calculate this value as the ratio of the right-hand side of the constraint (the "Solution" column of the tableau) to the corresponding coefficient of x_1. For example, for the first constraint's slack variable, S_1, the result is

$$\text{Intercept} = \frac{77}{7} = 11$$

The remaining intercepts can be calculated and listed as the last column of the tableau. Because 8 is the smallest positive intercept, it means that the second constraint line will be reached first as x_1 is increased. Hence, S_2 should be the leaving variable.

At this point, we have moved to point B ($x_2 = S_2 = 0$), and the new basic solution becomes

$$7x_1 + S_1 = 77$$
$$10x_1 = 80$$
$$x_1 + S_3 = 9$$
$$S_4 = 6$$

The solution of this system of equations effectively defines the values of the basic variables at point B: $x_1 = 8$, $S_1 = 21$, $S_3 = 1$, and $S_4 = 6$.

The tableau can be used to make the same calculation by employing the Gauss-Jordan method. Recall that the basic strategy behind Gauss-Jordan involved converting the pivot element to 1 and then eliminating the coefficients in the same column above and below the pivot element (recall Sec. 9.7).

For this example, the pivot row is S_2 (the leaving variable) and the pivot element is 10 (the coefficient of the entering variable, x_1). Dividing the row by 10 and replacing S_2 by x_1 gives

Basic	Z	x_1	x_2	S_1	S_2	S_3	S_4	Solution	Intercept
Z	1	−150	−175	0	0	0	0	0	
S_1	0	7	11	1	0	0	0	77	
x_1	0	1	0.8	0	0.1	0	0	8	
S_3	0	1	0	0	0	1	0	9	
S_4	0	0	1	0	0	0	1	6	

Next, the x_1 coefficients in the other rows can be eliminated. For example, for the objective function row, the pivot row is multiplied by −150 and the result subtracted from the first row to give

Z	x_1	x_2	S_1	S_2	S_3	S_4	Solution
1	−150	−175	0	0	0	0	0
−0	−(−150)	−(−120)	−0	−(−15)	0	0	−(−1200)
1	0	−55	0	15	0	0	1200

Similar operations can be performed on the remaining rows to give the new tableau:

Basic	Z	x_1	x_2	S_1	S_2	S_3	S_4	Solution	Intercept
Z	1	0	−55	0	15	0	0	1200	
S_1	0	0	5.4	1	−0.7	0	0	21	3.889
x_1	0	1	0.8	0	0.1	0	0	8	10
S_3	0	0	−0.8	0	−0.1	1	0	1	−1.25
S_4	0	0	1	0	0	0	1	6	6

Thus, the new tableau summarizes all the information for point B. This includes the fact that the move has increased the objective function to $Z = 1200$.

This tableau can then be used to chart our next, and in this case final, step. Only one more variable, x_2, has a negative value in the objective function, and it is therefore chosen as the entering variable. According to the intercept values (now calculated as the solution column over the coefficients in the x_2 column), the first constraint has the smallest positive value, and therefore, S_1 is selected as the leaving variable. Thus, the simplex method moves us from points B to C in Fig. 15.3. Finally, Gauss-Jordan elimination (recall Sec. 9.7) can be implemented to solve the simultaneous equations. The result is the final tableau:

Basic	Z	x_1	x_2	S_1	S_2	S_3	S_4	Solution
Z	1	0	0	10.1852	7.8704	0	0	1413.889
x_2	0	0	1	0.1852	−0.1296	0	0	3.889
x_1	0	1	0	−0.1481	0.2037	0	0	4.889
S_3	0	0	0	0.1481	−0.2037	1	0	4.111
S_4	0	0	0	−0.1852	0.1296	0	1	2.111

We know that the result is final because there are no negative coefficients remaining in the objective function row. The final solution is tabulated as $x_1 = 3.889$ and $x_2 = 4.889$, which give a maximum objective function value of $Z = 1413.889$. Further, because S_3 and S_4 are still in the basis, we know that the solution is limited by the first and second constraints.

15.2 NONLINEAR CONSTRAINED OPTIMIZATION

There are a number of approaches for handling nonlinear optimization problems in the presence of constraints. These can generally be divided into indirect and direct approaches (Rao 1996). A typical indirect approach uses so-called *penalty functions*. These involve including additional expressions to make the objective function less optimal as the solution approaches a constraint. Thus, the solution will be discouraged from violating constraints. Although such methods can be useful in some problems, they can become arduous when the problem involves many constraints.

The *generalized reduced gradient* (GRG) search method is one of the more popular of the direct methods (for details, see Fylstra et al. 1998; Lasdon et al. 1978; Lasdon and Smith 1992). It is, in fact, the nonlinear method used within the Excel Solver.

It first "reduces" the problem to an unconstrained optimization problem. It does this by solving a set of nonlinear equations for the basic variables in terms of the nonbasic variables. Then, the unconstrained problem is solved using approaches similar to those described in Chap. 14. First, a search direction is chosen along which an improvement in the objective function is sought. The default choice is a *quasi-Newton approach* (BFGS) that, as described in Chap. 14, requires storage of an approximation of the Hessian matrix. This approach performs very well for most cases. The *conjugate gradient method* is also available in Excel as an alternative for large problems. The Excel Solver has the nice feature that it automatically switches to the conjugate gradient method, depending on available storage. Once the search direction is established, a one-dimensional search is carried out along that direction using a variable step-size approach.

15.3   OPTIMIZATION WITH SOFTWARE PACKAGES

Software packages have great capabilities for optimization. In this section, we will give you an introduction to some of the more useful ones.

15.3.1 Excel for Linear Programming

There are a variety of software packages expressly designed to implement linear programming. However, because of its broad availability, we will focus on the Excel spreadsheet. This involves using the Solver option previously employed in Chap. 7 for root location.

   The manner in which Solver is used for linear programming is similar to our previous applications in that the data are entered into spreadsheet cells. The basic strategy is to arrive at a single cell that is to be optimized as a function of variations of other cells on the spreadsheet. The following example illustrates how this can be done for the gas-processing problem.

EXAMPLE 15.3

Using Excel's Solver for a Linear Programming Problem

Problem Statement.   Use Excel to solve the gas-processing problem we have been examining in this chapter.

Solution.   An Excel worksheet set up to calculate the pertinent values in the gas-processing problem is shown in Fig. 15.4. The unshaded cells are those containing numeric and labeling data. The shaded cells will hold quantities that are calculated based on other cells. Recognize that the cell to be maximized is D12, which contains the total profit. The cells to be varied are B4:C4, which hold the amounts of regular and premium gas produced.

FIGURE 15.4

Excel spreadsheet set up to use the Solver for linear programming.

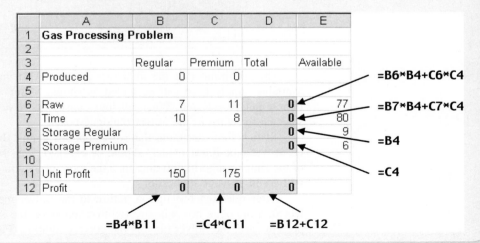

SOFTWARE

Once the spreadsheet is created, **Solver** is chosen from the Data tab (recall Sec. 7.7.1). At this point a dialogue box will be displayed, querying you for pertinent information. The pertinent cells of the Solver dialogue box are filled out as

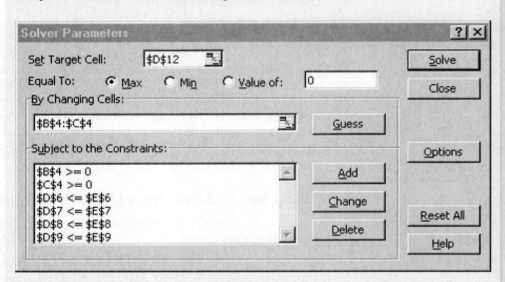

The constraints must be added one by one by selecting the Add button. This will open up a dialogue box that looks like

The constraint that the total raw gas (cell D6) must be less than or equal to the available supply (E6) can be added as shown. After adding each constraint, the Add button can be selected. When all four constraints have been entered, the OK button is selected to return to the Solver dialogue box.

Now, before execution, the Solver options button should be selected and the box labeled "Assume linear model" should be checked. This will make Excel employ a version of the simplex algorithm (rather than the more general nonlinear solver it usually uses) that will speed up your application.

After selecting this option, return to the Solver menu. When the OK button is selected, a dialogue box will open with a report on the success of the operation. For the present case, the Solver obtains the correct solution (Fig. 15.5)

	A	B	C	D	E
1	**Gas Processing Problem**				
2					
3		Regular	Premium	Total	Available
4	Produced	**4.888889**	**3.888889**		
5					
6	Raw	7	11	77	77
7	Time	10	8	80	80
8	Storage Regular			4.888889	9
9	Storage Premium			3.888889	6
10					
11	Unit Profit	150	175		
12	Profit	733.3333	680.5556	**1413.889**	

FIGURE 15.5
Excel spreadsheet showing solution to linear programming problem.

Beyond obtaining the solution, the Solver also provides some useful summary reports. We will explore these in the engineering applications described in Chap. 16.

15.3.2 Excel for Nonlinear Optimization

The manner in which Solver is used for nonlinear optimization is similar to our previous applications in that the data are entered into spreadsheet cells. Once again, the basic strategy is to arrive at a single cell that is to be optimized as a function of variations of other cells on the spreadsheet. The following example illustrates how this can be done for the parachutist problem we set up in the introduction to this part of the book (recall Example PT4.1).

EXAMPLE 15.4

Using Excel's Solver for Nonlinear Constrained Optimization

Problem Statement. Recall from Example PT4.1 that we developed a nonlinear constrained optimization to minimize the cost for a parachute drop into a refugee camp. Parameters for this problem are

Parameter	Symbol	Value	Unit
Total mass	M_t	2000	kg
Acceleration due to gravity	g	9.8	m/s^2
Cost coefficient (constant)	c_0	200	$
Cost coefficient (length)	c_1	56	$/m
Cost coefficient (area)	c_2	0.1	$/m^2$
Critical impact velocity	v_c	20	m/s
Area effect on drag	k_c	3	$kg/(s\ m^2)$
Initial drop height	z_0	500	m

SOFTWARE

Substituting these values into Eqs. (PT4.11) through (PT4.19) gives

Minimize $C = n(200 + 56\ell + 0.1A^2)$

subject to

$$v \leq 20$$
$$n \geq 1$$

where n is an integer and all other variables are real. In addition, the following quantities are defined:

$$A = 2\pi r^2$$
$$\ell = \sqrt{2}r$$
$$c = 3A$$
$$m = \frac{M_t}{n} \qquad \text{(E15.4.1)}$$
$$t = \text{root}\left[500 - \frac{9.8m}{c}t + \frac{9.8m^2}{c^2}\left(1 - e^{-(c/m)t}\right)\right] \qquad \text{(E15.4.2)}$$
$$v = \frac{9.8m}{c}\left(1 - e^{-(c/m)t}\right)$$

Use Excel to solve this problem for the design variables r and n that minimize cost C.

Solution. Before implementation of this problem on Excel, we must first deal with the problem of determining the root in the above formulation [Eq. (E15.4.2)]. One method might be to develop a macro to implement a root-location method such as bisection or the secant method. (Note that we will illustrate how this is done in the next chapter in Sec. 16.3.)

For the time being, an easier approach is possible by developing the following fixed-point iteration solution to Eq. (E15.4.2):

$$t_{i+1} = \left[500 + \frac{9.8m^2}{c^2}\left(1 - e^{-(c/m)t_i}\right)\right]\frac{c}{9.8m} \qquad \text{(E15.4.3)}$$

Thus, t can be adjusted until Eq. (E15.4.3) is satisfied. It can be shown that for the range of parameters used in the present problem, this formula always converges.

Now, how can this equation be solved on a spreadsheet? As shown below, two cells can be set up to hold a value for t and for the right-hand side of Eq. (E15.4.3) [that is, $f(t)$].

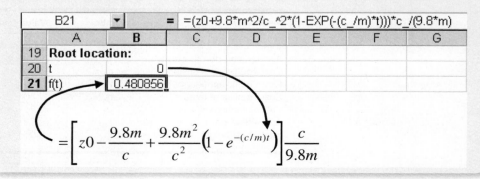

You can type Eq. (E15.4.3) into cell B21 so that it gets its time value from cell B20 and the other parameter values from cells elsewhere on the sheet (see below for how we set up the whole sheet). Then go to cell B20 and point its value to cell B21.

Once you enter these formulations, you will immediately get the error message: "Cannot resolve circular references" because B20 depends on B21 and vice versa. Now, go to the Tools/Options selections from the menu and select **Calculation.** From the calculation dialogue box, check off "iteration" and hit "OK." Immediately the spreadsheet will iterate these cells and the result will come out as

	A	B
19	**Root location:**	
20	t	10.2551
21	f(t)	10.25595

Thus, the cells will converge on the root. If you want to make it more precise, just strike the F9 key to make it iterate some more (the default is 100 iterations, which you can change if you wish).

An Excel worksheet to calculate the pertinent values can then be set up as shown in Fig. 15.6. The unshaded cells are those containing numeric and labeling data. The shaded

FIGURE 15.6
Excel spreadsheet set up for the nonlinear parachute optimization problem.

	A	B	C	D	E	F	G
1	**Parachute Optimization Problem**						
2							
3	**Parameters:**			**Design variables:**			
4	Mt	2000		r	1		
5	g	9.8		n	1		
6	cost1	200					
7	cost2	56		**Constraints:**			
8	cost3	0.1					
9	vc	20		**variable**		**type**	**limit**
10	kc	3		v	95.8786	<=	20
11	z0	500		n	1	>=	1
12							
13	**Calculated values:**			**Objective function:**			
14	A	6.283185					
15	l	1.414214		Cost	283.1438		
16	c	18.84956					
17	m	2000					
18							
19	**Root location:**						
20	t	10.26439					
21	f(t)	10.26439					

cells contain quantities that are calculated based on other cells. For example, the mass in B17 was computed with Eq. (E15.4.1) based on the values for M_t (B4) and n (E5). Note also that some cells are redundant. For example, cell E11 points back to cell E5. The information is repeated in cell E11 so that the structure of the constraints is evident from the sheet. Finally, recognize that the cell to be minimized is E15, which contains the total cost. The cells to be varied are E4:E5, which hold the radius and the number of parachutes.

Once the spreadsheet is created, Solver is chosen from the Data tab. At this point a dialogue box will be displayed, querying you for pertinent information. The pertinent cells of the Solver dialogue box would be filled out as

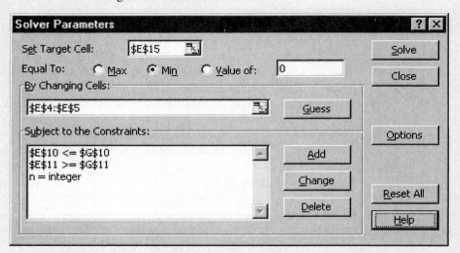

The constraints must be added one by one by selecting the Add button. This will open up a dialogue box that looks like

As shown, the constraint that the actual impact velocity (cell E10) must be less than or equal to the required velocity (G10) can be added. After adding each constraint, the Add button can be selected. Note that the down arrow allows you to choose among several types of constraints ($<=$, $>=$, $=$, and integer). Thus, we can force the number of parachutes (E5) to be an integer.

When all three constraints have been entered, the "OK" button is selected to return to the Solver Parameter box. When the "OK" button is selected, the Solver Results box will open with a report on the success of the operation. For the present case, the Solver obtains the correct solution as in Fig. 15.7.

	A	B	C	D	E	F	G
1	**Parachute Optimization Problem**						
2							
3	**Parameters:**			**Design variables:**			
4	Mt	2000		r	2.943652		
5	g	9.8		n	6		
6	cost1	200					
7	cost2	56		**Constraints:**			
8	cost3	0.1					
9	vc	20		**variable**		**type**	**limit**
10	kc	3		v	20	<=	20
11	z0	500		n	6	>=	1
12							
13	**Calculated values:**			**Objective function:**			
14	A	54.44435					
15	I	4.162953		Cost	**4377.264**		
16	c	163.333					
17	m	333.3333					
18							
19	**Root location:**						
20	t	27.04077					
21	f(t)	27.04077					

FIGURE 15.7
Excel spreadsheet showing the solution for the nonlinear parachute optimization problem.

Thus, we determine that the minimum cost of $4377.26 will occur if we break the load up into six parcels with a chute radius of 2.944 m. Beyond obtaining the solution, the Solver also provides some useful summary reports. We will explore these in the engineering applications described in Chap. 16.

15.3.3 MATLAB

As summarized in Table 15.1, MATLAB software has built-in functions to perform optimization. The following examples illustrates how they can be used.

TABLE 15.1 MATLAB functions to implement optimization.

Function	Description
fminbnd	Minimize function of one variable with bound constraints
fminsearch	Minimize function of several variables

EXAMPLE 15.5   **Using MATLAB for One-Dimensional Optimization**

Problem Statement.   Use the MATLAB `fminbnd` function to find the maximum of

$$f(x) = 2 \sin x - \frac{x^2}{2}$$

within the interval $x_l = 0$ and $x_u = 4$. Recall that in Chap. 13, we used several methods to solve this problem for $x = 1.4276$ and $f(x) = 1.7757$.

Solution.   First, we must create an M-file to hold the function.

```
function f=fx(x)
f = -(2*sin(x)-x^2/10)
```

Because we are interested in maximization, we enter the negative of the function. Then, we invoke the `fminbnd` function with

```
>> x=fminbnd('fx',0,4)
```

The result is

```
f =
   -1.7757

x =
     1.4275
```

Note that additional arguments can be included. One useful addition is to set optimization options such as error tolerance or maximum iterations. This is done with the `optimset` function, which was used previously in Example 7.6 and has the general format,

```
optimset('param₁',value₁,'param₂',value₂,...)
```

where $param_i$ is a parameter specifying the type of option and $value_i$ is the value assigned to that option. For example, if you wanted to set the tolerance at 1×10^{-2},

```
optimset('TolX',1e-2)
```

Thus, solving the present problem to a tolerance of 1×10^{-2} can be generated with

```
>> fminbnd('fx',0,4,optimset('TolX',1e-2))
```

with the result

```
f =
   -1.7757

ans =
     1.4270
```

A complete set of parameters can be found by invoking Help as in

```
>> Help optimset
```

MATLAB has a variety of capabilities for dealing with multidimensional functions. Recall from Chap. 13 that our visual image of a one-dimensional search was like a roller coaster. For two-dimensional cases, the image becomes that of mountains and valleys. As in the following example, MATLAB's graphic capabilities provide a handy means to visualize such functions.

EXAMPLE 15.6

Visualizing a Two-Dimensional Function

Problem Statement.   Use MATLAB's graphical capabilities to display the following function and visually estimate its minimum in the range $-2 \le x_1 \le 0$ and $0 \le x_2 \le 3$:

$$f(x_1, x_2) = 2 + x_1 - x_2 + 2x_1^2 + 2x_1x_2 + x_2^2$$

Solution.   The following script generates contour and mesh plots of the function:

```
x=linspace(-2,0,40);y=linspace(0,3,40);
[X,Y] = meshgrid(x,y);
Z=2+X-Y+2*X.^2+2*X.*Y+Y.^2;
subplot(1,2,1);
cs=contour(X,Y,Z);clabel(cs);
xlabel('x_1');ylabel('x_2');
title('(a) Contour plot');grid;
subplot(1,2,2);
cs=surfc(X,Y,Z);
zmin=floor(min(Z));
zmax=ceil(max(Z));
xlabel('x_1');ylabel('x_2');zlabel('f(x_1,x_2)');
title('(b) Mesh plot');
```

As displayed in Fig. 15.8, both plots indicate that the function has a minimum value of about $f(x_1, x_2) = 0$ to 1 located at about $x_1 = -1$ and $x_2 = 1.5$.

Standard MATLAB has a function fminsearch that can be used to determine the minimum of a multidimensional function. It is based on the Nelder-Mead method, which is a direct-search method that uses only function values (does not require derivatives) and handles nonsmooth objective functions. A simple expression of its syntax is

```
[xmin, fval] = fminsearch(function,x1,x2)
```

where *xmin* and *fval* are the location and value of the minimum, *function* is the name of the function being evaluated, and *x1* and *x2* are the bounds of the interval being searched.

FIGURE 15.8
(a) Contour and (b) mesh plots of a two-dimensional function.

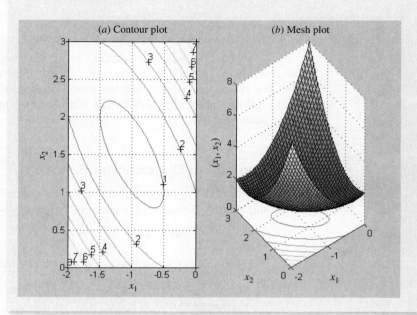

EXAMPLE 15.7 Using MATLAB for Multidimensional Optimization

Problem Statement.   Use the MATLAB fminsearch function to find the maximum for the simple function we just graphed in Example 15.6.

$$f(x_1, x_2) = 2 + x_1 - x_2 + 2x_1^2 + 2x_1x_2 + x_2^2$$

Employ initial guesses of $x_1 = -0.5$ and $x_2 = 0.5$.

Solution.   We can invoke the fminsearch function with

```
>> f=@(x) 2+x(1)-x(2)+2*x(1)^2+2*x(1)*x(2)+x(2)^2;
>> [x,fval]=fminsearch(f,[-0.5,0.5])

x =
    -1.0000     1.5000
fval =
     0.7500
```

Just as with fminbnd, arguments can be included in order to specify additional parameters of the optimization process. For example, the optimset function can be used to limit the maximum number of iterations

```
>> [x,fval]=fminsearch(f,[-0.5,0.5],optimset('MaxIter',2))
```

with the result

```
Exiting: Maximum number of iterations has been exceeded
        - increase MaxIter option.
        Current function value: 1.225625
x =
   -0.5000    0.5250
fval =
    1.2256
```

Thus, because we have set a very stringent limit on the iterations, the optimization terminates well before the maximum is reached.

15.3.4 Mathcad

Mathcad contains a numeric mode function called **Find** that can be used to solve up to 50 simultaneous nonlinear algebraic equations with inequality constraints. The use of this function for unconstrained applications was described in Part Two. If **Find** fails to locate a solution that satisfies the equations and constraints, it returns the error message "Did not find solution." However, Mathcad also contains a similar function called **Minerr**. This function gives solution results that minimize the errors in the constraints even when exact solutions cannot be found. This function solves equations and accommodates several constraints using the Levenberg-Marquardt method taken from the public-domain MINPACK algorithms developed and published by the Argonne National Laboratory.

Let's develop an example where **Find** is used to solve a system of nonlinear equations with constraints. Initial guesses of $x = -1$ and $y = 1$ are input using the definition symbol as shown in Fig. 15.9. The word "Given" then alerts Mathcad that what follows is a system of equations. Then we can enter the equations and the inequality constraint. Note that for this application, Mathcad requires the use of a symbolic equal sign (typed as [Ctrl]=) and > to separate the left and right sides of an equation. Now the vector consisting of xval and yval is computed using **Find**(x,y) and the values are shown using an equal sign.

A graph that displays the equations and constraints as well as the solution can be placed on the worksheet by clicking on the desired location. This places a red crosshair at that location. Then use the Insert/Graph/X-Y Plot pull-down menu to place an empty plot on the worksheet with placeholders for the expressions to be graphed and for the ranges of the x and y axes. Four variables are plotted on the y axis as shown: the top and bottom halves of the equation for the circle, the linear function, and a vertical line to represent the $x > 2$ constraint. In addition, the solution is included as a point. Once the graph has been created, you can use the Format/Graph/X-Y Plot pull-down menu to vary the type of graph; change the color, type, and weight of the trace of the function; and add titles, labels, and other features. The graph and the numerical values for xval and yval nicely portray the solution as the intersection of the circle and the line in the region where $x > 2$.

FIGURE 15.9
Mathcad screen for a nonlinear constrained optimization problem.

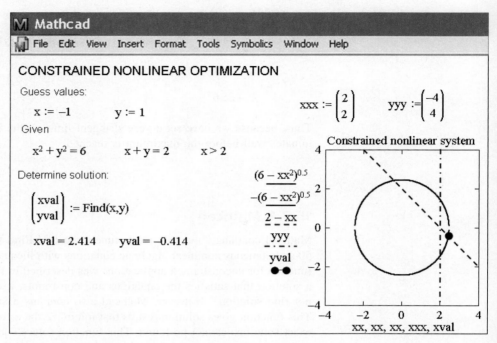

PROBLEMS

15.1 A company makes two types of products, A and B. These products are produced during a 40-hr work week and then shipped out at the end of the week. They require 20 and 5 kg of raw material per kg of product, respectively, and the company has access to 9500 kg of raw material per week. Only one product can be created at a time, with production times for each of 0.04 and 0.12 hr, respectively. The plant can only store 550 kg of total product per week. Finally, the company makes profits of $45 and $20 on each unit of A and B, respectively. Each unit of product is equivalent to 1 kg.
(a) Set up the linear programming problem to maximize profit.
(b) Solve the linear programming problem graphically.
(c) Solve the linear programming problem with the simplex method.
(d) Solve the problem with a software package.
(e) Evaluate which of the following options will raise profits the most: increasing raw material, storage, or production time.
15.2 Suppose that for Example 15.1, the gas-processing plant decides to produce a third grade of product with the following characteristics:

	Supreme
Raw gas	15 m^3/tonne
Production time	12 hr/tonne
Storage	5 tonnes
Profit	$250/tonne

In addition, suppose that a new source of raw gas has been discovered so that the total available is doubled, to 154 m^3/week.
(a) Set up the linear programming problem to maximize profit.
(b) Solve the linear programming problem with the simplex method.
(c) Solve the problem with a software package.
(d) Evaluate which of the following options will raise profits the most: increasing raw material, storage, or production time.
15.3 Consider the linear programming problem:

$$\text{Maximize } f(x, y) = 1.75x + 1.25y$$

subject to

$$1.2x + 2.25y \leq 14$$
$$x + 1.1y \leq 8$$
$$2.5x + y \leq 9$$
$$x \geq 0$$
$$y \geq 0$$

Obtain the solution:
(a) Graphically.
(b) Using the simplex method.
(c) Using an appropriate software package (for example, Excel, MATLAB, or Mathcad).

15.4 Consider the linear programming problem:

$$\text{Maximize } f(x, y) = 6x + 8y$$

subject to

$$5x + 2y \leq 40$$
$$6x + 6y \leq 60$$
$$2x + 4y \leq 32$$
$$x \geq 0$$
$$y \geq 0$$

Obtain the solution:
(a) Graphically.
(b) Using the simplex method.
(c) Using an appropriate software package (for example, Excel).

15.5 Use a software package (for example, Excel, MATLAB, Mathcad) to solve the following constrained nonlinear optimization problem:

$$\text{Maximize } f(x, y) = 1.2x + 2y - y^3$$

subject to

$$2x + y \leq 2$$
$$x \geq 0$$
$$y \geq 0$$

15.6 Use a software package (for example, Excel, MATLAB, or Mathcad) to solve the following constrained nonlinear optimization problem:

$$\text{Maximize } f(x, y) = 15x + 15y$$

subject to

$$x^2 + y^2 \leq 1$$
$$x + 2y \leq 2.1$$
$$x \geq 0$$
$$y \geq 0$$

15.7 Consider the following constrained nonlinear optimization problem:

$$\text{Minimize } f(x, y) = (x - 3)^2 + (y - 3)^2$$

subject to

$$x + 2y = 4$$

(a) Use a graphical approach to estimate the solution.
(b) Use a software package (for example, Excel) to obtain a more accurate estimate.

15.8 Use a software package to determine the maximum of

$$f(x, y) = 2xy + 1.5y - 1.25x^2 - 2y^2$$

15.9 Use a software package to determine the maximum of

$$f(x, y) = 3.5x + 2y + x^2 - x^4 - 3xy - 3y^2$$

15.10 Given the following function,

$$f(x, y) = -8x + x^2 + 12y + 4y^2 - 2xy$$

use a software package to determine the minimum:
(a) Graphically.
(b) Numerically.
(c) Substitute the result of **(b)** back into the function to determine the minimum $f(x, y)$.
(d) Determine the Hessian and its determinant, and substitute the result of part **(b)** back into the latter to verify that a minimum has been found.

15.11 You are asked to design a covered conical pit to store 50 m³ of waste liquid. Assume excavation costs at \$100/m³, side lining costs at \$50/m², and cover cost at \$25/m². Determine the dimensions of the pit that minimize cost **(a)** if the side slope is unconstrained and **(b)** if the side slope must be less than 45°.

15.12 An automobile company has two versions of the same model car for sale: a two-door coupe and the full-size four door.
(a) Graphically determine how many cars of each design should be produced to maximize profit and what that profit is.
(b) Solve the same problem with Excel.

	Two Door	Four Door	Availability
Profit	\$13,500/car	\$15,000/car	
Production time	15 h/car	20 h/car	8000 h/year
Storage	400 cars	350 cars	
Consumer demand	700/car	500/car	240,000 cars

15.13 Og is the leader of the surprisingly mathematically advanced, though technologically run-of-the-mill, Calm Waters caveman tribe. He must decide on the number of stone clubs and stone axes to be produced for the upcoming battle against the neighboring

Peaceful Sunset tribe. Experience has taught him that each club is good for, on the average, 0.45 kills and 0.65 maims, while each axe produces 0.70 kills and 0.35 maims. Production of a club requires 5.1 lb of stone and 2.1 man-hours of labor while an axe requires 3.2 lb of stone and 4.3 man-hours of labor. Og's tribe has 240 lb of stone available for weapons production, and a total of 200 man-hours of labor available before the expected time of this battle (that Og is sure will end war for all time). Og values a kill as worth two maims in quantifying the damage inflicted on the enemy, and he wishes to produce that mix of weapons that will maximize damage.

(a) Formulate this as a linear programming problem. Make sure to define your decision variables.

(b) Represent this problem graphically, making sure to identify all the feasible corner points and the infeasible corner points.

(c) Solve the problem graphically.

(d) Solve the problem using the computer.

15.14 Develop an M-file that is expressly designed to locate a maximum with the golden-section search algorithm. In other words, design the M-file so that it directly finds the maximum rather than finding the minimum of $-f(x)$. Test your program with the same problem as in Example 13.1. The function should have the following features:

- Iterate until the relative error falls below a stopping criterion or exceeds a maximum number of iterations.
- Return both the optimal x and $f(x)$.

15.15 Develop an M-file to locate a minimum with the golden-section search. Rather than using the standard stopping criteria (as in Fig. 13.5), determine the number of iterations needed to attain a desired tolerance.

15.16 Develop an M-file to implement parabolic interpolation to locate a minimum. Test your program with the same problem as in Example 13.2. The function should have the following features:

- Base it on two initial guesses, and have the program generate the third initial value at the midpoint of the interval.

- Check whether the guesses bracket a maximum. If not, the function should not implement the algorithm, but should return an error message.
- Iterate until the relative error falls below a stopping criterion or exceeds a maximum number of iterations.
- Return both the optimal x and $f(x)$.
- Use a bracketing approach (as in Example 13.2) to replace old values with new values.

15.17 The length of the longest ladder that can negotiate the corner depicted in Fig. P15.17 can be determined by computing the value of θ that minimizes the following function:

$$L(\theta) = \frac{w_1}{\sin \theta} + \frac{w_2}{\sin(\pi - \alpha - \theta)}$$

For the case where $w_1 = 1$ m and $w_2 = 2$ m, use a numerical method (including software) to develop a plot of L versus a range of α's from 45° to 135°.

FIGURE P15.17
A ladder negotiating a corner formed by two hallways.

Case Studies: Optimization

The purpose of this chapter is to use the numerical procedures discussed in Chaps. 13 through 15 to solve actual engineering problems involving optimization. These problems are important because engineers are often called upon to come up with the "best" solution to a problem. Because many of these cases involve complex systems and interactions, numerical methods and computers are often a necessity for developing optimal solutions.

The following applications are typical of those that are routinely encountered during upper-class and graduate studies. Furthermore, they are representative of problems you will address professionally. The problems are drawn from the major discipline areas of engineering: chemical/bio, civil/environmental, electrical, and mechanical/aerospace.

The first application, taken from *chemical/bio engineering,* deals with using nonlinear constrained optimization to design an optimal cylindrical tank. The Excel Solver is used to develop the solution.

Next, we use linear programming to assess a problem from *civil/environmental engineering:* minimizing the cost of waste treatment to meet water-quality objectives in a river. In this example, we introduce the notion of shadow prices and their use in assessing the sensitivity of a linear programming solution.

The third application, taken from *electrical engineering,* involves maximizing the power across a potentiometer in an electric circuit. The solution involves one-dimensional unconstrained optimization. Aside from solving the problem, we illustrate how the Visual Basic macro language allows access to the golden-section search algorithm within the context of the Excel environment.

Finally, the fourth application, taken from *mechanical/aerospace engineering,* involves determining the equilibrium position of a multispring system based on the minimum potential energy.

16.1 LEAST-COST DESIGN OF A TANK (CHEMICAL/BIO ENGINEERING)

Background.   Chemical engineers (as well as other specialists such as mechanical and civil engineers) often encounter the general problem of designing containers to transport liquids and gases. Suppose that you are asked to determine the dimensions of a small

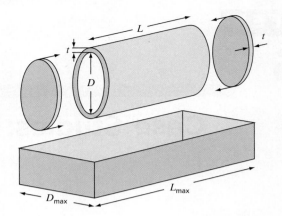

FIGURE 16.1
Parameters for determining the optimal dimensions of a cylindrical tank.

TABLE 16.1 Parameters for determining the optimal dimensions of a cylindrical tank used to transport toxic wastes.

Parameter	Symbol	Value	Units
Required volume	V_o	0.8	m³
Thickness	t	3	cm
Density	ρ	8000	kg/m³
Bed length	L_{max}	2	m
Bed width	D_{max}	1	m
Material cost	c_m	4.5	$/kg
Welding cost	c_w	20	$/m

cylindrical tank to transport toxic waste that is to be mounted on the back of a pickup truck. Your overall objective will be to minimize the cost of the tank. However, aside from cost, you must ensure that it holds the required amount of liquid and that it does not exceed the dimensions of the truck's bed. Note that because the tank will be carrying toxic waste, the tank thickness is specified by regulations.

A schematic of the tank and bed are shown in Fig. 16.1. As can be seen, the tank consists of a cylinder with two plates welded on each end.

The cost of the tank involves two components: (1) material expense, which is based on weight, and (2) welding expense based on length of weld. Note that the latter involves welding both the interior and the exterior seams where the plates connect with the cylinder. The data needed for the problem are summarized in Table 16.1.

Solution. The objective here is to construct a tank for a minimum cost. The cost is related to the design variables (length and diameter) as they affect the mass of the tank and the welding lengths. Further, the problem is constrained because the tank must (1) fit within the truck bed and (2) carry the required volume of material.

The cost consists of tank material and welding costs. Therefore, the objective function can be formulated as minimizing

$$C = c_m m + c_w \ell_w \tag{16.1}$$

where C = cost (\$), m = mass (kg), ℓ_w = weld length (m), and c_m and c_w = cost factors for mass (\$/kg) and weld length (\$/m), respectively.

Next, we will formulate how the mass and weld length are related to the dimensions of the drum. First, the mass can be calculated as the volume of material times its density. The volume of the material used to create the side walls (that is, the cylinder) can be computed as

$$V_{\text{cylinder}} = L\pi\left[\left(\frac{D}{2} + t\right)^2 - \left(\frac{D}{2}\right)^2\right]$$

For each circular end plate, it is

$$V_{\text{plate}} = \pi\left(\frac{D}{2} + t\right)^2 t$$

Thus, the mass is computed by

$$m = \rho\left\{L\pi\left[\left(\frac{D}{2} + t\right)^2 - \left(\frac{D}{2}\right)^2\right] + 2\pi\left(\frac{D}{2} + t\right)^2 t\right\} \tag{16.2}$$

where ρ = density (kg/m^3).

The weld length for attaching each plate is equal to the cylinder's inside and outside circumference. For the two plates, the total weld length would be

$$\ell_w = 2\left[2\pi\left(\frac{D}{2} + t\right) + 2\pi\frac{D}{2}\right] = 4\pi(D + t) \tag{16.3}$$

Given values for D and L (remember, thickness t is fixed by regulations), Eqs. (16.1) through (16.3) provide a means to compute cost. Also recognize that when Eqs. (16.2) and (16.3) are substituted into Eq. (16.1), the resulting objective function is nonlinear in the unknowns.

Next, we can formulate the constraints. First, we must compute how much volume can be held within the finished tank,

$$V = \frac{\pi D^2}{4} L$$

This value must be equal to the desired volume. Thus, one constraint is

$$\frac{\pi D^2 L}{4} = V_o$$

where V_o is the desired volume (m^3).

The remaining constraints deal with ensuring that the tank will fit within the dimensions of the truck bed,

$$L \leq L_{\text{max}} + 2t$$
$$D \leq D_{\text{max}} + 2t$$

The problem is now specified. Substituting the values from Table 16.1, it can be summarized as

$$\text{Maximize } C = 4.5m + 20\ell_w$$

subject to

$$\frac{\pi D^2 L}{4} = 0.8$$
$$L \le 2.06$$
$$D \le 1.06$$

where

$$m = 8000\left\{ L\pi\left[\left(\frac{D}{2} + 0.03\right)^2 - \left(\frac{D}{2}\right)^2 \right] + 2\pi\left(\frac{D}{2} + 0.03\right)^2 0.03 \right\}$$

and

$$\ell_w = 4\pi(D + 0.03)$$

The problem can now be solved in a number of ways. However, the simplest approach for a problem of this magnitude is to use a tool like the Excel Solver. The spreadsheet to accomplish this is shown in Fig. 16.2.

For the case shown, we enter the upper limits for D and L. For this case, the volume is more than required ($1.57 > 0.8$).

FIGURE 16.2
Excel spreadsheet set up to evaluate the cost of a tank subject to a volume requirement and size constraints.

	A	B	C	D	E	F	G
1	**Optimum tank design**						
2							
3	**Parameters:**			**Design variables:**			
4	V0	0.8		D	1		
5	t	0.03		L	2		
6	rho	8000					
7	Lmax	2		**Constraints:**			
8	Dmax	1		D	1	<=	1.06
9	cm	4.5		L	2	<=	2.06
10	cw	20		Vol	1.570796	=	0.8
11							
12	**Computed values:**			**Objective function:**			
13	m	1976.791		C	9154.425		
14	lw	12.94336					
15							
16	Vshell	0.19415					
17	Vends	0.052948					

Once the spreadsheet is created, Solver is chosen from the Data tab. At this point a dialogue box will be displayed, querying you for pertinent information. The pertinent cells of the Solver dialogue box would be filled out as

When the OK button is selected, a dialogue box will open with a report on the success of the operation. For the present case, the Solver obtains the correct solution, which is shown in Fig. 16.3. Notice that the optimal diameter is nudging up against the constraint of 1.06 m. Thus, if the required capacity of the tank were increased, we would run up against this constraint and the problem would reduce to a one-dimensional search for length.

FIGURE 16.3

Results of minimization. The price is reduced from $9154 to $5723 because of the smaller volume using dimensions of $D = 0.98$ m and $L = 1.05$ m.

	A	B	C	D	E	F	G
1	**Optimum tank design**						
2							
3	**Parameters:**			**Design variables:**			
4	V0	0.8		D	0.983184		
5	t	0.03		L	1.053732		
6	rho	8000					
7	Lmax	2		**Constraints:**			
8	Dmax	1		D	0.983184	<=	1.06
9	cm	4.5		L	1.053732	<=	2.06
10	cw	20		Vol	0.799999	=	0.8
11							
12	**Computed values:**			**Objective function:**			
13	m	1215.224		C	5723.148		
14	lw	12.73204					
15							
16	Vshell	0.100621					
17	Vends	0.051282					

16.2 LEAST-COST TREATMENT OF WASTEWATER (CIVIL/ENVIRONMENTAL ENGINEERING)

Background.   Wastewater discharges from big cities are often a major cause of river pollution. Figure 16.4 illustrates the type of system that an environmental engineer might confront. Several cities are located on a river and its tributary. Each generates pollution at a loading rate P that has units of milligrams per day (mg/d). The pollution loading is subject to waste treatment that results in a fractional removal x. Thus, the amount discharged to the river is the excess not removed by treatment,

$$W_i = (1 - x_i)P_i \tag{16.4}$$

where W_i = waste discharge from the ith city.

When the waste discharge enters the stream, it mixes with pollution from upstream sources. If complete mixing is assumed at the discharge point, the resulting concentration at the discharge point can be calculated by a simple mass balance,

$$c_i = \frac{W_i + Q_u c_u}{Q_i} \tag{16.5}$$

where Q_u = flow (L/d), c_u = concentration (mg/L) in the river immediately upstream of the discharge, and Q_i = flow downstream of the discharge point (L/d).

After the concentration at the mixing point is established, chemical and biological decomposition processes can remove some of the pollution as it flows downstream. For the present case, we will assume that this removal can be represented by a simple fractional reduction factor R.

Assuming that the headwaters (that is, the river system above cities 1 and 2) are pollution-free, the concentrations at the four nodes can be computed as

$$
\begin{aligned}
c_1 &= \frac{(1 - x_1)P_1}{Q_{13}} \\
c_2 &= \frac{(1 - x_2)P_2}{Q_{23}} \\
c_3 &= \frac{R_{13}Q_{13}c_1 + R_{23}Q_{23}c_2 + (1 - x_3)P_3}{Q_{34}} \\
c_4 &= \frac{R_{34}Q_{34}c_3 + (1 - x_4)P_4}{Q_{45}}
\end{aligned}
\tag{16.6}
$$

FIGURE 16.4

Four wastewater treatment plants discharging pollution to a river system. The river segments between the cities are labeled with circled numbers.

TABLE 16.2 Parameters for four wastewater treatment plants discharging pollution to a river system, along with the resulting concentration (c_i) for zero treatment. Flow, removal, and standards for the river segments are also listed.

City	P_i (mg/d)	d_i (10^{-6}/mg)	c_i (mg/L)	Segment	Q (L/d)	R	c_s (mg/L)
1	1.00×10^9	2	100	1–3	1.00×10^7	0.5	20
2	2.00×10^9	2	40	2–3	5.00×10^7	0.35	20
3	4.00×10^9	4	47.3	3–4	1.10×10^8	0.6	20
4	2.50×10^9	4	22.5	4–5	2.50×10^8		20

Next, it is recognized that the waste treatment costs a different amount, d_i (10^{-6}/mg) removed, at each of the facilities. Thus, the total cost of treatment (on a daily basis) can be calculated as

$$Z = d_1 P_1 x_1 + d_2 P_2 x_2 + d_3 P_3 x_3 + d_4 P_4 x_4 \qquad (16.7)$$

where Z is total daily cost of treatment ($1000/d).

The final piece in the "decision puzzle" involves environmental regulations. To protect the beneficial uses of the river (for example, boating, fisheries, bathing), regulations say that the pollution concentration must not exceed a water-quality standard of c_s.

Parameters for the river system in Fig. 16.4 are summarized in Table 16.2. Notice that there is a difference in treatment cost between the upstream (1 and 2) and the downstream cities (3 and 4) because of the outmoded nature of the downstream plants.

The concentration can be calculated with Eq. (16.6) and the result listed in the shaded column for the case where no waste treatment is implemented (that is, all the x's = 0). Notice that the standard of 20 mg/L is being violated at all mixing points.

Use linear programming to determine the treatment levels that meet the water-quality standards for the minimum cost. Also, evaluate the impact of making the standard more stringent below city 3. That is, redo the exercise, but with the standards for segments 3–4 and 4–5 lowered to 10 mg/L.

Solution.   All the factors outlined above can be combined into the following linear programming problem:

$$\text{Minimize } Z = d_1 P_1 x_1 + d_2 P_2 x_2 + d_3 P_3 x_3 + d_4 P_4 x_4 \qquad (16.8)$$

subject to the constraints

$$\frac{(1 - x_1)P_1}{Q_{13}} \leq c_{s1}$$

$$\frac{(1 - x_2)P_2}{Q_{23}} \leq c_{s2}$$

$$\frac{R_{13}Q_{13}c_1 + R_{23}Q_{23}c_2 + (1 - x_3)P_3}{Q_{34}} \leq c_{s3} \qquad (16.9)$$

$$\frac{R_{34}Q_{34}c_3 + (1 - x_4)P_4}{Q_{45}} \leq c_{s4}$$

$$0 \leq x_1, x_2, x_3, x_4 \leq 1 \qquad (16.10)$$

FIGURE 16.5
Excel spreadsheet set up to evaluate the cost of waste water treatment on a regulated river system. Column F contains the calculation of concentration according to Eq. (16.6). Cells F4 and H4 are highlighted to show the formulas used to calculate c_1 and treatment cost for city 1. In addition, highlighted cell H9 shows the formula [Eq. (16.8)] for total cost, which is to be minimized.

	A	B	C	D	E	F	G	H
1	Least-Cost Treatment of Wastewater							
2		Untreated	Treatment	Discharge	Unit cost	River	WQ	Treatment
3	City	P	x	W	d	conc	standard	cost
4	1	1.00E+09	0	1.00E+09	2.00E-06	**100.00**	20.00	**0.00**
5	2	2.00E+09	0	2.00E+09	2.00E-06	40.00	20.00	0.00
6	3	4.00E+09	0	4.00E+09	4.00E-06	47.27	20.00	0.00
7	4	2.50E+09	0	2.50E+09	4.00E-06	22.48	20.00	0.00
8		River	River					
9	Segment	flow	removal				Total	**0.00**
10	1-3	1.00E+07	0.5					
11	2-3	5.00E+07	0.35					
12	3-4	1.10E+08	0.6					
13	4-5	2.50E+08						

=D4/B10   =$B$4*$C$4*$E$4   =SUM(H4:H7)

Thus, the objective function is to minimize treatment cost [Eq. (16.8)] subject to the constraint that water-quality standards must be met for all parts of the system [Eq. (16.9)]. In addition, treatment cannot be negative or greater than 100% removal [Eq. (16.10)].

The problem can be solved using a variety of packages. For the present application, we use the Excel spreadsheet. As seen in Fig. 16.5, these data along with the concentration calculations can be set up nicely in the spreadsheet cells.

Once the spreadsheet is created, Solver is chosen from the Data tab. At this point a dialogue box will be displayed, querying you for pertinent information. The pertinent cells of the Solver dialogue box would be filled out as

Notice that not all the constraints are shown because the dialogue box displays only six constraints at a time.

	A	B	C	D	E	F	G	H
1	Least-Cost Treatment of Wastewater							
2		Untreated	Treatment	Discharge	Unit cost	River	WQ	Treament
3	City	P	x	W	d	conc	standard	cost
4	1	1.00E+09	0.80	2.00E+08	2.00E-06	20.00	20.00	1600.00
5	2	2.00E+09	0.50	1.00E+09	2.00E-06	20.00	20.00	2000.00
6	3	4.00E+09	0.5625	1.75E+09	4.00E-06	20.00	20.00	9000.00
7	4	2.50E+09	0.00	2.50E+09	4.00E-06	15.28	20.00	0.00
8		River	River					
9	Segment	flow	removal				Total	12600.00
10	1-3	1.00E+07	0.5					
11	2-3	5.00E+07	0.35					
12	3-4	1.10E+08	0.6					
13	4-5	2.50E+08						

FIGURE 16.6
Results of minimization. The water-quality standards are met at a cost of $12,600/day. Notice that despite the fact that no treatment is required for city 4, the concentration at its mixing point actually exceeds the standard.

When the OK button is selected, a dialogue box will open with a report on the success of the operation. For the present case, the Solver obtains the correct solution, which is shown in Fig. 16.6. Before accepting the solution (by selecting the OK button on the Solver Reports box), notice that three reports can be generated: Answer, Sensitivity, and Limits. Select the Sensitivity Report and then hit the OK button to accept the solution. The Solver will automatically generate a Sensitivity Report, as shown in Fig. 16.7.

Now let us examine the solution (Fig. 16.6). Notice that the standard will be met at all the mixing points. In fact, the concentration at city 4 will actually be less than the standard (16.28 mg/L), even though no treatment would be required for city 4.

As a final exercise, we can lower the standards for reaches 3–4 and 4–5 to 10 mg/L. Before doing this, we can examine the Sensitivity Report. For the present case, the key column of Fig. 16.7 is the one headed "Lagrange Multiplier" (aka the "shadow price"). The *shadow price* is a value that expresses the sensitivity of the objective function (in our case, cost) to a unit change of one of the constraints (water-quality standards). It therefore represents the additional cost that will be incurred by making the standards more stringent. For our example, it is revealing that the largest shadow price, $-\$440/\Delta c_{s3}$, occurs for one of the standard changes (that is, downstream from city 3) that we are contemplating. This tips us off that our modification will be costly.

This is confirmed when we rerun Solver with the new standards (that is, we lower cells G6 and G7 to 10). As seen in Table 16.3, the result is that treatment cost is increased from $12,600/day to $19,640/day. In addition, reducing the standard concentrations for the lower reaches means that city 4 must begin to treat its waste and city 3 must upgrade its treatment. Notice also that the treatment of the upstream cities is unaffected.

Microsoft Excel 11.0 Sensitivity Report
Worksheet: [Sec1602.xls]Sheet1
Report Created: 3/28/2005 11:37:39 AM

Adjustable Cells

Cell	Name	Final Value	Reduced Gradient
$C$4	x	0.80	0.00
$C$5	x	0.50	0.00
$C$6	x	0.5625	0.0000
$C$7	x	0.00	10000.00

Constraints

Cell	Name	Final Value	Lagrange Multiplier
$F$4	conc	20.00	0.00
$F$5	conc	20.00	-30.00
$F$6	conc	20.00	-440.00
$F$7	conc	15.28	0.00

FIGURE 16.7
Sensitivity Report for spread-sheet set up to evaluate the cost of waste water treatment on a regulated river system.

TABLE 16.3 Comparison of two scenarios involving the impact of different regulations on treatment costs.

Scenario 1: All $c_s = 20$			Scenario 2: Downstream $c_s = 10$		
City	x	c	City	x	c
1	0.8	20	1	0.8	20
2	0.5	20	2	0.5	20
3	0.5625	20	3	0.8375	10
4	0	15.28	4	0.264	10
Cost = $12,600			Cost = $19,640		

16.3 MAXIMUM POWER TRANSFER FOR A CIRCUIT (ELECTRICAL ENGINEERING)

Background. The simple resistor circuit in Fig. 16.8 contains three fixed resistors and one adjustable resistor. Adjustable resistors are called *potentiometers*. The values for the parameters are $V = 80$ V, $R_1 = 8$ Ω, $R_2 = 12$ Ω, and $R_3 = 10$ Ω. (*a*) Find the value of the adjustable resistance R_a that maximizes the power transfer across terminals 1 and 2. (*b*) Perform a sensitivity analysis to determine how the maximum power and the corresponding setting of the potentiometer (R_a) vary as V is varied over the range from 45 to 105 V.

FIGURE 16.8
A resistor circuit with an adjustable resistor, or potentiometer.

FIGURE 16.9
A plot of power transfer across terminals 1-2 from Fig. 16.8 as a function of the potentiometer resistance R_a.

Solution.   An expression for power for the circuit can be derived from Kirchhoff's laws as

$$P(R_a) = \frac{\left[\dfrac{VR_3 R_a}{R_1(R_a + R_2 + R_3) + R_3 R_a + R_3 R_2} \right]^2}{R_a} \tag{16.11}$$

Substituting the parameter values gives the plot shown in Fig. 16.9. Notice that a maximum power transfer occurs at a resistance of about 16 Ω.

We will solve this problem in two ways with the Excel spreadsheet. First, we will employ trial-and-error and the Solver option. Then, we will develop a Visual Basic macro program to perform the sensitivity analysis.

(a) An Excel spreadsheet to implement Eq. (16.11) is shown in Fig. 16.10. As indicated, Eq. (16.11) can be entered into cell B9. Then the value of R_a (cell B8) can be varied in a trial-and-error fashion until a minimum drag was determined. For this example, the result is a power of 30.03 W and a potentiometer setting of $R_a = 16.44$ Ω.

A superior approach involves using the Solver option from the spreadsheet's Data tab. When Solver is selected, a dialogue box will be displayed, querying you for pertinent information. The pertinent cells of the Solver dialogue box would be filled out as

Set target cell:	B9	
Equal to ● max ○ min ○ equal to		0
By changing cells	B8	

B9	▼	=	=(V*Res3*Ra/(Res1*(Ra+Res2+Res3)+Res3*Ra+Res3*Res2))^2/Ra

	A	B	C	D	E	F	G	H
1	**Maximum Power Transfer**							
2								
3	V	80						
4	Res1	8						
5	Res2	12						
6	Res3	10						
7								
8	Ra	16.444444						
9	P(Ra)	30.03003						
10								

FIGURE 16.10
Excel determination of maximum power across a potentiometer using trial-and-error.

When the OK button is selected, a dialogue box will open with a report on the success of the operation. For the present case, the Solver obtains the same correct solution shown in Fig. 16.10.

(b) Now, although the foregoing approach is excellent for a single evaluation, it is not convenient for cases where multiple optimizations would be employed. Such would be the case for the second part of this application, where we are interested in determining how the maximum power varies for different voltage settings. Of course, the Solver could be invoked multiple times for different parameter values, but this would be inefficient. A preferable course would involve developing a macro function to come up with the optimum.

Such a function is shown in Fig. 16.11. Notice how closely it resembles the golden-section-search pseudocode previously presented in Fig. 13.5. In addition, notice that a function must also be defined to compute power according to Eq. (16.11).

An Excel spreadsheet utilizing this macro to evaluate the sensitivity of the solution to voltage is given in Fig. 16.12. A column of values is set up that spans the range of V's (that is, from 45 to 105 V). A function call to the macro is written in cell B9 that references the adjacent value of V (the 45 in A9). In addition, the other parameters in the function argument are also included. Notice that, whereas the reference to V is relative, the references to the lower and upper guesses and the resistances are absolute (that is, they include a leading $). This was done so that when the formula is copied down, the absolute references stay fixed, whereas the relative reference corresponds to the voltage in the same row. A similar strategy is used to place Eq. (16.11) in cell C9.

When the formulas are copied downward, the result is as shown in Fig. 16.12. The maximum power can be plotted to visualize the impact of voltage variations. As seen in Fig. 16.13, the power increases with V.

The results for the corresponding potentiometer settings (R_a) are more interesting. The spreadsheet indicates that the same setting, 16.44 Ω, results in maximum power. Such a result might be difficult to intuit based on casual inspection of Eq. (16.11).

FIGURE 16.11

Excel macro written in Visual Basic to determine a maximum with the golden-section search.

```
Option Explicit

Function Golden(xlow, xhigh, R1, R2, R3, V)
Dim iter As Integer, maxit As Integer, ea As Double, es As Double
Dim fx As Double, xL As Double, xU As Double, d As Double, x1 as Double
Dim x2 As Double, f1 As Double, f2 As Double, xopt As Double
Const R As Double = (5 ^ 0.5 − 1) / 2
maxit = 50
es = 0.001
xL = xlow
xU = xhigh
iter = 1
d = R * (xU − xL)
x1 = xL + d
x2 = xU − d
f1 = f(x1, R1, R2, R3, V)
f2 = f(x2, R1, R2, R3, V)
If f1 > f2 Then
  xopt = x1
  fx = f1
Else
  xopt = x2
  fx = f2
End If
Do
  d = R * d
  If f1 > f2 Then
    xL = x2
    x2 = x1
    x1 = xL + d
    f2 = f1
    f1 = f(x1, R1, R2, R3, V)
  Else
    xU = x1
    x1 = x2
    x2 = xU − d
    f1 = f2
    f2 = f(x2, R1, R2, R3, V)
  End If
  iter = iter + 1
  If f1 > f2 Then
    xopt = x1
    fx =f1
  Else
    xopt = x2
    fx = f2
  End If
  If xopt <> 0 Then ea = (1 − R) * Abs((xU − xL) / xopt) * 100
  If ea <= es Or iter >= maxit Then Exit Do
Loop
Golden = xopt
End Function

Function f(Ra, R1, R2, R3, V)
f = (V * R3 * Ra / (R1 * (Ra + R2 + R3) + R3 * Ra + R3 * R2)) ^ 2 / Ra
END FUNCTION
```

	A	B	C	D
1	**Maximum Power Transfer**			
2				
3	R1	8		
4	R2	12		
5	R3	10		
6	Rmin	0.1		
7	Rmax	100		
8	V	Ra	P(Ra)	
9	45	**16.44444**	**9.501689**	
10	60	16.44444	16.89189	
11	75	16.44444	26.39358	
12	90	16.44444	38.00676	
13	105	16.44444	51.73142	

Call to Visual Basic macro function

= Golden($B$6,$B$7,$B$3,$B$4,$B$5,A9)

Power calculation

=(A9*$B$5*B9/($B$3*(B9+$B$4+$B$5)+$B$5*B9+$B$3*$B$4))^2/B9

FIGURE 16.12
Excel spreadsheet to implement a sensitivity analysis of the maximum power to variations of voltage. This routine accesses the macro function for golden-section search from Fig. 16.11.

FIGURE 16.13
Results of sensitivity analysis of the effect of voltage variations on maximum power.

16.4 EQUILIBRIUM AND MINIMUM POTENTIAL ENERGY (MECHANICAL/AEROSPACE ENGINEERING)

Background.   As shown in Fig. 16.14*a*, an unloaded spring can be attached to a wall mount. When a horizontal force is applied, the spring stretches. The displacement is related to the force by *Hooke's law*, $F = kx$. The *potential energy* of the deformed state consists of the difference between the strain energy of the spring and the work done by the force,

$$PE(x) = 0.5kx^2 - Fx \tag{16.12}$$

Equation (16.12) defines a parabola. Since the potential energy will be at a minimum at equilibrium, the solution for displacement can be viewed as a one-dimensional optimization

FIGURE 16.14
(a) An unloaded spring attached to a wall mount. (b) Application of a horizontal force stretches the spring where the relationship between force and displacement is described by Hooke's law.

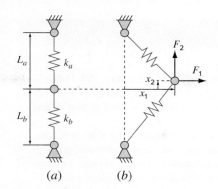

FIGURE 16.15
A two-spring system: (a) unloaded, and (b) loaded.

problem. Because this equation is so easy to differentiate, we can solve for the displacement as $x = F/k$. For example, if $k = 2$ N/cm and $F = 5$ N, $x = 5\text{N}/(2 \text{ N/cm})/5 = 2.5$ cm.

A more interesting two-dimensional case is shown in Fig. 16.15. In this system, there are two degrees of freedom in that the system can move both horizontally and vertically. In the same way that we approached the one-dimensional system, the equilibrium deformations are the values of x_1 and x_2 that minimize the potential energy,

$$PE(x_1, x_2) = 0.5k_a(\sqrt{x_1^2 + (L_a - x_2)^2} - L_a)^2$$
$$+ 0.5k_b(\sqrt{x_1^2 + (L_b + x_2)^2} - L_b)^2 - F_1x_1 - F_2x_2$$

(16.13)

If the parameters are $k_a = 9$ N/cm, $k_b = 2$ N/cm, $L_a = 10$ cm, $L_b = 10$ cm, $F_1 = 2$ N, and $F_2 = 4$ N, solve for the displacements and the potential energy.

Background.   We can use a variety of software tools to solve this problem. For example, using MATLAB, an M-file can be developed to hold the potential energy function,

```
function p=PE(x,ka,kb,La,Lb,F1,F2)
PEa=0.5*ka*(sqrt(x(1)^2+(La-x(2))^2)-La)^2;
PEb=0.5*kb*(sqrt(x(1)^2+(Lb+x(2))^2)-Lb)^2;
W=F1*x(1)+F2*x(2);
p=PEa+PEb-W;
```

The solution can then be obtained with the `fminsearch` function,

```
>> ka=9;kb=2;La=10;Lb=10;F1=2;F2=4;
>> [x,f]=fminsearch(@PE,[-0.5,0.5],[],ka,kb,La,Lb,F1,F2)
x =
    4.9523        1.2769
f =
   -9.6422
```

Thus, at equilibrium, the potential energy is -9.6422 N cm. The connecting point is located 4.9523 cm to the right and 1.2759 cm above its original position.

PROBLEMS

Chemical/Bio Engineering

16.1 Design the optimal cylindrical container (Fig. P16.1) that is open at one end and has walls of negligible thickness. The container is to hold 0.2 m³. Design it so that the areas of its bottom and sides are minimized.

16.2 **(a)** Design the optimal conical container (Fig. P16.2) that has a cover and has walls of negligible thickness. The container is to hold 0.2 m³. Design it so that the areas of its top and sides are minimized. **(b)** Repeat **(a)** but for a conical container without a cover.

16.3 Design the optimal cylindrical tank with dished ends (Fig. P16.3). The container is to hold 0.2 m³ and has walls of negligible thickness. Note that the area and volume of each of the dished ends can be computed with

$$A = \pi(h^2 + r^2) \qquad V = \frac{\pi h(h^2 + 3r^2)}{6}$$

(a) Design the tank so that its surface area is minimized. Interpret the result.

(b) Repeat part **(a)**, but add the constraint $L \geq 2h$.

FIGURE P16.3

FIGURE P16.1

A cylindrical container with no lid.

FIGURE P16.4

The specific growth rate of a yeast that produces an antibiotic versus the food concentration.

16.4 The specific growth rate g of a yeast that produces an antibiotic is a function of the food concentration c,

$$g = \frac{2c}{4 + 0.8c + c^2 + 0.2c^3}$$

As depicted in Fig. P16.4, growth goes to zero at very low concentrations due to food limitation. It also goes to zero at high concentrations due to toxicity effects. Find the value of c at which growth is a maximum.

16.5 A chemical plant makes three major products on a weekly basis. Each of these products requires a certain quantity of raw chemical and different production times, and yields different profits. The pertinent information is in Table P16.5. Note that there is sufficient warehouse space at the plant to store a total of 450 kg/week.

FIGURE P16.2

A conical container with a lid.

TABLE P16.5

	Product 1	Product 2	Product 3	Resource Availability
Raw chemical	6 kg/kg	4 kg/kg	12 kg/kg	2500 kg
Production time	0.05 hr/kg	0.1 hr/kg	0.2 hr/kg	55 hr/week
Profit	$30/kg	$30/kg	$35/kg	

(a) Set up a linear programming problem to maximize profit.
(b) Solve the linear programming problem with the simplex method.
(c) Solve the problem with a software package.
(d) Evaluate which of the following options will raise profits the most: increasing raw chemical, production time, or storage.

16.6 Recently chemical engineers have become involved in the area known as *waste minimization*. This involves the operation of a chemical plant so that impacts on the environment are minimized. Suppose a refinery develops a product $Z1$ made from two raw materials X and Y. The production of 1 metric tonne of the product involves 1 tonne of X and 2.5 tonnes of Y and produces 1 tonne of a liquid waste W. The engineers have come up with three alternative ways to handle the waste:

- Produce a tonne of a secondary product $Z2$ by adding an additional tonne of X to each tonne of W.
- Produce a tonne of another secondary product $Z3$ by adding an additional tonne of Y to each tonne of W.
- Treat the waste so that it is permissible to discharge it.

The products yield profits of $2000, −$75, and $250/tonne for $Z1$, $Z2$, and $Z3$, respectively. Note that producing $Z2$ actually creates a loss. The treatment process costs $300/tonne. In addition, the company has access to a limit of 7500 and 12,500 tonnes of X and Y, respectively, during the production period. Determine how much of the products and waste must be created in order to maximize profit.

16.7 A mixture of benzene and toluene are to be separated in a flash tank. At what temperature should the tank be operated to get the highest purity toluene in the liquid phase (maximizing x_T)? The pressure in the flash tank is 800 mm Hg. The units for *Antoine's equation* are mm Hg and °C for pressure and temperature, respectively:

$$x_B P_{\text{sat}B} + x_T P_{\text{sat}T} = P$$

$$\log_{10}(P_{\text{sat}B}) = 6.905 - \frac{1211}{T + 221}$$

$$\log_{10}(P_{\text{sat}T}) = 6.953 - \frac{1344}{T + 219}$$

16.8 A compound A will be converted into B in a stirred tank reactor. The product B and unreacted A are purified in a separation unit. Unreacted A is recycled to the reactor. A process engineer has found that the initial cost of the system is a function of the conversion x_A. Find the conversion that will result in the lowest-cost system. C is a proportionality constant.

$$\text{Cost} = C\left[\left(\frac{1}{(1 - x_A)^2}\right)^{0.6} + 6\left(\frac{1}{x_A}\right)^{0.6}\right]$$

16.9 In Prob. 16.8, only one reactor is used. If two reactors are used in series, the governing equation for the system changes. Find the conversions for both reactors (x_{A1} and x_{A2}) such that the total cost of the system is minimized.

$$\text{Cost} =$$

$$C\left[\left(\frac{x_{A1}}{x_{A2}(1 - x_{A1})^2}\right)^{0.6} + \left(\frac{1 - \left(\frac{x_{A1}}{x_{A2}}\right)}{(1 - x_{A2})^2}\right)^{0.6} + 6\left(\frac{1}{x_{A2}}\right)^{0.6}\right]$$

16.10 For the reaction

$$2A + B \Leftrightarrow C$$

equilibrium can be expressed as follows:

$$K = \frac{[C]}{[A]^2[B]} = \frac{[C]}{[A_0 - 2C]^2[B_0 - C]}$$

If $K = 2\ M^{-1}$, the initial concentration of A (A_0) can be varied. The initial concentration of B is fixed by the process, $B_0 = 100$. A costs $1/M and C sells for $10/M. What would be the optimum initial concentration of A to use such that the profits would be maximized?

16.11 A chemical plant requires 10^6 L/day of a solution. Three sources are available at different prices and supply rates. Each source also has a different concentration of an impurity that must be kept below a minimum level to prevent interference with the chemical. The data for the three sources are summarized in the following table. Determine the amount from each source to meet the requirements at the least cost.

	Source 1	Source 2	Source 3	Required
Cost ($/L)	0.50	1.00	1.20	minimize
Supply (10^5 L/day)	20	10	5	≥ 10
Concentration (mg/L)	135	100	75	≤ 100

FIGURE P16.12

FIGURE P16.15
A blood vessel branching at an angle θ from an artery.

FIGURE P16.13

16.12 You must design a triangular open channel to carry a waste stream from a chemical plant to a waste stabilization pond (Fig. P16.12). The mean velocity increases with the hydraulic radius $R_h = A/p$, where A is the cross-sectional area and p equals the wetted perimeter. Because the maximum flow rate corresponds to the maximum velocity, the optimal design amounts to minimizing the wetted perimeter. Determine the dimensions to minimize the wetted perimeter for a given cross-sectional area. Are the relative dimensions universal?

16.13 As an agricultural engineer, you must design a trapezoidal open channel to carry irrigation water (Fig. P16.13). Determine the optimal dimensions to minimize the wetted perimeter for a cross-sectional area of 100 m². Are the relative dimensions universal?

16.14 Find the optimal dimensions for a heated cylindrical tank designed to hold 10 m³ of fluid. The ends and sides cost \$200/m² and \$100/m², respectively. In addition, a coating is applied to the entire tank area at a cost of \$50/m².

16.15 As depicted in Fig. P16.15, a blood vessel branches from an artery at an angle θ. Under steady, laminar flow conditions, the resistance for blood flowing along the path ABC, R_{ABC}, can be computed based on the *Hagen-Poiseuille law* as

$$R_{\mathrm{ABC}} = K\left(\frac{L_1}{d_1^4} + \frac{L_2}{d_2^4}\right)$$

where K = a constant parameter, d_1 and d_2 = diameters of the artery and the vessel, respectively, and L_1 and L_2 = lengths of the artery and the vessel, respectively, which are related to the angle by

$$L_1 = L - H\cot\theta \qquad L_2 = H\csc\theta$$

Given $H = 2.5$ mm, $L = 10$ mm, $d_1 = 1$ mm, and $d_2 = 0.75$ mm, determine the angle that minimizes R_{ABC}.

16.16 You are asked to design a right circular cone (Fig. P16.16) to hold a specified volume of liquid while minimizing the amount of paper used to construct its surface area. The surface area and volume can be computed as a function of the cone's top radius, r (m), and height, h (m),

$$A = \pi r\sqrt{r^2 + h^2} \qquad V = \frac{\pi r^2 h}{3}$$

(a) Given $V = 160$ cm³, compute the height and radius that minimizes area, **(b)** compute the corresponding angle θ for this optimal solution, and **(c)** determine whether the result of **(b)** always holds.

FIGURE P16.16
A right circular cone.

Civil/Environmental Engineering
16.17 A finite-element model of a cantilever beam subject to loading and moments (Fig. P16.17) is given by optimizing

$$f(x, y) = 5x^2 - 5xy + 2.5y^2 - x - 1.5y$$

FIGURE P16.17
A cantilever beam.

where x = end displacement and y = end moment. Find the values of x and y that minimize $f(x, y)$.

16.18 Suppose that you are asked to design a column to support a compressive load P, as shown in Fig. P16.18a. The column has a cross section shaped as a thin-walled pipe, as shown in Fig. P16.18b.

The design variables are the mean pipe diameter d and the wall thickness t. The cost of the pipe is computed by

$$\text{Cost} = f(t, d) = c_1 W + c_2 d$$

where $c_1 = 4$ and $c_2 = 2$ are cost factors and W = weight of the pipe,

$$W = \pi dt H \rho$$

where ρ = density of the pipe material = 0.0025 kg/cm^3. The column must support the load under compressive stress and not buckle. Therefore,

Actual stress (σ) $\leq$ maximum compressive yield stress
$$= \sigma_y = 550 \text{ kg/cm}^2$$

Actual stress $\leq$ buckling stress

FIGURE P16.18
(a) A column supporting a compressive load P. (b) The column has a cross section shaped as a thin-walled pipe.

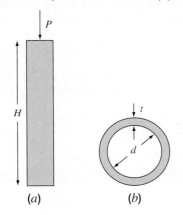

(a)　　(b)

The actual stress is given by

$$\sigma = \frac{P}{A} = \frac{P}{\pi dt}$$

The buckling stress can be shown to be

$$\sigma_b = \frac{\pi EI}{H^2 dt}$$

where E = modulus of elasticity and I = second moment of the area of the cross section. Calculus can be used to show that

$$I = \frac{\pi}{8} dt(d^2 + t^2)$$

Finally, diameters of available pipes are between d_1 and d_2 and thicknesses between t_1 and t_2. Develop and solve this problem by determining the values of d and t that minimize the cost. Note that $H = 275$ cm, $P = 2000$ kg, $E = 900{,}000$ kg/cm^2, $d_1 = 1$ cm, $d_2 = 10$ cm, $t_1 = 0.1$ cm, and $t_2 = 1$ cm.

16.19 The *Streeter-Phelps model* can be used to compute the dissolved oxygen concentration in a river below a point discharge of sewage (Fig. P16.19),

$$o = o_s - \frac{k_d L_o}{k_d + k_s - k_a}(e^{-k_a t} - e^{-(k_d + k_s)t}) - \frac{S_b}{k_a}(1 - e^{-k_a t})$$

(P16.19)

where o = dissolved oxygen concentration (mg/L), o_s = oxygen saturation concentration (mg/L), t = travel time (d), L_o = biochemical oxygen demand (BOD) concentration at the mixing point (mg/L), k_d = rate of decomposition of BOD (d^{-1}), k_s = rate of settling of BOD (d^{-1}), k_a = re-aeration rate (d^{-1}), and S_b = sediment oxygen demand (mg/L/d).

As indicated in Fig. P16.19, Eq. (P16.19) describes an oxygen "sag" that reaches a critical minimum level o_c some travel time t_c below the point discharge. This point is called "critical" because it

FIGURE P16.19
A dissolved oxygen "sag" below a point discharge of sewage into a river.

represents the location where biota that depend on oxygen (like fish) would be the most stressed. Determine the critical travel time and concentration, given the following values:

$$o_s = 10 \text{ mg/L} \quad k_d = 0.2 \text{ d}^{-1} \quad k_a = 0.8 \text{ d}^{-1}$$
$$k_s = 0.06 \text{ d}^{-1} \quad L_o = 50 \text{ mg/L} \quad S_b = 1 \text{ mg/L/d}$$

16.20 The two-dimensional distribution of pollutant concentration in a channel can be described by

$$c(x, y) = 7.7 + 0.15x + 0.22y - 0.05x^2$$
$$- 0.016y^2 - 0.007xy$$

Determine the exact location of the peak concentration given the function and the knowledge that the peak lies within the bounds $-10 \le x \le 10$ and $0 \le y \le 20$.

16.21 The flow Q (m³/s) in an open channel can be predicted with the *Manning equation*,

$$Q = \frac{1}{n} A_c R^{2/3} S^{1/2}$$

where n = Manning roughness coefficient (a dimensionless number used to parameterize the channel friction), A_c = cross-sectional area of the channel (m²), S = channel slope (dimensionless, meter drop per meter length), and R = hydraulic radius (m), which is related to more fundamental parameters by $R = A_c/P$, where P = wetted perimeter (m). As the name implies, the wetted perimeter is the length of the channel sides and bottom that is under water. For example, for a rectangular channel, it is defined as $P = B + 2H$, where H = depth (m). Suppose that you are using this formula to design a lined canal (note that farmers line canals to minimize leakage losses).
(a) Given the parameters $n = 0.03$, $S = 0.0004$, and $Q = 1$ m³/s, determine the values of B and H that minimize the wetted perimeter. Note that such a calculation would minimize cost if lining costs were much larger than excavation costs.
(b) Repeat part (a), but include the cost of excavation. To do this, minimize the following cost function

$$C = c_1 A_c + c_2 P$$

where c_1 is a cost factor for excavation = $100/m² and c_2 is a cost factor for lining = $50/m.
(c) Discuss the implications of your results.

16.22 A cylindrical beam carries a compression load $P = 3000$ kN. To prevent the beam from buckling, this load must be less than a critical load,

$$P_c = \frac{\pi^2 EI}{L^2}$$

where E = Young's modulus = 200×10^9 N/m², $I = \pi r^4/4$ (the area moment of inertia for a cylindrical beam of radius r), and L

is the beam length. If the volume of beam V cannot exceed 0.075 m³, find the largest height L that can be utilized and the corresponding radius.

16.23 The Splash River has a flow rate of 2×10^6 m³/d, of which up to 70% can be diverted into two channels where it flows through Splish County. These channels are used for transportation, irrigation, and electric power generation, with the latter two being sources of revenue. The transportation use requires a minimum diverted flow rate of 0.3×10^6 m³/d for Channel 1 and 0.2×10^6 m³/d for Channel 2. For political reasons it has been decided that the absolute difference between the flow rates in the two channels cannot exceed 40% of the total flow diverted into the channels. The Splish County Water Management Board has also limited maintenance costs for the channel system to be no more than 1.8×10^6 per year. Annual maintenance costs are estimated based on the daily flow rate. Channel 1 costs per year are estimated by multiplying $1.1 by the flow in m³/d; while for Channel 2 the multiplication factor is $1.4 per m³/d. Electric power production revenue is also estimated based on daily flow rate. For Channel 1 this is $4.0 per m³/d, while for Channel 2 it is $3.0 per m³/d. Annual revenue from irrigation is also estimated based on daily flow rate, but the flow rates must first be corrected for water loss in the channels previous to delivery for irrigation. This loss is 30% in Channel 1 and 20% in Channel 2. In both channels the revenue is $3.2 per m³/d. Determine the flows in the channels that maximize profit.

16.24 Determine the beam cross-sectional areas that result in the minimum weight for the truss we studied in Sec. 12.2 (Fig. 12.4). The critical buckling and maximum tensile strengths of compression and tension members are 10 and 20 ksi, respectively. The truss is to be constructed of steel (density = 3.5 lb/ft-in²). Note that the length of the horizontal member (2) is 50 ft. Also, recall that the stress in each member is equal to the force divided by cross-sectional area. Set up the problem as a linear programming problem. Obtain the solution graphically and with the Excel Solver.

16.25 Figure P16.25 shows a pinned-fixed beam subject to a uniform load. The equation for the resulting deflection is

$$y = -\frac{w}{48EI}(2x^4 - 3Lx^3 + L^3 x)$$

Develop a MATLAB script that uses `fminbnd` to (a) generate a labeled plot of deflection versus distance and (b) determine the

FIGURE P16.25

location and magnitude of the maximum deflection. Employ initial guesses of 0 and L and use `optimset` to display the iterations. Use the following parameter values in your computation (making sure that you use consistent units): $L = 400$ cm, $E = 52{,}000$ kN/cm^2, $I = 32{,}000$ cm^4, and $w = 4$ kN/cm.

16.26 As depicted in Fig. P16.26, a mobile fire hose projects a stream of water onto the roof of a building. At what angle, θ, and how far from the building, x_1, should the hose be placed in order to maximize the coverage of the roof, that is, to maximize $x_2 - x_1$? Note that the water velocity leaving the nozzle has a constant value of 3 m/s regardless of the angle, and the other parameter values are $h_1 = 0.06$ m, $h_2 = 0.2$ m, and $L = 0.12$ m. (Hint: The coverage is maximized for the trajectory that just clears the top front corner. That is, we want to choose an x_1 and θ that just clear the top corner while maximizing $x_2 - x_1$.)

16.27 Because many pollutants enter lakes (and other bodies of water for that matter) at their peripheries, an important water-quality problem involves modeling the distribution of contaminants in the vicinity of a waste discharge or a river. For a vertically well-mixed, constant-depth layer, the steady-state distribution of a pollutant reacting with first-order decay is represented by

$$0 = -U_x \frac{\partial c}{\partial x} + E\left(\frac{\partial^2 c}{\partial x^2} + \frac{\partial^2 c}{\partial y^2}\right) - kc$$

where the x and y axes are defined to be parallel and perpendicular to the shoreline, respectively (Fig. P16.27). The parameters and variables are U_x = the water velocity along the shoreline (m/d), c = concentration, E = the turbulent diffusion coefficient, and k = the first-order decay rate. For the case where a constant loading, W, enters at (0, 0), the solution for the concentration at any coordinate is given by

$$c = 2\left\{c(x, y) + \sum_{n=1}^{\infty}[c(x, y + 2nY) - c(x, y - 2nY)]\right\}$$

where

$$c(x, y) = \frac{W}{\pi H E}e^{U_x x/(2E)}K_0\left(\sqrt{(x^2 + y^2)\left[\frac{k}{E} + \left(\frac{U_x}{2E}\right)^2\right]}\right)$$

where Y = the width, H = depth, and K_0 = the modified Bessel function of the second kind. Develop a MATLAB script to generate a contour plot of concentration for a section of a lake with $Y = 4.8$ km and a length from $X = -2.4$ to 2.4 km using $\Delta x = \Delta y = 0.32$ km. Employ the following parameters for your calculation: $W = 1.2 \times 10^{12}$, $H = 20$, $E = 5 \times 10^6$, $U_x = 5 \times 10^3$, $k = 1$, and $n = 3$.

Electrical Engineering

16.28 A total charge Q is uniformly distributed around a ring-shaped conductor with radius a. A charge q is located at a distance

FIGURE P16.26

FIGURE P16.27
Plan view of a section of a lake with a point source of pollutant entering at the middle of the lower boundary.

x from the center of the ring (Fig. P16.28). The force exerted on the charge by the ring is given by

$$F = \frac{1}{4\pi e_0}\frac{qQx}{(x^2 + a^2)^{3/2}}$$

where $e_0 = 8.85 \times 10^{-12}$ C^2/(N m^2), $q = Q = 2 \times 10^{-5}$ C, and $a = 0.9$ m. Determine the distance x where the force is a maximum.

FIGURE P16.28

16.29 A system consists of two power plants that must deliver loads over a transmission network. The costs of generating power at plants 1 and 2 are given by

$$F_1 = 2p_1 + 2$$
$$F_2 = 10p_2$$

where p_1 and p_2 = power produced by plant 1 and plant 2, respectively. The losses of power due to transmission L are given by

$$L_1 = 0.2p_1 + 0.1p_2$$
$$L_2 = 0.2p_1 + 0.5p_2$$

The total demand for power is 30, and p_1 must not exceed 42. Determine the power generation needed to meet demand while minimizing cost using an optimization routine such as that found in, for example, Excel, MATLAB, or Mathcad software.

16.30 The torque transmitted to an induction motor is a function of the slip between the rotation of the stator field and the rotor speed s where slip is defined as

$$s = \frac{n - n_R}{n}$$

where n = revolutions per second of rotating stator speed and n_R = rotor speed. Kirchhoff's laws can be used to show that the torque (expressed in dimensionless form) and slip are related by

$$T = \frac{15(s - s^2)}{(1 - s)(4s^2 - 3s + 4)}$$

Figure P16.30 shows this function. Use a numerical method to determine the slip at which the maximum torque occurs.

FIGURE P16.30
Torque transmitted to an inductor as a function of slip.

16.31
(a) A computer equipment manufacturer produces scanners and printers. The resources needed for producing these devices and the corresponding profits are

Device	Capital ($/unit)	Labor (hr/unit)	Profit ($/unit)
Scanner	300	20	500
Printer	400	10	375

If there are $127,000 worth of capital and 4270 hr of labor available each day, how many of each device should be produced per day to maximize profit?
(b) Repeat the problem, but now assume that the profit for each printer sold P_p depends on the number of printers produced X_p, as in

$$P_p = 400 - X_p$$

16.32 A manufacturer provides specialized microchips. During the next 3 months, its sales, costs, and available time will be as follows:

	Month 1	Month 2	Month 3
Chips required	1000	2500	2200
Cost regular time ($/chip)	100	100	125
Cost overtime ($/chip)	110	120	125
Regular operation time (hr)	2400	2400	2400
Overtime (hr)	720	720	720

There are no chips in stock at the beginning of the first month. It takes 1.6 hr of production time to produce a chip and costs $5 to store a chip from one month to the next. Determine a production schedule that meets the demand requirements, does not exceed the monthly production time limitations, and minimizes cost. Note that no chips should be in stock at the end of the 3 months.

16.33 For the LCR circuit in Fig. P16.33, the charge on the capacitor can be computed with

$$q = q_0 e^{-Rt/(2L)} \cos \left[\sqrt{\frac{1}{LC} - \left(\frac{R}{2L} \right)^2} t \right]$$

Determine the time and magnitude of the minimum capacitor charge given the following parameter values: $q_0 = 10$ A, $L = 5$ H, $C = 10^{-4}$ F, and $R = 100 \ \Omega$.

FIGURE P16.33

Mechanical/Aerospace Engineering
16.34 The total drag on an airfoil can be estimated by

$$D = \underset{\text{Friction}}{0.01\sigma V^2} + \underset{\text{Lift}}{\frac{0.95}{\sigma} \left(\frac{W}{V} \right)^2}$$

FIGURE P16.34
Plot of drag versus velocity for an airfoil.

where D = drag, σ = ratio of air density between the flight altitude and sea level, W = weight, and V = velocity. As seen in Fig. P16.34, the two factors contributing to drag are affected differently as velocity increases. Whereas friction drag increases with velocity, the drag due to lift decreases. The combination of the two factors leads to a minimum drag.

(a) If $\sigma = 0.6$ and $W = 16,000$, determine the minimum drag and the velocity at which it occurs.

(b) In addition, develop a sensitivity analysis to determine how this optimum varies in response to a range of $W = 12,000$ to $20,000$ with $\sigma = 0.6$.

16.35 Roller bearings are subject to fatigue failure caused by large contact loads F (Fig. P16.35).

The problem of finding the location of the maximum stress along the x axis can be shown to be equivalent to maximizing the function

$$f(x) = \frac{0.4}{\sqrt{1 + x^2}} - \sqrt{1 + x^2}\left(1 - \frac{0.4}{1 + x^2}\right) + x$$

Find the value of x that maximizes $f(x)$.

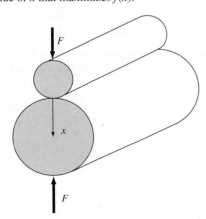

FIGURE P16.35
Roller bearings.

16.36 An aerospace company is developing a new fuel additive for commercial airliners. The additive is composed of three ingredients: X, Y, and Z. For peak performance, the total amount of additive must be at least 6 mL/L of fuel. For safety reasons, the sum of the highly flammable X and Y ingredients must not exceed 2.5 mL/L. In addition, the amount of the X ingredient must always be equal to or greater than the amount of Y, and the amount of Z must be greater than half of Y. If the cost per mL for the ingredients X, Y, and Z is \$0.05, \$0.025, and \$0.15, respectively, determine the minimum cost mixture for each liter of fuel.

16.37 A manufacturing firm produces four types of automobile parts. Each is first fabricated and then finished. The required worker hours and profit for each part are

	Part			
	A	B	C	D
Fabrication time (hr/100 units)	2.5	1.5	2.75	2
Finishing time (hr/100 units)	3.5	3	3	2
Profit (\$/100 units)	375	275	475	325

The capacities of the fabrication and finishing shops over the next month are 640 and 960 hours, respectively. Determine how many of each part should be produced in order to maximize profit.

16.38 In a similar fashion to the case study described in Sec. 16.4, develop the potential energy function for the system depicted in Fig. P16.38. Develop contour and surface plots in MATLAB. Minimize the potential energy function in order to determine the equilibrium displacements x_1 and x_2 given the

FIGURE P16.38
Two frictionless masses connected to a wall by a pair of linear elastic springs.

forcing function $F = 100$ N, and the parameter $k_a = 20$ and $k_b = 15$ N/m.

16.39 Recent interest in competitive and recreational cycling has meant that engineers have directed their skills toward the design and testing of mountain bikes (Fig. P16.39a). Suppose that you are given the task of predicting the horizontal and vertical

FIGURE P16.39
(a) A mountain bike along with (b) a free-body diagram for a part of the frame.

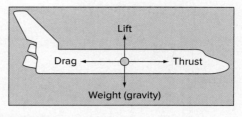

FIGURE P 16.40
The four major forces on a jet in steady, level flight.

displacement of a bike bracketing system in response to a force. Assume that the forces you must analyze can be simplified as depicted in Fig. P16.39b. You are interested in testing the response of the truss to a force exerted in any number of directions designated by the angle θ. The parameters for the problem are E = Young's modulus = 2×10^{11} Pa, A = cross-sectional area = 0.0001 m^2, w = width = 0.44 m, ℓ = length = 0.56 m, and h = height = 0.5 m. The displacements x and y can be found by determining the values that yield a minimum potential energy. Determine the displacements for a force of 10,000 N and a range of θ's from 0° (horizontal) to 90° (vertical).

16.40 For a jet in steady, level flight, thrust balances drag and lift balances weight (Fig. P16.40). Under these conditions, the optimal cruise velocity occurs when the ratio of drag force to velocity is minimized. The drag, C_D, can be computed as

$$C_D = C_{D0} + \frac{C_L^2}{\pi \cdot AR}$$

where C_{D0} = drag coefficient at zero lift, C_L = the lift coefficient, and AR = the aspect ratio. For steady, level flight, the lift coefficient can be computed as

$$C_L = \frac{2W}{\rho v^2 A}$$

where W = the jet's weight (N), ρ = air density (kg/m^3), v = velocity (m/s), and A = wing planform area (m^2). The drag force can then be computed as

$$F_D = W \frac{C_D}{C_L}$$

Use these formulas to determine the optimal steady cruise velocity for a 670 kN jet flying at 10 km above sea level. Employ the following parameters in your computation: A = 150 m^2, AR = 6.5, C_{D0} = 0.018, and ρ = 0.413 kg/m^3.

16.41 Develop a MATLAB script to generate a plot of the optimal velocity of the jet from Prob. 16.40 versus elevation above sea level. Employ a mass of 68,300 kg for the jet. Note that the gravitational acceleration at 45° latitude can be computed as a function of elevation with

$$g(h) = 9.8066 \left(\frac{r_e}{r_e + h} \right)^2$$

where $g(h)$ = gravitational acceleration (m/s^2) at elevation h (m) above sea level and r_e = Earth's mean radius (= 6.371 × 10^6 m). In addition, air density as a function of elevation can be calculated with

$$\rho(h) = -9.57926 \times 10^{-14}(h^3) + 4.71260 \times 10^{-9}(h^2)$$
$$-1.18951 \times 10^{-4}(h) + 1.22534$$

Employ the other parameters from Prob. 16.40, and design the plot for elevations ranging from h = 0 to 12 km above sea level.

EPILOGUE: PART FOUR

The epilogues of other parts of this book contain a discussion and a tabular summary of the trade-offs among various methods as well as important formulas and relationships. Most of the methods of this part are quite complicated and, consequently, cannot be summarized with simple formulas and tabular summaries. Therefore, we deviate somewhat here by providing the following narrative discussion of trade-offs and further references.

PT4.4 TRADE-OFFS

Chapter 13 dealt with finding the optimum of an unconstrained function of a single variable. The golden-section search method is a bracketing method requiring that an interval containing a single optimum be known. It has the advantage that it minimizes function evaluations and always converges. Parabolic interpolation also works best when implemented as a bracketing method, although it can also be programmed as an open method. However, in such cases, it may diverge. Both the golden-section search and parabolic interpolation do not require derivative evaluations. Thus, they are both appropriate methods when the bracket can be readily defined and function evaluations are costly.

Newton's method is an open method not requiring that an optimum be bracketed. It can be implemented in a closed-form representation when first and second derivatives can be determined analytically. It can also be implemented in a fashion similar to the secant method with finite-difference representations of the derivatives. Although Newton's method converges rapidly near the optimum, it is often divergent for poor guesses. Convergence is also dependent on the nature of the function.

Finally, hybrid approaches are available that orchestrate various methods to attain both reliability and efficiency. Brent's method does this by combining the reliable golden-section search with speedy parabolic interpolation.

Chapter 14 covered two general types of methods to solve multidimensional unconstrained optimization problems. Direct methods such as random searches and univariate searches do not require the evaluation of the function's derivatives and are often inefficient. However they also provide a tool to find global rather than local optima. Pattern search methods like Powell's method can be very efficient and also do not require derivative evaluation.

Gradient methods use either first or sometimes second derivatives to find the optimum. The method of steepest ascent/descent provides a reliable but sometimes slow approach. In contrast, Newton's method often converges rapidly when the starting point is in the vicinity of a root, but sometimes suffers from divergence. The Marquardt method uses the steepest descent method at a starting location far away from the optimum and switches to Newton's method near the optimum in an attempt to take advantage of the strengths of each method.

The Newton method can be computationally costly because it requires computation of both the gradient vector and the Hessian matrix. Quasi-Newton approaches attempt to circumvent these problems by using approximations to reduce the number of matrix evaluations (particularly the evaluation, storage, and inversion of the Hessian).

Research investigations continue today that explore the characteristics and respective strengths of various hybrid and tandem methods. Some examples are the Fletcher-Reeves conjugate gradient method and Davidon-Fletcher-Powell quasi-Newton methods.

Chapter 15 was devoted to constrained optimization. For linear problems, linear programming based on the simplex method provides an efficient means to obtain solutions. Approaches such as the generalized reduced gradient (GRG) method are available to solve nonlinear constrained problems.

Software packages include a wide variety of optimization capabilities. As described in Chap. 15, Excel, MATLAB software, and Mathcad all have built-in search capabilities that can be used for both one-dimensional and multidimensional problems routinely encountered in engineering and science.

PT4.5   ADDITIONAL REFERENCES

General overviews of optimization including some algorithms can be found in Press et al. (2007) and Moler (2004). For multidimensional problems, additional information can be found in Dennis and Schnabel (1996), Fletcher (1980, 1981), Gill et al. (1981), and Luenberger (1984).

In addition, there are a number of advanced methods that are well suited for specific problem contexts. For example, genetic algorithms use strategies inspired by evolutionary biology such as inheritance, mutation, and selection. Because they do not make assumptions regarding the underlying search space, such evolutionary algorithms are often useful for large problems with many local optima. Related techniques include simulated annealing and Tabu search. Hillier and Lieberman (2005) provide overviews of these and a number of other advanced techniques.

CURVE FITTING

PT5.1 MOTIVATION

Data are often given for discrete values along a continuum. However, you may require estimates at points between the discrete values. The present part of this book describes techniques to fit curves to such data to obtain intermediate estimates. In addition, you may require a simplified version of a complicated function. One way to do this is to compute values of the function at a number of discrete values along the range of interest. Then, a simpler function may be derived to fit these values. Both of these applications are known as *curve fitting*.

There are two general approaches for curve fitting that are distinguished from each other on the basis of the amount of error associated with these data. First, where these data exhibit a significant degree of error, or "noise," the strategy is to derive a single curve that represents the general trend of the data. Because any individual data point may be incorrect, we make no effort to intersect every point. Rather, the curve is designed to follow the pattern of the points taken as a group. One approach of this nature is called *least-squares regression* (Fig. PT5.1a).

Second, where these data are known to be very precise, the basic approach is to fit a curve or a series of curves that pass directly through each of the points. Such data usually originate from tables. Examples are values for the density of water or for the heat capacity of gases as a function of temperature. The estimation of values between well-known discrete points is called *interpolation* (Fig. PT5.1b and c).

PT5.1.1 Noncomputer Methods for Curve Fitting

The simplest method for fitting a curve to data is to plot the points and then sketch a line that visually conforms to these data. Although this is a valid option when quick estimates are required, the results are dependent on the subjective viewpoint of the person sketching the curve.

For example, Fig. PT5.1 shows sketches developed from the same set of data by three engineers. The first did not attempt to connect the points, but rather, characterized the general upward trend of these data with a straight line (Fig. PT5.1a). The second engineer used straight-line segments, or linear interpolation, to connect the points

447

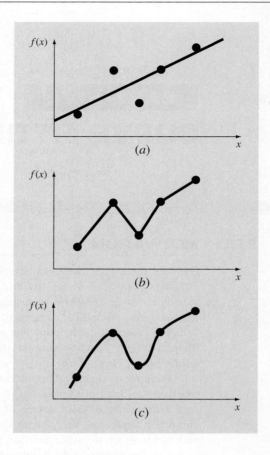

FIGURE PT5.1
Three attempts to fit a "best" curve through five data points. (a) Least-squares regression,
(b) linear interpolation, and (c) curvilinear interpolation.

(Fig. PT5.1b). This is a very common practice in engineering. If the values are truly
close to being linear or are spaced closely, such an approximation provides estimates that
are adequate for many engineering calculations. However, where the underlying relationship is highly curvilinear or the data are widely spaced, significant errors can be introduced by such linear interpolation. The third engineer used curves to try to capture the
meanderings suggested by these data (Fig. PT5.1c). A fourth or fifth engineer would
likely develop alternative fits. Obviously, our goal here is to develop systematic and
objective methods for the purpose of deriving such curves.

PT5.1.2 Curve Fitting and Engineering Practice

Your first exposure to curve fitting may have been to determine intermediate values from
tabulated data—for instance, from interest tables for engineering economics or from
steam tables for thermodynamics. Throughout the remainder of your career, you will
have frequent occasion to estimate intermediate values from such tables.

Although many of the widely used engineering properties have been tabulated, there are a great many more that are not available in this convenient form. Special cases and new problem contexts often require that you measure your own data and develop your own predictive relationships. Two types of applications are generally encountered when fitting experimental data: trend analysis and hypothesis testing.

Trend analysis is the process of using the pattern of the data to make predictions. For cases where the data are measured with high precision, you might utilize interpolating polynomials. Imprecise data are often analyzed with least-squares regression.

Trend analysis may be used to predict or forecast values of the dependent variable. This can involve extrapolation beyond the limits of the observed data or interpolation within the range of the data. All fields of engineering commonly involve problems of this type.

A second engineering application of experimental curve fitting is *hypothesis testing*. Here, an existing mathematical model is compared with measured data. If the model coefficients are unknown, it may be necessary to determine values that best fit the observed data. On the other hand, if estimates of the model coefficients are already available, it may be appropriate to compare predicted values of the model with observed values to test the adequacy of the model. Often, alternative models are compared and the "best" one is selected on the basis of empirical observations.

In addition to the above engineering applications, curve fitting is important in other numerical methods such as integration and the approximate solution of differential equations. Finally, curve-fitting techniques can be used to derive simple functions to approximate complicated functions.

PT5.2   MATHEMATICAL BACKGROUND

The prerequisite mathematical background for interpolation is found in the material on Taylor series expansions and finite divided differences introduced in Chap. 4. Least-squares regression requires additional information from the field of statistics. If you are familiar with the concepts of the mean, standard deviation, residual sum of the squares, normal distribution, and confidence intervals, feel free to skip the following pages and proceed directly to PT5.3. If you are unfamiliar with these concepts or are in need of a review, the following material is designed as a brief introduction to these topics.

PT5.2.1 Simple Statistics

Suppose that in the course of an engineering study, several measurements were made of a particular quantity. For example, Table PT5.1 contains 24 readings of the coefficient of thermal expansion of a structural steel. Taken at face value, these data provide

TABLE PT5.1 Measurements of the coefficient of thermal expansion of structural steel $[\times\ 10^{-6}\ \text{in}/(\text{in}\ {}^{\circ}\text{F})]$.

6.495	6.595	6.615	6.635	6.485	6.555
6.665	6.505	6.435	6.625	6.715	6.655
6.755	6.625	6.715	6.575	6.655	6.605
6.565	6.515	6.555	6.395	6.775	6.685

a limited amount of information—that is, that the values range from a minimum of 6.395 to a maximum of 6.775. Additional insight can be gained by summarizing these data in one or more well-chosen statistics that convey as much information as possible about specific characteristics of the data set. These descriptive statistics are most often selected to represent (1) the location of the center of the distribution of these data and (2) the degree of spread of the data set.

The most common location statistic is the arithmetic mean. The *arithmetic mean* ($\bar{y}$) of a sample is defined as the sum of the individual data points (y_i) divided by the number of points (n), or

$$\bar{y} = \frac{\Sigma y_i}{n} \tag{PT5.1}$$

where the summation (and all the succeeding summations in this introduction) is from $i = 1$ through n.

The most common measure of spread for a sample is the *standard deviation* (s_y) about the mean,

$$s_y = \sqrt{\frac{S_t}{n-1}} \tag{PT5.2}$$

where S_t is the total sum of the squares of the residuals between the data points and the mean, or

$$S_t = \Sigma (y_i - \bar{y})^2 \tag{PT5.3}$$

Thus, if the individual measurements are spread out widely around the mean, S_t (and, consequently, s_y) will be large. If they are grouped tightly, the standard deviation will be small. The spread can also be represented by the square of the standard deviation, which is called the *variance:*

$$s_y^2 = \frac{\Sigma (y_i - \bar{y})^2}{n-1} \tag{PT5.4}$$

Note that the denominator in both Eqs. (PT5.2) and (PT5.4) is $n - 1$. The quantity $n - 1$ is referred to as the degrees of freedom. Hence S_t and s_y are said to be based on $n - 1$ *degrees of freedom*. This nomenclature derives from the fact that the sum of the quantities upon which S_t is based (that is, $\bar{y} - y_1, \bar{y} - y_2, \ldots, \bar{y} - y_n$) is zero. Consequently, if $\bar{y}$ is known and $n - 1$ of the values are specified, the remaining value is fixed. Thus, only $n - 1$ of the values are said to be freely determined. Another justification for dividing by $n - 1$ is the fact that there is no such thing as the spread of a single data point. For the case where $n = 1$, Eqs. (PT5.2) and (PT5.4) yield a meaningless result of infinity.

It should be noted that an alternative, more convenient formula is available to compute the standard deviation,

$$s_y^2 = \frac{\Sigma y_i^2 - (\Sigma y_i)^2/n}{n-1}$$

This version does not require precomputation of $\bar{y}$ and yields the same result as Eq. (PT5.4).

A final statistic that has utility in quantifying the spread of data is the *coefficient of variation* (*c.v.*). This statistic is the ratio of the standard deviation to the mean. As such, it provides a normalized measure of the spread. It is often multiplied by 100 so that it can be expressed in the form of a percentage:

$$c.v. = \frac{s_y}{\bar{y}} 100\% \tag{PT5.5}$$

Notice that the coefficient of variation is similar in spirit to the percent relative error (ε_t) discussed in Sec. 3.3. That is, it is the ratio of a measure of error (s_y) to an estimate of the true value ($\bar{y}$).

EXAMPLE PT5.1

Simple Statistics of a Sample

Problem Statement.   Compute the mean, variance, standard deviation, and coefficient of variation for the data in Table PT5.1.

TABLE PT5.2 Computations for statistics for the readings of the coefficient of thermal expansion. The frequencies and bounds are developed to construct the histogram in Fig. PT5.2.

i	y_i	$(y_i - \bar{y})^2$	Frequency	Interval Lower Bound	Interval Upper Bound
1	6.395	0.042025	1	6.36	6.40
2	6.435	0.027225	1	6.40	6.44
3	6.485	0.013225			
4	6.495	0.011025	4	6.48	6.52
5	6.505	0.009025			
6	6.515	0.007225			
7	6.555	0.002025	2	6.52	6.56
8	6.555	0.002025			
9	6.565	0.001225			
10	6.575	0.000625	3	6.56	6.60
11	6.595	0.000025			
12	6.605	0.000025			
13	6.615	0.000225			
14	6.625	0.000625	5	6.60	6.64
15	6.625	0.000625			
16	6.635	0.001225			
17	6.655	0.003025			
18	6.655	0.003025	3	6.64	6.68
19	6.665	0.004225			
20	6.685	0.007225			
21	6.715	0.013225	3	6.68	6.72
22	6.715	0.013225			
23	6.755	0.024025	1	6.72	6.76
24	6.775	0.030625	1	6.76	6.80
Σ	158.4	0.217000			

Solution. These data are added (Table PT5.2), and the results are used to compute the mean [Eq. (PT5.1)]

$$\bar{y} = \frac{158.4}{24} = 6.6$$

As in Table PT5.2, the sum of the squares of the residuals is 0.217000, which can be used to compute the standard deviation [Eq. (PT5.2)]:

$$s_y = \sqrt{\frac{0.217000}{24 - 1}} = 0.097133$$

the variance [Eq. (PT5.4)]:

$$s_y^2 = 0.009435$$

and the coefficient of variation [Eq. (PT5.5)]:

$$c.v. = \frac{0.097133}{6.6}100\% = 1.47\%$$

PT5.2.2 The Normal Distribution

Another characteristic that bears on the present discussion is the *data distribution*—that is, the shape with which the given data are spread around the mean. A histogram provides a simple visual representation of the distribution. As seen in Table PT5.2, the histogram is constructed by sorting the measurements into intervals. The units of measurement are plotted on the abscissa and the frequency of occurrence of each interval is plotted on the ordinate. Thus, five of the measurements fall between 6.60 and 6.64. As in Fig. PT5.2, the histogram suggests that most of these data are grouped close to the mean value of 6.6.

If we have a very large set of data, the histogram often can be approximated by a smooth curve. The symmetric, bell-shaped curve superimposed on Fig. PT5.2 is one such characteristic shape—the *normal distribution*. Given enough additional measurements, the histogram for this particular case could eventually approach the normal distribution.

The concepts of the mean, standard deviation, residual sum of the squares, and normal distribution all have great relevance to engineering practice. A very simple example is their use to quantify the confidence that can be ascribed to a particular measurement. If a quantity is normally distributed, the range defined by $\bar{y} - s_y$ to $\bar{y} + s_y$ will encompass approximately 68 percent of the total measurements. Similarly, the range defined by $\bar{y} - 2s_y$ to $\bar{y} + 2s_y$ will encompass approximately 95 percent.

For example, for the data in Table PT5.1 ($\bar{y} = 6.6$ and $s_y = 0.097133$), we can make the statement that approximately 95 percent of the readings should fall between 6.405734 and 6.794266. If someone told us that they had measured a value of 7.35, we would suspect that the measurement might be erroneous. The following section elaborates on such evaluations.

PT5.2.3 Estimation of Confidence Intervals

As should be clear from the previous sections, one of the primary aims of statistics is to estimate the properties of a *population* based on a limited *sample* drawn from that

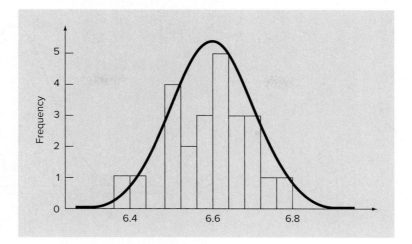

FIGURE PT5.2
A histogram used to depict the distribution of data. As the number of data points increases, the histogram could approach the smooth, bell-shaped curve called the normal distribution.

population. Clearly, it is impossible to measure the coefficient of thermal expansion for every piece of structural steel that has ever been produced. Consequently, as seen in Tables PT5.1 and PT5.2, we can randomly make a number of measurements and, on the basis of the sample, attempt to characterize the properties of the entire population.

Because we "infer" properties of the unknown population from a limited sample, the endeavor is called *statistical inference*. Because the results are often reported as estimates of the population parameters, the process is also referred to as *estimation*.

We have already shown how we estimate the central tendency (sample mean, $\bar{y}$) and spread (sample standard deviation and variance) of a limited sample. Now, we will briefly describe how we can attach probabilistic statements to the quality of these estimates. In particular, we will discuss how we can define a confidence interval around our estimate of the mean. We have chosen this particular topic because of its direct relevance to the regression models we will be describing in Chap. 17.

Note that in the following discussion, the nomenclature $\bar{y}$ and s_y refer to the sample mean and standard deviation, respectively. The nomenclature μ and σ refer to the population mean and standard deviation, respectively. The former are sometimes referred to as the "estimated" mean and standard deviation, whereas the latter are sometimes called the "true" mean and standard deviation.

An *interval estimator* gives the range of values within which the parameter is expected to lie with a given probability. Such intervals are described as being one-sided or two-sided. As the name implies, a *one-sided interval* expresses our confidence that the parameter estimate is less than or greater than the true value. In contrast, the *two-sided interval* deals with the more general proposition that the estimate agrees with the truth with no consideration to the sign of the discrepancy. Because it is more general, we will focus on the two-sided interval.

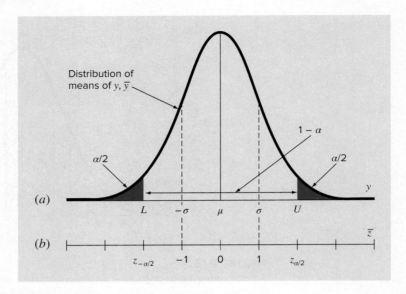

FIGURE PT5.3
A two-sided confidence interval. The abscissa scale in (a) is written in the natural units of the random variable y. The normalized version of the abscissa in (b) has the mean at the origin and scales the axis so that the standard deviation corresponds to a unit value.

A two-sided interval can be described by the statement

$$P\{L \le \mu \le U\} = 1 - \alpha$$

which reads, "the probability that the true mean of y, μ, falls within the bound from L to U is $1 - \alpha$." The quantity α is called the *significance level*. So the problem of defining a confidence interval reduces to estimating L and U. Although it is not absolutely necessary, it is customary to view the two-sided interval with the α probability distributed evenly as $\alpha/2$ in each tail of the distribution, as in Fig. PT5.3.

If the true variance of the distribution of y, σ^2, is known (which is not usually the case), statistical theory states that the sample mean $\bar{y}$ comes from a normal distribution with mean μ and variance σ^2/n (Box PT5.1). In the case illustrated in Fig. PT5.3, we really do not know μ. Therefore, we do not know where the normal curve is exactly located with respect to $\bar{y}$. To circumvent this dilemma, we compute a new quantity, the *standard normal estimate,*

$$\bar{z} = \frac{\bar{y} - \mu}{\sigma/\sqrt{n}} \tag{PT5.6}$$

which represents the normalized distance between $\bar{y}$ and μ. According to statistical theory, this quantity should be normally distributed with a mean of 0 and a variance of 1. Furthermore, the probability that $\bar{z}$ would fall within the unshaded region of Fig. PT5.3

Box PT5.1 A Little Statistics

Most engineers take several courses to become proficient at statistics. Because you may not have taken such a course yet, we would like to mention a few ideas that might make this present section more coherent.

As we have stated, the "game" of inferential statistics assumes that the random variable you are sampling, y, has a true mean (μ) and variance (σ^2). Further, in the present discussion, we also assume that it has a particular distribution: the normal distribution. The variance of this normal distribution has a finite value that specifies the "spread" of the normal distribution. If the variance is large, the distribution is broad. Conversely, if the variance is small, the distribution is narrow. Thus, the true variance quantifies the intrinsic uncertainty of the random variable.

In the game of statistics, we take a limited number of measurements of a random variable, called a sample. From this sample, we can compute an estimated mean ($\bar{y}$) and variance (s_y^2). The more measurements we take, the better the estimates approximate the true values. That is, as $n \to \infty$, $\bar{y} \to \mu$ and $s_y^2 \to \sigma^2$.

Suppose that we take n samples and compute an estimated mean $\bar{y}_1$. Then, we take another n samples and compute another, $\bar{y}_2$. We can keep repeating this process until we have generated a sample of means: $\bar{y}_1, \bar{y}_2, \bar{y}_3, \ldots, \bar{y}_m$, where m is large. We can then develop a histogram of these means and determine a "distribution of the means," as well as a "mean of the means" and a "standard deviation of the means." Now the question arises: Does this new distribution of means and its statistics behave in a predictable fashion?

There is an extremely important theorem known as the *Central Limit Theorem* that speaks directly to this question. It can be stated as

Let $y_1, y_2, \ldots, y_n$ be a random sample of size n from a distribution with mean μ and variance σ^2. Then, for large n, $\bar{y}$ is approximately normal with mean μ and variance σ^2/n. Furthermore, for large n, the random variable $(\bar{y} - \mu)/(\sigma/\sqrt{n})$ is approximately standard normal.

Thus, the theorem states the remarkable result that the distribution of means will always be normally distributed regardless of the underlying distribution of the random variables! It also yields the expected result that given a sufficiently large sample, the mean of the means should converge on the true population mean μ.

Further, the theorem says that as the sample size gets larger, the variance of the means should approach zero. This makes sense, because if n is small, our individual estimates of the mean should be poor and the variance of the means should be large. As n increases, our estimates of the mean will improve and hence their spread should shrink. The Central Limit Theorem neatly defines exactly how this shrinkage relates to both the true variance and the sample size, that is, as σ^2/n.

Finally, the theorem states the important result that we have given as Eq. (PT5.6). As is shown in this section, this result is the basis for constructing confidence intervals for the mean.

should be $1 - \alpha$. Therefore, the statement can be made that

$$\frac{\bar{y} - \mu}{\sigma/\sqrt{n}} < -z_{\alpha/2} \quad \text{or} \quad \frac{\bar{y} - \mu}{\sigma/\sqrt{n}} > z_{\alpha/2}$$

with a probability of α.

The quantity $z_{\alpha/2}$ is a standard normal random variable. This is the distance measured along the normalized axis above and below the mean that encompasses $1 - \alpha$ probability (Fig. PT5.3b). Values of $z_{\alpha/2}$ are tabulated in statistics books (for example, Milton and Arnold 2002). They can also be calculated using functions on software packages like Excel, MATLAB, and Mathcad. As an example, for $\alpha = 0.05$ (in other words, defining an interval encompassing 95%), $z_{\alpha/2}$ is equal to about 1.96. This means that an interval around the mean of width ±1.96 times the standard deviation will encompass approximately 95% of the distribution.

These results can be rearranged to yield

$$L \leq \mu \leq U$$

with a probability of $1 - \alpha$, where

$$L = \bar{y} - \frac{\sigma}{\sqrt{n}} z_{\alpha/2} \qquad U = \bar{y} + \frac{\sigma}{\sqrt{n}} z_{\alpha/2} \qquad \text{(PT5.7)}$$

Now, although the foregoing provides an estimate of L and U, it is based on knowledge of the true variance σ. For our case, we know only the estimated variance s_y. A straightforward alternative would be to develop a version of Eq. (PT5.6) based on s_y,

$$t = \frac{\bar{y} - \mu}{s_y/\sqrt{n}} \qquad \text{(PT5.8)}$$

Even when we sample from a normal distribution, this fraction will not be normally distributed, particularly when n is small. It was found by W. S. Gossett that the random variable defined by Eq. (PT5.8) follows the so-called Student-t distribution, or simply, t *distribution*. For this case,

$$L = \bar{y} - \frac{s_y}{\sqrt{n}} t_{\alpha/2,n-1} \qquad U = \bar{y} + \frac{s_y}{\sqrt{n}} t_{\alpha/2,n-1} \qquad \text{(PT5.9)}$$

where $t_{\alpha/2,n-1}$ is the standard random variable for the t distribution for a probability of $\alpha/2$. As was the case for $z_{\alpha/2}$, values are tabulated in statistics books and can also be calculated using software packages and libraries. For example, if $\alpha = 0.05$ and $n = 20$, $t_{\alpha/2,n-1} = 2.086$.

The t distribution can be thought of as a modification of the normal distribution that accounts for the fact that we have an imperfect estimate of the standard deviation. When n is small, it tends to be flatter than the normal (see Fig. PT5.4). Therefore, for small

FIGURE PT5.4
Comparison of the normal distribution with the t distribution for $n = 3$ and $n = 6$. Notice how the t distribution is generally flatter.

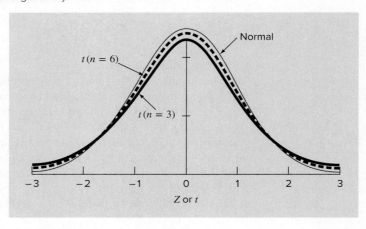

numbers of measurements, it yields wider and hence more conservative confidence intervals. As n grows larger, the t distribution converges on the normal.

Confidence Interval on the Mean

Problem Statement. Determine the mean and the corresponding 95% confidence interval for the data from Table PT5.1. Perform three estimates based on **(a)** the first 8, **(b)** the first 16, and **(c)** all 24 measurements.

Solution. **(a)** The mean and standard deviation for the first 8 values are

$$\bar{y} = \frac{52.72}{8} = 6.59 \qquad s_y = \sqrt{\frac{347.4814 - (52.72)^2/8}{8 - 1}} = 0.089921$$

The appropriate t statistic can be calculated as

$$t_{0.05/2,8-1} = t_{0.025,7} = 2.364623$$

which can be used to compute the interval:

$$L = 6.59 - \frac{0.089921}{\sqrt{8}}2.364623 = 6.5148$$

$$U = 6.59 + \frac{0.089921}{\sqrt{8}}2.364623 = 6.6652$$

or

$$6.5148 \le \mu \le 6.6652$$

FIGURE PT5.5
Estimates of the mean and 95% confidence intervals for samples of different size.

Thus, based on the first eight measurements, we conclude that there is a 95% probability that the true mean falls within the range 6.5148 to 6.6652.

The two other cases for **(b)** 16 values and **(c)** 24 values can be calculated in a similar fashion and the results tabulated along with case **(a)** as

n	$\bar{y}$	s_y	$t_{\alpha/2,n-1}$	L	U
8	6.5900	0.089921	2.364623	6.5148	6.6652
16	6.5794	0.095845	2.131451	6.5283	6.6304
24	6.6000	0.097133	2.068655	6.5590	6.6410

These results, which are also summarized in Fig. PT5.5, indicate the expected outcome that the confidence interval becomes more narrow as n increases. Thus, the more measurements we take, our estimate of the true value becomes more refined.

The above is just one simple example of how statistics can be used to make judgments regarding uncertain data. These concepts will also have direct relevance to our discussion of regression models. You can consult any basic statistics book (for example, Milton and Arnold 2002) to obtain additional information on the subject.

PT5.3 ORIENTATION

Before we proceed to numerical methods for curve fitting, some orientation might be helpful. The following is intended as an overview of the material discussed in Part Five. In addition, we have formulated some objectives to help focus your efforts when studying the material.

PT5.3.1 Scope and Preview

Figure PT5.6 provides a visual overview of the material to be covered in Part Five. *Chapter 17* is devoted to *least-squares regression*. We will first learn how to fit the "best" straight line through a set of uncertain data points. This technique is called *linear regression*. Besides discussing how to calculate the slope and intercept of this straight line, we also present quantitative and visual methods for evaluating the validity of the results.

In addition to fitting a straight line, we also present a general technique for fitting a "best" polynomial. Thus, you will learn to derive a parabolic, cubic, or higher-order polynomial that optimally fits uncertain data. Linear regression is a subset of this more general approach, which is called *polynomial regression*.

The next topic covered in Chap. 17 is *multiple linear regression*. It is designed for the case where the dependent variable y is a linear function of two or more independent variables $x_1, x_2, \ldots, x_m$. This approach has special utility for evaluating experimental data where the variable of interest is dependent on a number of different factors.

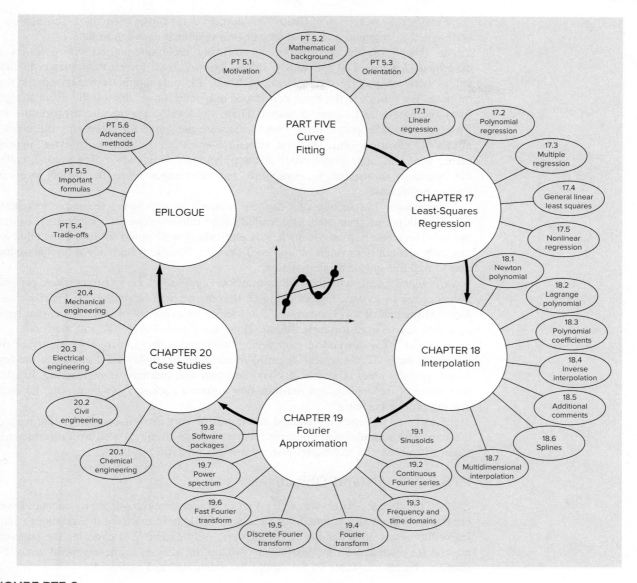

FIGURE PT5.6
Schematic of the organization of the material in Part Five: Curve Fitting.

After multiple regression, we illustrate how polynomial and multiple regression are both subsets of a *general linear least-squares model*. Among other things, this will allow us to introduce a concise matrix representation of regression and discuss its general statistical properties.

Finally, the last section of Chap. 17 is devoted to *nonlinear regression*. This approach is designed to compute a least-squares fit of a nonlinear equation to data.

In *Chap. 18,* the alternative curve-fitting technique called *interpolation* is described. As discussed previously, interpolation is used for estimating intermediate values between precise data points. In Chap. 18, polynomials are derived for this purpose. We introduce the basic concept of polynomial interpolation by using straight lines and parabolas to connect points. Then, we develop a generalized procedure for fitting an *n*th-order polynomial. Two formats are presented for expressing these polynomials in equation form. The first, called *Newton's interpolating polynomial,* is preferable when the appropriate order of the polynomial is unknown. The second, called the *Lagrange interpolating polynomial,* has advantages when the proper order is known beforehand.

The next-to-last section of Chap. 18 presents an alternative technique for fitting precise data points. This technique, called *spline interpolation,* fits polynomials to data but in a piecewise fashion. As such, it is particularly well-suited for fitting data that are generally smooth but exhibit abrupt local changes. Finally, we provide a brief introduction to multidimensional interpolation.

Chapter 19 deals with the Fourier transform approach to curve fitting where periodic functions are fit to data. Our emphasis in this chapter will be on the *fast Fourier transform*. At the end of this chapter, we also include an overview of several software packages that can be used for curve fitting. These are Excel, MATLAB, and Mathcad.

Chapter 20 is devoted to engineering applications that illustrate the utility of the numerical methods in engineering problem contexts. Examples are drawn from the four major specialty areas of chemical, civil, electrical, and mechanical engineering. In addition, some of the applications illustrate how software packages can be applied for engineering problem solving.

Finally, an epilogue is included at the end of Part Five. It contains a summary of the important formulas and concepts related to curve fitting as well as a discussion of trade-offs among the techniques and suggestions for future study.

PT5.3.2 Goals and Objectives

Study Objectives.   After completing Part Five, you should have greatly enhanced your capability to fit curves to data. In general, you should have mastered the techniques, have learned to assess the reliability of the answers, and be capable of choosing the preferred method (or methods) for any particular problem. In addition to these general goals, the specific concepts in Table PT5.3 should be assimilated and mastered.

Computer Objectives.   You have been provided with simple computer algorithms to implement the techniques discussed in Part Five. You may also have access to software packages and libraries. All have utility as learning tools.

Pseudocode algorithms are provided for most of the methods in Part Five. This information will allow you to expand your software library to include techniques beyond polynomial regression. For example, you may find it useful from a professional viewpoint to have software to implement multiple linear regression, Newton's interpolating polynomial, cubic spline interpolation, and the fast Fourier transform.

In addition, one of your most important goals should be to master several of the general-purpose software packages that are widely available. In particular, you should become adept at using these tools to implement numerical methods for engineering problem solving.

TABLE PT5.3 Specific study objectives for Part Five.

1. Understand the fundamental difference between regression and interpolation and realize why confusing the two could lead to serious problems.
2. Understand the derivation of linear least-squares regression and be able to assess the reliability of the fit using graphical and quantitative assessments.
3. Know how to linearize data by transformation.
4. Understand situations where polynomial, multiple, and nonlinear regression are appropriate.
5. Be able to recognize general linear models, understand the general matrix formulation of linear least squares, and know how to compute confidence intervals for parameters.
6. Understand that there is one and only one polynomial of degree n or less that passes exactly through $n + 1$ points.
7. Know how to derive the first-order Newton's interpolating polynomial.
8. Understand the analogy between Newton's polynomial and the Taylor series expansion and how it relates to the truncation error.
9. Recognize that the Newton and Lagrange equations are merely different formulations of the same interpolating polynomial and understand their respective advantages and disadvantages.
10. Realize that more accurate results are generally obtained if data used for interpolation are centered around and close to the unknown point.
11. Realize that data points do not have to be equally spaced or in any particular order for either the Newton or Lagrange polynomials.
12. Know why equispaced interpolation formulas have utility.
13. Recognize the liabilities and risks associated with extrapolation.
14. Understand why spline functions have utility for data with local areas of abrupt change.
15. Understand how interpolating polynomials can be applied in two dimensions.
16. Recognize how the Fourier series is used to fit data with periodic functions.
17. Understand the difference between the frequency and time domains.

Least-Squares Regression

Where substantial error is associated with data, polynomial interpolation is inappropriate and may yield unsatisfactory results when used to predict intermediate values. Experimental data are often of this type. For example, Fig. 17.1a shows seven experimentally derived data points exhibiting significant variability. Visual inspection of these data suggests a positive relationship between y and x. That is, the overall trend indicates that higher values of y are associated with higher values of x. Now, if a sixth-order interpolating polynomial is fitted to these data (Fig. 17.1b), it will pass exactly through all of the points. However, because of the variability in these data, the curve oscillates widely in the interval between the points. In particular, the interpolated values at $x = 1.5$ and $x = 6.5$ appear to be well beyond the range suggested by these data.

A more appropriate strategy for such cases is to derive an approximating function that fits the shape or general trend of the data without necessarily matching the individual points. Figure 17.1c illustrates how a straight line can be used to generally characterize the trend of these data without passing through any particular point.

One way to determine the line in Fig. 17.1c is to visually inspect the plotted data and then sketch a "best" line through the points. Although such "eyeball" approaches have commonsense appeal and are valid for "back-of-the-envelope" calculations, they are deficient because they are arbitrary. That is, unless the points define a perfect straight line (in which case, interpolation would be appropriate), different analysts would draw different lines.

To remove this subjectivity, some criterion must be devised to establish a basis for the fit. One way to do this is to derive a curve that minimizes the discrepancy between the data points and the curve. A technique for accomplishing this objective, called *least-squares regression,* will be discussed in the present chapter.

17.1 LINEAR REGRESSION

The simplest example of a least-squares approximation is fitting a straight line to a set of paired observations: (x_1, y_1), (x_2, y_2), . . . , (x_n, y_n). The mathematical expression for the straight line is

$$y = a_0 + a_1 x + e \tag{17.1}$$

FIGURE 17.1
(a) Data exhibiting significant error. (b) Polynomial fit oscillating beyond the range of the data. (c) More satisfactory result using the least-squares fit.

where a_0 and a_1 are coefficients representing the intercept and the slope, respectively, and e is the error, or residual, between the model and the observations, which can be represented by rearranging Eq. (17.1) as

$$e = y - a_0 - a_1 x$$

Thus, the error, or *residual,* is the discrepancy between the true value of y and the approximate value, $a_0 + a_1 x$, predicted by the linear equation.

17.1.1 Criteria for a "Best" Fit

One strategy for fitting a "best" line through the data would be to minimize the sum of the residual errors for all the available data, as in

$$\sum_{i=1}^{n} e_i = \sum_{i=1}^{n} (y_i - a_0 - a_1 x_i) \tag{17.2}$$

where n = total number of points. However, this is an inadequate criterion, as illustrated by Fig. 17.2a which depicts the fit of a straight line to two points. Obviously, the best

FIGURE 17.2
Examples of some criteria for "best fit" that are inadequate for regression: (a) minimizes the sum of the residuals, (b) minimizes the sum of the absolute values of the residuals, and (c) minimizes the maximum error of any individual point.

fit is the line connecting the points. However, any straight line passing through the mid-point of the connecting line (except a perfectly vertical line) results in a minimum value of Eq. (17.2) equal to zero because the errors cancel.

Therefore, another logical criterion might be to minimize the sum of the absolute values of the discrepancies, as in

$$\sum_{i=1}^{n} |e_i| = \sum_{i=1}^{n} |y_i - a_0 - a_1 x_i|$$

Figure 17.2*b* demonstrates why this criterion is also inadequate. For the four points shown, any straight line falling within the dashed lines will minimize the sum of the absolute values. Thus, this criterion also does not yield a unique best fit.

A third strategy for fitting a best line is the *minimax* criterion. In this technique, the line is chosen that minimizes the maximum distance that an individual point falls from the line. As depicted in Fig. 17.2*c*, this strategy is ill-suited for regression because it gives undue influence to an outlier, that is, a single point with a large error. It should be noted that the minimax principle is sometimes well-suited for fitting a simple function to a complicated function (Carnahan, Luther, and Wilkes 1969).

A strategy that overcomes the shortcomings of the aforementioned approaches is to minimize the sum of the squares of the residuals between the measured y and the y calculated with the linear model,

$$S_r = \sum_{i=1}^{n} e_i^2 = \sum_{i=1}^{n} (y_{i,\text{measured}} - y_{i,\text{model}})^2 = \sum_{i=1}^{n} (y_i - a_0 - a_1 x_i)^2 \qquad (17.3)$$

This criterion has a number of advantages, including the fact that it yields a unique line for a given set of data. Before discussing these properties, we will present a technique for determining the values of a_0 and a_1 that minimize the result of Eq. (17.3).

17.1.2   Least-Squares Fit of a Straight Line

To determine values for a_0 and a_1, Eq. (17.3) is differentiated with respect to each coefficient:

$$\frac{\partial S_r}{\partial a_0} = -2 \sum (y_i - a_0 - a_1 x_i)$$

$$\frac{\partial S_r}{\partial a_1} = -2 \sum [(y_i - a_0 - a_1 x_i) x_i]$$

Note that we have simplified the summation symbols; unless otherwise indicated, all summations are from $i = 1$ to n. Setting these derivatives equal to zero will result in a minimum S_r. If this is done, the equations can be expressed as

$$0 = \sum y_i - \sum a_0 - \sum a_1 x_i$$
$$0 = \sum y_i x_i - \sum a_0 x_i - \sum a_1 x_i^2$$

Now, realizing that $\Sigma a_0 = na_0$, we can express the equations as a set of two simultaneous linear equations with two unknowns (a_0 and a_1):

$$na_0 + \left(\sum x_i\right)a_1 = \sum y_i \tag{17.4}$$

$$\left(\sum x_i\right)a_0 + \left(\sum x_i^2\right)a_1 = \sum x_i y_i \tag{17.5}$$

These are called the *normal equations*. They can be solved simultaneously

$$a_1 = \frac{n\sum x_i y_i - \sum x_i \sum y_i}{n\sum x_i^2 - \left(\sum x_i\right)^2} \tag{17.6}$$

This result can then be used in conjunction with Eq. (17.4) to solve for

$$a_0 = \bar{y} - a_1\bar{x} \tag{17.7}$$

where $\bar{y}$ and $\bar{x}$ are the means of y and x, respectively.

EXAMPLE 17.1 | Linear Regression

Problem Statement.   Fit a straight line to the x and y values in the first two columns of Table 17.1.

Solution.   The following quantities can be computed:

$$n = 7 \qquad \sum x_i y_i = 119.5 \qquad \sum x_i^2 = 140$$

$$\sum x_i = 28 \qquad \bar{x} = \frac{28}{7} = 4$$

$$\sum y_i = 24 \qquad \bar{y} = \frac{24}{7} = 3.428571$$

Using Eqs. (17.6) and (17.7),

$$a_1 = \frac{7(119.5) - 28(24)}{7(140) - (28)^2} = 0.8392857$$

$$a_0 = 3.428571 - 0.8392857(4) = 0.07142857$$

TABLE 17.1 Computations for an error analysis of the linear fit.

x_i	y_i	$(y_i - \bar{y})$	$(y_i - a_0 - a_1 x_i)^2$
1	0.5	8.5765	0.1687
2	2.5	0.8622	0.5625
3	2.0	2.0408	0.3473
4	4.0	0.3265	0.3265
5	3.5	0.0051	0.5896
6	6.0	6.6122	0.7972
7	5.5	4.2908	0.1993
Σ	24.0	22.7143	2.9911

Therefore, the least-squares fit is

$$y = 0.07142857 + 0.8392857x$$

The line, along with the data, is shown in Fig. 17.1c.

17.1.3 Quantification of Error of Linear Regression

Any line other than the one computed in Example 17.1 results in a larger sum of the squares of the residuals. Thus, the line is unique and in terms of our chosen criterion is a "best" line through the points. A number of additional properties of this fit can be elucidated by examining more closely the way in which residuals were computed. Recall that the sum of the squares is defined as [Eq. (17.3)]

$$S_r = \sum_{i=1}^{n} e_i^2 = \sum_{i=1}^{n} (y_i - a_0 - a_1 x_i)^2 \qquad (17.8)$$

Notice the similarity between Eqs. (PT5.3) and (17.8). In the former case, the square of the residual represented the square of the discrepancy between the data and a single estimate of the measure of central tendency—the mean. In Eq. (17.8), the square of the residual represents the square of the vertical distance between the data and another measure of central tendency—the straight line (Fig. 17.3).

The analogy can be extended further for cases where (1) the spread of the points around the line is of similar magnitude along the entire range of the data and (2) the distribution of these points about the line is normal. It can be demonstrated that if these criteria are met, least-squares regression will provide the best (that is, the most likely) estimates of a_0 and a_1 (Draper and Smith 1981). This is called the *maximum likelihood*

FIGURE 17.3
The residual in linear regression represents the vertical distance between a data point and the straight line.

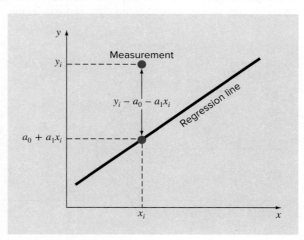

principle in statistics. In addition, if these criteria are met, a "standard deviation" for the regression line can be determined as [compare with Eq. (PT5.2)]

$$s_{y/x} = \sqrt{\frac{S_r}{n-2}}$$

(17.9)

where $s_{y/x}$ is called the *standard error of the estimate*. The subscript notation "y/x" designates that the error is for a predicted value of y corresponding to a particular value of x. Also, notice that we now divide by $n-2$ because two data-derived estimates—a_0 and a_1—were used to compute S_r; thus, we have lost two degrees of freedom. As in our discussion of the standard deviation in PT5.2.1, another justification for dividing by $n-2$ is that there is no such thing as the "spread of data" around a straight line connecting two points. Thus, for the case where $n=2$, Eq. (17.9) yields a meaningless result of infinity.

Just as was the case with the standard deviation, the standard error of the estimate quantifies the spread of the data. However, $s_{y/x}$ quantifies the spread *around the regression line,* as shown in Fig. 17.4b, in contrast to the original standard deviation s_y, which quantified the spread *around the mean* (Fig. 17.4a).

The above concepts can be used to quantify the "goodness" of our fit. This is particularly useful for comparison of several regressions (Fig. 17.5). To do this, we return to the original data and determine the *total sum of the squares* around the mean for the dependent variable (in our case, y). As was the case for Eq. (PT5.3), this quantity is designated S_t. This is the magnitude of the residual error associated with the dependent variable prior to regression. After performing the regression, we can compute S_r, the sum of the squares of the residuals around the regression line. This characterizes the residual error that remains after the regression. It is, therefore, sometimes called the unexplained

FIGURE 17.4
Regression data showing (*a*) the spread of the data around the mean of the dependent variable and (*b*) the spread of the data around the best-fit line. The reduction in the spread in going from (*a*) to (*b*), as indicated by the bell-shaped curves at the right, represents the improvement due to linear regression.

(*a*)   (*b*)

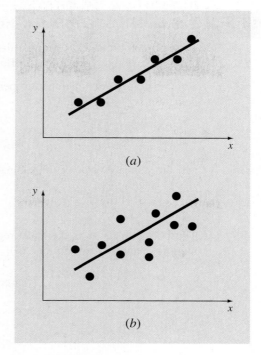

FIGURE 17.5
Examples of linear regression with (*a*) small and (*b*) large residual errors.

sum of the squares. The difference between the two quantities, $S_t - S_r$, quantifies the improvement, or error reduction, due to describing the data in terms of a straight line rather than as an average value. Because the magnitude of this quantity is scale-dependent, the difference is normalized to S_t to yield

$$r^2 = \frac{S_t - S_r}{S_t} \tag{17.10}$$

where r^2 is called the *coefficient of determination* and r is the *correlation coefficient* $(= \sqrt{r^2})$. For a perfect fit, $S_r = 0$ and $r = r^2 = 1$, signifying that the line explains 100 percent of the variability of the data. For $r = r^2 = 0$, $S_r = S_t$ and the fit represents no improvement. An alternative formulation for r that is more convenient for computer implementation is

$$r = \frac{n\Sigma x_i y_i - (\Sigma x_i)(\Sigma y_i)}{\sqrt{n\Sigma x_i^2 - (\Sigma x_i)^2}\sqrt{n\Sigma y_i^2 - (\Sigma y_i)^2}} \tag{17.11}$$

EXAMPLE 17.2 | **Estimation of Errors for the Linear Least-Squares Fit**

Problem Statement.   Compute the total standard deviation, the standard error of the estimate, and the correlation coefficient for the data in Example 17.1.

Solution.   The summations are performed and presented in Table 17.1. The standard deviation is [Eq. (PT5.2)]

$$s_y = \sqrt{\frac{22.7143}{7-1}} = 1.9457$$

and the standard error of the estimate is [Eq. (17.9)]

$$s_{y/x} = \sqrt{\frac{2.9911}{7-2}} = 0.7735$$

Thus, because $s_{y/x} < s_y$, the linear regression model has merit. The extent of the improvement is quantified by [Eq. (17.10)]

$$r^2 = \frac{22.7143 - 2.9911}{22.7143} = 0.868$$

or

$$r = \sqrt{0.868} = 0.932$$

These results indicate that 86.8% of the original uncertainty has been explained by the linear model.

Before proceeding to the computer program for linear regression, a word of caution is in order. Although the correlation coefficient provides a handy measure of goodness-of-fit, you should be careful not to ascribe more meaning to it than is warranted. Just because r is "close" to 1 does not mean that the fit is necessarily "good." For example, it is possible to obtain a relatively high value of r when the underlying relationship between y and x is not even linear. Draper and Smith (1981) provide guidance and additional material regarding assessment of results for linear regression. In addition, at the minimum, you should *always* inspect a plot of the data along with your regression curve. As described in the next section, software packages include such a capability.

17.1.4 Computer Program for Linear Regression

It is a relatively trivial matter to develop pseudocode for linear regression (Fig. 17.6). As mentioned above, a plotting option is critical to the effective use and interpretation of regression. Such capabilities are included in popular packages like MATLAB software and Excel. If your computer language has plotting capabilities, we recommend that you expand your program to include a plot of y versus x, showing both the data and the regression line. The inclusion of the capability will greatly enhance the utility of the program in problem-solving contexts.

```
SUB Regress(x, y, n, a1, a0, syx, r2)

    sumx = 0: sumxy = 0: st = 0
    sumy = 0: sumx2 = 0: sr = 0
    DOFOR i = 1, n
      sumx = sumx + xᵢ
      sumy = sumy + yᵢ
      sumxy = sumxy + xᵢ*yᵢ
      sumx2 = sumx2 + xᵢ*xᵢ
    END DO
    xm = sumx/n
    ym = sumy/n
    a1 = (n*sumxy - sumx*sumy)/(n*sumx2 - sumx*sumx)
    a0 = ym - a1*xm
    DOFOR i = 1, n
      st = st + (yᵢ - ym)²
      sr = sr + (yᵢ - a1*xᵢ - a0)²
    END DO
    syx = (sr/(n - 2))^0.5
    r2 = (st - sr)/st

END Regress
```

FIGURE 17.6
Algorithm for linear regression.

EXAMPLE 17.3

Linear Regression Using the Computer

Problem Statement. We can use software based on Fig. 17.6 to solve a hypothesis-testing problem associated with the falling parachutist discussed in Chap. 1. A theoretical mathematical model for the velocity of the parachutist was given as the following [Eq. (1.10)]:

$$v(t) = \frac{gm}{c}\left(1 - e^{(-c/m)t}\right)$$

where v = velocity (m/s), g = gravitational constant (9.8 m/s^2), m = mass of the parachutist, equal to 68.1 kg, and c = drag coefficient of 12.5 kg/s. The model predicts the velocity of the parachutist as a function of time, as described in Example 1.1.

An alternative empirical model for the velocity of the parachutist is given by

$$v(t) = \frac{gm}{c}\left(\frac{t}{3.75 + t}\right) \tag{E17.3.1}$$

Suppose that you would like to test and compare the adequacy of these two mathematical models. This might be accomplished by measuring the actual velocity of the

TABLE 17.2 Measured and calculated velocities for the falling parachutist.

Time, s	Measured v, m/s (a)	Model-calculated v, m/s [Eq. (1.10)] (b)	Model-calculated v, m/s [Eq. (E17.3.1)] (c)
1	10.00	8.953	11.240
2	16.30	16.405	18.570
3	23.00	22.607	23.729
4	27.50	27.769	27.556
5	31.00	32.065	30.509
6	35.60	35.641	32.855
7	39.00	38.617	34.766
8	41.50	41.095	36.351
9	42.90	43.156	37.687
10	45.00	44.872	38.829
11	46.00	46.301	39.816
12	45.50	47.490	40.678
13	46.00	48.479	41.437
14	49.00	49.303	42.110
15	50.00	49.988	42.712

parachutist at known values of time and comparing these results with the predicted velocities according to each model.

Such an experimental-data-collection program was implemented, and the results are listed in column (a) of Table 17.2. Computed velocities for each model are listed in columns (b) and (c).

Solution. The adequacy of the models can be tested by plotting the model-calculated velocity versus the measured velocity. Linear regression can be used to calculate the slope and the intercept of the plot. This line will have a slope of 1, an intercept of 0, and $r^2 = 1$ if the model matches the data perfectly. A significant deviation from these values can be used as an indication of the inadequacy of the model.

Figure 17.7a and b are plots of the line and data for the regressions of columns (b) and (c), respectively, versus column (a). For the first model [Eq. (1.10) as depicted in Fig. 17.7a],

$$v_{model} = -0.859 + 1.032v_{measure}$$

and for the second model [Eq. (E17.3.1) as depicted in Fig. 17.7b],

$$v_{model} = 5.776 + 0.752v_{measure}$$

These plots indicate that the linear regression between these data and each of the models is highly significant. Both models match the data with a correlation coefficient of greater than 0.99.

However, the model described by Eq. (1.10) conforms to our hypothesis test criteria much better than that described by Eq. (E17.3.1) because the slope and intercept are more nearly equal to 1 and 0, respectively. Thus, although each plot is well described by a straight line, Eq. (1.10) appears to be a better model than Eq. (E17.3.1).

FIGURE 17.7

(a) Results using linear regression to compare predictions computed with the theoretical model [Eq. (1.10)] versus measured values. (b) Results using linear regression to compare predictions computed with the empirical model [Eq. (E17.3.1)] versus measured values.

Model testing and selection are common and extremely important activities performed in all fields of engineering. The background material provided in this chapter, together with your software, should allow you to address many practical problems of this type.

There is one shortcoming with the analysis in Example 17.3. The example was unambiguous because the empirical model [Eq. (E17.3.1)] was clearly inferior to Eq. (1.10). Thus, the slope and intercept for the former were so much closer to the desired result of 1 and 0, that it was obvious which model was superior.

However, suppose that the slope were 0.85 and the intercept were 2. Obviously this would make the conclusion that the slope and intercept were 1 and 0 open to debate.

Clearly, rather than relying on a subjective judgment, it would be preferable to base such a conclusion on a quantitative criterion.

This can be done by computing confidence intervals for the model parameters in the same way that we developed confidence intervals for the mean in Sec. PT5.2.3. We will return to this topic later in this chapter.

17.1.5 Linearization of Nonlinear Relationships

Linear regression provides a powerful technique for fitting a best line to data. However, it is predicated on the fact that the relationship between the dependent and independent variables is linear. This is not always the case, and the first step in any regression analysis should be to plot and visually inspect the data to ascertain whether a linear model applies. For example, Fig. 17.8 shows some data whose plot is obviously curvilinear. In some cases, techniques such as polynomial regression, which is described in Sec. 17.2, are appropriate. For others, transformations can be used to express the data in a form that is compatible with linear regression.

One example is the *exponential model*,

$$y = \alpha_1 e^{\beta_1 x} \tag{17.12}$$

FIGURE 17.8
(*a*) Data that are ill-suited for linear least-squares regression. (*b*) Indication that a parabola is preferable.

where α_1 and β_1 are constants. This model is used in many fields of engineering to characterize quantities that increase (positive β_1) or decrease (negative β_1) at a rate that is directly proportional to their own magnitude. For example, population growth or radioactive decay can exhibit such behavior. As depicted in Fig. 17.9a, the equation represents a nonlinear relationship (for $\beta_1 \neq 0$) between y and x.

Another example of a nonlinear model is the simple power equation,

$$y = \alpha_2 x^{\beta_2} \qquad (17.13)$$

where α_2 and β_2 are constant coefficients. This model has wide applicability in all fields of engineering. As depicted in Fig. 17.9b, the graph of this equation (for $\beta_2 \neq 0$ or 1) is nonlinear.

FIGURE 17.9
(a) The exponential equation, (b) the power equation, and (c) the saturation-growth-rate equation. Parts (d), (e), and (f) are linearized versions of these equations that result from simple transformations.

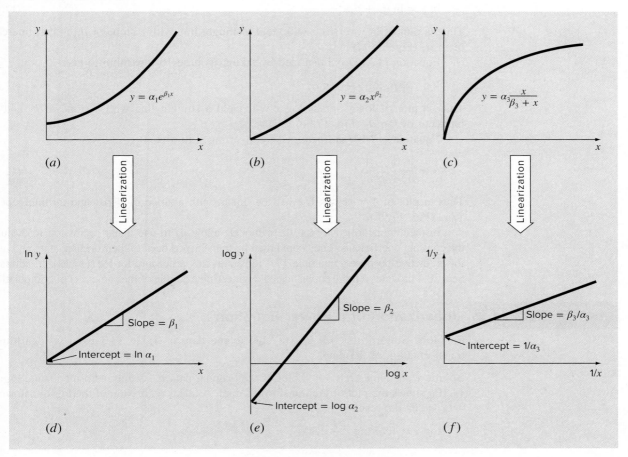

A third example of a nonlinear model is the saturation-growth-rate equation [recall Eq. (E17.3.1)],

$$y = \alpha_3 \frac{x}{\beta_3 + x} \tag{17.14}$$

where α_3 and β_3 are constant coefficients. This model, which is particularly well-suited for characterizing population growth rate under limiting conditions, also represents a nonlinear relationship between y and x (Fig. 17.9c) that levels off, or "saturates," as x increases.

Nonlinear regression techniques are available to fit these equations to experimental data directly. (Note that we will discuss nonlinear regression in Sec. 17.5.) However, a simpler alternative is to use mathematical manipulations to transform the equations into a linear form. Then, simple linear regression can be employed to fit the equations to data.

For example, Eq. (17.12) can be linearized by taking its natural logarithm to yield

$$\ln y = \ln \alpha_1 + \beta_1 x \ln e$$

But because $\ln e = 1$,

$$\ln y = \ln \alpha_1 + \beta_1 x \tag{17.15}$$

Thus, a plot of $\ln y$ versus x will yield a straight line with a slope of β_1 and an intercept of $\ln \alpha_1$ (Fig. 17.9d).

Equation (17.13) is linearized by taking its base-10 logarithm to give

$$\log y = \beta_2 \log x + \log \alpha_2 \tag{17.16}$$

Thus, a plot of $\log y$ versus $\log x$ will yield a straight line with a slope of β_2 and an intercept of $\log \alpha_2$ (Fig. 17.9e).

Equation (17.14) is linearized by inverting it to give

$$\frac{1}{y} = \frac{\beta_3}{\alpha_3} \frac{1}{x} + \frac{1}{\alpha_3} \tag{17.17}$$

Thus, a plot of $1/y$ versus $1/x$ will be linear, with a slope of β_3/α_3 and an intercept of $1/\alpha_3$ (Fig. 17.9f).

In their transformed forms, these models allow us to use linear regression to evaluate the constant coefficients. They could then be transformed back to their original state and used for predictive purposes. Example 17.4 illustrates this procedure for Eq. (17.13). In addition, Sec. 20.4 provides a more sophisticated engineering example of the same sort of computation.

EXAMPLE 17.4 **Linearization of a Power Equation**

Problem Statement. Fit Eq. (17.13) to the data in Table 17.3 using a logarithmic transformation of the data.

Solution. Figure 17.10a is a plot of the original data in its untransformed state. Figure 17.10b shows the plot of the transformed data. A linear regression of the log-transformed data yields the result

$$\log y = 1.75 \log x - 0.300$$

TABLE 17.3 Data to be fit to the power equation.

x	y	log x	log y
1	0.5	0	−0.301
2	1.7	0.301	0.226
3	3.4	0.477	0.534
4	5.7	0.602	0.753
5	8.4	0.699	0.922

FIGURE 17.10
(*a*) Plot of untransformed data with the power equation that fits these data. (*b*) Plot of transformed data used to determine the coefficients of the power equation.

(*a*)

(*b*)

Thus, the intercept, $\log \alpha_2$, equals -0.300, and therefore, by taking the antilogarithm, we get $\alpha_2 = 10^{-0.3} = 0.5$. The slope is $\beta_2 = 1.75$. Consequently, the power equation is

$$y = 0.5x^{1.75}$$

This curve, as plotted in Fig. 17.10a, indicates a good fit.

17.1.6 General Comments on Linear Regression

Before proceeding to curvilinear and multiple linear regression, we must emphasize the introductory nature of the foregoing material on linear regression. We have focused on the simple derivation and practical use of equations to fit data. You should be cognizant of the fact that there are theoretical aspects of regression that are of practical importance but are beyond the scope of this book. For example, some statistical assumptions that are inherent in the linear least-squares procedures are

1. Each x has a fixed value; it is not random and is known without error.
2. The y values are independent random variables and all have the same variance.
3. The y values for a given x must be normally distributed.

Such assumptions are relevant to the proper derivation and use of regression. For example, the first assumption means that (1) the x values must be error-free and (2) the regression of y versus x is not the same as x versus y (try Prob. 17.4 at the end of the chapter). You are urged to consult other references such as Draper and Smith (1981) to appreciate aspects and nuances of regression that are beyond the scope of this book.

17.2 POLYNOMIAL REGRESSION

In Sec. 17.1, a procedure was developed to derive the equation of a straight line using the least-squares criterion. Some engineering data, although exhibiting a marked pattern such as seen in Fig. 17.8, are poorly represented by a straight line. For these cases, a curve would be better suited to fit these data. As discussed in the previous section, one method to accomplish this objective is to use transformations. Another alternative is to fit polynomials to the data using *polynomial regression*.

The least-squares procedure can be readily extended to fit the data to a higher-order polynomial. For example, suppose that we fit a second-order polynomial or quadratic:

$$y = a_0 + a_1x + a_2x^2 + e$$

For this case the sum of the squares of the residuals is [compare with Eq. (17.3)]

$$S_r = \sum_{i=1}^{n} (y_i - a_0 - a_1x_i - a_2x_i^2)^2 \tag{17.18}$$

Following the procedure of the previous section, we take the derivative of Eq. (17.18) with respect to each of the unknown coefficients of the polynomial, as in

$$\frac{\partial S_r}{\partial a_0} = -2\sum (y_i - a_0 - a_1x_i - a_2x_i^2)$$

$$\frac{\partial S_r}{\partial a_1} = -2\sum x_i(y_i - a_0 - a_1 x_i - a_2 x_i^2)$$

$$\frac{\partial S_r}{\partial a_2} = -2\sum x_i^2(y_i - a_0 - a_1 x_i - a_2 x_i^2)$$

These equations can be set equal to zero and rearranged to develop the following set of normal equations:

$$(n)a_0 + \left(\sum x_i\right)a_1 + \left(\sum x_i^2\right)a_2 = \sum y_i$$

$$\left(\sum x_i\right)a_0 + \left(\sum x_i^2\right)a_1 + \left(\sum x_i^3\right)a_2 = \sum x_i y_i \qquad (17.19)$$

$$\left(\sum x_i^2\right)a_0 + \left(\sum x_i^3\right)a_1 + \left(\sum x_i^4\right)a_2 = \sum x_i^2 y_i$$

where all summations are from $i = 1$ through n. Note that the above three equations are linear and have three unknowns: a_0, a_1, and a_2. The coefficients of the unknowns can be calculated directly from the observed data.

For this case, we see that the problem of determining a least-squares second-order polynomial is equivalent to solving a system of three simultaneous linear equations. Techniques to solve such equations were discussed in Part Three.

The two-dimensional case can be easily extended to an mth-order polynomial:

$$y = a_0 + a_1 x + a_2 x^2 + \cdots + a_m x^m + e$$

The foregoing analysis can be easily extended to this more general case. Thus, we can recognize that determining the coefficients of an mth-order polynomial is equivalent to solving a system of $m + 1$ simultaneous linear equations. For this case, the standard error is formulated as

$$s_{y/x} = \sqrt{\frac{S_r}{n - (m + 1)}} \qquad (17.20)$$

This quantity is divided by $n - (m + 1)$ because $(m + 1)$ data-derived coefficients—a_0, $a_1, \ldots, a_m$—were used to compute S_r; thus, we have lost $m + 1$ degrees of freedom. In addition to the standard error, a coefficient of determination can also be computed for polynomial regression with Eq. (17.10).

EXAMPLE 17.5   **Polynomial Regression**

Problem Statement.   Fit a second-order polynomial to the data in the first two columns of Table 17.4.

Solution.   From the given data,

$$m = 2 \qquad \sum x_i = 15 \qquad \sum x_i^4 = 979$$
$$n = 6 \qquad \sum y_i = 152.6 \qquad \sum x_i y_i = 585.6$$
$$\bar{x} = 2.5 \qquad \sum x_i^2 = 55 \qquad \sum x_i^2 y_i = 2488.8$$
$$\bar{y} = 25.433 \qquad \sum x_i^3 = 225$$

TABLE 17.4 Computations for an error analysis of the quadratic least-squares fit.

x_i	y_i	$(y_i - \bar{y})^2$	$(y_i - a_0 - a_1 x_i - a_2 x_i^2)^2$
0	2.1	544.44	0.14332
1	7.7	314.47	1.00286
2	13.6	140.03	1.08158
3	27.2	3.12	0.80491
4	40.9	239.22	0.61951
5	61.1	1272.11	0.09439
Σ	152.6	2513.39	3.74657

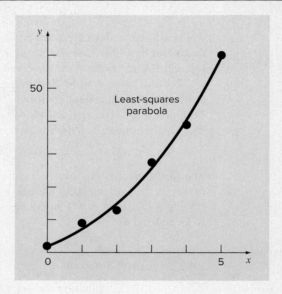

FIGURE 17.11
Fit of a second-order polynomial.

Therefore, the simultaneous linear equations are

$$\begin{bmatrix} 6 & 15 & 55 \\ 15 & 55 & 225 \\ 55 & 225 & 979 \end{bmatrix} \begin{Bmatrix} a_0 \\ a_1 \\ a_2 \end{Bmatrix} = \begin{Bmatrix} 152.6 \\ 585.6 \\ 2488.8 \end{Bmatrix}$$

Solving these equations through a technique such as Gauss elimination gives $a_0 = 2.47857$, $a_1 = 2.35929$, and $a_2 = 1.86071$. Therefore, the least-squares quadratic equation for this case is

$$y = 2.47857 + 2.35929x + 1.86071x^2$$

The standard error of the estimate based on the regression polynomial is [Eq. (17.20)]

$$s_{y/x} = \sqrt{\frac{3.74657}{6 - 3}} = 1.12$$

The coefficient of determination is

$$r^2 = \frac{2513.39 - 3.74657}{2513.39} = 0.99851$$

and the correlation coefficient is $r = 0.99925$.

These results indicate that 99.851% of the original uncertainty has been explained by the model. This result supports the conclusion that the quadratic equation represents an excellent fit, as is also evident from Fig. 17.11.

17.2.1 Algorithm for Polynomial Regression

An algorithm for polynomial regression is delineated in Fig. 17.12. Note that the primary task is the generation of the coefficients of the normal equations [Eq. (17.19)]. (Pseudocode for accomplishing this is presented in Fig. 17.13.) Then, techniques from Part Three can be applied to solve these simultaneous equations for the coefficients.

A potential problem associated with implementing polynomial regression on the computer is that the normal equations tend to be ill-conditioned. This is particularly true for higher-order versions. For these cases, the computed coefficients may be highly susceptible to round-off error, and consequently, the results can be inaccurate. Among other things, this problem is related to the structure of the normal equations and to the fact that for higher-order polynomials the normal equations can have very large and very small coefficients. This is because the coefficients are summations of the data raised to powers.

Although the strategies for mitigating round-off error discussed in Part Three, such as pivoting, can help to partially remedy this problem, a simpler alternative is to use a computer with higher precision. Fortunately, most practical problems are limited to lower-order polynomials for which round-off is usually negligible. In situations where higher-order versions are required, other alternatives are available for certain types of data. However, these techniques (such as orthogonal polynomials) are beyond the scope of this book. The reader should consult texts on regression, such as Draper and Smith (1981), for additional information regarding the problem and possible alternatives.

FIGURE 17.12
Algorithm for implementation of polynomial and multiple linear regression.

Step 1: Input order of polynomial to be fit, m.
Step 2: Input number of data points, n.
Step 3: If $n < m + 1$, print out an error message that regression is impossible and terminate the process. If $n \geq m + 1$, continue.
Step 4: Compute the elements of the normal equation in the form of an augmented matrix.
Step 5: Solve the augmented matrix for the coefficients $a_0, a_1, a_2, \ldots, a_m$ using an elimination method.
Step 6: Print out the coefficients.

```
DOFOR i = 1, order + 1
  DOFOR j = 1, i
    k = i + j - 2
    sum = 0
    DOFOR ℓ = 1, n
      sum = sum + xₗᵏ
    END DO
    aᵢ,ⱼ = sum
    aⱼ,ᵢ = sum
  END DO
  sum = 0
  DOFOR ℓ = 1, n
    sum = sum + yₗ · xₗⁱ⁻¹
  END DO
  aᵢ,order+2 = sum
END DO
```

FIGURE 17.13
Pseudocode to assemble the elements of the normal equations for polynomial regression.

17.3 MULTIPLE LINEAR REGRESSION

A useful extension of linear regression is the case where y is a linear function of two or more independent variables. For example, y might be a linear function of x_1 and x_2, as in

$$y = a_0 + a_1x_1 + a_2x_2 + e$$

Such an equation is particularly useful when fitting experimental data, where the variable being studied is often a function of two other variables. For this two-dimensional case, the regression "line" becomes a "plane" (Fig. 17.14).

FIGURE 17.14
Graphical depiction of multiple linear regression where y is a linear function of x_1 and x_2.

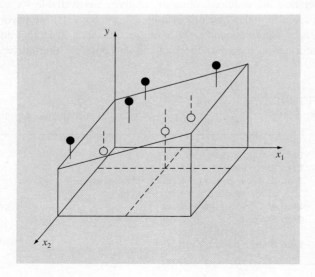

As with the previous cases, the "best" values of the coefficients are determined by setting up the sum of the squares of the residuals,

$$S_r = \sum_{i=1}^{n} (y_i - a_0 - a_1 x_{1i} - a_2 x_{2i})^2 \tag{17.21}$$

and differentiating with respect to each of the unknown coefficients,

$$\frac{\partial S_r}{\partial a_0} = -2 \sum (y_i - a_0 - a_1 x_{1i} - a_2 x_{2i})$$

$$\frac{\partial S_r}{\partial a_1} = -2 \sum x_{1i}(y_i - a_0 - a_1 x_{1i} - a_2 x_{2i})$$

$$\frac{\partial S_r}{\partial a_2} = -2 \sum x_{2i}(y_i - a_0 - a_1 x_{1i} - a_2 x_{2i})$$

The coefficients yielding the minimum sum of the squares of the residuals are obtained by setting the partial derivatives equal to zero and expressing the result in matrix form as

$$\begin{bmatrix} n & \sum x_{1i} & \sum x_{2i} \\ \sum x_{1i} & \sum x_{1i}^2 & \sum x_{1i}x_{2i} \\ \sum x_{2i} & \sum x_{1i}x_{2i} & \sum x_{2i}^2 \end{bmatrix} \begin{Bmatrix} a_0 \\ a_1 \\ a_2 \end{Bmatrix} = \begin{Bmatrix} \sum y_i \\ \sum x_{1i}y_i \\ \sum x_{2i}y_i \end{Bmatrix} \tag{17.22}$$

EXAMPLE 17.6   **Multiple Linear Regression**

Problem Statement.   The following data were calculated from the equation $y = 5 + 4x_1 - 3x_2$:

x_1	x_2	y
0	0	5
2	1	10
2.5	2	9
1	3	0
4	6	3
7	2	27

Use multiple linear regression to fit these data.

Solution.   The summations required to develop Eq. (17.22) are computed in Table 17.5. The result is

$$\begin{bmatrix} 6 & 16.5 & 14 \\ 16.5 & 76.25 & 48 \\ 14 & 48 & 54 \end{bmatrix} \begin{Bmatrix} a_0 \\ a_1 \\ a_2 \end{Bmatrix} = \begin{Bmatrix} 54 \\ 243.5 \\ 100 \end{Bmatrix}$$

This equation can be solved using a method such as Gauss elimination to give

$$a_0 = 5 \quad a_1 = 4 \quad a_2 = -3$$

which are consistent with the original equation from which these data were derived.

TABLE 17.5 Computations required to develop the normal equations for Example 17.6.

y	x_1	x_2	x_1^2	x_2^2	$x_1 x_2$	$x_1 y$	$x_2 y$
5	0	0	0	0	0	0	0
10	2	1	4	1	2	20	10
9	2.5	2	6.25	4	5	22.5	18
0	1	3	1	9	3	0	0
3	4	6	16	36	24	12	18
27	7	2	49	4	14	189	54
Σ 54	16.5	14	76.25	54	48	243.5	100

The foregoing two-dimensional case can be easily extended to m dimensions, as in

$$y = a_0 + a_1 x_1 + a_2 x_2 + \cdots + a_m x_m + e$$

where the standard error is formulated as

$$s_{y/x} = \sqrt{\frac{S_r}{n - (m + 1)}}$$

and the coefficient of determination is computed as in Eq. (17.10). An algorithm to set up the normal equations is listed in Fig. 17.15.

Although there may be certain cases where a variable is linearly related to two or more other variables, multiple linear regression has additional utility in the derivation of power equations of the general form

$$y = a_0 x_1^{a_1} x_2^{a_2} \cdots x_m^{a_m}$$

FIGURE 17.15
Pseudocode to assemble the elements of the normal equations for multiple regression. Note that aside from storing the independent variables in $x_{1,i}$, $x_{2,i}$, etc., 1's must be stored in $x_{0,i}$ for this algorithm to work.

```
DOFOR i = 1, order + 1
  DOFOR j = 1, i
    sum = 0
    DOFOR ℓ = 1, n
      sum = sum + x_{i-1,ℓ} · x_{j-1,ℓ}
    END DO
    a_{i,j} = sum
    a_{j,i} = sum
  END DO
  sum = 0
  DOFOR ℓ = 1, n
    sum = sum + y_ℓ · x_{i-1,ℓ}
  END DO
  a_{i,order+2} = sum
END DO
```

Such equations are extremely useful when fitting experimental data. To use multiple linear regression, the equation is transformed by taking its logarithm to yield

$$\log y = \log a_0 + a_1 \log x_1 + a_2 \log x_2 + \cdots + a_m \log x_m$$

This transformation is similar in spirit to the one used in Sec. 17.1.5 and Example 17.4 to fit a power equation when y was a function of a single variable x. Section 20.4 provides an example of such an application for two independent variables.

17.4   GENERAL LINEAR LEAST SQUARES

To this point, we have focused on the mechanics of obtaining least-squares fits of some simple functions to data. Before turning to nonlinear regression, there are several issues that we would like to discuss to enrich your understanding of the preceding material.

17.4.1 General Matrix Formulation for Linear Least Squares

In the preceding pages, we have introduced three types of regression: simple linear, polynomial, and multiple linear. In fact, all three belong to the following general linear least-squares model:

$$y = a_0 z_0 + a_1 z_1 + a_2 z_2 + \cdots + a_m z_m + e \tag{17.23}$$

where $z_0, z_1, \ldots, z_m$ are $m + 1$ basis functions. It can easily be seen how simple and multiple linear regression fall within this model—that is, $z_0 = 1, z_1 = x_1, z_2 = x_2, \ldots, z_m = x_m$. Further, polynomial regression is also included if the basis functions are simple monomials as in $z_0 = x^0 = 1, z_1 = x, z_2 = x^2, \ldots, z_m = x^m$.

Note that the terminology "linear" refers only to the model's dependence on its parameters—that is, the a's. As in the case of polynomial regression, the functions themselves can be highly nonlinear. For example, the z's can be sinusoids, as in

$$y = a_0 + a_1 \cos(\omega t) + a_2 \sin(\omega t)$$

Such a format is the basis of Fourier analysis described in Chap. 19.

On the other hand, a simple-looking model like

$$f(x) = a_0(1 - e^{-a_1 x})$$

is truly nonlinear because it cannot be manipulated into the format of Eq. (17.23). We will turn to such models at the end of this chapter.

For the time being, Eq. (17.23) can be expressed in matrix notation as

$$\{Y\} = [Z]\{A\} + \{E\} \tag{17.24}$$

where $[Z]$ is a matrix of the calculated values of the basis functions at the measured values of the independent variables,

$$[Z] = \begin{bmatrix} z_{01} & z_{11} & \cdots & z_{m1} \\ z_{02} & z_{12} & \cdots & z_{m2} \\ \cdot & \cdot & & \cdot \\ \cdot & \cdot & & \cdot \\ \cdot & \cdot & & \cdot \\ z_{0n} & z_{1n} & \cdots & z_{mn} \end{bmatrix}$$

where m is the number of variables in the model and n is the number of data points. Because $n \geq m + 1$, you should recognize that most of the time, $[Z]$ is not a square matrix.

The column vector $\{Y\}$ contains the observed values of the dependent variable

$$\{Y\}^T = \lfloor y_1 \quad y_2 \quad \cdots \quad y_n \rfloor$$

The column vector $\{A\}$ contains the unknown coefficients

$$\{A\}^T = \lfloor a_0 \quad a_1 \quad \cdots \quad a_m \rfloor$$

and the column vector $\{E\}$ contains the residuals

$$\{E\}^T = \lfloor e_1 \quad e_2 \quad \cdots \quad e_n \rfloor$$

As was done throughout this chapter, the sum of the squares of the residuals for this model can be defined as

$$S_r = \sum_{i=1}^{n} \left(y_i - \sum_{j=0}^{m} a_j z_{ji} \right)^2$$

This quantity can be minimized by taking its partial derivative with respect to each of the coefficients and setting the resulting equations equal to zero. The outcome of this process is the normal equations that can be expressed concisely in matrix form as

$$[[Z]^T[Z]]\{A\} = \{[Z]^T\{Y\}\} \tag{17.25}$$

It can be shown that Eq. (17.25) is, in fact, equivalent to the normal equations developed previously for simple linear, polynomial, and multiple linear regression.

Our primary motivation for the foregoing has been to illustrate the unity among the three approaches and to show how they can all be expressed simply in the same matrix notation. The matrix notation will also have relevance when we turn to nonlinear regression in the last section of this chapter.

From Eq. (PT3.6), recall that the matrix inverse can be employed to solve Eq. (17.25), as in

$$\{A\} = [[Z]^T[Z]]^{-1}\{[Z]^T\{Y\}\} \tag{17.26}$$

As we have learned in Part Three, this is an inefficient approach for solving a set of simultaneous equations. However, from a statistical perspective, there are a number of reasons why we might be interested in obtaining the inverse and examining its coefficients. These reasons will be discussed next.

17.4.2 Statistical Aspects of Least-Squares Theory

In Sec. PT5.2.1, we reviewed a number of descriptive statistics that can be used to describe a sample. These included the arithmetic mean, the standard deviation, and the variance.

Aside from yielding a solution for the regression coefficients, the matrix formulation of Eq. (17.26) provides estimates of their statistics. It can be shown (Draper and Smith 1981) that the diagonal and off-diagonal terms of the matrix $[[Z]^T[Z]]^{-1}$ give, respectively, the variances and the covariances[1] of the a's. If the diagonal elements of

---

[1]The covariance is a statistic that measures the dependency of one variable on another. Thus, $\text{cov}(x, y)$ indicates the dependency of x and y. For example, $\text{cov}(x, y) = 0$ would indicate that x and y are totally independent.

$[[Z]^T[Z]]^{-1}$ are designated as $z_{i,i}^{-1}$, then

$$\text{var}(a_{i-1}) = z_{i,i}^{-1} s_{y/x}^2 \tag{17.27}$$

and

$$\text{cov}(a_{i-1}, a_{j-1}) = z_{i,j}^{-1} s_{y/x}^2 \tag{17.28}$$

These statistics have a number of important applications. For our present purposes, we will illustrate how they can be used to develop confidence intervals for the intercept and slope.

Using an approach similar to that in Sec. PT5.2.3, it can be shown that lower and upper bounds on the intercept can be formulated as (see Milton and Arnold 2002, for details)

$$L = a_0 - t_{\alpha/2, n-2}\, s(a_0) \qquad U = a_0 + t_{\alpha/2, n-2}\, s(a_0) \tag{17.29}$$

where $s(a_j)$ = the standard error of coefficient $a_j = \sqrt{\text{var}(a_j)}$. In a similar manner, lower and upper bounds on the slope can be formulated as

$$L = a_1 - t_{\alpha/2, n-2}\, s(a_1) \qquad U = a_1 + t_{\alpha/2, n-2}\, s(a_1) \tag{17.30}$$

The following example illustrates how these intervals can be used to make quantitative inferences related to linear regression.

EXAMPLE 17.7   **Confidence Intervals for Linear Regression**

Problem Statement.   In Example 17.3, we used regression to develop the following relationship between measurements and model predictions:

$$y = -0.859 + 1.032x$$

where y = the model predictions and x = the measurements. We concluded that there was a good agreement between the two because the intercept was approximately equal to 0 and the slope approximately equal to 1. Recompute the regression but use the matrix approach to estimate standard errors for the parameters. Then employ these errors to develop confidence intervals, and use these to make a probabilistic statement regarding the goodness of fit.

Solution.   These data can be written in matrix format for simple linear regression as

$$[Z] = \begin{bmatrix} 1 & 10 \\ 1 & 16.3 \\ 1 & 23 \\ . & . \\ . & . \\ . & . \\ 1 & 50 \end{bmatrix} \qquad \{Y\} = \begin{Bmatrix} 8.953 \\ 16.405 \\ 22.607 \\ . \\ . \\ . \\ 49.988 \end{Bmatrix}$$

Matrix transposition and multiplication can then be used to generate the normal equations,

$$[[Z]^T[Z]] \qquad \{A\} = \{[Z]^T\{Y\}\}$$

$$\begin{bmatrix} 15 & 548.3 \\ 548.3 & 22191.21 \end{bmatrix} \begin{Bmatrix} a_0 \\ a_1 \end{Bmatrix} = \begin{Bmatrix} 552.741 \\ 22421.43 \end{Bmatrix}$$

Matrix inversion can be used to obtain the slope and intercept:

$$\{A\} = \quad [[Z]^T[Z]]^{-1} \quad \{[Z]^T\{Y\}\}$$

$$= \begin{bmatrix} 0.688414 & -0.01701 \\ -0.01701 & 0.000465 \end{bmatrix} \begin{Bmatrix} 552.741 \\ 22421.43 \end{Bmatrix} = \begin{Bmatrix} -0.85872 \\ 1.031592 \end{Bmatrix}$$

Thus, the intercept and the slope are determined as $a_0 = -0.85872$ and $a_1 = 1.031592$, respectively. These values in turn can be used to compute the standard error of the estimate as $s_{y/x} = 0.863403$. This value can be used along with the diagonal elements of the matrix inverse to calculate the standard errors of the coefficients,

$$s(a_0) = \sqrt{z_{11}^{-1}s_{y/x}^2} = \sqrt{0.688414(0.863403)^2} = 0.716372$$

$$s(a_1) = \sqrt{z_{22}^{-1}s_{y/x}^2} = \sqrt{0.000465(0.863403)^2} = 0.018625$$

The statistic, $t_{\alpha/2,n-1}$ needed for a 95% confidence interval with $n - 2 = 15 - 2 = 13$ degrees of freedom can be determined from a statistics table or using software. We used an Excel function, TINV, to come up with the proper value, as in

```
=TINV(0.05,13)
```

which yielded a value of 2.160368. Equations (17.29) and (17.30) can then be used to compute the confidence intervals as

$$a_0 = -0.85872 \pm 2.160368(0.716372)$$

$$= -0.85872 \pm 1.547627 = [-2.40634, 0.688912]$$

$$a_1 = 1.031592 \pm 2.160368(0.018625)$$

$$= 1.031592 \pm 0.040237 = [0.991355, 1.071828]$$

Notice that the desired values (0 for intercept and slope and 1 for the intercept) fall within the intervals. On the basis of this analysis we could make the following statement regarding the slope: We have strong grounds for believing that the slope of the true regression line lies within the interval from 0.991355 to 1.071828. Because 1 falls within this interval, we also have strong grounds for believing that the result supports the agreement between the measurements and the model. Because zero falls within the intercept interval, a similar statement can be made regarding the intercept.

As mentioned previously in Sec. 17.2.1, the normal equations are notoriously ill-conditioned. Hence, if solved with conventional techniques such as *LU* decomposition, the computed coefficients can be highly susceptible to round-off error. As a consequence, more sophisticated orthogonalization algorithms, such as *QR factorization,* are available to circumvent the problem. Because these techniques are beyond the scope of this book, the reader should consult texts on regression, such as Draper and Smith (1981), for additional information regarding the problem and possible alternatives. Moler (2004) also provides a nice discussion of the topic with emphasis on the numerical methods.

The foregoing is a limited introduction to the rich topic of statistical inference and its relationship to regression. There are many subleties that are beyond the scope of this book. Our primary motivation has been to illustrate the power of the matrix approach to general linear least squares. In addition, it should be noted that software packages such as Excel, MATLAB, and Mathcad can generate least-squares regression fits along with information relevant to inferential statistics. We will explore some of these capabilities when we describe these packages at the end of Chap. 19.

17.5 NONLINEAR REGRESSION

There are many cases in engineering where nonlinear models must be fit to data. In the present context, these models are defined as those that have a nonlinear dependence on their parameters. For example,

$$f(x) = a_0(1 - e^{-a_1 x}) + e \tag{17.31}$$

This equation cannot be manipulated so that it conforms to the general form of Eq. (17.23).

As with linear least squares, nonlinear regression is based on determining the values of the parameters that minimize the sum of the squares of the residuals. However, for the nonlinear case, the solution must proceed in an iterative fashion.

The *Gauss-Newton method* is one algorithm for minimizing the sum of the squares of the residuals between data and nonlinear equations. The key concept underlying the technique is that a Taylor series expansion is used to express the original nonlinear equation in an approximate, linear form. Then, least-squares theory can be used to obtain new estimates of the parameters that move in the direction of minimizing the residuals.

To illustrate how this is done, first the relationship between the nonlinear equation and the data can be expressed generally as

$$y_i = f(x_i; a_0, a_1, \ldots, a_m) + e_i$$

where y_i = a measured value of the dependent variable, $f(x_i; a_0, a_1, \ldots, a_m)$ = the equation that is a function of the independent variable x_i and a nonlinear function of the parameters $a_0, a_1, \ldots, a_m$, and e_i = a random error. For convenience, this model can be expressed in abbreviated form by omitting the parameters,

$$y_i = f(x_i) + e_i \tag{17.32}$$

The nonlinear model can be expanded in a Taylor series around the parameter values and curtailed after the first derivative. For example, for a two-parameter case,

$$f(x_i)_{j+1} = f(x_i)_j + \frac{\partial f(x_i)_j}{\partial a_0} \Delta a_0 + \frac{\partial f(x_i)_j}{\partial a_1} \Delta a_1 \tag{17.33}$$

where j = the initial guess, $j + 1$ = the prediction, $\Delta a_0 = a_{0,j+1} - a_{0,j}$, and $\Delta a_1 = a_{1,j+1} - a_{1,j}$. Thus, we have linearized the original model with respect to the parameters. Equation (17.33) can be substituted into Eq. (17.32) to yield

$$y_i - f(x_i)_j = \frac{\partial f(x_i)_j}{\partial a_0} \Delta a_0 + \frac{\partial f(x_i)_j}{\partial a_1} \Delta a_1 + e_i$$

or in matrix form [compare with Eq. (17.24)],

$$\{D\} = [Z_j]\{\Delta A\} + \{E\} \tag{17.34}$$

Here $[Z_j]$ is the matrix of partial derivatives of the function evaluated at the initial guess j,

$$[Z_j] = \begin{bmatrix} \partial f_1/\partial a_0 & \partial f_1/\partial a_1 \\ \partial f_2/\partial a_0 & \partial f_2/\partial a_1 \\ \cdot & \cdot \\ \cdot & \cdot \\ \cdot & \cdot \\ \partial f_n/\partial a_0 & \partial f_n/\partial a_1 \end{bmatrix}$$

where n = the number of data points and $\partial f_i/\partial a_k$ = the partial derivative of the function with respect to the kth parameter evaluated at the ith data point. The vector $\{D\}$ contains the differences between the measurements and the function values,

$$\{D\} = \begin{Bmatrix} y_1 - f(x_1) \\ y_2 - f(x_2) \\ \cdot \\ \cdot \\ \cdot \\ y_n - f(x_n) \end{Bmatrix}$$

and the vector $\{\Delta A\}$ contains the changes in the parameter values,

$$\{\Delta A\} = \begin{Bmatrix} \Delta a_0 \\ \Delta a_1 \\ \cdot \\ \cdot \\ \cdot \\ \Delta a_m \end{Bmatrix}$$

Applying linear least-squares theory to Eq. (17.34) results in the following normal equations [recall Eq. (17.25)]:

$$[[Z_j]^T[Z_j]]\{\Delta A\} = \{[Z_j]^T\{D\}\} \tag{17.35}$$

Thus, the approach consists of solving Eq. (17.35) for $\{\Delta A\}$, which can be employed to compute improved values for the parameters, as in

$$a_{0,j+1} = a_{0,j} + \Delta a_0$$

and

$$a_{1,j+1} = a_{1,j} + \Delta a_1$$

This procedure is repeated until the solution converges—that is, until

$$|\varepsilon_a|_k = \left| \frac{a_{k,j+1} - a_{k,j}}{a_{k,j+1}} \right| 100\% \tag{17.36}$$

falls below an acceptable stopping criterion.

EXAMPLE 17.8

Gauss-Newton Method

Problem Statement.   Fit the function $f(x; a_0, a_1) = a_0(1 - e^{-a_1 x})$ to these data:

x	0.25	0.75	1.25	1.75	2.25
y	0.28	0.57	0.68	0.74	0.79

Use initial guesses of $a_0 = 1.0$ and $a_1 = 1.0$ for the parameters. Note that for these guesses, the initial sum of the squares of the residuals is 0.0248.

Solution.   The partial derivatives of the function with respect to the parameters are

$$\frac{\partial f}{\partial a_0} = 1 - e^{-a_1 x} \tag{E17.8.1}$$

and

$$\frac{\partial f}{\partial a_1} = a_0 x e^{-a_1 x} \tag{E17.8.2}$$

Equations (E17.8.1) and (E17.8.2) can be used to evaluate the matrix:

$$[Z_0] = \begin{bmatrix} 0.2212 & 0.1947 \\ 0.5276 & 0.3543 \\ 0.7135 & 0.3581 \\ 0.8262 & 0.3041 \\ 0.8946 & 0.2371 \end{bmatrix}$$

This matrix multiplied by its transpose results in

$$[Z_0]^T [Z_0] = \begin{bmatrix} 2.3193 & 0.9489 \\ 0.9489 & 0.4404 \end{bmatrix}$$

which in turn can be inverted to yield

$$[[Z_0]^T [Z_0]]^{-1} = \begin{bmatrix} 3.6397 & -7.8421 \\ -7.8421 & 19.1678 \end{bmatrix}$$

The vector $\{D\}$ consists of the differences between the measurements and the model predictions,

$$\{D\} = \begin{Bmatrix} 0.28 - 0.2212 \\ 0.57 - 0.5276 \\ 0.68 - 0.7135 \\ 0.74 - 0.8262 \\ 0.79 - 0.8946 \end{Bmatrix} = \begin{Bmatrix} 0.0588 \\ 0.0424 \\ -0.0335 \\ -0.0862 \\ -0.1046 \end{Bmatrix}$$

It is multiplied by $[Z_0]^T$ to give

$$[Z_0]^T \{D\} = \begin{bmatrix} -0.1533 \\ -0.0365 \end{bmatrix}$$

The vector $\{\Delta A\}$ is then calculated by solving Eq. (17.35) for

$$\Delta A = \begin{Bmatrix} -0.2714 \\ 0.5019 \end{Bmatrix}$$

which can be added to the initial parameter guesses to yield

$$\begin{Bmatrix} a_0 \\ a_1 \end{Bmatrix} = \begin{Bmatrix} 1.0 \\ 1.0 \end{Bmatrix} + \begin{Bmatrix} -0.2714 \\ 0.5019 \end{Bmatrix} = \begin{Bmatrix} 0.7286 \\ 1.5019 \end{Bmatrix}$$

Thus, the improved estimates of the parameters are $a_0 = 0.7286$ and $a_1 = 1.5019$. The new parameters result in a sum of the squares of the residuals equal to 0.0242. Equation (17.36) can be used to compute ε_0 and ε_1 as 37% and 33%, respectively. The computation would then be repeated until these values fell below the prescribed stopping criterion. The final result is $a_0 = 0.79186$ and $a_1 = 1.6751$. These coefficients give a sum of the squares of the residuals of 0.000662.

A potential problem with the Gauss-Newton method as developed to this point is that the partial derivatives of the function may be difficult to evaluate. Consequently, many computer programs use difference equations to approximate the partial derivatives. One such equation is

$$\frac{\partial f_i}{\partial a_k} \cong \frac{f(x_i; a_0, \ldots, a_k + \delta a_k, \ldots, a_m) - f(x_i; a_0, \ldots, a_k, \ldots, a_m)}{\delta a_k} \qquad (17.37)$$

where δ = a small fractional perturbation.

The Gauss-Newton method has a number of other possible shortcomings:

1. It may converge slowly.
2. It may oscillate widely, that is, continually change directions.
3. It may not converge at all.

Modifications of the method (Booth and Peterson 1958; Hartley 1961) have been developed to remedy the shortcomings.

In addition, although there are several approaches expressly designed for regression, a more general approach is to use nonlinear optimization routines as described in Part Four. To do this, a guess for the parameters is made, and the sum of the squares of the residuals is computed. For example, for Eq. (17.31) it would be computed as

$$S_r = \sum_{i=1}^{n} [y_i - a_0(1 - e^{-a_1 x_i})]^2 \qquad (17.38)$$

Then, the parameters would be adjusted systematically to minimize S_r using search techniques of the type described previously in Chap. 14. We will illustrate how this is done when we describe software applications at the end of Chap. 19.

PROBLEMS

17.1 Given these data

0.90	1.42	1.30	1.55	1.63
1.32	1.35	1.47	1.95	1.66
1.96	1.47	1.92	1.35	1.05
1.85	1.74	1.65	1.78	1.71
2.29	1.82	2.06	2.14	1.27

determine **(a)** the mean, **(b)** the standard deviation, **(c)** the variance, **(d)** the coefficient of variation, and **(e)** the 95% confidence interval for the mean. **(f)** Construct a histogram using a range from 0.6 to 2.4 with intervals of 0.2.

17.2 Given these data

28.65	26.55	26.65	27.65	27.35	28.35	26.85
28.65	29.65	27.85	27.05	28.25	28.85	26.75
27.65	28.45	28.65	28.45	31.65	26.35	27.75
29.25	27.65	28.65	27.65	28.55	27.65	27.25

determine **(a)** the mean, **(b)** the standard deviation, **(c)** the variance, **(d)** the coefficient of variation, and **(e)** the 90% confidence interval for the mean. **(f)** Construct a histogram. Use a range from 26 to 32 with increments of 0.5. **(g)** Assuming that the distribution is normal and that your estimate of the standard deviation is valid, compute the range (that is, the lower and the upper values) that encompasses 68% of the readings. Determine whether this is a valid estimate for the data in this problem.

17.3 Use least-squares regression to fit a straight line to

x	0	2	4	6	9	11	12	15	17	19
y	4	6	7	6	9	8	8	10	12	12

Along with the slope and intercept, compute the standard error of the estimate and the correlation coefficient. Plot the data and the regression line. Then repeat the problem, but regress x versus y— that is, switch the variables. Interpret your results.

17.4 Use least-squares regression to fit a straight line to

x	6	7	11	15	17	21	23	29	29	37	39
y	29	21	29	14	21	15	7	7	13	0	3

Along with the slope and the intercept, compute the standard error of the estimate and the correlation coefficient. Plot the data and the regression line. If someone made an additional measurement of $x = 10$, $y = 10$, would you suspect, based on a visual assessment and the standard error, that the measurement was valid or faulty? Justify your conclusion.

17.5 Using the same approach as was employed to derive Eqs. (17.15) and (17.16), derive the least-squares fit of the following model:

$$y = a_1 x + e$$

That is, determine the slope that results in the least-squares fit for a straight line with a zero intercept. Fit the following data with this model and display the result graphically:

x	2	4	6	7	10	11	14	17	20
y	1	2	5	2	8	7	6	9	12

17.6 Use least-squares regression to fit a straight line to

x	1	2	3	4	5	6	7	8	9
y	1	1.5	2	3	4	5	8	10	13

(a) Along with the slope and intercept, compute the standard error of the estimate and the correlation coefficient. Plot the data and the straight line. Assess the fit.
(b) Recompute **(a)**, but use polynomial regression to fit a parabola to the data. Compare the results with those of **(a)**.

17.7 Fit the following data with **(a)** a saturation-growth-rate model, **(b)** a power equation, and **(c)** a parabola. In each case, plot the data and the equation.

x	0.75	2	3	4	6	8	8.5
y	1.2	1.95	2	2.4	2.5	2.7	2.6

17.8 Fit the following data with the power model ($y = ax^b$). Use the resulting power equation to predict y at $x = 9$:

x	2.5	3.5	5	6	7.5	10	12.5	15	17.5	20
y	13	11	8.5	8.2	7	6.2	5.2	4.8	4.6	4.3

17.9 Fit an exponential model to

x	0.4	0.8	1.2	1.6	2	2.3
y	800	980	1500	1945	2900	3600

Plot the data and the equation on both standard and semilogarithmic graph paper.

17.10 Rather than using the base-e exponential model [Eq. (17.12)], a common alternative is to use a base-10 model,

$$y = \alpha_5 10^{\beta_5 x}$$

When used for curve fitting, this equation yields identical results to the base-e version, but the value of the exponent parameter (β_5) will differ from that estimated with Eq. (17.12) (β_1). Use the base-10 version to solve Prob. 17.9. In addition, develop a formulation to relate β_1 to β_5.

17.11 Beyond the examples in Fig. 17.10, there are other models that can be linearized using transformations. For example,

$$y = \alpha_4 x e^{\beta_4 x}$$

Linearize this model and use it to estimate α_4 and β_4 based on the following data. Develop a plot of your fit along with the data.

x	0.1	0.2	0.4	0.6	0.9	1.3	1.5	1.7	1.8
y	0.75	1.25	1.45	1.25	0.85	0.55	0.35	0.28	0.18

17.12 An investigator has reported the data tabulated below for an experiment to determine the growth rate of bacteria k (per day) as a function of oxygen concentration c (mg/L). It is known that such data can be modeled by the following equation:

$$k = \frac{k_{\max}c^2}{c_s + c^2}$$

where c_s and $k_{\max}$ are parameters. Use a transformation to linearize this equation. Then use linear regression to estimate c_s and $k_{\max}$ and predict the growth rate at $c = 2$ mg/L.

c	0.5	0.8	1.5	2.5	4
k	1.1	2.4	5.3	7.6	8.9

17.13 An investigator has reported the data tabulated below. It is known that such data can be modeled by the following equation:

$$x = e^{(y-b)/a}$$

where a and b are parameters. Use a transformation to linearize this equation and then employ linear regression to determine a and b. Based on your analysis, predict y at $x = 2.6$.

x	1	2	3	4	5
y	1	3.8	5.9	7	8

17.14 It is known that the data tabulated below can be modeled by the following equation:

$$y = \left(\frac{a + \sqrt{x}}{b\sqrt{x}}\right)^2$$

Use a transformation to linearize this equation and then employ linear regression to determine the parameters a and b. Based on your analysis, predict y at $x = 1.6$.

x	0.5	1	2	3	4
y	10.4	5.8	3.3	2.4	2

17.15 The following model applies to third-order chemical reactions in batch reactors:

$$c = c_0 \frac{1}{\sqrt{1 + 2kc_0^2 t}}$$

where c = concentration (mg/L), c_0 = initial concentration (mg/L), k = reaction rate (L^2/(mg^2 d)), and t = time (d). Linearize this model and use it to estimate k and c_0 based on the following data.

Develop plots of your fit along with the data for both the original and the linearized equations.

t	0	0.5	1	1.5	2	3	4	5
c	3.26	2.09	1.62	1.48	1.17	1.06	0.9	0.85

17.16 The following data are provided:

x	1	2	3	4	5
y	2.2	2.8	3.6	4.5	5.5

You want to use least-squares regression to fit these data with the following model:

$$y = a + bx + \frac{c}{x}$$

Determine the coefficients by setting up and solving Eq. (17.25).

17.17 Given these data

x	5	10	15	20	25	30	35	40	45	50
y	17	24	31	33	37	37	40	40	42	41

use least-squares regression to fit **(a)** a straight line, **(b)** a power equation, **(c)** a saturation-growth-rate equation, and **(d)** a parabola. Plot the data along with all the curves. Is any one of the curves superior? If so, justify.

17.18 Fit a cubic equation to the following data:

x	3	4	5	7	8	9	11	12
y	1.6	3.6	4.4	3.4	2.2	2.8	3.8	4.6

Along with the coefficients, determine r^2 and $s_{y/x}$.

17.19 Use multiple linear regression to fit

x_1	0	1	1	2	2	3	3	4	4
x_2	0	1	2	1	2	1	2	1	2
y	15.1	17.9	12.7	25.6	20.5	35.1	29.7	45.4	40.2

Compute the coefficients, the standard error of the estimate, and the correlation coefficient.

17.20 Use multiple linear regression to fit

x_1	0	0	1	2	0	1	2	2	1
x_2	0	2	2	4	4	6	6	2	1
y	14	21	11	12	23	23	14	6	11

Compute the coefficients, the standard error of the estimate, and the correlation coefficient.

17.21 Use nonlinear regression to fit a parabola to the following data:

x	0.2	0.5	0.8	1.2	1.7	2	2.3
y	500	700	1000	1200	2200	2650	3750

17.22 Use nonlinear regression to fit a saturation-growth-rate equation to the data in Prob. 17.17.

17.23 Recompute the regression fits from Probs. **(a)** 17.3 and **(b)** 17.18 using the matrix approach. Estimate the standard errors and develop 90% confidence intervals for the coefficients.

17.24 Develop, debug, and test a program in either a high-level language or a macro language of your choice to implement linear regression. Among other things: **(a)** include statements to document the code and **(b)** determine the standard error and the coefficient of determination.

17.25 A material is tested for cyclic fatigue failure, whereby a stress, in MPa, is applied to the material and the number of cycles needed to cause failure is measured. The results are in the table below. When a log-log plot of stress versus cycles is generated, the data trend shows a linear relationship. Use least-squares regression to determine a best-fit equation for these data.

# of cycles	1	10	100	1000	10,000	100,000	1,000,000
Stress, MPa	1100	1000	925	800	625	550	420

17.26 The following data show the relationship between the viscosity of SAE 70 oil and temperature. After taking the log of the data, use linear regression to find the equation of the line that best fits the data and the r^2 value.

Temperature, °C	26.67	93.33	148.89	315.56
Viscosity, μ, N s/m^2	1.35	0.085	0.012	0.00075

17.27 The data below represent the bacterial growth in a liquid culture over a number of days.

Day	0	4	8	12	16	20
10^6 organisms	67	84	98	125	149	185

Find a best-fit equation to the data trend. Try several possibilities—linear, parabolic, and exponential. Use the software package of your choice to find the best equation to predict the amount of bacteria after 40 days.

17.28 The concentration of E. coli bacteria in a swimming area is monitored after a storm:

t (hr)	4	8	12	16	20	24
c (CFU/100 mL)	1600	1320	1000	890	650	560

The time is measured in hours following the end of the storm and the unit CFU is a "colony forming unit." Use these data to estimate **(a)** the concentration at the end of the storm ($t = 0$) and **(b)** the time at which the concentration will reach 200 CFU/100 mL. Note that your choice of model should be consistent with the fact that negative concentrations are impossible and that the bacteria concentration always decreases with time.

17.29 An object is suspended in a wind tunnel and the force measured for various levels of wind velocity. The results are tabulated below.

v, m/s	10	20	30	40	50	60	70	80
F, N	25	70	380	550	610	1220	830	1450

Use least-squares regression to fit these data with **(a)** a straight line, **(b)** a power equation based on log transformations, and **(c)** a power model based on nonlinear regression. Display the results graphically.

17.30 Fit a power model to the data from Prob. 17.29, but use natural logarithms to perform the transformations.

17.31 Derive the least-squares fit of the following model:

$$y = a_1 x + a_2 x^2 + e$$

That is, determine the coefficients that result in the least-squares fit for a second-order polynomial with a zero intercept. Test the approach by using it to fit the data from Prob. 17.29.

17.32 In Prob. 17.11 we used transformations to linearize and fit the following model:

$$y = \alpha_4 x e^{\beta_4 x}$$

Use nonlinear regression to estimate α_4 and β_4 based on the following data. Develop a plot of your fit along with the data.

x	0.1	0.2	0.4	0.6	0.9	1.3	1.5	1.7	1.8
y	0.75	1.25	1.45	1.25	0.85	0.55	0.35	0.28	0.18

17.33 Table P17.33 lists the 2015 world record times and holders for outdoor running. Note that all but the 100 m and the marathon (42,195 m) are run on oval tracks. Fit a power model for each gender and use it to predict the record time for a half marathon (21,097.5 m). Note that the actual records for the half marathon are 3503 s (Tadese) and 3909 s (Kiplagat) for men and women, respectively.

TABLE P17.33 World record times and holders for outdoor running in 2015

Event (m)	Men's Time (s)	Men's record holder	Women's Time (s)	Women's record holder
100	9.58	Bolt	10.49	Griffith-Joyner
200	19.19	Bolt	21.34	Griffith-Joyner
400	43.18	Johnson	47.60	Koch
800	100.90	Rudisha	113.28	Kratochvilova
1000	131.96	Ngeny	148.98	Masterkova
1500	206.00	El Guerrouj	230.07	Dibaba
2000	284.79	El Guerrouj	325.35	O'Sullivan
5000	757.40	Bekele	851.15	Dibaba
10,000	1577.53	Bekele	1771.78	Wang
20,000	3386.00	Gebrselassie	3926.60	Loroupe
42,195	7377.00	Kimetto	8125.00	Radcliffe

Interpolation

You will frequently have occasion to estimate intermediate values between precise data points. The most common method used for this purpose is polynomial interpolation. Recall that the general formula for an nth-order polynomial is

$$f(x) = a_0 + a_1x + a_2x^2 + \cdots + a_nx^n \tag{18.1}$$

For $n + 1$ data points, there is one and only one polynomial of order n that passes through all the points. For example, there is only one straight line (that is, a first-order polynomial) that connects two points (Fig. 18.1a). Similarly, only one parabola connects a set of three points (Fig. 18.1b). *Polynomial interpolation* consists of determining the unique nth-order polynomial that fits $n + 1$ data points. This polynomial then provides a formula to compute intermediate values.

Although there is one and only one nth-order polynomial that fits $n + 1$ points, there are a variety of mathematical formats in which this polynomial can be expressed. In this chapter, we will describe two alternatives that are well-suited for computer implementation: the Newton and the Lagrange polynomials.

FIGURE 18.1
Examples of interpolating polynomials: (*a*) first-order (linear) connecting two points, (*b*) second-order (quadratic or parabolic) connecting three points, and (*c*) third-order (cubic) connecting four points.

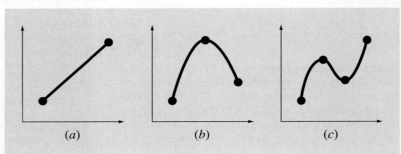

(*a*)　　(*b*)　　(*c*)

18.1   NEWTON'S DIVIDED-DIFFERENCE INTERPOLATING POLYNOMIALS

As stated above, there are a variety of alternative forms for expressing an interpolating polynomial. *Newton's divided-difference interpolating polynomial* is among the most popular and useful forms. Before presenting the general equation, we will introduce the first- and second-order versions because of their simple visual interpretation.

18.1.1 Linear Interpolation

The simplest form of interpolation is to connect two data points with a straight line. This technique, called *linear interpolation,* is depicted graphically in Fig. 18.2. From similar triangles,

$$\frac{f_1(x) - f(x_0)}{x - x_0} = \frac{f(x_1) - f(x_0)}{x_1 - x_0}$$

which can be rearranged to yield

$$f_1(x) = f(x_0) + \frac{f(x_1) - f(x_0)}{x_1 - x_0}(x - x_0) \tag{18.2}$$

which is a *linear-interpolation formula.* The notation $f_1(x)$ designates that this is a first-order interpolating polynomial. Notice that besides representing the slope of the line connecting the points, the term $[f(x_1) - f(x_0)]/(x_1 - x_0)$ is a finite-divided-difference

FIGURE 18.2
Graphical depiction of linear interpolation. The shaded areas indicate the similar triangles used to derive the linear-interpolation formula [Eq. (18.2)].

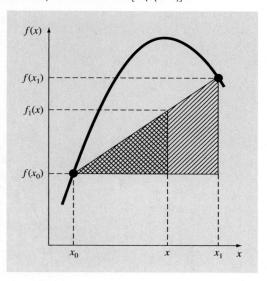

approximation of the first derivative [recall Eq. (4.17)]. In general, the smaller the interval between the data points, the better the approximation. This is due to the fact that, as the interval decreases, a continuous function will be better approximated by a straight line. This characteristic is demonstrated in the following example.

EXAMPLE 18.1

Linear Interpolation

Problem Statement.   Estimate the natural logarithm of 2 using linear interpolation. First, perform the computation by interpolating between $\ln 1 = 0$ and $\ln 6 = 1.791759$. Then, repeat the procedure, but use a smaller interval from $\ln 1$ to $\ln 4$ (1.386294). Note that the true value of $\ln 2$ is 0.6931472.

Solution.   We use Eq. (18.2) and a linear interpolation for $\ln 2$ from $x_0 = 1$ to $x_1 = 6$ to give

$$f_1(2) = 0 + \frac{1.791759 - 0}{6 - 1}(2 - 1) = 0.3583519$$

which represents an error of $\varepsilon_t = 48.3\%$. Using the smaller interval from $x_0 = 1$ to $x_1 = 4$ yields

$$f_1(2) = 0 + \frac{1.386294 - 0}{4 - 1}(2 - 1) = 0.4620981$$

Thus, using the shorter interval reduces the percent relative error to $\varepsilon_t = 33.3\%$. Both interpolations are shown in Fig. 18.3, along with the true function.

FIGURE 18.3
Two linear interpolations to estimate ln 2. Note how the smaller interval provides a better estimate.

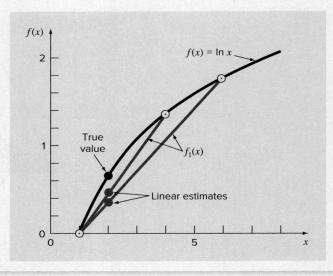

18.1.2 Quadratic Interpolation

The error in Example 18.1 resulted from our approximating a curve with a straight line. Consequently, a strategy for improving the estimate is to introduce some curvature into the line connecting the points. If three data points are available, this can be accomplished with a second-order polynomial (also called a quadratic polynomial, or a *parabola*). A particularly convenient form for this purpose is

$$f_2(x) = b_0 + b_1(x - x_0) + b_2(x - x_0)(x - x_1) \qquad (18.3)$$

Note that although Eq. (18.3) might seem to differ from the general polynomial [Eq. (18.1)], the two equations are equivalent. This can be shown by multiplying the terms in Eq. (18.3) to yield

$$f_2(x) = b_0 + b_1 x - b_1 x_0 + b_2 x^2 + b_2 x_0 x_1 - b_2 x x_0 - b_2 x x_1$$

or, collecting terms,

$$f_2(x) = a_0 + a_1 x + a_2 x^2$$

where

$$a_0 = b_0 - b_1 x_0 + b_2 x_0 x_1$$
$$a_1 = b_1 - b_2 x_0 - b_2 x_1$$
$$a_2 = b_2$$

Thus, Eqs. (18.1) and (18.3) are alternative, equivalent formulations of the unique second-order polynomial joining the three points.

A simple procedure can be used to determine the values of the coefficients. For b_0, Eq. (18.3) with $x = x_0$ can be used to compute

$$b_0 = f(x_0) \qquad (18.4)$$

Equation (18.4) can be substituted into Eq. (18.3), which can be evaluated at $x = x_1$ for

$$b_1 = \frac{f(x_1) - f(x_0)}{x_1 - x_0} \qquad (18.5)$$

Finally, Eqs. (18.4) and (18.5) can be substituted into Eq. (18.3), which can be evaluated at $x = x_2$ and solved (after some algebraic manipulations) for

$$b_2 = \frac{\dfrac{f(x_2) - f(x_1)}{x_2 - x_1} - \dfrac{f(x_1) - f(x_0)}{x_1 - x_0}}{x_2 - x_0} \qquad (18.6)$$

Notice that, as was the case with linear interpolation, b_1 still represents the slope of the line connecting points x_0 and x_1. Thus, the first two terms of Eq. (18.3) are equivalent to linear interpolation from x_0 to x_1, as specified previously in Eq. (18.2). The last term, $b_2(x - x_0)(x - x_1)$, introduces the second-order curvature into the formula.

Before illustrating how to use Eq. (18.3), we should examine the form of the coefficient b_2. It is very similar to the finite-divided-difference approximation of the second derivative introduced previously in Eq. (4.24). Thus, Eq. (18.3) is beginning to manifest a structure that is very similar to the Taylor series expansion. This observation will be explored further when we relate Newton's interpolating polynomials to the Taylor series in Sec. 18.1.4. But first, we will do an example that shows how Eq. (18.3) is used to interpolate among three points.

EXAMPLE 18.2

Quadratic Interpolation

Problem Statement. Fit a second-order polynomial to the three points used in Example 18.1:

$$x_0 = 1 \qquad f(x_0) = 0$$
$$x_1 = 4 \qquad f(x_1) = 1.386294$$
$$x_2 = 6 \qquad f(x_2) = 1.791759$$

Use the polynomial to evaluate ln 2.

Solution.   Applying Eq. (18.4) yields

$$b_0 = 0$$

Equation (18.5) yields

$$b_1 = \frac{1.386294 - 0}{4 - 1} = 0.4620981$$

and Eq. (18.6) gives

$$b_2 = \frac{\dfrac{1.791759 - 1.386294}{6 - 4} - 0.4620981}{6 - 1} = -0.0518731$$

FIGURE 18.4
The use of quadratic interpolation to estimate ln 2. The linear interpolation from $x = 1$ to 4 is also included for comparison.

Substituting these values into Eq. (18.3) yields the quadratic formula

$$f_2(x) = 0 + 0.4620981(x - 1) - 0.0518731(x - 1)(x - 4)$$

which can be evaluated at $x = 2$ to give

$$f_2(2) = 0.5658444$$

which represents a relative error of $\varepsilon_t = 18.4\%$. Thus, the curvature introduced by the quadratic formula (Fig. 18.4) improves the interpolation compared with the result obtained using straight lines in Example 18.1 and Fig. 18.3.

18.1.3 General Form of Newton's Interpolating Polynomials

The preceding analysis can be generalized to fit an nth-order polynomial to $n + 1$ data points. The nth-order polynomial is

$$f_n(x) = b_0 + b_1(x - x_0) + \cdots + b_n(x - x_0)(x - x_1) \cdots (x - x_{n-1}) \tag{18.7}$$

As was done previously with the linear and quadratic interpolations, data points can be used to evaluate the coefficients $b_0, b_1, \ldots, b_n$. For an nth-order polynomial, $n + 1$ data points are required: $[x_0, f(x_0)], [x_1, f(x_1)], \ldots, [x_n, f(x_n)]$. We use these data points and the following equations to evaluate the coefficients:

$$b_0 = f(x_0) \tag{18.8}$$

$$b_1 = f[x_1, x_0] \tag{18.9}$$

$$b_2 = f[x_2, x_1, x_0] \tag{18.10}$$

$$\vdots$$

$$b_n = f[x_n, x_{n-1}, \ldots, x_1, x_0] \tag{18.11}$$

where the bracketed function evaluations are finite divided differences. For example, the first finite divided difference is represented generally as

$$f[x_i, x_j] = \frac{f(x_i) - f(x_j)}{x_i - x_j} \tag{18.12}$$

The *second finite divided difference*, which represents the difference of two first divided differences, is expressed generally as

$$f[x_i, x_j, x_k] = \frac{f[x_i, x_j] - f[x_j, x_k]}{x_i - x_k} \tag{18.13}$$

Similarly, the *nth finite divided difference* is

$$f[x_n, x_{n-1}, \ldots, x_1, x_0] = \frac{f[x_n, x_{n-1}, \ldots, x_1] - f[x_{n-1}, x_{n-2}, \ldots, x_0]}{x_n - x_0} \tag{18.14}$$

i	x_i	$f(x_i)$	First	Second	Third
0	x_0	$f(x_0)$	$f[x_1, x_0]$	$f[x_2, x_1, x_0]$	$f[x_3, x_2, x_1, x_0]$
1	x_1	$f(x_1)$	$f[x_2, x_1]$	$f[x_3, x_2, x_1]$	
2	x_2	$f(x_2)$	$f[x_3, x_2]$		
3	x_3	$f(x_3)$			

FIGURE 18.5
Graphical depiction of the recursive nature of finite divided differences.

These differences can be used to evaluate the coefficients in Eqs. (18.8) through (18.11), which can then be substituted into Eq. (18.7) to yield the interpolating polynomial

$$f_n(x) = f(x_0) + (x - x_0)f[x_1, x_0] + (x - x_0)(x - x_1)f[x_2, x_1, x_0]$$
$$+ \cdots + (x - x_0)(x - x_1) \cdots (x - x_{n-1})f[x_n, x_{n-1}, \ldots, x_0] \qquad (18.15)$$

which is called *Newton's divided-difference interpolating polynomial*. It should be noted that it is not necessary that the data points used in Eq. (18.15) be equally spaced or that the abscissa values necessarily be in ascending order, as illustrated in the following example. Also, notice how Eqs. (18.12) through (18.14) are recursive—that is, higher-order differences are computed by taking differences of lower-order differences (Fig. 18.5). This property will be exploited when we develop an efficient computer program in Sec. 18.1.5 to implement the method.

EXAMPLE 18.3   Newton's Divided-Difference Interpolating Polynomials

Problem Statement.   In Example 18.2, data points at $x_0 = 1$, $x_1 = 4$, and $x_2 = 6$ were used to estimate ln 2 with a parabola. Now, adding a fourth point, $[x_3 = 5; f(x_3) = 1.609438]$, estimate ln 2 with a third-order Newton's interpolating polynomial.

Solution.   The third-order polynomial, Eq. (18.7) with $n = 3$, is

$$f_3(x) = b_0 + b_1(x - x_0) + b_2(x - x_0)(x - x_1) + b_3(x - x_0)(x - x_1)(x - x_2)$$

The first finite divided differences for the problem are [Eq. (18.12)]

$$f[x_1, x_0] = \frac{1.386294 - 0}{4 - 1} = 0.4620981$$

$$f[x_2, x_1] = \frac{1.791759 - 1.386294}{6 - 4} = 0.2027326$$

$$f[x_3, x_2] = \frac{1.609438 - 1.791759}{5 - 6} = 0.1823216$$

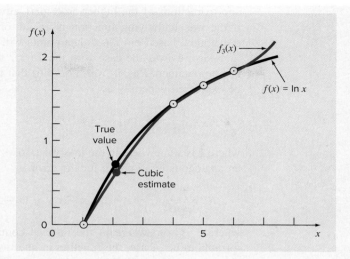

FIGURE 18.6
The use of cubic interpolation to estimate ln 2.

The second finite divided differences are [Eq. (18.13)]

$$f[x_2, x_1, x_0] = \frac{0.2027326 - 0.4620981}{6 - 1} = -0.05187311$$

$$f[x_3, x_2, x_1] = \frac{0.1823216 - 0.2027326}{5 - 4} = -0.02041100$$

The third finite divided difference is [Eq. (18.14) with $n = 3$]

$$f[x_3, x_2, x_1, x_0] = \frac{-0.02041100 - (-0.05187311)}{5 - 1} = 0.007865529$$

The results for $f[x_1, x_0]$, $f[x_2, x_1, x_0]$, and $f[x_3, x_2, x_1, x_0]$ represent the coefficients b_1, b_2, and b_3, respectively, of Eq. (18.7). With $b_0 = f(x_0) = 0.0$, Eq. (18.7) is

$$f_3(x) = 0 + 0.4620981(x - 1) - 0.05187311(x - 1)(x - 4)$$
$$+ 0.007865529(x - 1)(x - 4)(x - 6)$$

which can be used to evaluate $f_3(2) = 0.6287686$, which represents a relative error of $\varepsilon_t = 9.3\%$. The complete cubic polynomial is shown in Fig. 18.6.

18.1.4 Errors of Newton's Interpolating Polynomials

Notice that the structure of Eq. (18.15) is similar to the Taylor series expansion in the sense that terms are added sequentially to capture the higher-order behavior of the underlying function. These terms are finite divided differences and, thus, represent approximations of the higher-order derivatives. Consequently, as with the Taylor series, if the true underlying function is an nth-order polynomial, the nth-order interpolating polynomial based on $n + 1$ data points will yield exact results.

Also, as was the case with the Taylor series, a formulation for the truncation error can be obtained. Recall from Eq. (4.6) that the truncation error for the Taylor series can be expressed generally as

$$R_n = \frac{f^{(n+1)}(\xi)}{(n + 1)!}(x_{i+1} - x_i)^{n+1} \tag{4.6}$$

where ξ is somewhere in the interval x_i to x_{i+1}. For an nth-order interpolating polynomial, an analogous relationship for the error is

$$R_n = \frac{f^{(n+1)}(\xi)}{(n + 1)!}(x - x_0)(x - x_1) \cdots (x - x_n) \tag{18.16}$$

where ξ is somewhere in the interval containing the unknown and the data. For this formula to be of use, the function in question must be known and differentiable. This is not usually the case. Fortunately, an alternative formulation is available that does not require prior knowledge of the function. Rather, it uses a finite divided difference to approximate the $(n + 1)$th derivative,

$$R_n = f[x, x_n, x_{n-1}, \ldots , x_0](x - x_0)(x - x_1) \cdots (x - x_n) \tag{18.17}$$

where $f[x, x_n, x_{n-1}, \ldots , x_0]$ is the $(n + 1)$th finite divided difference. Because Eq. (18.17) contains the unknown $f(x)$, it cannot be solved for the error. However, if an additional data point $f(x_{n+1})$ is available, Eq. (18.17) can be used to estimate the error, as in

$$R_n \cong f[x_{n+1}, x_n, x_{n-1}, \ldots , x_0](x - x_0)(x - x_1) \cdots (x - x_n) \tag{18.18}$$

EXAMPLE 18.4 Error Estimation for Newton's Polynomial

Problem Statement.   Use Eq. (18.18) to estimate the error for the second-order polynomial interpolation of Example 18.2. Use the additional data point $f(x_3) = f(5) = 1.609438$ to obtain your results.

Solution.   Recall that in Example 18.2, the second-order interpolating polynomial provided an estimate of $f_2(2) = 0.5658444$, which represents an error of $0.6931472 - 0.5658444 = 0.1273028$. If we had not known the true value, as is most usually the case, Eq. (18.18), along with the additional value at x_3, could have been used to estimate the error, as in

$$R_2 = f[x_3, x_2, x_1, x_0](x - x_0)(x - x_1)(x - x_2)$$

or

$$R_2 = 0.007865529(x - 1)(x - 4)(x - 6)$$

where the value for the third-order finite divided difference is as computed previously in Example 18.3. This relationship can be evaluated at $x = 2$ to give

$$R_2 = 0.007865529(2 - 1)(2 - 4)(2 - 6) = 0.0629242$$

which is of the same order of magnitude as the true error.

From the previous example and from Eq. (18.18), it should be clear that the error estimate for the nth-order polynomial is equivalent to the difference between the $(n + 1)$th order and the nth-order prediction. That is,

$$R_n = f_{n+1}(x) - f_n(x) \tag{18.19}$$

In other words, the increment that is added to the nth-order case to create the $(n + 1)$th-order case [that is, Eq. (18.18)] is interpreted as an estimate of the nth-order error. This can be clearly seen by rearranging Eq. (18.19) to give

$$f_{n+1}(x) = f_n(x) + R_n$$

The validity of this approach is predicated on the fact that the series is strongly convergent. For such a situation, the $(n + 1)$th-order prediction should be much closer to the true value than the nth-order prediction. Consequently, Eq. (18.19) conforms to our standard definition of error as representing the difference between the truth and an approximation. However, note that whereas all other error estimates for iterative approaches introduced up to this point have been determined as a present prediction minus a previous one, Eq. (18.19) represents a future prediction minus a present one. This means that for a series that is converging rapidly, the error estimate of Eq. (18.19) could be less than the true error. This would represent a highly unattractive quality if the error estimate were being employed as a stopping criterion. However, as will be described in the following section, higher-order interpolating polynomials are highly sensitive to data errors—that is, they are very ill-conditioned. When employed for interpolation, they often yield predictions that diverge significantly from the true value. By "looking ahead" to sense errors, Eq. (18.19) is more sensitive to such divergence. As such, it is more valuable for the sort of exploratory data analysis for which Newton's polynomial is best-suited.

18.1.5 Computer Algorithm for Newton's Interpolating Polynomials

Three properties make Newton's interpolating polynomials extremely attractive for computer applications:

1. As in Eq. (18.7), higher-order versions can be developed sequentially by adding a single term to the next lower-order equation. This facilitates the evaluation of several different-order versions in the same program. Such a capability is especially valuable when the order of the polynomial is not known *a priori*. By adding new terms sequentially, we can determine when a point of diminishing returns is reached—that is, when addition of higher-order terms no longer significantly improves the estimate or in certain situations actually detracts from it. The error equations discussed below in (3) are useful in devising an objective criterion for identifying this point of diminishing terms.

2. The finite divided differences that constitute the coefficients of the polynomial [Eqs. (18.8) through (18.11)] can be computed efficiently. That is, as in Eq. (18.14) and Fig. 18.5, lower-order differences are used to compute higher-order differences. By utilizing this previously determined information, the coefficients can be computed efficiently. The algorithm in Fig. 18.7 contains such a scheme.

3. The error estimate [Eq. (18.18)] can be very simply incorporated into a computer algorithm because of the sequential way in which the prediction is built.

```
SUBROUTINE NewtInt(x, y, n, xi, yint, ea)
  LOCAL fdd_{n,n}
  DOFOR i = 0, n
    fdd_{i,0} = y_i
  END DO
  DOFOR j = 1, n
    DOFOR i = 0, n - j
      fdd_{i,j} = (fdd_{i+1,j-1} - fdd_{i,j-1})/(x_{i+j} - x_i)
    END DO
  END DO
  xterm = 1
  yint_0 = fdd_{0,0}
  DOFOR order = 1, n
    xterm = xterm * (xi - x_{order-1})
    yint2 = yint_{order-1} + fdd_{0,order} * xterm
    ea_{order-1} = yint2 - yint_{order-1}
    yint_{order} = yint2
  END order
END NewtInt
```

FIGURE 18.7
An algorithm for Newton's interpolating polynomial written in pseudocode.

All the above characteristics can be exploited and incorporated into a general algorithm for implementing Newton's polynomial (Fig. 18.7). Note that the algorithm consists of two parts: The first determines the coefficients from Eq. (18.7), and the second determines the predictions and their associated error. The utility of this algorithm is demonstrated in the following example.

EXAMPLE 18.5   Error Estimates to Determine the Appropriate Order of Interpolation

Problem Statement.   After incorporating the error [Eq. (18.18)], utilize the computer algorithm given in Fig. 18.7 and the following information to evaluate $f(x) = \ln x$ at $x = 2$:

x	f(x) = ln x
1	0
4	1.3862944
6	1.7917595
5	1.6094379
3	1.0986123
1.5	0.4054641
2.5	0.9162907
3.5	1.2527630

Solution.   The results of employing the algorithm in Fig. 18.7 to obtain a solution are shown in Fig. 18.8. The error estimates, along with the true error (based on the fact that $\ln 2 = 0.6931472$), are depicted in Fig. 18.9. Note that the estimated error and the true error are similar and that their agreement improves as the order increases. From these results, it can be concluded that the fifth-order version yields a good estimate and that higher-order terms do not significantly enhance the prediction.

This exercise also illustrates the importance of the positioning and ordering of the points. For example, up through the third-order estimate, the rate of improvement is slow because the points that are added (at $x = 4$, 6, and 5) are distant and on one side of the point in question at $x = 2$. The fourth-order estimate shows a somewhat greater improvement because the new point at $x = 3$ is closer to the unknown. However, the most dramatic decrease in the error is associated with the inclusion of the fifth-order term using the data point at $x = 1.5$. Not only is this point close to the unknown but it is also positioned on the opposite side from most of the other points. As a consequence, the error is reduced by almost an order of magnitude.

The significance of the positions and sequence of these data points can also be illustrated by using the same data to obtain an estimate for $\ln 2$ but considering the points in a different sequence. Figure 18.9 shows results for the case of reversing the order of the original data, that is, $x_0 = 3.5$, $x_1 = 2.5$, $x_3 = 1.5$, and so forth. Because the initial points for this case are closer to and spaced on either side of $\ln 2$, the error decreases much more rapidly than for the original situation. By the second-order term, the error has been reduced to less than $\varepsilon_t = 2\%$. Other combinations could be employed to obtain different rates of convergence.

FIGURE 18.8
The output of a program, based on the algorithm from Fig. 18.7, to evaluate ln 2.

```
NUMBER OF POINTS? 8
X( 0 ), y( 0 ) = ? 1,0
X( 1 ), y( 1 ) = ? 4,1.3862944
X( 2 ), y( 2 ) = ? 6,1.7917595
X( 3 ), y( 3 ) = ? 5,1.6094379
X( 4 ), y( 4 ) = ? 3,1.0986123
X( 5 ), y( 5 ) = ? 1.5,0.40546411
X( 6 ), y( 6 ) = ? 2.5,0.91629073
X( 7 ), y( 7 ) = ? 3.5,1.2527630

INTERPOLATION AT X = 2
ORDER          F(X)               ERROR
0              0.000000           0.462098
1              0.462098           0.103746
2              0.565844           0.062924
3              0.628769           0.046953
4              0.675722           0.021792
5              0.697514          -0.003616
6              0.693898          -0.000459
7              0.693439
```

FIGURE 18.9
Percent relative errors for the prediction of ln 2 as a function of the order of the interpolating polynomial.

The foregoing example illustrates the importance of the choice of base points. As should be intuitively obvious, the points should be centered around and as close as possible to the unknown. This observation is also supported by direct examination of the error equation [Eq. (18.17)]. If we assume that the finite divided difference does not vary markedly along the range of these data, the error is proportional to the product: $(x - x_0)(x - x_1) \cdots (x - x_n)$. Obviously, the closer the base points are to x, the smaller the magnitude of this product.

18.2 LAGRANGE INTERPOLATING POLYNOMIALS

The *Lagrange interpolating polynomial* is simply a reformulation of the Newton polynomial that avoids the computation of divided differences. It can be represented concisely as

$$f_n(x) = \sum_{i=0}^{n} L_i(x) f(x_i) \qquad (18.20)$$

where

$$L_i(x) = \prod_{\substack{j=0 \\ j \neq i}}^{n} \frac{x - x_j}{x_i - x_j}$$

(18.21)

and Π designates the "product of." For example, the linear version ($n = 1$) is

$$f_1(x) = \frac{x - x_1}{x_0 - x_1} f(x_0) + \frac{x - x_0}{x_1 - x_0} f(x_1)$$

(18.22)

and the second-order version is

$$f_2(x) = \frac{(x - x_1)(x - x_2)}{(x_0 - x_1)(x_0 - x_2)} f(x_0) + \frac{(x - x_0)(x - x_2)}{(x_1 - x_0)(x_1 - x_2)} f(x_1)$$
$$+ \frac{(x - x_0)(x - x_1)}{(x_2 - x_0)(x_2 - x_1)} f(x_2)$$

(18.23)

Equation (18.20) can be derived directly from Newton's polynomial (Box 18.1). However, the rationale underlying the Lagrange formulation can be grasped directly by realizing that each term $L_i(x)$ will be 1 at $x = x_i$ and 0 at all other sample points (Fig. 18.10). Thus, each product $L_i(x)f(x_i)$ takes on the value of $f(x_i)$ at the sample point x_i. Consequently, the summation of all the products designated by Eq. (18.20) is the unique nth-order polynomial that passes exactly through all $n + 1$ data points.

EXAMPLE 18.6   **Lagrange Interpolating Polynomials**

Problem Statement.   Use a Lagrange interpolating polynomial of the first and second order to evaluate ln 2 on the basis of the data given in Example 18.2:

$$x_0 = 1 \qquad f(x_0) = 0$$
$$x_1 = 4 \qquad f(x_1) = 1.386294$$
$$x_2 = 6 \qquad f(x_2) = 1.791760$$

Solution.   The first-order polynomial [Eq. (18.22)] can be used to obtain the estimate at $x = 2$,

$$f_1(2) = \frac{2 - 4}{1 - 4} 0 + \frac{2 - 1}{4 - 1} 1.386294 = 0.4620981$$

In a similar fashion, the second-order polynomial is developed as [Eq. (18.23)]

$$f_2(2) = \frac{(2 - 4)(2 - 6)}{(1 - 4)(1 - 6)} 0 + \frac{(2 - 1)(2 - 6)}{(4 - 1)(4 - 6)} 1.386294$$
$$+ \frac{(2 - 1)(2 - 4)}{(6 - 1)(6 - 4)} 1.791760 = 0.5658444$$

As expected, both these results agree with those previously obtained using Newton's interpolating polynomial.

Box 18.1   Derivation of the Lagrange Form Directly from Newton's Interpolating Polynomial

The Lagrange interpolating polynomial can be derived directly from Newton's formulation. We will do this for the first-order case only [Eq. (18.2)]. To derive the Lagrange form, we reformulate the divided differences. For example, the first finite divided difference,

$$f[x_1, x_0] = \frac{f(x_1) - f(x_0)}{x_1 - x_0} \qquad \text{(B18.1.1)}$$

can be reformulated as

$$f[x_1, x_0] = \frac{f(x_1)}{x_1 - x_0} + \frac{f(x_0)}{x_0 - x_1} \qquad \text{(B18.1.2)}$$

which is referred to as the *symmetric form*. Substituting Eq. (B18.1.2) into Eq. (18.2) yields

$$f_1(x) = f(x_0) + \frac{x - x_0}{x_1 - x_0} f(x_1) + \frac{x - x_0}{x_0 - x_1} f(x_0)$$

Finally, grouping similar terms and simplifying yields the Lagrange form,

$$f_1(x) = \frac{x - x_1}{x_0 - x_1} f(x_0) + \frac{x - x_0}{x_1 - x_0} f(x_1)$$

FIGURE 18.10
A visual depiction of the rationale behind the Lagrange polynomial. This figure shows a second-order case. Each of the three terms in Eq. (18.23) passes through one of the data points and is zero at the other two. The summation of the three terms must, therefore, be the unique second-order polynomial $f_2(x)$ that passes exactly through the three points.

```
FUNCTION Lagrng(x, y, n, xx)
  sum = 0
  DOFOR i = 0, n
    product = y_i
    DOFOR j = 0, n
      IF i ≠ j THEN
        product = product*(xx - x_j)/(x_i - x_j)
      ENDIF
    END DO
    sum = sum + product
  END DO
  Lagrng = sum
END Lagrng
```

FIGURE 18.11
Pseudocode to implement Lagrange interpolation. This algorithm is set up to compute a single nth-order prediction, where $n + 1$ is the number of data points.

Note that, as with Newton's method, the Lagrange version has an estimated error of [Eq. (18.17)]

$$R_n = f[x, x_n, x_{n-1}, \dots, x_0] \prod_{i=0}^{n} (x - x_i)$$

Thus, if an additional point is available at $x = x_{n+1}$, an error estimate can be obtained. However, because the finite divided differences are not employed as part of the Lagrange algorithm, this is rarely done.

Equations (18.20) and (18.21) can be very simply programmed for implementation on a computer. Figure 18.11 shows pseudocode that can be employed for this purpose.

In summary, for cases where the order of the polynomial is unknown, the Newton method has advantages because of the insight it provides into the behavior of the different-order formulas. In addition, the error estimate represented by Eq. (18.18) can usually be integrated easily into the Newton computation because the estimate employs a finite difference (Example 18.5). Thus, for exploratory computations, Newton's method is often preferable.

When only one interpolation is to be performed, the Lagrange and Newton formulations require comparable computational effort. However, the Lagrange version is somewhat easier to program. Because it does not require computation and storage of divided differences, the Lagrange form is often used when the order of the polynomial is known *a priori*.

| EXAMPLE 18.7 | **Lagrange Interpolation Using the Computer** |

Problem Statement.   We can use the algorithm from Fig. 18.11 to study a trend analysis problem associated with our now-familiar falling parachutist. Assume that we have developed instrumentation to measure the velocity of the parachutist. The measured data obtained for a particular test case are

Time, t (s)	Measured Velocity, v (cm/s)
1	800
3	2310
5	3090
7	3940
13	4755

Our problem is to estimate the velocity of the parachutist at $t = 10$ s to fill in the large gap in the measurements between $t = 7$ and $t = 13$ s. We are aware that the behavior of interpolating polynomials can be unexpected. Therefore, we will construct polynomials of orders 4, 3, 2, and 1 and compare the results.

Solution.   The Lagrange algorithm can be used to construct fourth-, third-, second-, and first-order interpolating polynomials.

The fourth-order polynomial and the input data can be plotted as shown in Fig. 18.12a. It is evident from this plot that the estimated value of y at $x = 10$ is higher than the overall trend of these data.

Figure 18.12b through d shows plots of the results of the computations for third-, second-, and first-order interpolating polynomials, respectively. It is noted that the lower the order, the lower the estimated value of the velocity at $t = 10$ s. The plots of the interpolating polynomials indicate that the higher-order polynomials tend to overshoot the trend of these data. This suggests that the first- or second-order versions are most appropriate for this particular trend analysis. It should be remembered, however, that because we are dealing with uncertain data, regression would actually be more appropriate.

FIGURE 18.12
Plots showing (a) fourth-order, (b) third-order, (c) second-order, and (d) first-order interpolations.

The preceding example illustrates that higher-order polynomials tend to be ill-conditioned; that is, they tend to be highly sensitive to round-off error. The same problem applies to higher-order polynomial regression. Double-precision arithmetic sometimes helps mitigate the problem. However, as the order increases, there will come a point at which round-off error will interfere with the ability to interpolate using the simple approaches covered to this point.

18.3 COEFFICIENTS OF AN INTERPOLATING POLYNOMIAL

Although both the Newton and the Lagrange polynomials are well-suited for determining intermediate values between points, they do not provide a convenient polynomial of the conventional form

$$f(x) = a_0 + a_1 x + a_2 x^2 + \cdots + a_n x^n \tag{18.24}$$

A straightforward method for computing the coefficients of this polynomial is based on the fact that $n + 1$ data points are required to determine the $n + 1$ coefficients. Thus, simultaneous linear algebraic equations can be used to calculate the a's. For example, suppose that you desired to compute the coefficients of the parabola

$$f(x) = a_0 + a_1 x + a_2 x^2 \tag{18.25}$$

Three data points are required: $[x_0, f(x_0)]$, $[x_1, f(x_1)]$, and $[x_2, f(x_2)]$. Each can be substituted into Eq. (18.25) to give

$$
\begin{aligned}
f(x_0) &= a_0 + a_1 x_0 + a_2 x_0^2 \\
f(x_1) &= a_0 + a_1 x_1 + a_2 x_1^2 \\
f(x_2) &= a_0 + a_1 x_2 + a_2 x_2^2
\end{aligned}
\tag{18.26}
$$

Thus, for this case, the x's are the knowns and the a's are the unknowns. Because there are the same number of equations as unknowns, Eq. (18.26) could be solved by an elimination method from Part Three.

It should be noted that the foregoing approach is not the most efficient method that is available to determine the coefficients of an interpolating polynomial. Press et al. (2007) provide a discussion and computer codes for more efficient approaches. Whatever technique is employed, a word of caution is in order. Systems such as Eq. (18.26) are notoriously ill-conditioned. Whether they are solved with an elimination method or with a more efficient algorithm, the resulting coefficients can be highly inaccurate, particularly for large n. When used for a subsequent interpolation, they often yield erroneous results.

In summary, if you are interested in determining an intermediate point, employ Newton or Lagrange interpolation. If you must determine an equation of the form of Eq. (18.24), limit yourself to lower-order polynomials and check your results carefully.

18.4 INVERSE INTERPOLATION

As the nomenclature implies, the $f(x)$ and x values in most interpolation contexts are the dependent and independent variables, respectively. As a consequence, the values of the x's are typically uniformly spaced. A simple example is a table of values derived for the function $f(x) = 1/x$,

x	1	2	3	4	5	6	7
$f(x)$	1	0.5	0.3333	0.25	0.2	0.1667	0.1429

Now suppose that you must use the same data, but you are given a value for $f(x)$ and must determine the corresponding value of x. For instance, for the data above, suppose that you were asked to determine the value of x that corresponded to $f(x) = 0.3$. For this case, because the function is available and easy to manipulate, the correct answer can be determined directly as $x = 1/0.3 = 3.3333$.

Such a problem is called *inverse interpolation*. For a more complicated case, you might be tempted to switch the $f(x)$ and x values [that is, merely plot x versus $f(x)$] and use an approach like Lagrange interpolation to determine the result. Unfortunately, when you reverse the variables, there is no guarantee that the values along the new abscissa [the $f(x)$'s] will be evenly spaced. In fact, in many cases, the values will be "telescoped." That is, they will have the appearance of a logarithmic scale with some adjacent points bunched together and others spread out widely. For example, for $f(x) = 1/x$ the result is

$f(x)$	0.1429	0.1667	0.2	0.25	0.3333	0.5	1
x	7	6	5	4	3	2	1

Such nonuniform spacing on the abscissa often leads to oscillations in the resulting interpolating polynomial. This can occur even for lower-order polynomials.

An alternative strategy is to fit an nth-order interpolating polynomial, $f_n(x)$, to the original data [that is, with $f(x)$ versus x]. In most cases, because the x's are evenly spaced, this polynomial will not be ill-conditioned. The answer to your problem then amounts to finding the value of x that makes this polynomial equal to the given $f(x)$. Thus, the interpolation problem reduces to a roots problem!

For example, for the problem outlined above, a simple approach would be to fit a quadratic polynomial to the three points: (2, 0.5), (3, 0.3333), and (4, 0.25). The result would be

$$f_2(x) = 1.08333 - 0.375x + 0.041667x^2$$

The answer to the inverse interpolation problem of finding the x corresponding to $f(x) = 0.3$ would therefore involve determining the root of

$$0.3 = 1.08333 - 0.375x + 0.041667x^2$$

For this simple case, the quadratic formula can be used to calculate

$$x = \frac{0.375 \pm \sqrt{(-0.375)^2 - 4(0.041667)0.78333}}{2(0.041667)} = \begin{matrix} 5.704158 \\ 3.295842 \end{matrix}$$

Thus, the second root, 3.296, is a good approximation of the true value, 3.333. If additional accuracy were desired, a third- or fourth-order polynomial along with one of the root location methods from Part Two could be employed.

18.5 ADDITIONAL COMMENTS

Before proceeding to the next section, we must mention two additional topics: interpolation with equally spaced data and extrapolation.

Because both the Newton and Lagrange polynomials are compatible with arbitrarily spaced data, you might wonder why we address the special case of equally spaced data (Box 18.2). Prior to the advent of digital computers, these techniques had great utility for interpolation from tables with equally spaced arguments. In fact, a computational framework known as a divided-difference table was developed to facilitate the implementation of these techniques. (Figure 18.5 is an example of such a table.)

However, because the formulas are subsets of the computer-compatible Newton and Lagrange schemes and because many tabular functions are available as library subroutines, the need for the equispaced versions has waned. In spite of this, we have included them at this point because of their relevance to later parts of this book. In particular, they are needed to derive numerical integration formulas that typically employ equispaced data (Chap. 21). Because the numerical integration formulas have relevance to the solution of ordinary differential equations, the material in Box 18.2 also has significance to Part Seven.

Extrapolation is the process of estimating a value of $f(x)$ that lies outside the range of the known base points, $x_0, x_1, \ldots, x_n$ (Fig. 18.13). In a previous section, we mentioned that the most accurate interpolation is usually obtained when the unknown lies near the center of the base points. Obviously, this is violated when the unknown lies outside the range, and consequently, the error in extrapolation can be very large. As depicted in Fig. 18.13, the open-ended nature of extrapolation represents a step into the unknown because the process extends the curve beyond the known region. As such, the true curve could easily diverge from the prediction. Extreme care should, therefore, be exercised whenever a case arises where one must extrapolate.

FIGURE 18.13

Illustration of the possible divergence of an extrapolated prediction. The extrapolation is based on fitting a parabola through the first three known points.

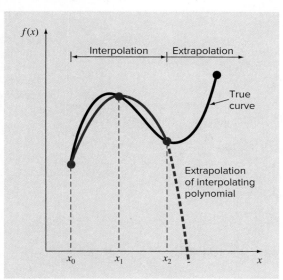

Box 18.2 Interpolation with Equally Spaced Data

If data are equally spaced and in ascending order, then the independent variable assumes values of

$$x_1 = x_0 + h$$
$$x_2 = x_0 + 2h$$
$$.$$
$$.$$
$$.$$
$$x_n = x_0 + nh$$

where h is the interval, or step size, between these data. On this basis, the finite divided differences can be expressed in concise form. For example, the second forward divided difference is

$$f[x_0, x_1, x_2] = \frac{\dfrac{f(x_2) - f(x_1)}{x_2 - x_1} - \dfrac{f(x_1) - f(x_0)}{x_1 - x_0}}{x_2 - x_0}$$

which can be expressed as

$$f[x_0, x_1, x_2] = \frac{f(x_2) - 2f(x_1) + f(x_0)}{2h^2} \qquad \text{(B18.2.1)}$$

because $x_1 - x_0 = x_2 - x_1 = (x_2 - x_0)/2 = h$. Now recall that the second forward difference is equal to [numerator of Eq. (4.24)]

$$\Delta^2 f(x_0) = f(x_2) - 2f(x_1) + f(x_0)$$

Therefore, Eq. (B18.2.1) can be represented as

$$f[x_0, x_1, x_2] = \frac{\Delta^2 f(x_0)}{2!h^2}$$

or, in general,

$$f[x_0, x_1, \dots, x_n] = \frac{\Delta^n f(x_0)}{n!h^n} \qquad \text{(B18.2.2)}$$

Using Eq. (B18.2.2), we can express Newton's interpolating polynomial [Eq. (18.15)] for the case of equispaced data as

$$f_n(x) = f(x_0) + \frac{\Delta f(x_0)}{h}(x - x_0)$$
$$+ \frac{\Delta^2 f(x_0)}{2!h^2}(x - x_0)(x - x_0 - h)$$
$$+ \dots + \frac{\Delta^n f(x_0)}{n!h^n}(x - x_0)(x - x_0 - h)$$
$$\dots [x - x_0 - (n-1)h] + R_n \qquad \text{(B18.2.3)}$$

where the remainder is the same as is given by Eq. (18.16). This equation is known as *Newton's formula,* or the *Newton-Gregory forward formula.* It can be simplified further by defining a new quantity, α:

$$\alpha = \frac{x - x_0}{h}$$

This definition can be used to develop the following simplified expressions for the terms in Eq. (B18.2.3):

$$x - x_0 = \alpha h$$
$$x - x_0 - h = \alpha h - h = h(\alpha - 1)$$
$$.$$
$$.$$
$$.$$
$$x - x_0 - (n-1)h = \alpha h - (n-1)h = h(\alpha - n + 1)$$

which can be substituted into Eq. (B18.2.3) to give

$$f_n(x) = f(x_0) + \Delta f(x_0)\alpha + \frac{\Delta^2 f(x_0)}{2!}\alpha(\alpha - 1)$$
$$+ \dots + \frac{\Delta^n f(x_0)}{n!}\alpha(\alpha - 1) \cdots (\alpha - n + 1) + R_n$$
$$\text{(B18.2.4)}$$

where

$$R_n = \frac{f^{(n+1)}(\xi)}{(n+1)!}h^{n+1}\alpha(\alpha - 1)(\alpha - 2) \cdots (\alpha - n)$$

This concise notation will have utility in our derivation and error analyses of the integration formulas in Chap. 21.

In addition to the forward formula, backward and central Newton-Gregory formulas are also available. Carnahan, Luther, and Wilkes (1969) can be consulted for further information regarding interpolation for equally spaced data.

18.6   SPLINE INTERPOLATION

In the previous sections, nth-order polynomials were used to interpolate between $n + 1$ data points. For example, for eight points, we can derive a perfect seventh-order polynomial. This curve would capture all the meanderings (at least up to and including seventh derivatives) suggested by the points. However, there are cases where these functions can lead to erroneous results because of round-off error and overshoot. An alternative approach is to apply lower-order polynomials to subsets of data points. Such connecting polynomials are called *spline functions*.

For example, third-order curves employed to connect each pair of data points are called *cubic splines*. These functions can be constructed so that the connections between adjacent cubic equations are visually smooth. On the surface, it would seem that the third-order approximation of the splines would be inferior to the seventh-order expression. You might wonder why a spline would ever be preferable.

Figure 18.14 illustrates a situation where a spline performs better than a higher-order polynomial. This is the case where a function is generally smooth but undergoes an abrupt change somewhere along the region of interest. The step increase depicted in Fig. 18.14 is an extreme example of such a change and serves to illustrate the point.

Figure 18.14*a* through *c* illustrates how higher-order polynomials tend to swing through wild oscillations in the vicinity of an abrupt change. In contrast, the spline also connects the points, but because it is limited to lower-order changes, the oscillations are kept to a minimum. As such, the spline usually provides a superior approximation of the behavior of functions that have local, abrupt changes.

The concept of the spline originated from the drafting technique of using a thin, flexible strip (called a *spline*) to draw smooth curves through a set of points. The process is depicted in Fig. 18.15 for a series of five pins (data points). In this technique, the drafter places paper over a wooden board and hammers nails or pins into the paper (and board) at the location of the data points. A smooth cubic curve results from interweaving the strip between the pins. Hence, the name "cubic spline" has been adopted for polynomials of this type.

In this section, simple linear functions will first be used to introduce some basic concepts and problems associated with spline interpolation. Then we derive an algorithm for fitting quadratic splines to data. Finally, we present material on the cubic spline, which is the most common and useful version in engineering practice.

18.6.1 Linear Splines

The notation used for splines is displayed in Fig. 18.16. For n data points ($i = 1, 2, \ldots, n$), there are $n - 1$ intervals. Each interval i has its own spline function, $s_i(x)$. For linear splines, each function is merely the straight line connecting the two points at each end of the interval, which is formulated as

$$s_i(x) = a_i + b_i(x - x_i) \tag{18.27}$$

where a_i is the intercept, which is defined as

$$a_i = f_i \tag{18.28}$$

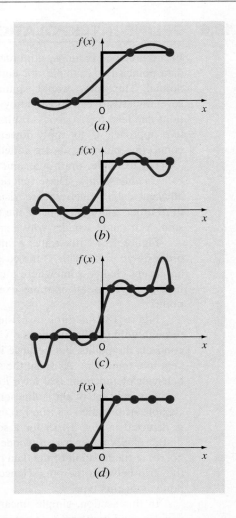

FIGURE 18.14
A visual representation of a situation where the splines are superior to higher-order interpolating polynomials. The function to be fit undergoes an abrupt increase at $x = 0$. Parts (a) through (c) indicate that the abrupt change induces oscillations in interpolating polynomials. In contrast, because it is limited to third-order curves with smooth transitions, a linear spline (d) provides a much more acceptable approximation.

and b_i is the slope of the straight line connecting the points:

$$b_i = \frac{f_{i+1} - f_i}{x_{i+1} - x_i} \tag{18.29}$$

where f_i is shorthand for $f(x_i)$. Substituting Eqs. (18.27) and (18.28) into Eq. (18.29) gives

$$s_i(x) = f_i + \frac{f_{i+1} - f_i}{x_{i+1} - x_i}(x - x_i) \tag{18.30}$$

FIGURE 18.15
The drafting technique of using a spline to draw smooth curves through a series of points. Notice how, at the end points, the spline straightens out. This is called a "natural" spline.

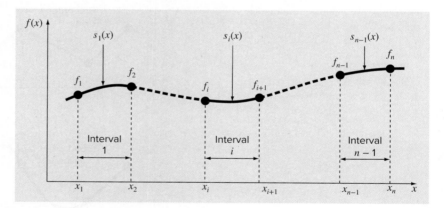

FIGURE 18.16
Notation used to derive splines. Notice that there are $n - 1$ intervals and n data points.

These equations can be used to evaluate the function at any point between x_1 and x_n by first locating the interval within which the point lies. Then the appropriate equation is used to determine the function value within the interval. Inspection of Eq. (18.30) indicates that the linear spline amounts to using Newton's first-order polynomial [Eq. (18.2)] to interpolate within each interval.

| EXAMPLE 18.8 | **First-Order Splines** |

Problem Statement. Fit the data in Table 18.1 with first-order splines. Evaluate the function at $x = 5$.

TABLE 18.1 Data to be fit with spline functions.

i	x_i	f_i
1	3.0	2.5
2	4.5	1.0
3	7.0	2.5
4	9.0	0.5

Solution.   The data can be substituted into Eq. (18.30) to generate the linear spline functions. For example, for the second interval from $x = 4.5$ to $x = 7$, the function is

$$s_2(x) = 1.0 + \frac{2.5 - 1.0}{7.0 - 4.5}(x - 4.5)$$

The equations for the other intervals can be generated, and the resulting first-order splines are plotted in Fig. 18.17a. The value at $x = 5$ is 1.3.

$$s_2(x) = 1.0 + \frac{2.5 - 1.0}{7.0 - 4.5}(5 - 4.5) = 1.3$$

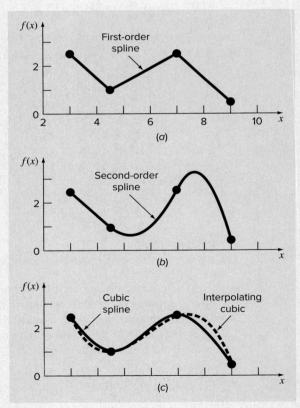

FIGURE 18.17
Spline fits of a set of four points. (a) Linear spline, (b) quadratic spline, and (c) cubic spline, with a cubic interpolating polynomial also plotted.

Visual inspection of Fig. 18.17a indicates that the primary disadvantage of first-order splines is that they are not smooth. In essence, at the data points where two splines meet (called a *knot*), the slope changes abruptly. In formal terms, the first derivative of the function is discontinuous at these points. This deficiency is overcome by using higher-order polynomial splines that ensure smoothness at the knots by equating derivatives at these points, as will be discussed subsequently.

18.6.2 Quadratic Splines

To ensure that the nth derivatives are continuous at the knots, a spline of at least $n + 1$ order must be used. Third-order polynomials or cubic splines that ensure continuous first and second derivatives are most frequently used in practice. Although third and higher derivatives can be discontinuous when cubic splines are used, they usually cannot be detected visually and consequently are ignored.

Because the derivation of cubic splines is somewhat involved, we have decided to first illustrate the concept of spline interpolation using second-order polynomials. These "quadratic splines" have continuous first derivatives at the knots. Although quadratic splines are not of practical importance, they serve nicely to demonstrate the general approach for developing higher-order splines.

The objective in quadratic splines is to derive a second-order polynomial for each interval between data points. The polynomial for each interval can be represented generally as

$$s_i(x) = a_i + b_i(x - x_i) + c_i(x - x_i)^2 \tag{18.31}$$

where the notation is as in Fig. 18.16. For n data points ($i = 1, 2, \ldots , n$), there are $n - 1$ intervals and, consequently, $3(n - 1)$ unknown constants (the a's, b's, and c's) to evaluate. Therefore, $3(n - 1)$ equations or conditions are required to evaluate the unknowns. These can be developed as follows:

1. The function must pass through all the points. This is called a *continuity condition.* It can be expressed mathematically as

$$f_i = a_i + b_i(x_i - x_i) + c_i(x_i - x_i)^2$$

which simplifies to

$$a_i = f_i \tag{18.32}$$

Therefore, the constant in each quadratic must be equal to the value of the dependent variable at the beginning of the interval. This result can be incorporated into Eq. (18.31):

$$s_i(x) = f_i + b_i(x - x_i) + c_i(x - x_i)^2$$

Note that because we have determined one of the coefficients, the number of conditions to be evaluated has now been reduced to $2(n - 1)$.

2. The function values of adjacent polynomials must be equal at the knots. This condition can be written for knot $i + 1$ as

$$f_i + b_i(x_{i+1} - x_i) + c_i(x_{i+1} - x_i)^2 = f_{i+1} + b_{i+1}(x_{i+1} - x_{i+1}) + c_{i+1}(x_{i+1} - x_{i+1})^2 \tag{18.33}$$

This equation can be simplified mathematically by defining the width of the ith interval as

$$h_i = x_{i+1} - x_i$$

Thus, Eq. (18.33) simplifies to

$$f_i + b_i h_i + c_i h_i^2 = f_{i+1} \tag{18.34}$$

This equation can be written for the nodes, $i = 1, \ldots, n - 1$. Since this amounts to $n - 1$ conditions, it means that there are $2(n - 1) - (n - 1) = n - 1$ remaining conditions.

3. The first derivatives at the interior nodes must be equal. This is an important condition, because it means that adjacent splines will be joined smoothly, rather than in the jagged fashion that we saw for the linear splines. Equation (18.31) can be differentiated to yield

$$s_i'(x) = b_i + 2c_i(x - x_i)$$

The equivalence of the derivatives at an interior node, $i + 1$, can therefore be written as

$$b_i + 2c_i h_i = b_{i+1} \tag{18.35}$$

Writing this equation for all the interior nodes amounts to $n - 2$ conditions. This means that there is $n - 1 - (n - 2) = 1$ remaining condition. Unless we have some additional information regarding the functions or their derivatives, we must make an arbitrary choice to successfully compute the constants. Although there are a number of different choices that can be made, we select the following condition.

4. Assume that the second derivative is zero at the first point. Because the second derivative of Eq. (18.31) is $2c_i$, this condition can be expressed mathematically as

$$c_1 = 0$$

The visual interpretation of this condition is that the first two points will be connected by a straight line.

EXAMPLE 18.9   **Quadratic Splines**

Problem Statement.   Fit quadratic splines to the same data employed in Example 18.8 (Table 18.1). Use the results to estimate the value of the function at $x = 5$.

Solution.   For the present problem, we have four data points and $n = 3$ intervals. Therefore, after applying the continuity condition and the zero second-derivative condition, this means that $2(4 - 1) - 1 = 5$ conditions are required. Equation (18.34) is written for $i = 1$ through 3 (with $c_1 = 0$) to give

$$f_1 + b_1 h_1 = f_2$$
$$f_2 + b_2 h_2 + c_2 h_2^2 = f_3$$
$$f_3 + b_3 h_3 + c_3 h_3^2 = f_4$$

Continuity of derivatives, Eq. (18.35), creates an additional $3 - 1 = 2$ conditions (again, recall that $c_1 = 0$):

$$b_1 = b_2$$
$$b_2 + 2c_2h_2 = b_3$$

The necessary function and interval width values are

$$
\begin{array}{ll}
f_1 = 2.5 & h_1 = 4.5 - 3.0 = 1.5 \\
f_2 = 1.0 & h_2 = 7.0 - 4.5 = 2.5 \\
f_3 = 2.5 & h_3 = 9.0 - 7.0 = 2.0 \\
f_4 = 0.5 &
\end{array}
$$

These values can be substituted into the conditions, which can be expressed in matrix form as

$$
\begin{bmatrix}
1.5 & 0 & 0 & 0 & 0 \\
0 & 2.5 & 6.25 & 0 & 0 \\
0 & 0 & 0 & 2 & 4 \\
1 & -1 & 0 & 0 & 0 \\
0 & 1 & 5 & -1 & 0
\end{bmatrix}
\begin{Bmatrix}
b_1 \\ b_2 \\ c_2 \\ b_3 \\ c_3
\end{Bmatrix}
=
\begin{Bmatrix}
-1.5 \\ 1.5 \\ -2 \\ 0 \\ 0
\end{Bmatrix}
$$

These equations can be solved with the results:

$$
\begin{array}{ll}
b_1 = -1 & \\
b_2 = -1 & c_2 = 0.64 \\
b_3 = 2.2 & c_3 = -1.6
\end{array}
$$

These results, along with the values for the a's [Eq. (18.32)], can be substituted into the original quadratic equations to develop the following quadratic splines for each interval:

$$s_1(x) = 2.5 - (x - 3)$$
$$s_2(x) = 1.0 - (x - 4.5) + 0.64(x - 4.5)^2$$
$$s_3(x) = 2.5 + 2.2(x - 7.0) - 1.6(x - 7.0)^2$$

Because $x = 5$ lies in the second interval, we use s_2 to make the prediction,

$$s_2(5) = 1.0 - (5 - 4.5) + 0.64(5 - 4.5)^2 = 0.66$$

The total quadratic spline fit is depicted in Fig. 18.17b. Notice that there are two shortcomings that detract from the fit: (1) the straight line connecting the first two points and (2) the spline for the last interval seems to swing too high. The cubic splines in the next section do not exhibit these shortcomings and, as a consequence, are better methods for spline interpolation.

18.6.3 Cubic Splines

As stated at the beginning of the previous section, cubic splines are most frequently used in practice. The shortcomings of linear and quadratic splines have already been discussed. Quartic or higher-order splines are not used because they tend to exhibit the instabilities inherent in higher-order polynomials. Cubic splines are preferred because they provide the simplest representation that exhibits the desired appearance of smoothness.

The objective with cubic splines is to derive a third-order polynomial for each interval between knots as represented generally by

$$s_i(x) = a_i + b_i(x - x_i) + c_i(x - x_i)^2 + d_i(x - x_i)^3 \tag{18.36}$$

Thus, for n data points ($i = 1, 2, \ldots, n$), there are $n - 1$ intervals and $4(n - 1)$ unknown coefficients to evaluate. Consequently, $4(n - 1)$ conditions are required for their evaluation.

The first conditions are identical to those used for the quadratic case. That is, they are set up so that the functions pass through the points and that the first derivatives at the knots are equal. In addition to these, conditions are developed to ensure that the second derivatives at the knots are also equal. This greatly enhances the fit's smoothness.

After these conditions are developed, two additional conditions are required to obtain the solution. This is a much nicer outcome than occurred for quadratic splines where we needed to specify a single condition. In that case, we had to arbitrarily specify a zero second derivative for the first interval, hence making the result asymmetric. For cubic splines, we are in the advantageous position of needing two additional conditions and can, therefore, apply them evenhandedly at both ends.

For cubic splines, these last two conditions can be formulated in several different ways. A very common approach is to assume that the second derivatives at the first and last knots are equal to zero. The visual interpretation of these conditions is that the function becomes a straight line at the end nodes. Specification of such an end condition leads to what is termed a "natural" spline. It is given this name because the drafting spline naturally behaves in this fashion (Fig. 18.15).

There are a variety of other end conditions that can be specified. Two of the more popular are the clamped condition and the not-a-knot conditions. We will describe these options in Sec. 18.6.5. For the following derivation, we will limit ourselves to natural splines.

Once the additional end conditions are specified, we would have the $4(n - 1)$ conditions needed to evaluate the $4(n - 1)$ unknown coefficients. Whereas it is certainly possible to develop cubic splines in this fashion, we will present an alternative approach that requires the solution of only $n - 1$ equations. Further, the simultaneous equations will be tridiagonal and hence can be solved very efficiently. Although the derivation of this approach is less straightforward than for quadratic splines, the gain in efficiency is well worth the effort.

18.6.4 Derivation of Cubic Splines

As was the case with quadratic splines, the first condition is that the spline must pass through all the data points:

$$f_i = a_i + b_i(x_i - x_i) + c_i(x_i - x_i)^2 + d_i(x_i - x_i)^3$$

which simplifies to

$$a_i = f_i \tag{18.37}$$

Therefore, the constant in each cubic must be equal to the value of the dependent variable at the beginning of the interval. This result can be incorporated into Eq. (18.36):

$$s_i(x) = f_i + b_i(x - x_i) + c_i(x - x_i)^2 + d_i(x - x_i)^3 \tag{18.38}$$

Next, we will apply the condition that each of the cubics must join at the knots. For knot $i + 1$, this can be represented as

$$f_i + b_i h_i + c_i h_i^2 + d_i h_i^3 = f_{i+1} \tag{18.39}$$

where

$$h_i = x_{i+1} - x_i$$

The first derivatives at the interior nodes must be equal. Equation (18.38) is differentiated to yield

$$s_i'(x) = b_i + 2c_i(x - x_i) + 3d_i(x - x_i)^2 \tag{18.40}$$

The equivalence of the derivatives at an interior node, $i + 1$, can therefore be written as

$$b_i + 2c_i h_i + 3d_i h_i^2 = b_{i+1} \tag{18.41}$$

The second derivatives at the interior nodes must also be equal. Equation (18.40) can be differentiated to yield

$$s_i''(x) = 2c_i + 6d_i(x - x_i) \tag{18.42}$$

The equivalence of the second derivatives at an interior node, $i + 1$, can therefore be written as

$$c_i + 3d_i h_i = c_{i+1} \tag{18.43}$$

Next, we can solve Eq. (18.43) for d_i:

$$d_i = \frac{c_{i+1} - c_i}{3h_i} \tag{18.44}$$

This can be substituted into Eq. (18.39) to give

$$f_i + b_i h_i + \frac{h_i^2}{3}(2c_i + c_{i+1}) = f_{i+1} \tag{18.45}$$

Equation (18.44) can also be substituted into Eq. (18.41) to give

$$b_{i+1} = b_i + h_i(c_i + c_{i+1}) \tag{18.46}$$

Equation (18.45) can be solved for

$$b_i = \frac{f_{i+1} - f_i}{h_i} - \frac{h_i}{3}(2c_i + c_{i+1}) \tag{18.47}$$

The index of this equation can be reduced by 1:

$$b_{i-1} = \frac{f_i - f_{i-1}}{h_{i-1}} - \frac{h_{i-1}}{3}(2c_{i-1} + c_i) \tag{18.48}$$

The index of Eq. (18.46) can also be reduced by 1:

$$b_i = b_{i-1} + h_{i-1}(c_{i-1} + c_i) \tag{18.49}$$

Equations (18.47) and (18.48) can be substituted into Eq. (18.49) and the result simplified to yield

$$h_{i-1}c_{i-1} + 2(h_{i-1} + h_i)c_i + h_ic_{i+1} = 3\frac{f_{i+1} - f_i}{h_i} - 3\frac{f_i - f_{i-1}}{h_{i-1}} \tag{18.50}$$

This equation can be made a little more concise by recognizing that the terms on the right-hand side are finite differences [recall Eq. (18.12)]:

$$f[x_i, x_j] = \frac{f_i - f_j}{x_i - x_j}$$

Therefore, Eq. (18.50) can be written as

$$h_{i-1}c_{i-1} + 2(h_{i-1} + h_i)c_i + h_ic_{i+1} = 3(f[x_{i+1}, x_i] - f[x_i, x_{i-1}]) \tag{18.51}$$

Equation (18.51) can be written for the interior knots, $i = 2, 3, \ldots, n - 2$, which results in $n - 3$ simultaneous tridiagonal equations with $n - 1$ unknown coefficients, $c_1, c_2, \ldots, c_{n-1}$. Therefore, if we have two additional conditions, we can solve for the c's. Once this is done, Eqs. (18.47) and (18.44) can be used to determine the remaining coefficients, b and d.

As stated previously, the two additional end conditions can be formulated in a number of ways. One common approach, the natural spline, assumes that the second derivatives at the end knots are equal to zero. To see how these can be integrated into the solution scheme, the second derivative at the first node [Eq. (18.42)] can be set to zero as in

$$s''_1(x_1) = 0 = 2c_1 + 6d_1(x_1 - x_1)$$

Thus, this condition amounts to setting c_1 equal to zero.

The same evaluation can be made at the last node:

$$s''_{n-1}(x_n) = 0 = 2c_{n-1} + 6d_{n-1}h_{n-1} \tag{18.52}$$

Recalling Eq. (18.43), we can conveniently define an extraneous parameter c_n; in which case Eq. (18.52) becomes

$$c_{n-1} + 3d_{n-1}h_{n-1} = c_n = 0$$

Thus, to impose a zero second derivative at the last node, we set $c_n = 0$.

The final equations can now be written in matrix form as

$$\begin{bmatrix} 1 \\ h_1 & 2(h_1 + h_2) & h_2 \\ & & h_{n-2} & 2(h_{n-2} + h_{n-1}) & h_{n-1} \\ & & & & 1 \end{bmatrix} \begin{Bmatrix} c_1 \\ c_2 \\ \vdots \\ c_{n-1} \\ c_n \end{Bmatrix}$$
$$= \begin{Bmatrix} 0 \\ 3(f[x_3, x_2] - f[x_2, x_1]) \\ \vdots \\ 3(f[x_n, x_{n-1}] - f[x_{n-1}, x_{n-2}]) \\ 0 \end{Bmatrix} \tag{18.53}$$

As shown, the system is tridiagonal and hence efficient to solve.

EXAMPLE 18.10   **Natural Cubic Splines**

Problem Statement.   Fit cubic splines to the same data used in Examples 18.8 and 18.9 (Table 18.1). Utilize the results to estimate the value of the function at $x = 5$.

Solution.   The first step is to employ Eq. (18.53) to generate the set of simultaneous equations that will be utilized to determine the c coefficients:

$$\begin{bmatrix} 1 & & & \\ h_1 & 2(h_1 + h_2) & h_2 & \\ & h_2 & 2(h_2 + h_3) & h_3 \\ & & & 1 \end{bmatrix} \begin{Bmatrix} c_1 \\ c_2 \\ c_3 \\ c_4 \end{Bmatrix} = \begin{Bmatrix} 0 \\ 3(f[x_3, x_2] - f[x_2, x_1]) \\ 3(f[x_4, x_3] - f[x_3, x_2]) \\ 0 \end{Bmatrix}$$

The necessary function and interval width values are

$$f_1 = 2.5 \qquad h_1 = 4.5 - 3.0 = 1.5$$
$$f_2 = 1.0 \qquad h_2 = 7.0 - 4.5 = 2.5$$
$$f_3 = 2.5 \qquad h_3 = 9.0 - 7.0 = 2.0$$
$$f_4 = 0.5$$

These can be substituted to yield

$$\begin{bmatrix} 1 & & & \\ 1.5 & 8 & 2.5 & \\ & 2.5 & 9 & 2 \\ & & & 1 \end{bmatrix} \begin{Bmatrix} c_1 \\ c_2 \\ c_3 \\ c_4 \end{Bmatrix} = \begin{Bmatrix} 0 \\ 4.8 \\ -4.8 \\ 0 \end{Bmatrix}$$

These equations can be solved using MATLAB with the results:

$$c_1 = 0 \qquad\qquad c_2 = 0.839543726$$
$$c_3 = -0.766539924 \qquad c_4 = 0$$

Equations (18.47) and (18.44) can be used to compute the b's and d's:

$$b_1 = -1.419771863 \qquad d_1 = 0.186565272$$
$$b_2 = -0.160456274 \qquad d_2 = -0.214144487$$
$$b_3 = 0.022053232 \qquad d_3 = 0.127756654$$

These results, along with the values for the a's [Eq. (18.37)], can be substituted into Eq. (18.36) to develop the following cubic splines for each interval:

$$s_1(x) = 2.5 - 1.419771863(x - 3) + 0.186565272(x - 3)^3$$
$$s_2(x) = 1.0 - 0.160456274(x - 4.5) + 0.839543726(x - 4.5)^2$$
$$\qquad - 0.214144487(x - 4.5)^3$$
$$s_3(x) = 2.5 + 0.022053232(x - 7.0) - 0.766539924(x - 7.0)^2$$
$$\qquad + 0.127756654(x - 7.0)^3$$

The three equations can then be employed to compute values within each interval. For example, the value at $x = 5$, which falls within the second interval, is calculated as

$$s_2(5) = 1.0 - 0.160456274(5 - 4.5) + 0.839543726(5 - 4.5)^2 - 0.214144487(5 - 4.5)^3$$
$$= 1.102889734.$$

The total cubic spline fit is depicted in Fig. 18.17c.

The results of Examples 18.8 through 18.10 are summarized in Fig. 18.17. Notice the progressive improvement of the fit as we move from linear to quadratic to cubic splines. We have also superimposed a cubic interpolating polynomial on Fig. 18.17c. Although the cubic spline consists of a series of third-order curves, the resulting fit differs from that obtained using the third-order polynomial. This is due to the fact that the natural spline requires zero second derivatives at the end knots, whereas the cubic polynomial has no such constraint.

18.6.5 End Conditions

Although its graphical basis is appealing, the natural spline is only one of several end conditions that can be specified for splines. Two of the most popular are

- *Clamped End Condition.* This option involves specifying the first derivatives at the first and last nodes. This is sometimes called a "clamped" spline because it is what occurs when you clamp the end of a drafting spline so that it has a desired slope. For example, if zero first derivatives are specified, the spline will level off or become horizontal at the ends.
- *"Not-a-Knot" End Condition.* A third alternative is to force continuity of the third derivative at the second and the next-to-last knots. Since the spline already specifies that the function value and its first and second derivatives are equal at these knots, specifying continuous third derivatives means that the same cubic functions will apply to each of the first and last two adjacent segments. Since the first internal knots no longer represent the junction of two different cubic functions, they are no longer true knots. Hence, this case is referred to as the *"not-a-knot" condition.* It has the additional property that for four points, it yields the same result as is obtained using an ordinary cubic interpolating polynomial of the sort described earlier in this chapter.

These conditions can be readily applied by using Eq. (18.51) for the interior knots, $i = 2, 3, \ldots, n - 2$, and using first (1) and last (n) equations as written in Table 18.2.

Figure 18.18 shows a comparison of the three end conditions as applied to fit the data from Table 18.1. The clamped case is set up so that the derivatives at the ends are equal to zero.

As expected, the spline fit for the clamped case levels off at the ends. In contrast, the natural and not-a-knot cases follow the trend of the data points more closely. Notice

TABLE 18.2 The first and last equations needed to specify some commonly used end conditions for cubic splines.

Condition	First and Last Equations
Natural	$c_1 = 0, \; c_n = 0$
Clamped (where f'_1 and f'_n are the specified first derivatives at the first and last nodes, respectively)	$2h_1 c_1 + h_1 c_2 = 3f\,[x_2, x_1] - 3f'_1$ $h_{n-1} c_{n-1} + 2h_{n-1} c_n = 3f'_n - 3f\,[x_n, x_{n-1}]$
Not-a-knot	$h_2 c_1 - (h_1 + h_2) c_2 + h_1 c_3 = 0$ $h_{n-1} c_{n-2} - (h_{n-2} + h_{n-1}) c_{n-1} + h_{n-2} c_n = 0$

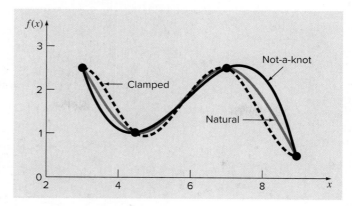

FIGURE 18.18

Comparison of the clamped (with zero first derivatives), not-a-knot, and natural splines for the data from Table 18.1.

how the natural spline tends to straighten out as would be expected because the second derivatives go to zero at the ends. Because it has nonzero second derivatives at the ends, the not-a-knot spline exhibits more curvature.

18.7 MULTIDIMENSIONAL INTERPOLATION

The interpolation methods for one-dimensional problems can be extended to multidimensional interpolation. In this section, we will describe the simplest case of two-dimensional interpolation in Cartesian coordinates.

18.7.1 Bilinear Interpolation

Two-dimensional interpolation deals with determining intermediate values for functions of two variables, $z = f(x_i, y_i)$. As depicted in Fig. 18.19, we have values at four points: $f(x_1, y_1)$, $f(x_2, y_1)$, $f(x_1, y_2)$, and $f(x_2, y_2)$. We want to interpolate between these points to estimate the value at an intermediate point $f(x_i, y_i)$. If we use a linear function, the result is a plane connecting the points as in Fig. 18.19. Such functions are called *bilinear*.

A simple approach for developing the bilinear function is depicted in Fig. 18.20. First, we can hold the y value fixed and apply one-dimensional linear interpolation in the x direction. Using the Lagrange form, the result at (x_i, y_1) is

$$f(x_i, y_1) = \frac{x_i - x_2}{x_1 - x_2} f(x_1, y_1) + \frac{x_i - x_1}{x_2 - x_1} f(x_2, y_1) \tag{18.54}$$

and at (x_i, y_2) is

$$f(x_i, y_2) = \frac{x_i - x_2}{x_1 - x_2} f(x_1, y_2) + \frac{x_i - x_1}{x_2 - x_1} f(x_2, y_2) \tag{18.55}$$

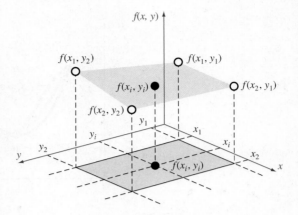

FIGURE 18.19
Graphical depiction of two-dimensional bilinear interpolation where an intermediate value (filled circle) is estimated based on four given values (open circles).

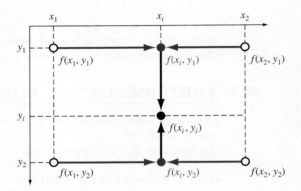

FIGURE 18.20
Two-dimensional bilinear interpolation can be implemented by first applying one-dimensional linear interpolation along the x dimension to determine values at x_i. These values can then be used to linearly interpolate along the y dimension to yield the final result at x_i, y_i.

These points can then be used to linearly interpolate along the y dimension to yield the final result,

$$f(x_i, y_i) = \frac{y_i - y_2}{y_1 - y_2} f(x_i, y_1) + \frac{y_i - y_1}{y_2 - y_1} f(x_i, y_2) \tag{18.56}$$

A single equation can be developed by substituting Eqs. (18.54) and (18.55) into Eq. (18.56) to give

$$f(x_i, y_i) = \frac{x_i - x_2}{x_1 - x_2} \frac{y_i - y_2}{y_1 - y_2} f(x_1, y_1) + \frac{x_i - x_1}{x_2 - x_1} \frac{y_i - y_2}{y_1 - y_2} f(x_2, y_1)$$

$$+ \frac{x_i - x_2}{x_1 - x_2} \frac{y_i - y_1}{y_2 - y_1} f(x_1, y_2) + \frac{x_i - x_1}{x_2 - x_1} \frac{y_i - y_1}{y_2 - y_1} f(x_2, y_2) \tag{18.57}$$

EXAMPLE 18.11

Bilinear Interpolation

Problem Statement.   Suppose you have measured temperatures at a number of coordinates on the surface of a rectangular heated plate:

$$T(2, 1) = 60 \qquad T(9, 1) = 57.5$$
$$T(2, 6) = 55 \qquad T(9, 6) = 70$$

Use bilinear interpolation to estimate the temperature at $x_i = 5.25$ and $y_i = 4.8$.

Solution.   Substituting these values into Eq. (18.57) gives

$$f(5.5, 4) = \frac{5.25 - 9}{2 - 9}\frac{4.8 - 6}{1 - 6}60 + \frac{5.25 - 2}{9 - 2}\frac{4.8 - 6}{1 - 6}57.5$$
$$+ \frac{5.25 - 9}{2 - 9}\frac{4.8 - 1}{6 - 1}55 + \frac{5.25 - 2}{9 - 2}\frac{4.8 - 1}{6 - 1}70 = 61.2143$$

Note that beyond the simple bilinear interpolation described in the foregoing example, higher-order polynomials and splines can also be used to interpolate in two dimensions. Further, these methods can be readily extended to three dimensions. We will return to this topic when we review software applications for interpolation at the end of Chap. 19.

PROBLEMS

18.1 The following data come from a table of values that were measured with high precision. Use a Newton interpolating polynomial to determine y at $x = 3.5$. Properly order *all the points* and then develop a divided difference table to compute the derivatives. Note that a polynomial will yield an exact value. Your solution should prove that your result is exact.

x	0	1	2.5	3	4.5	5	6
y	26	15.5	5.375	3.5	2.375	3.5	8

18.2 The following data were measured precisely:

t	2	2.1	2.2	2.7	3	3.4
z	6	7.752	10.256	36.576	66	125.168

(a) Use Newton interpolating polynomials to determine z at $t = 2.5$. Make sure that you order your points to attain the most accurate results. What do your results tell you regarding the order of the polynomial used to generate the data? **(b)** Use a third-order Lagrange interpolating polynomial to determine y at $t = 2.5$.

18.3 Estimate the natural logarithm of 10 using linear interpolation.
(a) Interpolate between $\ln 8 = 2.0794415$ and $\ln 12 = 2.4849066$.
(b) Interpolate between $\ln 9 = 2.1972246$ and $\ln 11 = 2.3978953$. For each of the interpolations, compute the percent relative error based on the true value.

18.4 Fit a second-order Newton's interpolating polynomial to estimate $\ln 10$ using the data from Prob. 18.3 at $x = 8$, 9, and 11. Compute the true percent relative error.

18.5 Fit a third-order Newton's interpolating polynomial to estimate $\ln 10$ using the data from Prob. 18.3.

18.6 Repeat Probs. 18.3 through 18.5 using the Lagrange polynomial.

18.7 Given these data:

x	1.6	2	2.5	3.2	4	4.5
f(x)	2	8	14	15	8	2

(a) Calculate $f(2.8)$ using Newton's interpolating polynomials of order 1 through 3. Choose the sequence of the points for your estimates to attain the best possible accuracy.
(b) Utilize Eq. (18.18) to estimate the error for each prediction.

18.8 Given these data

x	1	2	3	5	7	8
f(x)	3	6	19	99	291	444

Calculate $f(4)$ using Newton's interpolating polynomials of order 1 through 4. Choose your base points to attain good accuracy. What do your results indicate regarding the order of the polynomial used to generate the data in the table?

18.9 Repeat Prob. 18.8 using Lagrange polynomials of order 1 through 3.

18.10 The following data come from a table of values that were measured with high precision. Use the best numerical method (for this type of problem) to determine y at $x = 3.5$. Note that a polynomial will yield an exact value. Your solution should prove that your result is exact.

x	0	1.8	5	6	8.2	9.2	12
y	26	16.415	5.375	3.5	2.015	2.54	8

18.11 Use Newton's interpolating polynomial to determine y at $x = 3.5$ to the best possible accuracy. Compute the finite divided differences as in Fig. 18.5 and order your points to attain optimal accuracy and convergence.

x	0	1	2.5	3	4.5	5	6
y	2	5.4375	7.3516	7.5625	8.4453	9.1875	12

18.12 Use Newton's interpolating polynomial to determine y at $x = 8$ to the best possible accuracy. Compute the finite divided differences as in Fig. 18.5 and order your points to attain optimal accuracy and convergence.

x	0	1	2	5.5	11	13	16	18
y	0.5	3.134	5.3	9.9	10.2	9.35	7.2	6.2

18.13 Employ inverse interpolation using a cubic interpolating polynomial and bisection to determine the value of x that corresponds to $f(x) = 0.23$ for the following tabulated data:

x	2	3	4	5	6	7
y	0.5	0.3333	0.25	0.2	0.1667	0.1429

18.14 Employ inverse interpolation to determine the value of x that corresponds to $f(x) = 0.85$ for the following tabulated data:

x	0	1	2	3	4	5
f(x)	0	0.5	0.8	0.9	0.941176	0.961538

Note that the values in the table were generated with the function $f(x) = x^2/(1 + x^2)$.

(a) Determine the correct value analytically.
(b) Use cubic interpolation of x versus y.

(c) Use inverse interpolation with quadratic interpolation and the quadratic formula.
(d) Use inverse interpolation with cubic interpolation and bisection. For parts **(b)** through **(d)**, compute the true percent relative error.

18.15 Develop quadratic splines for the first five data points in Prob. 18.7 and predict $f(3.4)$ and $f(2.2)$.

18.16 Develop cubic splines for the data in Prob. 18.8, and **(a)** predict $f(4)$ and $f(2.5)$ and **(b)** verify that $f_2(3)$ and $f_3(3) = 19$.

18.17 Determine the coefficients of the parabola that passes through the last three points in Prob. 18.7.

18.18 Determine the coefficients of the cubic equation that passes through the first four points in Prob. 18.8.

18.19 Develop, debug, and test a program in either a high-level language or macro language of your choice to implement Newton's interpolating polynomial based on Fig. 18.7.

18.20 Test the program you developed in Prob. 18.19 by duplicating the computation from Example 18.5.

18.21 Use the program you developed in Prob. 18.19 to solve Probs. 18.3 through 18.5.

18.22 Use the program you developed in Prob. 18.19 to solve Probs. 18.7 and 18.8. Utilize all the data to develop first- through fifth-order polynomials. For both problems, plot the estimated error versus order.

18.23 Develop, debug, and test a program in either a high-level language or macro language of your choice to implement Lagrange interpolation. Base it on the pseudocode from Fig. 18.11. Test it by duplicating Example 18.7.

18.24 A useful application of Lagrange interpolation is called a *table look-up*. As the name implies, this involves "looking-up" an intermediate value from a table. To develop such an algorithm, the x and $f(x)$ values from the table are first stored in a pair of one-dimensional arrays. These values are then passed to a function along with the x value you wish to evaluate. The function then performs two tasks. First, it loops down through the table until it finds the interval within which the unknown lies. Then it applies a technique like Lagrange interpolation to determine the proper $f(x)$ value. Develop such a function using a cubic Lagrange polynomial to perform the interpolation. For intermediate intervals, this is a nice choice because the unknown will be located in the interval in the middle of the four points necessary to generate the cubic. For the first and last intervals, use a quadratic Lagrange polynomial. Also have your code detect when the user requests a value outside the range of x's. For such cases, the function should display an error message. Test your program for $f(x) = \ln x$ using data from $x = 1, 2, \ldots, 10$.

18.25 Develop, debug, and test a program in either a high-level language or macro language of your choice to implement cubic spline interpolation based on Sec. 18.6.3. Test the program by duplicating Example 18.10.

18.26 Use the software developed in Prob. 18.23 to fit cubic splines through the data in Probs. 18.7 and 18.8. For both cases, predict $f(2.25)$.

18.27 Use the portion of the given steam table for superheated H_2O at 200 MPa to **(a)** find the corresponding entropy s for a specific volume v of 0.108 m^3/kg with linear interpolation, **(b)** find the same corresponding entropy using quadratic interpolation, and **(c)** find the volume corresponding to an entropy of 6.6 using inverse interpolation.

v (m^3/kg)	0.10377	0.11144	0.1254
s (kJ/kg K)	6.4147	6.5453	6.7664

18.28 Runge's function is written as

$$f(x) = \frac{1}{1 + 25x^2}$$

(a) Develop a plot of this function for the interval from $x = -1$ to 1.
(b) Generate and plot the fourth-order Lagrange interpolating polynomial using equispaced function values corresponding to $x = -1, -0.5, 0, 0.5,$ and 1.
(c) Use the five points from **(b)** to estimate $f(0.8)$ with first-through fourth-order Newton interpolating polynomials.
(d) Generate and plot a cubic spline using the five points from **(b)**.
(e) Discuss your results.

18.29 The following is the *humps function* that is used to test a variety of numerical methods:

$$f(x) = \frac{1}{(x - 0.3)^2 + 0.01} + \frac{1}{(x - 0.9)^2 + 0.04} - 6$$

The humps function exhibits both flat and steep regions over a relatively short x range. Generate values of this function at intervals of 0.1 over the range from $x = 0$ to 1. Fit these data with a cubic spline and create a plot comparing the fit with the exact humps function.

18.30 The following data define the sea-level concentration of dissolved oxygen for fresh water as a function of temperature:

T (°C)	0	8	16	24	32	40
o (mg/L)	14.621	11.843	9.870	8.418	7.305	6.413

Estimate $o(27)$ using **(a)** linear interpolation, **(b)** Newton's interpolating polynomial, and **(c)** cubic splines. Note that the exact result is 7.986 mg/L.

18.31 Generate eight equally spaced points from the function

$$f(t) = \sin^2 t$$

for $t = 0$ to 2π. Fit these data with **(a)** a seventh-order interpolating polynomial and **(b)** a cubic spline.

18.32 Temperatures are measured at various points on a heated plate (Table P18.32). Estimate the temperature at **(a)** $x = 4$, $y = 3.2$, and **(b)** $x = 4.3$, $y = 2.7$.

TABLE P18.32 Temperature (°C) at various points on a square heated plate.

	$x = 0$	$x = 2$	$x = 4$	$x = 6$	$x = 8$
$y = 0$	100.00	90.00	80.00	70.00	60.00
$y = 2$	85.00	64.49	53.50	48.15	50.00
$y = 4$	70.00	48.90	38.43	35.03	40.00
$y = 6$	55.00	38.78	30.39	27.07	30.00
$y = 8$	40.00	35.00	30.00	25.00	20.00

18.33 Use the *precisely* measured data in Fig. P18.33 to estimate the value at $x = 2.2$ and $y = 1.5$

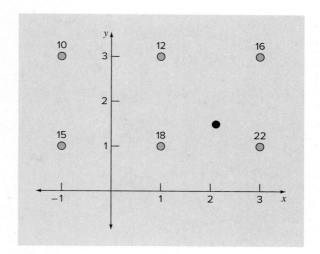

FIGURE P18.33

18.34 Bessel functions often arise in advanced engineering and scientific analyses such as the study of electric fields. These functions are usually not amenable to straightforward evaluation and, therefore, are often compiled in standard mathematical tables. For example,

x	1.8	2	2.2	2.4	2.6
$J_1(x)$	0.5815	0.5767	0.556	0.5202	0.4708

Estimate $J_1(2.1)$, using **(a)** an interpolating polynomial and **(b)** cubic splines. Note that the true value is 0.5683.

18.35 **(a)** Fit a cubic spline to the following data:

x	0	2	4	7	10	12
y	20	20	12	7	6	6

Determine the value of y at $x = 1.5$.

18.36 Runge's function is written as

$$f(x) = \frac{1}{1 + 25x^2}$$

Generate five equidistantly spaced values of this function over the interval $[1, 1]$. Fit this data with **(a)** a fourth-order polynomial, **(b)** a linear spline, and **(c)** a cubic spline. Present your results graphically.

18.37 Generate eight points from the function

$$f(t) = \sin^2 t$$

from $t = 0$ to 2π. Fit this data using a cubic spline. Develop a plot of the spline and the data as well as a plot of the absolute error ($E_t =$ approximation $-$ true).

CHAPTER 19

Fourier Approximation

To this point, our presentation of interpolation has emphasized standard polynomials—that is, linear combinations of the monomials $1, x, x^2, \ldots, x^m$ (Fig. 19.1a). We now turn to another class of functions that has immense importance in engineering. These are the trigonometric functions $1, \cos x, \cos 2x, \ldots, \cos nx, \sin x, \sin 2x, \ldots, \sin nx$ (Fig. 19.1b).

Engineers often deal with systems that oscillate or vibrate. As might be expected, trigonometric functions play a fundamental role in modeling such problem contexts.

FIGURE 19.1
The first five (a) monomials and (b) trigonometric functions. Note that for the intervals shown, both types of function range in value between −1 and 1. However, notice that the peak values for the monomials all occur at the extremes, whereas for the trigonometric functions the peaks are more uniformly distributed across the interval.

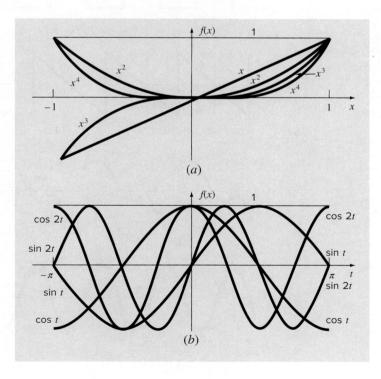

Fourier approximation represents a systematic framework for using trigonometric series for this purpose.

One of the hallmarks of a Fourier analysis is that it deals with both the time and the frequency domains. Because some engineers are not comfortable with the latter, we have devoted a large fraction of the subsequent material to a general overview of Fourier approximation. An important aspect of this overview will be to familiarize you with the frequency domain. This orientation is then followed by an introduction to numerical methods for computing discrete Fourier transforms.

19.1 CURVE FITTING WITH SINUSOIDAL FUNCTIONS

A periodic function $f(t)$ is one for which

$$f(t) = f(t + T) \tag{19.1}$$

FIGURE 19.2
Aside from trigonometric functions such as sines and cosines, periodic functions include waveforms such as (*a*) the square wave and (*b*) the sawtooth wave. Beyond these idealized forms, periodic signals in nature can be (*c*) nonideal and (*d*) contaminated by noise. The trigonometric functions can be used to represent and to analyze all these cases.

where T is a constant called the *period* that is the smallest value for which Eq. (19.1) holds. Common examples include waveforms such as square and sawtooth waves (Fig. 19.2). The most fundamental periodic functions are sinusoidal functions.

In the present discussion, we will use the term *sinusoid* to represent any waveform that can be described as a sine or cosine. There is no clear-cut convention for choosing either function, and in any case, the results will be identical. For this chapter, we will use the cosine, which is expressed generally as

$$f(t) = A_0 + C_1 \cos(\omega_0 t + \theta) \qquad (19.2)$$

Thus, four parameters serve to characterize the sinusoid (Fig. 19.3). The *mean value* A_0 sets the average height above the abscissa. The *amplitude* C_1 specifies the height of the

FIGURE 19.3

(*a*) A plot of the sinusoidal function $y(t) = A_0 + C_1 \cos(\omega_0 t + \theta)$. For this case, $A_0 = 1.7$, $C_1 = 1$, $\omega_0 = 2\pi/T = 2\pi/(1.5 \text{ s})$, and $\theta = \pi/3$ radians $= 1.0472$ ($= 0.25$ s). Other parameters used to describe the curve are the frequency $f = \omega_0/(2\pi)$, which for this case is 1 cycle/(1.5 s) and the period $T = 1.5$ s. (*b*) An alternative expression of the same curve is $y(t) = A_0 + A_1 \cos(\omega_0 t) + B_1 \sin(\omega_0 t)$. The three components of this function are depicted in (*b*), where $A_1 = 0.5$ and $B_1 = -0.866$. The summation of the three curves in (*b*) yields the single curve in (*a*).

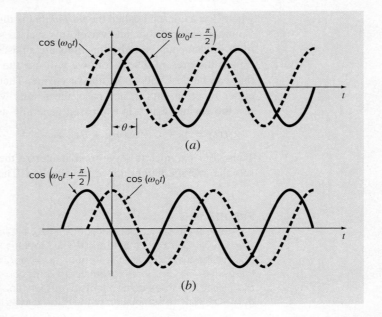

FIGURE 19.4
Graphical depictions of (a) a lagging phase angle and (b) a leading phase angle. Note that the lagging curve in (a) can be alternatively described as $\cos(\omega_0 t + 3\pi/2)$. In other words, if a curve lags by an angle of α, it can also be represented as leading by $2\pi - \alpha$.

oscillation. The *angular frequency* ω_0 characterizes how often the cycles occur. Finally, the phase angle, or *phase shift*, θ parameterizes the extent to which the sinusoid is shifted horizontally. It can be measured as the distance in radians from $t = 0$ to the point at which the cosine function begins a new cycle. As depicted in Fig. 19.4a, a negative value is referred to as a *lagging phase angle* because the curve $\cos(\omega_0 t - \theta)$ begins a new cycle θ radians after $\cos(\omega_0 t)$. Thus, $\cos(\omega_0 t - \theta)$ is said to lag $\cos(\omega_0 t)$. Conversely, as in Fig. 19.4b, a positive value is referred to as a leading phase angle.

Note that the angular frequency (in radians/time) is related to frequency f (in cycles/time) by

$$\omega_0 = 2\pi f \tag{19.3}$$

and frequency in turn is related to period T (in units of time) by

$$f = \frac{1}{T} \tag{19.4}$$

Although Eq. (19.2) is an adequate mathematical characterization of a sinusoid, it is awkward to work with from the standpoint of curve fitting because the phase shift is included in the argument of the cosine function. This deficiency can be overcome by invoking the trigonometric identity

$$C_1 \cos(\omega_0 t + \theta) = C_1[\cos(\omega_0 t)\cos(\theta) - \sin(\omega_0 t)\sin(\theta)] \tag{19.5}$$

Substituting Eq. (19.5) into Eq. (19.2) and collecting terms gives (Fig. 19.3b)

$$f(t) = A_0 + A_1 \cos(\omega_0 t) + B_1 \sin(\omega_0 t) \tag{19.6}$$

where

$$A_1 = C_1 \cos(\theta) \qquad B_1 = -C_1 \sin(\theta) \tag{19.7}$$

Dividing the two parts of Eq. (19.7) gives

$$\theta = \arctan\left(-\frac{B_1}{A_1}\right) \tag{19.8}$$

Note that if $A_1 < 0$, the angle θ should be incremented by π radians. Squaring and summing Eq. (19.7) leads to

$$C_1 = \sqrt{A_1^2 + B_1^2} \tag{19.9}$$

Thus, Eq. (19.6) represents an alternative formulation of Eq. (19.2) that still requires four parameters but that is cast in the format of a general linear model [recall Eq. (17.23)]. As we will discuss in the next section, it can be simply applied as the basis for a least-squares fit.

Before proceeding to the next section, however, we should stress that we could have employed a sine rather than a cosine as our fundamental model of Eq. (19.2). For example,

$$f(t) = A_0 + C_1 \sin(\omega_0 t + \delta)$$

could have been used. Simple relationships can be applied to convert between the two forms

$$\sin(\omega_0 t + \delta) = \cos\left(\omega_0 t + \delta - \frac{\pi}{2}\right)$$

and $\tag{19.10}$

$$\cos(\omega_0 t + \theta) = \sin\left(\omega_0 t + \theta + \frac{\pi}{2}\right)$$

In other words, $\theta = \delta - \pi/2$. The only important consideration is that one or the other format should be used consistently. Thus, we will use the cosine version throughout our discussion.

19.1.1 Least-Squares Fit of a Sinusoid

Equation (19.6) can be thought of as a linear least-squares model,

$$y = A_0 + A_1 \cos(\omega_0 t) + B_1 \sin(\omega_0 t) + e \tag{19.11}$$

which is just another example of the general model [recall Eq. (17.23)]

$$y = a_0 z_0 + a_1 z_1 + a_2 z_2 + \cdots + a_m z_m + e \tag{17.23}$$

where $z_0 = 1$, $z_1 = \cos(\omega_0 t)$, $z_2 = \sin(\omega_0 t)$, and all other z's $= 0$. Thus, our goal is to determine coefficient values that minimize

$$S_r = \sum_{i=1}^{N} \{y_i - [A_0 + A_1 \cos(\omega_0 t_i) + B_1 \sin(\omega_0 t_i)]\}^2$$

The normal equations to accomplish this minimization can be expressed in matrix form as [recall Eq. (17.25)]

$$
\begin{bmatrix}
N & \Sigma\cos(\omega_0 t) & \Sigma\sin(\omega_0 t) \\
\Sigma\cos(\omega_0 t) & \Sigma\cos^2(\omega_0 t) & \Sigma\cos(\omega_0 t)\sin(\omega_0 t) \\
\Sigma\sin(\omega_0 t) & \Sigma\cos(\omega_0 t)\sin(\omega_0 t) & \Sigma\sin^2(\omega_0 t)
\end{bmatrix}
\begin{Bmatrix} A_0 \\ A_1 \\ B_0 \end{Bmatrix}
$$

$$
= \begin{Bmatrix} \Sigma y \\ \Sigma y\cos(\omega_0 t) \\ \Sigma y\sin(\omega_0 t) \end{Bmatrix} \tag{19.12}
$$

These equations can be employed to solve for the unknown coefficients. However, rather than do this, we can examine the special case where there are N observations equispaced at intervals of Δt and with a total record length of $T = (N - 1)\,\Delta t$. For this situation, the following average values can be determined (see Prob. 19.1):

$$
\frac{\Sigma\sin(\omega_0 t)}{N} = 0 \qquad \frac{\Sigma\cos(\omega_0 t)}{N} = 0
$$

$$
\frac{\Sigma\sin^2(\omega_0 t)}{N} = \frac{1}{2} \qquad \frac{\Sigma\cos^2(\omega_0 t)}{N} = \frac{1}{2} \tag{19.13}
$$

$$
\frac{\Sigma\cos(\omega_0 t)\sin(\omega_0 t)}{N} = 0
$$

Thus, for equispaced points the normal equations become

$$
\begin{bmatrix}
N & 0 & 0 \\
0 & N/2 & 0 \\
0 & 0 & N/2
\end{bmatrix}
\begin{Bmatrix} A_0 \\ A_1 \\ B_1 \end{Bmatrix}
= \begin{Bmatrix} \Sigma y \\ \Sigma y\cos(\omega_0 t) \\ \Sigma y\sin(\omega_0 t) \end{Bmatrix}
$$

The inverse of a diagonal matrix is merely another diagonal matrix whose elements are the reciprocals of the original. Thus, the coefficients can be determined as

$$
\begin{Bmatrix} A_0 \\ A_1 \\ B_1 \end{Bmatrix}
= \begin{bmatrix}
1/N & 0 & 0 \\
0 & 2/N & 0 \\
0 & 0 & 2/N
\end{bmatrix}
\begin{Bmatrix} \Sigma y \\ \Sigma y\cos(\omega_0 t) \\ \Sigma y\sin(\omega_0 t) \end{Bmatrix}
$$

or

$$
A_0 = \frac{\Sigma y}{N} \tag{19.14}
$$

$$
A_1 = \frac{2}{N}\Sigma y\cos(\omega_0 t) \tag{19.15}
$$

$$
B_1 = \frac{2}{N}\Sigma y\sin(\omega_0 t) \tag{19.16}
$$

EXAMPLE 19.1 **Least-Squares Fit of a Sinusoid**

Problem Statement.   The curve in Fig. 19.3 is described by $y = 1.7 + \cos(4.189t + 1.0472)$. Generate 10 discrete values for this curve at intervals of $\Delta t = 0.15$ for the range

$t = 0$ to 1.35. Use this information to evaluate the coefficients of Eq. (19.11) by a least-squares fit.

Solution.   The data required to evaluate the coefficients with $\omega = 4.189$ are

t	y	$y \cos(\omega_0 t)$	$y \sin(\omega_0 t)$
0	2.200	2.200	0.000
0.15	1.595	1.291	0.938
0.30	1.031	0.319	0.980
0.45	0.722	−0.223	0.687
0.60	0.786	−0.636	0.462
0.75	1.200	−1.200	0.000
0.90	1.805	−1.460	−1.061
1.05	2.369	−0.732	−2.253
1.20	2.678	0.829	−2.547
1.35	2.614	2.114	−1.536
$\Sigma =$	17.000	2.502	−4.330

These results can be used to determine [Eqs. (19.14) through (19.16)]

$$A_0 = \frac{17.000}{10} = 1.7 \qquad A_1 = \frac{2}{10} 2.502 = 0.500 \qquad B_1 = \frac{2}{10}(-4.330) = -0.866$$

Thus, the least-squares fit is

$$y = 1.7 + 0.500 \cos(\omega_0 t) - 0.866 \sin(\omega_0 t)$$

The model can also be expressed in the format of Eq. (19.2) by calculating [Eq. (19.8)]

$$\theta = \arctan\left(-\frac{-0.866}{0.500}\right) = 1.0472$$

and [Eq. (19.9)]

$$C_1 = \sqrt{(0.5)^2 + (-0.866)^2} = 1.00$$

to give

$$y = 1.7 + \cos(\omega_0 t + 1.0472)$$

or alternatively, as a sine by using [Eq. (19.10)],

$$y = 1.7 + \sin(\omega_0 t + 2.618)$$

The foregoing analysis can be extended to the general model

$$f(t) = A_0 + A_1 \cos(\omega_0 t) + B_1 \sin(\omega_0 t) + A_2 \cos(2\omega_0 t) + B_2 \sin(2\omega_0 t)$$
$$+ \cdots + A_m \cos(m\omega_0 t) + B_m \sin(m\omega_0 t)$$

where, for equally spaced data, the coefficients can be evaluated by

$$A_0 = \frac{\Sigma y}{N}$$

$$\left.\begin{array}{l} A_j = \dfrac{2}{N}\Sigma y \cos(j\omega_0 t) \\[2mm] B_j = \dfrac{2}{N}\Sigma y \sin(j\omega_0 t) \end{array}\right\} \quad j = 1, 2, \ldots, m$$

Although these relationships can be used to fit data in the regression sense (that is, $N > 2m + 1$), an alternative application is to employ them for interpolation or collocation—that is, to use them for the case where the number of unknowns, $2m + 1$, is equal to the number of data points, N. This is the approach used in the continuous Fourier series, as described next.

19.2 CONTINUOUS FOURIER SERIES

In the course of studying heat-flow problems, Fourier showed that an arbitrary periodic function can be represented by an infinite series of sinusoids of harmonically related frequencies. For a function with period T, a continuous Fourier series can be written[1]

$$f(t) = a_0 + a_1 \cos(\omega_0 t) + b_1 \sin(\omega_0 t) + a_2 \cos(2\omega_0 t) + b_2 \sin(2\omega_0 t) + \cdots$$

or more concisely,

$$f(t) = a_0 + \sum_{k=1}^{\infty} [a_k \cos(k\omega_0 t) + b_k \sin(k\omega_0 t)] \tag{19.17}$$

where $\omega_0 = 2\pi/T$ is called the *fundamental frequency* and its constant multiples $2\omega_0$, $3\omega_0$, etc., are called *harmonics*. Thus, Eq. (19.17) expresses $f(t)$ as a linear combination of the basis functions: 1, $\cos(\omega_0 t)$, $\sin(\omega_0 t)$, $\cos(2\omega_0 t)$, $\sin(2\omega_0 t)$, ...

As described in Box 19.1, the coefficients of Eq. (19.17) can be computed via

$$a_k = \frac{2}{T} \int_0^T f(t) \cos(k\omega_0 t)\, dt \tag{19.18}$$

and

$$b_k = \frac{2}{T} \int_0^T f(t) \sin(k\omega_0 t)\, dt \tag{19.19}$$

for $k = 1, 2, \ldots$, and

$$a_0 = \frac{1}{T} \int_0^T f(t)\, dt \tag{19.20}$$

---

[1]The existence of the Fourier series is predicated on the Dirichlet conditions. These specify that the periodic function have a finite number of maxima and minima and that there be a finite number of jump discontinuities. In general, all physically derived periodic functions satisfy these conditions.

Box 19.1 Determination of the Coefficients of the Continuous Fourier Series

As was done for the discrete data of Sec. 19.1.1, the following relationships can be established:

$$\int_0^T \sin(k\omega_0 t)\, dt = \int_0^T \cos(k\omega_0 t)\, dt = 0 \qquad \text{(B19.1.1)}$$

$$\int_0^T \cos(k\omega_0 t) \sin(g\omega_0 t)\, dt = 0 \qquad \text{(B19.1.2)}$$

$$\int_0^T \sin(k\omega_0 t) \sin(g\omega_0 t)\, dt = 0 \qquad \text{(B19.1.3)}$$

$$\int_0^T \cos(k\omega_0 t) \cos(g\omega_0 t)\, dt = 0 \qquad \text{(B19.1.4)}$$

$$\int_0^T \sin^2(k\omega_0 t)\, dt = \int_0^T \cos^2(k\omega_0 t)\, dt = \frac{T}{2} \qquad \text{(B19.1.5)}$$

To evaluate its coefficients, each side of Eq. (19.17) can be integrated to give

$$\int_0^T f(t)\, dt = \int_0^T a_0\, dt + \int_0^T \sum_{k=1}^{\infty} [a_k \cos(k\omega_0 t) + b_k \sin(k\omega_0 t)]\, dt$$

Because every term in the summation is of the form of Eq. (B19.1.1), the equation becomes

$$\int_0^T f(t)\, dt = a_0 T$$

which can be solved for

$$a_0 = \frac{\int_0^T f(t)\, dt}{T}$$

Thus, a_0 is simply the average value of the function over the period.

To evaluate one of the cosine coefficients, for example, a_m, multiply Eq. (19.17) by $\cos(m\omega_0 t)$ and integrate to give

$$\int_0^T f(t) \cos(m\omega_0 t)\, dt = \int_0^T a_0 \cos(m\omega_0 t)\, dt$$

$$+ \int_0^T \sum_{k=1}^{\infty} a_k \cos(k\omega_0 t) \cos(m\omega_0 t)\, dt$$

$$+ \int_0^T \sum_{k=1}^{\infty} b_k \sin(k\omega_0 t) \cos(m\omega_0 t)\, dt \qquad \text{(B19.1.6)}$$

From Eqs. (B19.1.1), (B19.1.2), and (B19.1.4), we see that every term on the right-hand side is zero, with the exception of the case where $k = m$. This latter case can be evaluated by Eq. (B19.1.5) and, therefore, Eq. (B19.1.6) can be solved for a_m, or more generally [Eq. (19.18)],

$$a_k = \frac{2}{T} \int_0^T f(t) \cos(k\omega_0 t)\, dt$$

for $k = 1, 2, \ldots$.

In a similar fashion, Eq. (19.17) can be multiplied by $\sin(m\omega_0 t)$, integrated, and manipulated to yield Eq. (19.19).

EXAMPLE 19.2 Continuous Fourier Series Approximation

Problem Statement. Use the continuous Fourier series to approximate the square or rectangular wave function (Fig. 19.5)

$$f(t) = \begin{cases} -1 & -T/2 < t < -T/4 \\ 1 & -T/4 < t < T/4 \\ -1 & T/4 < t < T/2 \end{cases}$$

Solution. Because the average height of the wave is zero, a value of $a_0 = 0$ can be obtained directly. The remaining coefficients can be evaluated as [Eq. (19.18)]

$$a_k = \frac{2}{T} \int_{-T/2}^{T/2} f(t) \cos(k\omega_0 t)\, dt$$

$$= \frac{2}{T} \left[-\int_{-T/2}^{-T/4} \cos(k\omega_0 t)\, dt + \int_{-T/4}^{T/4} \cos(k\omega_0 t)\, dt - \int_{T/4}^{T/2} \cos(k\omega_0 t)\, dt \right]$$

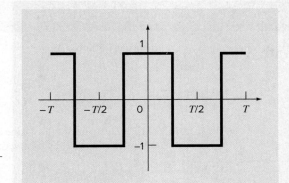

FIGURE 19.5
A square or rectangular wave-form with a height of 2 and a period $T = 2\pi/\omega_0$.

FIGURE 19.6
The Fourier series approximation of the square wave from Fig. 19.5. The series of plots shows the summation up to and including the (a) first, (b) second, and (c) third terms. The individual terms that were added at each stage are also shown.

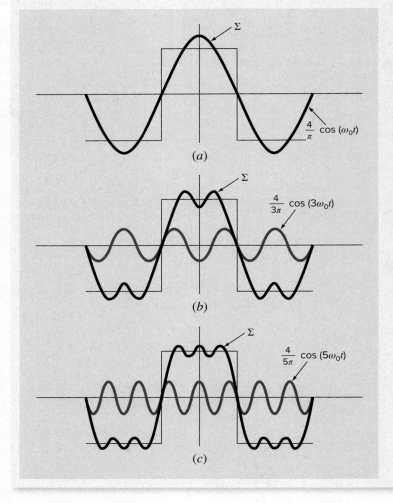

The integrals can be evaluated to give

$$a_k = \begin{cases} 4/(k\pi) & \text{for } k = 1, 5, 9, \ldots \\ -4/(k\pi) & \text{for } k = 3, 7, 11, \ldots \\ 0 & \text{for } k = \text{even integers} \end{cases}$$

Similarly, it can be determined that all the b's $= 0$. Therefore, the Fourier series approximation is

$$f(t) = \frac{4}{\pi}\cos(\omega_0 t) - \frac{4}{3\pi}\cos(3\omega_0 t) + \frac{4}{5\pi}\cos(5\omega_0 t) - \frac{4}{7\pi}\cos(7\omega_0 t) + \cdots$$

The results up to the first three terms are shown in Fig. 19.6.

It should be mentioned that the square wave in Fig. 19.5 is called an *even function* because $f(t) = f(-t)$. Another example of an even function is $\cos(t)$. It can be shown (Van Valkenburg 1974) that the b's in the Fourier series always equal zero for even functions. Note also that *odd functions* are those for which $f(t) = -f(-t)$. The function $\sin(t)$ is an odd function. For this case, the a's will equal zero.

In addition to the trigonometric format of Eq. (19.17), the Fourier series can be expressed in terms of exponential functions as (see Box 19.2 and App. A)

$$f(t) = \sum_{k=-\infty}^{\infty} \tilde{c}_k e^{ik\omega_0 t} \tag{19.21}$$

where $i = \sqrt{-1}$ and

$$\tilde{c}_k = \frac{1}{T}\int_{-T/2}^{T/2} f(t)e^{-ik\omega_0 t}\, dt \tag{19.22}$$

This alternative formulation will have utility throughout the remainder of the chapter.

19.3   FREQUENCY AND TIME DOMAINS

To this point, our discussion of Fourier approximation has been limited to the *time domain*. We have done this because most of us are fairly comfortable conceptualizing a function's behavior in this dimension. Although it is not as familiar, the *frequency domain* provides an alternative perspective for characterizing the behavior of oscillating functions.

Thus, just as amplitude can be plotted versus time, so also can it be plotted versus frequency. Both types of expression are depicted in Fig. 19.7a, where we have drawn a three-dimensional graph of a sinusoidal function,

$$f(t) = C_1 \cos\left(t + \frac{\pi}{2}\right)$$

In this plot, the magnitude or amplitude of the curve, $f(t)$, is the dependent variable and time t and frequency $f = \omega_0/2\pi$ are the independent variables. Thus, the amplitude and the time axes form a *time plane,* and the amplitude and the frequency axes form a

Box 19.2　Complex Form of the Fourier Series

The trigonometric form of the continuous Fourier series is

$$f(t) = a_0 + \sum_{k=1}^{\infty} [a_k \cos(k\omega_0 t) + b_k \sin(k\omega_0 t)] \qquad \text{(B19.2.1)}$$

From Euler's identity, the sine and cosine can be expressed in exponential form as

$$\sin x = \frac{e^{ix} - e^{-ix}}{2i} \qquad \text{(B19.2.2)}$$

$$\cos x = \frac{e^{ix} + e^{-ix}}{2} \qquad \text{(B19.2.3)}$$

which can be substituted into Eq. (B19.2.1) to give

$$f(t) = a_0 + \sum_{k=1}^{\infty} \left(e^{ik\omega_0 t} \frac{a_k - ib_k}{2} + e^{-ik\omega_0 t} \frac{a_k + ib_k}{2} \right) \qquad \text{(B19.2.4)}$$

because $1/i = -i$. We can define a number of constants:

$$\widetilde{c}_0 = a_0$$

$$\widetilde{c}_k = \frac{a_k - ib_k}{2} \qquad \text{(B19.2.5)}$$

$$\widetilde{c}_{-k} = \frac{a_{-k} - ib_{-k}}{2} = \frac{a_k + ib_k}{2}$$

where, because of the odd and even properties of the sine and cosine, $a_k = a_{-k}$ and $b_k = -b_{-k}$. Equation (B19.2.4) can, therefore, be re-expressed as

$$f(t) = \widetilde{c}_0 + \sum_{k=1}^{\infty} \widetilde{c}_k e^{ik\omega_0 t} + \sum_{k=1}^{\infty} \widetilde{c}_{-k} e^{-ik\omega_0 t}$$

or

$$f(t) = \sum_{k=0}^{\infty} \widetilde{c}_k e^{ik\omega_0 t} + \sum_{k=1}^{\infty} \widetilde{c}_{-k} e^{-ik\omega_0 t}$$

To simplify further, instead of summing the second series from 1 to ∞, perform the sum from -1 to $-\infty$,

$$f(t) = \sum_{k=0}^{\infty} \widetilde{c}_k e^{ik\omega_0 t} + \sum_{k=-1}^{-\infty} \widetilde{c}_k e^{ik\omega_0 t}$$

or

$$f(t) = \sum_{k=-\infty}^{\infty} \widetilde{c}_k e^{ik\omega_0 t} \qquad \text{(B19.2.6)}$$

where the summation includes a term for $k = 0$.

To evaluate the $\widetilde{c}_k$'s, Eqs. (19.18) and (19.19) can be substituted into Eq. (B19.2.5) to yield

$$\widetilde{c}_k = \frac{1}{T} \int_{-T/2}^{T/2} f(t) \cos(k\omega_0 t)\, dt - i \frac{1}{T} \int_{-T/2}^{T/2} f(t) \sin(k\omega_0 t)\, dt$$

Employing Eqs. (B19.2.2) and (B19.2.3) and simplifying gives

$$\widetilde{c}_k = \frac{1}{T} \int_{-T/2}^{T/2} f(t) e^{-ik\omega_0 t}\, dt \qquad \text{(B19.2.7)}$$

Therefore, Eqs. (B19.2.6) and (B19.2.7) are the complex versions of Eqs. (19.17) through (19.20). Note that App. A includes a summary of the interrelationships among all the formats of the Fourier series introduced in this chapter.

frequency plane. The sinusoid can, therefore, be conceived of as existing a distance $1/T$ out along the frequency axis and running parallel to the time axis. Consequently, when we speak about the behavior of the sinusoid in the time domain, we mean the projection of the curve onto the time plane (Fig. 19.7b). Similarly, the behavior in the frequency domain is merely its projection onto the frequency plane.

As in Fig. 19.7c, this projection is a measure of the sinusoid's maximum positive amplitude C_1. The full peak-to-peak swing is unnecessary because of the symmetry. Together with the location $1/T$ along the frequency axis, Fig. 19.7c defines the amplitude and frequency of the sinusoid. This is enough information to reproduce the shape and size of the curve in the time domain. However, one more parameter, namely, the phase angle, is required to position the curve relative to $t = 0$. Consequently, a phase diagram, as shown in Fig. 19.7d, must also be included. The phase angle is determined as the distance (in radians) from zero to the point at which the positive peak occurs. If the peak occurs after

FIGURE 19.7
(*a*) A depiction of how a sinusoid can be portrayed in the time and the frequency domains.
The time projection is reproduced in (*b*), whereas the amplitude-frequency projection is reproduced in (*c*). The phase-frequency projection is shown in (*d*).

zero, it is said to be delayed (recall our discussion of lags and leads in Sec. 19.1), and by
convention, the phase angle is given a negative sign. Conversely, a peak before zero is said
to be advanced and the phase angle is positive. Thus, for Fig. 19.7, the peak leads zero
and the phase angle is plotted as $+\pi/2$. Figure 19.8 depicts some other possibilities.

We can now see that Fig. 19.7*c* and *d* provide an alternative way to present or
summarize the pertinent features of the sinusoid in Fig. 19.7*a*. They are referred to as
line spectra. Admittedly, for a single sinusoid they are not very interesting. However,
when applied to a more complicated situation, say, a Fourier series, their true power and
value is revealed. For example, Fig. 19.9 shows the amplitude and phase line spectra for
the square-wave function from Example 19.2.

Such spectra provide information that would not be apparent from the time domain.
This can be seen by contrasting Figs. 19.6 and 19.9. Figure 19.6 presents two alternative
time-domain perspectives. The first, the original square wave, tells us nothing about

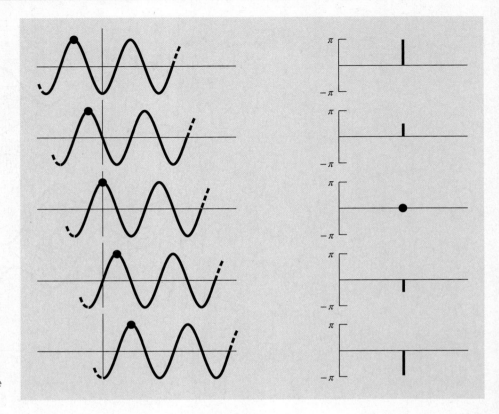

FIGURE 19.8
Various phases of a sinusoid showing the associated phase line spectra.

FIGURE 19.9
(a) Amplitude and (b) phase line spectra for the square wave from Fig. 19.5.

the sinusoids that comprise it. The alternative is to display these sinusoids—that is, $(4/\pi)\cos(\omega_0 t)$, $-(4/3\pi)\cos(3\omega_0 t)$, $(4/5\pi)\cos(5\omega_0 t)$, etc. This alternative does not provide an adequate visualization of the structure of these harmonics. In contrast, Fig. 19.9*a* and *b* provide a graphic display of this structure. As such, the line spectra represent "fingerprints" that can help us to characterize and understand a complicated waveform. They are particularly valuable for nonidealized cases, where they sometimes allow us to discern structure in otherwise obscure signals. In the next section, we will describe the Fourier transform that will allow us to extend such analyses to nonperiodic waveforms.

19.4 FOURIER INTEGRAL AND TRANSFORM

Although the Fourier series is a useful tool for investigating the spectrum of a periodic function, there are many waveforms that do not repeat themselves regularly. For example, a lightning bolt occurs only once (or at least it will be a long time until it occurs again), but it will cause interference with receivers operating on a broad range of frequencies—for example, TVs, radios, and shortwave receivers. Such evidence suggests that a nonrecurring signal such as that produced by lightning exhibits a continuous frequency spectrum. Because such phenomena are of great interest to engineers, an alternative to the Fourier series would be valuable for analyzing these aperiodic waveforms.

The *Fourier integral* is the primary tool available for this purpose. It can be derived from the exponential form of the Fourier series,

$$f(t) = \sum_{k=-\infty}^{\infty} \widetilde{c}_k e^{ik\omega_0 t} \tag{19.23}$$

where

$$\widetilde{c}_k = \frac{1}{T} \int_{-T/2}^{T/2} f(t) e^{-ik\omega_0 t}\, dt \tag{19.24}$$

and $\omega_0 = 2\pi/T$ and $k = 0, 1, 2, \ldots$.

The transition from a periodic to a nonperiodic function can be effected by allowing the period to approach infinity. In other words, as T becomes infinite, the function never repeats itself and thus becomes aperiodic. If this is allowed to occur, it can be demonstrated (for example, Van Valkenburg 1974; Hayt and Kemmerly 1986) that the Fourier series reduces to

$$f(t) = \frac{1}{2\pi} \int_{-\infty}^{\infty} F(i\omega_0) e^{i\omega_0 t}\, d\omega_0 \tag{19.25}$$

and the coefficients become a continuous function of the frequency variable ω, as in

$$F(i\omega_0) = \int_{-\infty}^{\infty} f(t) e^{-i\omega_0 t}\, dt \tag{19.26}$$

The function $F(i\omega_0)$, as defined by Eq. (19.26), is called the *Fourier integral* of $f(t)$. In addition, Eqs. (19.25) and (19.26) are collectively referred to as the *Fourier transform pair*. Thus, along with being called the Fourier integral, $F(i\omega_0)$ is also called the *Fourier transform* of $f(t)$. In the same spirit, $f(t)$, as defined by Eq. (19.25), is referred to as the

FIGURE 19.10

Illustration of how the discrete frequency spectrum of a Fourier series for a pulse train
(*a*) approaches a continuous frequency spectrum of a Fourier integral (*c*) as the period is
allowed to approach infinity.

inverse Fourier transform of $F(i\omega_0)$. Thus, the pair allows us to transform back and forth
between the time and the frequency domains for an aperiodic signal.

The distinction between the Fourier series and transform should now be quite clear.
The major difference is that each applies to a different class of functions—the series to
periodic and the transform to nonperiodic waveforms. Beyond this major distinction, the
two approaches differ in how they move between the time and the frequency domains.
The Fourier series converts a continuous, periodic time-domain function to frequency-
domain magnitudes at discrete frequencies. In contrast, the Fourier transform converts a
continuous time-domain function to a continuous frequency-domain function. Thus, the
discrete frequency spectrum generated by the Fourier series is analogous to a continuous
frequency spectrum generated by the Fourier transform.

The shift from a discrete to a continuous spectrum can be illustrated graphically. In
Fig. 19.10*a*, we can see a pulse train of rectangular waves with pulse widths equal to one-
half the period, along with its associated discrete spectrum. This is the same function as
was investigated previously in Example 19.2, with the exception that it is shifted vertically.

In Fig. 19.10*b*, a doubling of the pulse train's period has two effects on the spectrum.
First, two additional frequency lines are added on either side of the original components.
Second, the amplitudes of the components are reduced.

As the period is allowed to approach infinity, these effects continue as more and more spectral lines are packed together until the spacing between lines goes to zero. At the limit, the series converges on the continuous Fourier integral, depicted in Fig. 19.10c.

Now that we have introduced a way to analyze an aperiodic signal, we will take the final step in our development. In the next section, we will acknowledge the fact that a signal is rarely characterized as a continuous function of the sort needed to implement Eq. (19.26). Rather, these data are invariably in a discrete form. Thus, we will now show how to compute a Fourier transform for such discrete measurements.

19.5 DISCRETE FOURIER TRANSFORM (DFT)

In engineering, functions are often represented by finite sets of discrete values. Additionally, data are often collected in or converted to such a discrete format. As depicted in Fig. 19.11, an interval from 0 to t can be divided into N equispaced subintervals with widths of $\Delta t = T/N$. The subscript n is employed to designate the discrete times at which samples are taken. Thus, f_n designates a value of the continuous function $f(t)$ taken at t_n.

Note that the data points are specified at $n = 0, 1, 2, \ldots, N - 1$. A value is not included at $n = N$. (See Ramirez, 1985, for the rationale for excluding f_N.)

For the system in Fig. 19.11, a discrete Fourier transform (DFT) can be written as

$$F_k = \sum_{n=0}^{N-1} f_n e^{-ik\omega_0 n} \qquad \text{for } k = 0 \text{ to } N - 1 \tag{19.27}$$

FIGURE 19.11
The sampling points of the discrete Fourier series.

and the inverse Fourier transform as

$$f_n = \frac{1}{N}\sum_{k=0}^{N-1} F_k e^{ik\omega_0 n} \qquad \text{for } n = 0 \text{ to } N-1 \tag{19.28}$$

where $\omega_0 = 2\pi/N$.

Equations (19.27) and (19.28) represent the discrete analogs of Eqs. (19.25) and (19.26), respectively. As such, they can be employed to compute both a direct and an inverse Fourier transform for discrete data.

Before proceeding, several other aspects of the DFT bear mentioning. The highest frequency that can be measured in a signal, called the *Nyquist frequency*, is half the sampling frequency. Periodic variations that occur more rapidly than the shortest sampled time interval cannot be detected. The lowest frequency you can detect is the inverse of the total sample length.

As an example, suppose that you take 100 samples of data ($n = 100$ samples) at a sample frequency of $f_s = 1000$ Hz (i.e., 1000 samples per second). This means that the sample interval is

$$\Delta t = \frac{1}{f_s} = \frac{1}{1000 \text{ samples/s}} = 0.001 \text{ s/sample}$$

The total sample length is

$$t_n = \frac{n}{f_s} = \frac{100 \text{ samples}}{1000 \text{ samples/s}} = 0.1 \text{ s}$$

and the frequency increment is

$$\Delta f = \frac{f_s}{n} = \frac{1000 \text{ samples/s}}{100 \text{ samples}} = 10 \text{ Hz}$$

The Nyquist frequency is

$$f_{max} = 0.5 f_s = 0.5(1000 \text{ Hz}) = 500 \text{ Hz}$$

and the lowest detectable frequency is

$$f_{min} = \frac{1}{0.1 \text{ s}} = 10 \text{ Hz}$$

Thus, for this example, the DFT could detect signals with periods from $1/500 = 0.002$ s up to $1/10 = 0.1$ s.

EXAMPLE 19.3 Sampling Strategy to Detect a Range of Frequencies with the DFT

Problem Statement.   Suppose you want to determine the DFT for a simple sinusoid:

$$f(t) = 5 + \cos(2\pi(12.5)t) + \sin(2\pi(18.75)t)$$

If you generate eight equispaced points with $\Delta t = 0.02$ s, would the DFT be able to detect both the 12.5 and 18.75 Hz signals?

Solution.   The sampling frequency can be computed as

$$f_s = \frac{1}{\Delta t} = \frac{1}{0.02 \text{ s}} = 50 \text{ Hz}$$

The total sample length is

$$t_n = \frac{n}{f_s} = \frac{8 \text{ samples}}{50 \text{ samples/s}} = 0.16 \text{ s}$$

The Nyquist (or maximum detectable) frequency is

$$f_{max} = 0.5 f_s = 0.5(50 \text{ Hz}) = 25 \text{ Hz}$$

and the lowest detectable frequency is

$$f_{min} = \frac{1}{0.16 \text{ s}} = 6.25 \text{ Hz}$$

Thus, the analysis can detect signals with periods from $1/25 = 0.04$ s up to $1/6.25 = 0.16$ s and is able to detect both the 12.5 and 18.75 Hz signals.

Computer Algorithm for the DFT.   Note that the factor $1/N$ in Eq. (19.28) is merely a scale factor that can be included in either Eq. (19.27) or (19.28), but not both. For our computer algorithm, we will shift it to Eq. (19.27) so that the first coefficient F_0 (which is the analog of the continuous coefficient a_0) is equal to the arithmetic mean of the samples. Also, to develop an algorithm that can be implemented in languages that do not accommodate complex variables, we can use Euler's identity,

$$e^{\pm ia} = \cos a \pm i \sin a$$

to re-express Eqs. (19.27) and (19.28) as

$$F_k = \frac{1}{N} \sum_{n=0}^{N} [f_n \cos(k\omega_0 n) - i f_n \sin(k\omega_0 n)] \qquad (19.29)$$

and

$$f_n = \sum_{k=0}^{N-1} [F_k \cos(k\omega_0 n) - i F_k \sin(k\omega_0 n)] \qquad (19.30)$$

Pseudocode to implement Eq. (19.29) is listed in Fig. 19.12. This algorithm can be developed into a computer program to compute the DFT. The output from such a program is listed in Fig. 19.13 for the analysis of a cosine function.

FIGURE 19.12

Pseudocode for computing the DFT.

```
DOFOR k = 0, N - 1
  DOFOR n = 0, N - 1
    angle = kω₀n
    realₖ = realₖ + fₙ cos(angle)/N
    imaginaryₖ = imaginaryₖ - fₙ sin(angle)/N
  END DO
END DO
```

INDEX	f(t)	REAL	IMAGINARY
0	1.000	0.000	0.000
1	0.707	0.000	0.000
2	0.000	0.500	0.000
3	-0.707	0.000	0.000
4	-1.000	0.000	0.000
5	-0.707	0.000	0.000
6	0.000	0.000	0.000
7	0.707	0.000	0.000
8	1.000	0.000	0.000
9	0.707	0.000	0.000
10	0.000	0.000	0.000
11	-0.707	0.000	0.000
12	-1.000	0.000	0.000
13	-0.707	0.000	0.000
14	0.000	0.500	0.000
15	0.707	0.000	0.000

FIGURE 19.13
Output of a program based on the algorithm from Fig. 19.12 for the DFT of data generated by a cosine function $f(t) = \cos[2\pi(12.5)t]$ at 16 points with $\Delta t = 0.01$ s.

19.6 FAST FOURIER TRANSFORM (FFT)

Although the algorithm described in the previous section adequately calculates the DFT, it is computationally burdensome because N^2 operations are required. Consequently, for data samples of even moderate size, the direct determination of the DFT can be extremely time-consuming.

The *fast Fourier transform,* or *FFT,* is an algorithm that has been developed to compute the DFT in an extremely economical fashion. Its speed stems from the fact that it utilizes the results of previous computations to reduce the number of operations. In particular, it exploits the periodicity and symmetry of trigonometric functions to compute the transform with approximately $N \log_2 N$ operations (Fig. 19.14). Thus, for $N = 50$ samples, the FFT is about 10 times faster than the standard DFT. For $N = 1000$, it is about 100 times faster.

The first FFT algorithm was developed by Gauss in the early nineteenth century (Heideman et al. 1984). Other major contributions were made by Runge, Danielson, Lanczos, and others in the early twentieth century. However, because discrete transforms often took days to weeks to calculate by hand, they did not attract broad interest prior to the development of the modern digital computer.

In 1965, J. W. Cooley and J. W. Tukey published a key paper in which they outlined an algorithm for calculating the FFT. This scheme, which is similar to those of Gauss and other earlier investigators, is called the *Cooley-Tukey algorithm.* Today, there are a host of other approaches that are offshoots of this method.

The basic idea behind each of these algorithms is that a DFT of length N is decomposed, or "decimated," into successively smaller DFTs. There are a variety of different ways to implement this principle. For example, the Cooley-Tukey algorithm is a member of what are called *decimation-in-time* techniques. In the present section, we will describe

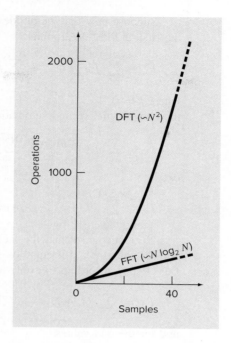

FIGURE 19.14
Plot of number of operations vs. sample size for the standard DFT and the FFT.

an alternative approach called the *Sande-Tukey algorithm*. This method is a member of another class of algorithms called *decimation-in-frequency* techniques. The distinction between the two classes will be discussed after we have elaborated on the method.

19.6.1 Sande-Tukey Algorithm

In the present case, N will be assumed to be an integral power of 2,

$$N = 2^M \tag{19.31}$$

where M is an integer. This constraint is introduced to simplify the resulting algorithm. Now, recall that the DFT can be generally represented as

$$F_k = \sum_{n=0}^{N-1} f_n e^{-i(2\pi/N)nk} \qquad \text{for } k = 0 \text{ to } N - 1 \tag{19.32}$$

where $2\pi/N = \omega_0$. Equation (19.32) can also be expressed as

$$F_k = \sum_{n=0}^{N-1} f_n W^{nk}$$

where W is a complex-valued weighting function defined as

$$W = e^{-i(2\pi/N)} \tag{19.33}$$

Suppose now that we divide the sample in half and express Eq. (19.32) in terms of the first and last $N/2$ points:

$$F_k = \sum_{n=0}^{(N/2)-1} f_n e^{-i(2\pi/N)kn} + \sum_{n=N/2}^{N-1} f_n e^{-i(2\pi/N)kn}$$

where $k = 0, 1, 2, \ldots, N - 1$. A new variable, $m = n - N/2$, can be created so that the range of the second summation is consistent with the first,

$$F_k = \sum_{n=0}^{(N/2)-1} f_n e^{-i(2\pi/N)kn} + \sum_{m=0}^{(N/2)-1} f_{m+N/2} e^{-i(2\pi/N)k(m+N/2)}$$

or

$$F_k = \sum_{n=0}^{(N/2)-1} (f_n + e^{-i\pi k} f_{n+N/2}) e^{-i2\pi kn/N} \qquad (19.34)$$

Next, recognize that the factor $e^{-i\pi k} = (-1)^k$. Thus, for even points it is equal to 1 and for odd points it is equal to -1. Therefore, the next step in the method is to separate Eq. (19.34) according to even values and odd values of k. For the even values,

$$F_{2k} = \sum_{n=0}^{(N/2)-1} (f_n + f_{n+N/2}) e^{-i2\pi(2k)n/N} = \sum_{n=0}^{(N/2)-1} (f_n + f_{n+N/2}) e^{-i2\pi kn/(N/2)}$$

and for the odd values,

$$F_{2k+1} = \sum_{n=0}^{(N/2)-1} (f_n - f_{n+N/2}) e^{-i2\pi(2k+1)n/N}$$

$$= \sum_{n=0}^{(N/2)-1} (f_n - f_{n+N/2}) e^{-i2\pi n/N} e^{-i2\pi kn/(N/2)}$$

for $k = 0, 1, 2, \ldots, (N/2) - 1$.

These equations can also be expressed in terms of Eq. (19.33). For the even values,

$$F_{2k} = \sum_{n=0}^{(N/2)-1} (f_n + f_{n+N/2}) W^{2kn}$$

and for the odd values,

$$F_{2k+1} = \sum_{n=0}^{(N/2)-1} (f_n - f_{n+N/2}) W^n W^{2kn}$$

Now, a key insight can be made. These even and odd expressions can be interpreted as being equal to the transforms of the $(N/2)$-length sequences,

$$g_n = f_n + f_{n+N/2} \qquad (19.35)$$

and

$$h_n = (f_n - f_{n+N/2}) W^n \qquad \text{for } n = 0, 1, 2, \ldots, (N/2) - 1 \qquad (19.36)$$

Thus, it directly follows that

$$\left. \begin{array}{l} F_{2k} = G_k \\ F_{2k+1} = H_k \end{array} \right\} \qquad \text{for } k = 0, 1, 2, \ldots, (N/2) - 1$$

In other words, one N-point computation has been replaced by two $(N/2)$-point computations. Because each of the latter requires approximately $(N/2)^2$ complex multiplications and additions, the approach produces a factor-of-2 savings—that is, N^2 versus $2(N/2)^2 = N^2/2$.

The scheme is depicted in Fig. 19.15 for $N = 8$. The DFT is computed by first forming the sequences g^n and h^n and then computing the $N/2$ DFTs to obtain the even- and odd-numbered transforms. The weights W^n are sometimes called *twiddle factors*.

Now it is clear that this "divide-and-conquer" approach can be repeated at the second stage. Thus, we can compute the $(N/4)$-point DFTs of the four $N/4$ sequences composed of the first and last $N/4$ points of Eqs. (19.35) and (19.36).

The strategy is continued to its inevitable conclusion when $N/2$ two-point DFTs are computed (Fig. 19.16). The total number of calculations for the entire computation is on the order of $N \log_2 N$. The contrast between this level of effort and that of the standard DFT (Fig. 19.14) illustrates why the *FFT* is so important.

Computer Algorithm.   It is a relatively straightforward proposition to express Fig. 19.16 as an algorithm. As was the case for the DFT algorithm of Fig. 19.12, we will use Euler's identity,

$$e^{\pm ia} = \cos a \pm i \sin a$$

to allow the algorithm to be implemented in languages that do not explicitly accommodate complex variables.

Close inspection of Fig. 19.16 indicates that its fundamental computational molecule is the so-called *butterfly network* depicted in Fig. 19.17a. Pseudocode to implement one of these molecules is shown in Fig. 19.17b.

Pseudocode for the FFT is listed in Fig. 19.18. The first part consists essentially of three nested loops to implement the computation embodied in Fig. 19.16. Note that the

FIGURE 19.15

Flow graph of the first stage in a decimation-in-frequency decomposition of an N-point DFT into two $(N/2)$-point DFTs for $N = 8$.

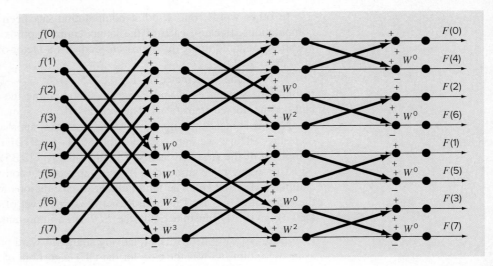

FIGURE 19.16
Flow graph of the complete decimation-in-frequency decomposition of an eight-point DFT.

FIGURE 19.17
(*a*) A butterfly network that represents the fundamental computation of Fig. 19.16.
(*b*) Pseudocode to implement (*a*).

real-valued data are originally stored in the array x. Also note that the outer loop steps through the M stages [recall Eq. (19.31)] of the flow graph.

After this first part is executed, the DFT will have been computed but in a scrambled order (see the right-hand side of Fig. 19.16). These Fourier coefficients can be unscrambled by a procedure called *bit reversal*. If the subscripts 0 through 7 are expressed in binary, the correct ordering can be obtained by reversing these bits (Fig. 19.19). The second part of the algorithm implements this procedure.

19.6.2 Cooley-Tukey Algorithm

Figure 19.20 shows a flow network to implement the Cooley-Tukey algorithm. For this case, the sample is initially divided into odd- and even-numbered points, and the final results are in correct order.

(a)
```
m = LOG(N)/LOG(2)
N2 = N
DOFOR k = 1, m
  N1 = N2
  N2 = N2/2
  angle = 0
  arg = 2π/N1
  DOFOR j = 0, N2 - 1
    c = cos(angle)
    s = -sin(angle)
    DOFOR i = j, N - 1, N1
      kk = i + N2
      xt = x(i) - x(kk)
      x(i) = x(i) + x(kk)
      yₜ = y(i) - y(kk)
      y(i) = y(i) + y(kk)
      x(kk) = xt * c - yt * s
      y(kk) = yt * c + xt * s
    END DO
    angle = (j + 1) * arg
  END DO
END DO
```

(b)
```
j = 0
DOFOR i = 0, N - 2
  IF (i < J) THEN
    xt = xⱼ
    xⱼ = xᵢ
    xᵢ = xt
    yt = yⱼ
    yⱼ = yᵢ
    yᵢ = yt
  END IF
  k = N/2
  DO
    IF (k ≥ j + 1) EXIT
    j = j - k
    k = k/2
  END DO
  j = j + k
END DO
DOFOR i = 0, N - 1
  x(i) = x(i)/N
  y(i) = y(i)/N
END DO
```

FIGURE 19.18

Pseudocode to implement a decimation-in-frequency FFT. Note that the pseudocode is composed of two parts: (a) the FFT itself and (b) a bit-reversal routine to unscramble the order of the resulting Fourier coefficients.

Scrambled Order (Decimal)		Scrambled Order (Binary)		Bit-Reversed Order (Binary)		Final Result (Decimal)
F(0)		F(000)		F(000)		F(0)
F(4)		F(100)		F(001)		F(1)
F(2)		F(010)		F(010)		F(2)
F(6)	⇒	F(110)	⇒	F(011)	⇒	F(3)
F(1)		F(001)		F(100)		F(4)
F(5)		F(101)		F(101)		F(5)
F(3)		F(011)		F(110)		F(6)
F(7)		F(111)		F(111)		F(7)

FIGURE 19.19

Depiction of the bit-reversal process.

This approach is called a *decimation in time*. It is the reverse of the Sande-Tukey algorithm described in the previous section. Although the two classes of method differ in organization, they both require only $N \log_2 N$ operations, which is the strength of the FFT approach.

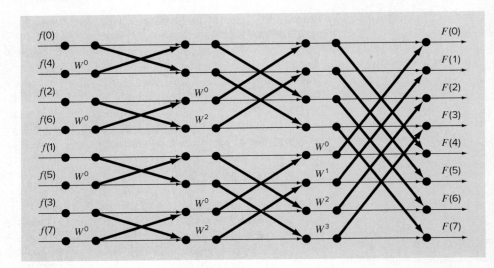

FIGURE 19.20
Flow graph of a decimation-in-time FFT of an eight-point DFT.

19.7 THE POWER SPECTRUM

The FFT has many engineering applications, ranging from vibration analysis of structures and mechanisms to signal processing. As described previously, amplitude and phase spectra provide a means to discern the underlying structure of seemingly random signals. Similarly, a useful analysis called a power spectrum can be developed from the Fourier transform.

As the name implies, the power spectrum derives from the analysis of the power output of electrical systems. In mathematical terms, the power of a periodic signal in the time domain can be defined as

$$P = \frac{1}{T} \int_{-T/2}^{T/2} f^2(t)\, dt \tag{19.37}$$

Now, another way to look at this information is to express it in the frequency domain by calculating the power associated with each frequency component. This information can be then displayed as a *power spectrum,* a plot of the power versus frequency.

If the Fourier series for $f(t)$ is

$$f(t) = \sum_{k=-\infty}^{\infty} F_k e^{ik\omega_0 t} \tag{19.38}$$

the following relation holds (see Gabel and Roberts, 1987, for details):

$$\frac{1}{T} \int_{-T/2}^{T/2} f^2(t)\, dt = \sum_{k=-\infty}^{\infty} |F_k|^2 \tag{19.39}$$

Thus, the power in $f(t)$ can be determined by adding together the squares of the Fourier coefficients, that is, the powers associated with the individual frequency components.

Now, remember that in this representation, the single real harmonic consists of both frequency components at $\pm k\omega_0$. We also know that the positive and negative coefficients are equal. Therefore, the power in $f_k(t)$, the kth real harmonic of $f(t)$, is

$$p_k = 2 \, |F_k|^2 \tag{19.40}$$

The power spectrum is the plot of p_k as a function of frequency $k\omega_0$. We will devote Sec. 20.3 to an engineering application involving the FFT and the power spectrum generated with software packages.

Additional Information. The foregoing has been a brief introduction to Fourier approximation and the FFT. Additional information on the former can be found in Van Valkenburg (1974), Chirlian (1969), and Hayt and Kemmerly (1986). References on the FFT include Davis and Rabinowitz (1975); Cooley, Lewis, and Welch (1977); and Brigham (1974). Nice introductions to both can be found in Ramirez (1985), Oppenheim and Schafer (1975), and Gabel and Roberts (1987).

19.8 CURVE FITTING WITH SOFTWARE PACKAGES

Software packages have great capabilities for curve fitting. In this section, we will give you a taste of some of the more useful ones.

19.8.1 Excel

In the present context, the most useful application of Excel is for regression analysis and, to a lesser extent, polynomial interpolation. Aside from a few built-in functions (see Table 19.1), there are two primary ways in which this capability can be implemented: the Trendline tool and the Data Analysis ToolPak.

The Trendline Tool. This tool allows a number of different trend models to be added to a chart. These models include linear, polynomial, logarithmic, exponential, power, and moving average fits. The following example illustrates how the Trendline command is invoked.

TABLE 19.1 Excel built-in functions related to regression fits of data.

Function	Description
FORECAST	Returns a value along a linear trend
GROWTH	Returns values along an exponential trend
INTERCEPT	Returns the intercept of the linear regression line
LINEST	Returns the parameters of a linear trend
LOGEST	Returns the parameters of an exponential trend
SLOPE	Returns the slope of the linear regression line
TREND	Returns values along a linear trend

SOFTWARE

EXAMPLE 19.4 **Using Excel's Trendline Command**

Problem Statement. You may have noticed that several of the fits available with Trendline were discussed previously in Chap. 17 (for example, linear, polynomial, exponential, and power). An additional capability is the logarithmic model,

$$y = a_0 + a_1 \log x$$

Fit the following data with this model using Excel's Trendline tool:

x	0.5	1	1.5	2	2.5	3	3.5	4	4.5	5	5.5
y	0.53	0.69	1.5	1.5	2	2.06	2.28	2.23	2.73	2.42	2.79

Solution. To invoke the Trendline tool, a chart relating a series of dependent and independent variables must be created. For the present case, we first create an xy-plot of the data.

Next, we can select the chart (by clicking on it) and the series (by positioning the mouse arrow on one of the values and right clicking). A menu will appear from which you can select **Add Trendline**.

At this point, a Format Trendline box opens where you can select the Trend/Regression Type as **Logarithmic**. In addition, select **Display Equation on chart** and **Display R-squared value on chart**. When the box is closed, the resulting fit along with r^2 is displayed as in Fig. 19.21.

The Trendline tool provides a handy way to fit a number of commonly used models to data. In addition, its inclusion of the **Polynomial** option means that it can also be used for polynomial interpolation. However, the fact that its statistical content is limited to r^2 means that it does not allow statistical inferences to be drawn regarding the model fit. The Data Analysis ToolPak described next provides a nice alternative where such inferences are necessary.

FIGURE 19.21
Fit of a logarithmic model to the data of Example 19.4.

$$y = 0.9846 \ln x + 1.0004$$
$$r^2 = 0.9444$$

The Data Analysis ToolPak. This Excel Add-in contains a comprehensive capability for curve fitting with general linear least squares. As previously described in Sec. 17.4, such models are of the general form

$$y = a_0 z_0 + a_1 z_1 + a_2 z_2 + \cdots + a_m z_m + e \tag{17.23}$$

where $z_0, z_1, \ldots, z_m$ are $m + 1$ different functions. The next example illustrates how such models can be fit with Excel.

EXAMPLE 19.5

Using Excel's Data Analysis ToolPak

Problem Statement.   The following data were collected for the slope, hydraulic radius, and velocity of water flowing in a canal:

S, m/m	0.0002	0.0002	0.0005	0.0005	0.001	0.001
R, m	0.2	0.5	0.2	0.5	0.2	0.5
U, m/s	0.25	0.5	0.4	0.75	0.5	1

There are theoretical reasons for believing that these data can be fit to a power model of the form

$$U = \alpha S^\sigma R^\rho$$

where α, σ, and ρ are empirically derived coefficients. There are also theoretical reasons for believing that σ and ρ should have values of approximately 0.5 and 0.667, respectively. Fit these data with Excel and evaluate whether your regression estimates contradict the expected values for the model coefficients.

Solution.   The logarithm of the power model is first used to convert it to the linear format of Eq. (17.23),

$$U = \log \alpha + \sigma \log S + \rho \log R$$

An Excel spreadsheet can be developed with both the original data and their common logarithms, as in the following:

	A	B	C	D	E	F
1	S	R	U	log(S)	log(R)	log(U)
2	0.0002	0.2	0.25	-3.69897	-0.69897	-0.60206
3	0.0002	0.5	0.5	-3.69897	-0.30103	-0.30103
4	0.0005	0.2	0.4	-3.30103	-0.69897	-0.39794
5	0.0005	0.5	0.75	-3.30103	-0.30103	-0.12494
6	0.001	0.2	0.5	-3	-0.69897	-0.30103
7	0.001	0.5	1	-3	-0.30103	0

=log(A2)

As shown, an efficient way to generate the logarithms is to type the formula to compute the first log(S). This formula can then be copied to the right and down to generate the other logarithms.

Because of its status as an "Add-in" on the version of Excel available at the time of this book's printing, the Data Analysis ToolPak must sometimes be loaded into Excel. To do this, choose File, Options, Add-Ins, Manage Excel Add-ins. Then, check the Analysis ToolPak box. The ToolPak should then be installed and a button to access it should appear on your Data tab.

After **Data Analysis** is selected from the Data menu, a Data Analysis menu will appear on the screen containing a large number of statistically oriented routines. Select **Regression** and a dialogue box will appear, prompting you for information on the regression. After making sure that the default selection **New Worksheet Ply** is selected, fill in F2:F7 for the *y* range and D2:E7 for the *x* range, and select OK. The following worksheet will be created:

	A	B	C	D	E	F	G
1	SUMMARY OUTPUT						
2							
3	*Regression Statistics*						
4	Multiple R	0.998353					
5	R Square	0.996708					
6	Adjusted R Square	0.994513					
7	Standard Error	0.015559					
8	Observations	6					
9							
10	ANOVA						
11		*df*	*SS*	*MS*	*F*	*Significance F*	
12	Regression	2	0.219867	0.10993	454.1106	0.0001889	
13	Residual	3	0.000726	0.00024			
14	Total	5	0.220593				
15							
16		*Coefficients*	*Standard Error*	*t Stat*	*P-value*	*Lower 95%*	*Upper 95%*
17	Intercept	**1.522452**	0.075932	20.05010	0.000271	1.2808009	1.7641028
18	X Variable 1	**0.433137**	0.022189	19.52030	0.000294	**0.362521**	**0.503752**
19	X Variable 2	**0.732993**	0.031924	22.96038	0.000181	**0.631395**	**0.834590**

Thus, the resulting fit is

$$\log U = 1.522 + 0.433 \log S + 0.733 \log R$$

or by taking the antilog,

$$U = 33.3 S^{0.433} R^{0.733}$$

Notice that 95% confidence intervals are generated for the coefficients. Thus, there is a 95% probability that the true slope exponent falls between 0.363 and 0.504, and the true hydraulic radius coefficient falls between 0.631 and 0.835. Thus, the fit does not contradict the theoretical exponents.

Finally, it should be noted that the Excel Solver tool can be used to perform *nonlinear regression* by directly minimizing the sum of the squares of the residuals between a nonlinear model prediction and data.

19.8.2 MATLAB

As summarized in Table 19.2, MATLAB software has a variety of built-in functions that span the total capabilities described in this part of the book. The following example illustrates how a few of them can be used.

TABLE 19.2 Some of the MATLAB functions to implement interpolation, regression, splines, and the FFT.

Function	Description
polyfit	Fit polynomial to data
interp1	1-D interpolation (1-D table lookup)
interp2	2-D interpolation (2-D table lookup)
spline	Cubic spline data interpolation
fft	Fast Fourier transform

EXAMPLE 19.6   **Using MATLAB for Curve Fitting**

Problem Statement.   Explore how MATLAB can be employed to fit curves to data. To do this, use the sine function to generate equally spaced $f(x)$ values from 0 to 10. Employ a step size of 1 so that the resulting characterization of the sine wave is sparse (Fig. 19.22). Then, fit it with (*a*) linear interpolation, (*b*) a fifth-order polynomial, and (*c*) a cubic spline.

Solution.

(**a**) The values for the independent and the dependent variables can be entered into vectors by

```
>> x=0:10;
>> y=sin(x);
```

A new, more finely spaced vector of independent variable values can be generated and stored in the vector **xi**,

```
>> xi=0:.25:10;
```

The MATLAB function interp1 can then be used to generate dependent variable values *yi* for all the *xi* values using linear interpolation. Both the original data (*x*, *y*) and the linearly interpolated values can be plotted together, as shown in the

FIGURE 19.22
Eleven points sampled from a sinusoid.

SOFTWARE

graph below:

```
>> yi=interp1(x,y,xi);
>> plot(x,y,'o',xi,yi)
```

(b) Next, the MATLAB `polyfit` function can be used to generate the coefficients of a fifth-order polynomial fit of the original sparse data,

```
>> p=polyfit(x,y,5)
p=
    0.0008   -0.0290   0.3542   -1.6854   2.5860   -0.0915
```

where the vector **p** holds the polynomial's coefficients. These can, in turn, be used to generate a new set of *yi* values, which can again be plotted along with the original sparse sample,

```
>> yi = polyval(p,xi);
>> plot(x,y,'o',xi,yi)
```

Thus, the polynomial captures the general pattern of the data, but misses most of the points.

(c) Finally, the MATLAB `spline` function can be used to fit a cubic spline to the original sparse data in the form of a new set of *yi* values, which can again be plotted along with the original sparse sample,

```
>> yi=spline(x,y,xi);
>> plot(x,y,'o',xi,yi)
```

It should be noted that MATLAB also has excellent capabilities to perform Fourier analysis. We devote Sec. 20.3 to an example of how this can be done.

MATLAB has two built-in functions for two- and three-dimensional piecewise interpolation: `interp2` and `interp3`. As you might expect from their names, these functions operate in a similar fashion to `interp1`. For example, a simple representation of the syntax of `interp2` is

```
zi = interp2(x, y, z, xi, yi, 'method')
```

where x and y = matrices containing the coordinates of the points at which the values in the matrix z are given, zi = a matrix containing the results of the interpolation as evaluated at the points in the matrices xi and yi, and $method$ = the desired method. Note that the methods are identical to those used by `interp1`; that is, `linear`, `nearest`, `spline`, and `cubic`.

As with `interp1`, if the $method$ argument is omitted, the default is linear interpolation. For example, `interp2` can be used to make the same evaluation as in Example 18.11 as

```
>> x=[2 9];
>> y=[1 6];
>> z=[60 57.5;55 70];
>> interp2(x,y,z,5.25,4.8)

ans =
   61.2143
```

19.8.3 Mathcad

Mathcad can perform a wide variety of statistical, curve-fitting, and data-smoothing tasks. These include relatively simple jobs like plotting histograms and calculating population statistic summaries such as mean, median, variance, standard deviations, and correlation coefficients. In addition, Mathcad contains a number of functions for performing regression. The **slope** and **intercept** functions return the slope and intercept of the least-squares regression fit line. The **regress** function is used for nth-order polynomial regression of a complete data set. The **loess** function performs localized nth-order polynomial regression over spans of the data that you can specify. The **interp** function can be used to return intermediate values of y from a regression fit for a given x point. The **regress** and **loess** functions can also perform multivariate polynomial regression. Mathcad also provides the **linfit** function that is used to model data with a linear combination of arbitrary functions. Finally, the **genfit** function is available for cases where model coefficients appear in arbitrary form. In this case, the more difficult nonlinear equations must be solved by iteration.

Mathcad also has considerable capabilities for interpolation. It can predict intermediate values by connecting known data points with either straight lines (linear interpolation) using the **linterp** function or with cubic spline interpolation using **cspline**, **pspline**, or **lspline**. These spline functions allow you to try different ways to deal with interpolation at the end points of the data. The **lspline** function generates a spline curve that is a straight line at the end points. The **pspline** function generates a spline curve that is a parabola at the end points. The **cspline** function generates a spline curve that is cubic at the end points. The **interp** function uses the curve-fitting results and returns an interpolated y value given an x value. In addition, you can perform two-dimensional cubic spline interpolation by passing a surface through a grid of points.

Let's do an example that shows how Mathcad is used to perform spline interpolation (Fig. 19.23). The data we will fit are simply some evenly spaced points sampled from a

FIGURE 19.23
Cubic spline interpolation with Mathcad.

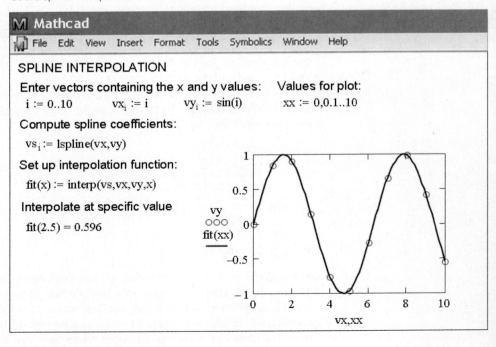

sinusoid. After generating these data, the definition symbol and the **lspline** function are used to compute the spline coefficients. Then, an interpolation function, fit, is developed with the **interp** function in order to generate interpolated values for specific values of x. Mathcad designed this sequence of operations so that the interpolating polynomials would not have to be recalculated every time an interpolation is desired. With the functions in place, you can then interpolate at any location using fit(x), as shown with $x = 2.5$. You can also construct a plot of these data along with the interpolated spline as shown in Fig. 19.23.

As another example of demonstrating some of Mathcad's curve fitting capabilities let's use the **fft** function for Fourier analysis as in Fig. 19.24. The first line uses the definition symbol to create i as a range variable. Next, x_i is formulated using the **rnd** Mathcad function to impart a random component to a sinusoidal signal. The graph of the signal can be placed on the worksheet by clicking to the desired location. This places a red crosshair at that location. Then use the Insert/Graph/X-Y Plot pull-down menu to place an empty plot on the worksheet with placeholders for the expressions to be graphed and for the ranges of the x and y axes. Simply type x_i in the placeholder on the y axis and i for the x-axis range. Mathcad does the rest to produce the first graph shown in Fig. 19.24. Once the graph has been created, you can use the Format/Graph/X-Y Plot pull-down menu to vary the type of graph; change the color, type, and weight of the trace of the function; and add titles, labels, and other features. Next, c is defined as **fft**(x). This function returns the Fourier transform of x. The result is a vector **c** of complex coefficients that represent values in the frequency domain. A plot of the magnitude of c_j is then constructed as described above.

FIGURE 19.24
FFT with Mathcad.

PROBLEMS

19.1 The average values of a function can be determined by

$$\overline{f(x)} = \frac{\int_0^x f(x)\,dx}{x}$$

Use this relationship to verify the results of Eq. (19.13).

19.2 The solar radiation for Tucson, Arizona, has been tabulated as

Time, mo	J	F	M	A	M	J	J	A	S	O	N	D
Radiation, W/m²	144	188	245	311	351	359	308	287	260	211	159	131

Assuming each month is 30 days long, fit a sinusoid to these data. Use the resulting equation to predict the radiation in mid-August.

19.3 The pH in a reactor varies sinusoidally over the course of a day. Use least-squares regression to fit Eq. (19.11) to the following data. Use your fit to determine the mean, amplitude, and time of maximum pH. Note that the period is 24 hr.

Time, hr	0	2	4	5	7	9	12	15	20	22	24
pH	7.6	7	7.1	6.5	7.4	7.2	8.9	8.8	8.9	7.9	7

FIGURE P19.4

A sawtooth wave.

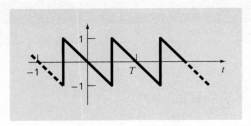

FIGURE P19.5

A triangular wave.

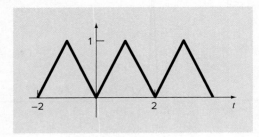

19.4 Use a continuous Fourier series to approximate the sawtooth wave in Fig. P19.4. Plot the first three terms along with the summation.

19.5 Use a continuous Fourier series to approximate the wave form in Fig. P19.5. Plot the first three terms along with the summation.

19.6 Construct amplitude and phase line spectra for Prob. 19.4.

19.7 Construct amplitude and phase line spectra for Prob. 19.5.

19.8 A half-wave rectifier can be characterized by

$$C_1 = \left[\frac{1}{\pi} + \frac{1}{2}\sin t - \frac{2}{3\pi}\cos 2t - \frac{2}{15\pi}\cos 4t \right.$$
$$\left. - \frac{2}{35\pi}\cos 6t - \cdots \right]$$

where C_1 is the amplitude of the wave. Plot the first four terms along with the summation.

19.9 Construct amplitude and phase line spectra for Prob. 19.8.

19.10 Develop a user-friendly program for the DFT based on the algorithm from Fig. 19.12. Test it by duplicating Fig. 19.13.

19.11 Use the program from Prob. 19.10 to compute a DFT for the triangular wave from Prob. 19.8. Sample the wave from $t = 0$ to $4T$. Use 32, 64, and 128 sample points. Time each run and plot execution versus N to verify Fig. 19.14.

19.12 Develop a user-friendly program for the FFT based on the algorithm from Fig. 19.18. Test it by duplicating Fig. 19.13.

19.13 Repeat Prob. 19.11 using the software you developed in Prob. 19.12.

19.14 An object is suspended in a wind tunnel and the force measured for various levels of wind velocity. The results are tabulated below. Use Excel's Trendline tool to fit a power equation to these data. Plot F versus v along with the power equation and r^2.

v, m/s	10	20	30	40	50	60	70	80
F, N	25	70	380	550	610	1220	830	1450

19.15 Use the Excel Data Analysis ToolPak to fit a regression polynomial to the following data for the dissolved oxygen concentration of fresh water versus temperature at sea level. Determine the order of polynomial necessary to match the precision of these data.

T, °C	0	8	16	24	32	40
o, mg/L	14.62	11.84	9.87	8.42	7.31	6.41

19.16 Use the Excel Data Analysis Toolpack to fit a straight line to the following data. Determine the 90% confidence interval for the intercept. If it encompasses zero, redo the regression, but with the intercept forced to be zero (this is an option on the **Regression** dialogue box).

x	2	4	6	8	10	12	14
y	6.5	7	13	17.8	19	25.8	26.9

19.17 **(a)** Use MATLAB to fit a cubic spline to the following data:

x	0	2	4	7	10	12
y	20	20	12	7	6	6

Determine the value of y at x = 1.5. **(b)** Repeat **(a)**, but with zero first derivatives at the end knots. Note that the MATLAB help facility describes how to prescribe end derivatives.

19.18 Use MATLAB to generate 64 points from the function

$$f(t) = \cos(10t) + \sin(3t)$$

from t = 0 to 2π. Add a random component to the signal with the function randn. Take an FFT of these values and plot the results.

19.19 In a fashion similar to the approach in Sec. 19.8.2, use MATLAB to fit the data from Prob. 19.15 using **(a)** linear interpolation, **(b)** a third-order regression polynomial, and **(c)** a spline. Use each approach to predict oxygen concentration at T = 10.

19.20 Runge's function is written as

$$f(x) = \frac{1}{1 + 25x^2}$$

Generate 11 equidistantly spaced values of this function over the interval [−1, 1]. Fit these data with **(a)** a tenth-order polynomial, **(b)** a linear spline, and **(c)** a cubic spline. Present your results graphically.

19.21 A dye is injected into the circulating blood volume to measure a patient's cardiac output, which is the volume flow rate of blood out of the left ventricle of the heart. In other words, cardiac output is the number of liters of blood your heart pumps in a minute. For a person at rest, the rate might be 5 or 6 liters per minute. If you are a trained marathon runner running a marathon, your cardiac output can be as high as 30 L/min. The data below show the response of an individual when 5 mg of dye was injected into the venous system.

Time (s)	2	6	9	12	15	18	20	24
Concentration (mg/L)	0	1.5	3.2	4.1	3.4	2	1	0

Fit a polynomial curve through the data points and use the function to approximate the patient's cardiac output, which can be calculated by:

$$\text{Cardiac output (in L/min)} = \frac{\text{amount of dye}}{\text{area under curve}}$$

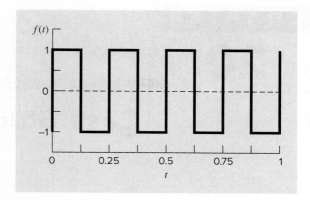

FIGURE P19.22

19.22 In electric circuits, it is common to see current behavior in the form of a square wave, as shown in Fig. P19.22. Solving for the Fourier series from

$$f(t) = \begin{cases} A_0 & 0 \le t \le T/2 \\ -A_0 & T/2 \le t \le T \end{cases}$$

we get the Fourier series

$$f(t) = \sum_{n=1}^{\infty} \left(\frac{4A_0}{(2n-1)\pi}\right) \sin\left(\frac{2\pi(2n-1)t}{T}\right)$$

Let $A_0 = 1$ and T = 0.25 s. Plot the first six terms of the Fourier series individually, as well as the sum of these six terms. Use a package such as Excel or MATLAB if possible.

19.23 Develop a plot of the following data with **(a)** a sixth-order interpolating polynomial, **(b)** a cubic spline, and **(c)** a cubic spline with zero end derivatives.

x	0	100	200	400	600	800	1000
f(x)	0	0.82436	1.00000	0.73576	0.40601	0.19915	0.09158

In each case, compare your plot with the following equation, which was used to generate these data

$$f(x) = \frac{x}{200} e^{-(x/200)+1}$$

19.24 Use the software package of your choice to compute the DFT for the following function:

$$f(t) = 1.5 + 1.8\cos(2\pi(12)t) + 0.8\sin(2\pi(20)t) - 1.25\cos(2\pi(28)t)$$

Take n = 64 samples with a sampling frequency of $f_s = 128$ samples/s.

19.25 Assuming 128 samples of data (n = 128 samples) with a total sample length of $t_n = 0.4$ s, compute **(a)** the sample frequency, f_s (sample/s); **(b)** the sample interval, Δt (s/sample); **(c)** the Nyquist frequency, f_{max} (Hz); and **(d)** the minimum frequency, f_{min} (Hz).

Case Studies: Curve Fitting

The purpose of this chapter is to use the numerical methods for curve fitting to solve some engineering problems. The first application, which is taken from chemical engineering, demonstrates how a nonlinear model can be linearized and fit to data using linear regression. The second application employs splines to study a problem that has relevance to the environmental area of civil engineering: heat and mass transport in a stratified lake.

The third application illustrates how a fast Fourier transform can be employed in electrical engineering to analyze a signal by determining its major harmonics. The final application demonstrates how multiple linear regression is used to analyze experimental data for a fluids problem taken from mechanical and aerospace engineering.

20.1 FITTING ENZYME KINETICS (CHEMICAL/BIO ENGINEERING)

Background. *Enzymes* act as catalysts to speed up the rate of chemical reactions in living cells. In most cases, they convert one chemical, the *substrate,* into another, the *product.* The *Michaelis-Menten* equation is commonly used to describe such reactions:

$$v = \frac{v_m[S]}{k_s + [S]} \tag{20.1}$$

where v = the initial reaction velocity, v_m = the maximum initial reaction velocity, $[S]$ = substrate concentration, and k_s = a half-saturation constant. As in Fig. 20.1, the equation describes a saturating relationship that levels off with increasing $[S]$. The graph also illustrates that the *half-saturation constant* corresponds to the substrate concentration at which the velocity is half the maximum.

Although the Michaelis-Menten model provides a nice starting point, it has been refined and extended to incorporate additional features of enzyme kinetics. One simple extension involves so-called *allosteric enzymes,* where the binding of a substrate molecule at one site leads to enhanced binding of subsequent molecules at other sites. For cases with two interacting bonding sites, the following second-order version often results in a better fit:

$$v = \frac{v_m[S]^2}{k_s^2 + [S]^2} \tag{20.2}$$

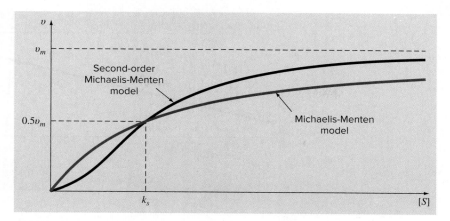

FIGURE 20.1
Two versions of the Michaelis-Menten model of enzyme kinetics.

This model also describes a saturating curve but, as depicted in Fig. 20.1, the squared concentrations tend to make the shape more *sigmoid,* or S-shaped.

Suppose that you are provided with the following data:

$[S]$	1.3	1.8	3	4.5	6	8	9
v	0.07	0.13	0.22	0.275	0.335	0.35	0.36

Employ linear regression to fit these data with linearized versions of Eqs. (20.1) and (20.2). Aside from estimating the model parameters, assess the validity of the fits with both statistical measures and graphs.

Solution.   Equation (20.1), which is in the format of the saturation-growth-rate model (Eq. 17.14), can be linearized by inverting it to give [recall Eq. (17.17)]

$$\frac{1}{v} = \frac{1}{v_m} + \frac{k_s}{v_m}\frac{1}{[S]}$$

If the model is valid, a plot of $1/v$ versus $1/[S]$ should yield a straight line with an intercept of $1/v_m$ and a slope of k_s/v_m. Using your own program or a software package, you should obtain an intercept of 0.1902 and a slope of 16.4022 with $r^2 = 0.9344$. The model coefficients can then be calculated as $v_m = 1/0.1902 = 5.2570$ and $k_s = 5.2570 \times 16.4022 = 86.2260$, and the best-fit model is

$$v = \frac{5.2570[S]}{86.2260 + [S]}$$

Although the high value of r^2 might lead you to believe that this result is acceptable, inspection of the coefficients might raise doubts. For example, the maximum velocity (5.2570) is much greater than the highest observed velocity (0.36). In addition, the half-saturation rate (86.2260) is much bigger than the maximum substrate concentration (9).

The problem is underscored when the fit is plotted along with the data. Figure 20.2*a* shows the transformed version. Although the straight line follows the upward trend, the

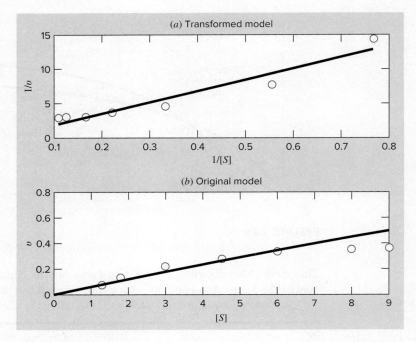

FIGURE 20.2
Plots of least-squares fit (line) of the Michaelis-Menten model along with data (points). The plot in (a) shows the transformed fit, and (b) shows how the fit looks when viewed in the untransformed, original form.

data clearly appear to be curved. When the original equation is plotted along with the data in the untransformed version (Fig. 20.2b), the fit is obviously unacceptable. The data are clearly leveling off at about 0.36 or 0.37. If this is correct, an eyeball estimate would suggest that v_m should be about 0.36, and k_s should be in the range of 2 to 3.

Beyond the visual evidence, the poorness of the fit is also reflected by statistics like the coefficient of determination. For the untransformed case, a much less acceptable result of $r^2 = 0.6406$ is obtained.

The foregoing analysis can be repeated for the second-order model. Equation (20.2) can also be linearized by inverting it to give

$$\frac{1}{v} = \frac{1}{v_m} + \frac{k_s^2}{v_m}\frac{1}{[S]^2}$$

If the model is valid, a plot of $1/v$ versus $1/[S]^2$ should yield a straight line with an intercept of $1/v_m$ and a slope of k_s^2/v_m. Using your own program or a software package, you should obtain an intercept of 2.4492 and a slope of 19.3760 with $r^2 = 0.9929$. The model coefficients can then be calculated as $v_m = 1/2.4492 = 0.4083$ and $k_s = \sqrt{0.4083(19.3760)} = 2.8127$, and the best-fit model is

$$v = \frac{0.4083[S]^2}{7.911 + [S]^2}$$

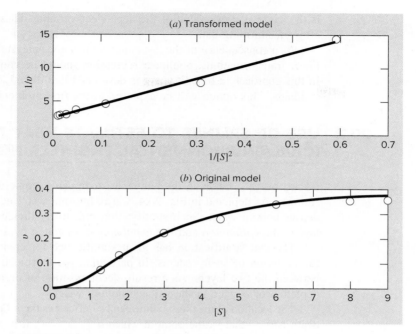

FIGURE 20.3
Plots of least-squares fit (line) of the second-order Michaelis-Menten model along with data (points). The plot in (a) shows the transformed fit, and (b) shows the untransformed, original form.

Although we know that a high r^2 does not guarantee a good fit, the fact that it is very high (0.9929) is promising. In addition, the parameter values also seem consistent with the trends in the data; that is, the v_m is slightly greater than the highest observed velocity and the half-saturation constant is lower than the maximum substrate concentration (9).

The adequacy of the fit can be assessed graphically. As seen in Fig. 20.3a, the transformed results appear linear. When the original equation is plotted along with the data in the untransformed version (Fig. 20.3b), the fit nicely follows the trend in the measurements. Beyond the graphs, the goodness of the fit is also reflected by the fact that the coefficient of determination for the untransformed case can be computed as $r^2 = 0.9896$.

Based on our analysis, we can conclude that the second-order model provides a good fit of this data set. This might suggest that we are dealing with an allosteric enzyme.

Beyond this specific result, there are a few other general conclusions that can be drawn from this case study. First, we should never solely rely on statistics such as r^2 as the sole basis of assessing goodness of fit. Second, regression equations should always be assessed graphically. And for cases where transformations are employed, a graph of the untransformed model and data should always be inspected.

Finally, although transformations may yield a decent fit of the transformed data, this does not always translate into an acceptable fit in the original format. The reason that this might occur is that minimizing squared residuals of transformed data is not the same as doing so for untransformed data. Linear regression assumes that the scatter of points around the best-fit line follows a Gaussian distribution, and that the standard deviation is the same at every value of the dependent variable. These assumptions are rarely true after transforming data.

As a consequence of the last conclusion, some analysts suggest that rather than using linear transformations, nonlinear regression should be employed to fit curvilinear data. In this approach, a best-fit curve is developed that directly minimizes the untransformed residuals. This option will be explored in the first problem at the end of this chapter.

20.2   USE OF SPLINES TO ESTIMATE HEAT TRANSFER (CIVIL/ENVIRONMENTAL ENGINEERING)

Background.   Lakes in the temperate zone can become thermally stratified during the summer. As depicted in Fig. 20.4, warm, buoyant water near the surface overlies colder, denser bottom water. Such stratification effectively divides the lake vertically into two layers: the *epilimnion* and the *hypolimnion* separated by a plane called the *thermocline*.

Thermal stratification has great significance for environmental engineers studying the pollution of such systems. In particular, the thermocline greatly diminishes mixing between the two layers. As a result, decomposition of organic matter can lead to severe depletion of oxygen in the isolated bottom water.

The location of the thermocline can be defined as the inflection point of the temperature-depth curve—that is, the point at which $d^2T/dx^2 = 0$. It is also the point at which the absolute value of the first derivative or gradient is a maximum. Use cubic splines to determine the thermocline depth for Platte Lake (Table 20.1). Also use the splines to determine the value of the gradient at the thermocline.

Solution.   The data are analyzed with a program that was developed using pseudocode based on Sec. 18.6.4. The results are displayed in Table 20.2, which lists the spline predictions along with first and second derivatives at intervals of 1 m down through the water column.

FIGURE 20.4
Temperature versus depth during summer for Platte Lake, Michigan.

The results are plotted in Fig. 20.5. Notice how the thermocline is clearly located at the depth where the gradient is highest (that is, the absolute value of the derivative is greatest) and the second derivative is zero. The depth is 11.35 m and the gradient at this point is $-1.61°C/m$.

TABLE 20.1 Temperature versus depth during summer for Platte Lake, Michigan.

T, °C	22.8	22.8	22.8	20.6	13.9	11.7	11.1	11.1
z, m	0	2.3	4.9	9.1	13.7	18.3	22.9	27.2

TABLE 20.2 Output of spline program based on pseudocode derived from Sec. 18.6.4.

Depth (m)	T(C)	dT/dz	d2T/dz2	Depth (m)	T(C)	dT/dz	d2T/dz2
0.	22.8000	−.0115	.0000	15.	12.7652	−.6518	.3004
1.	22.7907	−.0050	.0130	16.	12.2483	−.3973	.2086
2.	22.7944	.0146	.0261	17.	11.9400	−.2346	.1167
3.	22.8203	.0305	−.0085	18.	11.7484	−.1638	.0248
4.	22.8374	−.0055	−.0635	19.	11.5876	−.1599	.0045
5.	22.7909	−.0966	−.1199	20.	11.4316	−.1502	.0148
6.	22.6229	−.2508	−.1884	21.	11.2905	−.1303	.0251
7.	22.2665	−.4735	−.2569	22.	11.1745	−.1001	.0354
8.	21.6531	−.7646	−.3254	23.	11.0938	−.0596	.0436
9.	20.7144	−1.1242	−.3939	24.	11.0543	−.0212	.0332
10.	19.4118	−1.4524	−.2402	25.	11.0480	.0069	.0229
11.	17.8691	−1.6034	−.0618	26.	11.0646	.0245	.0125
12.	16.2646	−1.5759	.1166	27.	11.0936	.0318	.0021
13.	14.7766	−1.3702	.2950	28.	11.1000	.0000	.0000
14.	13.5825	−.9981	.3923				

FIGURE 20.5

Plots of (*a*) temperature, (*b*) gradient, and (*c*) second derivative versus depth (*m*) generated with the cubic spline program. The thermocline is located at the inflection point of the temperature-depth curve.

20.3 FOURIER ANALYSIS (ELECTRICAL ENGINEERING)

Background. Fourier analysis is used in many areas of engineering. However, it is extensively employed in electrical engineering applications such as signal processing.

In 1848, Johann Rudolph Wolf devised a method for quantifying solar activity by counting the number of individual spots and groups of spots on the sun's surface. He computed a quantity, now called a *Wolf sunspot number,* by adding 10 times the number of groups to the total count of individual spots. As in Fig. 20.6, the record of this number extends back to 1700. On the basis of the early historical records, Wolf determined the cycle's length to be 11.1 years.

Use a Fourier analysis to confirm this result by applying an FFT to the data from Fig. 20.6. Pinpoint the period by developing a power versus period plot.

Solution. The data for year and sunspot number were downloaded from the Web[1] and stored in a tab-delimited file: sunspot.dat. The file can be loaded into MATLAB software and the year and number information assigned to vectors of the same name,

```
>> load sunspot.dat
>> year=sunspot(:,1);number=sunspot(:,2);
```

Next, an FFT can be applied to the sunspot number vector,

```
>> y=fft(number);
```

After getting rid of the first harmonic, the length of the FFT is determined (n) and then the power and frequency are calculated,

```
>> y(1)=[ ];
>> n=length(y);
>> power=abs(y(1:n/2)).^2;
>> nyquist=1/2;
>> freq=(1:n/2)/(n/2)*nyquist;
```

FIGURE 20.6
Plot of Wolf sunspot number versus year.

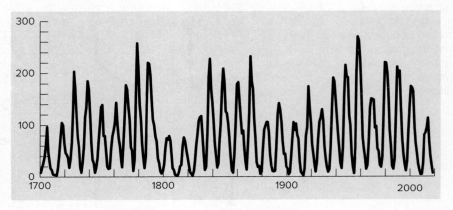

[1]At the time of this book's printing, the yearly mean total sunspot number data can be downloaded from http://www.sidc.be/silso/datafiles.

FIGURE 20.7
Power spectrum for Wolf sunspot numbers.

At this point, the power spectrum is a plot of power versus frequency. However, because period is more meaningful in the present context, we can determine the period and a power-period plot,

```
>> period=1./freq;
>> plot(period,power)
```

The result, as shown in Fig. 20.7, indicates a peak at about 11 years. The exact value can be computed with

```
>> index=find(power==max(power));
>> period(index)

ans=
    10.9655
```

20.4 ANALYSIS OF EXPERIMENTAL DATA (MECHANICAL/AEROSPACE ENGINEERING)

Background. Engineering design variables are often dependent on several independent variables. Often this functional dependence is best characterized by multivariate power equations. As discussed in Sec. 17.3, a multiple linear regression of log-transformed data provides a means to evaluate such relationships.

For example, a mechanical engineering study indicates that fluid flow through a pipe is related to pipe diameter and slope (Table 20.3). Use multiple linear regression to analyze these data. Then use the resulting model to predict the flow for a pipe with a diameter of 2.5 ft and a slope of 0.025 ft/ft.

Solution. The power equation to be evaluated is

$$Q = a_0 D^{a_1} S^{a_2} \tag{20.3}$$

TABLE 20.3 Experimental data for diameter, slope, and flow of circular concrete pipes.

Experiment	Diameter, ft	Slope, ft/ft	Flow, ft^3/s
1	1	0.001	1.4
2	2	0.001	8.3
3	3	0.001	24.2
4	1	0.01	4.7
5	2	0.01	28.9
6	3	0.01	84.0
7	1	0.05	11.1
8	2	0.05	69.0
9	3	0.05	200.0

where Q = flow (ft^3/s), S = slope (ft/ft), D = pipe diameter (ft), and a_0, a_1, and a_2 = coefficients. Taking the logarithm of this equation yields

$$\log Q = \log a_0 + a_1 \log D + a_2 \log S$$

In this form, the equation is suited for multiple linear regression because $\log Q$ is a linear function of $\log S$ and $\log D$. Using the logarithm (base 10) of the data in Table 20.3, we can generate the following normal equations expressed in matrix form [recall Eq. (17.22)]:

$$\begin{bmatrix} 9 & 2.334 & -18.903 \\ 2.334 & 0.954 & -4.903 \\ -18.903 & -4.903 & 44.079 \end{bmatrix} \begin{Bmatrix} \log a_0 \\ a_1 \\ a_2 \end{Bmatrix} = \begin{Bmatrix} 11.691 \\ 3.945 \\ -22.207 \end{Bmatrix}$$

This system can be solved using Gauss elimination to yield

$$\log a_0 = 1.7475$$
$$a_1 = 2.62$$
$$a_2 = 0.54$$

Since $\log a_0 = 1.7475$, then $a_0 = 10^{1.7475} = 55.9$ and Eq. (20.3) is

$$Q = 55.9 D^{2.62} S^{0.54} \tag{20.4}$$

Eq. (20.4) can be used to predict flow for the case where $D = 2.5$ ft and $S = 0.025$ ft/ft, as in

$$Q = 55.9(2.5)^{2.62}(0.025)S^{0.54} = 84.1 \text{ ft}^3/\text{s}$$

It should be noted that Eq. (20.4) can be used for other purposes besides computing flow. For example, the slope is related to head loss h_L and pipe length L by $S = h_L/L$. If this relationship is substituted into Eq. (20.4) and the resulting formula solved for h_L, the following equation can be developed:

$$h_L = \frac{L}{1721} Q^{1.85} D^{4.85}$$

This relationship is called the *Hazen-Williams equation.*

PROBLEMS

Chemical/Bio Engineering

20.1 Use nonlinear regression to fit the untransformed enzyme kinetics data from Sec. 20.1 to **(a)** the standard Michaelis-Menten model (Eq. 20.1), and **(b)** the second-order model (Eq. 20.2). Plot the resulting best-fit lines and the data and also calculate the coefficient of determination (r^2) and the standard error of the estimate ($s_{y/x}$) for both fits.

20.2 You perform experiments and determine the following values of heat capacity c at various temperatures T for a gas:

T	−50	−30	0	60	90	110
c	1270	1280	1350	1480	1580	1700

Use regression to determine a model to predict c as a function of T.

20.3 It is known that the tensile strength of a plastic increases as a function of the time it is heat-treated. The following data are collected:

Time, min	10	15	20	25	40	50	55	60	75
Tensile strength	5	20	18	40	33	54	70	60	78

(a) Fit a straight line to these data and use the equation to determine the tensile strength at a time of 32 min.

(b) Repeat the analysis but for a straight line with a zero intercept.

20.4 The following data were gathered to determine the relationship between pressure and temperature of a fixed volume of 1 kg of nitrogen. The volume is 10 m³.

T, °C	−40	0	40	80	120	160
p, N/m²	6900	8100	9300	10,500	11,700	12,900

Employ the ideal gas law $pV = nRT$ to determine R on the basis of these data. Note that for the law, T must be expressed in kelvins.

20.5 The specific volume of a superheated steam is listed in steam tables for various temperatures. For example, at a pressure of 3000 lb/in², absolute:

T, °F	700	720	740	760	780
v, ft³/lb$_\mathrm{m}$	0.0977	0.12184	0.14060	0.15509	0.16643

Determine v at $T = 750°F$.

20.6 A reactor is thermally stratified as in the following table:

Depth, m	0	0.5	1.0	1.5	2.0	2.5	3.0
Temperature, °C	70	68	55	22	13	11	10

As depicted in Fig. P20.6, the tank can be idealized as two zones separated by a strong temperature gradient, or thermocline. The depth of this gradient can be defined as the inflection point of the temperature-depth curve—that is, the point at which $d^2T/dz^2 = 0$.

FIGURE P20.6

At this depth, the heat flux from the surface to the bottom layer can be computed with Fourier's law,

$$J = -k\frac{dT}{dz}$$

Use a cubic spline fit of these data to determine the thermocline depth. If $k = 0.02$ cal/(s cm °C) compute the flux across this interface.

20.7 In Alzheimer's disease, the number of neurons in the cortex decreases as the disease progresses. The following data were taken to determine the number of neurotransmitter receptors left in a diseased brain. Free neurotransmitter ([F]) was incubated with tissue and the concentration that bound specifically to a receptor ([B]) was measured. When binding is specific to a receptor, the concentration bound is related to the free concentration by the following relationship:

$$[B] = \frac{B_{\max}[F]}{K + [F]}$$

Using the data below, determine the parameters that minimize the sum of the squares of the residuals. Also, compute r^2.

[F], nM	0.1	0.5	1	5	10	20	50
[B], nM	10.57	36.61	52.93	82.65	89.46	94.35	101.00

20.8 The following data were taken from a stirred tank reactor for the reaction $A \rightarrow B$. Use these data to determine the best possible estimates for k_{01} and E_1 for the following kinetic model,

$$-\frac{dA}{dt} = k_{01}e^{-E_1/(RT)}A$$

where R is the gas constant and equals 0.00198 Kcal/mol/K

$-dA/dt$ (moles/L/s)	460	960	2485	1600	1245
A (moles/L)	200	150	50	20	10
T (K)	280	320	450	500	550

20.9 Use the following set of pressure-volume data to find the best possible virial constants (A_1 and A_2) for the equation of state shown below. $R = 82.05$ mL atm/gmol K and $T = 303$ K.

$$\frac{PV}{RT} = 1 + \frac{A_1}{V} + \frac{A_2}{V^2}$$

P (atm)	0.985	1.108	1.363	1.631
V (mL)	25,000	22,200	18,000	15,000

20.10 Concentration data were taken at 15 time points for the polymerization reaction

$$xA + yB \rightarrow A_xB_y$$

We assume the reaction occurs via a complex mechanism consisting of many steps. Several models have been hypothesized and the sum of the squares of the residuals has been calculated for the fits of the models of these data. The results are shown below. Which model best describe these data (statistically)? Explain your choice.

	Model A	Model B	Model C
S_r	135	105	100
Number of model parameters fit	2	3	5

20.11 Below are data taken from a batch reactor of bacterial growth (after lag phase was over). The bacteria are allowed to grow as fast as possible for the first 2.5 hours, and then they are induced to produce a recombinant protein, the production of which slows the bacterial growth significantly. The theoretical growth of bacteria can be described by

$$\frac{dX}{dt} = \mu X$$

where X is the number of bacteria and μ is the specific growth rate of the bacteria during exponential growth. Based on these data, estimate the specific growth rate of the bacteria during the first 2 hours of growth and during the next 4 hours of growth.

Time, hr	0	1	2	3	4	5	6
[Cells], g/L	0.100	0.333	1.101	1.645	2.452	3.661	5.459

20.12 The molecular weight of a polymer can be determined from its viscosity by the following relationship:

$$[\eta] = KM_v^a$$

where $[\eta]$ is the intrinsic viscosity of the polymer, M_v is the viscosity averaged molecular weight, and K and a are constants specific for the polymer. The intrinsic viscosity is determined experimentally by determining the efflux time, or the time it takes for the polymer solution to flow between two etched lines in a capillary viscometer, at several different concentrations of dilute polymer and extrapolating to infinite dilution. A plot of

$$\frac{\frac{t}{t_0} - 1}{c} \text{ versus } c$$

should yield a straight line, with a y intercept equal to $[\eta]$. The concentration of the polymer solution is c, t is the efflux time of the polymer solution, and t_0 is the efflux time of the solvent without polymer. Using the data below of efflux times for dilute solutions of polystyrene in methyl ethyl ketone at 25°C and the constants $K = 3.9 \times 10^{-4}$ and $a = 0.58$, find the molecular weight of the polystyrene sample.

Polymer Concentration, g/dL	Efflux Time, s
0 (pure solvent)	83
0.04	89
0.06	95
0.08	104
0.10	114
0.12	126
0.14	139
0.16	155
0.20	191

20.13 On average, the surface area A of human beings is related to weight W and height H. Measurements on a number of individuals give the values of A in the following table:

H (cm)	182	180	179	187	189	194	195	193	200
W (kg)	74	88	94	78	84	98	76	86	96
A (m²)	1.92	2.11	2.15	2.02	2.09	2.31	2.02	2.16	2.31

Develop an equation to predict area as a function of height and weight. Use it to estimate the surface area for a 187-cm, 78-kg person.

20.14 Determine an equation to predict metabolism rate as a function of mass based on the following data:

Animal	Mass, kg	Metabolism, watts
Cow	400	270
Human	70	82
Sheep	45	50
Hen	2	4.8
Rat	0.3	1.45
Dove	0.16	0.97

20.15 Human blood behaves as a *Newtonian fluid* ($\tau = \mu\dot{\gamma}$; see Prob. 20.55) in the high shear rate region where $\dot{\gamma} > 100$. In the low shear rate region where $\dot{\gamma} < 50$, the red cells tend to aggregate into what are called rouleaux, which make the fluid behavior depart from Newtonian. This low shear rate region is called the Casson region, and there is a transition region between the two distinct flow regions. In the Casson region, as shear rate approaches zero, the shear stress goes to a finite value, similar to a

Bingham plastic, which is called the yield stress, τ_y, and this stress must be overcome in order to initiate flow in stagnant blood. Flow in the Casson region is usually plotted as the square root of shear rate versus the square root of shear stress, and it follows a straight-line relationship when plotted in this way. The Casson relationship is

$$\sqrt{\tau} = \sqrt{\tau_y} + K_c\sqrt{\dot{\gamma}}$$

where K_c = consistency index. In the table below are experimentally measured values of $\dot{\gamma}$ and τ from a single blood sample over the Casson and Newtonian flow regions.

$\dot{\gamma}$, 1/s	0.91	3.3	4.1	6.3	9.6	23	36	49	65	105	126	215	315	402
τ, N/m^2	0.059	0.15	0.19	0.27	0.39	0.87	1.33	1.65	2.11	3.44	4.12	7.02	10.21	13.01
Region				Casson				Transition				Newtonian		

Find the values of K_c and τ_y using linear regression in the Casson region, and find μ by fitting a straight line with zero intercept in the Newtonian region. Plot the two regression lines on a Casson plot ($\sqrt{\dot{\gamma}}$ versus $\sqrt{\tau}$) and extend the regression lines as dashed lines into adjoining regions; also include the data points in the plot.

20.16 Soft tissue follows an exponential deformation behavior in uniaxial tension while it is in the physiologic or normal range of elongation. This can be expressed as

$$\sigma = \frac{E_0}{a}(e^{a\varepsilon} - 1)$$

where σ = stress, ε = strain, and E_0 and a are material constants that are determined experimentally. To evaluate the two material constants, the above equation is differentiated with respect to ε. Doing so establishes the fundamental relationship for soft tissue:

$$\frac{d\sigma}{d\varepsilon} = E_0 + a\sigma$$

To evaluate E_0 and a, stress-strain data are plotted as $d\sigma/d\varepsilon$ versus σ and the intercept and slope of this plot are the two material constants, respectively.

The following table gives stress-strain data for heart chordae tendineae (small tendons used to hold heart valves closed during contraction of the heart muscle; these data are from loading the tissue, while different values are produced on unloading).

σ, 10^3 N/m^2	87.8	96.6	176	263	351	571	834	1229	1624	2107	2678	3380	4258
ε, 10^{-3} m/m	153	204	255	306	357	408	459	510	561	612	663	714	765

Calculate the derivative $d\sigma/d\varepsilon$ using finite differences. Plot these data and eliminate the data points near the zero points that appear not to follow the straight-line relationship. The error in these data comes from the inability of the instrumentation to read the small values in this region. Perform a regression analysis of the remaining data points to determine the values of E_0 and a.

Plot the stress versus strain data points along with the analytic curve expressed by the first equation. This will indicate how well the analytic curve matches these data.

Many times this does not work well because the value of E_0 is difficult to evaluate using this technique. To solve this problem E_0 is not used. A data point is selected, $(\bar{\sigma}, \bar{\varepsilon})$, that is in the middle of the regression analysis range. These values are substituted into the first equation and a value for E_0/a is determined and substituted into the first equation, which becomes

$$\sigma = \left(\frac{\bar{\sigma}}{e^{a\bar{\varepsilon}} - 1}\right)(e^{a\varepsilon} - 1)$$

Using this approach, experimental data that are well defined will produce a good match of the data points and the analytic curve. Use this new relationship and again plot the stress versus strain data points and the new analytic curve.

20.17 The thickness of the retina changes during certain eye diseases. One way to measure retinal thickness is to shine a low-energy laser at the retina and record the reflections on film. Because of the optical properties of the eye, the reflections from the front surface of the retina and the back surface of the retina will appear as two lines on the film separated by a distance. The distance between the lines on the film is proportional to the thickness of the retina. Below are data taken from a scanned film. Fit two Gaussian-shaped curves of arbitrary height and location to these data and determine the distance between the centers of the two peaks. A Gaussian curve has the form

$$f(x) = \frac{ke^{-k^2(x-a)^2}}{\sqrt{\pi}}$$

where k and a are constants that relate to the peak height and the center of the peak, respectively.

Position	Light Intensity	Position	Light Intensity
0.17	5.10	0.31	25.31
0.18	5.10	0.32	23.79
0.19	5.20	0.33	18.44
0.20	5.87	0.34	12.45
0.21	8.72	0.35	8.22
0.22	16.04	0.36	6.12
0.23	26.35	0.37	5.35
0.24	31.63	0.38	5.15
0.25	26.51	0.39	5.10
0.26	16.68	0.40	5.10
0.27	10.80	0.41	5.09
0.28	11.26	0.42	5.09
0.29	16.05	0.43	5.09
0.3	21.96	0.44	5.09

20.18 The data tabulated below were generated from an experiment starting with pure ammonium cyanate (NH_4OCN). It is known that such concentration changes can be modeled by the following equation:

$$c = \frac{c_0}{1 + kc_0 t}$$

where c_0 and k are parameters. Use a transformation to linearize this equation. Then employ linear regression to predict the concentration at $t = 160$ min.

t (min)	0	20	50	65	150
c (mole/L)	0.381	0.264	0.180	0.151	0.086

Civil/Environmental Engineering

20.19 The shear stresses, in kilopascals (kPa), of nine specimens taken at various depths in a clay stratum are listed below. Estimate the shear stress at a depth of 4.5 m.

Depth, m	1.9	3.1	4.2	5.1	5.8	6.9	8.1	9.3	10.0
Stress, kPa	14.4	28.7	19.2	43.1	33.5	52.7	71.8	62.2	76.6

20.20 A transportation engineering study was conducted to determine the proper design of bike lanes. Data were gathered on bike-lane widths and average distance between bikes and passing cars. The data from nine streets are

Distance, m	2.4	1.5	2.4	1.8	1.8	2.9	1.2	3	1.2
Lane width, m	2.9	2.1	2.3	2.1	1.8	2.7	1.5	2.9	1.5

(a) Plot these data.
(b) Fit a straight line to these data with linear regression. Add this line to the plot.
(c) If the minimum safe average distance between bikes and passing cars is considered to be 2 m, determine the corresponding minimum lane width.

20.21 The saturation concentration of dissolved oxygen in water as a function of temperature and chloride concentration is listed in Table P20.21. Use interpolation to estimate the dissolved oxygen level for $T = 18°C$ with chloride concentration = 10 g/L.

20.22 For the data in Table P20.21, use polynomial regression to derive a third-order predictive equation for dissolved oxygen concentration as a function of temperature for the case where chloride concentration is equal to 10 g/L. Use the equation to estimate the dissolved oxygen concentration for $T = 8°C$.

20.23 Use multiple linear regression to derive a predictive equation for dissolved oxygen concentration as a function of temperature and chloride concentration based on the data from Table P20.21. Use the equation to estimate the concentration of dissolved oxygen for a chloride concentration of 5 g/L at $T = 17°C$.

20.24 As compared to the models from Probs. 20.22 and 20.23, a somewhat more sophisticated model that accounts for the effect of both temperature and chloride concentration on dissolved oxygen saturation can be hypothesized as being of the form,

$$o_s = a_0 + f_3(T) + f_1(c)$$

That is, a constant plus a third-order polynomial in temperature and a linear relationship in chloride concentration are assumed to yield superior results. Use the general linear least-squares approach to fit this model to the data in Table P20.21. Use the resulting equation to estimate the dissolved oxygen concentration for a chloride concentration of 10 g/L at $T = 20°C$.

TABLE P20.21 Dissolved oxygen concentration in water as a function of temperature (°C) and chloride concentration (g/L).

| T, °C | Dissolved Oxygen (mg/L) for Temperature (°C) and Concentration of Chloride (g/L) | | |
	c = 0 g/L	c = 10 g/L	c = 20 g/L
0	14.6	12.9	11.4
5	12.8	11.3	10.3
10	11.3	10.1	8.96
15	10.1	9.03	8.08
20	9.09	8.17	7.35
25	8.26	7.46	6.73
30	7.56	6.85	6.20

20.25 In water-resources engineering, the sizing of reservoirs depends on accurate estimates of water flow in the river that is being impounded. For some rivers, long-term historical records of such flow data are difficult to obtain. In contrast, meteorological data on precipitation is often available for many years past. Therefore, it is often useful to determine a relationship between flow and precipitation. This relationship can then be used to estimate flows for years when only precipitation measurements were made. The following data are available for a river that is to be dammed:

Precipitation, cm	88.9	108.5	104.1	139.7	127	94	116.8	99.1
Flow, m³/s	14.6	16.7	15.3	23.2	19.5	16.1	18.1	16.6

(a) Plot these data.
(b) Fit a straight line to these data with linear regression. Superimpose this line on your plot.
(c) Use the best-fit line to predict the annual water flow if the precipitation is 120 cm.
(d) If the drainage area is 1100 km², estimate what fraction of the precipitation is lost via processes such as evaporation, deep groundwater infiltration, and consumptive use.

20.26 The concentration of total phosphorus (p in mg/m³) and chlorophyll a (c in mg/m³) for each of the Great Lakes in 1970 was

	p	c
Lake Superior	4.5	0.8
Lake Michigan	8.0	2.0
Lake Huron	5.5	1.2
Lake Erie:		
West basin	39.0	11.0
Central basin	19.5	4.4
East basin	17.5	3.8
Lake Ontario	21.0	5.5

The concentration of chlorophyll a indicates how much plant life is suspended in the water. As such, it indicates how unclear and unsightly the water appears. Use the tabulated data to determine the value of c as a function of p. Use this equation to predict the level of chlorophyll that can be expected if waste treatment is used to lower the phosphorus concentration of the west basin of Lake Erie to 10 mg/m³.

20.27 The vertical stress σ_z under the corner of a rectangular area subjected to a uniform load of intensity q is given by the solution of Boussinesq's equation:

$$\sigma = \frac{q}{4\pi}\left[\frac{2mn\sqrt{m^2+n^2+1}}{m^2+n^2+1+m^2n^2}\frac{m^2+n^2+2}{m^2+n^2+1} + \sin^{-1}\left(\frac{2mn\sqrt{m^2+n^2+1}}{m^2+n^2+1+m^2n^2}\right)\right]$$

Because this equation is inconvenient to solve manually, it has been reformulated as

$$\sigma_z = qf_z(m, n)$$

FIGURE P20.27

TABLE P20.27

m	n = 1.2	n = 1.4	n = 1.6
0.1	0.02926	0.03007	0.03058
0.2	0.05733	0.05894	0.05994
0.3	0.08323	0.08561	0.08709
0.4	0.10631	0.10941	0.11135
0.5	0.12626	0.13003	0.13241
0.6	0.14309	0.14749	0.15027
0.7	0.15703	0.16199	0.16515
0.8	0.16843	0.17389	0.17739

where $f_z(m, n)$ is called the influence value and m and n are dimensionless ratios, with $m = a/z$ and $n = b/z$ and a and b as defined in Fig. P20.27. Some results for influence values are given in Table P20.27. If $a = 4.6$ and $b = 14$, use a third-order interpolating polynomial to compute σ_z at a depth 10 m below the corner of a rectangular footing that is subject to a total load of 100 t (metric tons). Express your answer in tonnes per square meter. Note that q is equal to the load per area.

20.28 Three disease-carrying organisms decay exponentially in lake water according to the following model:

$$p(t) = Ae^{-1.5t} + Be^{-0.3t} + Ce^{-0.05t}$$

Estimate the initial population of each organism (A, B, and C) given the following measurements:

t, hr	0.5	1	2	3	4	5	6	7	9
p(t)	6.0	4.4	3.2	2.7	2.2	1.9	1.7	1.4	1.1

20.29 The mast of a sailboat has a cross-sectional area of 10.65 cm^2 and is constructed of an experimental aluminum alloy. Tests were performed to define the relationship between stress and strain. The test results are

Strain, cm/cm	0.0032	0.0045	0.0055	0.0016	0.0085	0.0005
Stress, N/cm^2	4970	5170	5500	3590	6900	1240

The stress caused by wind can be computed as F/A_c, where F = force in the mast and A_c = mast's cross-sectional area. This value can then be substituted into Hooke's law to determine the mast's deflection, $\Delta L = $ strain $\times L$, where $L = $ the mast's length. If the wind force is 25,000 N, use the given data to estimate the deflection of a 9-m mast.

20.30 Enzymatic reactions are used extensively to characterize biologically mediated reactions in environmental engineering. Proposed rate expressions for an enzymatic reaction are given below, where $[S]$ is the substrate concentration and v_0 is the initial rate of reaction. Which formula best fits the experimental data?

$$v_0 = k[S] \quad v_0 = \frac{k[S]}{K + [S]} \quad v_0 = \frac{k[S]^2}{K + [S]^2} \quad v_0 = \frac{k[S]^3}{K + [S]^3}$$

[S], M	Initial Rate, 10^{-6} M/s
0.01	6.3636×10^{-5}
0.05	7.9520×10^{-3}
0.1	6.3472×10^{-2}
0.5	6.0049
1	17.690
5	24.425
10	24.491
50	24.500
100	24.500

20.31 Environmental engineers dealing with the impacts of acid rain must determine the value of the ion product of water, K_w, as a function of temperature. Scientists have suggested the following equation to model this relationship:

$$-\log_{10} K_w = \frac{a}{T_a} + b \log_{10} T_a + cT_a + d$$

where T_a = absolute temperature (K), and a, b, c, and d are parameters. Employ the following data and regression to estimate the parameters:

T_a (K)	273.15	283.15	293.15	303.15	313.15
K_w	1.164×10^{-15}	2.950×10^{-15}	6.846×10^{-15}	1.467×10^{-14}	2.929×10^{-14}

20.32 The following data for water density (ρ) versus temperature (T) come from a table of values that were measured with high precision. Use inverse interpolation to determine the temperature corresponding to a density of 0.999245 g/cm^3. Base your estimate on a third-order interpolating polynomial using a tool like Excel or the MATLAB polyfit function to determine the polynomial's coefficients. Find the root with the Newton-Raphson method using an initial guess of $T = 14°C$.

T, °C	0	4	8	12	16
ρ, g/cm^3	0.99987	1.00000	0.99988	0.99952	0.99897

20.33 The following model is frequently used in environmental engineering to parameterize the effect of temperature T (°C) on biochemical reaction rates k (per day),

$$k = k_{20}\theta^{T-20}$$

where k_{20} and θ are parameters. Use a transformation to linearize this equation. Then employ linear regression to estimate k_{20} and θ and predict the reaction rate at $T = 17$°C.

T (°C)	6	12	18	24	30
k (per day)	0.14	0.20	0.31	0.46	0.69

20.34 As a member of Engineers Without Borders, you are working in a community that has contaminated drinking water. At $t = 0$, you add a disinfectant to a cistern that is contaminated with bacteria. You make the following measurements at several times thereafter:

t (hrs)	2	4	6	8	10
c (#/100 mL)	430	190	80	35	16

If the water is safe to drink when the concentration falls below 5 #/100 mL, estimate the time at which the concentration will fall below this limit.

Electrical Engineering

20.35 Perform the same computations as in Sec. 20.3, but analyze data generated with $f(t) = 4\cos(5t) - 7\sin(3t) + 6$.

20.36 You measure the voltage drop V across a resistor for a number of different values of current i. The results are

i	0.25	0.75	1.25	1.5	2.0
V	−0.45	−0.6	0.70	1.88	6.0

Use first- through fourth-order polynomial interpolation to estimate the voltage drop for $i = 1.15$. Interpret your results.

20.37 Duplicate the computation for Prob. 20.36, but use polynomial regression to derive best-fit equations of order 1 through 4 using all the data. Plot and evaluate your results.

20.38 The current in a wire is measured with great precision as a function of time:

t	0	0.1250	0.2500	0.3750	0.5000
i	0	6.24	7.75	4.85	0.0000

Determine i at $t = 0.23$.

20.39 The following data were taken from an experiment that measured the current in a wire for various imposed voltages:

V, V	2	3	4	5	7	10
i, A	5.2	7.8	10.7	13	19.3	27.5

(a) On the basis of a linear regression with these data, determine current for a voltage of 3.5 V. Plot the line and the data and evaluate the fit.

(b) Redo the regression and force the intercept to be zero.

20.40 It is known that the voltage drop across an inductor follows Faraday's law:

$$V_L = L\frac{di}{dt}$$

where V_L is the voltage drop (in volts), L is inductance (in henrys; 1 H = 1 V s/A), and i is current (in amperes). Employ the following data to estimate L:

di/dt, A/s	1	2	4	6	8	10
V_L, V	5.5	12.5	17.5	32	38	49

What is the meaning, if any, of the intercept of the regression equation derived from these data?

20.41 Ohm's law states that the voltage drop V across an ideal resistor is linearly proportional to the current i flowing through the resistor, as in $V = iR$, where R is the resistance. However, real resistors may not always obey Ohm's law. Suppose that you performed some very precise experiments to measure the voltage drop and corresponding current for a resistor. The results, as listed in Table P20.41, suggest a curvilinear relationship rather than the straight line represented by Ohm's law. In order to quantify this relationship, a curve must be fit to these data. Because of measurement error, regression would typically be the preferred method of curve fitting for analyzing such experimental data. However, the smoothness of the relationship, as well as the precision of the experimental methods, suggests that interpolation might be appropriate. Use Newton's interpolating polynomial to fit these data and compute V for $i = 0.10$. What is the order of the polynomial that was used to generate the data?

TABLE P20.41 Experimental data for voltage drop across a resistor subjected to various levels of current.

i	−2	−1	−0.5	0.5	1	2
V	−637	−96.5	−20.5	20.5	96.5	637

20.42 Repeat Prob. 20.41 but determine the coefficients of the polynomial (Sec. 18.4) that fit the data in Table P20.41.

20.43 An experiment is performed to determine the percent elongation of an electrical conducting material as a function of temperature. The resulting data are listed below. Predict the percent elongation for a temperature of 400°C.

Temperature, °C	200	250	300	375	425	475	600
% elongation	7.5	8.6	8.7	10	11.3	12.7	15.3

TABLE P20.49 Experimental values for elongation x and force F for the spring on an automobile suspension system.

Displacement, m	0.10	0.17	0.27	0.35	0.39	0.42	0.43	0.44
Force, 10^4 N	10	20	30	40	50	60	70	80

20.44 Bessel functions often arise in advanced engineering analyses such as the study of electric fields. These functions are usually not amenable to straightforward evaluation and, therefore, are often compiled in standard mathematical tables. For example,

x	1.8	2	2.2	2.4	2.6
$J_1(x)$	0.5815	0.5767	0.556	0.5202	0.4708

Estimate $J_1(2.1)$, **(a)** using an interpolating polynomial and **(b)** using cubic splines. Note that the true value is 0.568292.

20.45 The population (p) of a small community on the outskirts of a city grows rapidly over a 20-year period:

t	0	5	10	15	20
p	100	200	450	950	2000

As an engineer working for a utility company, you must forecast the population 5 years into the future in order to anticipate the demand for power. Employ an exponential model and linear regression to make this prediction.

Mechanical/Aerospace Engineering

20.46 Based on Table 20.3, use linear and quadratic interpolation to compute Q for $D = 1.23$ ft and $S = 0.001$ ft/ft. Compare your results with the value computed with the formula derived in Sec. 20.4.

20.47 Reproduce Sec. 20.4, but develop an equation to predict slope as a function of diameter and flow. Compare your results with the formula from Sec. 20.4 and discuss your results.

20.48 Dynamic viscosity of water μ (10^{-3} N s/m^2) is related to temperature T (°C) in the following manner:

T	0	5	10	20	30	40
μ	1.787	1.519	1.307	1.002	0.7975	0.6529

(a) Plot these data.
(b) Use interpolation to predict μ at $T = 7.5$°C.
(c) Use polynomial regression to fit a parabola to these data in order to make the same prediction.

20.49 *Hooke's law,* which holds when a spring is not stretched too far, signifies that the extension of the spring and the applied force are linearly related. The proportionality is parameterized by the spring constant k. A value for this parameter can be established

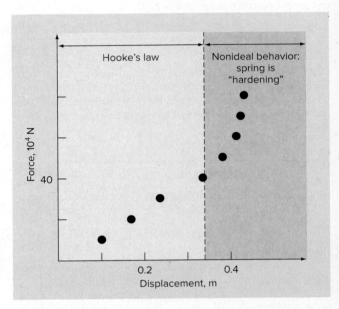

FIGURE P20.49
Plot of force (in 10^4 newtons) versus displacement (in meters) for the spring from the automobile suspension system.

experimentally by placing known weights onto the spring and measuring the resulting compression. Such data were contained in Table P20.49 and plotted in Fig. P20.49. Notice that above a weight of 40×10^4 N, the linear relationship between the force and displacement breaks down. This sort of behavior is typical of what is termed a "hardening spring." Employ linear regression to determine a value of k for the linear portion of this system. In addition, fit a nonlinear relationship to the nonlinear portion.

20.50 Repeat Prob. 20.49 but fit a power curve to all the data in Table P20.49. Comment on your results.

20.51 The distance required to stop an automobile consists of both thinking and braking components, each of which is a function of the car's speed. The following experimental data were collected to quantify this relationship. Develop a best-fit equation for both the thinking and braking components. Use these equations to estimate the total stopping distance for a car traveling at 110 km/hr.

Speed, km/hr	30	45	60	75	90	120
Thinking, m	5.6	8.5	11.1	14.5	16.7	22.4
Braking, m	5.0	12.3	21.0	32.9	47.6	84.7

20.52 An experiment is performed to define the relationship between applied stress and the time to fracture for a type of stainless steel. Eight different values of stress are applied, and the resulting data are

Applied stress x, kg/mm^2	5	10	15	20	25	30	35	40
Fracture time y, hr	40	30	25	40	18	20	22	15

Plot these data and then develop a best-fit equation to predict the fracture time for an applied stress of 20 kg/mm^2.

20.53 The acceleration due to gravity at an altitude y above the surface of the earth is given by

y, m	0	30,000	60,000	90,000	120,000
g, m/s^2	9.8100	9.7487	9.6879	9.6278	9.5682

Compute g at $y = 55,000$ m.

20.54 The creep rate $\dot{\varepsilon}$ is the time rate at which strain increases, and the stress data below were obtained from a testing procedure. Using a power-law fit,

$$\dot{\varepsilon} = B\sigma^m$$

find the value of B and m. Plot your results using a log-log scale.

Creep rate, min^{-1}	0.0004	0.0011	0.0021	0.0031
Stress, MPa	5.775	8.577	10.874	12.555

20.55 It is a common practice when examining a fluid's viscous behavior to plot the shear strain rate (velocity gradient),

$$\frac{dv}{dy} = \dot{\gamma}$$

on the abscissa versus shear stress (τ) on the ordinate. When a fluid has a straight-line behavior between these two variables, it is called a *Newtonian fluid*, and the resulting relationship is

$$\tau = \mu\dot{\gamma}$$

where μ is the fluid viscosity. Many common fluids follow this behavior, such as water, milk, and oil. Fluids that do not behave in this way are called *non-Newtonian*. Some examples of non-Newtonian fluids are shown in Fig. P20.55.

For *Bingham plastics*, there is a yield stress τ_y that must be overcome before flow will begin,

$$\tau = \tau_y + \mu\dot{\gamma}$$

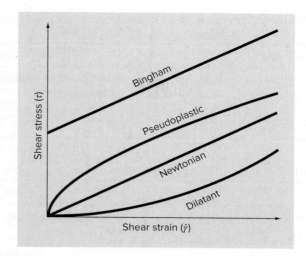

FIGURE P20.55

A common example of a Bingham plastic is toothpaste.

For *pseudoplastics*, or "*shear thinning*" fluids, the shear stress is raised to a power n less than one,

$$\tau = \mu\dot{\gamma}^n$$

Such fluids, such as yogurt, mayonnaise, and shampoo, exhibit a decrease in viscosity with increasing stress. Note that for cases where $n > 1$, called *dilatant* (or "*shear thickening*") *fluids*, viscosity actually increases with shear stress. Examples include starch in water and wet beach sand.

The following data show the relationship between the shear stress τ and the shear strain rate $\dot{\gamma}$ for a fluid. The yield stress τ_y is the amount of stress that must be exceeded before flow begins. Find the viscosity μ (slope), τ_y, and the r^2 value using a regression method. What is the type of fluid?

Stress τ, N/m^2	3.25	4.25	4.65	5.65	6.05
Shear strain rate $\dot{\gamma}$, 1/s	0.9	2.1	2.9	4.1	4.9

20.56 The relationship between stress τ and the shear strain rate $\dot{\gamma}$ for a pseudoplastic fluid (see Prob. 20.55) can be expressed by the equation $\tau = \mu\dot{\gamma}^n$. The following data come from a 0.5% hydroxyethylcellulose in water solution. Using a power-law fit, find the values of μ and n.

Shear strain rate $\dot{\gamma}$, 1/s	50	70	90	110	130
Stress τ, N/m^2	6.01	7.48	8.59	9.19	10.21

TABLE P20.59 Ideal specific heats, c_p (kJ/kg K), as a function of temperature for several gases.

Gas	250 K	300 K	350 K	450 K	550 K	650 K	800 K	900 K	1000 K
H_2	14.051	14.307	14.427	14.501	14.53	14.571	14.695	14.822	14.983
CO_2	0.791	0.846	0.895	0.978	1.046	1.102	1.169	1.204	1.234
O_2	0.913	0.918	0.928	0.956	0.988	1.017	1.054	1.074	1.09
N_2	1.039	1.039	1.041	1.049	1.065	1.086	1.121	1.145	1.167

TABLE P20.60 Temperatures (°C) at various points on a square heated plate.

	$x = 0$	$x = 2$	$x = 4$	$x = 6$	$x = 8$
$y = 0$	100.00	90.00	80.00	70.00	60.00
$y = 2$	85.00	64.49	53.50	48.15	50.00
$y = 4$	70.00	48.90	38.43	35.03	40.00
$y = 6$	55.00	38.78	30.39	27.07	30.00
$y = 8$	40.00	35.00	30.00	25.00	20.00

20.57 The velocity u of air flowing past a flat surface is measured at several distances y away from the surface. Fit a curve to these data assuming that the velocity is zero at the surface ($y = 0$). Use your result to determine the shear stress ($\mu \, du/dy$) at the surface. ($\mu = 1.8 \times 10^{-5}$ N s/m²)

y, m	0.002	0.006	0.012	0.018	0.024
u, m/s	0.287	0.899	1.915	3.048	4.299

20.58 *Andrade's equation* has been proposed as a model of the effect of temperature on viscosity,

$$\mu = De^{B/T_a}$$

where μ = dynamic viscosity of water (10^{-3} N s/m²), T_a = absolute temperature (K), and D and B are parameters. Fit this model to the data for water from Prob. 20.48.

20.59 Develop equations to fit the ideal specific heats, c_p (kJ/kg K), as a function of temperature T (K), for several gases as listed in Table P20.59.

20.60 Temperatures are measured at various points on a heated plate (Table P20.60). Estimate the temperature at **(a)** $x = 4$, $y = 3.2$, and **(b)** $x = 4.3$, $y = 2.7$.

20.61 The data below were obtained from a creep test performed at room temperature on a wire composed of 40% tin, 60% lead, and solid core solder. This was done by measuring the increase in strain over time while a constant load was applied to a test specimen. Using a linear regression method, find **(a)** the equation of the line that best fits these data and **(b)** the r^2 value. Plot your results. Does the line pass through the origin—that is, at time zero should there be any strain? If the line does not pass through the origin, force it to do so. Does this new line represent the data trend? Suggest a new equation that satisfies zero strain at zero time and also represents the data trend well.

Time, min	Strain, %	Time, min	Strain, %	Time, min	Strain, %
0.085	0.10	3.589	0.26	7.092	0.43
0.586	0.13	4.089	0.30	7.592	0.45
1.086	0.16	4.590	0.32	8.093	0.47
1.587	0.18	5.090	0.34	8.593	0.50
2.087	0.20	5.591	0.37	9.094	0.52
2.588	0.23	6.091	0.39	9.594	0.54
3.088	0.25	6.592	0.41	10.097	0.56

20.62 Use the following portion of the steam table for superheated H_2O at 200 MPa to (a) find the corresponding entropy s for a specific volume v of 0.108 m^3/kg with linear interpolation, (b) find the same corresponding entropy using quadratic interpolation, and (c) find the volume corresponding to an entropy of 6.6 using inverse interpolation.

v (m^3/kg)	0.10377	0.11144	0.12540
s (kJ/kg K)	6.4147	6.5453	6.7664

20.63 The U.S. Standard Atmosphere specifies atmospheric properties as a function of altitude above sea level. The following table shows selected values of temperature, pressure, and density

Altitude (km)	T (°C)	p (atm)	ρ (kg/m^3)
−0.5	18.4	1.0607	1.2850
2.5	−1.1	0.73702	0.95697
6	−23.8	0.46589	0.66015
11	−56.2	0.22394	0.36481
20	−56.3	0.054557	0.088911
28	−48.5	0.015946	0.025076
50	−2.3	7.8721×10^{-4}	1.0269×10^{-3}
60	−17.2	2.2165×10^{-4}	3.0588×10^{-4}
80	−92.3	1.0227×10^{-5}	1.9992×10^{-5}
90	−92.3	1.6216×10^{-6}	3.1703×10^{-6}

Develop a function to determine values of the three properties for a given altitude. If the user requests a value outside the range of altitudes, have the function display an error message and terminate the application.

20.64 Felix Baumgartner ascended to 39 km in a stratospheric balloon and made a free-fall jump, rushing toward earth at supersonic speeds before parachuting to the ground. As he fell, his drag coefficient changed primarily because the air density changed. Recall from Chap. 1 that the terminal velocity, $v_{terminal}$ (m/s), of a free-falling object can be computed as

$$v_{terminal} = \sqrt{\frac{gm}{c_d}}$$

where g = gravitational acceleration (m/s^2), m = mass (kg), and c_d = drag coefficient (kg/m). The drag coefficient can be computed as

$$c_d = 0.5\rho A C_d$$

where ρ = the fluid density (kg/m^3), A = the projected area (m^2), and C_d = a dimensionless drag coefficient. Note that the gravitational acceleration, g (m/s^2), can be related to elevation by

$$g = 9.806412 - 0.003039734z$$

where z = elevation above the earth's surface (km), and the density of air, ρ (kg/m^3), at various elevations can be tabulated as

z (km)	ρ (kg/m^3)	z (km)	ρ (kg/m^3)	z (km)	ρ (kg/m^3)
−1	1.347	6	0.6601	25	0.04008
0	1.225	7	0.5900	30	0.01841
1	1.112	8	0.5258	40	0.003996
2	1.007	9	0.4671	50	0.001027
3	0.9093	10	0.4135	60	0.0003097
4	0.8194	15	0.1948	70	8.283×10^{-5}
5	0.7364	20	0.08891	80	1.846×10^{-5}

Assume that $m = 80$ kg, $A = 0.55$ m^2, and $C_d = 1.1$. Develop a program to compute the terminal velocity versus elevation for a given value of elevation with a function based on splines to generate the required density. If the user requests a value outside the range of altitudes, have the function display an error message and terminate the application. Test your program for $z = −2$, 17, 62.5, 80, and 80.01 km.

EPILOGUE: PART FIVE

PT5.4 TRADE-OFFS

Table PT5.4 provides a summary of the trade-offs involved in curve fitting. The techniques are divided into two broad categories, depending on the uncertainty of the data. For imprecise measurements, regression is used to develop a "best" curve that fits the overall trend of the data without necessarily passing through any of the individual points. For precise measurements, interpolation is used to develop a curve that passes directly through each of the points.

All the regression methods are designed to fit functions that minimize the sum of the squares of the residuals between the data and the function. Such methods are termed least-squares regression. Linear least-squares regression is used for cases where a dependent and an independent variable are related to each other in a linear fashion. For situations where a dependent and an independent variable exhibit a curvilinear relationship, several options are available. In some cases, transformations can be used to linearize the relationship. In these instances, linear regression can be applied to the transformed variables to determine the best straight line. Alternatively, polynomial regression can be employed to fit a curve directly to the data.

Multiple linear regression is utilized when a dependent variable is a linear function of two or more independent variables. Logarithmic transformations can also be applied to this type of regression for some cases where the multiple dependency is curvilinear.

TABLE PT5.4 Comparison of the characteristics of alternative methods for curve fitting.

Method	Error Associated with Data	Match of Individual Data Points	Number of Points Matched Exactly	Programming Effort	Comments
Regression					
Linear regression	Large	Approximate	0	Easy	
Polynomial regression	Large	Approximate	0	Moderate	Round-off error becomes pronounced for higher-order versions
Multiple linear regression	Large	Approximate	0	Moderate	
Nonlinear regression	Large	Approximate	0	Difficult	
Interpolation					
Newton's divided-difference polynomials	Small	Exact	$n + 1$	Easy	Usually preferred for exploratory analyses
Lagrange polynomials	Small	Exact	$n + 1$	Easy	Usually preferred when order is known
Cubic splines	Small	Exact	Piecewise fit of data points	Moderate	First and second derivatives equal at knots

Polynomial regression and multiple linear regression (note that simple linear regression is a member of both) belong to a more general class of linear least-squares models. They are classified in this way because they are linear with respect to their coefficients. These models are typically implemented using linear algebraic systems that are sometimes ill-conditioned. However, in many engineering applications (that is, for lower-order fits), this does not come into play. For cases where it is a problem, alternative approaches are available. For example, a technique called orthogonal polynomials is available to perform polynomial regression (see Sec. PT5.6).

Equations that are not linear with respect to their coefficients are called nonlinear. Special regression techniques are available to fit such equations. These are approximate methods that start with initial parameter estimates and then iteratively home in on values that minimize the sum of the squares.

Polynomial interpolation is designed to fit a unique nth-order polynomial that passes exactly through $n + 1$ precise data points. This polynomial is presented in two alternative formats. Newton's divided-difference interpolating polynomial is ideally suited for those cases where the proper order of the polynomial is unknown. Newton's polynomial is appropriate for such situations because it is easily programmed in a format to compare results with different orders. In addition, an error estimate can be simply incorporated into the technique. Thus, you can compare and choose from results using several different-order polynomials.

The Lagrange interpolating polynomial is an alternative formulation that is appropriate when the order is known *a priori*. For these situations, the Lagrange version is somewhat simpler to program and does not require the computation and storage of finite divided differences.

Another approach to curve fitting is spline interpolation. This technique fits a low-order polynomial to each interval between data points. The fit is made smooth by setting the derivatives of adjacent polynomials to the same value at their connecting points. The cubic spline is the most common version. Splines are of great utility when fitting data that are generally smooth but exhibit local areas of abrupt change. Such data tend to induce wild oscillations in higher-order interpolating polynomials. Cubic splines are less prone to these oscillations because they are limited to third-order variations.

Beyond the one-dimensional case, interpolation can be implemented for multidimensional data. Both interpolating polynomials and splines can be used for this purpose. Software packages are available to expedite such applications.

The final method covered in this part of the book is Fourier approximation. This approach deals with using trigonometric functions to approximate waveforms. In contrast to the other techniques, the major emphasis of this approach is not to fit a curve to data points. Rather, the curve fit is employed to analyze the frequency characteristics of a signal. In particular, a fast Fourier transform is available to very efficiently transform a function from the time to the frequency domain to elucidate its underlying harmonic structure.

PT5.5   IMPORTANT RELATIONSHIPS AND FORMULAS

Table PT5.5 summarizes important information that was presented in Part Five. This table can be consulted to quickly access important relationships and formulas.

TABLE PT5.5 Summary of important information presented in Part Five.

Method	Formulation	Graphical Interpretation	Errors
Linear regression	$y = a_0 + a_1x$ where $a_1 = \dfrac{n\Sigma x_iy_i - \Sigma x_i \Sigma y_i}{n\Sigma x_i^2 - (\Sigma x_i)^2}$ $a_0 = \bar{y} - a_1\bar{x}$		$s_{y/x} = \sqrt{\dfrac{S_r}{n-2}}$ $r^2 = \dfrac{S_t - S_r}{S_t}$
Polynomial regression	$y = a_0 + a_1x + \cdots + a_mx^m$ (Evaluation of a's equivalent to solution of $m + 1$ linear algebraic equations)		$s_{y/x} = \sqrt{\dfrac{S_r}{n-(m+1)}}$ $r^2 = \dfrac{S_t - S_r}{S_t}$
Multiple linear regression	$y = a_0 + a_1x_1 + \cdots + a_mx_m$ (Evaluation of a's equivalent to solution of $m + 1$ linear algebraic equations)		$s_{y/x} = \sqrt{\dfrac{S_r}{n-(m+1)}}$ $r^2 = \dfrac{S_t - S_r}{S_t}$
Newton's divided-difference interpolating polynomial*	$f_2(x) = b_0 + b_1(x - x_0) + b_2(x - x_0)(x - x_1)$ where $b_0 = f(x_0)$ $b_1 = f[x_1, x_0]$ $b_2 = f[x_2, x_1, x_0]$		$R_2 = (x - x_0)(x - x_1)(x - x_2)\dfrac{f^{(3)}(\xi)}{6}$ or $R_2 = (x - x_0)(x - x_1)(x - x_2)f[x_3, x_2, x_1, x_0]$
Lagrange interpolating polynomial*	$f_2(x) = f(x_0)\left(\dfrac{x - x_1}{x_0 - x_1}\right)\left(\dfrac{x - x_2}{x_0 - x_2}\right)$ $+ f(x_1)\left(\dfrac{x - x_0}{x_1 - x_0}\right)\left(\dfrac{x - x_2}{x_1 - x_2}\right)$ $+ f(x_2)\left(\dfrac{x - x_0}{x_2 - x_0}\right)\left(\dfrac{x - x_1}{x_2 - x_1}\right)$		$R_2 = (x - x_0)(x - x_1)(x - x_2)\dfrac{f^{(3)}(\xi)}{6}$ or $R_2 = (x - x_0)(x - x_1)(x - x_2)f[x_3, x_2, x_1, x_0]$
Cubic splines	A cubic, $a_ix^3 + b_ix^2 + c_ix + d_i$ is fit to each interval between knots. First and second derivatives are equal at each knot		

*Note: For simplicity, second-order versions are shown.

PT5.6 ADVANCED METHODS AND ADDITIONAL REFERENCES

Although polynomial regression with normal equations is adequate for many engineering applications, there are problem contexts where its sensitivity to round-off error can represent a serious limitation. An alternative approach based on *orthogonal polynomials* can mitigate these effects. It should be noted that this approach does not yield a best-fit equation, but rather, yields individual predictions for given values of the independent variable. Information on orthogonal polynomials can be found in Shampine and Allen (1973) and Guest (1961).

Whereas the orthogonal polynomial technique is helpful for developing a polynomial regression, it does not represent a solution to the instability problem for the general linear regression model [Eq. (17.23)]. An alternative approach based on *single-value decomposition,* called the SVD method, is available for this purpose. Forsythe et al. (1977), Lawson and Hanson (1974), and Press et al. (2007) contain information on this approach.

In addition to the Gauss-Newton algorithm, there are a number of optimization methods that can be used to directly develop a least-squares fit for a nonlinear equation. These nonlinear regression techniques include Marquardt's and the steepest-descent methods (recall Part Four). General information on regression can be found in Draper and Smith (1981).

All the methods in Part Five have been couched in terms of fitting a curve to data points. In addition, you may also desire to fit a curve to another curve. The primary motivation for such *functional approximation* is to represent a complicated function by a simpler version that is easier to manipulate. One way to do this is to use the complicated function to generate a table of values. Then the techniques discussed in this part of the book can be used to fit polynomials to these discrete values.

An alternative approach is based on the *minimax principle* (recall Fig. 17.2c). This principle specifies that the coefficients of the approximating polynomial be chosen so that the maximum discrepancy is as small as possible. Thus, although the approximation may not be as good as that given by the Taylor series at the base point, it is generally better across the entire range of the fit. *Chebyshev economization* is an example of an approach for functional approximation based on such a strategy (Ralston and Rabinowitz 1978; Gerald and Wheatley 2004; and Carnahan, Luther, and Wilkes 1969).

An important area in curve fitting is the combining of splines with least-squares regression. Thus, a cubic spline is generated that does not intercept every point, but rather, minimizes the sum of the squares of the residuals between the data points and the spline curves. The approach involves using the so-called *B splines* as basis functions. These are so named because of their use as *basis* functions but also because of their characteristic bell shape. Such curves are consistent with a spline approach in that their value and their first and second derivatives have continuity at their extremes. Thus, continuity of $f(x)$ and its lower derivatives at the knots is ensured. Wold (1974), Prenter (1975), and Cheney and Kincaid (2008) present discussions of this approach.

In summary, the foregoing is intended to provide you with avenues for deeper exploration of the subject. Additionally, all the above references provide descriptions of the basic techniques covered in Part Five. We urge you to consult these alternative sources to broaden your understanding of numerical methods for curve fitting.

PART SIX

NUMERICAL DIFFERENTIATION AND INTEGRATION

PT6.1 MOTIVATION

Calculus is the mathematics of change. Because engineers must continuously deal with systems and processes that change, calculus is an essential tool of our profession. Standing at the heart of calculus are the related mathematical concepts of differentiation and integration.

According to the dictionary definition, to *differentiate* means "to mark off by differences; distinguish; . . . to perceive the difference in or between." Mathematically, the *derivative*, which serves as the fundamental vehicle for differentiation, represents the rate of change of a dependent variable with respect to an independent variable. As depicted in Fig. PT6.1, the mathematical definition of the derivative begins with a difference approximation:

$$\frac{\Delta y}{\Delta x} = \frac{f(x_i + \Delta x) - f(x_i)}{\Delta x} \tag{PT6.1}$$

where y and $f(x)$ are alternative representatives for the dependent variable and x is the independent variable. If Δx is allowed to approach zero, as occurs in moving from Fig. PT6.1a to c, the difference becomes a derivative

$$\frac{dy}{dx} = \lim_{\Delta x \to 0} \frac{f(x_i + \Delta x) - f(x_i)}{\Delta x}$$

where dy/dx [which can also be designated as y' or $f'(x_i)$] is the first derivative of y with respect to x evaluated at x_i. As seen in the visual depiction of Fig. PT6.1c, the derivative is the slope of the tangent to the curve at x_i.

The *second derivative* represents the derivative of the first derivative,

$$\frac{d^2y}{dx^2} = \frac{d}{dx}\left(\frac{dy}{dx}\right)$$

Thus, the second derivative tells us how fast the slope is changing. It is commonly referred to as the *curvature,* because a high value for the second derivative means high curvature.

Finally, *partial derivatives* are used for functions that depend on more than one variable. A partial derivative can be thought of as taking the derivative of the function at a point with all but one variable held constant. For example, given a function f that

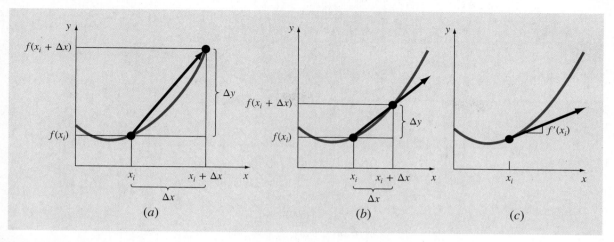

FIGURE PT6.1
The graphical definition of a derivative: As Δx approaches zero in going from (a) to (c), the difference approximation becomes a derivative.

depends on both x and y, the partial derivative of f with respect to x at an arbitrary point (x, y) is defined as

$$\frac{\partial f}{\partial x} = \lim_{\Delta x \to 0} \frac{f(x + \Delta x, y) - f(x, y)}{\Delta x}$$

Similarly, the partial derivative of f with respect to y is defined as

$$\frac{\partial f}{\partial y} = \lim_{\Delta y \to 0} \frac{f(x, y + \Delta y) - f(x, y)}{\Delta y}$$

To get an intuitive grasp of partial derivatives, recognize that a function that depends on two variables is a surface rather than a curve. Suppose you are mountain climbing and have access to a function, f, that yields elevation as a function of longitude (the east-west oriented x-axis) and latitude (the north-south oriented y-axis). If you stop at a particular point, (x_0, y_0), the slope to the east would be $\partial f(x_0, y_0)/\partial x$ and the slope to the north would be $\partial f(x_0, y_0)/\partial y$.

The inverse process to differentiation in calculus is integration. According to the dictionary definition, to *integrate* means "to bring together, as parts, into a whole; to unite; to indicate the total amount . . . ." Mathematically, integration is represented by

$$I = \int_a^b f(x)\, dx \qquad\qquad\qquad\qquad\qquad \text{(PT6.2)}$$

which stands for the integral of the function $f(x)$ with respect to the independent variable x, evaluated between the limits $x = a$ to $x = b$. The function $f(x)$ in Eq. (PT6.2) is referred to as the *integrand*.

As suggested by the dictionary definition, the "meaning" of Eq. (PT6.2) is the *total value*, or *summation*, of $f(x)\, dx$ over the range $x = a$ to b. In fact, the symbol $\int$ is actually

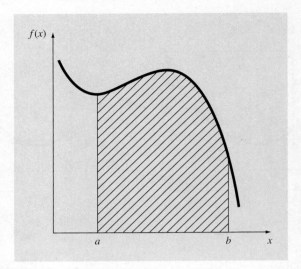

FIGURE PT6.2
Graphical representation of the integral of $f(x)$ between the limits $x = a$ to b. The integral is equivalent to the area under the curve.

a stylized capital S that is intended to signify the close connection between integration and summation.

Figure PT6.2 represents a graphical manifestation of the concept. For functions lying above the x axis, the integral expressed by Eq. (PT6.2) corresponds to the area under the curve of $f(x)$ between $x = a$ and b.[1]

As outlined above, the "marking off" or "discrimination" of differentiation and the "bringing together" of integration are closely linked processes that are, in fact, inversely related (Fig. PT6.3). For example, if we are given a function $y(t)$ that specifies an object's position as a function of time, differentiation provides a means to determine its velocity, as in (Fig. PT6.3a).

$$v(t) = \frac{d}{dt}y(t)$$

Conversely, if we are provided with velocity as a function of time, integration can be used to determine the object's position (Fig. PT6.3b),

$$y(t) = \int_{0}^{t} v(t)\, dt$$

Thus, we can make the general claim that the evaluation of the integral

$$I = \int_{a}^{b} f(x)\, dx$$

---

[1]It should be noted that the process represented by Eq. (PT6.2) and Fig. PT6.2 is called *definite integration*. There is another type called indefinite integration in which the limits a and b are unspecified. As will be discussed in Part Seven, *indefinite integration* deals with determining a function whose derivative is given.

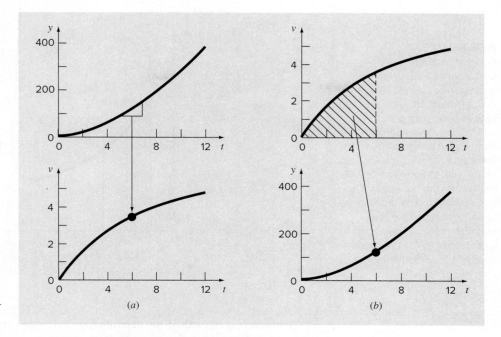

FIGURE PT6.3
The contrast between (*a*) differentiation and (*b*) integration.

is equivalent to solving the differential equation

$$\frac{dy}{dx} = f(x)$$

for $y(b)$ given the initial condition $y(a) = 0$.

Because of this close relationship, we have opted to devote this part of the book to both processes. Among other things, this will provide the opportunity to highlight their similarities and differences from a numerical perspective. In addition, our discussion will have relevance to the next parts of the book where we will cover differential equations.

PT6.1.1 Noncomputer Methods for Differentiation and Integration

The function to be differentiated or integrated will typically be in one of the following three forms:

1. A simple continuous function such as a polynomial, an exponential, or a trigonometric function.
2. A complicated continuous function that is difficult or impossible to differentiate or integrate directly.
3. A tabulated function, where values of x and $f(x)$ are given at a number of discrete points, as is often the case with experimental or field data.

In the first case, the derivative or integral of a simple function may be evaluated analytically using calculus. For the second case, analytical solutions are often impractical, and sometimes impossible, to obtain. In these instances, as well as in the third case of discrete data, approximate methods must be employed.

FIGURE PT6.4
Equal-area differentiation.
(a) Centered finite divided differences are used to estimate
the derivative for each interval
between the data points.
(b) The derivative estimates are
plotted as a bar graph. A
smooth curve is superimposed
on this plot to approximate the
area under the bar graph. This
is accomplished by drawing the
curve so that positive and
negative areas are balanced.
(c) Values of dy/dx can then be
read off the smooth curve.

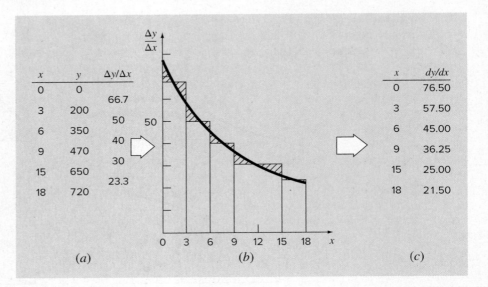

x	y	Δy/Δx
0	0	
		66.7
3	200	
		50
6	350	
		40
9	470	
		30
15	650	
		23.3
18	720	

x	dy/dx
0	76.50
3	57.50
6	45.00
9	36.25
15	25.00
18	21.50

(a)   (b)   (c)

A noncomputer method for determining derivatives from data is called *equal-area graphical differentiation*. In this method, the (x, y) data are tabulated and, for each interval, a simple divided difference Δy/Δx is employed to estimate the slope. Then these values are plotted as a stepped curve versus x (Fig. PT6.4). Next, a smooth curve is drawn that attempts to approximate the area under the stepped curve. That is, it is drawn so that, visually, the positive and negative areas are balanced. The rates at given values of x can then be read from the curve.

In the same spirit, visually oriented approaches were employed to integrate tabulated data and complicated functions in the precomputer era. A simple intuitive approach is to plot the function on a grid (Fig. PT6.5) and count the number of boxes that approximate the area. This number multiplied by the area of each box provides a rough estimate of the total area under the curve. This estimate can be refined, at the expense of additional effort, by using a finer grid.

Another commonsense approach is to divide the area into vertical segments, or strips, with a height equal to the function value at the midpoint of each strip (Fig. PT6.6). The areas of the rectangles can then be calculated and summed to estimate the total area. In this approach, it is assumed that the value at the midpoint provides a valid approximation of the average height of the function for each strip. As with the grid method, refined estimates are possible by using more (and thinner) strips to approximate the integral.

Although such simple approaches have utility for quick estimates, alternative numerical techniques are available for the same purpose. Not surprisingly, the simplest of these methods is similar in spirit to the noncomputer techniques.

For differentiation, the most fundamental numerical techniques use finite divided differences to estimate derivatives. For data with error, an alternative approach is to fit a smooth curve to those data with a technique such as least-squares regression and then differentiate this curve to obtain derivative estimates.

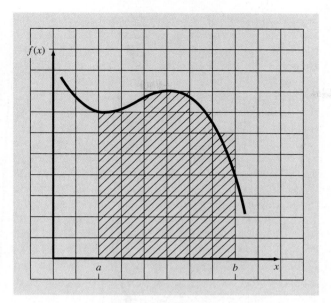

FIGURE PT6.5
The use of a grid to approximate an integral.

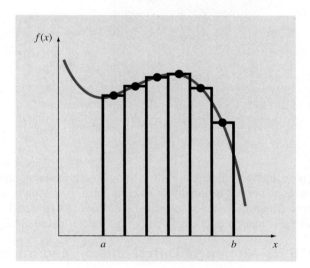

FIGURE PT6.6
The use of rectangles, or strips, to approximate the integral.

In a similar spirit, numerical integration, or *quadrature*, methods are available to obtain integrals. These methods, which are actually easier to implement than the grid approach, are similar in spirit to the strip method. That is, function heights are multiplied by strip widths and summed to estimate the integral. However, through clever choices of weighting factors, the resulting estimate can be made more accurate than that from the simple strip method.

As in the simple strip method, numerical integration and differentiation techniques utilize data at discrete points. Because tabulated information is already in such a form,

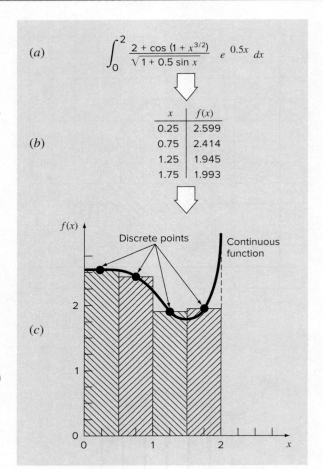

FIGURE PT6.7
Application of a numerical integration method: (*a*) A complicated, continuous function. (*b*) Table of discrete values of *f*(*x*) generated from the function. (*c*) Use of a numerical method (the strip method here) to estimate the integral on the basis of the discrete points. For a tabulated function, the data are already in tabular form (*b*); therefore, step (*a*) is unnecessary.

it is naturally compatible with many of the numerical approaches. Although continuous functions are not originally in discrete form, it is usually a simple proposition to use the given equation to generate a table of values. As depicted in Fig. PT6.7, the values from this table can then be evaluated with a numerical method.

PT6.1.2 Numerical Differentiation and Integration in Engineering

The differentiation and integration of a function has so many engineering applications that you were required to take differential and integral calculus in your first year at college. Many specific examples of such applications could be given for all fields of engineering.

Differentiation is commonplace in engineering because so much of our work involves characterizing the changes of variables in both time and space. In fact, many of the laws and other generalizations that figure so prominently in our work are based on the predictable ways in which change manifests itself in the physical world. A prime example is Newton's second law, which is not couched in terms of the position of an object but rather in terms of its change of position with respect to time.

Aside from such temporal examples, numerous laws governing the spatial behavior of variables are expressed in terms of derivatives. Among the most common of these are those laws involving potentials or gradients. For example, *Fourier's law of heat conduction* quantifies the observation that heat flows from regions of high to low temperature. For the one-dimensional case, this can be expressed mathematically as

$$\text{Heat flux} = -k' \frac{dT}{dx}$$

Thus, the derivative provides a measure of the intensity of the temperature change, or *gradient,* that drives the transfer of heat. Similar laws provide workable models in many other areas of engineering, including the modeling of fluid dynamics, mass transfer, chemical reaction kinetics, and electromagnetic flux. The ability to accurately estimate derivatives is an important facet of our capability to work effectively in these areas.

Just as accurate estimates of derivatives are important in engineering, the calculation of integrals is equally valuable. A number of examples relate directly to the idea of the integral as the area under a curve. Figure PT6.8 depicts a few cases where integration is used for this purpose.

Other common applications relate to the analogy between integration and summation. For example, a common application is to determine the mean of continuous functions. In Part Five, you were introduced to the concept of the mean of n discrete data points [recall Eq. (PT5.1)]:

$$\text{Mean} = \frac{\sum_{i=1}^{n} y_i}{n} \tag{PT6.3}$$

FIGURE PT6.8

Examples of how integration is used to evaluate areas in engineering applications. (*a*) A surveyor might need to know the area of a field bounded by a meandering stream and two roads. (*b*) A water-resource engineer might need to know the cross-sectional area of a river. (*c*) A structural engineer might need to determine the net force due to a nonuniform wind blowing against the side of a skyscraper.

(*a*)　　　　　(*b*)　　　　　(*c*)

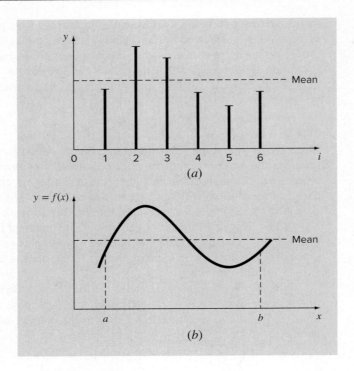

FIGURE PT6.9
An illustration of the mean for
(*a*) discrete and (*b*) continuous
data.

where y_i are individual measurements. The determination of the mean of discrete points is depicted in Fig. PT6.9*a*.

In contrast, suppose that y is a continuous function of an independent variable x, as depicted in Fig. PT6.9*b*. For this case, there are an infinite number of values between a and b. Just as Eq. (PT6.3) can be applied to determine the mean of the discrete readings, you might also be interested in computing the mean or average of the continuous function $y = f(x)$ for the interval from a to b. Integration is used for this purpose, as specified by the formula

$$\text{Mean} = \frac{\displaystyle\int_a^b f(x)\,dx}{b - a} \tag{PT6.4}$$

This formula has hundreds of engineering applications. For example, it is used to calculate the center of gravity of irregular objects in mechanical and civil engineering and to determine the root-mean-square current in electrical engineering.

Integrals are also employed by engineers to evaluate the total amount or quantity of a given physical variable. The integral may be evaluated over a line, an area, or a volume. For example, the total mass of chemical contained in a reactor is given as the product of the concentration of chemical and the reactor volume, or

$$\text{Mass} = \text{concentration} \times \text{volume}$$

where concentration has units of mass per volume. However, suppose that concentration varies from location to location within the reactor. In this case, it is necessary to sum the products of local concentrations c_i and corresponding elemental volumes ΔV_i:

$$\text{Mass} = \sum_{i=1}^{n} c_i \, \Delta V_i$$

where n is the number of discrete volumes. For the continuous case, where $c(x, y, z)$ is a known function and x, y, and z are independent variables designating position in Cartesian coordinates, integration can be used for the same purpose:

$$\text{Mass} = \iiint c(x, y, z) \, dx \, dy \, dz$$

or

$$\text{Mass} = \iiint_V c(V) \, dV$$

which is referred to as a *volume integral.* Notice the strong analogy between summation and integration.

Similar examples could be given in other fields of engineering. For example, the total rate of energy transfer across a plane where the flux (in calories per square centimeter per second) is a function of position is given by

$$\text{Heat transfer} = \iint_A \text{flux} \, dA$$

which is referred to as an *areal integral,* where A = area.

Similarly, for the one-dimensional case, the total mass of a variable-density rod with constant cross-sectional area is given by

$$m = A \int_0^L \rho(x) \, dx$$

where m = total weight (kg), L = length of the rod (m), $\rho(x)$ = known density (kg/m^3) as a function of length x (m), and A = cross-sectional area of the rod (m^2).

Finally, integrals are used to evaluate differential or rate equations. Suppose the velocity of a particle is a known continuous function of time $v(t)$,

$$\frac{dy}{dt} = v(t)$$

The total distance y traveled by this particle over a time t is given by (Fig. PT6.3b)

$$y = \int_0^t v(t) \, dt \tag{PT6.5}$$

These are just a few of the applications of differentiation and integration that you might face regularly in the pursuit of your profession. When the functions to be analyzed are simple, you will normally choose to evaluate them analytically. For example, in the falling parachutist problem, we determined the solution for velocity as a function of time [Eq. (1.10)]. This relationship could be substituted into Eq. (PT6.5), which could then be integrated easily to determine how far the parachutist fell over a time period t. For this case, the integral is simple to evaluate. However, it is difficult or impossible when the function is complicated, as is typically the case in more realistic examples. In addition, the underlying function is often unknown and defined only by measurement at discrete points. For both these cases, you must have the ability to obtain approximate values for derivatives and integrals using numerical techniques. Several such techniques will be discussed in this part of the book.

PT6.2　MATHEMATICAL BACKGROUND

In high school or during your first year of college, you were introduced to *differential* and *integral calculus.* There you learned techniques to obtain analytical or exact derivatives and integrals.

When we differentiate a function analytically, we generate a second function that can be used to compute the derivative for different values of the independent variable. General rules are available for this purpose. For example, in the case of the monomial

$$y = x^n$$

the following simple rule applies ($n \neq 0$):

$$\frac{dy}{dx} = nx^{n-1}$$

which is the expression of the more general rule for

$$y = u^n$$

where u = a function of x. For this equation, the derivative is computed via

$$\frac{dy}{dx} = nu^{n-1}\frac{du}{dx}$$

Two other useful formulas apply to the products and quotients of functions. For example, if the product of two functions of x (u and v) is represented as $y = uv$, then the derivative can be computed as

$$\frac{dy}{dx} = u\frac{dv}{dx} + v\frac{du}{dx}$$

For the division $y = u/v$, the derivative can be computed as

$$\frac{dy}{dx} = \frac{v\dfrac{du}{dx} - u\dfrac{dv}{dx}}{v^2}$$

Other useful formulas are summarized in Table PT6.1.

TABLE PT6.1 Some commonly used derivatives.

$$\frac{d}{dx} \sin x = \cos x \qquad\qquad \frac{d}{dx} \cot x = -\csc^2 x$$

$$\frac{d}{dx} \cos x = -\sin x \qquad\qquad \frac{d}{dx} \sec x = \sec x \tan x$$

$$\frac{d}{dx} \tan x = \sec^2 x \qquad\qquad \frac{d}{dx} \csc x = -\csc x \cot x$$

$$\frac{d}{dx} \ln x = \frac{1}{x} \qquad\qquad \frac{d}{dx} \log_a x = \frac{1}{x \ln a}$$

$$\frac{d}{dx} e^x = e^x \qquad\qquad \frac{d}{dx} a^x = a^x \ln a$$

Similar formulas are available for definite integration, which deals with determining an integral between specified limits, as in

$$I = \int_a^b f(x)\, dx \qquad\qquad\qquad\qquad (\text{PT6.6})$$

According to the *fundamental theorem of integral calculus*, Eq. (PT6.6) is evaluated as

$$\int_a^b f(x)\, dx = F(x)\big|_a^b$$

where $F(x)$ = the integral of $f(x)$—that is, any function such that $F'(x) = f(x)$. The nomenclature on the right-hand side stands for

$$F(x)\big|_a^b = F(b) - F(a) \qquad\qquad\qquad (\text{PT6.7})$$

An example of a definite integral is

$$I = \int_0^{0.8} (0.2 + 25x - 200x^2 + 675x^3 - 900x^4 + 400x^5)\, dx \qquad (\text{PT6.8})$$

For this case, the function is a simple polynomial that can be integrated analytically by evaluating each term according to the rule

$$\int_a^b x^n\, dx = \frac{x^{n+1}}{n+1}\bigg|_a^b \qquad\qquad\qquad (\text{PT6.9})$$

where n cannot equal -1. Applying this rule to each term in Eq. (PT6.8) yields

$$I = \left(0.2x + 12.5x^2 - \frac{200}{3}x^3 + 168.75x^4 - 180x^5 + \frac{400}{6}x^6\right)\bigg|_0^{0.8} \qquad (\text{PT6.10})$$

which can be evaluated according to Eq. (PT6.7) as $I = 1.6405333$. This value is equal to the area under the original polynomial [Eq. (PT6.8)] between $x = 0$ and 0.8.

TABLE PT6.2 Some simple integrals that are used in Part Six. The a and b in this table are constants and should not be confused with the limits of integration discussed in the text.

$$\int u \, dv = uv - \int v \, du$$

$$\int u^n \, du = \frac{u^{n+1}}{n+1} + C \qquad n \neq -1$$

$$\int a^{bx} \, dx = \frac{a^{bx}}{b \ln a} + C \qquad a > 0, a \neq 1$$

$$\int \frac{dx}{x} = \ln |x| + C \qquad x \neq 0$$

$$\int \sin (ax + b) \, dx = -\frac{1}{a} \cos (ax + b) + C$$

$$\int \cos (ax + b) \, dx = \frac{1}{a} \sin (ax + b) + C$$

$$\int \ln |x| \, dx = x \ln |x| - x + C$$

$$\int e^{ax} \, dx = \frac{e^{ax}}{a} + C$$

$$\int xe^{ax} \, dx = \frac{e^{ax}}{a^2} (ax - 1) + C$$

$$\int \frac{dx}{a + bx^2} = \frac{1}{\sqrt{ab}} \tan^{-1} \frac{\sqrt{ab}}{a} x + C$$

The foregoing integration depends on knowledge of the rule expressed by Eq. (PT6.9). Other functions follow different rules. These "rules" are all merely instances of *antidifferentiation,* that is, finding $F(x)$ so that $F'(x) = f(x)$. Consequently, analytical integration depends on prior knowledge of the answer. Such knowledge is acquired by training and experience. Many of the rules are summarized in handbooks and in tables of integrals. We list some commonly encountered integrals in Table PT6.2. However, many functions of practical importance are too complicated to be contained in such tables. One reason why the techniques in the present part of the book are so valuable is that they provide a means to evaluate relationships such as Eq. (PT6.8) without knowledge of the rules.

PT6.3   ORIENTATION

Before proceeding to the numerical methods for integration, some further orientation might be helpful. The following is intended as an overview of the material discussed in Part Six. In addition, we have formulated some objectives to help focus your efforts when studying the material.

PT6.3.1 Scope and Preview

Figure PT6.10 provides an overview of Part Six. *Chapter 21* is devoted to the most common approaches for numerical integration—the *Newton-Cotes formulas.* These

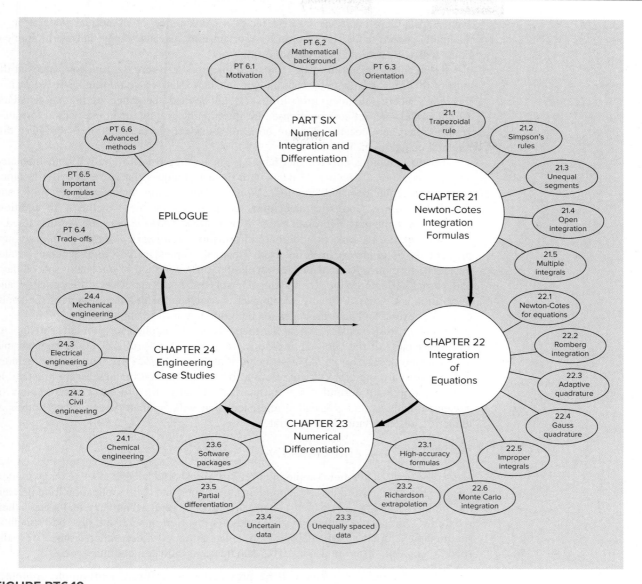

FIGURE PT6.10
Schematic of the organization of material in Part Six: Numerical Integration and Differentiation.

relationships are based on replacing a complicated function or tabulated data with a simple polynomial that is easy to integrate. Three of the most widely used Newton-Cotes formulas are discussed in detail: the *trapezoidal rule, Simpson's 1/3 rule,* and *Simpson's 3/8 rule.* All these formulas are designed for cases where the data to be integrated are evenly spaced. In addition, we also include a discussion of numerical integration of unequally spaced data. This is a very important topic because many real-world applications deal with data that are in this form.

All the above material relates to closed integration, where the function values at the ends of the limits of integration are known. In Chap. 21, we also present *open integration formulas,* where the integration limits extend beyond the range of the known data. Although they are not commonly used for definite integration, open integration formulas are presented here because they are utilized extensively in the solution of ordinary differential equations in Part Seven.

The formulations covered in Chap. 21 can be employed to analyze both tabulated data and equations. *Chapter 22* deals with three techniques that are expressly designed to integrate equations and functions: *Romberg integration, adaptive quadrature,* and *Gauss quadrature.* Computer algorithms are provided for these methods. In addition, methods for evaluating *improper integrals* and *Monte Carlo integration* are discussed.

In *Chap. 23,* we present additional information on *numerical differentiation* to supplement the introductory material from Chap. 4. Topics include high-accuracy finite-difference formulas, Richardson's extrapolation, differentiation of unequally spaced data, and partial differentiation. The effect of errors on both numerical differentiation and integration is discussed. Finally, the chapter is concluded with a description of the application of several software packages for integration and differentiation.

Chapter 24 demonstrates how the methods can be applied for problem solving. As with other parts of the book, applications are drawn from various fields of engineering.

A review section, or *epilogue,* is included at the end of Part Six. This review includes a discussion of trade-offs that are relevant to implementation in engineering practice. In addition, important formulas are summarized. Finally, we present a short review of advanced methods and alternative references that will facilitate your further studies of numerical differentiation and integration.

PT6.3.2 Goals and Objectives

Study Objectives.   After completing Part Six, you should be able to solve many numerical integration and differentiation problems and appreciate their application for engineering problem solving. You should strive to master several techniques and assess their reliability. You should understand the trade-offs involved in selecting the "best" method (or methods) for any particular problem. In addition to these general objectives, the specific concepts listed in Table PT6.3 should be assimilated and mastered.

Computer Objectives.   You will be provided with software and simple computer algorithms to implement the techniques discussed in Part Six. All have utility as learning tools.

Algorithms are provided for most of the other methods in Part Six. This information will allow you to expand your software library to include techniques beyond the trapezoidal rule. For example, you may find it useful from a professional viewpoint to have software to implement numerical integration and differentiation of unequally spaced data. You may also want to develop your own software for Simpson's rules, Romberg integration, adaptive integration, and Gauss quadrature, which are usually more efficient and accurate than the trapezoidal rule.

TABLE PT6.3 Specific study objectives for Part Six.

1. Understand the derivation of the Newton-Cotes formulas; know how to derive the trapezoidal rule and how to set up the derivation of both of Simpson's rules; recognize that the trapezoidal and Simpson's 1/3 and 3/8 rules represent the areas under first-, second-, and third-order polynomials, respectively.
2. Know the formulas and error equations for (*a*) the trapezoidal rule, (*b*) the multiple-application trapezoidal rule, (*c*) Simpson's 1/3 rule, (*d*) Simpson's 3/8 rule, and (*e*) the multiple-application Simpson's rule. Be able to choose the "best" among these formulas for any particular problem context.
3. Recognize that Simpson's 1/3 rule is fourth-order accurate even though it is based on only three points; realize that all the even-segment–odd-point Newton-Cotes formulas have similar enhanced accuracy.
4. Know how to evaluate the integral and derivative of unequally spaced data.
5. Recognize the difference between open and closed integration formulas.
6. Understand how to evaluate multiple integrals numerically.
7. Understand the theoretical basis of Richardson extrapolation and how it is applied in the Romberg integration algorithm and for numerical differentiation.
8. Understand the fundamental difference between Newton-Cotes and Gauss quadrature formulas.
9. Recognize why Romberg integration, adaptive quadrature, and Gauss quadrature all have utility when integrating equations (as opposed to tabular or discrete data).
10. Know how open integration formulas are employed to evaluate improper integrals.
11. Understand how to use random numbers to implement simple Monte Carlo integration.
12. Understand the application of high-accuracy numerical differentiation formulas.
13. Know how to differentiate unequally spaced data.
14. Recognize the differing effects of data error on the processes of numerical integration and differentiation.

Finally, one of your most important goals should be to master several of the general-purpose software packages that are widely available. In particular, you should become adept at using these tools to implement numerical methods for engineering problem solving.

Newton-Cotes Integration Formulas

The *Newton-Cotes formulas* are the most common numerical integration schemes. They are based on the strategy of replacing a complicated function or tabulated data with an approximating function that is easy to integrate:

$$I = \int_a^b f(x)\,dx \cong \int_a^b f_n(x)\,dx \tag{21.1}$$

where $f_n(x)$ is a polynomial of the form

$$f_n(x) = a_0 + a_1 x + \cdots + a_{n-1} x^{n-1} + a_n x^n$$

and n is the order of the polynomial. For example, in Fig. 21.1a, a first-order polynomial (a straight line) is used as an approximation. In Fig. 21.1b, a parabola is employed for the same purpose.

The integral can also be approximated using a series of polynomials applied piece-wise to the function or data over segments of constant length. For example, in Fig. 21.2,

FIGURE 21.1
The approximation of an integral by the area under (*a*) a single straight line and (*b*) a single parabola.

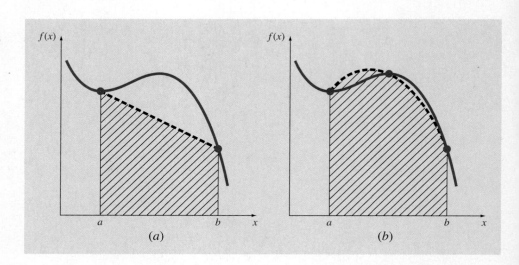

three straight-line segments are used to approximate the integral. Higher-order polynomials can be utilized for the same purpose. With this background, we now recognize that the "strip method" in Fig. PT6.6 employed a series of zero-order polynomials (that is, constants) to approximate the integral.

Closed and open forms of the Newton-Cotes formulas are available. The *closed forms* are those where the data points at the beginning and end of the limits of integration are known (Fig. 21.3a). The *open forms* have integration limits that extend beyond the range of the data (Fig. 21.3b). In this sense, they are akin to extrapolation as discussed in Sec. 18.5. Open Newton-Cotes formulas are not generally used for definite integration.

FIGURE 21.2

The approximation of an integral by the area under three straight-line segments.

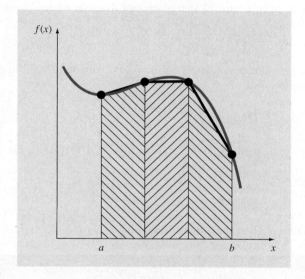

FIGURE 21.3

The difference between (a) closed and (b) open integration formulas.

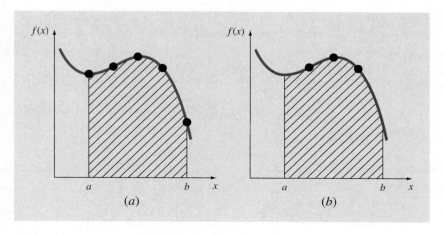

However, they are utilized for evaluating improper integrals and for the solution of ordinary differential equations. This chapter emphasizes the closed forms. However, material on open Newton-Cotes formulas is briefly introduced at the end of this chapter.

21.1 THE TRAPEZOIDAL RULE

The *trapezoidal rule* is the first of the Newton-Cotes closed integration formulas. It corresponds to the case where the polynomial in Eq. (21.1) is first-order:

$$I = \int_a^b f(x)\, dx \cong \int_a^b f_1(x)\, dx$$

Recall from Chap. 18 that a straight line can be represented as [Eq. (18.2)]

$$f_1(x) = f(a) + \frac{f(b) - f(a)}{b - a}(x - a) \tag{21.2}$$

The area under this straight line is an estimate of the integral of $f(x)$ between the limits a and b:

$$I = \int_a^b \left[f(a) + \frac{f(b) - f(a)}{b - a}(x - a) \right] dx$$

The result of the integration (see Box 21.1 for details) is

$$I = (b - a)\frac{f(a) + f(b)}{2} \tag{21.3}$$

which is called the *trapezoidal rule*.

Box 21.1 Derivation of Trapezoidal Rule

Before integration, Eq. (21.2) can be expressed as

$$f_1(x) = \frac{f(b) - f(a)}{b - a}x + f(a) - \frac{af(b) - af(a)}{b - a}$$

Grouping the last two terms gives

$$f_1(x) = \frac{f(b) - f(a)}{b - a}x + \frac{bf(a) - af(a) - af(b) + af(a)}{b - a}$$

or

$$f_1(x) = \frac{f(b) - f(a)}{b - a}x + \frac{bf(a) - af(b)}{b - a}$$

which can be integrated between $x = a$ and $x = b$ to yield

$$I = \frac{f(b) - f(a)}{b - a}\frac{x^2}{2} + \frac{bf(a) - af(b)}{b - a}x \Bigg|_a^b$$

This result can be evaluated to give

$$I = \frac{f(b) - f(a)}{b - a}\frac{(b^2 - a^2)}{2} + \frac{bf(a) - af(b)}{b - a}(b - a)$$

Now, since $b^2 - a^2 = (b - a)(b + a)$,

$$I = [f(b) - f(a)]\frac{b + a}{2} + bf(a) - af(b)$$

Multiplying and collecting terms yields

$$I = (b - a)\frac{f(a) + f(b)}{2}$$

which is the formula for the trapezoidal rule.

Geometrically, the trapezoidal rule is equivalent to approximating the area of the trapezoid under the straight line connecting $f(a)$ and $f(b)$ in Fig. 21.4. Recall from geometry that the formula for computing the area of a trapezoid is the height times the average of the bases (Fig. 21.5a). In our case, the concept is the same but the trapezoid is on its side (Fig. 21.5b). Therefore, the integral estimate can be represented as

$$I \cong \text{width} \times \text{average height} \tag{21.4}$$

FIGURE 21.4
Graphical depiction of the trapezoidal rule.

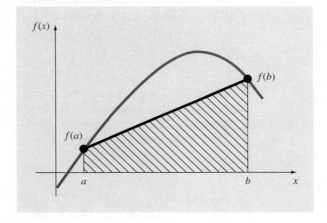

FIGURE 21.5
(*a*) The formula for computing the area of a trapezoid: height times the average of the bases.
(*b*) For the trapezoidal rule, the concept is the same but the trapezoid is on its side.

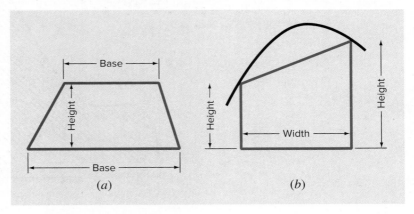

or

$$I \cong (b - a) \times \text{average height} \tag{21.5}$$

where, for the trapezoidal rule, the average height is the average of the function values at the end points, or $[f(a) + f(b)]/2$.

All the Newton-Cotes closed formulas can be expressed in the general format of Eq. (21.5). In fact, they differ only with respect to the formulation of the average height.

21.1.1 Error of the Trapezoidal Rule

When we employ the integral under a straight-line segment to approximate the integral under a curve, we obviously can incur an error that may be substantial (Fig. 21.6). An estimate for the local truncation error of a single application of the trapezoidal rule is (Box. 21.2)

$$E_t = -\frac{1}{12} f''(\xi)(b - a)^3 \tag{21.6}$$

where ξ lies somewhere in the interval from a to b. Equation (21.6) indicates that if the function being integrated is linear, the result from the trapezoidal rule will be exact. Otherwise, for functions with second- and higher-order derivatives (that is, with curvature), some error can occur.

FIGURE 21.6
Graphical depiction of the use of a single application of the trapezoidal rule to approximate the integral of $f(x) = 0.2 + 25x - 200x^2 + 675x^3 - 900x^4 + 400x^5$ from $x = 0$ to 0.8.

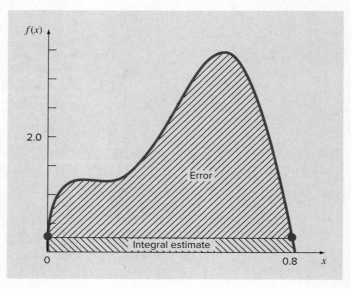

Box 21.2 Derivation and Error Estimate of the Trapezoidal Rule

An alternative derivation of the trapezoidal rule is possible by integrating the forward Newton-Gregory interpolating polynomial. Recall that for the first-order version with an error term, the integral would be (Box 18.2)

$$I = \int_a^b \left[f(a) + \Delta f(a)\alpha + \frac{f''(\xi)}{2}\alpha(\alpha - 1)h^2 \right] dx \quad (B21.2.1)$$

To simplify the analysis, realize that because $\alpha = (x - a)/h$,

$$dx = h\, d\alpha$$

Inasmuch as $h = b - a$ (for the one-segment trapezoidal rule), the limits of integration a and b correspond to 0 and 1, respectively. Therefore, Eq. (B21.2.1) can be expressed as

$$I = h \int_0^1 \left[f(a) + \Delta f(a)\alpha + \frac{f''(\xi)}{2}\alpha(\alpha - 1)h^2 \right] d\alpha$$

If it is assumed that, for small h, the term $f''(\xi)$ is approximately constant, this equation can be integrated:

$$I = h \left[\alpha f(a) + \frac{\alpha^2}{2}\Delta f(a) + \left(\frac{\alpha^3}{6} - \frac{\alpha^2}{4} \right)f''(\xi)h^2 \right]_0^1$$

and evaluated as

$$I = h \left[f(a) + \frac{\Delta f(a)}{2} \right] - \frac{1}{12}f''(\xi)h^3$$

Because $\Delta f(a) = f(b) - f(a)$, the result can be written as

$$I = h\underbrace{\frac{f(a) + f(b)}{2}}_{\text{Trapezoidal rule}} - \underbrace{\frac{1}{12}f''(\xi)h^3}_{\text{Truncation error}}$$

Thus, the first term is the trapezoidal rule and the second is an approximation for the error.

EXAMPLE 21.1   Single Application of the Trapezoidal Rule

Problem Statement.   Use Eq. (21.3) to numerically integrate

$$f(x) = 0.2 + 25x - 200x^2 + 675x^3 - 900x^4 + 400x^5$$

from $a = 0$ to $b = 0.8$. Recall from Sec. PT6.2 that the exact value of the integral can be determined analytically to be 1.640533.

Solution.   The function values

$$f(0) = 0.2$$
$$f(0.8) = 0.232$$

can be substituted into Eq. (21.3) to yield

$$I \cong 0.8\frac{0.2 + 0.232}{2} = 0.1728$$

which represents an error of

$$E_t = 1.640533 - 0.1728 = 1.467733$$

which corresponds to a percent relative error of $\varepsilon_t = 89.5\%$. The reason for this large error is evident from the graphical depiction in Fig. 21.6. Notice that the area under the straight line neglects a significant portion of the integral lying above the line.

In actual situations, we would have no foreknowledge of the true value. Therefore, an approximate error estimate is required. To obtain this estimate, the function's second derivative over the interval can be computed by differentiating the original function twice to give

$$f''(x) = -400 + 4050x - 10{,}800x^2 + 8000x^3$$

The average value of the second derivative can be computed using Eq. (PT6.4):

$$\bar{f}''(x) = \frac{\int_0^{0.8}(-400 + 4050x - 10{,}800x^2 + 8000x^3)\,dx}{0.8 - 0} = -60$$

which can be substituted into Eq. (21.6) to yield

$$E_a = -\frac{1}{12}(-60)(0.8)^3 = 2.56$$

which is of the same order of magnitude and sign as the true error. A discrepancy does exist, however, because of the fact that for an interval of this size, the average second derivative is not necessarily an accurate approximation of $f''(\xi)$. Thus, we denote that the error is approximate by using the notation E_a, rather than using E_t.

21.1.2 The Multiple-Application Trapezoidal Rule

One way to improve the accuracy of the trapezoidal rule is to divide the integration interval from a to b into a number of segments and apply the method to each segment (Fig. 21.7). The areas of individual segments can then be added to yield the integral for the entire interval. The resulting equations are called *multiple-application,* or *composite, integration formulas.*

Figure 21.8 shows the general format and nomenclature we will use to characterize multiple-application integrals. There are $n + 1$ equally spaced base points ($x_0, x_1, x_2, \ldots, x_n$). Consequently, there are n segments of equal width:

$$h = \frac{b - a}{n} \tag{21.7}$$

If a and b are designated as x_0 and x_n, respectively, the total integral can be represented as

$$I = \int_{x_0}^{x_1} f(x)\,dx + \int_{x_1}^{x_2} f(x)\,dx + \cdots + \int_{x_{n-1}}^{x_n} f(x)\,dx$$

Substituting the trapezoidal rule for each integral yields

$$I = h\frac{f(x_0) + f(x_1)}{2} + h\frac{f(x_1) + f(x_2)}{2} + \cdots + h\frac{f(x_{n-1}) + f(x_n)}{2} \tag{21.8}$$

or, grouping terms,

$$I = \frac{h}{2}\left[f(x_0) + 2\sum_{i=1}^{n-1}f(x_i) + f(x_n)\right] \tag{21.9}$$

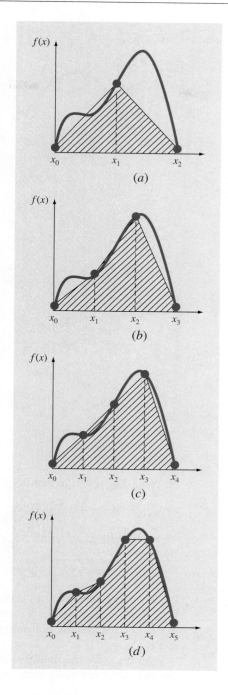

FIGURE 21.7
Illustration of the multiple-application trapezoidal rule. (*a*) Two segments, (*b*) three segments, (*c*) four segments, and (*d*) five segments.

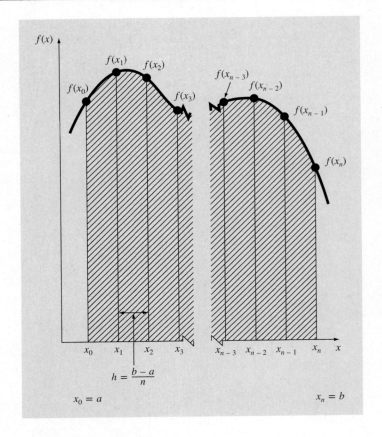

or, using Eq. (21.7) to express Eq. (21.9) in the general form of Eq. (21.5),

$$I = (b - a) \underbrace{\vphantom{\frac{f(x_0) + 2 \sum_{i=1}^{n-1} f(x_i) + f(x_n)}{2n}}}_{\text{Width}} \underbrace{\frac{f(x_0) + 2 \sum_{i=1}^{n-1} f(x_i) + f(x_n)}{2n}}_{\text{Average height}} \tag{21.10}$$

Because the summation of the coefficients of $f(x)$ in the numerator divided by $2n$ is equal to 1, the average height represents a weighted average of the function values. According to Eq. (21.10), the interior points are given twice the weight of the two end points $f(x_0)$ and $f(x_n)$.

An error for the multiple-application trapezoidal rule can be obtained by summing the individual errors for each segment to give

$$E_t = -\frac{(b - a)^3}{12n^3} \sum_{i=1}^{n} f''(\xi_i) \tag{21.11}$$

where $f''(\xi_i)$ is the second derivative at a point ξ_i located in segment i. This result can be simplified by estimating the mean or average value of the second derivative for the entire interval as [Eq. (PT6.3)]:

$$\bar{f}'' \cong \frac{\displaystyle\sum_{i=1}^{n} f''(\xi_i)}{n} \tag{21.12}$$

Therefore, $\Sigma f''(\xi_i) \cong n\bar{f}''$ and Eq. (21.11) can be rewritten as

$$E_a = -\frac{(b-a)^3}{12n^2} \bar{f}'' \tag{21.13}$$

Thus, if the number of segments is doubled, the truncation error will be quartered. Note that Eq. (21.13) gives an approximate error because of the approximate nature of Eq. (21.12).

EXAMPLE 21.2 **Multiple-Application Trapezoidal Rule**

Problem Statement.   Use the two-segment trapezoidal rule to estimate the integral of

$$f(x) = 0.2 + 25x - 200x^2 + 675x^3 - 900x^4 + 400x^5$$

from $a = 0$ to $b = 0.8$. Employ Eq. (21.13) to estimate the error. Recall that the correct value for the integral is 1.640533.

Solution.   For $n = 2$ ($h = 0.4$):

$$f(0) = 0.2 \qquad f(0.4) = 2.456 \qquad f(0.8) = 0.232$$

$$I = 0.8 \frac{0.2 + 2(2.456) + 0.232}{4} = 1.0688$$

$$E_t = 1.640533 - 1.0688 = 0.57173 \qquad \varepsilon_t = 34.9\%$$

$$E_a = -\frac{0.8^3}{12(2)^2}(-60) = 0.64$$

where -60 is the average second derivative determined previously in Example 21.1.

The results of the previous example, along with three- through ten-segment applications of the trapezoidal rule, are summarized in Table 21.1. Notice how the error decreases as the number of segments increases. However, also notice that the rate of decrease is gradual. This is because the error is inversely related to the square of n [Eq. (21.13)]. Therefore, doubling the number of segments quarters the error. In subsequent sections we develop higher-order formulas that are more accurate and that converge more quickly on the true integral as the segments are increased. However, before investigating these formulas, we will first discuss computer algorithms to implement the trapezoidal rule.

21.1.3 Computer Algorithms for the Trapezoidal Rule

Two simple algorithms for the trapezoidal rule are listed in Fig. 21.9. The first (Fig. 21.9a) is for the single-segment version. The second (Fig. 21.9b) is for the multiple-segment

TABLE 21.1 Results for multiple-application trapezoidal rule to estimate the integral of $f(x) = 0.2 + 25x - 200x^2 + 675x^3 - 900x^4 + 400x^5$ from $x = 0$ to 0.8. The exact value is 1.640533.

n	h	I	ε_t (%)
2	0.4	1.0688	34.9
3	0.2667	1.3695	16.5
4	0.2	1.4848	9.5
5	0.16	1.5399	6.1
6	0.1333	1.5703	4.3
7	0.1143	1.5887	3.2
8	0.1	1.6008	2.4
9	0.0889	1.6091	1.9
10	0.08	1.6150	1.6

(a) Single-segment

```
FUNCTION Trap(h, f0, f1)
  Trap = h * (f0 + f1)/2
END Trap
```

(b) Multiple-segment

```
FUNCTION Trapm(h, n, f)
  sum = f0
  DOFOR i = 1, n - 1
    sum = sum + 2 * fi
  END DO
  sum = sum + fn
  Trapm = h * sum / 2
END Trapm
```

FIGURE 21.9
Algorithms for the (a) single-segment and (b) multiple-segment trapezoidal rule.

version with a constant segment width. Note that both are designed for data that are in tabulated form. A general program should have the capability to evaluate known functions or equations as well. We will illustrate how functions are handled in Chap. 22.

EXAMPLE 21.3

Evaluating Integrals with the Computer

Problem Statement.   Use software based on Fig. 21.9b to solve a problem related to our friend, the falling parachutist. As you recall from Example 1.1, the velocity of the parachutist is given as the following function of time:

$$v(t) = \frac{gm}{c}\left(1 - e^{-(c/m)t}\right) \tag{E21.3.1}$$

where v = velocity (m/s); g = the gravitational constant, 9.8 m/s^2; m = mass of the parachutist, equal to 68.1 kg; and c = the drag coefficient, 12.5 kg/s. The model predicts the velocity of the parachutist as a function of time, as described in Example 1.1.

Suppose we would like to know how far the parachutist has fallen after a certain time t. This distance is given by [Eq. (PT6.5)],

$$d = \int_0^t v(t)\, dt$$

where d is the distance in meters. Substituting Eq. (E21.3.1),

$$d = \frac{gm}{c} \int_0^t \left(1 - e^{-(c/m)t}\right) dt$$

Use your software to determine this integral with the multiple-segment trapezoidal rule using different numbers of segments. Note that performing the integration analytically and substituting known parameter values results in an exact value of $d = 289.43515$ m.

Solution.   For the case where $n = 10$ segments, a calculated integral of 288.7491 is obtained. Thus, we have evaluated the integral to three significant digits of accuracy. Results for other numbers of segments can be readily generated.

Segments	Segment Size	Estimated d, m	ε_t (%)
10	1.0	288.7491	0.237
20	0.5	289.2636	0.0593
50	0.2	289.4076	9.5×10^{-3}
100	0.1	289.4282	2.4×10^{-3}
200	0.05	289.4336	5.4×10^{-4}
500	0.02	289.4348	1.2×10^{-4}
1000	0.01	289.4360	-3.0×10^{-4}
2000	0.005	289.4369	-5.9×10^{-4}
5000	0.002	289.4337	5.2×10^{-4}
10,000	0.001	289.4317	1.2×10^{-3}

Up to about 500 segments, the multiple-application trapezoidal rule attains excellent accuracy. However, notice how the error changes sign and begins to increase in absolute value beyond the 500-segment case. The 10,000-segment case actually seems to be diverging from the true value. This is due to the intrusion of round-off error because of the great number of computations for this many segments. Thus, the level of precision is limited, and we would never reach the exact result of 289.4351 obtained analytically. This limitation and ways to overcome it will be discussed in further detail in Chap. 22.

Three major conclusions can be drawn from the Example 21.3:

- For individual applications with nicely behaved functions, the multiple-segment trapezoidal rule is just fine for attaining the type of accuracy required in many engineering applications.
- If high accuracy is required, the multiple-segment trapezoidal rule demands a great deal of computational effort. Although this effort may be negligible for a single application, it could be very important when (*a*) numerous integrals are being evaluated or (*b*) where the function itself is time-consuming to evaluate. For such cases, more efficient approaches (like those in the remainder of this chapter and the next) may be necessary.

- Finally, round-off errors can limit our ability to determine integrals. This is due both to the machine precision as well as to the numerous computations involved in simple techniques like the multiple-segment trapezoidal rule.

We now turn to one way in which efficiency is improved: by using higher-order polynomials to approximate the integral.

21.2   SIMPSON'S RULES

Aside from applying the trapezoidal rule with finer segmentation, another way to obtain a more accurate estimate of an integral is to use higher-order polynomials to connect the points. For example, if there is an extra point midway between $f(a)$ and $f(b)$, the three points can be connected with a parabola (Fig. 21.10a). If there are two points equally spaced between $f(a)$ and $f(b)$, the four points can be connected with a third-order polynomial (Fig. 21.10b). The formulas that result from taking the integrals under these polynomials are called *Simpson's rules*.

21.2.1 Simpson's 1/3 Rule

Simpson's 1/3 rule results when a second-order interpolating polynomial is substituted into Eq. (21.1):

$$I = \int_a^b f(x)\, dx \cong \int_a^b f_2(x)\, dx$$

If a and b are designated as x_0 and x_2 and $f_2(x)$ is represented by a second-order Lagrange polynomial [Eq. (18.23)], the integral becomes

$$I = \int_{x_0}^{x_2} \left[\frac{(x - x_1)(x - x_2)}{(x_0 - x_1)(x_0 - x_2)} f(x_0) + \frac{(x - x_0)(x - x_2)}{(x_1 - x_0)(x_1 - x_2)} f(x_1) \right.$$
$$\left. + \frac{(x - x_0)(x - x_1)}{(x_2 - x_0)(x_2 - x_1)} f(x_2) \right] dx$$

FIGURE 21.10

(a) Graphical depiction of Simpson's 1/3 rule: It consists of taking the area under a parabola connecting three points. (b) Graphical depiction of Simpson's 3/8 rule: It consists of taking the area under a cubic equation connecting four points.

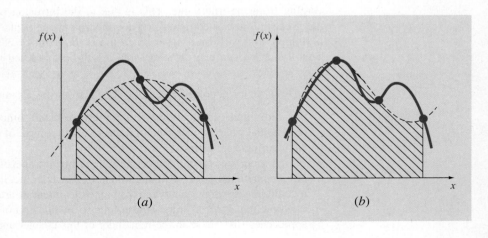

After integration and algebraic manipulation, the following formula results:

$$I \cong \frac{h}{3}\left[f(x_0) + 4f(x_1) + f(x_2)\right] \tag{21.14}$$

where, for this case, $h = (b - a)/2$. This equation is known as *Simpson's 1/3 rule*. It is the second Newton-Cotes closed integration formula. The label "1/3" stems from the fact that h is divided by 3 in Eq. (21.14). An alternative derivation is shown in Box 21.3 where the Newton-Gregory polynomial is integrated to obtain the same formula.

Simpson's 1/3 rule can also be expressed using the format of Eq. (21.5):

$$I \cong \underbrace{(b - a)}_{\text{Width}} \underbrace{\frac{f(x_0) + 4f(x_1) + f(x_2)}{6}}_{\text{Average height}} \tag{21.15}$$

Box 21.3 Derivation and Error Estimate of Simpson's 1/3 Rule

As was done in Box 21.2 for the trapezoidal rule, Simpson's 1/3 rule can be derived by integrating the forward Newton-Gregory interpolating polynomial (Box 18.2):

$$I = \int_{x_0}^{x_2}\left[f(x_0) + \Delta f(x_0)\alpha + \frac{\Delta^2 f(x_0)}{2}\alpha(\alpha - 1)\right.$$
$$+ \frac{\Delta^3 f(x_0)}{6}\alpha(\alpha - 1)(\alpha - 2)$$
$$\left.+ \frac{f^{(4)}(\xi)}{24}\alpha(\alpha - 1)(\alpha - 2)(\alpha - 3)h^4\right]dx$$

Notice that we have written the polynomial up to the fourth-order term rather than the third-order term as would be expected. The reason for this will be apparent shortly. Also notice that the limits of integration are from x_0 to x_2. Therefore, when the simplifying substitutions are made (recall Box 21.2), the integral is from $\alpha = 0$ to 2:

$$I = h\int_0^2\left[f(x_0) + \Delta f(x_0)\alpha + \frac{\Delta^2 f(x_0)}{2}\alpha(\alpha - 1)\right.$$
$$+ \frac{\Delta^3 f(x_0)}{6}\alpha(\alpha - 1)(\alpha - 2)$$
$$\left.+ \frac{f^{(4)}(\xi)}{24}\alpha(\alpha - 1)(\alpha - 2)(\alpha - 3)h^4\right]d\alpha$$

which can be integrated to yield

$$I = h\left[\alpha f(x_0) + \frac{\alpha^2}{2}\Delta f(x_0) + \left(\frac{\alpha^3}{6} - \frac{\alpha^2}{4}\right)\Delta^2 f(x_0)\right.$$
$$+ \left(\frac{\alpha^4}{24} - \frac{\alpha^3}{6} + \frac{\alpha^2}{6}\right)\Delta^3 f(x_0)$$
$$\left.+ \left(\frac{\alpha^5}{120} - \frac{\alpha^4}{16} + \frac{11\alpha^3}{72} - \frac{\alpha^2}{8}\right)f^{(4)}(\xi)h^4\right]_0^2$$

and evaluated for the limits to give

$$I = h\left[2f(x_0) + 2\Delta f(x_0) + \frac{\Delta^2 f(x_0)}{3}\right.$$
$$\left.+ (0)\Delta^3 f(x_0) - \frac{1}{90}f^{(4)}(\xi)h^4\right] \tag{B21.3.1}$$

Notice the significant result that the coefficient of the third divided difference is zero. Because $\Delta f(x_0) = f(x_1) - f(x_0)$ and $\Delta^2 f(x_0) = f(x_2) - 2f(x_1) + f(x_0)$, Eq. (B21.3.1) can be rewritten as

$$I = \underbrace{\frac{h}{3}\left[f(x_0) + 4f(x_1) + f(x_2)\right]}_{\text{Simpson's 1/3}} - \underbrace{\frac{1}{90}f^{(4)}(\xi)h^5}_{\text{Truncation error}}$$

Thus, the first term is Simpson's 1/3 rule and the second is the truncation error. Because the third divided difference dropped out, we obtain the significant result that the formula is third-order accurate.

where $a = x_0$, $b = x_2$, and $x_1 =$ the point midway between a and b, which is given by $(b + a)/2$. Notice that, according to Eq. (21.15), the middle point is weighted by two-thirds and the two end points by one-sixth.

It can be shown that a single-segment application of Simpson's 1/3 rule has a truncation error of (Box 21.3)

$$E_t = -\frac{1}{90}h^5 f^{(4)}(\xi)$$

or, because $h = (b - a)/2$,

$$E_t = -\frac{(b - a)^5}{2880}f^{(4)}(\xi) \tag{21.16}$$

where ξ lies somewhere in the interval from a to b. Thus, Simpson's 1/3 rule is more accurate than the trapezoidal rule. However, comparison with Eq. (21.6) indicates that it is more accurate than expected. Rather than being proportional to the third derivative, the error is proportional to the fourth derivative. This is because, as shown in Box 21.3, the coefficient of the third-order term goes to zero during the integration of the interpolating polynomial. Consequently, Simpson's 1/3 rule is third-order accurate even though it is based on only three points. In other words, it yields exact results for cubic polynomials even though it is derived from a parabola!

EXAMPLE 21.4

Single Application of Simpson's 1/3 Rule

Problem Statement. Use Eq. (21.15) to integrate

$$f(x) = 0.2 + 25x - 200x^2 + 675x^3 - 900x^4 + 400x^5$$

from $a = 0$ to $b = 0.8$. Recall that the exact value is 1.640533.

Solution.

$$f(0) = 0.2 \qquad f(0.4) = 2.456 \qquad f(0.8) = 0.232$$

Therefore, Eq. (21.15) can be used to compute

$$I \cong 0.8\frac{0.2 + 4(2.456) + 0.232}{6} = 1.367467$$

which represents an exact error of

$$E_t = 1.640533 - 1.367467 = 0.2730667 \qquad \varepsilon_t = 16.6\%$$

which is approximately five times more accurate than for a single application of the trapezoidal rule (Example 21.1).

The estimated error is [Eq. (21.16)]

$$E_a = -\frac{(0.8)^5}{2880}(-2400) = 0.2730667$$

where -2400 is the average fourth derivative for the interval, as obtained using Eq. (PT6.4). As was the case in Example 21.1, the error is approximate (E_a) because the average fourth derivative is not an exact estimate of $f^{(4)}(\xi)$. However, because this case deals with a fifth-order polynomial, the result matches.

21.2.2 The Multiple-Application Simpson's 1/3 Rule

Just as with the trapezoidal rule, Simpson's rule can be improved by dividing the integration interval into a number of segments of equal width (Fig. 21.11):

$$h = \frac{b - a}{n} \tag{21.17}$$

The total integral can be represented as

$$I = \int_{x_0}^{x_2} f(x)\, dx + \int_{x_2}^{x_4} f(x)\, dx + \cdots + \int_{x_{n-2}}^{x_n} f(x)\, dx$$

Substituting Simpson's 1/3 rule for the individual integrals yields

$$I \cong 2h\frac{f(x_0) + 4f(x_1) + f(x_2)}{6} + 2h\frac{f(x_2) + 4f(x_3) + f(x_4)}{6}$$

$$+ \cdots + 2h\frac{f(x_{n-2}) + 4f(x_{n-1}) + f(x_n)}{6}$$

or, combining terms and using Eq. (21.17),

$$I \cong \underbrace{(b - a)}_{\text{Width}} \underbrace{\frac{f(x_0) + 4\sum\limits_{i=1,3,5}^{n-1} f(x_i) + 2\sum\limits_{j=2,4,6}^{n-2} f(x_j) + f(x_n)}{3n}}_{\text{Average height}} \tag{21.18}$$

FIGURE 21.11

Graphical representation of the multiple application of Simpson's 1/3 rule. Note that the method can be employed only if the number of segments is even.

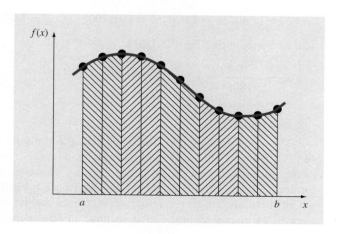

Notice that, as illustrated in Fig. 21.11, an even number of segments must be utilized to implement the method. In addition, the coefficients 4 and 2 in Eq. (21.18) might seem peculiar at first glance. However, they follow naturally from Simpson's 1/3 rule. The odd points represent the middle term for each application and hence carry the weight of 4 from Eq. (21.15). The even points are common to adjacent applications and hence are counted twice.

An error estimate for the multiple-application Simpson's rule is obtained in the same fashion as for the trapezoidal rule by summing the individual errors for the segments and averaging the derivative to yield

$$E_a = -\frac{(b-a)^5}{180n^4} \bar{f}^{(4)} \qquad\qquad (21.19)$$

where $\bar{f}^{(4)}$ is the average fourth derivative for the interval.

EXAMPLE 21.5   **Multiple-Application Version of Simpson's 1/3 Rule**

Problem Statement.   Use Eq. (21.18) with $n = 4$ to estimate the integral of

$$f(x) = 0.2 + 25x - 200x^2 + 675x^3 - 900x^4 + 400x^5$$

from $a = 0$ to $b = 0.8$. Recall that the exact value is 1.640533.

Solution.   For $n = 4$ ($h = 0.2$):

$$f(0) = 0.2 \qquad f(0.2) = 1.288$$
$$f(0.4) = 2.456 \qquad f(0.6) = 3.464$$
$$f(0.8) = 0.232$$

From Eq. (21.18),

$$I = 0.8 \frac{0.2 + 4(1.288 + 3.464) + 2(2.456) + 0.232}{12} = 1.623467$$

$$E_t = 1.640533 - 1.623467 = 0.017067 \qquad \varepsilon_t = 1.04\%$$

The estimated error [Eq. (21.19)] is

$$E_a = -\frac{(0.8)^5}{180(4)^4}(-2400) = 0.017067$$

The previous example illustrates that the multiple-application version of Simpson's 1/3 rule yields very accurate results. For this reason, it is considered superior to the trapezoidal rule for most applications. However, as mentioned previously, it is limited to cases where the values are equispaced. Further, it is limited to situations where there are an even number of segments and an odd number of points. Consequently, as discussed in the next section, an odd-segment–even-point formula known as Simpson's 3/8 rule is used in conjunction with the 1/3 rule to permit evaluation of both even and odd numbers of segments.

21.2.3 Simpson's 3/8 Rule

In a similar manner to the derivation of the trapezoidal and Simpson's 1/3 rule, a third-order Lagrange polynomial can be fit to four points and integrated:

$$I = \int_a^b f(x)\, dx \cong \int_a^b f_3(x)\, dx$$

to yield

$$I \cong \frac{3h}{8}[f(x_0) + 3f(x_1) + 3f(x_2) + f(x_3)]$$

where $h = (b - a)/3$. This equation is called *Simpson's 3/8 rule* because h is multiplied by 3/8. It is the third Newton-Cotes closed integration formula. The 3/8 rule can also be expressed in the form of Eq. (21.5):

$$I \cong \underbrace{(b - a)}_{\text{Width}} \underbrace{\frac{f(x_0) + 3f(x_1) + 3f(x_2) + f(x_3)}{8}}_{\text{Average height}} \tag{21.20}$$

Thus, the two interior points are given weights of three-eighths, whereas the end points are weighted with one-eighth. Simpson's 3/8 rule has an error of

$$E_t = -\frac{3}{80}h^5 f^{(4)}(\xi)$$

or, because $h = (b - a)/3$,

$$E_t = -\frac{(b - a)^5}{6480} f^{(4)}(\xi) \tag{21.21}$$

Because the denominator of Eq. (21.21) is larger than that of Eq. (21.16), the 3/8 rule is somewhat more accurate than the 1/3 rule.

Simpson's 1/3 rule is usually the method of preference because it attains third-order accuracy with three points rather than the four points required for the 3/8 version. However, the 3/8 rule has utility when the number of segments is odd. For instance, in Example 21.5 we used Simpson's rule to integrate the given function for four segments. Suppose that you desired an estimate for five segments. One option would be to use a multiple-application version of the trapezoidal rule as was done in Examples 21.2 and 21.3. This may not be advisable, however, because of the large truncation error associated with this method. An alternative would be to apply Simpson's 1/3 rule to the first two segments and Simpson's 3/8 rule to the last three (Fig. 21.12). In this way, you could obtain an estimate with third-order accuracy across the entire interval.

FIGURE 21.12
Illustration of how Simpson's 1/3 and 3/8 rules can be applied in tandem to handle multiple applications with odd numbers of intervals.

EXAMPLE 21.6

Simpson's 3/8 Rule

Problem Statement.

(a) Use Simpson's 3/8 rule to integrate

$$f(x) = 0.2 + 25x - 200x^2 + 675x^3 - 900x^4 + 400x^5$$

from $a = 0$ to $b = 0.8$.

(b) Use it in conjunction with Simpson's 1/3 rule to integrate the same function for five segments.

Solution.

(a) A single application of Simpson's 3/8 rule requires four equally spaced points:

$$f(0) = 0.2 \qquad\qquad f(0.2667) = 1.432724$$

$$f(0.5333) = 3.487177 \qquad\qquad f(0.8) = 0.232$$

Using Eq. (21.20),

$$I \cong 0.8 \frac{0.2 + 3(1.432724 + 3.487177) + 0.232}{8} = 1.519170$$

$$E_t = 1.640533 - 1.519170 = 0.1213630 \qquad \varepsilon_t = 7.4\%$$

$$E_a = -\frac{(0.8)^5}{6480}(-2400) = 0.1213630$$

(b) The data needed for a five-segment application ($h = 0.16$) is

$$f(0) = 0.2 \qquad f(0.16) = 1.296919$$
$$f(0.32) = 1.743393 \qquad f(0.48) = 3.186015$$
$$f(0.64) = 3.181929 \qquad f(0.80) = 0.232$$

The integral for the first two segments is obtained using Simpson's 1/3 rule:

$$I \cong 0.32 \frac{0.2 + 4(1.296919) + 1.743393}{6} = 0.3803237$$

For the last three segments, the 3/8 rule can be used to obtain

$$I \cong 0.48 \frac{1.743393 + 3(3.186015 + 3.181929) + 0.232}{8} = 1.264754$$

The total integral is computed by summing the two results:

$$I = 0.3803237 + 1.264753 = 1.645077$$
$$E_t = 1.640533 - 1.645077 = -0.00454383 \qquad \varepsilon_t = -0.28\%$$

21.2.4 Computer Algorithms for Simpson's Rules

Pseudocodes for a number of forms of Simpson's rule are outlined in Fig. 21.13. Note that all are designed for data that are in tabulated form. A general program should have the capability to evaluate known functions or equations as well. We will illustrate how functions are handled in Chap. 22.

Notice that the program in Fig. 21.13d is set up so that either an even or odd number of segments may be used. For the even case, Simpson's 1/3 rule is applied to each pair of segments, and the results are summed to compute the total integral. For the odd case, Simpson's 3/8 rule is applied to the last three segments, and the 1/3 rule is applied to all the previous segments.

21.2.5 Higher-Order Newton-Cotes Closed Formulas

As noted previously, the trapezoidal rule and both of Simpson's rules are members of a family of integrating equations known as the Newton-Cotes closed integration formulas. Some of the formulas are summarized in Table 21.2 along with their truncation-error estimates.

Notice that, as was the case with Simpson's 1/3 and 3/8 rules, the five- and six-point formulas have errors of the same order. This general characteristic holds for the higher-point formulas and leads to the result that the even-segment–odd-point formulas (for example, 1/3 rule and *Boole's rule*) are usually the methods of preference.

However, it must also be stressed that, in engineering practice, the higher-order (that is, greater than four-point) formulas are rarely used. Simpson's rules are sufficient for most applications. Accuracy can be improved by using the multiple-application version. Furthermore, when the function is known and high accuracy is required, methods such

(a)
```
FUNCTION Simp13(h, f0, f1, f2)
  Simp13 = 2*h* (f0+4*f1+f2) / 6
END Simp13
```

(b)
```
FUNCTION Simp38(h, f0, f1, f2, f3)
  Simp38 = 3*h* (f0+3*(f1+f2)+f3) / 8
END Simp38
```

(c)
```
FUNCTION Simp13m(h, n, f)
  sum = f(0)
  DOFOR i = 1, n − 2, 2
    sum = sum + 4 * f_i +2 * f_{i+1}
  END DO
  sum = sum + 4 * f_{n-1} + f_n
  Simp13m = h * sum / 3
END Simp13m
```

(d)
```
FUNCTION SimpInt(a,b,n,f)
  h = (b − a) / n
  IF n = 1 THEN
    sum = Trap(h, f_{n-1}, f_n)
  ELSE
    m = n
    odd = n / 2 − INT(n / 2)
    IF odd > 0 AND n > 1 THEN
      sum = sum+Simp38(h, f_{n-3}, f_{n-2}, f_{n-1}, f_n)
      m = n −3
    END IF
    IF m > 1 THEN
      sum = sum + Simp13m(h,m,f)
    END IF
  END IF
  SimpInt = sum
END SimpInt
```

FIGURE 21.13
Pseudocode for Simpson's rules. (a) Single-application Simpson's 1/3 rule, (b) single-application Simpson's 3/8 rule, (c) multiple-application Simpson's 1/3 rule, and (d) multiple-application Simpson's rule for both odd and even number of segments. Note that for all cases, n must be ≥ 1.

TABLE 21.2 Newton-Cotes closed integration formulas. The formulas are presented in the format of Eq. (21.5) so that the weighting of the data points to estimate the average height is apparent. The step size is given by $h = (b − a)/n$.

Segments (n)	Points	Name	Formula	Truncation Error
1	2	Trapezoidal rule	$(b - a)\dfrac{f(x_0) + f(x_1)}{2}$	$-(1/12)h^3 f''(\xi)$
2	3	Simpson's 1/3 rule	$(b - a)\dfrac{f(x_0) + 4f(x_1) + f(x_2)}{6}$	$-(1/90)h^5 f^{(4)}(\xi)$
3	4	Simpson's 3/8 rule	$(b - a)\dfrac{f(x_0) + 3f(x_1) + 3f(x_2) + f(x_3)}{8}$	$-(3/80)h^5 f^{(4)}(\xi)$
4	5	Boole's rule	$(b - a)\dfrac{7f(x_0) + 32f(x_1) + 12f(x_2) + 32f(x_3) + 7f(x_4)}{90}$	$-(8/945)h^7 f^{(6)}(\xi)$
5	6		$(b - a)\dfrac{19f(x_0) + 75f(x_1) + 50f(x_2) + 50f(x_3) + 75f(x_4) + 19f(x_5)}{288}$	$-(275/12{,}096)h^7 f^{(6)}(\xi)$

as Romberg integration or Gauss quadrature, described in Chap. 22, offer viable and attractive alternatives.

21.3 INTEGRATION WITH UNEQUAL SEGMENTS

To this point, all formulas for numerical integration have been based on equally spaced data points. In practice, there are many situations where this assumption does not hold and we must deal with unequal-sized segments. For example, experimentally derived data are often of this type. For these cases, one method is to apply the trapezoidal rule to each segment and sum the results:

$$I = h_1 \frac{f(x_0) + f(x_1)}{2} + h_2 \frac{f(x_1) + f(x_2)}{2} + \cdots + h_n \frac{f(x_{n-1}) + f(x_n)}{2} \qquad (21.22)$$

where h_i = the width of segment i. Note that this was the approach used for the multiple-application trapezoidal rule. The only difference between Eqs. (21.8) and (21.22) is that the h's in the former are constant. Consequently, Eq. (21.8) could be simplified by grouping terms to yield Eq. (21.9). Although this simplification cannot be applied to Eq. (21.22), a computer program can be easily developed to accommodate unequal-sized segments. Before describing such an algorithm, we will illustrate in the following example how Eq. (21.22) is applied to evaluate an integral.

EXAMPLE 21.7

Trapezoidal Rule with Unequal Segments

Problem Statement.   The information in Table 21.3 was generated using the same polynomial employed in Example 21.1. Use Eq. (21.22) to determine the integral for these data. Recall that the correct answer is 1.640533.

Solution.   Applying Eq. (21.22) to the data in Table 21.3 yields

$$I = 0.12 \frac{1.309729 + 0.2}{2} + 0.10 \frac{1.305241 + 1.309729}{2} + \cdots + 0.10 \frac{0.232 + 2.363}{2}$$

$$= 0.090584 + 0.130749 + \cdots + 0.12975 = 1.594801$$

which represents an absolute percent relative error of $\varepsilon_t = 2.8\%$.

TABLE 21.3 Data for $f(x) = 0.2 + 25x - 200x^2 + 675x^3 - 900x^4 + 400x^5$, with unequally spaced values of x.

x	f(x)	x	f(x)
0.0	0.200000	0.44	2.842985
0.12	1.309729	0.54	3.507297
0.22	1.305241	0.64	3.181929
0.32	1.743393	0.70	2.363000
0.36	2.074903	0.80	0.232000
0.40	2.456000		

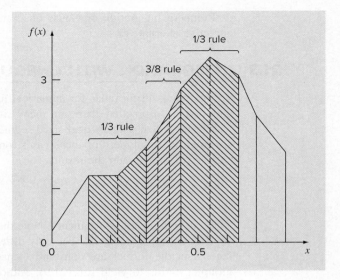

FIGURE 21.14
Use of the trapezoidal rule to determine the integral of unevenly spaced data. Notice how the shaded segments could be evaluated with Simpson's rule to attain higher accuracy.

The data from Example 21.7 are depicted in Fig. 21.14. Notice that some adjacent segments are of equal width and, consequently, could have been evaluated using Simpson's rules. This usually leads to more accurate results, as illustrated by the following example.

EXAMPLE 21.8	**Inclusion of Simpson's Rules in the Evaluation of Uneven Data**

Problem Statement.   Recompute the integral for the data in Table 21.3, but use Simpson's rules for those segments where they are appropriate.

Solution.   The first segment is evaluated with the trapezoidal rule:

$$I = 0.12 \frac{1.309729 + 0.2}{2} = 0.09058376$$

Because the next two segments from $x = 0.12$ to 0.32 are of equal length, their integral can be computed with Simpson's 1/3 rule:

$$I = 0.2 \frac{1.743393 + 4(1.305241) + 1.309729}{6} = 0.2758029$$

The next three segments are also equal and, as such, may be evaluated with the 3/8 rule to give $I = 0.2726863$. Similarly, the 1/3 rule can be applied to the two segments from $x = 0.44$ to 0.64 to yield $I = 0.6684701$. Finally, the last two segments, which are of unequal length, can be evaluated with the trapezoidal rule to give values of 0.1663479 and 0.1297500, respectively. The area of these individual segments can be summed to yield a total integral of 1.603641. This represents an error of $\varepsilon_t = 2.2\%$, which is superior to the result using the trapezoidal rule in Example 21.7.

Computer Program for Unequally Spaced Data. It is a fairly simple proposition to program Eq. (21.22). Such an algorithm is listed in Fig. 21.15a.

However, as demonstrated in Example 21.8, the approach is enhanced if it implements Simpson's rules wherever possible. For this reason, we have developed a second algorithm that incorporates this capability. As depicted in Fig 21.15b, the algorithm checks the lengths of adjacent segments. If two consecutive segments are of equal length, Simpson's 1/3 rule is applied. If three are equal, the 3/8 rule is used. When adjacent segments are of unequal length, the trapezoidal rule is implemented.

FIGURE 21.15

Pseudocode for integrating unequally spaced data. (a) Trapezoidal rule and (b) combination of Simpson's and trapezoidal rules.

(a)
```
FUNCTION Trapun(x, y, n)
  LOCAL i, sum
  sum = 0
  DOFOR i = 1, n
    sum = sum + (xᵢ − xᵢ₋₁)*(yᵢ₋₁ + yᵢ)/2
  END DO
  Trapun = sum
END Trapun
```

(b)
```
FUNCTION Uneven(n,x,f)
  h = x₁ − x₀
  k = 1
  sum = 0.
  DOFOR j = 1, n
    hf = xⱼ₊₁ − xⱼ
    IF ABS (h − hf) < .000001 THEN
      IF k = 3 THEN
        sum = sum + Simp13(h,fⱼ₋₃,fⱼ₋₂,fⱼ₋₁)
        k = k − 1
      ELSE
        k = k + 1
      END IF
    ELSE
      IF k = 1 THEN
        sum = sum + Trap(h,fⱼ₋₁,fⱼ)
      ELSE
        IF k = 2 THEN
          sum = sum + Simp13(h,fⱼ₋₂,fⱼ₋₁,fⱼ)
        ELSE
          sum = sum + Simp38(h,fⱼ₋₃,fⱼ₋₂,fⱼ₋₁,fⱼ)
        END IF
        k = 1
      END IF
    END IF
    h = hf
  END DO
  Uneven = sum
END Uneven
```

Thus, not only does it allow evaluation of unequally spaced data, but if equally spaced information is used, it reduces to using Simpson's rules. As such, it represents a basic, all-purpose algorithm for the determination of the integral of tabulated data.

21.4 OPEN INTEGRATION FORMULAS

Recall from Fig 21.3b that open integration formulas have limits that extend beyond the range of the data. Table 21.4 summarizes the Newton-Cotes open integration formulas. The formulas are expressed in the form of Eq. (21.5) so that the weighting factors are evident. As with the closed versions, successive pairs of the formulas have the same-order error. The even-segment–odd-point formulas are usually the methods of preference because they require fewer points to attain the same accuracy as the odd-segment–even-point formulas.

The open formulas are not often used for definite integration. However, as discussed in Chap. 22, they have utility for analyzing improper integrals. In addition, they will have relevance to our discussion of multistep methods for solving ordinary differential equations in Chap. 26.

TABLE 21.4 Newton-Cotes open integration formulas. The formulas are presented in the format of Eq. (21.5) so that the weighting of the data points to estimate the average height is apparent. The step size is given by $h = (b - a)/n$.

Segments (n)	Points	Name	Formula	Truncation Error
2	1	Midpoint method	$(b - a) f(x_1)$	$(1/3)h^3 f''(\xi)$
3	2		$(b - a) \dfrac{f(x_1) + f(x_2)}{2}$	$(3/4)h^3 f''(\xi)$
4	3		$(b - a) \dfrac{2f(x_1) - f(x_2) + 2f(x_3)}{3}$	$(14/45)h^5 f^{(4)}(\xi)$
5	4		$(b - a) \dfrac{11f(x_1) + f(x_2) + f(x_3) + 11f(x_4)}{24}$	$(95/144)h^5 f^{(4)}(\xi)$
6	5		$(b - a) \dfrac{11f(x_1) - 14f(x_2) + 26f(x_3) - 14f(x_4) + 11f(x_5)}{20}$	$(41/140)h^7 f^{(6)}(\xi)$

21.5 MULTIPLE INTEGRALS

Multiple integrals are widely used in engineering. For example, a general equation to compute the average of a two-dimensional function can be written as (recall Eq. PT6.4)

$$\bar{f} = \frac{\displaystyle\int_c^d \left(\int_a^b f(x, y)dx \right) dy}{(d - c)(b - a)} \tag{21.23}$$

The numerator is called a double integral.

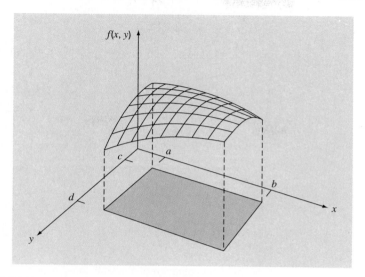

FIGURE 21.16
Double integral as the area under the function surface.

The techniques discussed in this chapter (and the following chapter) can be readily employed to evaluate multiple integrals. A simple example would be to take the double integral of a function over a rectangular area (Fig. 21.16).

Recall from calculus that such integrals can be computed as iterated integrals,

$$\int_c^d \left(\int_a^b f(x, y)\, dx \right) dy = \int_a^b \left(\int_c^d f(x, y)\, dy \right) dx \tag{21.24}$$

Thus, the integral in one of the dimensions is evaluated first. The result of this first integration is integrated in the second dimension. Equation (21.24) states that the order of integration is not important.

A numerical double integral would be based on the same idea. First, methods like the multiple-segment trapezoidal or Simpson's rule would be applied in the first dimension with each value of the second dimension held constant. Then the method would be applied to integrate the second dimension. The approach is illustrated in the following example.

EXAMPLE 21.9   **Using a Double Integral to Determine Average Temperature**

Problem Statement.   Suppose that the temperature of a rectangular heated plate is described by the following function:

$$T(x, y) = 2xy + 2x - x^2 - 2y^2 + 72$$

If the plate is 8-m long (x dimension) and 6-m wide (y dimension), compute the average temperature.

FIGURE 21.17

Numerical evaluation of a double integral using the two-segment trapezoidal rule.

Solution.   First, let us merely use two-segment applications of the trapezoidal rule in each dimension. The temperatures at the necessary x and y values are depicted in Fig. 21.17. Note that a simple average of these values is 47.33. The function can also be evaluated analytically to yield a result of 58.66667.

To make the same evaluation numerically, the trapezoidal rule is first implemented along the x dimension for each y value. These values are then integrated along the y dimension to give the final result of 2544. Dividing this by the area yields the average temperature as $2544/(6 \times 8) = 53$.

Now we can apply a single-segment Simpson's 1/3 rule in the same fashion. This results in an integral value of 2816 and an average of 58.66667, which is exact. Why does this occur? Recall that Simpson's 1/3 rule yielded perfect results for cubic polynomials. Since the highest-order term in the function is second-order, the same exact result occurs for the present case.

For higher-order algebraic functions as well as transcendental functions, it would be necessary to use multisegment applications to attain accurate integral estimates. In addition, Chap. 22 introduces techniques that are more efficient than the Newton-Cotes formulas for evaluating integrals of given functions. These often provide a superior means to implement the numerical integrations for multiple integrals.

PROBLEMS

21.1 Evaluate the following integral:

$$\int_0^{\pi/2} (8 + 4 \cos x)\, dx$$

(a) analytically and then by **(b)** single application of the trapezoidal rule; **(c)** multiple-application trapezoidal rule, with $n = 2$ and 4; **(d)** single application of Simpson's 1/3 rule; **(e)** multiple-application

Simpson's 1/3 rule, with $n = 4$; **(f)** single application of Simpson's 3/8 rule; and **(g)** multiple-application Simpson's rule, with $n = 5$. For each of the numerical estimates **(b)** through **(g)**, determine the percent relative error based on **(a)**.

21.2 Evaluate the following integral:

$$\int_0^3 (1 - e^{-x})\, dx$$

(a) analytically and then by **(b)** single application of the trapezoidal rule; **(c)** multiple-application trapezoidal rule, with $n = 2$ and 4; **(d)** single application of Simpson's 1/3 rule; **(e)** multiple-application Simpson's 1/3 rule, with $n = 4$; **(f)** single application of Simpson's 3/8 rule; and **(g)** multiple-application Simpson's rule, with $n = 5$. For each of the numerical estimates **(b)** through **(g)**, determine the percent relative error based on **(a)**.

21.3 Evaluate the following integral:

$$\int_{-2}^{4} (1 - x - 4x^3 + 2x^5)\, dx$$

(a) analytically and then by **(b)** single application of the trapezoidal rule; **(c)** composite trapezoidal rule, with $n = 2$ and 4; **(d)** single application of Simpson's 1/3 rule; **(e)** Simpson's 3/8 rule; and **(f)** Boole's rule. For each of the numerical estimates **(b)** through **(f)** determine the percent relative error based on **(a)**.

21.4 Integrate the following function analytically and using the trapezoidal rule, with $n = 1, 2, 3,$ and 4:

$$\int_{1}^{2} (x + 1/x)^2\, dx$$

Use the analytical solution to compute true percent relative errors to evaluate the accuracy of the trapezoidal approximations.

21.5 Integrate the following function both analytically and using Simpson's rules, with $n = 4$ and 5. Discuss the results.

$$\int_{-3}^{5} (4x - 3)^3\, dx$$

21.6 Integrate the following function both analytically and numerically. Use both the trapezoidal and Simpson's 1/3 rules to numerically integrate the function. For both cases, use the multiple-application version, with $n = 4$. Compute percent relative errors for the numerical results.

$$\int_{0}^{3} x^2 e^x\, dx$$

21.7 Integrate the following function both analytically and numerically. For the numerical evaluations use **(a)** a single application of the trapezoidal rule, **(b)** Simpson's 1/3 rule, **(c)** Simpson's 3/8 rule, **(d)** Boole's rule, **(e)** the midpoint method, **(f)** the 3-segment–2-point open integration formula, and **(g)** the 4-segment–3-point open integration formula. Compute percent relative errors for the numerical results.

$$\int_{0}^{1} 15^{2x}\, dx$$

21.8 Integrate the following function both analytically and numerically. For the numerical evaluations use **(a)** single application of the trapezoidal rule; **(b)** Simpson's 1/3 rule; **(c)** Simpson's 3/8 rule; **(d)** multiple application of Simpson's rules, with $n = 5$; **(e)** Boole's

rule; **(f)** the midpoint method; **(g)** the 3-segment–2-point open integration formula; and **(h)** the 4-segment–3-point open integration formula.

$$\int_{0}^{3} (5 + 3 \cos x)\, dx$$

Compute percent relative errors for the numerical results.

21.9 Suppose that the upward force of air resistance on a falling object is proportional to the square of the velocity. For this case, the velocity can be computed as

$$v(t) = \sqrt{\frac{gm}{c_d}} \tanh\left(\sqrt{\frac{gc_d}{m}}\, t\right)$$

where c_d = a second-order drag coefficient. **(a)** If $g = 9.81$ m/s^2, $m = 68.1$ kg, and $c_d = 0.25$ kg/m, use analytical integration to determine how far the object falls in 10 s. **(b)** Make the same evaluation, but evaluate the integral with the multiple-segment trapezoidal rule. Use a sufficiently high n that you get three significant digits of accuracy.

21.10 Evaluate the integral of the following tabular data with **(a)** the trapezoidal rule and **(b)** Simpson's rules:

x	0	0.1	0.2	0.3	0.4	0.5
f(x)	1	8	4	3.5	6	1

21.11 Evaluate the integral of the following tabular data with **(a)** the trapezoidal rule and **(b)** Simpson's rules:

x	−2	0	2	4	6	8	10
f(x)	35	5	−10	2	5	3	20

21.12 Determine the mean value of the function

$$f(x) = -46 + 45x - 13.8x^2 + 1.71x^3 - 0.0729x^4$$

between $x = 2$ and 10 by **(a)** graphing the function and visually estimating the mean value, **(b)** using Eq. (PT6.4) and the analytical evaluation of the integral, and **(c)** using Eq. (PT6.4) and a five-segment version of Simpson's rule to estimate the integral. Calculate the relative percent error.

21.13 The function $f(x) = 2e^{-1.5x}$ can be used to generate the following table of unequally spaced data:

x	0	0.05	0.15	0.25	0.35	0.475	0.6
f(x)	2	1.8555	1.5970	1.3746	1.1831	0.9808	0.8131

Evaluate the integral from $a = 0$ to $b = 0.6$ using **(a)** analytical means, **(b)** the trapezoidal rule, and **(c)** a combination of the trapezoidal and Simpson's rules; employ Simpson's rules wherever possible to obtain the highest accuracy. For **(b)** and **(c)**, compute the percent relative error (ε_t).

21.14 Evaluate the following double integral:

$$\int_{-2}^{2} \int_{0}^{4} (x^2 - 3y^2 + xy^3)\, dx\, dy$$

(a) analytically; (b) using a multiple-application trapezoidal rule, with $n = 2$; and (c) using single applications of Simpson's 1/3 rule. For (b) and (c), compute the percent relative error (ε_t).

21.15 Evaluate the following triple integral (a) analytically and (b) using single applications of Simpson's 1/3 rule. For (b) compute the percent relative error (ε_t).

$$\int_{-4}^{4} \int_{0}^{6} \int_{-1}^{3} (x^3 - 2yz) \, dx \, dy \, dz$$

21.16 Develop a user-friendly computer program for the multiple-application trapezoidal rule based on Fig. 21.9. Test your program by duplicating the computation from Example 21.2.

21.17 Develop a user-friendly computer program for the multiple-application version of Simpson's rule based on Fig. 21.13c. Test it by duplicating the computations from Example 21.5.

21.18 Develop a user-friendly computer program for integrating unequally spaced data based on Fig. 21.15b. Test it by duplicating the computation from Example 21.8.

21.19 The following data were collected for a cross section of a river (y = distance from bank, H = depth, and U = velocity):

y, m	0	1	3	5	7	8	9	10
H, m	0	1	1.5	3	3.5	3.2	2	0
U, m/s	0	0.1	0.12	0.2	0.25	0.3	0.15	0

Use numerical integration to compute the (a) average depth, (b) cross-sectional area, (c) average velocity, and (d) the flow rate. Note that the cross-sectional area (A_c) and the flow rate (Q) can be computed as

$$A_c = \int_{0}^{y} H(y) \, dy \qquad Q = \int_{0}^{y} H(y)U(y) \, dy$$

21.20 The outflow concentration from a reactor is measured at a number of times over a 24-hr period:

t, hr	0	1	5.5	10	12	14	16	18	20	24
c, mg/L	1	1.5	2.3	2.1	4	5	5.5	5	3	1.2

The flow rate for the outflow in m^3/s can be computed with the following equation:

$$Q(t) = 20 + 10 \sin\left(\frac{2\pi}{24}(t - 10)\right)$$

Use the best numerical integration method to determine the flow-weighted average concentration leaving the reactor over the 24-hr period,

$$\bar{c} = \frac{\int_{0}^{t} Q(t)c(t) \, dt}{\int_{0}^{t} Q(t) \, dt}$$

21.21 An 11-m beam is subjected to a load, and the shear force is given by the equation

$$V(x) = 5 + 0.25x^2$$

where V is the shear force and x is length in distance along the beam. We know that $V = dM/dx$, and M is the bending moment. Integration yields the relationship

$$M = M_0 + \int_{0}^{x} V \, dx$$

If M_0 is zero and $x = 11$, calculate M using (a) analytical integration, (b) multiple-application trapezoidal rule, and (c) multiple-application Simpson's rules. For (b) and (c), use 1-m increments.

21.22 The work produced by a constant temperature, pressure-volume thermodynamic process can be computed as

$$W = \int p \, dV$$

where W is work, p is pressure, and V is volume. Using a combination of the trapezoidal rule, Simpson's 1/3 rule, and Simpson's 3/8 rule, use the following data to compute the work in kJ (kJ = kN m):

Pressure (kPa)	336	294.4	266.4	260.8	260.5	249.6	193.6	165.6
Volume (m^3)	0.5	2	3	4	6	8	10	11

21.23 Determine the distance traveled for the following data:

t, min	1	2	3.25	4.5	6	7	8	9	9.5	10
v, m/s	5	6	5.5	7	8.5	8	6	7	7	5

(a) Use the trapezoidal rule, (b) use the best combination of the trapezoidal and Simpson's rules, and (c) analytically integrate second- and third-order polynomials determined by regression.

21.24 The total mass of a variable-density rod is given by

$$m = \int_{0}^{L} \rho(x)A_c(x) \, dx$$

where m = mass, $\rho(x)$ = density, $A_c(x)$ = cross-sectional area, x = distance along the rod, and L = the total length of the rod. The following data have been measured for a 10-m length rod. Determine the mass in kilograms to the best possible accuracy.

x, m	0	2	3	4	6	8	10
ρ, g/cm^3	4.00	3.95	3.89	3.80	3.60	3.41	3.30
A_c, cm^2	100	103	106	110	120	133	150

21.25 A transportation engineering study requires that you determine the number of cars that pass through an intersection during morning rush hour. You stand at the side of the road and count the number of cars that pass every 4 minutes at several times as tabulated below. Use the best numerical method to determine **(a)** the total number of cars that pass between 7:30 and 9:15, and **(b)** the rate of cars going through the intersection per minute. (*Hint:* Be careful with units.)

Time (hr)	7:30	7:45	8:00	8:15	8:45	9:15
Rate (cars per 4 min)	18	24	14	24	21	9

21.26 Determine the average value for the data in Fig. P21.26. Perform the integral needed for the average in the order shown by the following equation:

$$I = \int_{x_0}^{x_n} \left[\int_{y_0}^{y_m} f(x, y) \, dy \right] dx$$

FIGURE P21.26

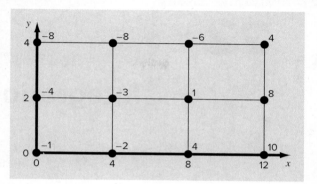

Integration of Equations

In the introduction to Part Six, we noted that functions to be integrated numerically will typically be of two forms: a table of values or a function. The form of the data has an important influence on the approaches that can be used to evaluate the integral. For tabulated information, you are limited by the number of points that are given. In contrast, if the function is available, you can generate as many values of $f(x)$ as are required to attain acceptable accuracy (recall Fig. PT6.7).

This chapter is devoted to three techniques that are expressly designed to analyze cases where the function is given. Each capitalizes on the ability to generate function values to develop efficient schemes for numerical integration. The first is based on *Richardson extrapolation,* which is a method for combining two numerical integral estimates to obtain a third, more accurate value. The computational algorithm for implementing Richardson extrapolation in a highly efficient manner is called *Romberg integration.* This technique is recursive and can be used to generate an integral estimate within a prespecified error tolerance.

The second method, *adaptive integration,* is based on dividing the integration interval into successively more refined subintervals in a recursive fashion. Thus, more refined spacing is employed where the function varies rapidly and coarser spacing used where the function varies slowly in order to attain a desired global accuracy with the least computational effort.

The third method is called *Gauss quadrature.* Recall that, in the last chapter, values of $f(x)$ for the Newton-Cotes formulas were determined at specified values of x. For example, if we used the trapezoidal rule to determine an integral, we were constrained to take the weighted average of $f(x)$ at the ends of the interval. Gauss-quadrature formulas employ x values that are positioned between a and b in such a manner that a much more accurate integral estimate results.

In addition to these three standard techniques, we devote a section to the evaluation of *improper integrals.* In this discussion, we focus on integrals with infinite limits and show how a change of variable and open integration formulas prove useful for such cases. Finally, we discuss Monte Carlo integration, a technique that employs random numbers to evaluate definite integrals.

22.1 NEWTON-COTES ALGORITHMS FOR EQUATIONS

In Chap. 21, we presented algorithms for multiple-application versions of the trapezoidal rule and Simpson's rules. Although these pseudocodes can certainly be used to analyze

(a)

```
FUNCTION TrapEq (n, a, b)
  h = (b - a) / n
  x = a
  sum = f(x)
  DOFOR i = 1, n - 1
    x = x + h
    sum = sum + 2 * f(x)
  END DO
  sum = sum + f(b)
  TrapEq = (b - a) * sum / (2 * n)
END TrapEq
```

(b)

```
FUNCTION SimpEq (n, a, b)
  h = (b - a) / n
  x = a
  sum = f(x)
  DOFOR i = 1, n - 2, 2
    x = x + h
    sum = sum + 4 * f(x)
    x = x + h
    sum = sum + 2 * f(x)
  END DO
  x = x + h
  sum = sum + 4 * f(x)
  sum = sum + f(b)
  SimpEq = (b - a) * sum /(3 * n)
END SimpEq
```

FIGURE 22.1

Algorithms for multiple applications of the (*a*) trapezoidal and (*b*) Simpson's 1/3 rules, where the function is available.

equations, in our effort to make them compatible with either data or functions, they could not exploit the convenience of the latter.

Figure 22.1 shows pseudocodes that are expressly designed for cases where the function is analytical. In particular, notice that neither the independent nor the dependent variable values are passed into the function via its argument as was the case for the codes in Chap. 21. For the independent variable x, the integration interval (a, b) and the number of segments are passed. This information is then employed to generate equispaced values of x within the function. For the dependent variable, the function values in Fig. 22.1 are computed using calls to the function being analyzed, $f(x)$.

We developed single-precision programs based on these pseudocodes to analyze the effort involved and the errors incurred as we progressively used more segments to estimate the integral of a simple function. For an analytical function, the error equations [Eqs. (21.13) and (21.19)] indicate that increasing the number of segments n will result in more accurate integral estimates. This observation is borne out by Fig. 22.2, which is a plot of true error versus n for the integral of $f(x) = 0.2 + 25x - 200x^2 + 675x^3 - 900x^4 + 400x^5$. Notice how the error drops as n increases. However, also notice that at large values of n, the error starts to increase as round-off errors begin to dominate. Also observe that a very large number of function evaluations (and, hence, computational effort) is required to attain high levels of accuracy. As a consequence of these shortcomings, the multiple-application trapezoidal rule and Simpson's rules are sometimes inadequate for problem contexts where high efficiency and low errors are needed.

22.2 ROMBERG INTEGRATION

Romberg integration is one technique that is designed to attain efficient numerical integrals of functions. It is quite similar to the techniques discussed in Chap. 21 in the sense that it is based on successive application of the trapezoidal rule. However, through mathematical manipulations, superior results are attained for less effort.

22.2.1 Richardson Extrapolation

Recall that, in Sec. 10.3.3, we used iterative refinement to improve the solution of a set of simultaneous linear equations. Error-correction techniques are also available to improve the results of numerical integration on the basis of the integral estimates themselves. Generally called *Richardson extrapolation,* these methods use two estimates of an integral to compute a third, more accurate approximation.

The estimate and error associated with a multiple-application trapezoidal rule can be represented generally as

$$I = I(h) + E(h)$$

where I = the exact value of the integral, $I(h)$ = the approximation from an n-segment application of the trapezoidal rule with step size $h = (b - a)/n$, and $E(h)$ = the truncation error. If we make two separate estimates using step sizes of h_1 and h_2 and have exact values for the error,

$$I(h_1) + E(h_1) = I(h_2) + E(h_2) \tag{22.1}$$

FIGURE 22.2

Absolute value of the true percent relative error versus number of segments for the determination of the integral of $f(x) = 0.2 + 25x - 200x^2 + 675x^3 - 900x^4 + 400x^5$, evaluated from $a = 0$ to $b = 0.8$ using the multiple-application trapezoidal rule and the multiple-application Simpson's 1/3 rule. Note that both results indicate that for a large number of segments, round-off errors limit precision.

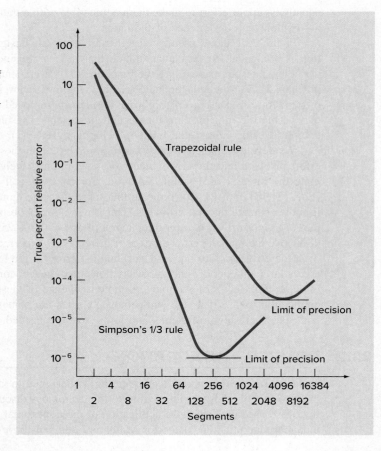

Now recall that the error of the multiple-application trapezoidal rule can be represented approximately by Eq. (21.13) [with $n = (b - a)/h$]:

$$E \cong -\frac{b - a}{12} h^2 \bar{f}''$$

(22.2)

If it is assumed that $\bar{f}''$ is constant regardless of step size, Eq. (22.2) can be used to determine that the ratio of the two errors will be

$$\frac{E(h_1)}{E(h_2)} \cong \frac{h_1^2}{h_2^2}$$

(22.3)

This calculation has the important effect of removing the term $\bar{f}''$ from the computation. In so doing, we have made it possible to utilize the information embodied by Eq. (22.2) without prior knowledge of the function's second derivative. To do this, we rearrange Eq. (22.3) to give

$$E(h_1) \cong E(h_2)\left(\frac{h_1}{h_2}\right)^2$$

which can be substituted into Eq. (22.1):

$$I(h_1) + E(h_2)\left(\frac{h_1}{h_2}\right)^2 \cong I(h_2) + E(h_2)$$

which can be solved for

$$E(h_2) \cong \frac{I(h_1) - I(h_2)}{1 - (h_1/h_2)^2}$$

Thus, we have developed an estimate of the truncation error in terms of the integral estimates and their step sizes. This estimate can then be substituted into

$$I = I(h_2) + E(h_2)$$

to yield an improved estimate of the integral:

$$I \cong I(h_2) + \frac{1}{(h_1/h_2)^2 - 1}[I(h_2) - I(h_1)]$$

(22.4)

It can be shown (Ralston and Rabinowitz 1978) that the error of this estimate is $O(h^4)$. Thus, we have combined two trapezoidal rule estimates of $O(h^2)$ to yield a new estimate of $O(h^4)$. For the special case where the interval is halved ($h_2 = h_1/2$), this equation becomes

$$I \cong I(h_2) + \frac{1}{2^2 - 1}[I(h_2) - I(h_1)]$$

or, collecting terms,

$$I \cong \frac{4}{3}I(h_2) - \frac{1}{3}I(h_1)$$

(22.5)

EXAMPLE 22.1   Error Corrections of the Trapezoidal Rule

Problem Statement.   In the previous chapter (Example 21.1 and Table 21.1), we used a variety of numerical integration methods to evaluate the integral of $f(x) = 0.2 + 25x -$

$200x^2 + 675x^3 - 900x^4 + 400x^5$ from $a = 0$ to $b = 0.8$. For example, single and multiple applications of the trapezoidal rule yielded the following results:

Segments	h	Integral	ε_t, %
1	0.8	0.1728	89.5
2	0.4	1.0688	34.9
4	0.2	1.4848	9.5

Use this information along with Eq. (22.5) to compute improved estimates of the integral.

Solution. The estimates for one and two segments can be combined to yield

$$I \cong \frac{4}{3}(1.0688) - \frac{1}{3}(0.1728) = 1.367467$$

The error of the improved integral is $E_t = 1.640533 - 1.367467 = 0.273067$ ($\varepsilon_t = 16.6\%$), which is superior to the estimates upon which it was based.

In the same manner, the estimates for two and four segments can be combined to give

$$I \cong \frac{4}{3}(1.4848) - \frac{1}{3}(1.0688) = 1.623467$$

which represents an error of $E_t = 1.640533 - 1.623467 = 0.017067$ ($\varepsilon_t = 1.0\%$).

Equation (22.4) provides a way to combine two applications of the trapezoidal rule with error $O(h^2)$ to compute a third estimate with error $O(h^4)$. This approach is a subset of a more general method for combining integrals to obtain improved estimates. For instance, in Example 22.1, we computed two improved integrals of $O(h^4)$ on the basis of three trapezoidal rule estimates. These two improved estimates can, in turn, be combined to yield an even better value with $O(h^6)$. For the special case where the original trapezoidal estimates are based on successive halving of the step size, the equation used for $O(h^6)$ accuracy is

$$I \cong \frac{16}{15}I_m - \frac{1}{15}I_l \tag{22.6}$$

where I_m and I_l are the more and less accurate estimates, respectively. Similarly, two $O(h^6)$ results can be combined to compute an integral that is $O(h^8)$ using

$$I \cong \frac{64}{63}I_m - \frac{1}{63}I_l \tag{22.7}$$

EXAMPLE 22.2 **Higher-Order Error Correction of Integral Estimates**

Problem Statement. In Example 22.1, we used Richardson extrapolation to compute two integral estimates of $O(h^4)$. Utilize Eq. (22.6) to combine these estimates to compute an integral with $O(h^6)$.

Solution. The two integral estimates of $O(h^4)$ obtained in Example 22.1 were 1.367467 and 1.623467. These values can be substituted into Eq. (22.6) to yield

$$I = \frac{16}{15}(1.623467) - \frac{1}{15}(1.367467) = 1.640533$$

which is the correct answer to the seven significant figures that are carried in this example.

22.2.2 The Romberg Integration Algorithm

Notice that the coefficients in each of the extrapolation equations [Eqs. (22.5), (22.6), and (22.7)] add up to 1. Thus, they represent weighting factors that, as accuracy increases, place relatively greater weight on the superior integral estimate. These formulations can be expressed in a general form that is well-suited for computer implementation:

$$I_{j,k} \cong \frac{4^{k-1} I_{j+1,k-1} - I_{j,k-1}}{4^{k-1} - 1} \tag{22.8}$$

where $I_{j+1,k-1}$ and $I_{j,k-1}$ = the more and less accurate integrals, respectively, and $I_{j,k}$ = the improved integral. The index k signifies the level of the integration, where $k = 1$ corresponds to the original trapezoidal rule estimates, $k = 2$ corresponds to $O(h^4)$, $k = 3$ to $O(h^6)$, and so forth. The index j is used to distinguish between the more $(j + 1)$ and the less (j) accurate estimates. For example, for $k = 2$ and $j = 1$, Eq. (22.8) becomes

$$I_{1,2} \cong \frac{4I_{2,1} - I_{1,1}}{3}$$

which is equivalent to Eq. (22.5).

The general form represented by Eq. (22.8) is attributed to Romberg, and its systematic application to evaluate integrals is known as *Romberg integration.* Figure 22.3 is a

FIGURE 22.3
Graphical depiction of the sequence of integral estimates generated using Romberg integration. (*a*) First iteration. (*b*) Second iteration. (*c*) Third iteration.

graphical depiction of the sequence of integral estimates generated using this approach. Each matrix corresponds to a single iteration. The first column contains the trapezoidal rule evaluations that are designated $I_{j,1}$, where $j = 1$ is for a single-segment application (step size is $b - a$), $j = 2$ is for a two-segment application [step size is $(b - a)/2$], $j = 3$ is for a four-segment application [step size is $(b - a)/4$], and so forth. The other columns of the matrix are generated by systematically applying Eq. (22.8) to obtain successively better estimates of the integral.

For example, the first iteration (Fig. 22.3a) involves computing the one- and two-segment trapezoidal rule estimates ($I_{1,1}$ and $I_{2,1}$). Equation (22.8) is then used to compute the element $I_{1,2} = 1.367467$, which has an error of $O(h^4)$.

Now, we must check to determine whether this result is adequate for our needs. As in other approximate methods in this book, a termination, or stopping, criterion is required to assess the accuracy of the results. One method that can be employed for the present purposes is [Eq. (3.5)]

$$|\varepsilon_a| = \left| \frac{I_{1,k} - I_{2,k-1}}{I_{1,k}} \right| 100\% \tag{22.9}$$

where ε_a = an estimate of the percent relative error. Thus, as was done previously in other iterative processes, we compare the new estimate with a previous value. When the change between the old and new values as represented by ε_a is below a prespecified error criterion ε_s, the computation is terminated. For Fig. 22.3a, this evaluation indicates an 21.8 percent change over the course of the first iteration.

The object of the second iteration (Fig. 22.3b) is to obtain the $O(h^6)$ estimate—$I_{1,3}$. To do this, an additional trapezoidal rule estimate, $I_{3,1} = 1.4848$, is determined. Then it is combined with $I_{2,1}$ using Eq. (22.8) to generate $I_{2,2} = 1.623467$. The result is, in turn, combined with $I_{1,2}$ to yield $I_{1,3} = 1.640533$. Equation (22.9) can be applied to determine that this result represents a change of 1.0 percent when compared with the previous result $I_{1,2}$.

The third iteration (Fig. 22.3c) continues the process in the same fashion. In this case, a trapezoidal estimate is added to the first column, and then Eq. (22.8) is applied to compute successively more accurate integrals along the lower diagonal. After only three iterations, because we are evaluating a fifth-order polynomial, the result ($I_{1,4} = 1.640533$) is exact.

Romberg integration is more efficient than the trapezoidal rule and Simpson's rules discussed in Chap. 21. For example, for determination of the integral as shown in Fig. 22.1, Simpson's 1/3 rule would require a 256-segment application to yield an estimate of 1.640533. Finer approximations would not be possible because of round-off error. In contrast, Romberg integration yields an exact result (to seven significant figures) based on combining one-, two-, four-, and eight-segment trapezoidal rules; that is, with only 15 function evaluations!

Figure 22.4 presents pseudocode for Romberg integration. By using loops, this algorithm implements the method in an efficient manner. Romberg integration is designed for cases where the function to be integrated is known. This is because knowledge of the function permits the evaluations required for the initial implementations of the trapezoidal rule. Tabulated data are rarely in the form needed to make the necessary successive halvings.

```
FUNCTION Romberg (a, b, maxit, es)
  LOCAL I(10, 10)
  n = 1
  I₁,₁ = TrapEq(n, a, b)
  iter = 0
  DO
    iter = iter + 1
    n = 2^iter
    I_iter+1,1 = TrapEq(n, a, b)
    DOFOR k = 2, iter + 1
      j = 2 + iter - k
      I_j,k = (4^(k-1) * I_j+1,k-1 - I_j,k-1) / (4^(k-1) - 1)
    END DO
    ea = ABS((I₁,iter+1 - I₂,iter) / I₁,iter+1) * 100
    IF (iter ≥ maxit OR ea ≤ es) EXIT
  END DO
  Romberg = I₁,iter+1
END Romberg
```

FIGURE 22.4
Pseudocode for Romberg integration that uses the equal-size-segment version of the trapezoidal rule from Fig. 22.1.

22.3 ADAPTIVE QUADRATURE

Although Romberg integration is more efficient than the composite Simpson's 1/3 rule, both use equally spaced points. This global perspective ignores the fact that many functions have regions of high variability along with other sections where change is gradual.

Adaptive quadrature methods remedy this situation by adjusting the step size so that small intervals are used in regions of rapid variations and larger intervals are used where the function changes gradually. Most of these techniques are based on applying the composite Simpson's 1/3 rule to subintervals in a fashion that is very similar to the way in which the composite trapezoidal rule was used in Richardson extrapolation. That is, the 1/3 rule is applied at two levels of refinement and the difference between these two levels is used to estimate the truncation error. If the truncation error is acceptable, no further refinement is required and the integral estimate for the subinterval is deemed acceptable. If the error estimate is too large, the step size is refined and the process repeated until the error falls to acceptable levels. The total integral is then computed as the summation of the integral estimates for the subintervals.

The theoretical basis of the approach can be illustrated for an interval $x = a$ to $x = b$ with a width of $h_1 = b - a$. A first estimate of the integral can be estimated with Simpson's 1/3 rule,

$$I(h_1) = \frac{h_1}{6}(f(a) + 4f(c) + f(b)) \tag{22.10}$$

where $c = (a + b)/2$.

As in Richardson extrapolation, a more refined estimate can be obtained by halving the step size. That is, by applying the multiple-application Simpson's 1/3 rule with $n = 4$,

$$I(h_2) = \frac{h_2}{6}(f(a) + 4f(d) + 2f(c) + 4f(e) + f(b)) \tag{22.11}$$

where $d = (a + c)/2$, $e = (c + b)/2$, and $h_2 = h_1/2$.

Because both $I(h_1)$ and $I(h_2)$ are estimates of the same integral, their difference provides a measure of the error. That is,

$$E \cong I(h_2) - I(h_1) \tag{22.12}$$

In addition, the estimate and error associated with either application can be represented generally as

$$I = I(h) + E(h) \tag{22.13}$$

where I = the exact value of the integral, $I(h)$ = the approximation from an n-segment application of the Simpson's 1/3 rule with step size $h = (b - a)/n$, and $E(h)$ = the corresponding truncation error.

Using an approach similar to Richardson extrapolation, we can derive an estimate for the error of the more refined estimate, $I(h_2)$, as a function of the difference between the two integral estimates,

$$E(h_2) = \frac{1}{15}[I(h_2) - I(h_1)] \tag{22.14}$$

The error can then be added to $I(h_2)$ to generate an even better estimate

$$I = I(h_2) + \frac{1}{15}[I(h_2) - I(h_1)] \tag{22.15}$$

This result is equivalent to *Boole's rule.*

The equations developed above can now be combined into an efficient algorithm. Figure 22.5 presents pseudocode for such an algorithm that is based on a MATLAB software M-file developed by Cleve Moler (2004).

The function consists of a main calling function, `quadapt`, along with a recursive function, `qstep`, that actually performs the integration. As set up in Fig. 22.5, both `qadapt` and `qstep` must have access to another function, `f`, that evaluates the integrand.

The main calling function, `quadapt`, is passed the integration limits, a and b. After setting the tolerance, the function evaluations required for the initial application of Simpson's 1/3 rule (Eq. 22.10) are computed. These values along with the integration limits are then passed to `qstep`. Within `qstep`, the remaining step sizes and function values are determined and the two integral estimates (Eqs. 22.10 and 22.11) are computed.

At this point, the error is estimated as the absolute difference between the integral estimates. Depending on the value of the error, two things can then happen:

1) If the error is less than or equal to the tolerance, Boole's rule is generated, the function terminates and the result is returned.
2) If the error is larger than the tolerance, `qstep` is invoked twice to evaluate each of the two subintervals of the current call.

```
FUNCTION quadapt(a, b)                    (main calling function)
tol = 0.000001
c = (a + b)/2                                    (initialization)
fa = f(a)
fc = f(c)
fb = f(b)
quadapt = qstep(a, b, tol, fa, fc, fb)
END quadapt

FUNCTION qstep(a, b, tol, fa, fc, fb) (recursive function)
h1 = b - a
h2 = h1/2
c = (a + b)/2
fd = f((a + c)/2)
fe = f((c + b)/2)
I1 = h1/6 * (fa + 4 * fc + fb)            (Simpson's 1/3 rule)
I2 = h2/6 * (fa + 4 * fd + 2 * fc + 4 * fe + fb)
IF |I2 - I1| ≤ tol THEN       (terminate after Boole's rule)
  I = I2 + (I2 - I1)/15
ELSE                            (recursive calls if needed)
  Ia = qstep(a, c, tol, fa, fd, fc)
  Ib = qstep(c, b, tol, fc, fe, fb)
  I = Ia + Ib
END IF
qstep = I
END qstep
```

FIGURE 22.5

Pseudocode for simplified
adaptive quadrature algorithm
based on a MATLAB M-file
presented in Moler (2004).

The two recursive calls in the second step represent the real beauty of this algorithm. They just keep subdividing until the tolerance is met. Once this occurs, their results are passed back up the recursive path, combining with the other integral estimates along the way. The process ends when the final call is satisfied and the total integral is evaluated and returned to the main calling function.

It should be stressed that the algorithm in Fig. 22.5 is a stripped-down version of the quad function that is the professional quadrature function employed in MATLAB. Thus, it does not guard against failure such as cases where integrals do not exist. Nevertheless, it works just fine for many applications, and certainly serves to illustrate how adaptive quadrature works.

22.4   GAUSS QUADRATURE

In Chap. 21, we studied the group of numerical integration or quadrature formulas known as the Newton-Cotes equations. A characteristic of these formulas (with the exception of the special case of Sec. 21.3) was that the integral estimate was based on evenly spaced function values. Consequently, the location of the base points used in these equations was predetermined or fixed.

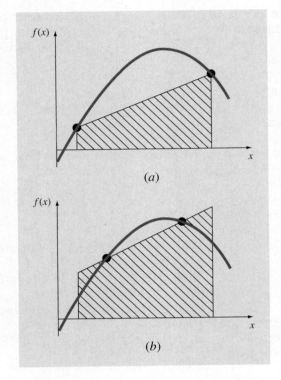

FIGURE 22.6
(a) Graphical depiction of the trapezoidal rule as the area under the straight line joining fixed end points. (b) An improved integral estimate obtained by taking the area under the straight line passing through two intermediate points. By positioning these points wisely, the positive and negative errors are balanced, and an improved integral estimate results.

For example, as depicted in Fig. 22.6a, the trapezoidal rule is based on taking the area under the straight line connecting the function values at the ends of the integration interval. The formula that is used to compute this area is

$$I \cong (b - a)\frac{f(a) + f(b)}{2} \tag{22.16}$$

where a and b = the limits of integration and $b - a$ = the width of the integration interval. Because the trapezoidal rule must pass through the end points, there are cases such as Fig. 22.6a where the formula results in a large error.

Now, suppose that the constraint of fixed base points was removed and we were free to evaluate the area under a straight line joining any two points on the curve. By positioning these points wisely, we could define a straight line that would balance the positive and negative errors. Hence, as in Fig. 22.6b, we would arrive at an improved estimate of the integral.

Gauss quadrature is the name for one class of techniques to implement such a strategy. The particular Gauss quadrature formulas described in this section are called *Gauss-Legendre* formulas. Before describing the approach, we will show how numerical integration formulas such as the trapezoidal rule can be derived using the method of undetermined coefficients. This method will then be employed to develop the Gauss-Legendre formulas.

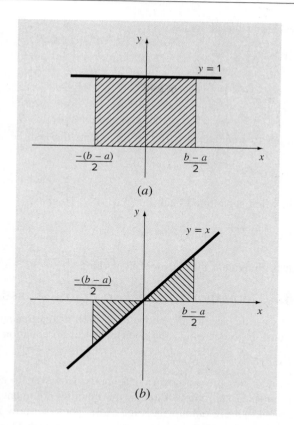

FIGURE 22.7

Two integrals that should be evaluated exactly by the trapezoidal rule: (a) a constant and (b) a straight line.

22.4.1 Method of Undetermined Coefficients

In Chap. 21, we derived the trapezoidal rule by integrating a linear interpolating polynomial and by geometrical reasoning. The *method of undetermined coefficients* offers a third approach that also has utility in deriving other integration techniques such as Gauss quadrature.

To illustrate the approach, Eq. (22.16) is expressed as

$$I \cong c_0 f(a) + c_1 f(b) \tag{22.17}$$

where the c's = constants. Now realize that the trapezoidal rule should yield exact results when the function being integrated is a constant or a straight line. Two simple equations that represent these cases are $y = 1$ and $y = x$. Both are illustrated in Fig. 22.7. Thus, the following equalities should hold:

$$c_0 + c_1 = \int_{-(b-a)/2}^{(b-a)/2} 1\, dx$$

and

$$-c_0 \frac{b-a}{2} + c_1 \frac{b-a}{2} = \int_{-(b-a)/2}^{(b-a)/2} x\, dx$$

or, evaluating the integrals,

$$c_0 + c_1 = b - a$$

and

$$-c_0 \frac{b - a}{2} + c_1 \frac{b - a}{2} = 0$$

These are two equations with two unknowns that can be solved for

$$c_0 = c_1 = \frac{b - a}{2}$$

which, when substituted back into Eq. (22.17), gives

$$I = \frac{b - a}{2} f(a) + \frac{b - a}{2} f(b)$$

which is equivalent to the trapezoidal rule.

22.4.2 Derivation of the Two-Point Gauss-Legendre Formula

Just as was the case for the above derivation of the trapezoidal rule, the object of Gauss quadrature is to determine the coefficients of an equation of the form

$$I \cong c_0 f(x_0) + c_1 f(x_1) \tag{22.18}$$

where the c's = the unknown coefficients. However, in contrast to the trapezoidal rule that used fixed end points a and b, the function arguments x_0 and x_1 are not fixed at the end points, but are unknowns (Fig. 22.8). Thus, we now have a total of four unknowns that must be evaluated, and consequently, we require four conditions to determine them exactly.

FIGURE 22.8
Graphical depiction of the unknown variables x_0 and x_1 for integration by Gauss quadrature.

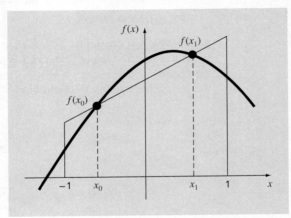

Just as for the trapezoidal rule, we can obtain two of these conditions by assuming that Eq. (22.18) fits the integral of a constant and a linear function exactly. Then, to arrive at the other two conditions, we merely extend this reasoning by assuming that it also fits the integral of a parabolic ($y = x^2$) and a cubic ($y = x^3$) function. By doing this, we determine all four unknowns and in the bargain derive a linear two-point integration formula that is exact for cubics. The four equations to be solved are

$$c_0 f(x_0) + c_1 f(x_1) = \int_{-1}^{1} 1\, dx = 2 \tag{22.19}$$

$$c_0 f(x_0) + c_1 f(x_1) = \int_{-1}^{1} x\, dx = 0 \tag{22.20}$$

$$c_0 f(x_0) + c_1 f(x_1) = \int_{-1}^{1} x^2\, dx = \frac{2}{3} \tag{22.21}$$

$$c_0 f(x_0) + c_1 f(x_1) = \int_{-1}^{1} x^3\, dx = 0 \tag{22.22}$$

Equations (22.19) through (22.22) can be solved simultaneously for

$$c_0 = c_1 = 1$$

$$x_0 = -\frac{1}{\sqrt{3}} = -0.5773503 \ldots$$

$$x_1 = \frac{1}{\sqrt{3}} = 0.5773503 \ldots$$

which can be substituted into Eq. (22.18) to yield the two-point Gauss-Legendre formula

$$I \cong f\left(\frac{-1}{\sqrt{3}}\right) + f\left(\frac{1}{\sqrt{3}}\right) \tag{22.23}$$

Thus, we arrive at the interesting result that the simple addition of the function values at $x = 1/\sqrt{3}$ and $-1/\sqrt{3}$ yields an integral estimate that is third-order accurate.

Notice that the integration limits in Eqs. (22.19) through (22.22) are from -1 to 1. This was done to simplify the mathematics and to make the formulation as general as possible. A simple change of variable can be used to translate other limits of integration into this form. This is accomplished by assuming that a new variable x_d is related to the original variable x in a linear fashion, as in

$$x = a_0 + a_1 x_d \tag{22.24}$$

If the lower limit, $x = a$, corresponds to $x_d = -1$, these values can be substituted into Eq. (22.24) to yield

$$a = a_0 + a_1(-1) \tag{22.25}$$

Similarly, the upper limit, $x = b$, corresponds to $x_d = 1$, to give

$$b = a_0 + a_1(1) \tag{22.26}$$

Equations (22.25) and (22.26) can be solved simultaneously for

$$a_0 = \frac{b+a}{2} \tag{22.27}$$

and

$$a_1 = \frac{b-a}{2} \tag{22.28}$$

which can be substituted into Eq. (22.24) to yield

$$x = \frac{(b+a) + (b-a)x_d}{2} \tag{22.29}$$

This equation can be differentiated to give

$$dx = \frac{b-a}{2} dx_d \tag{22.30}$$

Equations (22.29) and (22.30) can be substituted for x and dx, respectively, in the equation to be integrated. These substitutions effectively transform the integration interval without changing the value of the integral. The following example illustrates how this is done in practice.

EXAMPLE 22.3

Two-Point Gauss-Legendre Formula

Problem Statement. Use Eq. (22.23) to evaluate the integral of

$$f(x) = 0.2 + 25x - 200x^2 + 675x^3 - 900x^4 + 400x^5$$

between the limits $x = 0$ to 0.8. Recall that this was the same problem that we solved in Chap. 21 using a variety of Newton-Cotes formulations. The exact value of the integral is 1.640533.

Solution. Before integrating the function, we must perform a change of variable so that the limits are from -1 to $+1$. To do this, we substitute $a = 0$ and $b = 0.8$ into Eq. (22.29) to yield

$$x = 0.4 + 0.4x_d$$

The derivative of this relationship is [Eq. (22.30)]

$$dx = 0.4\,dx_d$$

Both of these can be substituted into the original equation to yield

$$\int_0^{0.8} (0.2 + 25x - 200x^2 + 675x^3 - 900x^4 + 400x^5)\,dx$$

$$= \int_{-1}^{1} [0.2 + 25(0.4 + 0.4x_d) - 200(0.4 + 0.4x_d)^2 + 675(0.4 + 0.4x_d)^3$$

$$- 900(0.4 + 0.4x_d)^4 + 400(0.4 + 0.4x_d)^5]0.4\,dx_d$$

Therefore, the right-hand side is in the form that is suitable for evaluation using Gauss quadrature. The transformed function can be evaluated at $-1/\sqrt{3}$ to be equal to 0.516741 and at $1/\sqrt{3}$ to be equal to 1.305837. Therefore, the integral according to Eq. (22.23) is

$$I \cong 0.516741 + 1.305837 = 1.822578$$

which represents a percent relative error of -11.1 percent. This result is comparable in magnitude to a four-segment application of the trapezoidal rule (Table 21.1) or a single application of Simpson's 1/3 and 3/8 rules (Examples 21.4 and 21.6). This latter result is to be expected because Simpson's rules are also third-order accurate. However, because of the clever choice of base points, Gauss quadrature attains this accuracy on the basis of only two function evaluations.

22.4.3 Higher-Point Formulas

Beyond the two-point formula described in the previous section, higher-point versions can be developed in the general form

$$I \cong c_0 f(x_0) + c_1 f(x_1) + \cdots + c_{n-1} f(x_{n-1}) \tag{22.31}$$

where n = the number of points. Values for c's and x's for up to and including the six-point formula are summarized in Table 22.1.

TABLE 22.1 Weighting factors c and function arguments x used in Gauss-Legendre formulas.

Points	Weighting Factors	Function Arguments	Truncation Error
2	$c_0 = 1.0000000$ $c_1 = 1.0000000$	$x_0 = -0.577350269$ $x_1 = 0.577350269$	$\cong f^{(4)}(\xi)$
3	$c_0 = 0.5555556$ $c_1 = 0.8888889$ $c_2 = 0.5555556$	$x_0 = -0.774596669$ $x_1 = 0.0$ $x_2 = 0.774596669$	$\cong f^{(6)}(\xi)$
4	$c_0 = 0.3478548$ $c_1 = 0.6521452$ $c_2 = 0.6521452$ $c_3 = 0.3478548$	$x_0 = -0.861136312$ $x_1 = -0.339981044$ $x_2 = 0.339981044$ $x_3 = 0.861136312$	$\cong f^{(8)}(\xi)$
5	$c_0 = 0.2369269$ $c_1 = 0.4786287$ $c_2 = 0.5688889$ $c_3 = 0.4786287$ $c_4 = 0.2369269$	$x_0 = -0.906179846$ $x_1 = -0.538469310$ $x_2 = 0.0$ $x_3 = 0.538469310$ $x_4 = 0.906179846$	$\cong f^{(10)}(\xi)$
6	$c_0 = 0.1713245$ $c_1 = 0.3607616$ $c_2 = 0.4679139$ $c_3 = 0.4679139$ $c_4 = 0.3607616$ $c_5 = 0.1713245$	$x_0 = -0.932469514$ $x_1 = -0.661209386$ $x_2 = -0.238619186$ $x_3 = 0.238619186$ $x_4 = 0.661209386$ $x_5 = 0.932469514$	$\cong f^{(12)}(\xi)$

EXAMPLE 22.4	**Three-Point Gauss-Legendre Formula**

Problem Statement. Use the three-point formula from Table 22.1 to estimate the integral for the same function as in Example 22.3.

Solution. According to Table 22.1, the three-point formula is

$$I = 0.5555556\, f(-0.7745967) + 0.8888889\, f(0) + 0.5555556\, f(0.7745967)$$

which is equal to

$$I = 0.2813013 + 0.8732444 + 0.4859876 = 1.640533$$

which is exact.

Because Gauss quadrature requires function evaluations at nonuniformly spaced points within the integration interval, it is not appropriate for cases where the function is unknown. Thus, it is not suited for engineering problems that deal with tabulated data. However, where the function is known, its efficiency can be a decided advantage. This is particularly true when numerous integral evaluations must be performed.

EXAMPLE 22.5	**Applying Gauss Quadrature to the Falling Parachutist Problem**

Problem Statement. In Example 21.3, we used the multiple-application trapezoidal rule to evaluate

$$d = \frac{gm}{c} \int_0^{10} [1 - e^{-(c/m)t}]\, dt$$

where $g = 9.8$, $c = 12.5$, and $m = 68.1$. The exact value of the integral was determined by calculus to be 289.4351. Recall that the best estimate obtained using a 500-segment trapezoidal rule was 289.4348 with an $|\varepsilon_t| \cong 1.15 \times 10^{-4}$ percent. Repeat this computation using Gauss quadrature.

Solution. After modifying the function, the following results are obtained:

Two-point estimate = 290.0145

Three-point estimate = 289.4393

Four-point estimate = 289.4352

Five-point estimate = 289.4351

Six-point estimate = 289.4351

Thus, the five- and six-point estimates yield results that are exact to seven significant figures.

22.4.4 Error Analysis for Gauss Quadrature

The error for the Gauss-Legendre formulas is specified generally by (Carnahan et al. 1969)

$$E_t = \frac{2^{2n+3}[(n+1)!]^4}{(2n+3)[(2n+2)!]^3} f^{(2n+2)}(\xi) \tag{22.32}$$

where n = the number of points minus one and $f^{(2n+2)}(\xi)$ = the $(2n + 2)$th derivative of the function after the change of variable with ξ located somewhere on the interval from -1 to 1. Comparison of Eq. (22.32) with Table 21.2 indicates the superiority of Gauss quadrature to Newton-Cotes formulas, provided the higher-order derivatives do not increase substantially with increasing n. Problem 22.8 at the end of this chapter illustrates a case where the Gauss-Legendre formulas perform poorly. In these situations, the multiple-application Simpson's rule or Romberg integration would be preferable. However, for many functions confronted in engineering practice, Gauss quadrature provides an efficient means for evaluating integrals.

22.5 IMPROPER INTEGRALS

To this point, we have dealt exclusively with integrals having finite limits and bounded integrands. Although these types are commonplace in engineering, there will be times when improper integrals must be evaluated. In this section, we will focus on one type of improper integral—that is, one with a lower limit of $-\infty$ or an upper limit of $+\infty$.

Such integrals usually can be evaluated by making a change of variable that transforms the infinite range to one that is finite. The following identity serves this purpose and works for any function that decreases toward zero at least as fast as $1/x^2$ as x approaches infinity:

$$\int_a^b f(x) \, dx = \int_{1/b}^{1/a} \frac{1}{t^2} f\left(\frac{1}{t}\right) dt \tag{22.33}$$

for $ab > 0$. Therefore, it applies when a is positive and b is ∞ or when a is $-\infty$ and b is negative. For cases where the limits are from $-\infty$ to a positive value or from a negative value to ∞, the integral can be implemented in two steps. For example,

$$\int_{-\infty}^b f(x) \, dx = \int_{-\infty}^{-A} f(x) \, dx + \int_{-A}^b f(x) \, dx \tag{22.34}$$

where $-A$ is chosen as a sufficiently large negative value so that the function has begun to approach zero asymptotically at least as fast as $1/x^2$. After the integral has been divided into two parts, the first can be evaluated with Eq. (22.33) and the second with a Newton-Cotes closed formula such as Simpson's 1/3 rule.

One problem with using Eq. (22.33) to evaluate an integral is that the transformed function will be singular at one of the limits. The open integration formulas can be used to circumvent this dilemma as they allow evaluation of the integral without employing data at the end points of the integration interval. To allow the maximum flexibility, a multiple-application version of one of the open formulas from Table 21.4 is required.

Multiple-application versions of the open formulas can be concocted by using closed formulas for the interior segments and open formulas for the ends. For example, the multiple-segment trapezoidal rule and the midpoint rule can be combined to give

$$\int_{x_0}^{x_n} f(x) \, dx = h\left[\frac{3}{2} f(x_1) + \sum_{i=2}^{n-2} f(x_i) + \frac{3}{2} f(x_{n-1}) \right]$$

FIGURE 22.9
Placement of data points relative to integration limits for the extended midpoint rule.

In addition, semiopen formulas can be developed for cases where one or the other end of the interval is closed. For example, a formula that is open at the lower limit and closed at the upper limit is given as

$$\int_{x_0}^{x_n} f(x)\, dx = h\left[\frac{3}{2}f(x_1) + \sum_{i=2}^{n-1} f(x_i) + \frac{1}{2}f(x_n)\right]$$

Although these relationships can be used, a preferred formula is (Press et al. 2007)

$$\int_{x_0}^{x_n} f(x)\, dx = h[f(x_{1/2}) + f(x_{3/2}) + \cdots + f(x_{n-3/2}) + f(x_{n-1/2})] \qquad (22.35)$$

which is called the *extended midpoint rule*. Notice that this formula is based on limits of integration that are $h/2$ after and before the first and last data points (Fig. 22.9).

EXAMPLE 22.6   **Evaluation of an Improper Integral**

Problem Statement.   The *cumulative normal distribution* is an important formula in statistics (see Fig. 22.10):

$$N(x) = \int_{-\infty}^{x} \frac{1}{\sqrt{2\pi}} e^{-x^2/2}\, dx \qquad (E22.6.1)$$

where $x = (y - \bar{y})/s_y$ is called the *normalized standard deviate*. It represents a change of variable to scale the normal distribution so that it is centered on zero and the distance along the abscissa is measured in multiples of the standard deviation (Fig. 22.10b).

Equation (E22.6.1) represents the probability that an event will be less than x. For example, if $x = 1$, Eq. (E22.6.1) can be used to determine that the probability that an event will occur that is less than one standard deviation above the mean is $N(1) = 0.8413$. In other words, if 100 events occur, approximately 84 will be less than the mean plus one standard deviation. Because Eq. (E22.6.1) cannot be evaluated in a simple functional form, it is solved numerically and listed in statistical tables. Use Eq. (22.34) in conjunction with Simpson's 1/3 rule and the extended midpoint rule to determine $N(1)$ numerically.

Solution.   Equation (E22.6.1) can be reexpressed in terms of Eq. (22.34) as

$$N(x) = \frac{1}{\sqrt{2\pi}}\left(\int_{-\infty}^{-2} e^{-x^2/2}\, dx + \int_{-2}^{1} e^{-x^2/2}\, dx\right)$$

The first integral can be evaluated by applying Eq. (22.33) to give

$$\int_{-\infty}^{-2} e^{-x^2/2}\, dx = \int_{-1/2}^{0} \frac{1}{t^2} e^{-1/(2t^2)}\, dt$$

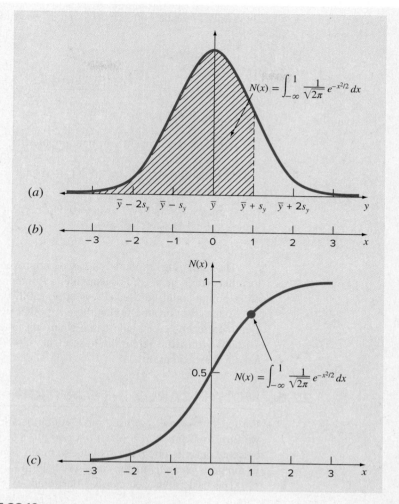

FIGURE 22.10
(a) The normal distribution, (b) the transformed abscissa in terms of the standardized normal deviate, and (c) the cumulative normal distribution. The shaded area in (a) and the point in (c) represent the probability that a random event will be less than the mean plus one standard deviation.

Then the extended midpoint rule with $h = 1/8$ can be employed to estimate

$$\int_{-1/2}^{0} \frac{1}{t^2} e^{-1/(2t^2)} \, dt \cong \frac{1}{8} [f(x_{-7/16}) + f(x_{-5/16}) + f(x_{-3/16}) + f(x_{-1/16})]$$

$$= \frac{1}{8}[0.3833 + 0.0612 + 0 + 0] = 0.0556$$

Simpson's 1/3 rule with $h = 0.5$ can be used to estimate the second integral as

$$\int_{-2}^{1} e^{-x^2/2} dx$$

$$= [1 - (-2)]\frac{0.1353 + 4(0.3247 + 0.8825 + 0.8825) + 2(0.6065 + 1) + 0.6065}{3(6)}$$

$$= 2.0523$$

Therefore, the final result can be computed as

$$N(1) \cong \frac{1}{\sqrt{2\pi}}(0.0556 + 2.0523) = 0.8409$$

which represents an error of $\varepsilon_t = 0.046$ percent.

The foregoing computation can be improved in a number of ways. First, higher-order formulas could be used. For example, a Romberg integration could be employed. Second, more points could be used. Press et al. (2007) explore both options in depth.

Aside from infinite limits, there are other ways in which an integral can be improper. Common examples include cases where the integral is singular at either the limits or at a point within the integral. Press et al. (2007) provide a nice discussion of ways to handle these situations.

22.6 MONTE CARLO INTEGRATION

Earlier in this book, we used random numbers for optimization (recall Sec. 14.1.1). Now, we turn to *Monte Carlo (MC) integration*, a technique that employs random numbers to evaluate definite integrals. As with so much of integration, a visual explanation grounded in integrals as areas under curves is helpful.

The basic approach can be illustrated for the standard one-dimensional definite integral,

$$I = \int_{a}^{b} f(x)\, dx$$

As depicted in Fig. 22.11, because of its similarity to the game of darts, this method can be called *dartboard Monte Carlo integration*. The function to be integrated, $f(x)$, is contained within a rectangular target, with the rectangle's width spanning the integration interval, and its height equal to or greater than the function's maximum value over the integration interval. Just like aimlessly throwing darts at a target, random values of x and y can be generated within the whole rectangle and then identified as being either below (shaded symbols) or above (unshaded) the function. The integral estimate corresponds to the product of the rectangle's area times the ratio of the filled to the total number of symbols. The more darts that are thrown, the better the integral estimate.

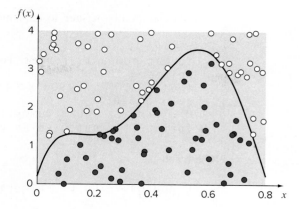

FIGURE 22.11
Graphical depiction of the "dartboard" version of Monte Carlo integration for the polynomial $f(x) = 400x^5 - 900x^4 + 675x^3 - 200x^2 + 25x + 0.2$ between $a = 0$ and $b = 0.8$. For this example, 100 samples were generated with 51 landing between the function and the abscissa. Thus, the resulting value of the integral is $0.51 \times 0.8 \times 4 = 1.6320$, as compared with the exact value of 1.6405333.

| EXAMPLE 22.7 | Dartboard Monte Carlo Integration |

Problem Statement.   Develop computer code for the dartboard MC integration method for the special case of a function that is positive across the whole integration interval but can be greater than or equal to zero at the limits. Use it to estimate the same definite integral evaluated in Fig. 22.11,

$$I = \int_0^{0.8} (400x^5 - 900x^4 + 675x^3 - 200x^2 + 25x + 0.2)\, dx$$

where the exact value of the integral is 1.640533. Test your code for $n = 100{,}000$.

Solution.   The following MATLAB function can be developed to implement the approach:

```
function Q = MonteCarloDartDummy(func,n,a,b)
ymx=@(x) -func(x);
[xmax,ymax]=fminbnd(ymx,a,b);
ymax=-ymax;
TotalArea=ymax*(b-a);
count = 0;
for i = 1:n
  x=a+(b-a)*rand(1); y=ymax*rand(1);
  if y<func(x), count = count + 1; end
end
Q = count/n*TotalArea;
```

A script can then be written to implement the function for this example:

```
clear,clc,
format compact, format long
points = 100000;
f=@(x) 400*x.^5-900*x.^4+675*x.^3-200*x.^2+25*x+0.2;
a=0; b=0.8;
Q = MonteCarloDartDummy(f,points,a,b);
fprintf('Computed value of integral is %8.7f.\n',Q);
```

The result is

```
Computed value of integral is 1.6373760.
```

Because each toss is random, every time this code is executed, the result will be some-what different.

Whereas the simple dartboard version described above certainly offers a straightfor-ward illustration of the general idea behind MC integration, it converges very slowly. In general, the convergence rates of most simple MC integration approaches are proportional to $1/\sqrt{n}$. This means that four times more samples are needed to halve the error.

In summary, MC integration has the advantage of simplicity and is especially advan-tageous for multidimensional integrals. Further, it is well-suited for parallel processing as each processor can make a random trial. It has disadvantages related to its slow conver-gence rate and the difficulty of evaluating the error of the results. Additional information on Monte Carlo methods, including descriptions of advanced MC integration techniques that improve the performance, can be found elsewhere (Hammersley and Handscomb 1964; Press et al. 2007; Newman and Barkema 1999; Robert and Casella 2004).

PROBLEMS

22.1 Use order of h^8 Romberg integration to evaluate

$$\int_0^3 xe^{2x}\,dx$$

Compare ε_a and ε_t.

22.2 Use Romberg integration to evaluate

$$I = \int_1^2 \left(x + \frac{1}{x}\right)^2 dx$$

to an accuracy of $\varepsilon_s = 0.5\%$ based on Eq. (22.9). Your results should be presented in the form of Fig. 22.3. Use the analytical solution of the integral to determine the percent relative error of the result ob-tained with Romberg integration. Check that ε_t is less than the stop-ping criterion ε_s.

22.3 Use Romberg integration to evaluate

$$\int_0^2 \frac{e^x \sin x}{1 + x^2}\,dx$$

to an accuracy of $\varepsilon_s = 0.5\%$. Your results should be presented in the form of Fig. 22.3.
22.4 Obtain an estimate of the integral from Prob. 22.2, but using two-, three-, and four-point Gauss-Legendre formulas. Compute ε_t for each case on the basis of the analytical solution.
22.5 Obtain an estimate of the integral from Prob. 22.1, but using two-, three-, and four-point Gauss-Legendre formulas. Compute ε_t for each case on the basis of the analytical solution.
22.6 Obtain an estimate of the integral from Prob. 22.3 using the five-point Gauss-Legendre formula.

22.7 Perform the computation in Examples 21.3 and 22.5 for the falling parachutist, but use Romberg integration ($\varepsilon_s = 0.05\%$).

22.8 Employ two- through six-point Gauss-Legendre formulas to solve

$$\int_{-3}^{3} \frac{1}{1 + x^2}\, dx$$

Interpret your results in light of Eq. (22.32).

22.9 Use numerical integration to evaluate the following:

(a) $\displaystyle\int_{2}^{\infty} \frac{dx}{x(x + 2)}$ **(b)** $\displaystyle\int_{0}^{\infty} e^{-y} \sin^2 y\, dy$

(c) $\displaystyle\int_{0}^{\infty} \frac{1}{(1 + y^2)(1 + y^2/2)}\, dy$ **(d)** $\displaystyle\int_{-2}^{\infty} y e^{-y}\, dy$

(e) $\displaystyle\int_{0}^{\infty} \frac{1}{\sqrt{2\pi}} e^{-x^2/2}\, dx$

Note that **(e)** is the normal distribution (recall Fig. 22.10).

22.10 Develop a user-friendly computer program for the multiple-segment **(a)** trapezoidal and **(b)** Simpson's 1/3 rule based on Fig. 22.1. Test it by integrating

$$\int_{0}^{1} x^{0.1}(1.2 - x)(1 - e^{20(x-1)})\, dx$$

Use the true value of 0.602298 to compute ε_t for $n = 4$.

22.11 Develop a user-friendly computer program for Romberg integration based on Fig. 22.4. Test it by duplicating the results of Examples 22.3 and 22.4 and the function in Prob. 22.10.

22.12 Develop a user-friendly computer program for adaptive quadrature based on Fig. 22.5. Test it by solving Prob. 22.10.

22.13 Develop a user-friendly computer program for Gauss quadrature. Test it by duplicating the results of Examples 22.3 and 22.4 and the function in Prob. 22.10.

22.14 There is no closed form solution for the error function,

$$\text{erf}(a) = \frac{2}{\sqrt{\pi}} \int_{0}^{a} e^{-x^2}\, dx$$

Use the two-point Gauss quadrature approach to estimate erf(1.5). Note that the exact value is 0.966105.

22.15 The amount of mass transported via a pipe over a period of time can be computed as

$$M = \int_{t_1}^{t_2} Q(t)c(t)\, dt$$

where M = mass (mg), t_1 = the initial time (min), t_2 = the final time (min), $Q(t)$ = flow rate (m^3/min), and $c(t)$ = concentration (mg/m^3). The following functional representations define the temporal variations in flow and concentration:

$$Q(t) = 8 + 5\cos^2(0.4t)$$

$$c(t) = 5e^{-0.5t} + 3e^{0.15t}$$

Determine the mass transported between $t_1 = 2$ and $t_2 = 8$ min with Romberg integration to a tolerance of 0.1%.

22.16 The depths of a river H are measured at equally spaced distances across a channel as tabulated below. The river's cross-sectional area can be determined by integration as in

$$A_c = \int_{0}^{x} H(x)\, dx$$

Use Romberg integration to perform the integration to a stopping criterion of 1%.

x, m	0	2	4	6	8	10	12	14	16
H, m	0	1.8	2	2	2.5	2.6	2.25	1.12	0

22.17 Recall that the velocity of the freefalling parachutist with linear drag can be computed analytically as

$$v(t) = \frac{gm}{c}\left(1 - e^{-(c/m)t}\right)$$

where $v(t)$ = velocity (m/s), t = time (s), $g = 9.81$ m/s^2, m = mass (kg), c = linear drag coefficient (kg/s). Use Romberg integration to compute how far the jumper travels during the first 8 seconds of free fall given $m = 80$ kg and $c = 10$ kg/s. Compute the answer to $\varepsilon_s = 1\%$.

22.18 Prove that Eq. (22.15) is equivalent to Boole's rule.

22.19 Whereas the "dartboard" version described in Sec. 22.6 certainly offered a straightforward illustration of the general idea behind MC integration, it converges slowly and inefficiently. An equally simple alternative, which we will call *rectangular MC integration*, is based on this algorithm:

- Repeat the following three steps $i = 1$ to n times:
  Generate a random point, x_i, along the abscissa within the integration interval, a to b.
  Compute the function value, $f(x_i)$, at that point.
  Compute the area of a rectangle, $A_i = f(x_i) \times (b - a)$.
- The integral is then computed as the average of the areas:

$$I = \frac{\sum_{i=1}^{n} A_i}{n}$$

Because each iteration requires a single random variable along a single dimension (x), this algorithm is more efficient than the dartboard version. Develop a computer code for the rectangle MC integration method. Use it to estimate the same definite integral evaluated with the dartboard method in Fig. 22.11 and Example 22.7,

$$I = \int_{0}^{0.8} (400x^5 - 900x^4 + 675x^3 - 200x^2 + 25x + 0.2)\, dx$$

for which the exact value is 1.640533.

22.20 Using the rectangular Monte Carlo algorithm described in Prob. 22.19, estimate the integral

$$I = \int_{1.5}^{15} (-0.0125x^3 + 0.115x^2 + 1.18x - 6.3)\,dx$$

Employ $n = 20{,}000$ and compare your result with the analytical solution.

22.21 Develop computer code for dartboard MC integration for cases where the function can be positive, negative, or zero across the integration interval. Use your code to estimate the derivative of the integral from Prob. 22.20.

22.22 Another example of MC integration is its use to evaluate π. If a square is drawn circumscribing a circle (Fig. P22.22), the ratio of the two areas is equal to $\pi/4$. If we randomly generate n_s points within the square and then count the number of points falling within the circle, n_c, $\pi \cong 4 \times (n_c/n_s)$. Note that because the circle and square are centered on the origin, we can obtain the same result by limiting the analysis to the first quadrant. If the following condition tests true, $x_2 + y_2 \le R$, can conclude that the trial is within the circle. Use this approach to estimate π for 5 cases: $n = 10^3, 10^4, 10^5, 10^6,$ and 10^7. Compute the absolute % relative error, $|\varepsilon_t|$, for each case. Develop a log-log plot of the error versus n in order to evaluate the dependency of the error on the number of iterations.

22.23 Evaluate the following integral:

$$I = \int_0^4 x^3\,dx$$

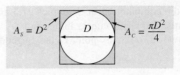

FIGURE P22.22

(a) Analytically
(b) Using Romberg integration with a stopping criterion of $\varepsilon_s = 0.01\%$
(c) Using three-point Gauss quadrature
(d) Using dartboard Monte Carlo integration with $n = 10{,}000$
For **(b)** through **(d)**, compute the absolute % relative error.

22.24 The *humps function* defines a curve with two maxima (peaks) of unequal height,

$$f(x) = \frac{1}{(x - 0.3)^2 + 0.01} + \frac{1}{(x - 0.9)^2 + 0.04} - 6$$

Estimate the definite integral of this function between the limits $a = 0$ and $b = 2$ with

(a) Romberg integration with a stopping criterion of $\varepsilon_s = 0.01\%$
(b) Three-point Gauss quadrature
(c) Monte Carlo integration with $n = 10{,}000$, using the software you developed in Prob. 22.19 or 22.21.

Numerical Differentiation

We have already introduced the notion of numerical differentiation in Chap. 4. Recall that we employed Taylor series expansions to derive finite-divided-difference approximations of derivatives. In Chap. 4, we developed forward, backward, and centered difference approximations of first and higher derivatives. Recall that, at best, these estimates had errors that were $O(h^2)$—that is, their errors were proportional to the square of the step size. This level of accuracy is due to the number of terms of the Taylor series that were retained during the derivation of these formulas. We will now illustrate how to develop more accurate formulas by retaining more terms.

23.1 HIGH-ACCURACY DIFFERENTIATION FORMULAS

As noted above, high-accuracy divided-difference formulas can be generated by including additional terms from the Taylor series expansion. For example, the forward Taylor series expansion can be written as [Eq. (4.21)]

$$f(x_{i+1}) = f(x_i) + f'(x_i)h + \frac{f''(x_i)}{2}h^2 + \cdots \tag{23.1}$$

which can be solved for

$$f'(x_i) = \frac{f(x_{i+1}) - f(x_i)}{h} - \frac{f''(x_i)}{2}h + O(h^2) \tag{23.2}$$

In Chap. 4, we truncated this result by excluding the second- and higher-derivative terms and were thus left with a final result of

$$f'(x_i) = \frac{f(x_{i+1}) - f(x_i)}{h} + O(h) \tag{23.3}$$

In contrast to this approach, we now retain the second-derivative term by substituting the following approximation of the second derivative [recall Eq. (4.24)]

$$f''(x_i) = \frac{f(x_{i+2}) - 2f(x_{i+1}) + f(x_i)}{h^2} + O(h) \tag{23.4}$$

667

into Eq. (23.2) to yield

$$f'(x_i) = \frac{f(x_{i+1}) - f(x_i)}{h} - \frac{f(x_{i+2}) - 2f(x_{i+1}) + f(x_i)}{2h^2} h + O(h^2)$$

or, after collecting terms,

$$f'(x_i) = \frac{-f(x_{i+2}) + 4f(x_{i+1}) - 3f(x_i)}{2h} + O(h^2) \tag{23.5}$$

Notice that inclusion of the second-derivative term has improved the accuracy to $O(h^2)$. Similar improved versions can be developed for the backward and centered formulas as well as for the approximations of the higher derivatives. The formulas are summarized in Figs. 23.1 through 23.3 along with all the results from Chap. 4. The following example illustrates the utility of these formulas for estimating derivatives.

FIGURE 23.1
Forward finite-divided-difference formulas: Two versions are presented for each derivative. The latter version incorporates more terms of the Taylor series expansion and is, consequently, more accurate.

First Derivative   Error

$$f'(x_i) = \frac{f(x_{i+1}) - f(x_i)}{h} \qquad O(h)$$

$$f'(x_i) = \frac{-f(x_{i+2}) + 4f(x_{i+1}) - 3f(x_i)}{2h} \qquad O(h^2)$$

Second Derivative

$$f''(x_i) = \frac{f(x_{i+2}) - 2f(x_{i+1}) + f(x_i)}{h^2} \qquad O(h)$$

$$f''(x_i) = \frac{-f(x_{i+3}) + 4f(x_{i+2}) - 5f(x_{i+1}) + 2f(x_i)}{h^2} \qquad O(h^2)$$

Third Derivative

$$f'''(x_i) = \frac{f(x_{i+3}) - 3f(x_{i+2}) + 3f(x_{i+1}) - f(x_i)}{h^3} \qquad O(h)$$

$$f'''(x_i) = \frac{-3f(x_{i+4}) + 14f(x_{i+3}) - 24f(x_{i+2}) + 18f(x_{i+1}) - 5f(x_i)}{2h^3} \qquad O(h^2)$$

Fourth Derivative

$$f''''(x_i) = \frac{f(x_{i+4}) - 4f(x_{i+3}) + 6f(x_{i+2}) - 4f(x_{i+1}) + f(x_i)}{h^4} \qquad O(h)$$

$$f''''(x_i) = \frac{-2f(x_{i+5}) + 11f(x_{i+4}) - 24f(x_{i+3}) + 26f(x_{i+2}) - 14f(x_{i+1}) + 3f(x_i)}{h^4} \qquad O(h^2)$$

First Derivative

$$f'(x_i) = \frac{f(x_i) - f(x_{i-1})}{h}$$

Error

$O(h)$

$$f'(x_i) = \frac{3f(x_i) - 4f(x_{i-1}) + f(x_{i-2})}{2h}$$

$O(h^2)$

Second Derivative

$$f''(x_i) = \frac{f(x_i) - 2f(x_{i-1}) + f(x_{i-2})}{h^2}$$

$O(h)$

$$f''(x_i) = \frac{2f(x_i) - 5f(x_{i-1}) + 4f(x_{i-2}) - f(x_{i-3})}{h^2}$$

$O(h^2)$

Third Derivative

$$f'''(x_i) = \frac{f(x_i) - 3f(x_{i-1}) + 3f(x_{i-2}) - f(x_{i-3})}{h^3}$$

$O(h)$

$$f'''(x_i) = \frac{5f(x_i) - 18f(x_{i-1}) + 24f(x_{i-2}) - 14f(x_{i-3}) + 3f(x_{i-4})}{2h^3}$$

$O(h^2)$

Fourth Derivative

$$f''''(x_i) = \frac{f(x_i) - 4f(x_{i-1}) + 6f(x_{i-2}) - 4f(x_{i-3}) + f(x_{i-4})}{h^4}$$

$O(h)$

$$f''''(x_i) = \frac{3f(x_i) - 14f(x_{i-1}) + 26f(x_{i-2}) - 24f(x_{i-3}) + 11f(x_{i-4}) - 2f(x_{i-5})}{h^4}$$

$O(h^2)$

FIGURE 23.2
Backward finite-divided-difference formulas: Two versions are presented for each derivative. The latter version incorporates more terms of the Taylor series expansion and is, consequently, more accurate.

First Derivative

$$f'(x_i) = \frac{f(x_{i+1}) - f(x_{i-1})}{2h}$$

Error

$O(h^2)$

$$f'(x_i) = \frac{-f(x_{i+2}) + 8f(x_{i+1}) - 8f(x_{i-1}) + f(x_{i-2})}{12h}$$

$O(h^4)$

Second Derivative

$$f''(x_i) = \frac{f(x_{i+1}) - 2f(x_i) + f(x_{i-1})}{h^2}$$

$O(h^2)$

$$f''(x_i) = \frac{-f(x_{i+2}) + 16f(x_{i+1}) - 30f(x_i) + 16f(x_{i-1}) - f(x_{i-2})}{12h^2}$$

$O(h^4)$

Third Derivative

$$f'''(x_i) = \frac{f(x_{i+2}) - 2f(x_{i+1}) + 2f(x_{i-1}) - f(x_{i-2})}{2h^3}$$

$O(h^2)$

$$f'''(x_i) = \frac{-f(x_{i+3}) + 8f(x_{i+2}) - 13f(x_{i+1}) + 13f(x_{i-1}) - 8f(x_{i-2}) + f(x_{i-3})}{8h^3}$$

$O(h^4)$

Fourth Derivative

$$f''''(x_i) = \frac{f(x_{i+2}) - 4f(x_{i+1}) + 6f(x_i) - 4f(x_{i-1}) + f(x_{i-2})}{h^4}$$

$O(h^2)$

$$f''''(x_i) = \frac{-f(x_{i+3}) + 12f(x_{i+2}) - 39f(x_{i+1}) + 56f(x_i) - 39f(x_{i-1}) + 12f(x_{i-2}) - f(x_{i-3})}{6h^4}$$

$O(h^4)$

FIGURE 23.3
Centered finite-divided-difference formulas: Two versions are presented for each derivative. The latter version incorporates more terms of the Taylor series expansion and is, consequently, more accurate.

EXAMPLE 23.1   **High-Accuracy Differentiation Formulas**

Problem Statement.   Recall that in Example 4.4 we estimated the derivative of

$$f(x) = -0.1x^4 - 0.15x^3 - 0.5x^2 - 0.25x + 1.2$$

at $x = 0.5$ using finite divided differences and a step size of $h = 0.25$,

	Forward $O(h)$	Backward $O(h)$	Centered $O(h^2)$
Estimate	−1.155	−0.714	−0.934
ε_t (%)	−26.5	21.7	−2.4

where the errors were computed on the basis of the true value of −0.9125. Repeat this computation, but employ the high-accuracy formulas from Figs. 23.1 through 23.3.

Solution.   The data needed for this example are

$$
\begin{aligned}
x_{i-2} &= 0 & f(x_{i-2}) &= 1.2 \\
x_{i-1} &= 0.25 & f(x_{i-1}) &= 1.1035156 \\
x_i &= 0.5 & f(x_i) &= 0.925 \\
x_{i+1} &= 0.75 & f(x_{i+1}) &= 0.6363281 \\
x_{i+2} &= 1 & f(x_{i+2}) &= 0.2
\end{aligned}
$$

The forward difference of accuracy $O(h^2)$ is computed as (Fig. 23.1)

$$f'(0.5) = \frac{-0.2 + 4(0.6363281) - 3(0.925)}{2(0.25)} = -0.859375 \qquad \varepsilon_t = 5.82\%$$

The backward difference of accuracy $O(h^2)$ is computed as (Fig. 23.2)

$$f'(0.5) = \frac{3(0.925) - 4(1.1035156) + 1.2}{2(0.25)} = -0.878125 \qquad \varepsilon_t = 3.77\%$$

The centered difference of accuracy $O(h^4)$ is computed as (Fig. 23.3)

$$f'(0.5) = \frac{-0.2 + 8(0.6363281) - 8(1.1035156) + 1.2}{12(0.25)} = -0.9125 \qquad \varepsilon_t = 0\%$$

As expected, the errors for the forward and backward differences are considerably more accurate than the results from Example 4.4. However, surprisingly, the centered difference yields a perfect result. This is because the formulas based on the Taylor series are equivalent to passing polynomials through the data points.

23.2 RICHARDSON EXTRAPOLATION

To this point, we have seen that there are two ways to improve derivative estimates when employing finite divided differences: (1) decrease the step size or (2) use a higher-order formula that employs more points. A third approach, based on Richardson extrapolation, uses two derivative estimates to compute a third, more accurate approximation.

Recall from Sec. 22.2.1 that Richardson extrapolation provided a means to obtain an improved integral estimate I with the formula [Eq. (22.4)]

$$I \cong I(h_2) + \frac{1}{(h_1/h_2)^2 - 1}[I(h_2) - I(h_1)] \tag{23.6}$$

where $I(h_1)$ and $I(h_2)$ are integral estimates using two step sizes, h_1 and h_2. Because of its convenience when expressed as a computer algorithm, this formula is usually written for the case where $h_2 = h_1/2$, as in

$$I \cong \frac{4}{3}I(h_2) - \frac{1}{3}I(h_1) \tag{23.7}$$

In a similar fashion, Eq. (23.7) can be written for derivatives as

$$D \cong \frac{4}{3}D(h_2) - \frac{1}{3}D(h_1) \tag{23.8}$$

For centered difference approximations with $O(h^2)$, the application of this formula will yield a new derivative estimate of $O(h^4)$.

EXAMPLE 23.2   **Richardson Extrapolation**

Problem Statement.   Using the same function as in Example 23.1, estimate the first derivative at $x = 0.5$ employing step sizes of $h_1 = 0.5$ and $h_2 = 0.25$. Then use Eq. (23.8) to compute an improved estimate with Richardson extrapolation. Recall that the true value is -0.9125.

Solution.   The first-derivative estimates can be computed with centered differences as

$$D(0.5) = \frac{0.2 - 1.2}{1} = -1.0 \qquad \varepsilon_t = -9.6\%$$

and

$$D(0.25) = \frac{0.6363281 - 1.1035156}{0.5} = -0.934375 \qquad \varepsilon_t = -2.4\%$$

The improved estimate can be determined by applying Eq. (23.8) to give

$$D = \frac{4}{3}(-0.934375) - \frac{1}{3}(-1) = -0.9125$$

which for the present case is a perfect result.

The previous example yielded a perfect result because the function being analyzed was a fourth-order polynomial. The perfect outcome was due to the fact that Richardson extrapolation is actually equivalent to fitting a higher-order polynomial to the data and then evaluating the derivatives by centered divided differences. Thus, the present case matched the derivative of the fourth-order polynomial precisely. For most other functions, of course, this would not occur and our derivative estimate would be improved but not perfect. Consequently, as was the case for the application of Richardson extrapolation, the approach can be applied iteratively using a Romberg algorithm until the result falls below an acceptable error criterion.

23.3 DERIVATIVES OF UNEQUALLY SPACED DATA

The approaches discussed to this point are primarily designed to determine the derivative of a given function. For the finite-divided-difference approximations of Sec. 23.1, these data had to be evenly spaced. For the Richardson extrapolation technique of Sec. 23.2, these data had to be evenly spaced and generated for successively halved intervals. Such control of data spacing is usually available only in cases where we can use a function to generate a table of values.

In contrast, empirically derived information—that is, data from experiments or field studies—is often collected at unequal intervals. Such information cannot be analyzed with the techniques discussed to this point.

One way to handle nonequispaced data is to fit a second-order Lagrange interpolating polynomial [recall Eq. (18.23)] to each set of three adjacent points. Remember that this polynomial does not require that the points be equispaced. The second-order polynomial can be differentiated analytically to give

$$f'(x) = f(x_{i-1}) \frac{2x - x_i - x_{i+1}}{(x_{i-1} - x_i)(x_{i-1} - x_{i+1})} + f(x_i) \frac{2x - x_{i-1} - x_{i+1}}{(x_i - x_{i-1})(x_i - x_{i+1})}$$

$$+ f(x_{i+1}) \frac{2x - x_{i-1} - x_i}{(x_{i+1} - x_{i-1})(x_{i+1} - x_i)} \tag{23.9}$$

where x is the value at which you want to estimate the derivative. Although this equation is certainly more complicated than the first-derivative approximations from Figs. 23.1 through 23.3, it has some important advantages. First, it can be used to estimate the derivative anywhere within the range prescribed by the three points. Second, the points themselves do not have to be equally spaced. Third, the derivative estimate is of the same accuracy as the centered difference [Eq. (4.22)]. In fact, for equispaced points, Eq. (23.9) evaluated at $x = x_i$ reduces to Eq. (4.22).

EXAMPLE 23.3   **Differentiating Unequally Spaced Data**

Problem Statement.   As in Fig. 23.4, a temperature gradient can be measured down into the soil. The heat flux at the soil-air interface can be computed with Fourier's law,

$$q(z = 0) = -k\rho C \left. \frac{dT}{dz} \right|_{z=0}$$

where q = heat flux (W/m^2), k = coefficient of thermal diffusivity in soil ($\cong 3.5 \times 10^{-7}$ m^2/s), ρ = soil density ($\cong 1800$ kg/m^3), and C = soil specific heat ($\cong 840$ J/(kg °C)). Note that a positive value for flux means that heat is transferred from the air to the soil. Use numerical differentiation to evaluate the gradient at the soil-air interface and employ this estimate to determine the heat flux into the ground.

FIGURE 23.4
Temperature versus depth into the soil.

Solution.   Equation (23.9) can be used to calculate the derivative as

$$f'(x) = 13.5\frac{2(0) - 1.25 - 3.75}{(0 - 1.25)(0 - 3.75)} + 12\frac{2(0) - 0 - 3.75}{(1.25 - 0)(1.25 - 3.75)}$$

$$+ 10\frac{2(0) - 0 - 1.25}{(3.75 - 0)(3.75 - 1.25)}$$

$$= -14.4 + 14.4 - 1.333333 = -1.333333°C/cm$$

which can be used to compute (note that 1 W = 1 J/s),

$$q(z = 0) = -3.5 \times 10^{-7}\frac{m^2}{s}\left(1800\frac{kg}{m^3}\right)\left(840\frac{J}{kg\,°C}\right)\left(-133.3333\frac{°C}{m}\right)$$

$$= 70.56 \text{ W/m}^2$$

23.4 DERIVATIVES AND INTEGRALS FOR DATA WITH ERRORS

Aside from unequal spacing, another problem related to differentiating empirical data is that it usually includes measurement error. A shortcoming of numerical differentiation is that it tends to amplify errors in the data. Figure 23.5a shows smooth, error-free data that when numerically differentiated yield a smooth result (Fig. 23.5c). In contrast, Fig. 23.5b uses the same data, but with some points raised and some lowered slightly. This minor modification is barely apparent from Fig. 23.5b. However, the resulting effect in Fig. 23.5d is significant because the process of differentiation amplifies errors.

As might be expected, the primary approach for determining derivatives for imprecise data is to use least-squares regression to fit a smooth, differentiable function to the data. In the absence of any other information, a lower-order polynomial regression might be a good first choice. Obviously, if the true functional relationship between the dependent and independent variable is known, this relationship should form the basis for the least-squares fit.

FIGURE 23.5
Illustration of how small data errors are amplified by numerical differentiation: (a) data with no error, (b) data modified slightly, (c) the resulting numerical differentiation of curve (a), and (d) the resulting differentiation of curve (b), manifesting increased variability. In contrast, the reverse operation of integration [moving from (d) to (b) by taking the area under (d)] tends to attenuate or smooth data errors.

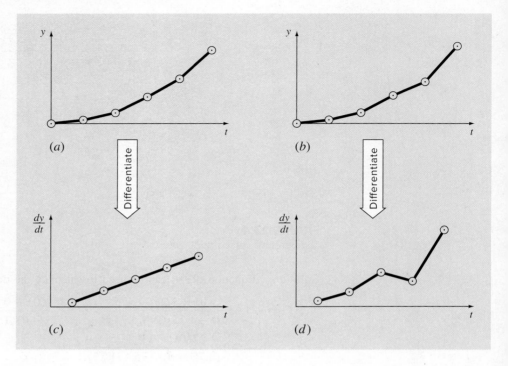

23.4.1 Differentiation versus Integration of Uncertain Data

Just as curve-fitting techniques like regression can be used to differentiate uncertain data, a similar process can be employed for integration. However, because of the difference in stability between differentiation and integration, this is rarely done.

As depicted in Fig. 23.5, differentiation tends to be unstable—that is, it amplifies errors. In contrast, the fact that integration is a summing process tends to make it very forgiving with regard to uncertain data. In essence, as points are summed to form an integral, random positive and negative errors tend to cancel out. In contrast, because differentiation is subtractive, random positive and negative errors tend to add.

23.5 PARTIAL DERIVATIVES

Partial derivatives along a single dimension are computed in the same fashion as ordinary derivatives. For example, suppose that we want to determine to partial derivatives for a two-dimensional function, $f(x, y)$. For equally spaced data, the partial first derivatives can be approximated with centered differences,

$$\frac{\partial f}{\partial x} = \frac{f(x + \Delta x, y) - f(x - \Delta x, y)}{2\Delta x} \tag{23.10}$$

$$\frac{\partial f}{\partial y} = \frac{f(x, y + \Delta y) - f(x, y - \Delta y)}{2\Delta y} \tag{23.11}$$

All the other formulas and approaches discussed to this point can be applied to evaluate partial derivatives in a similar fashion.

For higher-order derivatives, we might want to differentiate a function with respect to two or more different variables. The result is called a *mixed partial derivative*. For example, we might want to take the partial derivative of $f(x, y)$ with respect to both independent variables

$$\frac{\partial^2 f}{\partial x \partial y} = \frac{\partial}{\partial x}\left(\frac{\partial f}{\partial y}\right) \tag{23.12}$$

To develop a finite-difference approximation, we can first form a difference in x of the partial derivatives in y,

$$\frac{\partial^2 f}{\partial x \partial y} = \frac{\dfrac{\partial f}{\partial y}(x + \Delta x, y) - \dfrac{\partial f}{\partial y}(x - \Delta x, y)}{2\Delta x} \tag{23.13}$$

Then, we can use finite differences to evaluate each of the partials in y,

$$\frac{\partial^2 f}{\partial x \partial y} = \frac{\dfrac{f(x + \Delta x, y + \Delta y) - f(x + \Delta x, y - \Delta y)}{2\Delta y} - \dfrac{f(x - \Delta x, y + \Delta y) - f(x - \Delta x, y - \Delta y)}{2\Delta y}}{2\Delta x} \tag{23.14}$$

Collecting terms yields the final result:

$$\frac{\partial^2 f}{\partial x \partial y} = \frac{f(x + \Delta x, y + \Delta y) - f(x + \Delta x, y - \Delta y) - f(x - \Delta x, y + \Delta y) + f(x - \Delta x, y - \Delta y)}{4\Delta x \Delta y} \tag{23.15}$$

23.6 NUMERICAL INTEGRATION/DIFFERENTIATION WITH SOFTWARE PACKAGES

Software packages have great capabilities for numerical integration and differentiation. In this section, we will give you a taste of some of the more useful ones.

23.6.1 MATLAB

MATLAB software has a variety of built-in functions that allow functions and data to be integrated and differentiated (Table 23.1). In this section, we will illustrate some of these capabilities.

MATLAB can integrate both discrete data and functions. For example, `trapz` computes the integral of discrete values using the multiple-application trapezoidal rule. A simple representation of its syntax is

```
q = trapz(x, y)
```

where the two vectors, x and y, hold the independent and dependent variables, respectively, and q holds the resulting integral. MATLAB also has another function, `cumtrapz`, that computes the cumulative integral. For this case, the result is a vector whose elements $q(k)$ hold the integral from $x(1)$ to $x(k)$.

TABLE 23.1 MATLAB functions to implement (a) integration and (b) differentiation.

Function	Description
(a) Integration:	
cumtrapz	Cumulative trapezoidal numerical integration
dblquad	Numerically evaluate double integral
polyint	Integrate polynomial analytically
quad	Numerically evaluate integral, adaptive Simpson quadrature
quadgk	Numerically evaluate integral, adaptive Gauss-Kronrod quadrature
quadl	Numerically evaluate integral, adaptive Lobatto quadrature
quadv	Vectorized quadrature
trapz	Trapezoidal numerical integration
triplequad	Numerically evaluate triple integral
(b) Differentiation:	
del2	Discrete Laplacian
diff	Differences and approximate derivatives
gradient	Numerical gradient
polyder	Polynomial derivative

When the integrand is available in functional form, quad generates the definite integral using adaptive quadrature. A simple representation of its syntax is

$$q = \text{quad}(fun, a, b)$$

where fun is the function to be integrated and a and b are the integration limits.

EXAMPLE 23.4   **Using Numerical Integration to Compute Distance from Velocity**

Problem Statement.   As described in Sec. PT6.1, integration can be used to compute the distance traveled, $y(t)$, of an object based on its velocity, $v(t)$, as in,

$$y(t) = \int_0^t v(t)\, dt \tag{E23.4.1}$$

Recall from Sec. 1.1, that the velocity of a free-falling parachutist, subject to linear drag and with zero initial velocity, can be computed with

$$v(t) = \frac{gm}{c}\left(1 - e^{-(c/m)t}\right) \tag{E23.4.2}$$

If we substitute, Eq. (E23.4.2) into Eq. (E23.4.1), the result can be integrated analytically, with the initial condition, $y(0) = 0$, to yield

$$y(t) = \frac{gm}{c}t - \frac{gm^2}{c^2}\left(1 - e^{-(c/m)t}\right)$$

This result can be used to compute that a 70-kg parachutist with a drag coefficient of 12.5 kg/s will fall 799.73 m over a 20-s period.

Use MATLAB functions to perform the same integration numerically. In addition, develop a plot of the analytical and computed distances along with velocity on the same graph.

Solution. We can first use Eq. (E23.4.2) to generate some unequally spaced times and velocities. We can then round these velocities so that they are more like measured values; that is, they are not exact,

```
>> format short g
>> t=[0 1 2 3 4.3 7 12 16];
>> g=9.81;m=70;c=12.5;
>> v=round(g*m/c*(1-exp(-c/m*t)));
```

The total distance can then be computed as

```
>> y=trapz(t,v)

y =
   789.6
```

Thus, after 20 seconds, the jumper has fallen 789.6 m, which is reasonably close to the exact, analytical solution of 799.73 m.

If we desire the cumulative distance traveled at each time, cumtrapz can be employed to compute,

```
>> yc=cumtrapz(t,v)

yc =
   0   4.5   17   36.5   70.3   162.1   379.6   579.6   789.6
```

A graph of the numerical and analytical solutions along with both the exact and rounded velocities is generated with the following commands,

```
>> ta=linspace (t(1), t(length(t)));
>> ya=g*m/c*ta-g*m^2/c^2*(1-exp(-c/m*ta));
>> plot (ta, ya, t, yc, 'o')
>> title ('Distance versus time')
>> xlabel ('t (s)'), ylabel ('x (m)')
>> legend ('analytical', 'numerical')
```

As in Fig. 23.6, the numerical and analytical results match fairly well.

Finally, the quad function can be used to evaluate the integral with adaptive quadrature

```
>> va=@(t) g*m/c*(1-exp(-c/m*t));
>> yq=quad(va,t(1),t(length(t)))

yq =
   799.73
```

This result is identical to the analytical solution to within the 5 significant digits displayed.

As listed in Table 23.1*b*, MATLAB has a number of built-in functions for evaluating derivatives including the diff and gradient functions. When it is passed a one-dimensional vector of length n, the diff function returns a vector of length $n - 1$

FIGURE 23.6
Plot of distance versus time.
The line was computed with
the analytical solution,
whereas the points were determined numerically with the
cumtrapz function.

containing the differences between adjacent elements. These can then be employed to determine finite-difference approximations of first derivatives.

The gradient function also returns differences. However, it does so in a manner that is more compatible with evaluating derivatives at the values themselves rather than in the intervals between values. A simple representation of its syntax is

    fx = gradient(f)

where f = a one-dimensional vector of length n and fx is a vector of length n containing differences based on f. Just as with the diff function, the first value returned is the difference between the first and second value. However, for the intermediate values, a centered difference based on the adjacent values is returned,

$$\text{diff}_i = \frac{f_{i+1} - f_{i-1}}{2}$$

The last value is then computed as the difference between the final two values. Hence, the results correspond to using centered differences for all the intermediate values, with forward and backward differences at the ends.

Note that the spacing between points is assumed to be equal to 1. If the vector represents equally spaced data, the following version divides all the results by the interval and hence returns the actual values of the derivatives,

    fx = gradient(f, h)

where h = the spacing between points.

EXAMPLE 23.5 | **Using diff and gradient for Differentiation**

Problem Statement.   Explore how the MATLAB's diff and gradient functions can be employed to differentiate the function $f(x) = 0.2 + 25x - 200x^2 + 675x^3 - 900x^4 +$

$400x^5$ from $x = 0$ to 0.8. Compare your results with the exact solution: $f'(x) = 25 - 400x^2 + 2025x^2 - 3600x^3 + 2000x^4$.

Solution.   We can first express $f(x)$ as an anonymous function

```
>> f=@(x) 0.2+25*x-200*x.^2+675*x.^3-900*x.^4+400*x.^5;
```

We then generate a series of equally spaced values of the independent and dependent variables,

```
>> x=0:0.1:0.8;
>> y=f(x);
```

The diff function is then used to determine the differences between adjacent elements of each vector. For example,

```
>> format short g
>> diff(x)

 0.1000  0.1000  0.1000  0.1000  0.1000  0.1000  0.1000  0.1000
```

As expected, the result represents the differences between each pair of elements of x. To compute divided-difference approximations of the derivative, we merely perform a vector division of the y differences by the x differences by entering

```
>> d=diff(y)./diff(x)

 10.89  -0.01  3.19  8.49  8.69  1.39  -11.01  -21.31
```

Note that because we are using equally spaced values, after generating the x values, we could have simply performed the above computation concisely as

```
>> d=diff(f(x))/0.1;
```

The vector d now contains derivative estimates corresponding to the midpoints between adjacent elements. Therefore, in order to develop a plot of our results, we must first generate a vector holding the x values for the midpoint of each interval

```
>> n=length(x);
>> xm=(x(1:n-1)+x(2:n))./2;
```

We can compute values for the analytical derivative at a finer level of resolution to include on the plot for comparison.

```
>> xa=0:.01:.8;
>> ya=25-400*xa+3*675*xa.^2-4*900*xa.^3+5*400*xa.^4;
```

A plot of the numerical and analytical estimates is then generated with

```
subplot (1, 2, 1), plot (xm, d, 'o', xa, ya)
xlabel ('x'), ylabel ('y')
legend ('numerical', 'analytical'),title ('(a) diff')
```

As displayed in Fig. 23.7a, the results of the numerical approximation compare favorably with the exact, analytical solution for this case.

FIGURE 23.7
Comparison of the exact derivative (line) with numerical estimates (dots) computed with MATLAB's (*a*) diff, and (*b*) gradient functions.

We can also use the gradient function to determine the derivatives as

```
>> dy=gradient(y,0.1)
dy = 10.89  5.44  1.59  5.84  8.59  5.04  -4.81  -16.16  -21.31
```

As was done for the diff function, we can also display both the numerical and analytical estimates on a plot,

```
>> subplot(1,2,2), plot(x,dy,'o',xa,ya)
>> xlabel('x')
>> legend('numerical','analytical'),title('(b)gradient')
```

The results (Fig. 23.7*b*) are not as accurate as those obtained with the diff function (Fig. 23.7*a*). This is due to the fact that gradient employs intervals that are two times (0.2) as wide as for those used for diff (0.1).

Beyond one-dimensional vectors, the gradient function is particularly well-suited for determining the partial derivatives of matrices. For example, for a two-dimensional matrix, the function can be invoked as

```
[fx, fy] = gradient(f, h)
```

where f is a two-dimensional array, fx corresponds to the differences in the x (column) direction and fy corresponds to the differences in the y (row) direction, and $h =$ the spacing between points. If *h* is omitted, the spacing between points in both dimensions is assumed to be 1. In Sec. 31.4.2, we will illustrate how this capability can be used to visualize vector fields.

23.6.2 Mathcad

Mathcad has operators that perform numerical integration and differentiation. These operators employ and look like the same traditional mathematical symbols you have used since high school or your first semester of college.

The integration operator uses a sequence of trapezoidal rule evaluations of the integral and the Romberg algorithm. Iterations are performed until successive results vary by less than a specified tolerance. The derivative operator uses a similar method to compute derivatives between order 0 and order 5. This operator creates a table of approximations based on divided-difference calculations of the derivative using various orders and step sizes. Extrapolation techniques are used to estimate values in a manner resembling Richardson's method.

Figure 23.8 shows a Mathcad example where f(x) is created using the definition symbol (:=), and then the integral is calculated over a range from x = 0 to x = 0.8. In this case, we used the simple polynomial we evaluated throughout Chap. 21. Note that the range as defined by the variables a and b is input with the definition symbol.

Figure 23.9 shows a Mathcad example where a function f(x) is created with the definition symbol (:=) and then first and third derivatives are calculated at a point where x = −6. Note that the location of the point and the order of the derivative are input with the definition symbol.

FIGURE 23.8

Mathcad screen to determine the integral of a polynomial with Romberg integration.

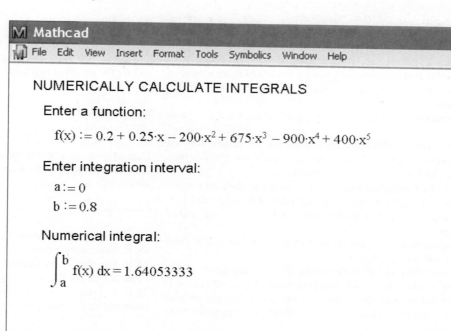

NUMERICALLY CALCULATE INTEGRALS

Enter a function:

$$f(x) := 0.2 + 0.25 \cdot x - 200 \cdot x^2 + 675 \cdot x^3 - 900 \cdot x^4 + 400 \cdot x^5$$

Enter integration interval:

$$a := 0$$
$$b := 0.8$$

Numerical integral:

$$\int_a^b f(x)\, dx = 1.64053333$$

M Mathcad

File   Edit   View   Insert   Format   Tools   Symbolics   Window   Help

NUMERICALLY CALCULATE DERIVATIVES

Enter a function:

$f(x) := 2 \cdot x + 3 + \cos(x)^2$

Enter a point to evaluate derivative:

$x := -6$

Compute the first derivative:

$\dfrac{d}{dx} f(x) := 1.46342708$

Compute the third derivative:

$n := 3$

$\dfrac{d^n}{dx^n} f(x) := 2.14629167$

FIGURE 23.9
Mathcad screen to implement numerical differentiation.

PROBLEMS

23.1 Compute forward and backward difference approximations of $O(h)$ and $O(h^2)$, and central difference approximations of $O(h^2)$ and $O(h^4)$ for the first derivative of $y = \sin x$ at $x = \pi/4$ using a value of $h = \pi/12$. Estimate the true percent relative error, ε_t, for each approximation.

23.2 Repeat Prob. 23.1, but for $y = \log x$ evaluated at $x = 25$ with $h = 2$.

23.3 Use centered difference approximations to estimate the first and second derivatives of $y = e^x$ at $x = 2$ for $h = 0.1$. Employ both $O(h^2)$ and $O(h^4)$ formulas for your estimates.

23.4 Use Richardson extrapolation to estimate the first derivative of $y = \cos x$ at $x = \pi/4$ using step sizes of $h_1 = \pi/3$ and $h_2 = \pi/6$. Employ centered differences of $O(h^2)$ for the initial estimates.

23.5 Repeat Prob. 23.4, but for the first derivative of $\ln x$ at $x = 5$ using $h_1 = 2$ and $h_2 = 1$.

23.6 Employ Eq. (23.9) to determine the first derivative of $y = 2x^4 - 6x^3 - 14x - 8$ at $x = 0$ based on values at $x_0 = -0.5$, $x_1 = 1$, and $x_2 = 2$. Compare this result with the true value and with an estimate obtained using a centered difference approximation based on $h = 1$.

23.7 Prove that for equispaced data points, Eq. (23.9) reduces to Eq. (4.22) at $x = x_i$.

23.8 Compute the first-order central difference approximations of $O(h^4)$ for each of the following functions at the specified location and for the specified step size:

(a) $y = x^3 + 3x - 15$	at $x = 0$,	$h = 0.25$
(b) $y = x^2 \cos x$	at $x = 0.5$,	$h = 0.1$
(c) $y = \tan(x/3)$	at $x = 2$,	$h = 0.5$
(d) $y = \sin(0.5\sqrt{x})/x$	at $x = 1$,	$h = 0.2$
(e) $y = e^x + x$	at $x = 3$,	$h = 0.2$

Compare your results with the analytical solutions.

23.9 The following data were collected for the distance traveled versus time for a rocket:

t, s	0	25	50	75	100	125
y, km	0	32	58	78	92	100

Use numerical differentiation to estimate the rocket's velocity and acceleration at each time.

23.10 Develop a user-friendly program to apply a Romberg algorithm to estimate the derivative of a given function.

23.11 Develop a user-friendly program to obtain first-derivative estimates for unequally spaced data. Test it with the following data:

x	1	1.5	1.6	2.5	3.5
$f(x)$	0.6767	0.3734	0.3261	0.08422	0.01596

where $f(x) = 5e^{-2x}x$. Compare your results with the true derivatives.

23.12 The following data are provided for the velocity of an object as a function of time,

t, s	0	4	8	12	16	20	24	28	32	36
v, m/s	0	34.7	61.8	82.8	99.2	112.0	121.9	129.7	135.7	140.4

(a) Using the best numerical method available, how far does the object travel from $t = 0$ to 28 s?

(b) Using the best numerical method available, what is the object's acceleration at $t = 28$ s?

(c) Using the best numerical method available, what is the object's acceleration at $t = 0$ s?

23.13 Recall that for the falling parachutist problem, the velocity is given by

$$v(t) = \frac{gm}{c}\left(1 - e^{-(c/m)t}\right) \qquad (P23.13.1)$$

and the distance traveled can be obtained by

$$d(t) = \frac{gm}{c}\int_0^t \left(1 - e^{-(c/m)t}\right)dt \qquad (P23.13.2)$$

Given $g = 9.81$, $m = 70$, and $c = 12$,

(a) Use MATLAB or Mathcad to integrate Eq. (P23.13.1) from $t = 0$ to 10.

(b) Analytically integrate Eq. (P23.13.2) with the initial condition that $d = 0$ at $t = 0$. Evaluate the result at $t = 10$ to confirm **(a)**.

(c) Use MATLAB or Mathcad to differentiate Eq. (P23.13.1) at $t = 10$.

(d) Analytically differentiate Eq. (P23.13.1) at $t = 10$ to confirm **(c)**.

23.14 The normal distribution is defined as

$$f(x) = \frac{1}{\sqrt{2\pi}}e^{-x^2/2}$$

(a) Use MATLAB or Mathcad to integrate this function from $x = -1$ to 1 and from -2 to 2.

(b) Use MATLAB or Mathcad to determine the inflection points of this function.

23.15 The following data were generated from the normal distribution:

x	−2	−1.5	−1	−0.5	0	0.5	1	1.5	2
f(x)	0.05399	0.12952	0.24197	0.35207	0.39894	0.35207	0.24197	0.12952	0.05399

(a) Use MATLAB to integrate these data from $x = -1$ to 1 and -2 to 2 with the `trap` function.

(b) Use MATLAB to estimate the inflection points of these data.

23.16 Evaluate $\partial f/\partial x$, $\partial f/\partial y$, and $\partial f/(\partial x\partial y)$ for the following function at $x = y = 1$ **(a)** analytically and **(b)** numerically with $\Delta x = \Delta y = 0.0001$,

$$f(x, y) = 3xy + 3x - x^3 - 3y^3$$

23.17 Evaluate the following integral with MATLAB,

$$\int_0^{2\pi} \frac{\sin t}{t}\,dt$$

using both the `quad` and `quadl` functions. To learn more about `quadl`, type `help quadl` at the MATLAB prompt.

23.18 Use the `diff` function in MATLAB and compute the finite-difference approximation to the first and second derivative at each x-value in the table below, excluding the two end points. Use finite-difference approximations that are second-order correct, $O(\Delta x^2)$.

x	0	1	2	3	4	5	6	7	8	9	10
y	1.5	2.1	3.3	4.8	7.0	6.6	8.7	7.5	9.0	10.9	10

23.19 The objective of this problem is to compare second-order accurate forward, backward, and centered finite-difference approximations of the first derivative of a function to the actual value of the derivative. This will be done for

$$f(x) = e^{-2x} - x$$

(a) Use calculus to determine the correct value of the derivative at $x = 2$.

(b) To evaluate the centered finite-difference approximations, start with $x = 0.5$. Thus, for the first evaluation, the x values for the centered difference approximation will be $x = 2 \pm 0.5$ or $x = 1.5$ and 2.5. Then, decrease in increments of 0.01 down to a minimum value of $\Delta x = 0.01$.

(c) Repeat part **(b)** for the second-order forward and backward differences. (Note that these can be done at the same time that the centered difference is computed in the loop.)

(d) Plot the results of **(b)** and **(c)** versus x. Include the exact result on the plot for comparison.

23.20 Use a Taylor series expansion to derive a centered finite-difference approximation to the third derivative that is second-order accurate. To do this, you will have to use four different expansions for the points x_{i-2}, x_{i-1}, x_{i+1}, and x_{i+2}. In each case, the expansion will be around the point x_i. The interval Δx will be used in each case of $i - 1$ and $i + 1$, and $2\Delta x$ will be used in each case of $i - 2$ and $i + 2$. The four equations must then be combined in a way to eliminate the first and second derivatives. Carry enough terms along in each expansion to evaluate the first term that will be truncated to determine the order of the approximation.

23.21 Use the following data to find the velocity and acceleration at $t = 10$ seconds:

Time, t, s	0	2	4	6	8	10	12	14	16
Position, x, m	0	0.7	1.8	3.4	5.1	6.3	7.5	8.0	8.4

Use second-order correct (a) centered finite-difference, (b) forward finite-difference, and (c) backward finite-difference methods.

23.22 A plane is being tracked by radar, and data are taken every second in polar coordinates θ and r:

t, s	200	202	204	206	208	210
θ, rad	0.75	0.72	0.70	0.68	0.67	0.66
r, m	5120	5370	5560	5800	6030	6240

At 206 s, use the centered finite difference (second-order correct) to find the vector expressions for velocity $\vec{v}$, and acceleration $\vec{a}$. The velocity and acceleration given in polar coordinates are

$$\vec{v} = \dot{r}\vec{e}_r + r\dot{\theta}\vec{e}_\theta \quad \text{and} \quad \vec{a} = (\ddot{r} - r\dot{\theta}^2)\vec{e}_r + (r\ddot{\theta} + 2\dot{r}\dot{\theta})\vec{e}_\theta$$

23.23 Develop an Excel/VBA macro program to read in adjacent columns of x and y values from a worksheet. Evaluate the derivatives at each point using Eq. (23.9), and display the results in a third column adjacent to the x and y values on the spreadsheet. Test your program by applying it to evaluate the velocities for the time–position values from Prob. 23.21.

23.24 Use regression to estimate the acceleration at each time for the following data with second-, third-, and fourth-order polynomials. Plot the results.

t	1	2	3.3	4.5	6	7	8	8.6	9.3	10
v	10	12	11	14	17	16	12	14	13.5	10

23.25 You have to measure the flow rate of water through a small pipe. In order to do it, you place a bucket at the pipe's outlet and measure the volume in the bucket as a function of time as tabulated below. Estimate the flow rate at $t = 6.7$ s.

Time, s	0	1	5	8
Volume, cm^3	0	1	8	16.4

23.26 The velocity v (m/s) of air flowing past a flat surface is measured at several distances y (m) away from the surface. Determine the shear stress τ (N/m^2) at the surface ($y = 0$), using *Newton's viscosity law*,

$$\tau = \mu \frac{dv}{dy}$$

Assume this value of dynamic viscosity: $\mu = 1.8 \times 10^{-5}$ N s/m^2.

y, m	0	0.002	0.006	0.012	0.018	0.024
v, m/s	0	0.287	0.899	1.915	3.048	4.299

23.27 Chemical reactions often follow the model:

$$\frac{dc}{dt} = -kc^n$$

where c = concentration, t = time, k = reaction rate, and n = reaction order. Given values of c and dc/dt, k and n can be evaluated by a linear regression of the logarithm of this equation:

$$\log\left(-\frac{dc}{dt}\right) = \log k + n \log c$$

Use this approach along with the following data to estimate k and n:

t	10	20	30	40	50	60
c	3.5	2.5	1.8	1.3	0.9	0.6

23.28 The velocity profile of a fluid in a circular pipe can be represented as

$$v = 10\left(1 - \frac{r}{r_0}\right)^{1/n}$$

where v = velocity, r = radial distance measured out from the pipe's centerline, r_0 = the pipe's radius, and n = a parameter. Determine the flow in the pipe if $r_0 = 0.75$ and $n = 7$ using (a) Romberg integration to a tolerance of 0.1%, (b) the two-point Gauss-Legendre formula, and (c) the MATLAB quad function. Note that flow is equal to velocity times area.

23.29 The amount of mass transported via a pipe over a period of time can be computed as

$$M = \int_{t_1}^{t_2} Q(t)c(t)\, dt$$

where M = mass (mg), t_1 = the initial time (min), t_2 = the final time (min), $Q(t)$ = flow rate (m^3/min), and $c(t)$ = concentration (mg/m^3). The following functional representations define the temporal variations in flow and concentration,

$$Q(t) = 9 + 4\cos^2(0.4t)$$
$$c(t) = 5e^{-0.5t} + 2e^{0.15t}$$

Determine the mass transported between $t_1 = 2$ and $t_2 = 8$ min using (a) Romberg integration to a tolerance of 0.1% and (b) the MATLAB quad function.

23.30 The velocity (m/s) of an object at time t (in seconds) is given by

$$v = \frac{2t}{\sqrt{1 + t^2}}$$

Using Richardson extrapolation, find the acceleration of the particle at time $t = 5$ s using $h = 0.5$ and 0.25. Employ the exact solution to compute the true percent relative error of each estimate.

Case Studies: Numerical Integration and Differentiation

The purpose of this chapter is to apply the methods of numerical integration and differentiation discussed in Part Six to practical engineering problems. Two situations are most frequently encountered. In the first case, the function under study can be expressed in analytic form but is too complicated to be readily evaluated using the methods of calculus. Numerical methods are applied to situations of this type by using the analytic expression to generate a table of arguments and function values. In the second case, the function to be evaluated is inherently tabular in nature. This type of function usually represents a series of measurements, observations, or some other empirical information. Data for either case are directly compatible with several schemes discussed in this part of the book.

Section 24.1, which deals with heat calculations from chemical engineering, involves equations. In this application, an analytic function is integrated numerically to determine the heat required to raise the temperature of a material.

Sections 24.2 and 24.3 also involve functions that are available in equation form. Section 24.2, which is taken from civil engineering, uses numerical integration to determine the total wind force acting on the mast of a racing sailboat. Section 24.3 determines the root-mean-square current for an electric circuit. This example is used to demonstrate the utility of Romberg integration and Gauss quadrature.

Section 24.4 focuses on the analysis of tabular information to determine the work required to move a block. Although this application has a direct connection with mechanical engineering, it is germane to all other areas of engineering. Among other things, we use this example to illustrate the integration of unequally spaced data.

24.1 INTEGRATION TO DETERMINE THE TOTAL QUANTITY OF HEAT (CHEMICAL/BIO ENGINEERING)

Background.   Heat calculations are employed routinely in chemical and bio engineering as well as in many other fields of engineering. This application provides a simple but useful example of such computations.

One problem that is often encountered is the determination of the quantity of heat required to raise the temperature of a material. The characteristic that is needed to carry out this computation is the heat capacity, c. This parameter represents the quantity of heat required to raise a unit mass by a unit temperature. If c is constant over the range of temperatures being examined, the required heat ΔH (in calories) can be calculated by

$$\Delta H = mc\ \Delta T \tag{24.1}$$

where c has units of cal/(g °C), m = mass (g), and ΔT = change in temperature (°C). For example, the amount of heat required to raise 20 g of water from 5 to 10°C is equal to

$$\Delta H = 20(1)(10 - 5) = 100\ \text{cal}$$

where the heat capacity of water is approximately 1 cal/(g °C). Such a computation is adequate when ΔT is small. However, for large ranges of temperature, the heat capacity is not constant and, in fact, varies as a function of temperature. For example, the heat capacity of a material could increase with temperature according to a relationship such as

$$c(T) = 0.132 + 1.56 \times 10^{-4}T + 2.64 \times 10^{-7}T^2 \tag{24.2}$$

Here you are asked to compute the heat required to raise 1000 g of this material from -100 to 200°C.

Solution.   Equation (PT6.4) provides a way to calculate the average value of $c(T)$:

$$\bar{c}(T) = \frac{\displaystyle\int_{T_1}^{T_2} c(T)\, dT}{T_2 - T_1} \tag{24.3}$$

which can be substituted into Eq. (24.1) to yield

$$\Delta H = m \int_{T_1}^{T_2} \bar{c}(T)\, dT \tag{24.4}$$

where $\Delta T = T_2 - T_1$. Now because, for the present case, $c(T)$ is a simple quadratic, ΔH can be determined analytically. Equation (24.2) is substituted into Eq. (24.4) and the result integrated to yield an exact value of $\Delta H = 42{,}732$ cal. It is useful and instructive to compare this result with those from the numerical methods developed in Chap. 21. To accomplish this, it is necessary to generate a table of values of c for various values of T:

T, °C	c, cal/(g °C)
−100	0.11904
−50	0.12486
0	0.13200
50	0.14046
100	0.15024
150	0.16134
200	0.17376

These points can be used in conjunction with a six-segment Simpson's 1/3 rule to compute an integral estimate of 42,732. This result can be substituted into Eq. (24.4) to yield a value of $\Delta H = 42{,}732$ cal, which agrees exactly with the analytical solution. This exact agreement would occur no matter how many segments were used. This is to be expected because c is a quadratic function and Simpson's rule is exact for polynomials of the third order or less (see Sec. 21.2.1).

TABLE 24.1 Results using the trapezoidal rule with various step sizes.

Step Size, °C	ΔH	ε_t (%)
300	96,048	125
150	43,029	0.7
100	42,864	0.3
50	42,765	0.07
25	42,740	0.018
10	42,733.3	<0.01
5	42,732.3	<0.01
1	42,732.01	<0.01
0.05	42,732.00003	<0.01

The results using the trapezoidal rule are listed in Table 24.1. It is seen that the trapezoidal rule is also capable of estimating the total heat very accurately. However, a small step (<10°C) is required for five-place accuracy. The same calculation can also be implemented with software. For example, MATLAB software yields

```
>> m=1000;
>> DH=m*quad(@(T) 0.132+1.56e-4*T+2.64e-7*T.^2,-100,200)
DH =
    42732
```

24.2 EFFECTIVE FORCE ON THE MAST OF A RACING SAILBOAT (CIVIL/ENVIRONMENTAL ENGINEERING)

Background.   A cross section of a racing sailboat is shown in Fig. 24.1a. Wind force (f) exerted per foot of mast from the sails varies as a function of distance above the

FIGURE 24.1
(a) Cross section of a racing sailboat. (b) Wind force f exerted per foot of mast as a function of distance z above the deck of the boat.

$\theta = \tan^{-1}(3/30)$
$= 0.0996687$

$F = 1480.6$ lb

$d = 13.05$ ft

T

0

3 ft

H

V

FIGURE 24.2
Free-body diagram of the forces exerted on the mast of a sailboat.

deck of the boat (z), as in Fig. 24.1b. Calculate the tensile force T in the left mast support cable, assuming that the right support cable is completely slack and the mast joins the deck in a manner that transmits horizontal or vertical forces but no moments. Assume that the mast remains vertical.

Solution.   To proceed with this problem, it is required that the distributed force f be converted to an equivalent total force F and that its effective location above the deck d be calculated (Fig. 24.2). This computation is complicated by the fact that the force exerted per foot of mast varies with the distance above the deck. The total force exerted on the mast can be expressed as the integral of a continuous function:

$$F = \int_0^{30} 200\left(\frac{z}{5+z}\right) e^{-2z/30}\, dz \qquad (24.5)$$

This nonlinear integral is difficult to evaluate analytically. Therefore, it is convenient to employ numerical approaches such as Simpson's rule and the trapezoidal rule for this problem. This is accomplished by calculating $f(z)$ for various values of z and then using Eq. (21.10) or (21.18). For example, Table 24.2 has values of $f(z)$ for a step size of 3 ft that provide data for Simpson's 1/3 rule or the trapezoidal rule. Results for several step sizes are given in Table 24.3. It is observed that both methods give a value of $F = 1480.6$ lb as the step size becomes small. In this case, step sizes of 0.05 ft for the trapezoidal rule and 0.5 ft for Simpson's rule provide good results.

TABLE 24.2 Values of $f(z)$ for a step size of 3 ft that provide data for the trapezoidal rule and Simpson's 1/3 rule.

z, ft	0	3	6	9	12	15	18	21	24	27	30
$f(z)$, lb/ft	0	61.40	73.13	70.56	63.43	55.18	47.14	39.83	33.42	27.89	23.20

TABLE 24.3 Values of F computed on the basis of various versions of the trapezoidal rule and Simpson's 1/3 rule.

Technique	Step Size, ft	Segments	F, lb
Trapezoidal rule	15	2	1001.7
	10	3	1222.3
	6	5	1372.3
	3	10	1450.8
	1	30	1477.1
	0.5	60	1479.7
	0.25	120	1480.3
	0.1	300	1480.5
	0.05	600	1480.6
Simpson's 1/3 rule	15	2	1219.6
	5	6	1462.9
	3	10	1476.9
	1	30	1480.5
	0.5	60	1480.6

The effective line of action of d (Fig. 24.2) can be calculated by evaluation of the integral

$$d = \frac{\int_0^{30} zf(z)\,dz}{\int_0^{30} f(z)\,dz} \tag{24.6}$$

or

$$d = \frac{\int_0^{30} 200z[z/(5+z)]e^{-2z/30}\,dz}{1480.6} \tag{24.7}$$

This integral can be evaluated using methods similar to the above. For example, Simpson's 1/3 rule with a step size of 0.5 gives $d = 19{,}326.9/1480.6 = 13.05$ ft.

With F and d known from numerical methods, a free-body diagram is used to develop force and moment balance equations. This free-body diagram is shown in Fig. 24.2. Summing forces in the horizontal and vertical direction and taking moments about point 0 gives

$$\Sigma F_H = 0 = F - T\sin\theta - H \tag{24.8}$$
$$\Sigma F_V = 0 = V - T\cos\theta \tag{24.9}$$
$$\Sigma M_0 = 0 = 3V - Fd \tag{24.10}$$

where T = the tension in the cable and H and V = the unknown reactions on the mast transmitted by the deck. The direction, as well as the magnitude, of H and V is unknown. Equation (24.10) can be solved directly for V because F and d are known:

$$V = \frac{Fd}{3} = \frac{(1480.6)(13.05)}{3} = 6440.6 \text{ lb}$$

Therefore, from Eq. (24.9),

$$T = \frac{V}{\cos\theta} = \frac{6440.6}{0.995} = 6473 \text{ lb}$$

and from Eq. (24.8),

$$H = F - T\sin\theta = 1480.6 - (6473)(0.0995) = 836.54 \text{ lb}$$

These forces now enable you to proceed with other aspects of the structural design of the boat such as the cables and the deck support system for the mast. This problem illustrates nicely two uses of numerical integration that may be encountered during the engineering design of structures. It is seen that both the trapezoidal rule and Simpson's 1/3 rule are easy to apply and are practical problem-solving tools. Simpson's 1/3 rule is more accurate than the trapezoidal rule for the same step size and thus may often be preferred.

24.3 ROOT-MEAN-SQUARE CURRENT BY NUMERICAL INTEGRATION (ELECTRICAL ENGINEERING)

Background. The average value of an oscillating electric current over one period may be zero. For example, suppose that the current is described by a simple sinusoid: $i(t) = \sin(2\pi t/T)$, where T is the period. The average value of this function can be determined

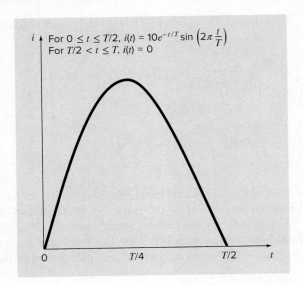

For $0 \le t \le T/2$, $i(t) = 10e^{-t/T}\sin\left(2\pi\frac{t}{T}\right)$
For $T/2 < t \le T$, $i(t) = 0$

FIGURE 24.3
A periodically varying electric current.

by the following equation:

$$i = \frac{\int_0^T \sin\left(\frac{2\pi t}{T}\right) dt}{T - 0} = \frac{-\cos(2\pi) + \cos 0}{T} = 0$$

Despite the fact that the net result is zero, such current is capable of performing work and generating heat. Therefore, electrical engineers often characterize such current by

$$I_{RMS} = \sqrt{\frac{1}{T}\int_0^T i^2(t)\,dt} \qquad (24.11)$$

where $i(t)$ = the instantaneous current. Calculate the RMS, or root-mean-square, current of the waveform shown in Fig. 24.3 using the trapezoidal rule, Simpson's 1/3 rule, Romberg integration, and Gauss quadrature for $T = 1$ s.

Solution. Integral estimates for various applications of the trapezoidal rule and Simpson's 1/3 rule are listed in Table 24.4. Notice that Simpson's rule is more accurate than the trapezoidal rule.

The exact value for the integral is 15.41261. This result is obtained using a 128-segment trapezoidal rule or a 32-segment Simpson's rule. The same estimate is also determined using Romberg integration (Fig. 24.4).

In addition, Gauss quadrature can be used to make the same estimate. The determination of the root-mean-square current involves the evaluation of the integral ($T = 1$),

$$I = \int_0^{1/2} (10e^{-t}\sin 2\pi t)^2\,dt \qquad (24.12)$$

TABLE 24.4 Values for the integral calculated using various numerical schemes. The percent relative error ε_t is based on a true value of 15.41261.

Technique	Segments	Integral	ε_t (%)
Trapezoidal rule	1	0.0	100
	2	15.16327	1.62
	4	15.40143	0.0725
	8	15.41196	4.21×10^{-3}
	16	15.41257	2.59×10^{-4}
	32	15.41261	1.62×10^{-5}
	64	15.41261	1.30×10^{-6}
	128	15.41261	0
Simpson's 1/3 rule	2	20.21769	−31.2
	4	15.48082	−0.443
	8	15.41547	−0.0186
	16	15.41277	1.06×10^{-3}
	32	15.41261	0

FIGURE 24.4
Result of using Romberg integration to estimate the RMS current.

$O(h^2)$	$O(h^4)$	$O(h^6)$	$O(h^8)$	$O(h^{10})$	$O(h^{12})$
0	20.21769	15.16503	15.41502	15.41261	15.41261
15.16327	15.48082	15.41111	15.41262	15.41261	
15.40143	15.41547	15.41225	15.41261		
15.41196	15.41277	15.41261			
15.41257	15.41262				
15.41261					

First, a change in variable is performed by applying Eqs. (22.29) and (22.30) to yield

$$t = \frac{1}{4} + \frac{1}{4}t_d \qquad dt = \frac{1}{4}dt_d$$

These relationships can be substituted into Eq. (24.12) to yield

$$I = \int_{-1}^{1} \left[10e^{-[1/4+(1/4)t_d]} \sin 2\pi \left(\frac{1}{4} + \frac{1}{4}t_d \right) \right]^2 \frac{1}{4}\,dt \tag{24.13}$$

For the two-point Gauss-Legendre formula, this function is evaluated at $t_d = -1/\sqrt{3}$ and $1/\sqrt{3}$, with the results being 7.684096 and 4.313728, respectively. These values can be substituted into Eq. (22.23) to yield an integral estimate of 11.99782, which represents an error of $\varepsilon_t = 22.1\%$.

The three-point formula is (Table 22.1)

$$I = 0.5555556(1.237449) + 0.8888889(15.16327) + 0.5555556(2.684915)$$
$$= 15.65755 \qquad |\varepsilon_t| = 1.6$$

The results of using the higher-point formulas are summarized in Table 24.5.

TABLE 24.5 Results of using various-point Gauss quadrature formulas to approximate the integral.

Points	Estimate	ε_t (%)
2	11.9978243	22.1
3	15.6575502	−1.59
4	15.4058023	4.42×10^{-2}
5	15.4126391	-2.01×10^{-4}
6	15.4126109	-1.82×10^{-5}

The integral estimate of 15.41261 can be substituted into Eq. (24.12) to compute an I_{RMS} of 3.925890 A. This result could then be employed to guide other aspects of the design and operation of the circuit.

24.4 NUMERICAL INTEGRATION TO COMPUTE WORK (MECHANICAL/AEROSPACE ENGINEERING)

Background.   Many engineering problems involve the calculation of work. The general formula is

Work = force × distance

When you were introduced to this concept in high school physics, simple applications were presented using forces that remained constant throughout the displacement. For example, if a force of 10 lb was used to pull a block a distance of 15 ft, the work would be calculated as 150 ft lb.

Although such a simple computation is useful for introducing the concept, realistic problem settings are usually more complex. For example, suppose that the force varies during the course of the calculation. In such cases, the work equation is re-expressed as

$$W = \int_{x_0}^{x_n} F(x) \, dx \qquad (24.14)$$

where W = work (ft lb), x_0 and x_n = the initial and final positions, respectively, and $F(x)$ a force that varies as a function of position. If $F(x)$ is easy to integrate, Eq. (24.14) can be evaluated analytically. However, in a realistic problem setting, the force might not be expressed in such a manner. In fact, when analyzing measured data, the force might be available only in tabular form. For such cases, numerical integration is the only viable option for the evaluation.

Further complexity is introduced if the angle between the force and the direction of movement also varies as a function of position (Fig. 24.5). The work equation can be modified further to account for this effect, as in

$$W = \int_{x_0}^{x_n} F(x)\cos\left[\theta(x)\right] dx \qquad (24.15)$$

Again, if $F(x)$ and $\theta(x)$ are simple functions, Eq. (24.15) might be solved analytically. However, as in Fig. 24.5, it is more likely that the functional relationship is complicated. For this situation, numerical methods provide the only alternative for determining the integral.

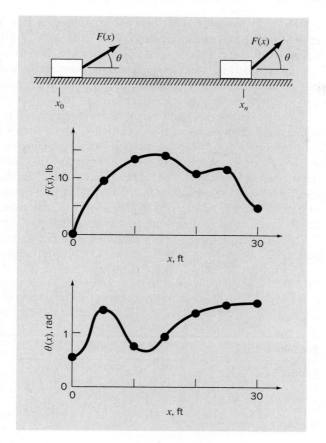

FIGURE 24.5
The case of a variable force acting on a block. For this case, the angle, as well as the magnitude, of the force varies.

TABLE 24.6 Data for force $F(x)$ and angle $\theta(x)$ as a function of position x.

x, ft	$F(x)$, lb	θ, rad	$F(x) \cos \theta$
0	0.0	0.50	0.0000
5	9.0	1.40	1.5297
10	13.0	0.75	9.5120
15	14.0	0.90	8.7025
20	10.5	1.30	2.8087
25	12.0	1.48	1.0881
30	5.0	1.50	0.3537

Suppose that you have to perform the computation for the situation depicted in Fig. 24.5. Although the figure shows the continuous values for $F(x)$ and $\theta(x)$, assume that, because of experimental constraints, you are provided with only discrete measurements at $x = 5$-ft intervals (Table 24.6). Use single- and multiple-application versions of the trapezoidal rule and Simpson's 1/3 and 3/8 rules to compute work for this data.

Solution. The results of the analysis are summarized in Table 24.7. A percent relative error ε_t was computed in reference to a true value of the integral of 129.52 that was estimated on the basis of values taken from Fig. 24.5 at 1-ft intervals.

The results are interesting because the most accurate outcome occurs for the simple two-segment trapezoidal rule. More refined estimates using more segments, as well as Simpson's rules, yield less accurate results.

The reason for this apparently counterintuitive result is that the coarse spacing of the points is not adequate to capture the variations of the forces and angles. This is particularly evident in Fig. 24.6, where we have plotted the continuous curve for the product of $F(x)$ and $\cos [\theta(x)]$. Notice how the use of seven points to characterize the continuously varying function misses the two peaks at $x = 2.5$ and 12.5 ft. The omission of these two points effectively limits the accuracy of the numerical integration estimates in Table 24.7. The fact that the two-segment trapezoidal rule yields the most accurate result is due to the chance positioning of the points for this particular problem (Fig. 24.7).

The conclusion to be drawn from Fig. 24.6 is that an adequate number of measurements must be made to accurately compute integrals. For the present case, if data were

TABLE 24.7 Estimates of work calculated using the trapezoidal rule and Simpson's rules. The percent relative error ε_t was computed in reference to a true value of the integral (129.52 ft lb) that was estimated on the basis of values at 1-ft intervals.

Technique	Segments	Work	ε_t, %
Trapezoidal	1	5.31	95.9
	2	133.19	2.84
	3	124.98	3.51
	6	119.09	8.05
Simpson's 1/3 rule	2	175.82	−35.75
	6	117.13	9.57
Simpson's 3/8 rule	3	139.93	−8.04

FIGURE 24.6
A continuous plot of $F(x) \cos [\theta(x)]$ versus position with the seven discrete points used to develop the numerical integration estimates in Table 24.7. Notice how the use of seven points to characterize this continuously varying function misses two peaks at $x = 2.5$ and 12.5 ft.

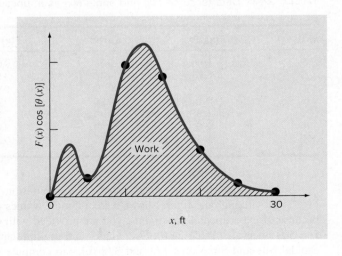

FIGURE 24.7
Graphical depiction of why the two-segment trapezoidal rule yields a good estimate of the integral for this particular case. By chance, the use of two trapezoids happens to lead to an even balance between positive and negative errors.

FIGURE 24.8
The unequal segmentation scheme that results from the inclusion of two additional points at $x = 2.5$ and 12.5 in the data in Table 24.6. The numerical integration formulas applied to each set of segments are shown.

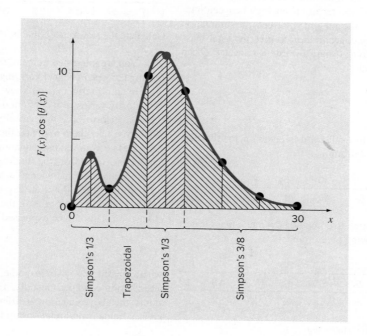

available at $F(2.5) \cos [\theta(2.5)] = 4.3500$ and $F(12.5) \cos [\theta(12.5)] = 11.3600$, we could determine an integral estimate using the algorithm for unequally spaced data described previously in Sec. 21.3. Figure 24.8 illustrates the unequal segmentation for this case. Including the two additional points yields an improved integral estimate of 126.9 ($\varepsilon_t = 2.02\%$). Thus, the inclusion of the additional data would incorporate the peaks that were missed previously and, as a consequence, lead to better results.

PROBLEMS

Chemical/Bio Engineering

24.1 Perform the same computation as in Sec. 24.1, but compute the amount of heat required to raise the temperature of 1200 g of the material from -150 to $100°C$. Use Simpson's rule for your computation, with values of T at $50°C$ increments.

24.2 Repeat Prob. 24.1, but use Romberg integration to $\varepsilon_s = 0.01\%$.

24.3 Repeat Prob. 24.1, but use a two- and a three-point Gauss-Legendre formula. Interpret your results.

24.4 Integration provides a means to compute how much mass enters or leaves a reactor over a specified time period, as in

$$M = \int_{t_1}^{t_2} Qc \, dt$$

where t_1 and t_2 = the initial and final times, respectively. This formula makes intuitive sense if you recall the analogy between integration and summation. Thus, the integral represents the summation of the product of flow times concentration to give the total mass entering or leaving from t_1 to t_2. If the flow rate is constant, Q can be moved outside the integral:

$$M = Q \int_{t_1}^{t_2} c \, dt \qquad (P24.4.1)$$

Use numerical integration to evaluate this equation for the data listed below. Note that $Q = 4$ m³/min.

t, min	0	10	20	30	35	40	45	50
c, mg/m³	10	36	55	52	41	37	33	34

24.5 Use numerical integration to compute how much mass leaves a reactor based on the following measurements.

t, min	0	10	20	30	35	40	45	50
Q, m³/min	4	4.7	5.2	5.0	4.6	4.3	4.3	4.9
c, mg/m³	10	35	55	52	42	37	32	34

24.6 *Fick's first diffusion law* states that

$$\text{Mass flux} = -D \frac{dc}{dx} \qquad (P24.6.1)$$

where mass flux = the quantity of mass that passes across a unit area per unit time (g/cm²/s), D = a diffusion coefficient (cm²/s), c = concentration, and x = distance (cm). An environmental engineer measures the following concentration of a pollutant in the

sediments underlying a lake ($x = 0$ at the sediment-water interface and increases downward):

x, cm	0	1	3
c, 10^{-6} g/cm³	0.06	0.32	0.6

Use the best numerical differentiation technique available to estimate the derivative at $x = 0$. Employ this estimate in conjunction with Eq. (P24.6.1) to compute the mass flux of pollutant out of the sediments and into the overlying waters ($D = 1.6 \times 10^{-6}$ cm²/s). For a lake with 4×10^6 m² of sediments, how much pollutant would be transported into the lake over a year's time?

24.7 The following data were collected when a large oil tanker was loading:

t, min	0	10	20	30	45	60	75
V, 10^6 barrels	0.4	0.7	0.8	0.9	1.05	1.2	1.35

Calculate the flow rate Q (that is, dV/dt) for each time with an error of $O(h^2)$.

24.8 You are interested in measuring the fluid velocity in a narrow rectangular open channel carrying petroleum waste between locations in an oil refinery. You know that, because of bottom friction, the velocity varies with depth in the channel. If your technician has time to perform only two velocity measurements, at what depths would you take them to obtain the best estimate of the average velocity? State your recommendation in terms of the percent of total depth d measured from the fluid surface. For example, measuring at the top would be 0%d, whereas at the very bottom would be 100%d.

24.9 Soft tissue follows an exponential deformation behavior in uniaxial tension while it is in the physiologic or normal range of elongation. This can be expressed as

$$\sigma = \frac{E_o}{a}(e^{a\varepsilon} - 1)$$

where σ = stress, ε = strain, and E_o and a are material constants that are determined experimentally. To evaluate the two material constants, the above equation is differentiated with respect to ε, which gives a fundamental relationship for soft tissue:

$$\frac{d\sigma}{d\varepsilon} = E_o + a\sigma$$

To evaluate E_o and a, stress-strain data are used to plot $d\sigma/d\varepsilon$ versus σ and the slope and intercept of this plot are the two material constants, respectively. The table contains stress-strain data for heart chordae tendineae (small tendons that hold heart valves closed during contraction of the heart muscle). These are data from loading the tissue; different curves are produced on unloading.

$\sigma \times 10^3$ N/m^2	87.8	96.6	176	263	350	569	833	1227	1623	2105	2677	3378	4257
$\varepsilon \times 10^{-3}$ m/m	153	198	270	320	355	410	460	512	562	614	664	716	766

(a) Calculate the derivative $d\sigma/d\varepsilon$ using finite differences that are second-order accurate. Plot the data and eliminate the data points near the zero points that appear not to follow the straight-line relationship. The error in these data comes from the inability of the instrumentation to read the small values in this region. Perform a regression analysis of the remaining data points to determine the values of E_o and a. Plot the stress versus strain data points along with the analytic curve expressed by the first equation. This will indicate how well the analytic curve matches these data.

(b) Often the previous analysis does not work well because the value of E_o is difficult to evaluate. To solve this problem, E_o is not used. A data point is selected, $(\bar{\sigma}, \bar{\varepsilon})$, that is in the middle of the range used for the regression analysis. These values are substituted into the first equation, and a value for E_o/a is determined and substituted into the first equation:

$$\sigma = \left(\frac{\bar{\sigma}}{e^{a\bar{\varepsilon}} - 1}\right)(e^{a\varepsilon} - 1)$$

Using this approach, experimental data that are well defined will produce a good match of the data points and the analytic curve. Use this new relationship and again plot the stress versus strain data points and the new analytic curve.

24.10 The standard technique for determining cardiac output is the indicator dilution method developed by Hamilton. One end of a small catheter is inserted into the radial artery and the other end is connected to a densitometer, which can automatically record the concentration of the dye in the blood. A known amount of dye, 5.6 mg, is injected rapidly, and the following data are obtained:

Time, s	Concentration, mg/L	Time, s	Concentration, mg/L
5	0	21	2.3
7	0.1	23	1.1
9	0.11	25	0.9
11	0.4	27	1.75
13	4.1	29	2.06
15	9.1	31	2.25
17	8	33	2.32
19	4.2	35	2.43

Plotting the above data results in the dye dilution curve in Fig. P24.10a. The concentration reaches a maximum value at about 15 seconds and then falls off, followed by a rise due to the recirculation of dye. The curve is replotted on a semilog graph in Fig. P24.10b. Notice that a straight line approximates the descending limb of the

FIGURE P24.10

(a)   (b)

dilution curve. In order to separate out the recirculation effect, analysts extend the straight-line portion. The cardiac output can then be calculated from

$$C = \frac{M}{A} \times 60 \text{ s/min}$$

where C = cardiac output [L/min], M = amount of injected dye (mg), and A = area under the curve with the linear correction. Calculate the cardiac output of this patient using the trapezoidal rule with a step size of 2 s.

24.11 Glaucoma is the second leading cause of vision loss worldwide. High intraocular pressure (pressure inside the eye) almost always accompanies vision loss. It is postulated that the high pressure damages a subset of cells in the eye that are responsible for vision. One investigator theorizes that the relationship between vision loss and pressure can be described by

$$VL = A \exp\left(k \int_{25}^{t} (P - 13)\, dt \right)$$

where VL is percent vision loss, P is intraocular pressure (mm Hg), t is time (years), and k and A are constants. Using the data (Table P24.11) from three patients, estimate the constants k and A.

collected the following data on the mass flux of insulin being delivered through the patch (and skin) as a function of time:

Flux, mg/cm^2/h	Time, h	Flux, mg/cm^2/h	Time, h
15	0	8	5
14	1	5	10
12	2	2.5	15
11	3	2	20
9	4	1	24

Remember that mass flux is flow rate through an area, or $(1/A)$ dm/dt. Provide your best possible estimate for the amount of drug delivered through the skin in 24 hours using a 12 cm^2 patch.

24.13 *Videoangiography* is used to measure blood flow and determine the status of circulatory function. In order to quantify the videoangiograms, blood vessel diameter and blood velocity are needed such that total blood flow is determined. Table P24.13 provides data for the densitometric profile taken from a videoangiogram of a blood vessel. One way to determine consistently where the edge of the blood vessel is from the angiogram is to determine where the first derivative of the profile is an extreme value. Using

TABLE P24.11 Data for three patients related to glaucoma

Patient	A		B		C	
Age at diagnosis	65		43		80	
VL	60		40		30	
Pressure vs. Age	**Age, years**	**P, mm Hg**	**Age, years**	**P, mm Hg**	**Age, years**	**P, mm Hg**
	25	13	25	11	25	13
	40	15	40	30	40	14
	50	22	41	32	50	15
	60	23	42	33	60	17
	65	24	43	35	80	19

24.12 One of your colleagues has designed a new transdermal patch to deliver insulin through the skin to diabetic patients in a controlled way, eliminating the need for painful injections. She has

the data provided, find the boundaries of the blood vessel and estimate the blood vessel diameter. Use both $O(h^2)$ and $O(h^4)$ centered difference formulas and compare the results.

TABLE P24.13 Densitometric profile data taken from a videoangiogram of a blood vessel.

Distance	Density	Distance	Density	Distance	Density	Distance	Density
0	26.013	28	38.273	56	39.124	84	37.331
4	26.955	32	39.103	60	38.813	88	35.980
8	26.351	36	39.025	64	38.925	92	31.936
12	28.343	40	39.432	68	38.804	96	28.843
16	31.100	44	39.163	72	38.806	100	26.309
20	34.667	48	38.920	76	38.666	104	26.146
24	37.251	52	38.631	80	38.658		

FIGURE P24.19
A stream cross section.

24.14 The *respiratory quotient* is a dimensionless number used to estimate the basal metabolic rate of microorganisms. It is computed as the ratio of carbon dioxide (CO_2) produced to the oxygen (O_2) consumed by the organisms during the breakdown of organic compounds. In addition, the rates can be integrated to determine the total amounts of CO_2 produced and O_2 consumed over the course of the experiment. The following cumulative quantities of the two gases were measured during an experiment in which bacteria broke down fats.

Time (hr)	0	1	2	3	4	5	6	7	8	9	10
CO_2 generated (g)	0	75	151	225	297	367	436	502	568	634	701
O_2 consumed (g)	0	76	149	224	301	381	463	547	631	715	801

Thus, at the end of the experiment, the *respiratory quotient* was 701/801.

(a) Use numerical differentiation to determine the instantaneous rates at each hour. Develop a plot of these versus time. Use $O(h^2)$ centered differences for the intermediate times, and forward and backward differences of order $O(h^2)$ for the initial and final times, respectively.

(b) Numerically integrate the results of **(a)** using both multiple-application trapezoidal and Simpson's 1/3 rule. Compare these results to the totals at $t = 10$ hr.

24.15 The pressure p and volume v of a given mass of gas are connected by the relation

$$(p + a/v^2)(v - b)$$

where a, b, and k are constants. Express p in terms of v, and write a script to compute the work done by the gas in expanding from an initial volume to a final volume. Test your solution with $a = 0.01$, $b = 0.001$, initial pressure and volume = 100 kPa and 1 m^3, respectively, and final volume = 2 m^3.

Civil/Environmental Engineering

24.16 Perform the same computation as in Sec. 24.2, but use $O(h^8)$ Romberg integration to evaluate the integral.

24.17 Perform the same computation as in Sec. 24.2, but use Gauss quadrature to evaluate the integral.

24.18 As in Sec. 24.2, compute F using the trapezoidal rule and Simpson's 1/3 and Simpson's 3/8 rules but use the following force. Divide the mast into 5-ft intervals.

$$F = \int_0^{30} \frac{250z}{6+z} e^{-z/10} \, dz$$

24.19 Stream cross-sectional areas (A) are required for a number of tasks in water resources engineering, including flood forecasting and reservoir designing. Unless electronic sounding devices are available to obtain continuous profiles of the channel bottom, the engineer must rely on discrete depth measurements to compute A. An example of a typical stream cross section is shown in Fig. P24.19. The data points represent locations where a boat was anchored and depth readings taken. Use two trapezoidal rule applications ($h = 4$ and 2 m) and Simpson's 1/3 rule ($h = 2$ m) to estimate the cross-sectional area from these data.

24.20 As described in Prob. 24.19, the cross-sectional area of a channel can be computed as

$$A_c = \int_0^B H(y) \, dy$$

where B = the total channel width (m), H = the depth (m), and y = distance from the bank (m). In a similar fashion, the average flow Q (m³/s) can be computed as

$$Q = \int_0^B U(y)H(y)\,dy$$

where U = water velocity (m/s). Use these relationships and a numerical method to determine A_c and Q for the following data:

y, m	0	2	4	5	6	9
H, m	0.5	1.3	1.25	1.7	1	0.25
U, m/s	0.03	0.06	0.05	0.12	0.11	0.02

24.21 The following relationships can be used to analyze uniform beams subject to distributed loads,

$$\frac{dy}{dx} = \theta(x) \quad \frac{d\theta}{dx} = \frac{M(x)}{EI} \quad \frac{dM}{dx} = V(x) \quad \frac{dV}{dx} = -w(x)$$

where x = distance along beam (m), y = deflection (m), $\theta(x)$ = slope (m/m), E = modulus of elasticity (Pa = N/m²), I = moment of inertia (m⁴), $M(x)$ = moment (N m), $V(x)$ = shear (N), and $w(x)$ = distributed load (N/m). For the case of a linearly increasing load (recall Fig. P8.21), the slope can be computed analytically as

$$\theta(x) = \frac{w_0}{120EIL}(-5x^4 + 6L^2x^2 - L^4) \tag{P24.21.1}$$

Employ **(a)** numerical integration to compute the deflection (in m) and **(b)** numerical differentiation to compute the moment (in N m) and shear (in N). Base your numerical calculations on values of the slope computed with Eq. P24.21.1 at equally spaced intervals of $\Delta x = 0.125$ m along a 3-m beam. Use the following parameter values in your computation: $E = 200$ GPa, $I = 0.0003$ m⁴, and $w_0 = 2.5$ kN/cm. In addition, the deflections at the ends of the beam are set at $y(0) = y(L) = 0$. Be careful of units.

24.22 You measure the following deflections along the length of a simply supported uniform beam (see Prob. 24.21)

x, m	0	0.375	0.75	1.125	1.5	1.875	2.25	2.625	3
y, cm	0	−0.2571	−0.9484	−1.9689	−3.2262	−4.6414	−6.1503	−7.7051	−9.275

Employ numerical differentiation to compute the slope, the moment (in N m), the shear (in N) and the distributed load (in N/m). Use the following parameter values in your computation: $E = 200$ GPa, and $I = 0.0003$ m⁴.

24.23 A transportation engineering study requires the calculation of the total number of cars that pass through an intersection over a 24-h period. An individual visits the intersection at various times during the course of a day and counts the number of cars that pass through the intersection in a minute. Utilize the data summarized in Table P24.23, to estimate the total number of cars that pass through the intersection per day. (Be careful of units.)

24.24 A wind force distributed against the side of a skyscraper is measured as

Height, l, m	0	30	60	90	120	150	180	210	240
Force, $F(l)$, N/m	0	340	1200	1600	2700	3100	3200	3500	3800

Compute the net force and the line of action due to this distributed wind force.

24.25 Water exerts pressure on the upstream face of a dam as shown in Fig. P24.25. The pressure can be characterized by

$$p(z) = \rho g(D - z) \tag{P24.25.1}$$

where $p(z)$ = pressure in pascals (or N/m²) exerted at an elevation z meters above the reservoir bottom; ρ = density of water, which for this problem is assumed to be a constant 10^3 kg/m³; g = acceleration due to gravity (9.8 m/s²); and D = elevation (in m) of the water surface above the reservoir bottom. According to Eq. (P24.25.1), pressure increases linearly with depth, as depicted in Fig. P24.25a. Omitting atmospheric pressure (because it works against both sides of the dam face and essentially cancels out), the total force f_t can be determined by multiplying pressure times the area of the dam face (as shown in Fig. P24.25b). Because both pressure and area vary with elevation, the total force is obtained by evaluating

$$f_t = \int_0^D \rho g w(z)(D - z)\,dz$$

where $w(z)$ = width of the dam face (m) at elevation z (Fig. P24.25b). The line of action can also be obtained by evaluating

$$d = \frac{\displaystyle\int_0^D \rho g z w(z)(D - z)\,dz}{\displaystyle\int_0^D \rho g w(z)(D - z)\,dz}$$

Use Simpson's rule to compute f_t and d. Check the results with your computer program for the trapezoidal rule.

24.26 To estimate the size of a new dam, you have to determine the total volume of water (m³) that flows down a river in a year's time. You have available the following long-term average data for the river:

Date	Mid-Jan.	Mid-Feb.	Mid-Mar.	Mid-Apr.	Mid-June	Mid-Sept.	Mid-Oct.	Mid-Nov.	Mid-Dec.
Flow, m³/s	30	38	82	125	95	20	22	24	35

TABLE P24.23 Traffic flow rate (cars/min) for an intersection measured at various times within a 24-h period.

Time	Rate	Time	Rate	Time	Rate
12:00 midnight	2	9:00 A.M.	11	6:00 P.M.	20
2:00 A.M.	2	10:30 A.M.	4	7:00 P.M.	10
4:00 A.M.	0	11:30 A.M.	11	8:00 P.M.	8
5:00 A.M.	2	12:30 P.M.	12	9:00 P.M.	10
6:00 A.M.	6	2:00 P.M.	8	10:00 P.M.	8
7:00 A.M.	7	4:00 P.M.	7	11:00 P.M.	7
8:00 A.M.	23	5:00 P.M.	26	12:00 midnight	3

(a)   (b)

FIGURE P24.25
Water exerting pressure on the upstream face of a dam: (a) side view showing force increasing linearly with depth; (b) front view showing width of dam in meters.

Determine the total volume. Be careful of units, and take care to make a proper estimate of flow at the end points.

24.27 The data listed in the following table give hourly measurements of heat flux q (cal/cm²/h) at the surface of a solar collector. As an architectural engineer, you must estimate the total heat absorbed by a 150,000-cm² collector panel during a 14-h period. The panel has an absorption efficiency e_{ab} of 45%. The total heat absorbed is given by

$$h = e_{ab} \int_0^t qA \, dt$$

where A = area and q = heat flux.

t	0	2	4	6	8	10	12	14
q	0.10	5.32	7.80	8.00	8.03	6.27	3.54	0.20

24.28 The heat flux q is the quantity of heat flowing through a unit area of a material per unit time. It can be computed with Fourier's law,

$$J = -k \frac{dT}{dx}$$

where J has units of J/m²/s or W/m²; k is a coefficient of thermal conductivity that parameterizes the heat-conducting properties of

the material and has units of W/(°C m); T = temperature (°C); and x = distance (m) along the path of heat flow. Fourier's law is used routinely by architectural engineers to determine heat flow through walls. The following temperatures are measured from the surface ($x = 0$) into a stone wall:

x, cm	0	0.08	0.16
T, °C	20	17	15

If the flux at $x = 0$ is 60 W/m², compute k.

24.29 The horizontal surface area A_s (m²) of a lake at a particular depth can be computed from volume by differentiation,

$$A_s(z) = -\frac{dV}{dz}(z)$$

where V = volume (m³) and z = depth (m) as measured from the surface down to the bottom. The average concentration $\bar{c}$ (g/m³) of a substance that varies with depth can be computed by integration

$$\bar{c} = \frac{\displaystyle\int_0^Z c(z)A_s(z) \, dz}{\displaystyle\int_0^Z A_s(z) \, dz}$$

where Z = the total depth (m). Determine the average concentration based on the following data:

z, m	0	4	8	12	16
V, 10^6 m^3	9.8175	5.1051	1.9635	0.3927	0.0000
c, g/m^3	10.2	8.5	7.4	5.2	4.1

24.30 During a storm, a high wind blows along one side of a rectangular skyscraper as depicted in Fig. P24.30. Use the best lower-order Newton-Cotes formulas (trapezoidal, Simpsons 1/3 and 3/8

rules) to determine **(a)** the force on the building in newtons and **(b)** the line of force in meters.

24.31 A useful application of integration is to find the arc length of a curve. If a curve in Cartesian (x-y) coordinates is given by the function $y = f(x)$, the length of the curve between $x = a$ and b can be computed as

$$L = \int_a^b \sqrt{1 + [f'(x)]^2}\, dx$$

As depicted in Fig. P24.31, a two-dimensional roller coaster track's x-y coordinates are defined by the following function:

$$f(x) = \frac{1}{(0.0025x - 0.3)^2 + 0.01} + \frac{1}{(0.0025x - 0.9)^2 + 0.04} - 1.14$$

After differentiating this equation, use numerical integration to estimate the length of track from $a = 0$ to $b = 600$

24.32 A *catenary cable* is one that is hung between two points not in the same vertical line. As depicted in Fig. P24.32, it is subject to no loads other than its own weight. Thus, its weight (N/m) acts as a uniform load per unit length along the cable. The following formula provides a way to calculate its x-y trajectory,

$$y = \frac{T_H}{w} \cosh\left(\frac{w}{T_H}x\right) + y_0 - \frac{T_H}{w}$$

where T_H = the horizontal tension force at the lowest point (N), w = the downward force of the cable due to its weight expressed as a uniform load per unit length (N/m), and y_0 = height above the x axis of the lowest point (m). Given the following parameter values: $w = 10$ N/m, $y_0 = 5$ m, and $T_H = 1.266 \times 10^{-2}$ N,

(a) Develop a plot of y from $x = -50$ to 100 m.

(b) As described in Prob. 24.31, employ integration to determine the length of the cable. Do this by generating y values at intervals of $\Delta x = 2.5$ m. Then use numerical differentiation and integration to determine the length.

FIGURE P24.30

FIGURE P24.31

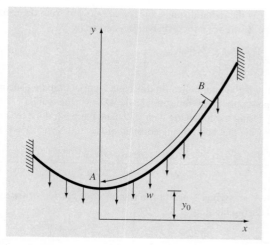

FIGURE P24.32
A catenary cable.

Electrical Engineering

24.33 Perform the same computation as in Sec. 24.3, but for the current as specified by

$$i(t) = 5e^{-1.25t} \sin 2\pi t \quad \text{for } 0 \le t \le T/2$$
$$i(t) = 0 \quad \text{for } T/2 < t \le T$$

where $T = 1$ s. Use five-point Gauss quadrature to estimate the integral.

24.34 Repeat Prob. 24.33, but use five applications of Simpson's 1/3 rule.

24.35 Repeat Prob. 24.33, but use Romberg integration to $\varepsilon_s = 1\%$.

24.36 Faraday's law characterizes the voltage drop across an inductor as

$$V_L = L\frac{di}{dt}$$

where V_L = voltage drop (V), L = inductance (in henrys; 1 H = 1 V s/A), i = current (A), and t = time (s). Determine the voltage drop as a function of time from the following data for an inductance of 4 H.

t	0	0.1	0.2	0.3	0.5	0.7
i	0	0.16	0.32	0.56	0.84	2.0

24.37 Based on Faraday's law (Prob. 24.36), use the following voltage data to estimate the inductance in henrys if a current of 2 A is passed through the inductor over 400 milliseconds.

t, ms	0	10	20	40	60	80	120	180	280	400
V, volts	0	18	29	44	49	46	35	26	15	7

24.38 Suppose that the current through a resistor is described by the function

$$i(t) = (60 - t)^2 + (60 - t)\sin(\sqrt{t})$$

and the resistance is a function of the current,

$$R = 10i + 2i^{2/3}$$

Compute the average voltage over $t = 0$ to 60 s using the multiple-segment Simpson's 1/3 rule.

24.39 If a capacitor initially holds no charge, the voltage across it as a function of time can be computed as

$$V(t) = \frac{1}{C}\int_0^t i(t)\,dt$$

If $C = 10^{-5}$ farad, use the following current data to develop a plot of voltage versus time:

t, s	0	0.2	0.4	0.6	0.8	1	1.2
i, 10^{-3} A	0.2	0.3683	0.3819	0.2282	0.0486	0.0082	0.1441

Mechanical/Aerospace Engineering

24.40 Perform the same computation as in Sec. 24.4, but use the following equations:

$$F(x) = 1.6x - 0.045x^2$$
$$\theta(x) = 0.8 + 0.125x - 0.009x^2 + 0.0002x^3$$

Use 4-, 8-, and 16-segment trapezoidal rules to compute the integral.

24.41 Repeat Prob. 24.40, but use **(a)** Simpson's 1/3 rule, **(b)** Romberg integration to $\varepsilon_s = 0.5\%$, and **(c)** Gauss quadrature.

24.42 Compute work as described in Sec. 24.4, but use the following equations for $F(x)$ and $\theta(x)$:

$$F(x) = 1.6x - 0.045x^2$$
$$\theta(x) = -0.00055x^3 + 0.0123x^2 + 0.13x$$

The force is in newtons and the angle is in radians. Perform the integration from $x = 0$ to 30 m.

24.43 As was done in Sec. 24.4, determine the work performed if a constant force of 1 N applied at an angle θ results in the following displacements. Use the MATLAB function cumtrapz to determine the cumulative work and plot the result versus θ.

x, m	0	1	2.7	3.8	3.7	3	1.4
θ, deg	0	30	60	90	120	150	180

24.44 The work done on an object is equal to the force times the distance moved in the direction of the force. The velocity of an object in the direction of a force is given by

$$v = 4t \quad 0 \le t \le 4$$
$$v = 16 + (4 - t)^2 \quad 4 \le t \le 14$$

where v = m/s. Employ the multiple-application Simpson's rule to determine the work if a constant force of 200 N is applied for all t.

24.45 The rate of cooling of a body (Fig. P24.45) can be expressed as

$$\frac{dT}{dt} = -k(T - T_a)$$

where T = temperature of the body (°C), T_a = temperature of the surrounding medium (°C), and k = a proportionality constant (per minute). Thus, this equation (called *Newton's law of cooling*) specifies that the rate of cooling is proportional to the difference in the temperatures of the body and of the surrounding medium. If a metal ball heated to 80°C is dropped into water that is held constant at $T_a = 20$°C, the temperature of the ball changes, as in

Time, min	0	5	10	15	20	25
T, °C	80	44.5	30.0	24.1	21.7	20.7

Utilize numerical differentiation to determine dT/dt at each value of time. Plot dT/dt versus $T - T_a$ and employ linear regression to evaluate k.

FIGURE P24.45

24.46 A rod subject to an axial load (Fig. P24.46a) will be deformed, as shown in the stress-strain curve in Fig. P24.46b. The area under the curve from zero stress out to the point of rupture is called the *modulus of toughness* of the material. It provides a measure of the energy per unit volume required to cause the material to rupture. As such, it is representative of the material's ability to withstand an impact load. Use numerical integration to compute the modulus of toughness for the stress-strain curve seen in Fig. P24.46b.

24.47 If the velocity distribution of a fluid flowing through a pipe is known (Fig. P24.47), the flow rate Q (that is, the volume of water passing through the pipe per unit time) can be computed by $Q = \int v \, dA$, where v is the velocity and A is the pipe's cross-sectional area. (To grasp the meaning of this relationship physically, recall the close connection between summation and integration.) For a circular pipe, $A = \pi r^2$ and $dA = 2\pi r \, dr$. Therefore,

$$Q = \int_0^r v(2\pi r) \, dr$$

where r is the radial distance measured outward from the center of the pipe. If the velocity distribution is given by

$$v = 2\left(1 - \frac{r}{r_0}\right)^{1/6}$$

where r_0 is the total radius (in this case, 3 cm), compute Q using the multiple-application trapezoidal rule. Discuss the results.

24.48 Using the following data, calculate the work done by stretching a spring that has a spring constant of $k = 300$ N/m to $x = 0.35$ m:

F, 10^3 N	0	0.01	0.028	0.046	0.063	0.082	0.11	0.13
x, m	0	0.05	0.10	0.15	0.20	0.25	0.30	0.35

24.49 A jet fighter's position on an aircraft carrier's runway was timed during landing:

t, s	0	0.52	1.04	1.75	2.37	3.25	3.83
x, m	153	185	208	249	261	271	273

where x is the distance from the end of the carrier. Estimate **(a)** velocity (dx/dt) and **(b)** acceleration (dv/dt) using numerical differentiation.

24.50 Employ the multiple-application Simpson's rule to evaluate the vertical distance traveled by a rocket if the vertical velocity is given by

$$\begin{aligned} v &= 11t^2 - 5t & 0 \le t \le 10 \\ v &= 1100 - 5t & 10 \le t \le 20 \\ v &= 50t + 2(t - 20)^2 & 20 \le t \le 30 \end{aligned}$$

In addition, use numerical differentiation to develop graphs of the acceleration (dv/dt) and the *jerk* (d^2v/dt^2) versus time for $t = 0$ to 30 s. Note that the jerk is very important because it is highly correlated with injuries such as whiplash.

24.51 The upward velocity of a rocket can be computed by the following formula:

$$v = u \ln\left(\frac{m_0}{m_0 - qt}\right) - gt$$

where v = upward velocity, u = velocity at which fuel is expelled relative to the rocket, m_0 = initial mass of the rocket at time $t = 0$, q = fuel consumption rate, and g = downward acceleration due to gravity (assumed constant = 9.8 m/s^2). If $u = 1800$ m/s, $m_0 = 160{,}000$ kg, and $q = 2500$ kg/s, use six-segment trapezoidal and Simpson's 1/3 rule, six-point Gauss quadrature, and $O(h^8)$ Romberg integration to determine how high the rocket will fly in 30 s. In addition, use numerical differentiation to generate a graph of acceleration as a function of time.

24.52 Referring to the data from Problem 20.61, find the strain rate using finite-difference methods. Use second-order accurate derivative approximations and plot your results. Looking at the

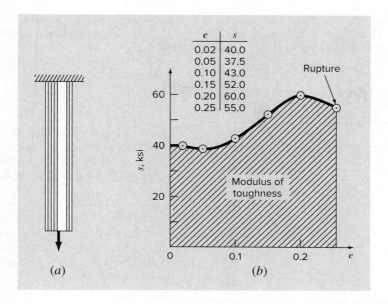

FIGURE P24.46

(a) A rod under axial loading and (b) the resulting stress-strain curve where stress is in kips per square inch (10^3 lb/in^2) and strain is dimensionless.

FIGURE P24.47

graph, it is apparent that there is some experimental startup error. Find the mean and standard deviation of the strain rate after eliminating the data points representing the experimental startup error.

24.53 Fully developed flow moving through a 40-cm-diameter pipe has the following velocity profile:

Radius, r, cm	0.0	2.5	5.0	7.5	10.0	12.5	15.0	17.5	20.0
Velocity, v, m/s	0.914	0.890	0.847	0.795	0.719	0.543	0.427	0.204	0

Find the volume flow rate Q using the relationship $Q = \int_0^R 2\pi r v \, dr$, where r is the radial axis of the pipe, R is the radius of the pipe, and v is the velocity. Solve the problem using two different approaches.

(a) Fit a polynomial curve to the velocity data and integrate analytically.

(b) Use multiple-application Simpson's 1/3 rule to integrate.

(c) Find the percent error using the integral of the polynomial fit as the more correct value.

24.54 Fully developed flow of a Bingham plastic fluid moving through a 12-in diameter pipe has the given velocity profile. The flow of a Bingham fluid does not shear the center core, producing plug flow in the region around the centerline.

Radius, r, in	0	1	2	3	4	5	6
Velocity, v, ft/s	5.00	5.00	4.62	4.01	3.42	1.69	0.00

Find the total volume flow rate Q using the relationship $Q = \int_{r_1}^{r_2} 2\pi r v \, dr + v_c A_c$, where r is the radial axis of the pipe, v is the velocity, v_c is the velocity at the core, and A_c is the cross-sectional area of the plug. Solve the problem using two different approaches.

(a) Fit a polynomial curve to the noncore data and integrate.

(b) Use multiple-application Simpson's rule to integrate.

(c) Find the percent error using the integral of the polynomial fit as the more correct value.

24.55 The enthalpy of a real gas is a function of pressure as described in Eq. (P24.55.1). These data were taken for a real fluid. Estimate the enthalpy of the fluid at 400 K and 50 atm (evaluate the integral from 0.1 atm to 50 atm).

$$H = \int_0^P \left(V - T \left(\frac{\partial V}{\partial T} \right)_P \right) dP \qquad \text{(P24.55.1)}$$

P, atm	V, L		
	T = 350 K	T = 400 K	T = 450 K
0.1	220	250	282.5
5	4.1	4.7	5.23
10	2.2	2.5	2.7
20	1.35	1.49	1.55
25	1.1	1.2	1.24
30	0.90	0.99	1.03
40	0.68	0.75	0.78
45	0.61	0.675	0.7
50	0.54	0.6	0.62

24.56 Given the data below, find the isothermal work done on the gas as it is compressed from 23 L to 3 L (remember that $W = -\int_{V_1}^{V_2} P \, dV$).

V, L	3	8	13	18	23
P, atm	12.5	3.5	1.8	1.4	1.2

(a) Find the work performed on the gas numerically, using the 1-, 2-, and 4-segment trapezoidal rule.

(b) Compute the ratios of the errors in these estimates and relate them to the error analysis of the multiple-application trapezoidal rule discussed in Chap. 21.

24.57 The *Rosin-Rammler-Bennet (RRB) equation* is used to describe size distribution in fine dust. $F(x)$ represents the cumulative mass of dust particles of diameter x and smaller; x' and n' are constants equal to 30 μm and 1.44, respectively. The mass density distribution $f(x)$, or the mass of dust particles of a diameter x, is found by taking the derivative of the cumulative distribution:

$$F(x) = 1 - e^{-(x/x')^{n'}} \qquad f(x) = \frac{dF(x)}{dx}$$

(a) Numerically calculate the mass density distribution $f(x)$ and graph both $f(x)$ and the cumulative distribution $F(x)$.

(b) Using your results from part (a), calculate the mode size of the mass density distribution—that is, the size at which the derivative of $f(x)$ is equal to zero.

(c) Find the surface area per mass of the dust S_m (cm²/g) using

$$S_m = \frac{6}{\rho} \int_{d_{min}}^{\infty} \frac{f(x)}{x} dx$$

The equation is valid only for spherical particles. Assume a density $\rho = 1$ g cm^{-3} and a minimum diameter of dust included in the distribution d_{min} of 1 μm.

24.58 For fluid flow over a surface, the heat flux to the surface can be computed as

$$J = -k \frac{dT}{dy}$$

where J = heat flux (W/m²), k = thermal conductivity (W/m K), T = temperature (K), and y = distance normal to the surface (m). The following measurements are made for air flowing over a flat plate that is 200 cm long and 50 cm wide:

y, cm	0	1	3	5
T, K	900	480	270	200

If $k = 0.028$ J/s m K, (a) determine the flux at the surface and (b) the heat transfer in watts. Note that 1 J = 1 W s.

24.59 The pressure gradient for laminar flow through a constant radius tube is given by

$$\frac{dp}{dx} = -\frac{8\mu \, Q}{\pi r^4}$$

where p = pressure (N/m²), x = distance along the tube's centerline (m), μ = dynamic viscosity (N s/m²), Q = flow (m³/s), and r = radius (m).

(a) Determine the pressure drop for a 10-cm length tube for a viscous liquid ($\mu = 0.005$ N s/m², density = $\rho = 1 \times 10^3$ kg/m³) with a flow of 10×10^{-6} m³/s and the following varying radii along the tube's length:

x, cm	0	2	4	5	6	7	10
r, mm	2	1.35	1.34	1.6	1.58	1.42	2

(b) Compare your result with the pressure drop that would have occurred if the tube had a constant radius equal to the average radius.

(c) Determine the average Reynolds number for the tube to verify that flow is truly laminar (Re = $\rho v D / \mu < 2100$, where v = velocity).

24.60 Velocity data for air are collected at different radii from the centerline of a circular 16-cm-diameter pipe as tabulated below:

r, cm	0	1.60	3.20	4.80	6.40	7.47	7.87	7.95	8
v, m/s	10	9.69	9.30	8.77	7.95	6.79	5.57	4.89	0

Use numerical integration to determine the mass flow rate, which can be computed as

$$\int_0^R \rho v 2\pi r \, dr$$

where ρ = density (= 1.2 kg/m³). Express your results in kg/s.

24.61 The following data are provided for the velocity of an object as a function of time,

t, s	0	4	8	12	16	20	24	28	30
v, m/s	0	18	31	42	50	56	61	65	70

(a) Limiting yourself to the trapezoidal rule and Simpson's 1/3 and 3/8 rules, make the best estimate of how far the object travels from $t = 0$ to 30 s.

(b) Employ the results of **(a)** to compute the average velocity.

24.62 The total mass of a variable-density rod is given by

$$m = \int_0^L \rho(x) A_c(x)\, dx$$

where m = mass, $\rho(x)$ = density, $A_c(x)$ = cross-sectional area, and x = distance along the rod. The following data have been measured for a 10-m length rod:

x (m)	0	2	3	4	6	8	10
ρ (g/cm³)	4.00	3.95	3.89	3.80	3.60	3.41	3.30
A_c (cm²)	100	103	106	110	120	133	150

Determine the mass *in grams* to the best possible accuracy limiting yourself to the trapezoidal rule and Simpson's 1/3 and 3/8 rules.

24.63 A gas that is expanding in an engine cylinder follows the law

$$PV^{1.3} = c$$

where c is a constant. The initial pressure is 2550 kPa, and the final pressure is 210 kPa. If the volume at the end of expansion is 0.75 m³, compute the work done by the gas.

EPILOGUE: PART SIX

PT6.4 TRADE-OFFS

Table PT6.4 provides a summary of the trade-offs involved in numerical integration or quadrature. Most of these methods are based on the simple physical interpretation of an integral as the area under a curve. These techniques are designed to evaluate the integral of two different cases: (1) a mathematical function and (2) discrete data in tabular form.

The Newton-Cotes formulas are the primary methods discussed in Chap. 21. They are applicable to both continuous and discrete functions. Both closed and open versions of these formulas are available. The open forms, which have integration limits that extend beyond the range of the data, are rarely used for the evaluation of definite integrals. However, they have utility for the solution of ordinary differential equations and for evaluating improper integrals.

The closed Newton-Cotes formulas are based on replacing a mathematical function or tabulated data by an interpolating polynomial that is easy to integrate. The simplest version is the trapezoidal rule, which is based on taking the area below a straight line joining adjacent values of the function. One way to improve the accuracy of the trapezoidal rule is to divide the integration interval from a to b into a number of segments and apply the method to each segment.

Aside from applying the trapezoidal rule with finer segmentation, another way to obtain a more accurate estimate of the integral is to use higher-order polynomials to

TABLE PT6.4 Comparison of the characteristics of alternative methods for numerical integration. The comparisons are based on general experience and do not account for the behavior of special functions.

Method	Data Points Required for One Application	Data Points Required for n Applications	Truncation Error	Application	Programming Effort	Comments
Trapezoidal rule	2	$n + 1$	$\cong h^3 f''(\xi)$	Wide	Easy	
Simpson's 1/3 rule	3	$2n + 1$	$\cong h^5 f^{(4)}(\xi)$	Wide	Easy	
Simpson's rule (1/3 and 3/8)	3 or 4	≥ 3	$\cong h^5 f^{(4)}(\xi)$	Wide	Easy	
Higher-order Newton-Cotes	≥ 5	N/A	$\cong h^7 f^{(6)}(\xi)$	Rare	Easy	
Romberg integration	3			Requires $f(x)$ be known	Moderate	Inappropriate for tabular data
Gauss quadrature	≥ 2	N/A		Requires $f(x)$ be known	Easy	Inappropriate for tabular data
Monte Carlo Integration	N/A	N/A	N/A	Requires $f(x)$ be known	Moderate	Inappropriate for tabular data

connect the points. If a quadratic equation is employed, the result is Simpson's 1/3 rule. If a cubic equation is used, the result is Simpson's 3/8 rule. Because they are much more accurate than the trapezoidal rule, these formulas are usually preferred and multiple-application versions are available. For situations with an even number of segments, the multiple application of the 1/3 rule is recommended. For an odd number of segments, the 3/8 rule can be applied to the last three segments and the 1/3 rule to the remaining segments.

Higher-order Newton-Cotes formulas are also available. However, they are rarely used in practice. Where high accuracy is required, Romberg integration, adaptive quadrature, and Gauss quadrature methods may be applied. It should be noted that these approaches are usually of practical value only in cases where the function is available. Finally, Monte Carlo integration provides a means to estimate integrals using large numbers of random samples of function values. These techniques are ill-suited for tabulated data.

PT6.5   IMPORTANT RELATIONSHIPS AND FORMULAS

Table PT6.5 summarizes important formulas presented in Part Six. This table can be consulted to quickly access important relationships and formulas.

PT6.6   ADVANCED METHODS AND ADDITIONAL REFERENCES

Although we have reviewed a number of numerical integration techniques, there are other methods that have utility in engineering practice. For example, adaptive schemes for solving ordinary differential equations can be used to evaluate complicated integrals, as will be discussed in Chap. 25.

Another method for obtaining integrals is to fit *cubic splines* to the data. The resulting cubic equations can be integrated easily (Forsythe et al. 1977). A similar approach is also sometimes used for differentiation. Finally, aside from the Gauss-Legendre formulas discussed in Sec. 22.4, there are a variety of other quadrature formulas. Carnahan, Luther, and Wilkes (1969) and Ralston and Rabinowitz (1978) summarize many of these approaches.

In summary, the foregoing is intended to provide you with avenues for deeper exploration of the subject. Additionally, all the above references describe basic techniques covered in Part Six. We urge you to consult these alternative sources to broaden your understanding of numerical methods for integration.

TABLE PT6.5 Summary of important formulas presented in Part Six.

Method	Formulation	Graphic Interpretations	Error
Trapezoidal rule	$I \cong (b - a) \dfrac{f(a) + f(b)}{2}$		$-\dfrac{(b - a)^3}{12} f''(\xi)$
Multiple-application trapezoidal rule	$I \cong (b - a) \dfrac{f(x_0) + 2 \sum\limits_{i=1}^{n-1} f(x_i) + f(x_n)}{2n}$		$-\dfrac{(b - a)^3}{12n^2} \bar{f}''$
Simpson's 1/3 rule	$I \cong (b - a) \dfrac{f(x_0) + 4f(x_1) + f(x_2)}{6}$		$-\dfrac{(b - a)^5}{2880} f^{(4)}(\xi)$
Multiple-application Simpson's 1/3 rule	$I \cong (b - a) \dfrac{f(x_0) + 4 \sum\limits_{i=1,3}^{n-1} f(x_i) + 2 \sum\limits_{j=2,4}^{n-2} f(x_j) + f(x_n)}{3n}$		$-\dfrac{(b - a)^5}{180n^4} f^{(4)}$
Simpson's 3/8 rule	$I \cong (b - a) \dfrac{f(x_0) + 3f(x_1) + 3f(x_2) + f(x_3)}{8}$		$-\dfrac{(b - a)^5}{6480} f^{(4)}(\xi)$
Romberg integration	$I_{j,k} = \dfrac{4^{k-1} I_{j+1,\, k-1} - I_{j,\, k-1}}{4^{k-1} - 1}$		$O(h^{2k})$
Gauss quadrature	$I \cong c_0 f(x_0) + c_1 f(x_1) + \cdots + c_{n-1} f(x_{n-1})$		$\cong f^{(2n+2)}(\xi)$

PART SEVEN

ORDINARY DIFFERENTIAL EQUATIONS

PT7.1 MOTIVATION

In the first chapter of this book, we derived the following equation based on Newton's second law to compute the velocity v of a falling parachutist as a function of time t [recall Eq. (1.9)]:

$$\frac{dv}{dt} = g - \frac{c}{m}v \qquad\qquad (PT7.1)$$

where g is the gravitational constant, m is the mass, and c is a drag coefficient. Such equations, which are composed of an unknown function and its derivatives, are called *differential equations.* Equation (PT7.1) is sometimes referred to as a *rate equation* because it expresses the rate of change of a variable as a function of variables and parameters. Such equations play a fundamental role in engineering because many physical phenomena are best formulated mathematically in terms of their rate of change.

In Eq. (PT7.1), the quantity being differentiated, v, is called the *dependent variable.* The quantity with respect to which v is differentiated, t, is called the *independent variable.* When the function involves one independent variable, the equation is called an *ordinary differential equation* (or *ODE*). This is in contrast to a *partial differential equation* (or *PDE*) that involves two or more independent variables.

Differential equations are also classified as to their order. For example, Eq. (PT7.1) is called a *first-order equation* because the highest derivative is a first derivative. A *second-order equation* would include a second derivative. For example, the equation describing the position x of a mass-spring system with damping is the second-order equation,

$$m\frac{d^2x}{dt^2} + c\frac{dx}{dt} + kx = 0 \qquad\qquad (PT7.2)$$

where c is a damping coefficient and k is a spring constant. Similarly, an nth-order equation would include an nth derivative.

Higher-order equations can be reduced to a system of first-order equations. For Eq. (PT7.2), this is done by defining a new variable y, where

$$y = \frac{dx}{dt} \tag{PT7.3}$$

which itself can be differentiated to yield

$$\frac{dy}{dt} = \frac{d^2x}{dt^2} \tag{PT7.4}$$

Equations (PT7.3) and (PT7.4) can then be substituted into Eq. (PT7.2) to give

$$m\frac{dy}{dt} + cy + kx = 0 \tag{PT7.5}$$

or

$$\frac{dy}{dt} = -\frac{cy + kx}{m} \tag{PT7.6}$$

Thus, Eqs. (PT7.3) and (PT7.6) are a pair of first-order equations that are equivalent to the original second-order equation. Because other nth-order differential equations can be similarly reduced, this part of the book focuses on the solution of first-order equations. Some of the engineering applications in Chap. 28 deal with the solution of second-order ODEs by reduction to a pair of first-order equations.

PT7.1.1 Noncomputer Methods for Solving ODEs

Without computers, ODEs are usually solved with analytical integration techniques. For example, Eq. (PT7.1) could be multiplied by dt and integrated to yield

$$v = \int \left(g - \frac{c}{m}v \right) dt \tag{PT7.7}$$

The right-hand side of this equation is called an *indefinite integral* because the limits of integration are unspecified. This is in contrast to the definite integrals discussed previously in Part Six [compare Eq. (PT7.7) with Eq. (PT6.6)].

An analytical solution for Eq. (PT7.7) is obtained if the indefinite integral can be evaluated exactly in equation form. For example, recall that for the falling parachutist problem, Eq. (PT7.7) was solved analytically by Eq. (1.10) (assuming $v = 0$ at $t = 0$):

$$v(t) = \frac{gm}{c}(1 - e^{-(c/m)t}) \tag{1.10}$$

The mechanics of deriving such analytical solutions will be discussed in Sec. PT7.2. For the time being, the important fact is that exact solutions for many ODEs of practical importance are not available. As is true for most situations discussed in other parts of this book, numerical methods offer the only viable alternative for these cases. Because these numerical methods usually require computers, engineers in the precomputer era were somewhat limited in the scope of their investigations.

FIGURE PT7.1
The swinging pedulum.

One very important method that engineers and applied mathematicians developed to overcome this dilemma was *linearization*. A linear ordinary differential equation is one that fits the general form

$$a_n(x)y^{(n)} + \cdots + a_1(x)y' + a_0(x)y = f(x) \tag{PT7.8}$$

where $y^{(n)}$ is the nth derivative of y with respect to x and the a's and f's are specified functions of x. This equation is called *linear* because there are no products or nonlinear functions of the dependent variable y and its derivatives. The practical importance of linear ODEs is that they can be solved analytically. In contrast, most nonlinear equations cannot be solved exactly. Thus, in the precomputer era, one tactic for solving nonlinear equations was to linearize them.

A simple example is the application of ODEs to predict the motion of a swinging pendulum (Fig. PT7.1). In a manner similar to the derivation of the falling parachutist problem, Newton's second law can be used to develop the following differential equation (see Sec. 28.4 for the complete derivation):

$$\frac{d^2\theta}{dt^2} + \frac{g}{l}\sin\theta = 0 \tag{PT7.9}$$

where θ is the angle of displacement of the pendulum, g is the gravitational constant, and l is the pendulum length. This equation is nonlinear because of the term $\sin\theta$. One way to obtain an analytical solution is to realize that for small displacements of the pendulum from equilibrium (that is, for small values of θ),

$$\sin\theta \cong \theta \tag{PT7.10}$$

Thus, if it is assumed that we are interested only in cases where θ is small, Eq. (PT7.10) can be substituted into Eq. (PT7.9) to give

$$\frac{d^2\theta}{dt^2} + \frac{g}{l}\theta = 0 \tag{PT7.11}$$

We have, therefore, transformed Eq. (PT7.9) into a linear form that is easy to solve analytically.

Although linearization remains a very valuable tool for engineering problem solving, there are cases where it cannot be invoked. For example, suppose that we were interested in studying the behavior of the pendulum for large displacements from equilibrium. In such instances, numerical methods offer a viable option for obtaining solutions. Today, the widespread availability of computers places this option within reach of all practicing engineers.

PT7.1.2 ODEs and Engineering Practice

The fundamental laws of physics, mechanics, electricity, and thermodynamics are usually based on empirical observations that explain variations in physical properties and states of systems. Rather than describing the states of physical systems directly, the laws are usually couched in terms of spatial and temporal changes.

Several examples are listed in Table PT7.1. These laws define mechanisms of change. When combined with continuity laws for energy, mass, or momentum, differential equations

TABLE PT7.1 Examples of fundamental laws that are written in terms of the rate of change of variables (t = time and x = position).

Law	Mathematical Expression	Variables and Parameters
Newton's second law of motion	$\dfrac{dv}{dt} = \dfrac{F}{m}$	Velocity (v), force (F), and mass (m)
Fourier's heat law	$q = -k' \dfrac{dT}{dx}$	Heat flux (q), thermal conductivity (k') and temperature (T)
Fick's law of diffusion	$J = -D \dfrac{dc}{dx}$	Mass flux (J), diffusion coefficient (D), and concentration (c)
Faraday's law (voltage drop across an inductor)	$\Delta V_L = L \dfrac{di}{dt}$	Voltage drop (ΔV_L), inductance (L), and current (i)

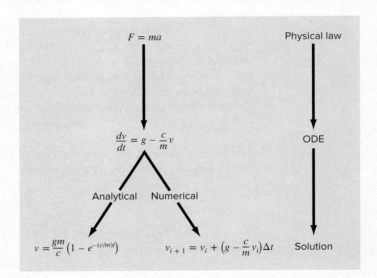

FIGURE PT7.2
The sequence of events in the application of ODEs for engineering problem solving. The example shown is the velocity of a falling parachutist.

result. Subsequent integration of these differential equations results in mathematical functions that describe the spatial and temporal state of a system in terms of energy, mass, or velocity variations.

The falling parachutist problem introduced in Chap. 1 is an example of the derivation of an ordinary differential equation from a fundamental law. Recall that Newton's second law was used to develop an ODE describing the rate of change of velocity of a falling parachutist. By integrating this relationship, we obtained an equation to predict fall velocity as a function of time (Fig. PT7.2). This equation could be utilized in a number of different ways, including design purposes.

In fact, such mathematical relationships are the basis of the solution for a great number of engineering problems. However, as described in the previous section, many of the differential equations of practical significance cannot be solved using the analytical methods of calculus. Thus, the methods discussed in the following chapters are extremely important in all fields of engineering.

PT7.2   MATHEMATICAL BACKGROUND

A solution of an ordinary differential equation is a specific function of the independent variable and parameters that satisfies the original differential equation. To illustrate this concept, let us start with a given function,

$$y = -0.5x^4 + 4x^3 - 10x^2 + 8.5x + 1 \qquad (PT7.12)$$

which is a fourth-order polynomial (Fig. PT7.3a). Now, if we differentiate Eq. (PT7.12), we obtain an ODE:

$$\frac{dy}{dx} = -2x^3 + 12x^2 - 20x + 8.5 \qquad (PT7.13)$$

This equation also describes the behavior of the polynomial, but in a manner different from Eq. (PT7.12). Rather than explicitly representing the values of y for each value of x, Eq. (PT7.13) gives the rate of change of y with respect to x (that is, the slope) at every value of x. Figure PT7.3 shows both the function and the derivative plotted versus x. Notice how the zero values of the derivatives correspond to the point at which the original function is flat—that is, has a zero slope. Also, the maximum absolute values of the derivatives are at the ends of the interval where the slopes of the function are greatest.

Although, as just demonstrated, we can determine a differential equation given the original function, the object here is to determine the original function given the differential equation. The original function then represents the solution. For the present case, we can determine this solution analytically by integrating Eq. (PT7.13):

$$y = \int (-2x^3 + 12x^2 - 20x + 8.5)\, dx$$

Applying the integration rule (recall Table PT6.2),

$$\int u^n\, du = \frac{u^{n+1}}{n+1} + C \qquad n \neq -1$$

to each term of the equation gives the solution,

$$y = -0.5x^4 + 4x^3 - 10x^2 + 8.5x + C \qquad (PT7.14)$$

which is identical to the original function with one notable exception. In the course of differentiating and then integrating, we lost the constant value of 1 in the original equation and gained the value C. This C is called a *constant of integration*. The fact that such an arbitrary constant appears indicates that the solution is not unique. In fact, it is but one of an infinite number of possible functions (corresponding to an infinite number of

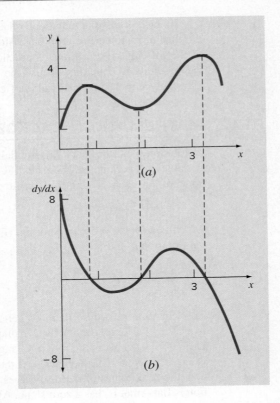

FIGURE PT7.3
Plots of (a) y versus x and (b) dy/dx versus x for the function
$y = -0.5x^4 + 4x^3 - 10x^2 + 8.5x + 1$.

possible values of C) that satisfy the differential equation. For example, Fig. PT7.4 shows six possible functions that satisfy Eq. (PT7.14).

Therefore, to specify the solution completely, a differential equation is usually accompanied by *auxiliary conditions*. For first-order ODEs, a type of auxiliary condition called an *initial value* is required to determine the constant and obtain a unique solution. For example, Eq. (PT7.13) could be accompanied by the initial condition that at $x = 0$, $y = 1$. These values could be substituted into Eq. (PT7.14):

$$1 = -0.5(0)^4 + 4(0)^3 - 10(0)^2 + 8.5(0) + C \qquad \text{(PT7.15)}$$

to determine $C = 1$. Therefore, the unique solution that satisfies both the differential equation and the specified initial condition is obtained by substituting $C = 1$ into Eq. (PT7.14) to yield

$$y = -0.5x^4 + 4x^3 - 10x^2 + 8.5x + 1 \qquad \text{(PT7.16)}$$

Thus, we have "pinned down" Eq. (PT7.14) by forcing it to pass through the initial condition, and in so doing, we have developed a unique solution to the ODE and have come full circle to the original function [Eq. (PT7.12)].

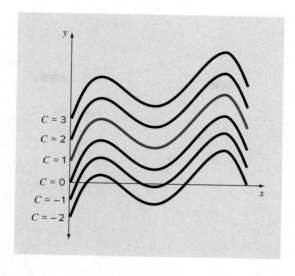

FIGURE PT7.4

Six possible solutions for the integral of $-2x^3 + 12x^2 - 20x + 8.5$. Each conforms to a different value of the constant of integration C.

Initial conditions usually have very tangible interpretations for differential equations derived from physical problem settings. For example, in the falling parachutist problem, the initial condition was reflective of the physical fact that at time zero the vertical velocity was zero. If the parachutist had already been in vertical motion at time zero, the solution would have been modified to account for this initial velocity.

When dealing with an nth-order differential equation, n conditions are required to obtain a unique solution. If all conditions are specified at the same value of the independent variable (for example, at x or $t = 0$), then the problem is called an *initial-value problem*. This is in contrast to *boundary-value problems* where specification of conditions occurs at different values of the independent variable. Chapters 25 and 26 will focus on initial-value problems. Boundary-value problems are covered in Chap. 27 along with eigenvalues.

PT7.3   ORIENTATION

Before proceeding to numerical methods for solving ordinary differential equations, some orientation might be helpful. The following material is intended to provide you with an overview of the material discussed in Part Seven. In addition, we have formulated objectives to focus your studies of the subject area.

PT7.3.1 Scope and Preview

Figure PT7.5 provides an overview of Part Seven. Two broad categories of numerical methods for initial-value problems will be discussed in this part of this book. One-step methods, which are covered in Chap. 25, permit the calculation of y_{i+1}, given the

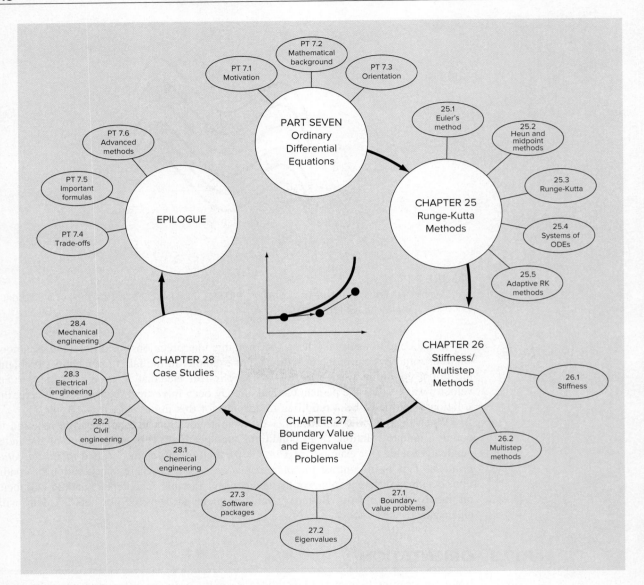

FIGURE PT7.5
Schematic representation of the organization of Part Seven: Ordinary Differential Equations.

differential equation and y_i. Multistep methods, which are covered in Chap. 26, require additional values of y other than at i.

With all but a minor exception, the *one-step methods* in *Chap. 25* belong to what are called Runge-Kutta techniques. Although the chapter might have been organized around this theoretical notion, we have opted for a more graphical, intuitive approach to introduce the methods. Thus, we begin the chapter with *Euler's method,* which has a very straightforward graphical interpretation. Then, we use visually oriented arguments to develop two improved versions of Euler's method—the *Heun* and the *midpoint* techniques. After this introduction, we formally develop the concept of *Runge-Kutta* (or *RK*) approaches and demonstrate how the foregoing techniques are actually first- and second-order RK methods. This is followed by a discussion of the higher-order RK formulations that are frequently used for engineering problem solving. In addition, we cover the application of one-step methods to *systems of ODEs*. Finally, the chapter ends with a discussion of *adaptive RK methods* that automatically adjust the step size in response to the truncation error of the computation.

Chapter 26 starts with a description of *stiff ODEs.* These are both individual and systems of ODEs that have both fast and slow components to their solution. We introduce the idea of an *implicit solution* technique as one commonly used remedy for this problem.

Next, we discuss *multistep methods.* These algorithms retain information of previous steps to more effectively capture the trajectory of the solution. They also yield the truncation error estimates that can be used to implement step-size control. In this section, we initially take a visual, intuitive approach by using a simple method—the *non-self-starting Heun*—to introduce all the essential features of the multistep approaches.

In *Chap. 27* we turn to *boundary-value* and *eigenvalue* problems. For the former, we introduce both *shooting* and *finite-difference methods*. For the latter, we discuss several approaches, including the *polynomial* and the *power methods*. Finally, the chapter concludes with a description of the application of several *software packages* and *libraries* for solution of ODEs and eigenvalues.

Chapter 28 is devoted to applications from all the fields of engineering. Finally, a short review section is included at the end of Part Seven. This epilogue summarizes and compares the important formulas and concepts related to ODEs. The comparison includes a discussion of trade-offs that are relevant to their implementation in engineering practice. The epilogue also summarizes important formulas and includes references for advanced topics.

PT7.3.2 Goals and Objectives

Study Objectives.   After completing Part Seven, you should have greatly enhanced your capability to confront and solve ordinary differential equations and eigenvalue problems. General study goals should include mastering the techniques, having the capability to assess the reliability of the answers, and being able to choose the "best" method (or methods) for any particular problem. In addition to these general objectives, the specific study objectives in Table PT7.2 should be mastered.

Computer Objectives.   Algorithms are provided for many of the methods in Part Seven. This information will allow you to expand your software library. For example,

TABLE PT7.2 Specific study objectives for Part Seven.

1. Understand the visual representations of Euler's, Heun's, and the midpoint methods.
2. Know the relationship of Euler's method to the Taylor series expansion and the insight it provides regarding the error of the method.
3. Understand the difference between local and global truncation errors and how they relate to the choice of a numerical method for a particular problem.
4. Know the order and the step-size dependency of the global truncation errors for all the methods described in Part Seven; understand how these errors bear on the accuracy of the techniques.
5. Understand the basis of predictor-corrector methods; in particular, realize that the efficiency of the corrector is highly dependent on the accuracy of the predictor.
6. Know the general form of the Runge-Kutta methods; understand the derivation of the second-order RK method and how it relates to the Taylor series expansion; realize that there are an infinite number of possible versions for second- and higher-order RK methods.
7. Know how to apply any of the RK methods to systems of equations; be able to reduce an nth-order ODE to a system of n first-order ODEs.
8. Recognize the type of problem context where step-size adjustment is important.
9. Understand how adaptive step-size control is integrated into a fourth-order RK method.
10. Recognize how the combination of slow and fast components makes an equation or a system of equations stiff.
11. Understand the distinction between explicit and implicit solution schemes for ODEs; in particular, recognize how the latter (1) ameliorates the stiffness problem and (2) complicates the solution mechanics.
12. Understand the difference between initial-value and boundary-value problems.
13. Know the difference between multistep and one-step methods; realize that all multistep methods are predictor-correctors but that not all predictor-correctors are multistep methods.
14. Understand the connection between integration formulas and predictor-corrector methods.
15. Recognize the fundamental difference between Newton-Cotes and Adams integration formulas.
16. Know the rationale behind the polynomial and the power methods for determining eigenvalues; in particular, recognize their strengths and limitations.
17. Understand how Hoteller's deflation allows the power method to be used to compute intermediate eigenvalues.
18. Know how to use software packages and/or libraries to integrate ODEs and evaluate eigenvalues.

you may find it useful from a professional viewpoint to have software that employs the fourth-order Runge-Kutta method for more than five equations and to solve ODEs with an adaptive step-size approach.

In addition, one of your most important goals should be to master several of the general-purpose software packages that are widely available. In particular, you should become adept at using these tools to implement numerical methods for engineering problem solving.

Runge-Kutta Methods

This chapter is devoted to solving ordinary differential equations of the form

$$\frac{dy}{dx} = f(x, y)$$

In Chap. 1, we used a numerical method to solve such an equation for the velocity of the falling parachutist. Recall that the method was of the general form

New value = old value + slope × step size

or, in mathematical terms,

$$y_{i+1} = y_i + \phi h \tag{25.1}$$

According to this equation, the slope estimate of ϕ is used to extrapolate from an old value y_i to a new value y_{i+1} over a distance h (Fig. 25.1). This formula can be applied step by step to compute out into the future and, hence, trace out the trajectory of the solution.

FIGURE 25.1
Graphical depiction of a one-step method.

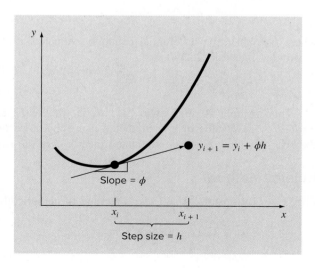

$$y_{i+1} = y_i + \phi h$$

Slope = ϕ

x_i x_{i+1} x

Step size = h

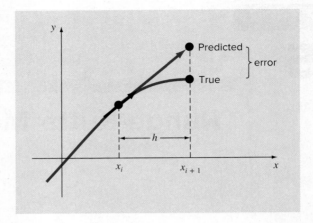

FIGURE 25.2
Euler's method.

All one-step methods can be expressed in this general form, with the only difference being the manner in which the slope is estimated. As in the falling parachutist problem, the simplest approach is to use the differential equation to estimate the slope in the form of the first derivative at x_i. In other words, the slope at the beginning of the interval is taken as an approximation of the average slope over the whole interval. This approach, called *Euler's method,* is discussed in the first part of this chapter. This is followed by other one-step methods that employ alternative slope estimates that result in more accurate predictions. All these techniques are generally called *Runge-Kutta methods.*

25.1 EULER'S METHOD

The first derivative provides a direct estimate of the slope at x_i (Fig. 25.2):

$$\phi = f(x_i, y_i)$$

where $f(x_i, y_i)$ is the differential equation evaluated at x_i and y_i. This estimate can be substituted into Eq. (25.1):

$$y_{i+1} = y_i + f(x_i, y_i)h \tag{25.2}$$

This formula is referred to as *Euler's* (or the *Euler-Cauchy* or the *point-slope*) *method.* A new value of y is predicted using the slope (equal to the first derivative at the original value of x) to extrapolate linearly over the step size h (Fig. 25.2).

EXAMPLE 25.1   **Euler's Method**

Problem Statement.   Use Euler's method to numerically integrate Eq. (PT7.13):

$$\frac{dy}{dx} = -2x^3 + 12x^2 - 20x + 8.5$$

from $x = 0$ to $x = 4$ with a step size of 0.5. The initial condition at $x = 0$ is $y = 1$. Recall that the exact solution is given by Eq. (PT7.16):

$$y = -0.5x^4 + 4x^3 - 10x^2 + 8.5x + 1$$

Solution. Equation (25.2) can be used to implement Euler's method:

$$y(0.5) = y(0) + f(0, 1)0.5$$

where $y(0) = 1$ and the slope estimate at $x = 0$ is

$$f(0, 1) = -2(0)^3 + 12(0)^2 - 20(0) + 8.5 = 8.5$$

Therefore,

$$y(0.5) = 1.0 + 8.5(0.5) = 5.25$$

The true solution at $x = 0.5$ is

$$y = -0.5(0.5)^4 + 4(0.5)^3 - 10(0.5)^2 + 8.5(0.5) + 1 = 3.21875$$

Thus, the error is

$$E_t = \text{true} - \text{approximate} = 3.21875 - 5.25 = -2.03125$$

or, expressed as percent relative error, $\varepsilon_t = -63.1\%$. For the second step,

$$\begin{aligned} y(1) &= y(0.5) + f(0.5, 5.25)0.5 \\ &= 5.25 + [-2(0.5)^3 + 12(0.5)^2 - 20(0.5) + 8.5]0.5 \\ &= 5.875 \end{aligned}$$

The true solution at $x = 1.0$ is 3.0, and therefore, the percent relative error is -95.8%. The computation is repeated, and the results are compiled in Table 25.1 and Fig. 25.3.

TABLE 25.1 Comparison of true and approximate values of the integral of $y' = -2x^3 + 12x^2 - 20x + 8.5$, with the initial condition that $y = 1$ at $x = 0$. The approximate values were computed using Euler's method with a step size of 0.5. The local error refers to the error incurred over a single step. It is calculated with a Taylor series expansion as in Example 25.2. The global error is the total discrepancy due to past as well as present steps.

x	y_{true}	y_{Euler}	Percent Relative Error Global	Local
0.0	1.00000	1.00000		
0.5	3.21875	5.25000	−63.1	−63.1
1.0	3.00000	5.87500	−95.8	−28.1
1.5	2.21875	5.12500	−131.0	−1.4
2.0	2.00000	4.50000	−125.0	20.3
2.5	2.71875	4.75000	−74.7	17.2
3.0	4.00000	5.87500	−46.9	3.9
3.5	4.71875	7.12500	−51.0	−11.3
4.0	3.00000	7.00000	−133.3	−53.1

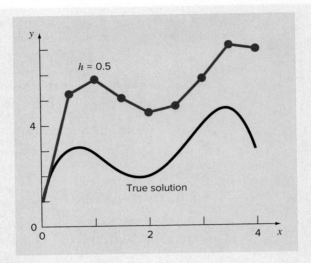

FIGURE 25.3
Comparison of the true solution with a numerical solution using Euler's method for the integral of $y' = -2x^3 + 12x^2 - 20x + 8.5$ from $x = 0$ to $x = 4$ with a step size of 0.5. The initial condition at $x = 0$ is $y = 1$.

Note that, although the computation captures the general trend of the true solution, the error is considerable. As discussed in the next section, this error can be reduced by using a smaller step size.

The preceding example uses a simple polynomial for the differential equation to facilitate the error analyses that follow. Thus,

$$\frac{dy}{dx} = f(x)$$

Obviously, a more general (and more common) case involves ODEs that depend on both x and y,

$$\frac{dy}{dx} = f(x, y)$$

As we progress through this part of the text, our examples will increasingly involve ODEs that depend on both the independent and the dependent variables.

25.1.1 Error Analysis for Euler's Method

The numerical solution of ODEs involves two types of error (recall Chaps. 3 and 4):

1. *Truncation,* or discretization, errors caused by the nature of the techniques employed to approximate values of y.

2. *Round-off* errors caused by the limited numbers of significant digits that can be retained by a computer.

The truncation errors are composed of two parts. The first is a *local truncation error* that results from an application of the method in question over a single step. The second is a *propagated truncation error* that results from the approximations produced during the previous steps. The sum of the two is the *total*, or *global, truncation error*.

Insight into the magnitude and properties of the truncation error can be gained by deriving Euler's method directly from the Taylor series expansion. To do this, realize that the differential equation being integrated will be of the general form

$$y' = f(x, y) \tag{25.3}$$

where $y' = dy/dx$ and x and y are the independent and the dependent variables, respectively. If the solution—that is, the function describing the behavior of y—has continuous derivatives, it can be represented by a Taylor series expansion about a starting value (x_i, y_i), as in [recall Eq. (4.7)]

$$y_{i+1} = y_i + y_i' h + \frac{y_i''}{2!} h^2 + \cdots + \frac{y_i^{(n)}}{n!} h^n + R_n \tag{25.4}$$

where $h = x_{i+1} - x_i$ and R_n = the remainder term, defined as

$$R_n = \frac{y^{(n+1)}(\xi)}{(n+1)!} h^{n+1} \tag{25.5}$$

where ξ lies somewhere in the interval from x_i to x_{i+1}. An alternative form can be developed by substituting Eq. (25.3) into Eqs. (25.4) and (25.5) to yield

$$y_{i+1} = y_i + f(x_i, y_i)h + \frac{f'(x_i, y_i)}{2!} h^2 + \cdots + \frac{f^{(n-1)}(x_i, y_i)}{n!} h^n + O(h^{n+1}) \tag{25.6}$$

where $O(h^{n+1})$ specifies that the local truncation error is proportional to the step size raised to the $(n + 1)$th power.

By comparing Eqs. (25.2) and (25.6), it can be seen that Euler's method corresponds to the Taylor series up to and including the term $f(x_i, y_i)h$. Additionally, the comparison indicates that a truncation error occurs because we approximate the true solution using a finite number of terms from the Taylor series. We thus truncate, or leave out, a part of the true solution. For example, the truncation error in Euler's method is attributable to the remaining terms in the Taylor series expansion that were not included in Eq. (25.2). Subtracting Eq. (25.2) from Eq. (25.6) yields

$$E_t = \frac{f'(x_i, y_i)}{2!} h^2 + \cdots + O(h^{n+1}) \tag{25.7}$$

where E_t = the true local truncation error. For sufficiently small h, the errors in the terms in Eq. (25.7) usually decrease as the order increases (recall Example 4.2 and the accompanying discussion), and the result is often represented as

$$E_a = \frac{f'(x_i, y_i)}{2!} h^2 \tag{25.8}$$

or

$$E_a = O(h^2) \qquad (25.9)$$

where E_a = the approximate local truncation error.

EXAMPLE 25.2   **Taylor Series Estimate for the Error of Euler's Method**

Problem Statement.   Use Eq. (25.7) to estimate the error of the first step of Example 25.1. Also use it to determine the error due to each higher-order term of the Taylor series expansion.

Solution.   Because we are dealing with a polynomial, we can use the Taylor series to obtain exact estimates of the errors in Euler's method. Equation (25.7) can be written as

$$E_t = \frac{f'(x_i, y_i)}{2!} h^2 + \frac{f''(x_i, y_i)}{3!} h^3 + \frac{f^{(3)}(x_i, y_i)}{4!} h^4 \qquad (E25.2.1)$$

where $f'(x_i, y_i)$ = the first derivative of the differential equation (that is, the second derivative of the solution). For the present case, this is

$$f'(x_i, y_i) = -6x^2 + 24x - 20 \qquad (E25.2.2)$$

and $f''(x_i, y_i)$ is the second derivative of the ODE,

$$f''(x_i, y_i) = -12x + 24 \qquad (E25.2.3)$$

and $f^{(3)}(x_i, y_i)$ is the third derivative of the ODE,

$$f^{(3)}(x_i, y_i) = -12 \qquad (E25.2.4)$$

We can omit additional terms (that is, fourth derivatives and higher) from Eq. (E25.2.1) because for this particular case they equal zero. It should be noted that for other functions (for example, transcendental functions such as sinusoids or exponentials) this would not necessarily be true, and higher-order terms would have nonzero values. However, for the present case, Eqs. (E25.2.1) through (E25.2.4) completely define the truncation error for a single application of Euler's method.

For example, the error due to truncation of the second-order term can be calculated as

$$E_{t,2} = \frac{-6(0.0)^2 + 24(0.0) - 20}{2}(0.5)^2 = -2.5 \qquad (E25.2.5)$$

For the third-order term:

$$E_{t,3} = \frac{-12(0.0) + 24}{6}(0.5)^3 = 0.5$$

and for the fourth-order term:

$$E_{t,4} = \frac{-12}{24}(0.5)^4 = -0.03125$$

These three results can be added to yield the total truncation error:

$$E_t = E_{t,2} + E_{t,3} + E_{t,4} = -2.5 + 0.5 - 0.03125 = -2.03125$$

which is exactly the error that was incurred in the initial step of Example 25.1. Note that $E_{t, 2} > E_{t, 3} > E_{t, 4}$, which supports the approximation represented by Eq. (25.8).

As illustrated in Example 25.2, the Taylor series provides a means of quantifying the error in Euler's method. However, there are limitations associated with its use for this purpose:

1. The Taylor series provides only an estimate of the local truncation error—that is, the error created during a single step of the method. It does not provide a measure of the propagated and, hence, the global, truncation error. In Table 25.1, we have included the local and global truncation errors for Example 25.1. The local error was computed for each time step with Eq. (25.2) but using the true value of y_i (the second column of the table) to compute each y_{i+1} rather than the approximate value (the third column), as is done in the Euler method. As expected, the average absolute local truncation error (25%) is less than the average global error (90%). The only reason that we can make these exact error calculations is that we know the true value *a priori*. Such would not be the case in an actual problem. Consequently, as discussed below, you must usually apply techniques such as Euler's method using a number of different step sizes to obtain an indirect estimate of the errors involved.
2. As mentioned above, in actual problems we usually deal with functions that are more complicated than simple polynomials. Consequently, the derivatives that are needed to evaluate the Taylor series expansion would not always be easy to obtain.

Although these limitations preclude exact error analysis for most practical problems, the Taylor series still provides valuable insight into the behavior of Euler's method. According to Eq. (25.9), we see that the local error is proportional to the square of the step size and the first derivative of the differential equation. It can also be demonstrated that the global truncation error is $O(h)$; that is, it is proportional to the step size (Carnahan et al. 1969). These observations lead to some useful conclusions:

1. The error can be reduced by decreasing the step size.
2. The method will provide error-free predictions if the underlying function (that is, the solution of the differential equation) is linear, because for a straight line the second derivative would be zero.

This latter conclusion makes intuitive sense because Euler's method uses straight-line segments to approximate the solution. Hence, Euler's method is referred to as a *first-order method*.

It should also be noted that this general pattern holds for the higher-order one-step methods described in the following pages. That is, an nth-order method will yield perfect results if the underlying solution is an nth-order polynomial. Further, the local truncation error will be $O(h^{n+1})$ and the global truncation error $O(h^n)$.

EXAMPLE 25.3 Effect of Reduced Step Size on Euler's Method

Problem Statement. Repeat the computation of Example 25.1 but use a step size of 0.25.

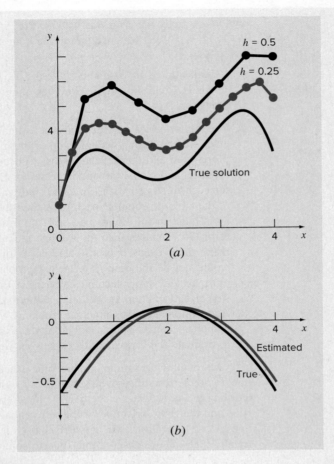

FIGURE 25.4
(a) Comparison of two numerical solutions with Euler's method using step sizes of 0.5 and 0.25. (b) Comparison of true and estimated local truncation error for the case where the step size is 0.5. Note that the "estimated" error is based on Eq. (E25.2.5).

Solution. The computation is repeated, and the results are compiled in Fig. 25.4a. Halving the step size reduces the absolute value of the average global error to 40% and the absolute value of the local error to 6.4%. This is compared to global and local errors for Example 25.1 of 90% and 24.8%, respectively. Thus, as expected, the local error is quartered and the global error is halved.

Also, notice how the local error changes sign for intermediate values along the range. This is due primarily to the fact that the first derivative of the differential equation is a parabola that changes sign [recall Eq. (E25.2.2) and see Fig. 25.4b]. Because the local error is proportional to this function, the net effect of the oscillation in sign is to keep the global error from continuously growing as the calculation proceeds. Thus, from $x = 0$ to $x = 1.25$, the local errors are all negative, and consequently, the global error increases over this interval. In the intermediate section of the range, positive local errors begin to reduce the global error. Near the end, the process is reversed and the global error again inflates. If the local error continuously changes sign over the computation interval, the net effect is usually to reduce the global error. However, where the local errors are of the same sign, the numerical solution may diverge farther and farther from the true solution as the computation proceeds. Such results are said to be *unstable*.

The effect of further step-size reductions on the global truncation error of Euler's method is illustrated in Fig. 25.5. This plot shows the absolute percent relative error at $x = 5$ as a function of step size for the problem we have been examining in Examples 25.1 through 25.3. Notice that even when h is reduced to 0.001, the error still exceeds 0.1%. Because this step size corresponds to 5000 steps to proceed from $x = 0$ to $x = 5$, the plot suggests that a first-order technique such as Euler's method demands great computational effort to obtain acceptable error levels. Later in this chapter, we present higher-order techniques that attain much better accuracy for the same computational effort. However, it should be noted that, despite its inefficiency, the simplicity of Euler's method makes it an extremely attractive option for many engineering problems. Because it is very easy to program, the technique is particularly useful for quick analyses. In the next section, a computer algorithm for Euler's method is developed.

FIGURE 25.5

Effect of step size on the global truncation error of Euler's method for the integral of $y' = -2x^3 + 12x^2 - 20x + 8.5$. The plot shows the absolute percent relative global error at $x = 5$ as a function of step size.

25.1.2 Algorithm for Euler's Method

Algorithms for one-step techniques such as Euler's method are extremely simple to program. As specified previously at the beginning of this chapter, all one-step methods have the general form

$$\text{New value} = \text{old value} + \text{slope} \times \text{step size} \tag{25.10}$$

The only way in which the methods differ is in the calculation of the slope.

Suppose that you want to perform the simple calculation outlined in Table 25.1. That is, you would like to use Euler's method to integrate $y' = -2x^3 + 12x^2 - 20x + 8.5$, with the initial condition that $y = 1$ at $x = 0$. You would like to integrate out to $x = 4$ using a step size of 0.5, and display all the results. A simple pseudocode to accomplish this task could be written as in Fig. 25.6.

Although this program will "do the job" of duplicating the results of Table 25.1, it is not very well designed. First, and foremost, it is not very modular. Although this is not very important for such a small program, it would be critical if we desired to modify and improve the algorithm.

Further, there are a number of issues related to the way we have set up the iterations. For example, suppose that the step size were to be made very small to obtain better accuracy. In such cases, because every computed value is displayed, the number of output values might be very large. Further, the algorithm is predicated on the assumption that the calculation interval is evenly divisible by the step size. Finally, the accumulation of x in the line $x = x + dx$ can be subject to quantizing errors of the sort previously discussed

FIGURE 25.6
Pseudocode for a "dumb" version of Euler's method.

```
'set integration range
xi = 0
xf = 4
'initialize variables
x = xi
y = 1
'set step size and determine
'number of calculation steps
dx = 0.5
nc = (xf - xi)/dx
'output initial condition
PRINT x, y
'loop to implement Euler's method
'and display results
DOFOR i = 1, nc
    dydx = -2x^3 + 12x^2 - 20x + 8.5
    y = y + dydx · dx
    x = x + dx
    PRINT x, y
END DO
```

in Sec. 3.4.1. For example, if dx were changed to 0.01 and standard IEEE floating point representation were used (about seven significant digits), the result at the end of the calculation would be 3.999997 rather than 4. For $dx = 0.001$, it would be 3.999892!

A much more modular algorithm that avoids these difficulties is displayed in Fig. 25.7. The algorithm does not output all calculated values. Rather, the user specifies an output interval, $xout$, that dictates the interval at which calculated results are stored in arrays, xp_m and yp_m. These values are stored in arrays so that they can be output in a variety of ways after the computation is completed (for example, printed, graphed, or written to a file).

The Driver Program takes big output steps and calls an Integrator routine that takes finer calculation steps. Note that the loops controlling both large and small steps exit on logical conditions. Thus, the intervals do not have to be evenly divisible by the step sizes.

The Integrator routine then calls an Euler routine that takes a single step with Euler's method. The Euler routine calls a Derivative routine that calculates the derivative value.

Whereas such modularization might seem like overkill for the present case, it will greatly facilitate modifying the program in later sections. For example, although the program in Fig. 25.7 is specifically designed to implement Euler's method, the Euler module is the only part that is method-specific. Thus, all that is required to apply this algorithm to the other one-step methods is to modify this routine.

FIGURE 25.7
Pseudocode for an "improved" modular version of Euler's method.

(a) Main or "Driver" Program

```
Assign values for
y = initial value dependent variable
xi = initial value independent variable
xf = final value independent variable
dx = calculation step size
xout = output interval

x = xi
m = 0
xp_m = x
yp_m = y
DO
    xend = x + xout
    IF (xend > xf) THEN xend = xf
    h = dx
    CALL Integrator(x, y, h, xend)
    m = m + 1
    xp_m = x
    yp_m = y
    IF (x ≥ xf) EXIT
END DO
DISPLAY RESULTS
END
```

(b) Routine to Take One Output Step

```
SUB Integrator(x, y, h, xend)
    DO
        IF (xend − x < h) THEN h = xend − x
        CALL Euler(x, y, h, ynew)
        y = ynew
        IF (x ≥ xend) EXIT
    END DO
END SUB
```

(c) Euler's Method for a Single ODE

```
SUB Euler(x, y, h, ynew)
    CALL Derivs(x, y, dydx)
    ynew = y + dydx * h
    x = x + h
END SUB
```

(d) Routine to Determine Derivative

```
SUB Derivs(x, y, dydx)
    dydx = ...
END SUB
```

EXAMPLE 25.4   Solving ODEs with the Computer

Problem Statement. A computer program can be developed from the pseudocode in Fig. 25.7. We can use this software to solve another problem associated with the falling parachutist. You recall from Part One that our mathematical model for the velocity was based on Newton's second law in the form

$$\frac{dv}{dt} = g - \frac{c}{m}v \qquad (E25.4.1)$$

This differential equation was solved both analytically (Example 1.1) and numerically using Euler's method (Example 1.2). These solutions were for the case where $g = 9.81$, $c = 12.5$, $m = 68.1$, and $v = 0$ at $t = 0$.

The objective of the present example is to repeat these numerical computations employing a more complicated model for the velocity based on a more complete mathematical description of the drag force caused by wind resistance. This model is given by

$$\frac{dv}{dt} = g - \frac{c}{m}\left[v + a\left(\frac{v}{v_{max}}\right)^{b}\right] \qquad (E25.4.2)$$

where g, m, and c are the same as for Eq. (E25.4.1), and a, b, and v_{max} are empirical constants, which for this case are equal to 8.3, 2.2, and 46, respectively. Note that this model is more capable of accurately fitting empirical measurements of drag force versus velocity than is the simple linear model of Example 1.1. However, this increased flexibility is gained at the expense of evaluating three coefficients rather than one. Furthermore, the resulting mathematical model is more difficult to solve analytically. In this case, Euler's method provides a convenient alternative to obtain an approximate numerical solution.

Solution. The results for both the linear and nonlinear model are displayed in Fig. 25.8 with an integration step size of 0.1 s. The plot in Fig. 25.8 also shows an overlay of the solution of the linear model for comparison purposes.

FIGURE 25.8
Graphical results for the solution of the nonlinear ODE [Eq. (E25.4.2)]. Notice that the plot also shows the solution for the linear model [Eq. (E25.4.1)] for comparative purposes.

The results of the two simulations indicate how increasing the complexity of the formulation of the drag force affects the velocity of the parachutist. In this case, the terminal velocity is lowered because of resistance caused by the higher-order terms in Eq. (E25.4.2).

Alternative models could be tested in a similar fashion. The combination of a computer-generated solution makes this an easy and efficient task. This convenience should allow you to devote more of your time to considering creative alternatives and holistic aspects of the problem rather than to performing tedious manual computations.

25.1.3 Higher-Order Taylor Series Methods

One way to reduce the error of Euler's method would be to include higher-order terms of the Taylor series expansion in the solution. For example, including the second-order term from Eq. (25.6) yields

$$y_{i+1} = y_i + f(x_i, y_i)h + \frac{f'(x_i, y_i)}{2!}h^2 \qquad (25.11)$$

with a local truncation error of

$$E_a = \frac{f''(x_i, y_i)}{6}h^3$$

Although the incorporation of higher-order terms is simple enough to implement for polynomials, their inclusion is not so trivial when the ODE is more complicated. In particular, ODEs that are a function of both the dependent and independent variable require chain-rule differentiation. For example, the first derivative of $f(x, y)$ is

$$f'(x_i, y_i) = \frac{\partial f(x, y)}{\partial x} + \frac{\partial f(x, y)}{\partial y}\frac{dy}{dx}$$

The second derivative is

$$f''(x_i, y_i) = \frac{\partial[\partial f/\partial x + (\partial f/\partial y)(dy/dx)]}{\partial x} + \frac{\partial[\partial f/\partial x + (\partial f/\partial y)(dy/dx)]}{\partial y}\frac{dy}{dx}$$

Higher-order derivatives become increasingly more complicated.

Consequently, as described in the following sections, alternative one-step methods have been developed. These schemes are comparable in performance to the higher-order Taylor-series approaches but require only the calculation of first derivatives.

25.2   IMPROVEMENTS OF EULER'S METHOD

A fundamental source of error in Euler's method is that the derivative at the beginning of the interval is assumed to apply across the entire interval. Two simple modifications are available to help circumvent this shortcoming. As will be demonstrated in Sec. 25.3, both modifications actually belong to a larger class of solution techniques called Runge-Kutta methods. However, because they have a very straightforward graphical interpretation, we will present them prior to their formal derivation as Runge-Kutta methods.

25.2.1 Heun's Method

One method to improve the estimate of the slope involves the determination of two derivatives for the interval—one at the initial point and another at the end point. The two derivatives are then averaged to obtain an improved estimate of the slope for the entire interval. This approach, called *Heun's method,* is depicted graphically in Fig. 25.9.

Recall that in Euler's method, the slope at the beginning of an interval,

$$y_i' = f(x_i, y_i) \tag{25.12}$$

is used to extrapolate linearly to y_{i+1}:

$$y_{i+1}^0 = y_i + f(x_i, y_i)h \tag{25.13}$$

For the standard Euler method we would stop at this point. However, in Heun's method the y_{i+1}^0 calculated in Eq. (25.13) is not the final answer, but an intermediate prediction. This is why we have distinguished it with a superscript 0. Equation (25.13) is called a

FIGURE 25.9
Graphical depiction of Heun's method. (*a*) Predictor and (*b*) corrector.

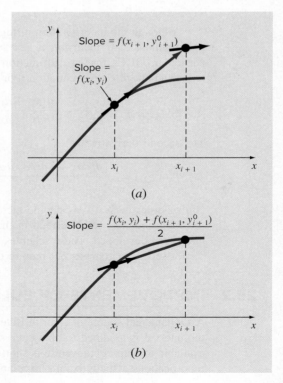

predictor equation. It provides an estimate of y_{i+1} that allows the calculation of an estimated slope at the end of the interval:

$$y'_{i+1} = f(x_{i+1}, y^0_{i+1}) \tag{25.14}$$

Thus, the two slopes [Eqs. (25.12) and (25.14)] can be combined to obtain an average slope for the interval:

$$\bar{y}' = \frac{y'_i + y'_{i+1}}{2} = \frac{f(x_i, y_i) + f(x_{i+1}, y^0_{i+1})}{2}$$

This average slope is then used to extrapolate linearly from y_i to y_{i+1} using Euler's method:

$$y_{i+1} = y_i + \frac{f(x_i, y_i) + f(x_{i+1}, y^0_{i+1})}{2} h$$

This is called a *corrector equation.*

The Heun method is a *predictor-corrector approach.* All the multistep methods to be discussed subsequently in Chap. 26 are of this type. The Heun method is the only one-step predictor-corrector method described in this book. As derived above, it can be expressed concisely as

Predictor (Fig. 25.9a): $y^0_{i+1} = y_i + f(x_i, y_i)h$ (25.15)

Corrector (Fig. 25.9b): $y_{i+1} = y_i + \dfrac{f(x_i, y_i) + f(x_{i+1}, y^0_{i+1})}{2} h$ (25.16)

Note that because Eq. (25.16) has y_{i+1} on both sides of the equal sign, it can be applied in an iterative fashion. That is, an old estimate can be used repeatedly to provide an improved estimate of y_{i+1}. The process is depicted in Fig. 25.10. It should be understood that

FIGURE 25.10

Graphical representation of iterating the corrector of Heun's method to obtain an improved estimate.

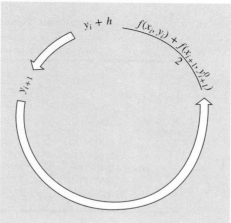

this iterative process does not necessarily converge on the true answer but will converge on an estimate with a finite truncation error, as demonstrated in the following example.

As with similar iterative methods discussed in previous sections of the book, a termination criterion for convergence of the corrector is provided by [recall Eq. (3.5)]

$$|\varepsilon_a| = \left| \frac{y_{i+1}^j - y_{i+1}^{j-1}}{y_{i+1}^j} \right| 100\% \qquad (25.17)$$

where y_{i+1}^{j-1} and y_{i+1}^j are the result from the prior and the present iteration of the corrector, respectively.

EXAMPLE 25.5

Heun's Method

Problem Statement.   Use Heun's method to integrate $y' = 4e^{0.8x} - 0.5y$ from $x = 0$ to $x = 4$ with a step size of 1. The initial condition at $x = 0$ is $y = 2$.

Solution.   Before solving the problem numerically, we can use calculus to determine the following analytical solution:

$$y = \frac{4}{1.3}\left(e^{0.8x} - e^{-0.5x}\right) + 2e^{-0.5x} \qquad (E25.5.1)$$

This formula can be used to generate the true solution values in Table 25.2.

First, the slope at (x_0, y_0) is calculated as

$$y_0' = 4e^0 - 0.5(2) = 3$$

This result is quite different from the actual average slope for the interval from 0 to 1.0, which is equal to 4.1946, as calculated from the differential equation using Eq. (PT6.4).

The numerical solution is obtained by using the predictor [Eq. (25.15)] to obtain an estimate of y at 1.0:

$$y_1^0 = 2 + 3(1) = 5$$

TABLE 25.2 Comparison of true and approximate values of the integral of $y' = 4e^{0.8x} - 0.5y$, with the initial condition that $y = 2$ at $x = 0$. The approximate values were computed using the Heun method with a step size of 1. Two cases, corresponding to different numbers of corrector iterations, are shown, along with the absolute percent relative error.

| | | Iterations of Heun's Method | | | |
| | | 1 | | 15 | |
x	y_{true}	y_{Heun}	$|\varepsilon_t|$ (%)	y_{Heun}	$|\varepsilon_t|$ (%)
0	2.0000000	2.0000000	0.00	2.0000000	0.00
1	6.1946314	6.7010819	8.18	6.3608655	2.68
2	14.8439219	16.3197819	9.94	15.3022367	3.09
3	33.6771718	37.1992489	10.46	34.7432761	3.17
4	75.3389626	83.3377674	10.62	77.7350962	3.18

Note that this is the result that would be obtained by the standard Euler method. The true value in Table 25.2 shows that it corresponds to a percent relative error of 19.3 percent.

Now, to improve the estimate for y_{i+1}, we use the value y_1^0 to predict the slope at the end of the interval,

$$y_1' = f(x_1, y_1^0) = 4e^{0.8(1)} - 0.5(5) = 6.402164$$

which can be combined with the initial slope to yield an average slope over the interval from $x = 0$ to 1,

$$y' = \frac{3 + 6.402164}{2} = 4.701082$$

which is closer to the true average slope of 4.1946. This result can then be substituted into the corrector [Eq. (25.16)] to give the prediction at $x = 1$,

$$y_1 = 2 + 4.701082(1) = 6.701082$$

which represents a percent relative error of -8.18 percent. Thus, the Heun method without iteration of the corrector reduces the absolute value of the error by a factor of 2.4 as compared with Euler's method.

Now this estimate can be used to refine or correct the prediction of y_1 by substituting the new result back into the right-hand side of Eq. (25.16):

$$y_1 = 2 + \frac{[3 + 4e^{0.8(1)} - 0.5(6.701082)]}{2} 1 = 6.275811$$

which represents an absolute percent relative error of 1.31 percent. This result, in turn, can be substituted back into Eq. (25.16) to further correct:

$$y_1 = 2 + \frac{[3 + 4e^{0.8(1)} - 0.5(6.275811)]}{2} 1 = 6.382129$$

which represents an $|\varepsilon_t|$ of 3.03%. Notice how the errors can grow as the iterations proceed. Such increases can occur, especially for large step sizes, and they prevent us from drawing the general conclusion that an additional iteration will always improve the result. However, for a sufficiently small step size, the iterations should eventually converge on a single value. For our case, 6.360865, which represents a relative error of 2.68 percent, is attained after 15 iterations. Table 25.2 shows results for the remainder of the computation using the method with 1 and 15 iterations per step.

In the previous example, the derivative is a function of both the dependent variable y and the independent variable x. For cases such as polynomials, where the ODE is solely a function of the independent variable, the predictor step [Eq. (25.16)] is not required and the corrector is applied only once for each iteration. For such cases, the technique is expressed concisely as

$$y_{i+1} = y_i + \frac{f(x_i) + f(x_{i+1})}{2} h \qquad (25.18)$$

Notice the similarity between the right-hand side of Eq. (25.18) and the trapezoidal rule [Eq. (21.3)]. The connection between the two methods can be formally demonstrated by starting with the ordinary differential equation

$$\frac{dy}{dx} = f(x)$$

This equation can be solved for y by integration:

$$\int_{y_i}^{y_{i+1}} dy = \int_{x_i}^{x_{i+1}} f(x)\,dx \tag{25.19}$$

which yields

$$y_{i+1} - y_i = \int_{x_i}^{x_{i+1}} f(x)\,dx \tag{25.20}$$

or

$$y_{i+1} = y_i + \int_{x_i}^{x_{i+1}} f(x)\,dx \tag{25.21}$$

Now, recall from Chap. 21 that the trapezoidal rule [Eq. (21.3)] is defined as

$$\int_{x_i}^{x_{i+1}} f(x)\,dx \cong \frac{f(x_i) + f(x_{i+1})}{2} h \tag{25.22}$$

where $h = x_{i+1} - x_i$. Substituting Eq. (25.22) into Eq. (25.21) yields

$$y_{i+1} = y_i + \frac{f(x_i) + f(x_{i+1})}{2} h \tag{25.23}$$

which is equivalent to Eq. (25.18).

Because Eq. (25.23) is a direct expression of the trapezoidal rule, the local truncation error is given by [recall Eq. (21.6)]

$$E_t = -\frac{f''(\xi)}{12} h^3 \tag{25.24}$$

where ξ is between x_i and x_{i+1}. Thus, the method is second order because the second derivative of the ODE is zero when the true solution is a quadratic. In addition, the local and global errors are $O(h^3)$ and $O(h^2)$, respectively. Therefore, decreasing the step size decreases the error at a faster rate than for Euler's method. Figure 25.11, which shows the result of using Heun's method to solve the polynomial from Example 25.1, demonstrates this behavior.

25.2.2 The Midpoint (or Improved Polygon) Method

Figure 25.12 illustrates another simple modification of Euler's method. Called the *midpoint method* (or the *improved polygon* or the *modified Euler*), this technique uses Euler's method to predict a value of y at the midpoint of the interval (Fig. 25.12a):

$$y_{i+1/2} = y_i + f(x_i, y_i)\frac{h}{2} \tag{25.25}$$

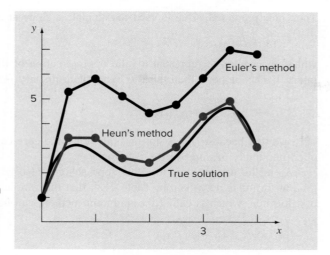

FIGURE 25.11
Comparison of the true solution with a numerical solution using Euler's and Heun's methods for the integral of $y' = -2x^3 + 12x^2 - 20x + 8.5$.

FIGURE 25.12
Graphical depiction of the midpoint method.
(a) Eq. (25.25) and
(b) Eq. (25.27).

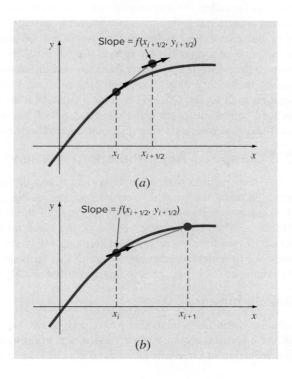

Then, this predicted value is used to calculate a slope at the midpoint:

$$y'_{i+1/2} = f(x_{i+1/2}, y_{i+1/2}) \tag{25.26}$$

which is assumed to represent a valid approximation of the average slope for the entire interval. This slope is then used to extrapolate linearly from x_i to x_{i+1} (Fig. 25.12b):

$$y_{i+1} = y_i + f(x_{i+1/2}, y_{i+1/2})h \tag{25.27}$$

Observe that because y_{i+1} is not on both sides, the corrector [Eq. (25.27)] cannot be applied iteratively to improve the solution.

As in the previous section, this approach can also be linked to Newton-Cotes integration formulas. Recall from Table 21.4, that the simplest Newton-Cotes open integration formula, which is called the midpoint method, can be represented as

$$\int_a^b f(x)\,dx \cong (b-a)f(x_1)$$

where x_1 is the midpoint of the interval (a, b). Using the nomenclature for the present case, it can be expressed as

$$\int_{x_i}^{x_{i+1}} f(x)\,dx \cong hf(x_{i+1/2})$$

Substitution of this formula into Eq. (25.21) yields Eq. (25.27). Thus, just as the Heun method can be called the trapezoidal rule, the *midpoint method* gets its name from the underlying integration formula upon which it is based.

The midpoint method is superior to Euler's method because it utilizes a slope estimate at the midpoint of the prediction interval. Recall from our discussion of numerical differentiation in Sec. 4.1.3 that centered finite divided differences are better approximations of derivatives than either forward or backward versions. In the same sense, a centered approximation such as Eq. (25.26) has a local truncation error of $O(h^2)$ in comparison with the forward approximation of Euler's method, which has an error of $O(h)$. Consequently, the local and global errors of the midpoint method are $O(h^3)$ and $O(h^2)$, respectively.

25.2.3 Computer Algorithms for Heun and Midpoint Methods

Both the Heun method with a single corrector and the midpoint method can be easily programmed using the general structure depicted in Fig. 25.7. As in Fig. 25.13a and b, simple routines can be written to take the place of the Euler routine in Fig. 25.7.

However, when the iterative version of the Heun method is to be implemented, the modifications are a bit more involved (although they are still localized within a single module). We have developed pseudocode for this purpose in Fig. 25.13c. This algorithm can be combined with Fig. 25.7 to develop software for the iterative Heun method.

25.2.4 Summary

By tinkering with Euler's method, we have derived two new second-order techniques. Even though these versions require more computational effort to determine the slope, the accompanying reduction in error will allow us to conclude in a subsequent section (Sec. 25.3.4)

(a) Simple Heun without Iteration

```
SUB Heun(x, y, h, ynew)
  CALL Derivs(x, y, dy1dx)
  ye = y + dy1dx · h
  CALL Derivs(x + h, ye, dy2dx)
  Slope = (dy1dx + dy2dx)/2
  ynew = y + Slope · h
  x = x + h
END SUB
```

(b) Midpoint Method

```
SUB Midpoint(x, y, h, ynew)
  CALL Derivs(x, y, dydx)
  ym = y + dydx · h/2
  CALL Derivs(x + h/2, ym, dymdx)
  ynew = y + dymdx · h
  x = x + h
END SUB
```

(c) Heun with Iteration

```
SUB HeunIter(x, y, h, ynew)
  es = 0.01
  maxit = 20
  CALL Derivs(x, y, dy1dx)
  ye = y + dy1dx · h
  iter = 0
  DO
    yeold = ye
    CALL Derivs(x + h, ye, dy2dx)
    slope = (dy1dx + dy2dx)/2
    ye = y + slope · h
    iter = iter + 1
```

$$ea = \left| \frac{ye - yeold}{ye} \right| 100\%$$

```
    IF (ea ≤ es OR iter > maxit) EXIT
  END DO
  ynew = ye
  x = x + h
END SUB
```

FIGURE 25.13
Pseudocode to implement the (a) simple Heun, (b) midpoint, and (c) iterative Heun methods.

that the improved accuracy is usually worth the effort. Although there are certain cases where easily programmable techniques such as Euler's method can be applied to advantage, the Heun and midpoint methods are generally superior and should be implemented if they are consistent with the problem objectives.

As noted at the beginning of this section, the Heun (without iterations), the midpoint method, and in fact, the Euler technique itself are versions of a broader class of one-step approaches called Runge-Kutta methods. We now turn to a formal derivation of these techniques.

25.3 RUNGE-KUTTA METHODS

Runge-Kutta (RK) methods achieve the accuracy of a Taylor series approach without requiring the calculation of higher derivatives. Many variations exist but all can be cast in the generalized form of Eq. (25.1):

$$y_{i+1} = y_i + \phi(x_i, y_i, h)h \tag{25.28}$$

where $\phi(x_i, y_i, h)$ is called an *increment function*, which can be interpreted as a representative slope over the interval. The increment function can be written in general form as

$$\phi = a_1 k_1 + a_2 k_2 + \cdots + a_n k_n \tag{25.29}$$

where the a's are constants and the k's are

$$k_1 = f(x_i, y_i) \qquad\qquad\qquad (25.29a)$$

$$k_2 = f(x_i + p_1h, y_i + q_{11}k_1h) \qquad\qquad (25.29b)$$

$$k_3 = f(x_i + p_2h, y_i + q_{21}k_1h + q_{22}k_2h) \qquad\qquad (25.29c)$$

$$\cdot$$
$$\cdot$$
$$\cdot$$

$$k_n = f(x_i + p_{n-1}h, y_i + q_{n-1,\,1}k_1h + q_{n-1,\,2}k_2h + \cdots + q_{n-1,\,n-1}k_{n-1}h) \qquad (25.29d)$$

where the p's and q's are constants. Notice that the k's are a series of recurrence relationships. That is, k_1 appears in the equation for k_2, which appears in the equation for k_3, and so forth. Because each k is a functional evaluation, this recurrence makes RK methods efficient for computer calculations.

Various types of Runge-Kutta methods can be devised by employing different numbers of terms in the increment function as specified by n. Note that the first-order RK method with $n = 1$ is, in fact, Euler's method. Once n is chosen, values for the a's, p's, and q's are evaluated by setting Eq. (25.28) equal to terms in a Taylor series expansion (Box 25.1). Thus, at least for the lower-order versions, the number of terms, n, usually represents the order of the approach. For example, in the next section, second-order RK methods use an increment function with two terms ($n = 2$). These second-order methods will be exact if the solution to the differential equation is quadratic. In addition, because terms with h^3 and higher are dropped during the derivation, the local truncation error is $O(h^3)$ and the global error is $O(h^2)$. In subsequent sections, the third- and fourth-order RK methods ($n = 3$ and 4, respectively) are developed. For these cases, the global truncation errors are $O(h^3)$ and $O(h^4)$, respectively.

25.3.1 Second-Order Runge-Kutta Methods

The second-order version of Eq. (25.28) is

$$y_{i+1} = y_i + (a_1k_1 + a_2k_2)h \qquad\qquad\qquad (25.30)$$

where

$$k_1 = f(x_i, y_i) \qquad\qquad\qquad (25.30a)$$

$$k_2 = f(x_i + p_1h, y_i + q_{11}k_1h) \qquad\qquad (25.30b)$$

As described in Box 25.1, values for a_1, a_2, p_1, and q_{11} are evaluated by setting Eq. (25.30) equal to a Taylor series expansion to the second-order term. By doing this, we derive three equations to evaluate the four unknown constants. The three equations are

$$a_1 + a_2 = 1 \qquad\qquad\qquad (25.31)$$

$$a_2p_1 = \frac{1}{2} \qquad\qquad\qquad (25.32)$$

$$a_2q_{11} = \frac{1}{2} \qquad\qquad\qquad (25.33)$$

Box 25.1 Derivation of the Second-Order Runge-Kutta Methods

The second-order version of Eq. (25.28) is

$$y_{i+1} = y_i + (a_1 k_1 + a_2 k_2)h \tag{B25.1.1}$$

where

$$k_1 = f(x_i, y_i) \tag{B25.1.2}$$

and

$$k_2 = f(x_i + p_1 h, y_i + q_{11}k_1 h) \tag{B25.1.3}$$

To use Eq. (B25.1.1), we have to determine values for the constants a_1, a_2, p_1, and q_{11}. To do this, we recall that the second-order Taylor series for y_{i+1} in terms of y_i and $f(x_i, y_i)$ is written as [Eq. (25.11)]

$$y_{i+1} = y_i + f(x_i, y_i)h + \frac{f'(x_i, y_i)}{2!}h^2 \tag{B25.1.4}$$

where $f'(x_i, y_i)$ must be determined by chain-rule differentiation (Sec. 25.1.3):

$$f'(x_i, y_i) = \frac{\partial f(x, y)}{\partial x} + \frac{\partial f(x, y)}{\partial y}\frac{dy}{dx} \tag{B25.1.5}$$

Substituting Eq. (B25.1.5) into (B25.1.4) gives

$$y_{i+1} = y_i + f(x_i, y_i)h + \left(\frac{\partial f}{\partial x} + \frac{\partial f}{\partial y}\frac{dy}{dx}\right)\frac{h^2}{2!} \tag{B25.1.6}$$

The basic strategy underlying Runge-Kutta methods is to use algebraic manipulations to solve for values of a_1, a_2, p_1, and q_{11} that make Eqs. (B25.1.1) and (B25.1.6) equivalent.

To do this, we first use a Taylor series to expand Eq. (B25.1.3). The Taylor series for a two-variable function is defined as [recall Eq. (4.26)]

$$g(x + r, y + s) = g(x, y) + r\frac{\partial g}{\partial x} + s\frac{\partial g}{\partial y} + \cdots$$

Applying this method to expand Eq. (B25.1.3) gives

$$f(x_i + p_1 h, y_i + q_{11}k_1 h) = f(x_i, y_i) + p_1 h\frac{\partial f}{\partial x}$$
$$+ q_{11}k_1 h\frac{\partial f}{\partial y} + O(h^2)$$

This result can be substituted along with Eq. (B25.1.2) into Eq. (B25.1.1) to yield

$$y_{i+1} = y_i + a_1 hf(x_i, y_i) + a_2 hf(x_i, y_i) + a_2 p_1 h^2\frac{\partial f}{\partial x}$$
$$+ a_2 q_{11}h^2 f(x_i, y_i)\frac{\partial f}{\partial y} + O(h^3)$$

or, by collecting terms,

$$y_{i+1} = y_i + [a_1 f(x_i, y_i) + a_2 f(x_i, y_i)]h$$
$$+ \left[a_2 p_1\frac{\partial f}{\partial x} + a_2 q_{11}f(x_i, y_i)\frac{\partial f}{\partial y}\right]h^2 + O(h^3) \tag{B25.1.7}$$

Now, comparing like terms in Eqs. (B25.1.6) and (B25.1.7), we determine that for the two equations to be equivalent, the following must hold:

$$a_1 + a_2 = 1$$
$$a_2 p_1 = \frac{1}{2}$$
$$a_2 q_{11} = \frac{1}{2}$$

These three simultaneous equations contain the four unknown constants. Because there is one more unknown than the number of equations, there is no unique set of constants that satisfy the equations. However, by assuming a value for one of the constants, we can determine the other three. Consequently, there is a family of second-order methods rather than a single version.

Because we have three equations with four unknowns, we must assume a value of one of the unknowns to determine the other three. Suppose that we specify a value for a_2. Then Eqs. (25.31) through (25.33) can be solved simultaneously for

$$a_1 = 1 - a_2 \tag{25.34}$$

$$p_1 = q_{11} = \frac{1}{2a_2} \tag{25.35}$$

Because we can choose an infinite number of values for a_2, there are an infinite number of second-order RK methods. Every version would yield exactly the same results if the solution to the ODE were quadratic, linear, or a constant. However, they yield different results when (as is typically the case) the solution is more complicated. We present three of the most commonly used and preferred versions:

Heun Method with a Single Corrector ($a_2 = 1/2$). If a_2 is assumed to be $1/2$, Eqs. (25.34) and (25.35) can be solved for $a_1 = 1/2$ and $p_1 = q_{11} = 1$. These parameters, when substituted into Eq. (25.30), yield

$$y_{i+1} = y_i + \left(\frac{1}{2}k_1 + \frac{1}{2}k_2 \right)h \qquad (25.36)$$

where

$$k_1 = f(x_i, y_i) \qquad (25.36a)$$
$$k_2 = f(x_i + h, y_i + k_1 h) \qquad (25.36b)$$

Note that k_1 is the slope at the beginning of the interval and k_2 is the slope at the end of the interval. Consequently, this second-order Runge-Kutta method is actually Heun's technique without iteration.

The Midpoint Method ($a_2 = 1$). If a_2 is assumed to be 1, then $a_1 = 0$, $p_1 = q_{11} = 1/2$, and Eq. (25.30) becomes

$$y_{i+1} = y_i + k_2 h \qquad (25.37)$$

where

$$k_1 = f(x_i, y_i) \qquad (25.37a)$$
$$k_2 = f\left(x_i + \frac{1}{2}h, y_i + \frac{1}{2}k_1 h \right) \qquad (25.37b)$$

This is the midpoint method.

Ralston's Method ($a_2 = 2/3$). Ralston (1962) and Ralston and Rabinowitz (1978) determined that choosing $a_2 = 2/3$ provides a minimum bound on the truncation error for the second-order RK algorithms. For this version, $a_1 = 1/3$ and $p_1 = q_{11} = 3/4$ and we have

$$y_{i+1} = y_i + \left(\frac{1}{3}k_1 + \frac{2}{3}k_2 \right)h \qquad (25.38)$$

where

$$k_1 = f(x_i, y_i) \qquad (25.38a)$$
$$k_2 = f\left(x_i + \frac{3}{4}h, y_i + \frac{3}{4}k_1 h \right) \qquad (25.38b)$$

EXAMPLE 25.6

Comparison of Various Second-Order RK Schemes

Problem Statement.   Use the midpoint method [Eq. (25.37)] and Ralston's method [Eq. (25.38)] to numerically integrate Eq. (PT7.13),

$$f(x, y) = -2x^3 + 12x^2 - 20x + 8.5$$

from $x = 0$ to $x = 4$ using a step size of 0.5. The initial condition at $x = 0$ is $y = 1$. Compare the results with the values obtained using another second-order RK algorithm, that is, the Heun method without corrector iteration (Table 25.3).

Solution.   The first step in the midpoint method is to use Eq. (25.37a) to compute

$$k_1 = -2(0)^3 + 12(0)^2 - 20(0) + 8.5 = 8.5$$

However, because the ODE is a function of x only, this result has no bearing on the second step—the use of Eq. (25.37b) to compute

$$k_2 = -2(0.25)^3 + 12(0.25)^2 - 20(0.25) + 8.5 = 4.21875$$

Notice that this estimate of the slope is much closer to the average value for the interval (4.4375) than the slope at the beginning of the interval (8.5) that would have been used for Euler's approach. The slope at the midpoint can then be substituted into Eq. (25.37) to predict

$$y(0.5) = 1 + 4.21875(0.5) = 3.109375 \qquad \varepsilon_t = 3.4\%$$

The computation is repeated, and the results are summarized in Fig. 25.14 and Table 25.3.

FIGURE 25.14
Comparison of the true solution with numerical solutions using three second-order RK methods and Euler's method.

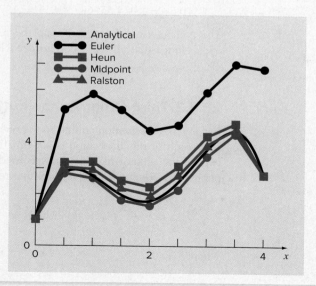

TABLE 25.3 Comparison of true and approximate values of the integral of $y' = -2x^3 + 12x^2 - 20x + 8.5$, with the initial condition that $y = 1$ at $x = 0$. The approximate values were computed using three versions of second-order RK methods with a step size of 0.5.

x	y_{true}	Heun		Midpoint		Second-Order Ralston RK	
		y	$|\varepsilon_t|$ (%)	y	$|\varepsilon_t|$ (%)	y	$|\varepsilon_t|$ (%)
0.0	1.00000	1.00000	0	1.00000	0	1.00000	0
0.5	3.21875	3.43750	6.8	3.109375	3.4	3.277344	1.8
1.0	3.00000	3.37500	12.5	2.81250	6.3	3.101563	3.4
1.5	2.21875	2.68750	21.1	1.984375	10.6	2.347656	5.8
2.0	2.00000	2.50000	25.0	1.75	12.5	2.140625	7.0
2.5	2.71875	3.18750	17.2	2.484375	8.6	2.855469	5.0
3.0	4.00000	4.37500	9.4	3.81250	4.7	4.117188	2.9
3.5	4.71875	4.93750	4.6	4.609375	2.3	4.800781	1.7
4.0	3.00000	3.00000	0	3	0	3.031250	1.0

For Ralston's method, k_1 for the first interval also equals 8.5 and [Eq. (25.38b)]

$$k_2 = -2(0.375)^3 + 12(0.375)^2 - 20(0.375) + 8.5 = 2.58203125$$

The average slope is computed by

$$\phi = \frac{1}{3}(8.5) + \frac{2}{3}(2.58203125) = 4.5546875$$

which can be used to predict

$$y(0.5) = 1 + 4.5546875(0.5) = 3.27734375 \qquad \varepsilon_t = -1.82\%$$

The computation is repeated, and the results are summarized in Fig. 25.14 and Table 25.3. Notice that all the second-order RK methods are superior to Euler's method.

25.3.2 Third-Order Runge-Kutta Methods

For $n = 3$, a derivation similar to the one for the second-order method can be performed. The result of this derivation is six equations with eight unknowns. Therefore, values for two of the unknowns must be specified *a priori* in order to determine the remaining parameters. One common version that results is

$$y_{i+1} = y_i + \frac{1}{6}(k_1 + 4k_2 + k_3)h \qquad (25.39)$$

where

$$k_1 = f(x_i, y_i) \qquad (25.39a)$$

$$k_2 = f\left(x_i + \frac{1}{2}h, y_i + \frac{1}{2}k_1h\right) \tag{25.39b}$$

$$k_3 = f(x_i + h, y_i - k_1h + 2k_2h) \tag{25.39c}$$

Note that if the derivative is a function of x only, this third-order method reduces to Simpson's 1/3 rule. Ralston (1962) and Ralston and Rabinowitz (1978) have developed an alternative version that provides a minimum bound on the truncation error. In any case, the third-order RK methods have local and global errors of $O(h^4)$ and $O(h^3)$, respectively, and yield exact results when the solution is a cubic. When dealing with polynomials, Eq. (25.39) will also be exact when the differential equation is cubic and the solution is quartic. This is because Simpson's 1/3 rule provides exact integral estimates for cubics (recall Box 21.3).

25.3.3 Fourth-Order Runge-Kutta Methods

The most popular RK methods are fourth order. As with the second-order approaches, there are an infinite number of versions. The following is the most commonly used form, and we therefore call it the *classical fourth-order RK method:*

$$y_{i+1} = y_i + \frac{1}{6}(k_1 + 2k_2 + 2k_3 + k_4)h \tag{25.40}$$

where

$$k_1 = f(x_i, y_i) \tag{25.40a}$$

$$k_2 = f\left(x_i + \frac{1}{2}h, y_i + \frac{1}{2}k_1h\right) \tag{25.40b}$$

FIGURE 25.15
Graphical depiction of the slope estimates comprising the fourth-order RK method.

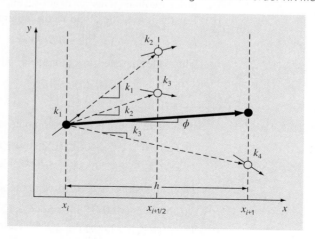

$$k_3 = f\left(x_i + \frac{1}{2}h, y_i + \frac{1}{2}k_2h\right) \tag{25.40c}$$

$$k_4 = f(x_i + h, y_i + k_3h) \tag{25.40d}$$

Notice that for ODEs that are a function of x alone, the classical fourth-order RK method is similar to Simpson's 1/3 rule. In addition, the fourth-order RK method is similar to the Heun approach in that multiple estimates of the slope are developed in order to come up with an improved average slope for the interval. As depicted in Fig. 25.15, each of the k's represents a slope. Equation (25.40) then represents a weighted average of these to arrive at the improved slope.

EXAMPLE 25.7

Classical Fourth-Order RK Method

Problem Statement.

(a) Use the classical fourth-order RK method [Eq. (25.40)] to integrate

$$f(x, y) = -2x^3 + 12x^2 - 20x + 8.5$$

using a step size of $h = 0.5$ and an initial condition of $y = 1$ at $x = 0$.

(b) Similarly, integrate

$$f(x, y) = 4e^{0.8x} - 0.5y$$

using $h = 0.5$ with $y(0) = 2$ from $x = 0$ to 0.5.

Solution.

(a) Equations (25.40a) through (25.40d) are used to compute $k_1 = 8.5$, $k_2 = 4.21875$, $k_3 = 4.21875$ and $k_4 = 1.25$, which are substituted into Eq. (25.40) to yield

$$y(0.5) = 1 + \left\{\frac{1}{6}[8.5 + 2(4.21875) + 2(4.21875) + 1.25]\right\}0.5$$

$$= 3.21875$$

which is exact. Thus, because the true solution is a quartic [Eq. (PT7.16)], the fourth-order method gives an exact result.

(b) For this case, the slope at the beginning of the interval is computed as

$$k_1 = f(0, 2) = 4e^{0.8(0)} - 0.5(2) = 3$$

This value is used to compute a value of y and a slope at the midpoint,

$$y(0.25) = 2 + 3(0.25) = 2.75$$

$$k_2 = f(0.25, 2.75) = 4e^{0.8(0.25)} - 0.5(2.75) = 3.510611$$

This slope in turn is used to compute another value of y and another slope at the midpoint,

$$y(0.25) = 2 + 3.510611(0.25) = 2.877653$$

$$k_3 = f(0.25, 2.877653) = 4e^{0.8(0.25)} - 0.5(2.877653) = 3.446785$$

Next, this slope is used to compute a value of y and a slope at the end of the interval,

$$y(0.5) = 2 + 3.071785(0.5) = 3.723392$$

$$k_4 = f(0.5, 3.723392) = 4e^{0.8(0.5)} - 0.5(3.723392) = 4.105603$$

Finally, the four slope estimates are combined to yield an average slope. This average slope is then used to make the final prediction at the end of the interval.

$$\phi = \frac{1}{6}[3 + 2(3.510611) + 2(3.446785) + 4.105603] = 3.503399$$

$$y(0.5) = 2 + 3.503399(0.5) = 3.751699$$

which compares favorably with the true solution of 3.751521.

25.3.4 Higher-Order Runge-Kutta Methods

Where more accurate results are required, *Butcher's* (1964) *fifth-order RK method* is recommended:

$$y_{i+1} = y_i + \frac{1}{90}(7k_1 + 32k_3 + 12k_4 + 32k_5 + 7k_6)h \tag{25.41}$$

where

$$k_1 = f(x_i, y_i) \tag{25.41a}$$

$$k_2 = f\left(x_i + \frac{1}{4}h, y_i + \frac{1}{4}k_1h\right) \tag{25.41b}$$

$$k_3 = f\left(x_i + \frac{1}{4}h, y_i + \frac{1}{8}k_1h + \frac{1}{8}k_2h\right) \tag{25.41c}$$

$$k_4 = f\left(x_i + \frac{1}{2}h, y_i - \frac{1}{2}k_2h + k_3h\right) \tag{25.41d}$$

$$k_5 = f\left(x_i + \frac{3}{4}h, y_i + \frac{3}{16}k_1h + \frac{9}{16}k_4h\right) \tag{25.41e}$$

$$k_6 = f\left(x_i + h, y_i - \frac{3}{7}k_1h + \frac{2}{7}k_2h + \frac{12}{7}k_3h - \frac{12}{7}k_4h + \frac{8}{7}k_5h\right) \tag{25.41f}$$

Note the similarity between Butcher's method and Boole's rule in Table 21.2. Higher-order RK formulas such as Butcher's method are available, but in general, beyond fourth-order methods the gain in accuracy is offset by the added computational effort and complexity.

EXAMPLE 25.8 **Comparison of Runge-Kutta Methods**

Problem Statement.   Use first- through fifth-order RK methods to solve

$$f(x, y) = 4e^{0.8x} - 0.5y$$

with $y(0) = 2$ from $x = 0$ to $x = 4$ with various step sizes. Compare the accuracy of the various methods for the result at $x = 4$ based on the exact answer of $y(4) = 75.33896$.

Solution.   The computation is performed using Euler's, the noniterative Heun, the third-order RK [Eq. (25.39)], the classical fourth-order RK, and Butcher's fifth-order RK methods. The results are presented in Fig. 25.16, where we have plotted the absolute

FIGURE 25.16
Comparison of percent relative error versus computational effort for first- through fifth-order RK methods.

value of the percent relative error versus the computational effort. This latter quantity is equivalent to the number of function evaluations required to attain the result, as in

$$\text{Effort} = n_f \frac{b - a}{h} \qquad\qquad\qquad \text{(E25.8.1)}$$

where n_f = the number of function evaluations involved in the particular RK computation. For orders ≤ 4, n_f is equal to the order of the method. However, note that Butcher's fifth-order technique requires six function evaluations [Eq. (25.41a) through (25.41f)]. The quantity $(b - a)/h$ is the total integration interval divided by the step size—that is, it is the number of applications of the RK technique required to obtain the result. Thus, because the function evaluations are usually the primary time-consuming steps, Eq. (E25.8.1) provides a rough measure of the run time required to attain the answer.

Inspection of Fig. 25.16 leads to a number of conclusions: first, that the higher-order methods attain better accuracy for the same computational effort and, second, that the gain in accuracy for the additional effort tends to diminish after a point. (Notice that the curves drop rapidly at first and then tend to level off.)

Example 25.8 and Fig. 25.16 might lead one to conclude that higher-order RK techniques are always the preferred methods. However, other factors such as programming

```
SUB RK4(x, y, h, ynew)
  CALL Derivs(x, y, k1)
  ym = y + k1 · h/2
  CALL Derivs(x + h/2, ym, k2)
  ym = y + k2 · h/2
  CALL Derivs(x + h/2, ym, k3)
  ye = y + k3 · h
  CALL Derivs(x + h, ye, k4)
  slope = (k1 + 2(k2 + k3) + k4)/6
  ynew = y + slope · h
  x = x + h
END SUB
```

FIGURE 25.17
Pseudocode to determine a single step of the fourth-order RK method.

costs and the accuracy requirements of the problem also must be considered when choosing a solution technique. Such trade-offs will be explored in detail in the engineering applications in Chap. 28 and in the epilogue for Part Seven.

25.3.5 Computer Algorithms for Runge-Kutta Methods

As with all the methods covered in this chapter, the RK techniques fit nicely into the general algorithm embodied in Fig. 25.7. Figure 25.17 presents pseudocode to determine the slope for the classic fourth-order RK method [Eq. (25.40)]. Subroutines to compute slopes for all the other versions can be easily programmed in a similar fashion.

25.4 SYSTEMS OF EQUATIONS

Many practical problems in engineering and science require the solution of a system of simultaneous ordinary differential equations rather than a single equation. Such systems may be represented generally as

$$\frac{dy_1}{dx} = f_1(x, y_1, y_2, \ldots, y_n)$$

$$\frac{dy_2}{dx} = f_2(x, y_1, y_2, \ldots, y_n)$$

$$\cdot$$
$$\cdot$$
$$\cdot$$

$$\frac{dy_n}{dx} = f_n(x, y_1, y_2, \ldots, y_n) \tag{25.42}$$

The solution of such a system requires that n initial conditions be known at the starting value of x.

25.4.1 Euler's Method

All the methods discussed in this chapter for single equations can be extended to the system shown above. Engineering applications can involve thousands of simultaneous equations. In each case, the procedure for solving a system of equations simply involves applying the one-step technique for every equation at each step before proceeding to the next step. This is best illustrated by the following example for the simple Euler's method.

EXAMPLE 25.9   Solving Systems of ODEs Using Euler's Method

Problem Statement.   Solve the following set of differential equations using Euler's method, assuming that at $x = 0$, $y_1 = 4$ and $y_2 = 6$. Integrate to $x = 2$ with a step size of 0.5.

$$\frac{dy_1}{dx} = -0.5y_1 \qquad \frac{dy_2}{dx} = 4 - 0.3y_2 - 0.1y_1$$

Solution.   Euler's method is implemented for each variable as in Eq. (25.2):

$$y_1(0.5) = 4 + [-0.5(4)]0.5 = 3$$
$$y_2(0.5) = 6 + [4 - 0.3(6) - 0.1(4)]0.5 = 6.9$$

Note that $y_1(0) = 4$ is used in the second equation rather than $y_1(0.5) = 3$ computed with the first equation. Proceeding in a like manner gives

x	y_1	y_2
0	4	6
0.5	3	6.9
1.0	2.25	7.715
1.5	1.6875	8.44525
2.0	1.265625	9.094087

25.4.2 Runge-Kutta Methods

Note that any of the higher-order RK methods in this chapter can be applied to systems of equations. However, care must be taken in determining the slopes. Figure 25.15 is helpful in visualizing the proper way to do this for the fourth-order method. That is, we first develop slopes for all variables at the initial value. These slopes (a set of k_1's) are then used to make predictions of the dependent variable at the midpoint of the interval. These midpoint values are in turn used to compute a set of slopes at the midpoint (the k_2's). These new slopes are then taken back to the starting point to make another set of midpoint predictions that lead to new slope predictions at the midpoint (the k_3's). These are then employed to make predictions at the end of the interval that are used to develop slopes at the end of the interval (the k_4's). Finally, the k's are combined into a set of increment functions [as in Eq. (25.40)] and brought back to the beginning to make the final prediction. The following example illustrates the approach.

EXAMPLE 25.10

Solving Systems of ODEs Using the Fourth-Order RK Method

Problem Statement.   Use the fourth-order RK method to solve the ODEs from Example 25.9.

Solution.   First, we must solve for all the slopes at the beginning of the interval:

$$k_{1,1} = f_1(0, 4, 6) = -0.5(4) = -2$$
$$k_{1,2} = f_2(0, 4, 6) = 4 - 0.3(6) - 0.1(4) = 1.8$$

where $k_{i,j}$ is the ith value of k for the jth dependent variable. Next, we must calculate the first values of y_1 and y_2 at the midpoint:

$$y_1 + k_{1,1}\frac{h}{2} = 4 + (-2)\frac{0.5}{2} = 3.5$$

$$y_2 + k_{1,2}\frac{h}{2} = 6 + (1.8)\frac{0.5}{2} = 6.45$$

which can be used to compute the first set of midpoint slopes,

$$k_{2,1} = f_1(0.25, 3.5, 6.45) = -1.75$$
$$k_{2,2} = f_2(0.25, 3.5, 6.45) = 1.715$$

These are used to determine the second set of midpoint predictions,

$$y_1 + k_{2,1}\frac{h}{2} = 4 + (-1.75)\frac{0.5}{2} = 3.5625$$

$$y_2 + k_{2,2}\frac{h}{2} = 6 + (1.715)\frac{0.5}{2} = 6.42875$$

which can be used to compute the second set of midpoint slopes,

$$k_{3,1} = f_1(0.25, 3.5625, 6.42875) = -1.78125$$
$$k_{3,2} = f_2(0.25, 3.5625, 6.42875) = 1.715125$$

These are used to determine the predictions at the end of the interval,

$$y_1 + k_{3,1}h = 4 + (-1.78125)(0.5) = 3.109375$$
$$y_2 + k_{3,2}h = 6 + (1.715125)(0.5) = 6.857563$$

which can be used to compute the endpoint slopes,

$$k_{4,1} = f_1(0.5, 3.109375, 6.857563) = -1.554688$$
$$k_{4,2} = f_2(0.5, 3.109375, 6.857563) = 1.631794$$

The values of k can then be used to compute [Eq. (25.40)]:

$$y_1(0.5) = 4 + \frac{1}{6}[-2 + 2(-1.75 - 1.78125) - 1.554688]0.5 = 3.115234$$

$$y_2(0.5) = 6 + \frac{1}{6}[1.8 + 2(1.715 + 1.715125) + 1.631794]0.5 = 6.857670$$

Proceeding in a like manner for the remaining steps yields

x	y_1	y_2
0	4	6
0.5	3.115234	6.857670
1.0	2.426171	7.632106
1.5	1.889523	8.326886
2.0	1.471577	8.946865

25.4.3 Computer Algorithm for Solving Systems of ODEs

The computer code for solving a single ODE with Euler's method (Fig. 25.7) can be easily extended to systems of equations. The modifications include the following:

1. Inputting the number of equations, n.
2. Inputting the initial values for each of the n dependent variables.
3. Modifying the algorithm so that it computes slopes for each of the dependent variables.
4. Including additional equations to compute derivative values for each of the ODEs.
5. Including loops to compute a new value for each dependent variable.

Such an algorithm is outlined in Fig. 25.18 for the fourth-order RK method. Notice how similar it is in structure and organization to Fig. 25.7. Most of the differences relate to

1. The fact that there are n equations.
2. The added detail of the fourth-order RK method.

EXAMPLE 25.11 **Solving Systems of ODEs with the Computer**

Problem Statement.   A computer program to implement the fourth-order RK method for systems can be easily developed based on Fig. 25.18. Such software makes it convenient to compare different models of a physical system. For example, a linear model for a swinging pendulum is given by [recall Eq. (PT7.11)]

$$\frac{dy_1}{dx} = y_2 \qquad \frac{dy_2}{dx} = -16.1y_1$$

where y_1 and y_2 = angular displacement and velocity, respectively. A nonlinear model of the same system is [recall Eq. (PT7.9)]

$$\frac{dy_3}{dx} = y_4 \qquad \frac{dy_4}{dx} = -16.1\sin(y_3)$$

where y_3 and y_4 = angular displacement and velocity for the nonlinear case. Solve these systems for two cases: **(a)** a small initial displacement ($y_1 = y_3 = 0.1$ radian; $y_2 = y_4 = 0$) and **(b)** a large displacement ($y_1 = y_3 = \pi/4 = 0.785398$ radian; $y_2 = y_4 = 0$).

(a) Main or "Driver" Program

```
Assign values for
n = number of equations
yi = initial values of n dependent
     variables
xi = initial value independent
     variable
xf = final value independent variable
dx = calculation step size
xout = output interval


x = xi
m = 0
xp_m = x
DOFOR i = 1, n
  yp_{i,m} = yi_i
  y_i = yi_i
END DO
DO
  xend = x + xout
  IF (xend > xf) THEN xend = xf
  h = dx
  CALL Integrator(x, y, n, h, xend)
  m = m + 1
  xp_m = x
  DOFOR i = 1, n
    yp_{i,m} = y_i
  END DO
  IF (x ≥ xf) EXIT
END DO
DISPLAY RESULTS
END
```

(b) Routine to Take One Output Step

```
SUB Integrator(x, y, n, h, xend)
  DO
    IF (xend − x < h) THEN h = xend − x
    CALL RK4(x, y, n, h)
    IF (x ≥ xend) EXIT
  END DO
END SUB
```

(c) Fourth-Order RK Method for a System of ODEs

```
SUB RK4(x, y, n, h)
  CALL Derivs(x, y, k1)
  DOFOR i = 1, n
    ym_i = y_i + k1_i * h / 2
  END DO
  CALL Derivs(x + h / 2, ym, k2)
  DOFOR i = 1, n
    ym_i = y_i + k2_i * h / 2
  END DO
  CALL Derivs(x + h / 2, ym, k3)
  DOFOR i = 1, n
    ye_i = y_i + k3_i * h
  END DO
  CALL Derivs(x + h, ye, k4)
  DOFOR i = 1, n
    slope_i = (k1_i + 2*(k2_i+k3_i)+k4_i)/6
    y_i = y_i + slope_i * h
  END DO
  x = x + h
END SUB
```

(d) Routine to Determine Derivatives

```
SUB Derivs(x, y, dy)
  dy_1 = ...
  dy_2 = ...
END SUB
```

FIGURE 25.18
Pseudocode for the fourth-order RK method for systems.

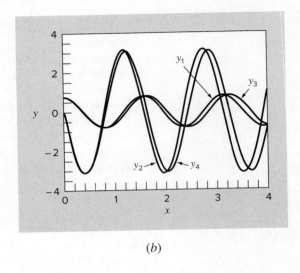

(a)   (b)

FIGURE 25.19
Solutions obtained with a computer program for the fourth-order RK method. The plots represent solutions for both linear and nonlinear pendulums with (a) small and (b) large initial displacements.

Solution.

(a) The calculated results for the linear and nonlinear models are almost identical (Fig. 25.19a). This is as expected because when the initial displacement is small, $\sin \theta \cong \theta$.

(b) When the initial displacement is $\pi/4 = 0.785398$, the solutions are much different and the difference is magnified as time becomes larger and larger (Fig. 25.19b). This is expected because the assumption that $\sin \theta = \theta$ is poor when theta is large.

25.5 ADAPTIVE RUNGE-KUTTA METHODS

To this point, we have presented methods for solving ODEs that employ a constant step size. For a significant number of problems, this can represent a serious limitation. For example, suppose that we are integrating an ODE with a solution of the type depicted in Fig. 25.20. For most of the range, the solution changes gradually. Such behavior suggests that a fairly large step size could be employed to obtain adequate results. However, for a localized region from $x = 1.75$ to $x = 2.25$, the solution undergoes an abrupt change. The practical consequence of dealing with such functions is that a very small step size would be required to accurately capture the impulsive behavior. If a constant step-size algorithm were employed, the smaller step size required for the region of abrupt change would have to be applied to the entire computation. As a consequence, a much smaller step size than necessary—and, therefore, many more calculations—would be wasted on the regions of gradual change.

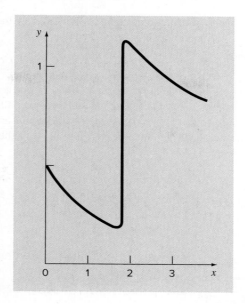

FIGURE 25.20
An example of a solution of an ODE that exhibits an abrupt change. Automatic step-size adjustment has great advantages for such cases.

Algorithms that automatically adjust the step size can avoid such overkill and hence be of great advantage. Because they "adapt" to the solution's trajectory, they are said to have *adaptive step-size control*. Implementation of such approaches requires that an estimate of the local truncation error be obtained at each step. This error estimate can then serve as a basis for either lengthening or decreasing the step size.

Before proceeding, we should mention that aside from solving ODEs, the methods described in this chapter can also be used to evaluate definite integrals. As mentioned previously in the introduction to Part Six, the evaluation of the integral

$$I = \int_a^b f(x)\,dx$$

is equivalent to solving the differential equation

$$\frac{dy}{dx} = f(x)$$

for $y(b)$ given the initial condition $y(a) = 0$. Thus, the following techniques can be employed to efficiently evaluate definite integrals involving functions that are generally smooth but exhibit regions of abrupt change.

There are two primary approaches to incorporate adaptive step-size control into one-step methods. In the first, the error is estimated as the difference between two predictions using the same-order RK method but with different step sizes. In the second, the local truncation error is estimated as the difference between two predictions using different-order RK methods.

25.5.1 Adaptive RK or Step-Halving Method

Step halving (also called *adaptive RK*) involves taking each step twice, once as a full step and independently as two half steps. The difference in the two results represents an estimate of the local truncation error. If y_1 designates the single-step prediction and y_2 designates the prediction using the two half steps, the error Δ can be represented as

$$\Delta = y_2 - y_1 \tag{25.43}$$

In addition to providing a criterion for step-size control, Eq. (25.43) can also be used to correct the y_2 prediction. For the fourth-order RK version, the correction is

$$y_2 \leftarrow y_2 + \frac{\Delta}{15} \tag{25.44}$$

This estimate is fifth-order accurate.

EXAMPLE 25.12   **Adaptive Fourth-Order RK Method**

Problem Statement.   Use the adaptive fourth-order RK method to integrate $y' = 4e^{0.8x} - 0.5y$ from $x = 0$ to 2 using $h = 2$ and an initial condition of $y(0) = 2$. This is the same differential equation that was solved previously in Example 25.5. Recall that the true solutions is $y(2) = 14.84392$.

Solution.   The single prediction with a step of h is computed as

$$y(2) = 2 + \frac{1}{6}[3 + 2(6.40216 + 4.70108) + 14.11105]2 = 15.10584$$

The two half-step predictions are

$$y(1) = 2 + \frac{1}{6}[3 + 2(4.21730 + 3.91297) + 5.945681]1 = 6.20104$$

and

$$y(2) = 6.20104 + \frac{1}{6}[5.80164 + 2(8.72954 + 7.99756) + 12.71283]1 = 14.86249$$

Therefore, the approximate error is

$$E_a = \frac{14.86249 - 15.10584}{15} = -0.01622$$

which compares favorably with the true error of

$$E_t = 14.84392 - 14.86249 = -0.01857$$

The error estimate can also be used to correct the prediction:

$$y(2) = 14.86249 - 0.01622 = 14.84627$$

which has an $E_t = -0.00235$.

25.5.2 Runge-Kutta Fehlberg

Aside from step halving as a strategy to adjust step size, an alternative approach for obtaining an error estimate involves computing two RK predictions of different order. The results can then be subtracted to obtain an estimate of the local truncation error. One shortcoming of this approach is that it greatly increases the computational overhead. For example, computing fourth- and fifth-order predictions amounts to a total of 10 function evaluations per step. The *Runge-Kutta Fehlberg*, or *embedded RK*, *method* cleverly circumvents this problem by using a fifth-order RK method that employs the function evaluations from the accompanying fourth-order RK method. Thus, the approach yields the error estimate on the basis of only six function evaluations!

For the present case, we use the following fourth-order estimate:

$$y_{i+1} = y_i + \left(\frac{37}{378}k_1 + \frac{250}{621}k_3 + \frac{125}{594}k_4 + \frac{512}{1771}k_6 \right) h \tag{25.45}$$

along with the fifth-order formula:

$$y_{i+1} = y_i + \left(\frac{2825}{27,648}k_1 + \frac{18,575}{48,384}k_3 + \frac{13,525}{55,296}k_4 + \frac{277}{14,336}k_5 + \frac{1}{4}k_6 \right) h \tag{25.46}$$

where

$$k_1 = f(x_i, y_i)$$

$$k_2 = f\left(x_i + \frac{1}{5}h, y_i + \frac{1}{5}k_1 h \right)$$

$$k_3 = f\left(x_i + \frac{3}{10}h, y_i + \frac{3}{40}k_1 h + \frac{9}{40}k_2 h \right)$$

$$k_4 = f\left(x_i + \frac{3}{5}h, y_i + \frac{3}{10}k_1 h - \frac{9}{10}k_2 h + \frac{6}{5}k_3 h \right)$$

$$k_5 = f\left(x_i + h, y_i - \frac{11}{54}k_1 h + \frac{5}{2}k_2 h - \frac{70}{27}k_3 h + \frac{35}{27}k_4 h \right)$$

$$k_6 = f\left(x_i + \frac{7}{8}h, y_i + \frac{1631}{55,296}k_1 h + \frac{175}{512}k_2 h + \frac{575}{13,824}k_3 h + \frac{44,275}{110,592}k_4 h \right.$$
$$\left. + \frac{253}{4096}k_5 h \right)$$

Thus, the ODE can be solved with Eq. (25.46) and the error estimated as the difference of the fifth- and fourth-order estimates. It should be noted that the particular coefficients used above were developed by Cash and Karp (1990). Therefore, it is sometimes called the *Cash-Karp RK method*.

EXAMPLE 25.13 **Runge-Kutta Fehlberg Method**

Problem Statement.   Use the Cash-Karp version of the Runge-Kutta Fehlberg approach to perform the same calculation as in Example 25.12 from $x = 0$ to 2 using $h = 2$.

Solution. The calculation of the k's can be summarized in the following table:

	x	y	$f(x, y)$
k_1	0	2	3
k_2	0.4	3.2	3.908511
k_3	0.6	4.20883	4.359883
k_4	1.2	7.228398	6.832587
k_5	2	15.42765	12.09831
k_6	1.75	12.17686	10.13237

These can then be used to compute the fourth-order prediction:

$$y_1 = 2 + \left(\frac{37}{378}3 + \frac{250}{621}4.359883 + \frac{125}{594}6.832587 + \frac{512}{1771}10.13237 \right)2 = 14.83192$$

along with a fifth-order formula:

$$y_1 = 2 + \left(\frac{2825}{27,648}3 + \frac{18,575}{48,384}4.359883 + \frac{13,525}{55,296}6.832587 \right.$$
$$\left. + \frac{227}{14,336}12.09831 + \frac{1}{4}10.13237 \right)2 = 14.83677$$

The error estimate is obtained by subtracting these two equations to give

$$E_a = 14.83677 - 14.83192 = 0.004842$$

25.5.3 Step-Size Control

Now that we have developed ways to estimate the local truncation error, it can be used to adjust the step size. In general, the strategy is to increase the step size if the error is too small and decrease it if the error is too large. Press et al. (2007) have suggested the following criterion to accomplish this:

$$h_{\text{new}} = h_{\text{present}} \left| \frac{\Delta_{\text{new}}}{\Delta_{\text{present}}} \right|^\alpha \tag{25.47}$$

where h_{present} and $h_{\text{new}} =$ the present and the new step sizes, respectively, $\Delta_{\text{present}} =$ the computed present accuracy, $\Delta_{\text{new}} =$ the desired accuracy, and $\alpha =$ a constant power that is equal to 0.2 when the step size is increased (that is, when $\Delta_{\text{present}} \leq \Delta_{\text{new}}$) and 0.25 when the step size is decreased ($\Delta_{\text{present}} > \Delta_{\text{new}}$).

The key parameter in Eq. (25.47) is obviously Δ_{new} because it is your vehicle for specifying the desired accuracy. One way to do this would be to relate Δ_{new} to a relative error level. Although this works well when only positive values occur, it can cause problems for solutions that pass through zero. For example, you might be simulating an oscillating function that repeatedly passes through zero but is bounded by maximum absolute values. For such a case, you might want these maximum values to figure in the desired accuracy.

A more general way to handle such cases is to determine Δ_{new} as

$$\Delta_{new} = \varepsilon y_{scale}$$

where ε = an overall tolerance level. Your choice of y_{scale} will then determine how the error is scaled. For example, if $y_{scale} = y$, the accuracy will be couched in terms of fractional relative errors. If you are dealing with a case where you desire constant errors relative to a prescribed maximum bound, set y_{scale} equal to that bound. A trick suggested by Press et al. (2007) to obtain the constant relative errors except very near zero crossings is

$$y_{scale} = |y| + \left| h\frac{dy}{dx} \right|$$

This is the version we will use in our algorithm.

25.5.4 Computer Algorithm

Figures 25.21 and 25.22 outline pseudocode to implement the Cash-Karp version of the Runge-Kutta Fehlberg algorithm. This algorithm is patterned after a more detailed implementation by Press et al. (2007) for systems of ODEs.

Figure 25.21 implements a single step of the Cash-Karp routine [that is, Eqs. (25.45) and (25.46)]. Figure 25.22 outlines a general driver program along with a subroutine that actually adapts the step size.

FIGURE 25.21
Pseudocode for a single step of the Cash-Karp RK method.

```
SUBROUTINE RKkc(y,dy,x,h,yout,yerr)
PARAMETER (a2=0.2,a3=0.3,a4=0.6,a5=1.,a6=0.875,
  b21=0.2,b31=3./40.,b32=9./40.,b41=0.3,b42=-0.9,
  b43=1.2,b51=-11./54.,b52=2.5,b53=-70./27.,
  b54=35./27.,b61=1631./55296.,b62=175./512.,
  b63=575./13824.,b64=44275./110592.,b65=253./4096.,
  c1=37./378.,c3=250./621.,c4=125./594.,
  c6=512./1771.,dc1=c1-2825./27648.,
  dc3=c3-18575./48384.,dc4=c4-13525./55296.,
  dc5=-277./14336.,dc6=c6-0.25)
ytemp=y+b21*h*dy
CALL Derivs(x+a2*h,ytemp,k2)
ytemp=y+h*(b31*dy+b32*k2)
CALL Derivs(x+a3*h,ytemp,k3)
ytemp=y+h*(b41*dy+b42*k2+b43*k3)
CALL Derivs(x+a4*h,ytemp,k4)
ytemp=y+h*(b51*dy+b52*k2+b53*k3+b54*k4)
CALL Derivs(x+a5*h,ytemp,k5)
ytemp=y+h*(b61*dy+b62*k2+b63*k3+b64*k4+b65*k5)
CALL Derivs(x+a6*h,ytemp,k6)
yout=y+h*(c1*dy+c3*k3+c4*k4+c6*k6)
yerr=h*(dc1*dy+dc3*k3+dc4*k4+dc5*k5+dc6*k6)
END RKkc
```

(a) Driver Program

```
INPUT xi, xf, yi
maxstep=100
hi=.5; tiny = 1. × 10⁻³⁰
eps=0.00005
print *, xi,yi
x=xi
y=yi
h=hi
istep=0
DO
  IF (istep > maxstep AND x ≤ xf) EXIT
  istep=istep+1
  CALL Derivs(x,y,dy)
  yscal=ABS(y)+ABS(h*dy)+tiny
  IF (x+h>xf) THEN h=xf-x
  CALL Adapt(x,y,dy,h,yscal,eps,hnxt)
  PRINT x,y
  h=hnxt
END DO
END
```

(b) Adaptive Step Routine

```
SUB Adapt(x,y,dy,htry,yscal,eps,hnxt)
PARAMETER (safety=0.9, econ=1.89e-4)
h=htry
DO
  CALL RKkc(y,dy,x,h,ytemp,yerr)
  emax=abs(yerr/yscal/eps)
  IF emax ≤ 1 EXIT
  htemp=safety*h*emax⁻⁰·²⁵
  h=max(abs(htemp),0.25*abs(h))
  xnew=x+h
  IF xnew=x THEN pause
END DO
IF emax > econ THEN
  hnxt=safety*emax⁻·²*h
ELSE
  hnxt=4.*h
END IF
x=x+h
y=ytemp
END Adapt
```

FIGURE 25.22
Pseudocode for a (a) driver program and an (b) adaptive step routine to solve a single ODE.

EXAMPLE 25.14

Computer Application of an Adaptive Fourth-Order RK Scheme

Problem Statement.   The adaptive RK method is well-suited for the following ordinary differential equation:

$$\frac{dy}{dx} + 0.6y = 10e^{-(x-2)^2/[2(0.075)^2]} \tag{E25.14.1}$$

Notice for the initial condition, $y(0) = 0.5$, the general solution is

$$y = 0.5e^{-0.6x} \tag{E25.14.2}$$

which is a smooth curve that gradually approaches zero as x increases. In contrast, the particular solution undergoes an abrupt transition in the vicinity of $x = 2$ due to the nature of the forcing function (Fig. 25.23a). Use a standard fourth-order RK scheme to solve Eq. (E25.14.1) from $x = 0$ to 4. Then employ the adaptive scheme described in this section to perform the same computation.

Solution.   First, the classical fourth-order scheme is used to compute the solid curve in Fig. 25.23b. For this computation, a step size of 0.1 is used so that $4/(0.1) = 40$ applications of the technique are made. Then, the calculation is repeated with a step size of 0.05 for a total of 80 applications. The major discrepancy between the two results occurs in the region from 1.8 to 2.0. The magnitude of the discrepancy is about 0.1 to 0.2 percent.

FIGURE 25.23

(a) A bell-shaped forcing function that induces an abrupt change in the solution of an ODE [Eq. (E25.14.1)]. (b) The solution. The points indicate the predictions of an adaptive step-size routine.

Next, the algorithm in Figs. 25.21 and 25.22 is developed into a computer program and used to solve the same problem. An initial step size of 0.5 and an $\varepsilon = 0.00005$ were chosen. The results were superimposed on Fig. 25.23b. Notice how large steps are taken in the regions of gradual change. Then, in the vicinity of $x = 2$, the steps are decreased to accommodate the abrupt nature of the forcing function.

The utility of an adaptive integration scheme obviously depends on the nature of the functions being modeled. It is particularly advantageous for those solutions with long smooth stretches and short regions of abrupt change. In addition, it has utility in those situations where the correct step size is not known *a priori*. For these cases, an adaptive routine will "feel" its way through the solution while keeping the results within the desired tolerance. Thus, it will tiptoe through the regions of abrupt change and step out briskly when the variations become more gradual.

PROBLEMS

25.1 Solve the following initial-value problem over the interval from $t = 0$ to 2 where $y(0) = 1$. Display all your results on the same graph.

$$\frac{dy}{dt} = yt^3 - 1.5y$$

(a) Analytically.
(b) Euler's method with $h = 0.5$ and 0.25.
(c) Midpoint method with $h = 0.5$.
(d) Fourth-order RK method with $h = 0.5$.

25.2 Solve the following problem over the interval from $x = 0$ to 1 using a step size of 0.25, where $y(0) = 1$. Display all your results on the same graph.

$$\frac{dy}{dx} = (1 + 2t)\sqrt{x}$$

(a) Analytically.
(b) Euler's method.
(c) Heun's method without iteration.
(d) Ralston's method.
(e) Fourth-order RK method.

25.3 Use the **(a)** Euler and **(b)** Heun (without iteration) methods to solve

$$\frac{d^2y}{dt^2} - t + y = 0$$

where $y(0) = 2$ and $y'(0) = 0$. Solve from $x = 0$ to 4 using $h = 0.1$. Compare the methods by plotting the solutions.

25.4 Solve the following problem with the fourth-order RK method:

$$\frac{d^2y}{dx^2} + 0.5\frac{dy}{dx} + 7y = 0$$

where $y(0) = 4$ and $y'(0) = 0$. Solve from $x = 0$ to 5 with $h = 0.5$. Plot your results.

25.5 Solve from $t = 0$ to 3 with $h = 0.1$ using **(a)** Heun (without corrector) and **(b)** Ralston's second-order RK method:

$$\frac{dy}{dt} = y\sin^3(t) \qquad y(0) = 1$$

25.6 Solve the following problem numerically from $t = 0$ to 3:

$$\frac{dy}{dt} = -y + t^2 \qquad y(0) = 1$$

Use the third-order RK method with a step size of 0.5.

25.7 Use **(a)** Euler's and **(b)** the fourth-order RK method to solve

$$\frac{dy}{dt} = -2y + 4e^{-t}$$

$$\frac{dz}{dt} = -\frac{yz^2}{3}$$

over the range $t = 0$ to 1 using a step size of 0.2 with $y(0) = 2$ and $z(0) = 4$.

25.8 Compute the first step of Example 25.14 using the adaptive fourth-order RK method with $h = 0.5$. Verify whether step-size adjustment is in order.

25.9 If $\varepsilon = 0.001$, determine whether step-size adjustment is required for Example 25.12.

25.10 Use the RK Fehlberg approach to perform the same calculation as in Example 25.12 from $x = 0$ to 1 with $h = 1$.

25.11 Write a computer program based on Fig. 25.7. Among other things, place documentation statements throughout the program to identify what each section is intended to accomplish.

25.12 Test the program you developed in Prob. 25.11 by duplicating the computations from Examples 25.1 and 25.4.

25.13 Develop a user-friendly program for the Heun method with an iterative corrector. Test the program by duplicating the results in Table 25.2.

25.14 Develop a user-friendly computer program for the classical fourth-order RK method. Test the program by duplicating Example 25.7.

25.15 Develop a user-friendly computer program for systems of equations using the fourth-order RK method. Use this program to duplicate the computation in Example 25.10.

25.16 The motion of a damped spring-mass system (Fig. P25.16) is described by the following ordinary differential equation:

$$m\frac{d^2x}{dt^2} + c\frac{dx}{dt} + kx = 0$$

where $x =$ displacement from equilibrium position (m), $t =$ time (s), $m = 20$-kg mass, and $c =$ the damping coefficient (N s/m). The damping coefficient c takes on three values: 5 (underdamped), 40 (critically damped), and 200 (overdamped). The spring constant $k = 20$ N/m. The initial velocity is zero, and the initial displacement $x = 1$ m. Solve this equation using a numerical method over the time period $0 \le t \le 15$ s. Plot the displacement versus time for each of the three values of the damping coefficient on the same curve.

FIGURE P25.16

25.17 If water is drained from a vertical cylindrical tank by opening a valve at the base, the water will flow fast when the tank is full and slow down as it continues to drain. As it turns out, the rate at which the water level drops is

$$\frac{dy}{dt} = -k\sqrt{y}$$

where k is a constant depending on the shape of the hole and the cross-sectional area of the tank and drain hole. The depth of the water y is measured in meters and the time t in minutes. If $k = 0.05$, determine how long it takes the tank to drain if the fluid level is initially 3 m. Solve by applying Euler's equation and writing a computer program or using Excel. Use a step of 0.5 minute.

25.18 The following is an initial-value, second-order differential equation:

$$\frac{d^2x}{dt^2} + (5x)\frac{dx}{dt} + (x + 7)\sin(\omega t) = 0$$

where

$$\frac{dx}{dt}(0) = 1.5 \quad \text{and} \quad x(0) = 6$$

Note that $\omega = 1$. Decompose the equation into two first-order differential equations. After the decomposition, solve the system from $t = 0$ to 15 and plot the results.

25.19 Assuming that drag is proportional to the square of velocity, we can model the velocity of a falling object like a parachutist with the following differential equation:

$$\frac{dv}{dt} = g - \frac{c_d}{m}v^2$$

where v is velocity (m/s), t = time (s), g is the acceleration due to gravity (9.81 m/s^2), c_d = a second-order drag coefficient (kg/m), and m = mass (kg). Solve for the velocity and distance fallen by a 90-kg object with a drag coefficient of 0.25 kg/m. If the initial height is 1 km, determine when the object hits the ground. Obtain your solution with (a) Euler's method and (b) the fourth-order RK method.

25.20 A spherical tank has a circular orifice in its bottom through which liquid flows out (Fig. P25.20). The flow rate through the hole can be estimated as

$$Q_{out} = CA\sqrt{2gH}$$

where Q_{out} = outflow (m^3/s), C = an empirically derived coefficient, A = the area of the orifice (m^2), g = the gravitational constant (= 9.81 m/s^2), and H = the depth of liquid in the tank. Use one of the numerical methods described in this chapter to determine how long it will take for water to flow out of a 3-m-diameter tank if the water's initial depth is 2.8 m. Note that the orifice has a diameter of 3 cm and $C = 0.6$.

FIGURE P25.20
A spherical tank.

25.21 The logistic model is used to simulate population, as in

$$\frac{dp}{dt} = k_{gm}(1 - p/p_{max})p$$

where p = population, k_{gm} = the maximum growth rate under unlimited conditions, and p_{max} = the carrying capacity. Simulate the world's population from 1950 to 2000 using one of the numerical methods described in this chapter. Employ the following initial conditions and parameter values for your simulation: p_0 (in 1950) = 2555 million people, $k_{gm} = 0.026$/yr, and $p_{max} = 12{,}000$ million people. Have the function generate output corresponding to the dates for the following measured population data. Develop a plot of your simulation along with these data.

t	1950	1960	1970	1980	1990	2000
p	2555	3040	3710	4455	5275	6080

25.22 Suppose that a projectile is launched upward from the earth's surface. Assume that the only force acting on the object is the downward force of gravity. Under these conditions, a force balance can be used to derive

$$\frac{dv}{dt} = -g(0)\frac{R^2}{(R + x)^2}$$

where v = upward velocity (m/s), t = time (s), x = altitude (m) measured upward from the earth's surface, $g(0)$ = the gravitational acceleration at the earth's surface ($\cong 9.81$ m/s^2), and R = the earth's radius ($\cong 6.37 \times 10^6$ m). Recognizing that $dx/dt = v$, use Euler's method to determine the maximum height that would be obtained if $v(t = 0) = 1400$ m/s.

25.23 The *humps function* exhibits both flat and steep regions over a relatively short x region:

$$f(x) = \frac{1}{(x - 0.3)^2 + 0.01} + \frac{1}{(x - 0.9)^2 + 0.04} - 6$$

Determine the value of the definite integral of this function between $x = 0$ and 1 using an adaptive RK method.

25.24 Given the initial conditions $y(0) = 1$ and $y'(0) = 0$, solve the following initial-value problem from $t = 0$ to 4:

$$\frac{d^2y}{dt^2} + 4y = 0$$

Obtain your solutions with **(a)** Euler's method and **(b)** the fourth-order RK method. In both cases, use a step size of 0.125. Plot both solutions on the same graph along with the exact solution $y = \cos 2t$.

25.25 Use the following differential equations to compute the velocity and position of a soccer ball that is kicked straight up in the air with an initial velocity of 40 m/s:

$$\frac{dy}{dt} = v$$

$$\frac{dv}{dt} = -g - \frac{c_d}{m}v|v|$$

where y = upward distance (m), t = time (s), v = upward velocity (m/s), g = gravitational constant (= 9.81 m/s²), c_d = drag coefficient (kg/m), and m = mass (kg). Note that the drag coefficient is related to more fundamental parameters by

$$c_d = \frac{1}{2}\rho A C_d$$

where ρ = air density (kg/m³), A = area (m²), and C_d = the dimensionless drag coefficient. Use the following parameter values for your calculation: $d = 22$ cm, $m = 0.4$ kg, $\rho = 1.3$ kg/m³, and $C_d = 0.52$.

25.26 Three linked bungee jumpers are depicted in Fig. P25.26. If the bungee cords are idealized as linear springs (i.e., governed by Hooke's law), the following differential equations based on force balances can be developed:

$$m_1\frac{d^2x_1}{dt^2} = m_1g + k_2(x_2 - x_1) - k_1x_1$$

$$m_2\frac{d^2x_2}{dt^2} = m_2g + k_3(x_3 - x_2) + k_2(x_1 - x_2)$$

$$m_3\frac{d^2x_3}{dt^2} = m_3g + k_3(x_2 - x_3)$$

where m_i = the mass of jumper i (kg), k_j = the spring constant for cord j (N/m), x_i = the displacement of jumper i measured downward

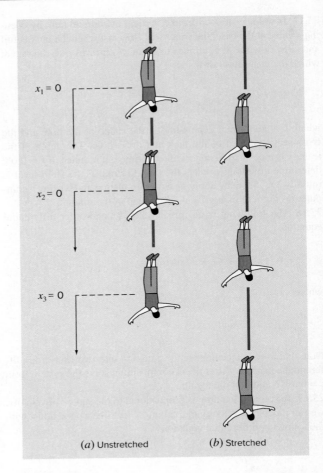

FIGURE P25.26
Three individuals connected by bungee cords.

from the jumper's equilibrium position (m), and g = gravitational acceleration (9.81 m/s²). Solve these equations for the positions and velocities of the three jumpers given the initial conditions that all positions and velocities are zero at $t = 0$. Use the following parameters for your calculations: $m_1 = 60$ kg, $m_2 = 72$ kg, $m_3 = 84$ kg, $k_1 = k_3 = 60$, and $k_2 = 120$ (N/m).

Stiffness and Multistep Methods

This chapter covers two areas. First, we describe *stiff ODEs*. These are both individual and systems of ODEs that have both fast and slow components to their solution. We introduce the idea of an *implicit solution technique* as one commonly used remedy for this problem. Then we discuss *multistep methods*. These algorithms retain information from previous steps to more effectively capture the trajectory of the solution. They also yield the truncation error estimates that can be used to implement adaptive step-size control.

26.1 STIFFNESS

Stiffness is a special problem that can arise in the solution of ordinary differential equations. A *stiff system* is one involving rapidly changing components together with slowly changing ones. In many cases, the rapidly varying components are ephemeral transients that die away quickly, after which the solution becomes dominated by the slowly varying components. Although the transient phenomena exist for only a short part of the integration interval, they can dictate the time step for the entire solution.

Both individual and systems of ODEs can be stiff. An example of a single stiff ODE is

$$\frac{dy}{dt} = -1000y + 3000 - 2000e^{-t} \tag{26.1}$$

If $y(0) = 0$, the analytical solution can be developed as

$$y = 3 - 0.998e^{-1000t} - 2.002e^{-t} \tag{26.2}$$

As in Fig. 26.1, the solution is initially dominated by the fast exponential term (e^{-1000t}). After a short period ($t < 0.005$), this transient dies out and the solution becomes dictated by the slow exponential (e^{-t}).

Insight into the step size required for stability of such a solution can be gained by examining the homogeneous part of Eq. (26.1),

$$\frac{dy}{dt} = -ay \tag{26.3}$$

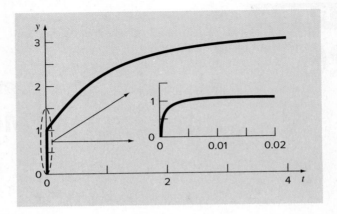

FIGURE 26.1
Plot of a stiff solution of a single ODE. Although the solution appears to start at 1, there is actually a fast transient from $y = 0$ to 1 that occurs in less than 0.005 time unit. This transient is perceptible only when the response is viewed on the finer timescale in the inset.

If $y(0) = y_0$, calculus can be used to determine the solution as

$$y = y_0 e^{-at}$$

Thus, the solution starts at y_0 and asymptotically approaches zero.

Euler's method can be used to solve the same problem numerically:

$$y_{i+1} = y_i + \frac{dy_i}{dt} h$$

Substituting Eq. (26.3) gives

$$y_{i+1} = y_i - ay_i h$$

or

$$y_{i+1} = y_i(1 - ah) \tag{26.4}$$

The stability of this formula clearly depends on the step size h. That is, $|1 - ah|$ must be less than 1. Thus, if $h > 2/a$, $|y_i| \to \infty$ as $i \to \infty$.

For the fast transient part of Eq. (26.2), this criterion can be used to show that the step size to maintain stability must be $< 2/1000 = 0.002$. In addition, it should be noted that, whereas this criterion maintains stability (that is, a bounded solution), an even smaller step size would be required to obtain an accurate solution. Thus, although the transient occurs for only a small fraction of the integration interval, it controls the maximum allowable step size.

Superficially, you might suppose that the adaptive step-size routines described at the end of the last chapter might offer a solution for this dilemma. You might think that they would use small steps during the rapid transients and large steps otherwise. However, this is not the case, because the stability requirement will necessitate using very small steps throughout the entire solution.

Rather than using explicit approaches, implicit methods offer an alternative remedy. Such representations are called *implicit* because the unknown appears on both sides of the equation. An implicit form of Euler's method can be developed by evaluating the derivative at the future time,

$$y_{i+1} = y_i + \frac{dy_{i+1}}{dt}h$$

This is called the *backward*, or *implicit, Euler's method*. Substituting Eq. (26.3) yields

$$y_{i+1} = y_i - ay_{i+1}h$$

which can be solved for

$$y_{i+1} = \frac{y_i}{1 + ah} \tag{26.5}$$

For this case, regardless of the size of the step, $|y_i| \to 0$ as $i \to \infty$. Hence, the approach is called *unconditionally stable*.

EXAMPLE 26.1 **Explicit and Implicit Euler**

Problem Statement. Use both the explicit and implicit Euler methods to solve

$$\frac{dy}{dt} = -1000y + 3000 - 2000e^{-t}$$

where $y(0) = 0$. **(a)** Use the explicit Euler with step sizes of 0.0005 and 0.0015 to solve for y between $t = 0$ and 0.006. **(b)** Use the implicit Euler with a step size of 0.05 to solve for y between 0 and 0.4.

Solution.

(a) For this problem, the explicit Euler's method is

$$y_{i+1} = y_i + (-1000y_i + 3000 - 2000e^{-t_i})h$$

The result for $h = 0.0005$ is displayed in Fig. 26.2a along with the analytical solution. Although it exhibits some truncation error, the result captures the general shape of the analytical solution. In contrast, when the step size is increased to a value just below the stability limit ($h = 0.0015$), the solution manifests oscillations. Using $h > 0.002$ would result in a totally unstable solution; that is, it would go to infinity as the solution progressed.

(b) The implicit Euler's method is

$$y_{i+1} = y_i + (-1000y_{i+1} + 3000 - 2000e^{-t_{i+1}})h$$

Now because the ODE is linear, we can rearrange this equation so that y_{i+1} is isolated on the left-hand side,

$$y_{i+1} = \frac{y_i + 3000h - 2000he^{-t_{i+1}}}{1 + 1000h}$$

The result for $h = 0.05$ is displayed in Fig. 26.2b along with the analytical solution. Notice that even though we have used a much bigger step size than the one that

FIGURE 26.2
Solution of a "stiff" ODE with (*a*) the explicit and (*b*) implicit Euler methods.

induced instability for the explicit Euler, the numerical solution tracks nicely on the analytical result.

Systems of ODEs can also be stiff. An example is

$$\frac{dy_1}{dt} = -5y_1 + 3y_2 \tag{26.6a}$$

$$\frac{dy_2}{dt} = 100y_1 - 301y_2 \tag{26.6b}$$

For the initial conditions $y_1(0) = 52.29$ and $y_2(0) = 83.82$, the exact solution is

$$y_1 = 52.96e^{-3.9899t} - 0.67e^{-302.0101t} \tag{26.7a}$$

$$y_2 = 17.83e^{-3.9899t} + 65.99e^{-302.0101t} \tag{26.7b}$$

Note that the exponents are negative and differ by about two orders of magnitude. As with the single equation, it is the large exponents that respond rapidly and are at the heart of the system's stiffness.

An implicit Euler's method for the example system can be formulated as

$$y_{1,i+1} = y_{1,i} + (-5y_{1,i+1} + 3y_{2,i+1})h \tag{26.8a}$$

$$y_{2,i+1} = y_{2,i} + (100y_{1,i+1} - 301y_{2,i+1})h \tag{26.8b}$$

Collecting terms gives

$$(1 + 5h)y_{1,i+1} - 3hy_{2,i+1} = y_{1,i} \tag{26.9a}$$

$$-100hy_{1,i+1} + (1 + 301h)y_{2,i+1} = y_{2,i} \tag{26.9b}$$

Thus, we can see that the problem consists of solving a set of simultaneous equations for each time step.

For nonlinear ODEs, the solution becomes even more difficult since it involves solving a system of nonlinear simultaneous equations (recall Sec. 6.6). Thus, although stability is gained through implicit approaches, a price is paid in the form of added solution complexity.

The implicit Euler method is unconditionally stable and only first-order accurate. It is also possible to develop in a similar manner a second-order accurate implicit trapezoidal rule integration scheme for stiff systems. It is usually desirable to have higher-order methods. The Adams-Moulton formulas described later in this chapter can also be used to devise higher-order implicit methods. However, the stability limits of such approaches are very stringent when applied to stiff systems. Gear (1971) developed a special series of implicit schemes that have much larger stability limits based on backward difference formulas. Extensive efforts have been made to develop software to efficiently implement Gear's method. As a result, this is probably the most widely used method to solve stiff systems. In addition, Rosenbrock and others (see Press et al. 2007) have proposed implicit Runge-Kutta algorithms where the k terms appear implicitly. These methods have good stability characteristics and are quite suitable for solving systems of stiff ordinary differential equations.

26.2 MULTISTEP METHODS

The one-step methods described in the previous sections utilize information at a single point x_i to predict a value of the dependent variable y_{i+1} at a future point x_{i+1} (Fig. 26.3a). Alternative approaches, called *multistep methods* (Fig. 26.3b), are based on the insight that, once the computation has begun, valuable information from previous points is at our command. The curvature of the lines connecting these previous values provides information regarding the trajectory of the solution. The multistep methods explored in this chapter exploit this information to solve ODEs. Before describing the higher-order versions, we will present a simple second-order method that serves to demonstrate the general characteristics of multistep approaches.

26.2.1 The Non-Self-Starting Heun Method

Recall that the Heun approach uses *Euler's method* as a *predictor* [Eq. (25.15)]:

$$y_{i+1}^0 = y_i + f(x_i, y_i)h \tag{26.10}$$

FIGURE 26.3
Graphical depiction of the fundamental difference between (*a*) one-step and (*b*) multistep methods for solving ODEs.

and the *trapezoidal rule* as a *corrector* [Eq. (25.16)]:

$$y_{i+1} = y_i + \frac{f(x_i, y_i) + f(x_{i+1}, y_{i+1}^0)}{2} h \tag{26.11}$$

Thus, the predictor and the corrector have local truncation errors of $O(h^2)$ and $O(h^3)$, respectively. This suggests that the predictor is the weak link in the method because it has the greatest error. This weakness is significant because the efficiency of the iterative corrector step depends on the accuracy of the initial prediction. Consequently, one way to improve Heun's method is to develop a predictor that has a local error of $O(h^3)$. This can be accomplished by using Euler's method and the slope at y_i, and extra information from a previous point y_{i-1}, as in

$$y_{i+1}^0 = y_{i-1} + f(x_i, y_i)2h \tag{26.12}$$

Notice that Eq. (26.12) attains $O(h^3)$ at the expense of employing a larger step size, $2h$. In addition, note that Eq. (26.12) is not self-starting because it involves a previous value of the dependent variable y_{i-1}. Such a value would not be available in a typical initial-value problem. Because of this fact, Eqs. (26.11) and (26.12) are called the *non-self-starting Heun method*.

As depicted in Fig. 26.4, the derivative estimate in Eq. (26.12) is now located at the midpoint rather than at the beginning of the interval over which the prediction is made. As demonstrated subsequently, this centering improves the error of the predictor to $O(h^3)$. However, before proceeding to a formal derivation of the non-self-starting Heun, we will summarize the method and express it using a slightly modified nomenclature:

Predictor: $y_{i+1}^0 = y_{i-1}^m + f(x_i, y_i^m)2h$ (26.13)

Corrector: $y_{i+1}^j = y_i^m + \dfrac{f(x_i, y_i^m) + f(x_{i+1}, y_{i+1}^{j-1})}{2} h$

(for $j = 1, 2, \ldots, m$)   (26.14)

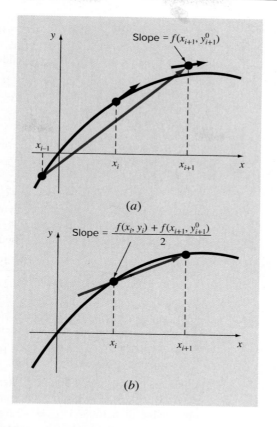

FIGURE 26.4
A graphical depiction of the non-self-starting Heun method. (*a*) The midpoint method that is used as a predictor. (*b*) The trapezoidal rule that is employed as a corrector.

where the superscripts have been added to denote that the corrector is applied iteratively from $j = 1$ to m to obtain refined solutions. Note that y_i^m and y_{i-1}^m are the final results of the corrector iterations at the previous time steps. The iterations are terminated at any time step on the basis of the stopping criterion,

$$|\varepsilon_a| = \left| \frac{y_{i+1}^j - y_{i+1}^{j-1}}{y_{i+1}^j} \right| 100\%$$
(26.15)

When ε_a is less than a prespecified error tolerance ε_s, the iterations are terminated. At this point, $j = m$. The use of Eqs. (26.13) through (26.15) to solve an ODE is demonstrated in the following example.

EXAMPLE 26.2 **Non-Self-Starting Heun Method**

Problem Statement. Use the non-self-starting Heun method to perform the same computations as were performed previously in Example 25.5 using Heun's method. That is, integrate $y' = 4e^{0.8x} - 0.5y$ from $x = 0$ to $x = 4$ using a step size of 1.0. As with Example 25.5, the initial condition at $x = 0$ is $y = 2$. However, because we are now dealing with a multistep method, we require the additional information that y is equal to -0.3929953 at $x = -1$.

Solution.　The predictor [Eq. (26.13)] is used to extrapolate linearly from $x = -1$ to $x = 1$:

$$y_1^0 = -0.3929953 + [4e^{0.8(0)} - 0.5(2)]2 = 5.607005$$

The corrector [Eq. (26.14)] is then used to compute the value:

$$y_1^1 = 2 + \frac{4e^{0.8(0)} - 0.5(2) + 4e^{0.8(1)} - 0.5(5.607005)}{2}1 = 6.549331$$

which represents a percent relative error of -5.73% (true value = 6.194631). This error is somewhat smaller than the value of -8.18% incurred with the self-starting Heun.

Now, Eq. (26.14) can be applied iteratively to improve the solution:

$$y_1^2 = 2 + \frac{3 + 4e^{0.8(1)} - 0.5(6.549331)}{2}1 = 6.313749$$

which represents an ε_t of -1.92%. An approximate estimate of the error can also be determined using Eq. (26.15):

$$|\varepsilon_a| = \left| \frac{6.313749 - 6.549331}{6.313749} \right| 100\% = 3.7\%$$

Equation (26.14) can be applied iteratively until ε_a falls below a prespecified value of ε_s. As was the case with the Heun method (recall Example 25.5), the iterations converge on a value of 6.360865 ($\varepsilon_t = -2.68\%$). However, because the initial predictor value is more accurate, the multistep method converges at a somewhat faster rate.

For the second step, the predictor is

$$y_2^0 = 2 + [4e^{0.8(1)} - 0.5(6.360865)]2 = 13.44346 \qquad \varepsilon_t = 9.43\%$$

which is superior to the prediction of 12.08260 ($\varepsilon_t = 18\%$) that was computed with the original Heun method. The first corrector yields 15.76693 ($\varepsilon_t = 6.8\%$), and subsequent iterations converge on the same result as was obtained with the self-starting Heun method: 15.30224 ($\varepsilon_t = -3.1\%$). As with the previous step, the rate of convergence of the corrector is somewhat improved because of the better initial prediction.

Derivation and Error Analysis of Predictor-Corrector Formulas.　We have just employed graphical concepts to derive the non-self-starting Heun. We will now show how the same equations can be derived mathematically. This derivation is particularly interesting because it ties together ideas from curve fitting, numerical integration, and ODEs. The exercise is also useful because it provides a simple procedure for developing higher-order multistep methods and estimating their errors.

The derivation is based on solving the general ODE,

$$\frac{dy}{dx} = f(x, y)$$

This equation can be solved by multiplying both sides by dx and integrating between limits at i and $i + 1$:

$$\int_{y_i}^{y_{i+1}} dy = \int_{x_i}^{x_{i+1}} f(x, y) \, dx$$

The left side can be integrated and evaluated using [recall Eq. (25.21)]

$$y_{i+1} = y_i + \int_{x_i}^{x_{i+1}} f(x, y) \, dx \tag{26.16}$$

Equation (26.16) represents a solution to the ODE if the integral can be evaluated. That is, it provides a means to compute a new value of the dependent variable y_{i+1} on the basis of a prior value y_i and the differential equation.

Numerical integration formulas such as those developed in Chap. 21 provide one way to make this evaluation. For example, the trapezoidal rule [Eq. (21.3)] can be used to evaluate the integral, as in

$$\int_{x_i}^{x_{i+1}} f(x, y) \, dx = \frac{f(x_i, y_i) + f(x_{i+1}, y_{i+1})}{2} h \tag{26.17}$$

where $h = x_{i+1} - x_i$ is the step size. Substituting Eq. (26.17) into Eq. (26.16) yields

$$y_{i+1} = y_i + \frac{f(x_i, y_i) + f(x_{i+1}, y_{i+1})}{2} h$$

which is the corrector for the Heun method. Because this equation is based on the trapezoidal rule, the truncation error can be taken directly from Table 21.2,

$$E_c = -\frac{1}{12} h^3 y^{(3)}(\xi_c) = -\frac{1}{12} h^3 f''(\xi_c) \tag{26.18}$$

where the subscript c designates that this is the error of the corrector.

A similar approach can be used to derive the predictor. For this case, the integration limits are from $i - 1$ to $i + 1$:

$$\int_{y_{i-1}}^{y_{i+1}} dy = \int_{x_{i-1}}^{x_{i+1}} f(x, y) \, dx$$

which can be integrated and rearranged to yield

$$y_{i+1} = y_{i-1} + \int_{x_{i-1}}^{x_{i+1}} f(x, y) \, dx \tag{26.19}$$

Now, rather than using a closed formula from Table 21.2, we can use the first Newton-Cotes open integration formula (see Table 21.4) to evaluate the integral, as in

$$\int_{x_{i-1}}^{x_{i+1}} f(x, y) \, dx = 2hf(x_i, y_i) \tag{26.20}$$

which is called the *midpoint method*. Substituting Eq. (26.20) into Eq. (26.19) yields

$$y_{i+1} = y_{i-1} + 2hf(x_i, y_i)$$

which is the predictor for the non-self-starting Heun. As with the corrector, the local truncation error can be taken directly from Table 21.4:

$$E_p = \frac{1}{3}h^3 y^{(3)}(\xi_p) = \frac{1}{3}h^3 f''(\xi_p) \tag{26.21}$$

where the subscript p designates that this is the error of the predictor.

Thus, the predictor and the corrector for the non-self-starting Heun method have truncation errors of the same order. Aside from upgrading the accuracy of the predictor, this fact has additional benefits related to error analysis, as elaborated in the next section.

Error Estimates. If the predictor and the corrector of a multistep method are of the same order, the local truncation error may be estimated during the course of a computation. This is a tremendous advantage because it establishes a criterion for adjustment of the step size.

The local truncation error for the predictor is estimated by Eq. (26.21). This error estimate can be combined with the estimate of y_{i+1} from the predictor step to yield [recall our basic definition of Eq. (3.1)]

$$\text{True value} = y_{i+1}^0 + \frac{1}{3}h^3 y^{(3)}(\xi_p) \tag{26.22}$$

Using a similar approach, the error estimate for the corrector [Eq. (26.18)] can be combined with the corrector result y_{i+1} to give

$$\text{True value} = y_{i+1}^m - \frac{1}{12}h^3 y^{(3)}(\xi_c) \tag{26.23}$$

Equation (26.22) can be subtracted from Eq. (26.23) to yield

$$0 = y_{i+1}^m - y_{i+1}^0 - \frac{5}{12}h^3 y^{(3)}(\xi) \tag{26.24}$$

where ξ is now between x_{i-1} and x_{i+1}. Now, dividing Eq. (26.24) by 5 and rearranging the result gives

$$\frac{y_{i+1}^0 - y_{i+1}^m}{5} = -\frac{1}{12}h^3 y^{(3)}(\xi) \tag{26.25}$$

Notice that the right-hand sides of Eqs. (26.18) and (26.25) are identical, with the exception of the argument of the third derivative. If the third derivative does not vary appreciably over the interval in question, we can assume that the right-hand sides are equal, and therefore, the left-hand sides should also be equivalent, as in

$$E_c = -\frac{y_{i+1}^0 - y_{i+1}^m}{5} \tag{26.26}$$

Thus, we have arrived at a relationship that can be used to estimate the per-step truncation error on the basis of two quantities—the predictor (y_{i+1}^0) and the corrector (y_{i+1}^m)—that are routine by-products of the computation.

EXAMPLE 26.3

Estimate of Per-Step Truncation Error

Problem Statement. Use Eq. (26.26) to estimate the per-step truncation error of Example 26.2. Note that the true values at $x = 1$ and 2 are 6.194631 and 14.84392, respectively.

Solution. At $x_{i+1} = 1$, the predictor gives 5.607005 and the corrector yields 6.360865. These values can be substituted into Eq. (26.26) to give

$$E_c = -\frac{6.360865 - 5.607005}{5} = -0.1507722$$

which compares well with the exact error,

$$E_t = 6.194631 - 6.360865 = -0.1662341$$

At $x_{i+1} = 2$, the predictor gives 13.44346 and the corrector yields 15.30224, which can be used to compute

$$E_c = -\frac{15.30224 - 13.44346}{5} = -0.3717550$$

which also compares favorably with the exact error, $E_t = 14.84392 - 15.30224 = -0.4583148$.

The ease with which the error can be estimated using Eq. (26.26) provides a rational basis for step-size adjustment during the course of a computation. For example, if Eq. (26.26) indicates that the error is greater than an acceptable level, the step size can be decreased.

Modifiers. Before discussing computer algorithms, we must note two other ways in which the non-self-starting Heun method can be made more accurate and efficient. First, you should realize that besides providing a criterion for step-size adjustment, Eq. (26.26) represents a numerical estimate of the discrepancy between the final corrected value at each step y_{i+1} and the true value. Thus, it can be added directly to y_{i+1} to refine the estimate further:

$$y_{i+1}^m \leftarrow y_{i+1}^m - \frac{y_{i+1}^m - y_{i+1}^0}{5} \tag{26.27}$$

Equation (26.27) is called a *corrector modifier*. (The symbol $\leftarrow$ is read "is replaced by.") The left-hand side is the modified value of y_{i+1}^m.

A second improvement, one that relates more to program efficiency, is a *predictor modifier*, which is designed to adjust the predictor result so that it is closer to the final convergent value of the corrector. This is advantageous because, as noted previously at the beginning of this section, the number of iterations of the corrector is highly dependent on the accuracy of the initial prediction. Consequently, if the prediction is modified properly, we might reduce the number of iterations required to converge on the ultimate value of the corrector.

Such a modifier can be derived simply by assuming that the third derivative is relatively constant from step to step. Therefore, using the result of the previous step at i, Eq. (26.25) can be solved for

$$h^3 y^{(3)}(\xi) = -\frac{12}{5}(y_i^0 - y_i^m)$$

(26.28)

which, assuming that $y^{(3)}(\xi) \cong y^{(3)}(\xi_p)$, can be substituted into Eq. (26.21) to give

$$E_p = \frac{4}{5}(y_i^m - y_i^0)$$

(26.29)

which can then be used to modify the predictor result:

$$y_{i+1}^0 \leftarrow y_{i+1}^0 + \frac{4}{5}(y_i^m - y_i^0)$$

(26.30)

EXAMPLE 26.4 **Effect of Modifiers on Predictor-Corrector Results**

Problem Statement.   Recompute Example 26.3 using both modifiers.

Solution.   As in Example 26.3, the initial predictor result is 5.607005. Because the predictor modifier [Eq. (26.30)] requires values from a previous iteration, it cannot be employed to improve this initial result. However, Eq. (26.27) can be used to modify the corrected value of 6.360865 ($\varepsilon_t = -2.684\%$), as in

$$y_1^m = 6.360865 - \frac{6.360865 - 5.607005}{5} = 6.210093$$

which represents an $\varepsilon_t = -0.25\%$. Thus, the error is reduced by over an order of magnitude. For the next iteration, the predictor [Eq. (26.13)] is used to compute

$$y_2^0 = 2 + [4e^{0.8(0)} - 0.5(6.210093)] 2 = 13.59423 \qquad \varepsilon_t = 8.42\%$$

which is about half the error of the predictor for the second iteration of Example 26.3 ($\varepsilon_t = 18.6\%$). This improvement occurs because we are using a superior estimate of y (6.210093 as opposed to 6.360865) in the predictor. In other words, the propagated and global errors are reduced by the inclusion of the corrector modifier.

Now because we have information from the prior iteration, Eq. (26.30) can be employed to modify the predictor, as in

$$y_2^0 = 13.59423 + \frac{4}{5}(6.360865 - 5.607005) = 14.19732 \qquad \varepsilon_t = -4.36\%$$

which, again, halves the error.

This modification has no effect on the final outcome of the subsequent corrector step. Regardless of whether the unmodified or modified predictors are used, the corrector will ultimately converge on the same answer. However, because the rate or efficiency of convergence depends on the accuracy of the initial prediction, the modification can reduce the number of iterations required for convergence.

Implementing the corrector yields a result of 15.21178 ($\varepsilon_t = -2.48\%$), which represents an improvement over Example 26.3 because of the reduction of global error. Finally, this result can be modified using Eq. (26.27):

$$y_2^m = 15.21178 - \frac{15.21178 - 13.59423}{5} = 14.88827 \qquad \varepsilon_t = -0.30\%$$

Again, the error has been reduced by an order of magnitude.

As in the previous example, the addition of the modifiers increases both the efficiency and accuracy of multistep methods. In particular, the corrector modifier effectively increases the order of the technique. Thus, the non-self-starting Heun with modifiers is third order rather than second order as is the case for the unmodified version. However, it should be noted that there are situations where the corrector modifier will affect the stability of the corrector iteration process. As a consequence, the modifier is not included in the algorithm for the non-self-starting Heun delineated in Fig. 26.5. Nevertheless, the corrector modifier can still have utility for step-size control, as discussed next.

FIGURE 26.5

The sequence of formulas used to implement the non-self-starting Heun method. Note that the corrector error estimates can be used to modify the corrector. However, because this can affect the corrector's stability, the modifier is not included in this algorithm. The corrector error estimate is included because of its utility for step-size adjustment.

Predictor:

$$y_{i+1}^0 = y_{i-1}^m + f(x_i, y_i^m)2h$$

(Save result as $y_{i+1,u}^0 = y_{i+1}^0$ where the subscript u designates that the variable is unmodified.)

Predictor Modifier:

$$y_{i+1}^0 \leftarrow y_{i+1,u}^0 + \frac{4}{5}(y_{i,u}^m - y_{i,u}^0)$$

Corrector:

$$y_{i+1}^j = y_i^m + \frac{f(x_i, y_i^m) + f(x_{i+1}, y_{i+1}^{j-1})}{2}h \qquad \text{(for } j = 1 \text{ to maximum iterations } m)$$

Error Check:

$$|\varepsilon_a| = \left| \frac{y_{i+1}^j - y_{i+1}^{j-1}}{y_{i+1}^j} \right| 100\%$$

(If $|\varepsilon_a| >$ error criterion, set $j = j + 1$ and repeat corrector; if $\varepsilon_a \leq$ error criterion, save result as $y_{i+1,u}^m = y_{i+1}^m$.)

Corrector Error Estimate:

$$E_c = -\frac{1}{5}(y_{i+1,u}^m - y_{i+1,u}^0)$$

(If computation is to continue, set $i = i + 1$ and return to predictor.)

26.2.2 Step-Size Control and Computer Programs

Constant Step Size.   It is relatively simple to develop a constant step-size version of the non-self-starting Heun method. About the only complication is that a one-step method is required to generate the extra point to start the computation.

Additionally, because a constant step size is employed, a value for h must be chosen prior to the computation. In general, experience indicates that an optimal step size should be small enough to ensure convergence within two iterations of the corrector (Hull and Creemer 1963). In addition, it must be small enough to yield a sufficiently small truncation error. At the same time, the step size should be as large as possible to minimize run-time cost and round-off error. As with other methods for ODEs, the only practical way to assess the magnitude of the global error is to compare the results for the same problem but with a halved step size.

Variable Step Size.   Two criteria are typically used to decide whether a change in step size is warranted. First, if the value from Eq. (26.26) is greater than some prespecified error criterion, the step size is decreased. Second, the step size is chosen so that the convergence criterion of the corrector is satisfied in two iterations. This criterion is intended to account for the trade-off between the rate of convergence and the total number of steps in the calculation. For smaller values of h, convergence will be more rapid but more steps are required. For larger h, convergence is slower but fewer steps result. Experience (Hull and Creemer 1963) suggests that the total steps will be minimized if h is chosen so that the corrector converges within two iterations. Therefore, if over two iterations are required, the step size is decreased, and if less than two iterations are required, the step size is increased.

Although the above strategy specifies when step-size modifications are in order, it does not indicate how the step size should be changed. This is a critical question because multistep methods by definition require several points to compute a new point. Once the step size is changed, a new set of points must be determined. One approach is to restart the computation and use the one-step method to generate a new set of starting points.

A more efficient strategy that makes use of existing information is to increase and decrease by doubling and halving the step size. As depicted in Fig. 26.6b, if a sufficient number of previous values have been generated, increasing the step size by doubling is a relatively straightforward task (Fig. 26.6c). All that is necessary is to keep track of subscripts so that old values of x and y become the appropriate new values. Halving the step size is somewhat more complicated because some of the new values will be unavailable (Fig. 26.6a). However, interpolating polynomials of the type developed in Chap. 18 can be used to determine these intermediate values.

In any event, the decision to incorporate step-size control represents a trade-off between initial investment in program complexity versus the long-term return because of increased efficiency. Obviously, the magnitude and importance of the problem itself will have a strong bearing on this trade-off. Fortunately, several software packages and libraries have multistep routines that you can use to obtain solutions without having to program them from scratch. We will mention some of these when we review packages and libraries at the end of Chap. 27.

26.2.3 Integration Formulas

The non-self-starting Heun method is characteristic of most multistep methods. It employs an open integration formula (the midpoint method) to make an initial estimate.

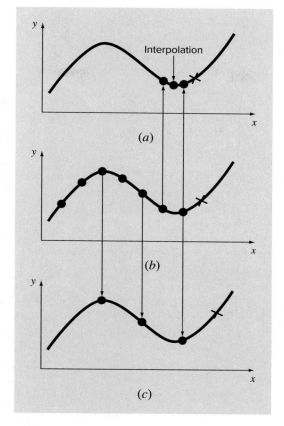

FIGURE 26.6
A plot indicating how a halving-doubling strategy allows the use of (*b*) previously calculated values for a third-order multistep method. (*a*) Halving; (*c*) doubling.

This predictor step requires a previous data point. Then, a closed integration formula (the trapezoidal rule) is applied iteratively to improve the solution.

It should be obvious that a strategy for improving multistep methods would be to use higher-order integration formulas as predictors and correctors. For example, the higher-order Newton-Cotes formulas developed in Chap. 21 could be used for this purpose.

Before describing these higher-order methods, we will review the most common integration formulas upon which they are based. As mentioned above, the first of these are the Newton-Cotes formulas. However, there is a second class called the Adams formulas that we will also review and that are often preferred. As depicted in Fig. 26.7, the fundamental difference between the Newton-Cotes and Adams formulas relates to the manner in which the integral is applied to obtain the solution. As depicted in Fig. 26.7*a*, the Newton-Cotes formulas estimate the integral over an interval spanning several points. This integral is then used to project from the beginning of the interval to the end. In contrast, the Adams formulas (Fig. 26.7*b*) use a set of points from an interval to estimate the integral solely for the last segment in the interval. This integral is then used to project across this last segment.

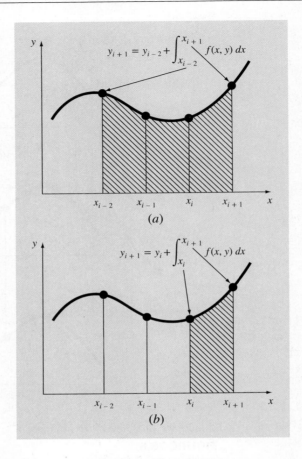

FIGURE 26.7
Illustration of the fundamental difference between the Newton-Cotes and Adams integration formulas. (*a*) The Newton-Cotes formulas use a series of points to obtain an integral estimate over a number of segments. The estimate is then used to project across the entire range. (*b*) The Adams formulas use a series of points to obtain an integral estimate for a single segment. The estimate is then used to project across the segment.

Newton-Cotes Formulas. Some of the most common formulas for solving ordinary differential equations are based on fitting an *n*th-degree interpolating polynomial to $n + 1$ known values of y and then using this equation to compute the integral. As discussed previously in Chap. 21, the Newton-Cotes integration formulas are based on such an approach. These formulas are of two types: open and closed forms.

Open Formulas. For n equally spaced data points, the open formulas can be expressed in the form of a solution of an ODE, as was done previously for Eq. (26.19). The general equation for this purpose is

$$y_{i+1} = y_{i-n} + \int_{x_{i-n}}^{x_{i+1}} f_n(x)\, dx \qquad (26.31)$$

where $f_n(x)$ is an nth-order interpolating polynomial. The evaluation of the integral employs the nth-order Newton-Cotes open integration formula (Table 21.4). For example, if $n = 1$,

$$y_{i+1} = y_{i-1} + 2hf_i \tag{26.32}$$

where f_i is an abbreviation for $f(x_i, y_i)$—that is, the differential equation evaluated at x_i and y_i. Equation (26.32) is referred to as the midpoint method and was used previously as the predictor in the non-self-starting Heun method. For $n = 2$,

$$y_{i+1} = y_{i-2} + \frac{3h}{2}(f_i + f_{i-1})$$

and for $n = 3$,

$$y_{i+1} = y_{i-3} + \frac{4h}{3}(2f_i - f_{i-1} + 2f_{i-2}) \tag{26.33}$$

Equation (26.33) is depicted graphically in Fig. 26.8a.

Closed Formulas. The closed form can be expressed generally as

$$y_{i+1} = y_{i-n+1} + \int_{x_{i-n+1}}^{x_{i+1}} f_n(x)\, dx \tag{26.34}$$

where the integral is approximated by an nth-order Newton-Cotes closed integration formula (Table 21.2). For example, for $n = 1$,

$$y_{i+1} = y_i + \frac{h}{2}(f_i + f_{i+1})$$

which is equivalent to the trapezoidal rule. For $n = 2$,

$$y_{i+1} = y_{i-1} + \frac{h}{3}(f_{i-1} + 4f_i + f_{i+1}) \tag{26.35}$$

which is equivalent to Simpson's 1/3 rule. Equation (26.35) is depicted in Fig. 26.8b.

Adams Formulas. The other types of integration formulas that can be used to solve ODEs are the Adams formulas. Many popular computer algorithms for multistep solution of ODEs are based on these methods.

Open Formulas (Adams-Bashforth). The Adams formulas can be derived in a variety of ways. One technique is to write a forward Taylor series expansion around x_i:

$$y_{i+1} = y_i + f_i h + \frac{f_i'}{2}h^2 + \frac{f_i''}{6}h^3 + \cdots$$

which can also be written as

$$y_{i+1} = y_i + h\left(f_i + \frac{h}{2}f_i' + \frac{h^2}{3}f_i'' + \cdots\right) \tag{26.36}$$

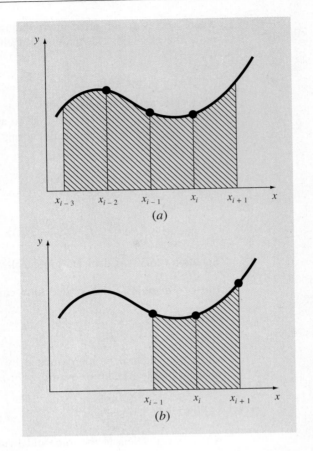

FIGURE 26.8
Graphical depiction of open and closed Newton-Cotes integration formulas. (*a*) The third open formula [Eq. (26.33)] and (*b*) Simpson's 1/3 rule [Eq. (26.35)].

Recall from Sec. 4.1.3 that a backward difference can be used to approximate the derivative:

$$f'_i = \frac{f_i - f_{i-1}}{h} + \frac{f''_i}{2}h + O(h^2)$$

which can be substituted into Eq. (26.36) to yield

$$y_{i+1} = y_i + h\left\{f_i + \frac{h}{2}\left[\frac{f_i - f_{i-1}}{h} + \frac{f''_i}{2}h + O(h^2)\right] + \frac{h^2}{6}f''_i + \cdots\right\}$$

or, after collecting terms,

$$y_{i+1} = y_i + h\left(\frac{3}{2}f_i - \frac{1}{2}f_{i-1}\right) + \frac{5}{12}h^3f''_i + O(h^4) \tag{26.37}$$

TABLE 26.1 Coefficients and truncation error for Adams-Bashforth predictors.

Order	β_0	β_1	β_2	β_3	β_4	β_5	Local Truncation Error
1	1						$\frac{1}{2}h^2 f'(\xi)$
2	3/2	−1/2					$\frac{5}{12}h^3 f''(\xi)$
3	23/12	−16/12	5/12				$\frac{9}{24}h^4 f^{(3)}(\xi)$
4	55/24	−59/24	37/24	−9/24			$\frac{251}{720}h^5 f^{(4)}(\xi)$
5	1901/720	−2774/720	2616/720	−1274/720	251/720		$\frac{475}{1440}h^6 f^{(5)}(\xi)$
6	4277/720	−7923/720	9982/720	−7298/720	2877/720	−475/720	$\frac{19{,}087}{60{,}480}h^7 f^{(6)}(\xi)$

This formula is called the *second-order open Adams formula*. Open Adams formulas are also referred to as *Adams-Bashforth formulas*. Consequently, Eq. (26.37) is sometimes called the *second Adams-Bashforth formula*.

Higher-order Adams-Bashforth formulas can be developed by substituting higher-difference approximations into Eq. (26.36). The *n*th-order open Adams formula can be represented generally as

$$y_{i+1} = y_i + h \sum_{k=0}^{n-1} \beta_k f_{i-k} + O(h^{n+1}) \tag{26.38}$$

The coefficients β_k are compiled in Table 26.1. The fourth-order version is depicted in Fig. 26.9a. Notice that the first-order version is Euler's method.

Closed Formulas (Adams-Moulton). A backward Taylor series around x_{i+1} can be written as

$$y_i = y_{i+1} - f_{i+1}h + \frac{f'_{i+1}}{2}h^2 - \frac{f''_{i+1}}{3}h^3 + \cdots$$

Solving for y_{i+1} yields

$$y_{i+1} = y_i + h\left(f_{i+1} - \frac{h}{2}f'_{i+1} + \frac{h^2}{6}f''_{i+1} + \cdots \right) \tag{26.39}$$

A difference can be used to approximate the first derivative:

$$f'_{i+1} = \frac{f_{i+1} - f_i}{h} + \frac{f''_{i+1}}{2}h + O(h^2)$$

FIGURE 26.9
Graphical depiction of open and closed Adams integration formulas. (*a*) The fourth Adams-Bashforth open formula and (*b*) the fourth Adams-Moulton closed formula.

which can be substituted into Eq. (26.39), and collecting terms gives

$$y_{i+1} = y_i + h\left(\frac{1}{2}f_{i+1} + \frac{1}{2}f_i\right) - \frac{1}{12}h^3 f''_{i+1} - O(h^4)$$

This formula is called the *second-order closed Adams formula*, or the *second Adams-Moulton formula*. Also, notice that it is the trapezoidal rule.

The *n*th-order closed Adams formula can be written generally as

$$y_{i+1} = y_i + h\sum_{k=0}^{n-1} \beta_k f_{i+1-k} + O(h^{n+1})$$

The coefficients β_k are listed in Table 26.2. The fourth-order method is depicted in Fig. 26.9*b*.

TABLE 26.2 Coefficients and truncation error for Adams-Moulton correctors.

Order	β_0	β_1	β_2	β_3	β_4	β_5	Local Truncation Error
2	1/2	1/2					$-\dfrac{1}{12}h^3 f''(\xi)$
3	5/12	8/12	−1/12				$-\dfrac{1}{24}h^4 f^{(3)}(\xi)$
4	9/24	19/24	−5/24	1/24			$-\dfrac{19}{720}h^5 f^{(4)}(\xi)$
5	251/720	646/720	−264/720	106/720	−19/720		$-\dfrac{27}{1440}h^6 f^{(5)}(\xi)$
6	475/1440	1427/1440	−798/1440	482/1440	−173/1440	27/1440	$-\dfrac{863}{60,480}h^7 f^{(6)}(\xi)$

26.2.4 Higher-Order Multistep Methods

Now that we have formally developed the Newton-Cotes and Adams integration formulas, we can use them to derive higher-order multistep methods. As was the case with the non-self-starting Heun method, the integration formulas are applied in tandem as predictor-corrector methods. In addition, if the open and closed formulas have local truncation

---

Box 26.1   Derivation of General Relationships for Modifiers

The relationship between the true value, the approximation, and the error of a predictor can be represented generally as

$$\text{True value} = y_{i+1}^0 + \frac{\eta_p}{\delta_p}h^{n+1}y^{(n+1)}(\xi_p) \qquad (B26.1.1)$$

where η_p and δ_p = the numerator and denominator, respectively, of the constant of the truncation error for either an open Newton-Cotes (Table 21.4) or an Adams-Bashforth (Table 26.1) predictor, and n is the order.

A similar relationship can be developed for the corrector:

$$\text{True value} = y_{i+1}^m - \frac{\eta_c}{\delta_c}h^{n+1}y^{(n+1)}(\xi_c) \qquad (B26.1.2)$$

where η_c and δ_c = the numerator and denominator, respectively, of the constant of the truncation error for either a closed Newton-Cotes (Table 21.2) or an Adams-Moulton (Table 26.2) corrector. As was done in the derivation of Eq. (26.24), Eq. (B26.1.1) can be subtracted from Eq. (B26.1.2) to yield

$$0 = y_{i+1}^m - y_{i+1}^0 - \frac{\eta_c + \eta_p\delta_c/\delta_p}{\delta_c}h^{n+1}y^{(n+1)}(\xi) \qquad (B26.1.3)$$

Now, dividing the equation by $\eta_c + \eta_p\delta_c/\delta_p$, multiplying the last term by δ_p/δ_p, and rearranging provides an estimate of the local

truncation error of the corrector:

$$E_c \cong -\frac{\eta_c\delta_p}{\eta_c\delta_p + \eta_p\delta_c}(y_{i+1}^m - y_{i+1}^0) \qquad (B26.1.4)$$

For the predictor modifier, Eq. (B26.1.3) can be solved at the previous step for

$$h^n y^{(n+1)}(\xi) = -\frac{\delta_c\delta_p}{\eta_c\delta_p + \eta_p\delta_c}(y_i^0 - y_i^m)$$

which can be substituted into the error term of Eq. (B26.1.1) to yield

$$E_p = \frac{\eta_p\delta_c}{\eta_c\delta_p + \eta_p\delta_c}(y_i^m - y_i^0) \qquad (B26.1.5)$$

Equations (B26.1.4) and (B26.1.5) are general versions of modifiers that can be used to improve multistep algorithms. For example, Milne's method has $\eta_p = 14$, $\delta_p = 45$, $\eta_c = 1$, $\delta_c = 90$. Substituting these values into Eqs. (B26.1.4) and (B26.1.5) yields Eqs. (26.43) and (26.42), respectively. Similar modifiers can be developed for other pairs of open and closed formulas that have local truncation errors of the same order.

errors of the same order, modifiers of the type listed in Fig. 26.5 can be incorporated to improve accuracy and allow step-size control. Box 26.1 provides general equations for these modifiers. In this section, we present two of the most common higher-order multistep approaches: Milne's method and the fourth-order Adams method.

Milne's Method.   Milne's method is the most common multistep method based on Newton-Cotes integration formulas. It uses the three-point Newton-Cotes open formula as a predictor:

$$y_{i+1}^0 = y_{i-3}^m + \frac{4h}{3}(2f_i^m - f_{i-1}^m + 2f_{i-2}^m) \tag{26.40}$$

and the three-point Newton-Cotes closed formula (Simpson's 1/3 rule) as a corrector:

$$y_{i+1}^j = y_{i-1}^m + \frac{h}{3}(f_{i-1}^m + 4f_i^m + f_{i+1}^{j-1}) \tag{26.41}$$

where j is an index representing the number of iterations of the modifier. The predictor and corrector modifiers for Milne's method can be developed from the formulas in Box 26.1 and the error coefficients in Tables 21.2 and 21.4:

$$E_p = \frac{28}{29}(y_i^m - y_i^0) \tag{26.42}$$

$$E_c \cong -\frac{1}{29}(y_{i+1}^m - y_{i+1}^0) \tag{26.43}$$

EXAMPLE 26.5   Milne's Method

Problem Statement.   Use Milne's method to integrate $y' = 4e^{0.8x} - 0.5y$ from $x = 0$ to $x = 4$ using a step size of 1. The initial condition at $x = 0$ is $y = 2$. Because we are dealing with a multistep method, previous points are required. In an actual application, a one-step method such as a fourth-order RK would be used to compute the required points. For the present example, we will use the analytical solution [recall Eq. (E25.5.1) from Example 25.5] to compute exact values at $x_{i-3} = -3$, $x_{i-2} = -2$, and $x_{i-1} = -1$: $y_{i-3} = -4.547302$, $y_{i-2} = -2.306160$, and $y_{i-1} = -0.3929953$, respectively.

Solution.   The predictor [Eq. (26.40)] is used to calculate a value at $x = 1$:

$$y_1^0 = -4.54730 + \frac{4(1)}{3}[2(3) - 1.99381 + 2(1.96067)] = 6.02272 \qquad \varepsilon_t = 2.8\%$$

The corrector [Eq. (26.41)] is then employed to compute

$$y_1^1 = -0.3929953 + \frac{1}{3}[1.99381 + 4(3) + 5.890802] = 6.235210 \qquad \varepsilon_t = -0.66\%$$

This result can be substituted back into Eq. (26.41) to iteratively correct the estimate. This process converges on a final corrected value of 6.204855 ($\varepsilon_t = -0.17\%$).

This value is more accurate than the comparable estimate of 6.360865 ($\varepsilon_t = -2.68\%$) obtained previously with the non-self-starting Heun method (Examples 26.2 through 26.4). The results for the remaining steps are $y(2) = 14.86031$ ($\varepsilon_t = -0.11\%$), $y(3) = 33.72426$ ($\varepsilon_t = -0.14\%$), and $y(4) = 75.43295$ ($\varepsilon_t = -0.12\%$).

As in the previous example, Milne's method usually yields results of high accuracy. However, there are certain cases where it performs poorly (see Ralston and Rabinowitz, 1978). Before elaborating on these cases, we will describe another higher-order multistep approach—the fourth-order Adams method.

Fourth-Order Adams Method. A popular multistep method based on the Adams integration formulas uses the fourth-order Adams-Bashforth formula (Table 26.1) as the predictor:

$$y_{i+1}^0 = y_i^m + h\left(\frac{55}{24}f_i^m - \frac{59}{24}f_{i-1}^m + \frac{37}{24}f_{i-2}^m - \frac{9}{24}f_{i-3}^m\right) \tag{26.44}$$

and the fourth-order Adams-Moulton formula (Table 26.2) as the corrector:

$$y_{i+1}^j = y_i^m + h\left(\frac{9}{24}f_{i+1}^{j-1} + \frac{19}{24}f_i^m - \frac{5}{24}f_{i-1}^m + \frac{1}{24}f_{i-2}^m\right) \tag{26.45}$$

The predictor and corrector modifiers for the fourth-order Adams method can be developed from the formulas in Box 26.1 and the error coefficients in Tables 26.1 and 26.2 as

$$E_p = \frac{251}{270}(y_i^m - y_i^0) \tag{26.46}$$

$$E_c = -\frac{19}{270}(y_{i+1}^m - y_{i+1}^0) \tag{26.47}$$

EXAMPLE 26.6

Fourth-Order Adams Method

Problem Statement. Use the fourth-order Adams method to solve the same problem as in Example 26.5.

Solution. The predictor [Eq. (26.44)] is used to compute a value at $x = 1$:

$$y_1^0 = 2 + 1\left(\frac{55}{24}3 - \frac{59}{24}1.993814 + \frac{37}{24}1.960667 - \frac{9}{24}2.6365228\right) = 6.007539$$

$$\varepsilon_t = 3.1\%$$

which is comparable to but somewhat less accurate than the result using the Milne method. The corrector [Eq. (26.45)] is then employed to calculate

$$y_1^1 = 2 + 1\left(\frac{9}{24}5.898394 + \frac{19}{24}3 - \frac{5}{24}1.993814 + \frac{1}{24}1.960666\right) = 6.253214$$

$$\varepsilon_t = -0.96\%$$

which again is comparable to but less accurate than the result using Milne's method. This result can be substituted back into Eq. (26.45) to iteratively correct the estimate. The process converges on a final corrected value of 6.214424 ($\varepsilon_t = 0.32\%$), which is an accurate result but again somewhat inferior to that obtained with the Milne method.

Stability of Multistep Methods.   The superior accuracy of the Milne method exhibited in Examples 26.5 and 26.6 would be anticipated on the basis of the error terms for the predictors [Eqs. (26.42) and (26.46)] and the correctors [Eqs. (26.43) and (26.47)]. The coefficients for the Milne method, 14/45 and 1/90, are smaller than for the fourth-order Adams, 251/720 and 19/720. Additionally, the Milne method employs fewer function evaluations to attain these higher accuracies. At face value, these results might lead to the conclusion that the Milne method is superior and, therefore, preferable to the fourth-order Adams. Although this conclusion holds for many cases, there are instances where the Milne method performs unacceptably. Such behavior is exhibited in the following example.

EXAMPLE 26.7

Stability of Milne's and Fourth-Order Adams Methods

Problem Statement.   Employ Milne's and the fourth-order Adams methods to solve

$$\frac{dy}{dx} = -y$$

with the initial condition that $y = 1$ at $x = 0$. Solve this equation from $x = 0$ to $x = 10$ using a step size of $h = 0.5$. Note that the analytical solution is $y = e^{-x}$.

Solution.   The results, as summarized in Fig. 26.10, indicate problems with Milne's method. Shortly after the onset of the computation, the errors begin to grow and oscillate

FIGURE 26.10
Graphical depiction of the instability of Milne's method.

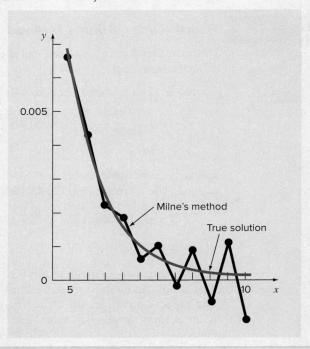

in sign. By $x = 10$, the relative error has inflated to 2831% and the predicted value itself has started to oscillate in sign.

In contrast, the results for the Adams method would be much more acceptable. Although the error also grows, it would do so at a slow rate. Additionally, the discrepancies would not exhibit the wild swings in sign exhibited by the Milne method.

The unacceptable behavior manifested in the previous example by the Milne method is referred to as *instability*. Although it does not always occur, its possibility leads to the conclusion that Milne's approach should be avoided. Thus, the fourth-order Adams method is normally preferred.

The instability of Milne's method is due to the corrector. Consequently, attempts have been made to rectify the shortcoming by developing stable correctors. One commonly used alternative that employs this approach is *Hamming's method,* which uses the Milne predictor and a stable corrector:

$$y_{i+1}^j = \frac{9y_i^m - y_{i-2}^m + 3h(y_{i+1}^{j-1} + 2f_i^m - f_{i-1}^m)}{8}$$

which has the following local truncation error:

$$E_c = \frac{1}{40}h^5 y^{(4)}(\xi_c)$$

Hamming's method also includes modifiers of the form

$$E_p = \frac{9}{121}(y_i^m - y_i^0)$$

$$E_c = -\frac{112}{121}(y_{i+1}^m - y_{i+1}^0)$$

The reader can obtain additional information on this and other multistep methods elsewhere (Hamming 1973; Lapidus and Seinfield 1971).

PROBLEMS

26.1 Given

$$\frac{dy}{dx} = -200{,}000y + 200{,}000e^{-x} - e^{-x}$$

(a) Estimate the step size required to maintain stability using the explicit Euler method.
(b) If $y(0) = 0$, use the implicit Euler to obtain a solution from $t = 0$ to 2 using a step size of 0.1.

26.2 Given

$$\frac{dy}{dt} = 30(\sin t - y) + 3 \cos t$$

If $y(0) = 1$, use the implicit Euler to obtain a solution from $t = 0$ to 4 using a step size of 0.4.

26.3 Given

$$\frac{dx_1}{dt} = 999x_1 + 1999x_2$$

$$\frac{dx_2}{dt} = -1000x_1 - 2000x_2$$

If $x_1(0) = x_2(0) = 1$, obtain a solution from $t = 0$ to 0.2 using a step size of 0.05 with the **(a)** explicit and **(b)** implicit Euler methods.

26.4 Solve the following initial-value problem over the interval from $t = 2$ to 3:

$$\frac{dy}{dt} = -0.5y + e^{-t}$$

Use the non-self-starting Heun method with a step size of 0.5 and initial conditions of $y(1.5) = 5.222138$ and $y(2.0) = 4.143883$. Iterate the corrector to $\varepsilon_s = 0.1\%$. Compute the true percent relative errors ε_t for your results based on the analytical solution.

26.5 Repeat Prob. 26.4, but use the fourth-order Adams method. [Note: $y(0.5) = 8.132548$ and $y(1.0) = 6.542609$.] Iterate the corrector to $\varepsilon_s = 0.01\%$.

26.6 Solve the following initial-value problem from $t = 4$ to 5:

$$\frac{dy}{dt} = -\frac{2y}{t}$$

Use a step size of 0.5 and initial values of $y(2.5) = 0.48$, $y(3) = 0.333333$, $y(3.5) = 0.244898$, and $y(4) = 0.1875$. Obtain your solutions using the following techniques: **(a)** the non-self-starting Heun method ($\varepsilon_s = 1\%$), and **(b)** the fourth-order Adams method ($\varepsilon_s = 0.01\%$). [Note: The exact answers obtained analytically are $y(4.5) = 0.148148$ and $y(5) = 0.12$.] Compute the true percent relative errors ε_t for your results.

26.7 Solve the following initial-value problem from $x = 0$ to $x = 0.75$:

$$\frac{dy}{dx} = yx^2 - y$$

Use the non-self-starting Heun method with a step size of 0.25. If $y(0) = 1$, employ the fourth-order RK method with a step size of 0.25 to predict the starting value at $y(0.25)$.

26.8 Solve the following initial-value problem from $t = 1.5$ to $t = 2.5$:

$$\frac{dy}{dt} = \frac{-2y}{1 + t}$$

Use the fourth-order Adams method. Employ a step size of 0.5 and the fourth-order RK method to predict the start-up values if $y(0) = 2$.

26.9 Develop a program for the implicit Euler method for a single linear ODE. Test it by duplicating Prob. 26.1*b*.

26.10 Develop a program for the implicit Euler method for a pair of linear ODEs. Test it by solving Eq. (26.6).

26.11 Develop a user-friendly program for the non-self-starting Heun method with a predictor modifier. Employ a fourth-order RK method to compute starter values. Test the program by duplicating Example 26.4.

26.12 Use the program developed in Prob. 26.11 to solve Prob. 26.7.

FIGURE P26.13

26.13 Consider the thin rod of length l moving in the x-y plane as shown in Fig. P26.13. The rod is fixed with a pin on one end and a mass at the other. Note that $g = 9.81$ m/s^2 and $l = 0.5$ m. This system can be solved using

$$\ddot{\theta} - \frac{g}{l}\theta = 0$$

Let $\theta = 0$ and $\dot{\theta}(0) = 0.25$ rad/s. Solve using any method studied in this chapter. Plot the angle versus time and the angular velocity versus time. (Hint: Decompose the second-order ODE.)

26.14 Given the first-order ODE

$$\frac{dx}{dt} = -700x - 1000e^{-t}$$

$$x(t = 0) = 4$$

Solve this stiff differential equation using a numerical method over the time period $0 \leq t \leq 5$. Also solve analytically and plot the analytic and numerical solution for both the fast transient and slow transition phase of the timescale.

26.15 The following second-order ODE is considered to be stiff:

$$\frac{d^2y}{dx^2} = -1001\frac{dy}{dx} - 1000y$$

Solve this differential equation **(a)** analytically and **(b)** numerically for $x = 0$ to 5. For **(b)** use an implicit approach with $h = 0.5$. Note that the initial conditions are $y(0) = 1$ and $y'(0) = 0$. Display both results graphically.

26.16 Solve the following differential equation from $t = 0$ to 1:

$$\frac{dy}{dt} = -10y$$

with the initial condition $y(0) = 1$. Use the following techniques to obtain your solutions: **(a)** analytically, **(b)** the explicit Euler method, and **(c)** the implicit Euler method. For **(b)** and **(c)**, use $h = 0.1$ and 0.2. Plot your results.

Boundary-Value and Eigenvalue Problems

Recall from our discussion at the beginning of Part Seven that an ordinary differential equation is accompanied by auxiliary conditions. These conditions are used to evaluate the constants of integration that result during the solution of the equation. For an nth-order equation, n conditions are required. If all the conditions are specified at the same value of the independent variable, then we are dealing with an *initial-value problem* (Fig. 27.1a). To this point, the material in Part Seven has been devoted to this type of problem.

FIGURE 27.1

Initial-value versus boundary-value problems. (a) An initial-value problem where all the conditions are specified at the same value of the independent variable. (b) A boundary-value problem where the conditions are specified at different values of the independent variable.

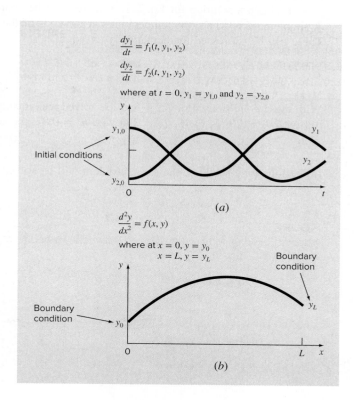

In contrast, there is another application for which the conditions are not known at a single point, but rather, are known at different values of the independent variable. Because these values are often specified at the extreme points or boundaries of a system, these applications are customarily referred to as *boundary-value problems* (Fig. 27.1*b*). A variety of significant engineering applications fall within this class. In this chapter, we discuss two general approaches for obtaining their solution: the shooting method and the finite-difference approach. Additionally, we present techniques to approach a special type of boundary-value problem: the determination of eigenvalues. Of course, eigenvalues also have many applications beyond those involving boundary-value problems.

27.1 GENERAL METHODS FOR BOUNDARY-VALUE PROBLEMS

The conservation of heat can be used to develop a heat balance for a long, thin rod (Fig. 27.2). If the rod is not insulated along its length and the system is at a steady state, the equation that results is

$$\frac{d^2T}{dx^2} + h'(T_a - T) = 0 \tag{27.1}$$

where h' is a heat transfer coefficient (m^{-2}) that parameterizes the rate of heat dissipation to the surrounding air and T_a is the temperature of the surrounding air (°C).

To obtain a solution for Eq. (27.1), there must be appropriate boundary conditions. A simple case is where the temperatures at the ends of the rod are held at fixed values. These can be expressed mathematically as

$$T(0) = T_1$$
$$T(L) = T_2$$

With these conditions, Eq. (27.1) can be solved analytically using calculus. For a 10-m rod with $T_a = 20$, $T_1 = 40$, $T_2 = 200$, and $h' = 0.01$, the solution is

$$T = 73.4523e^{0.1x} - 53.4523e^{-0.1x} + 20 \tag{27.2}$$

In the following sections, the same problem will be solved using numerical approaches.

FIGURE 27.2
A noninsulated uniform rod positioned between two bodies of constant but different temperature. For this case $T_1 > T_2$ and $T_2 > T_a$.

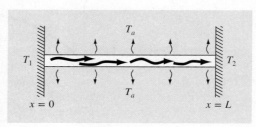

27.1.1 The Shooting Method

The *shooting method* is based on converting the boundary-value problem into an equivalent initial-value problem. A trial-and-error approach is then implemented to solve the initial-value version. The approach can be illustrated by an example.

EXAMPLE 27.1 **The Shooting Method**

Problem Statement.   Use the shooting method to solve Eq. (27.1) for a 10-m rod with $h' = 0.01$ m^{-2}, $T_a = 20$, and the boundary conditions

$$T(0) = 40 \qquad T(10) = 200$$

Solution.   Using the same approach as was employed to transform Eq. (PT7.2) into Eqs. (PT7.3) through (PT7.6), the second-order equation can be expressed as two first-order ODEs:

$$\frac{dT}{dx} = z \tag{E27.1.1}$$

$$\frac{dz}{dx} = h'(T - T_a) \tag{E27.1.2}$$

To solve these equations, we require an initial value for z. For the shooting method, we guess a value—say, $z(0) = 10$. The solution is then obtained by integrating Eq. (E27.1.1) and (E27.1.2) simultaneously. For example, using a fourth-order RK method with a step size of 2, we obtain a value at the end of the interval of $T(10) = 168.3797$ (Fig. 27.3a), which differs from the boundary condition of $T(10) = 200$. Therefore, we make another guess, $z(0) = 20$, and perform the computation again. This time, the result of $T(10) = 285.8980$ is obtained (Fig. 27.3b).

Now, because the original ODE is linear, the values

$$z(0) = 10 \qquad T(10) = 168.3797$$

and

$$z(0) = 20 \qquad T(10) = 285.8980$$

are linearly related. As such, they can be used to compute the value of $z(0)$ that yields $T(10) = 200$. A linear interpolation formula [recall Eq. (18.2)] can be employed for this purpose:

$$z(0) = 10 + \frac{20 - 10}{285.8980 - 168.3797}(200 - 168.3797) = 12.6907$$

This value can then be used to determine the correct solution, as depicted in Fig. 27.3c.

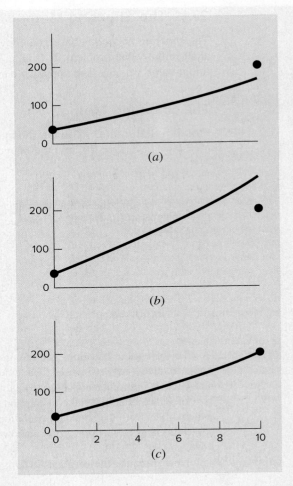

FIGURE 27.3
The shooting method: (*a*) the first "shot," (*b*) the second "shot," and (*c*) the final exact "hit."

Nonlinear Two-Point Problems.   For nonlinear boundary-value problems, linear interpolation or extrapolation through two solution points will not necessarily result in an accurate estimate of the required boundary condition to attain an exact solution. An alternative is to perform three applications of the shooting method and use a quadratic interpolating polynomial to estimate the proper boundary condition. However, it is unlikely that such an approach would yield the exact answer, and additional iterations would be necessary to obtain the solution.

Another approach for a nonlinear problem involves recasting it as a roots problem. Recall that the general form of a roots problem is to find the value of *x* that makes the function $f(x) = 0$. Now, let us use Example 27.1 to understand how the shooting method can be recast in this form.

First, recognize that the solution of the pair of differential equations is also a "function" in the sense that we guess a condition at the left-hand end of the rod, z_0, and the integration yields a prediction of the temperature at the right-hand end, T_{10}. Thus, we can think of the integration as

$$T_{10} = f(z_0)$$

That is, it represents a process whereby a guess of z_0 yields a prediction of T_{10}. Viewed in this way, we can see that what we desire is the value of z_0 that yields a specific value of T_{10}. If, as in the example, we desire $T_{10} = 200$, the problem can be posed as

$$200 = f(z_0)$$

By bringing the goal of 200 over to the right-hand side of the equation, we generate a new function, $g(z_0)$, that represents the difference between what we have, $f(z_0)$, and what we want, 200.

$$g(z_0) = f(z_0) - 200$$

If we drive this new function to zero, we will obtain the solution. The next example illustrates the approach.

EXAMPLE 27.2   **The Shooting Method for Nonlinear Problems**

Problem Statement.   Although it served our purposes for proving a simple boundary-value problem, our model for the rod in Eq. (27.1) was not very realistic. For one thing, such a rod would lose heat by mechanisms such as radiation that are nonlinear.

Suppose that the following nonlinear ODE is used to simulate the temperature of the heated rod:

$$\frac{d^2T}{dx^2} + h''(T_a - T)^4 = 0$$

where $h'' = 5 \times 10^{-8}$. Now, although it is still not a very good representation of heat transfer, this equation is straightforward enough to allow us to illustrate how the shooting method can be used to solve a two-point nonlinear boundary-value problem. The remaining problem conditions are as specified in Example 27.1.

Solution.   The second-order equation can be expressed as two first-order ODEs:

$$\frac{dT}{dx} = z$$

$$\frac{dz}{dx} = h''(T - T_a)^4$$

Now, these equations can be integrated using any of the methods described in Chaps. 25 and 26. We used the constant step-size version of the fourth-order RK approach described in Chap. 25. We implemented this approach as an Excel macro function written in Visual

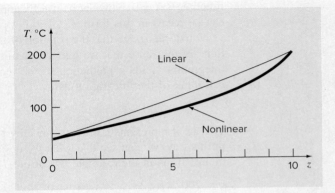

FIGURE 27.4
The result of using the shooting method to solve a nonlinear problem.

BASIC. The function integrated the equations based on an initial guess for $z(0)$ and returned the temperature at $x = 10$. The difference between this value and the goal of 200 was then placed in a spreadsheet cell. The Excel Solver was then invoked to adjust the value of $z(0)$ until the difference was driven to zero.

The result is shown in Fig. 27.4 along with the original linear case. As might be expected, the nonlinear case is curved more than the linear model. This is due to the quartic term in the heat transfer relationship.

The shooting method can become arduous for higher-order equations, where the necessity to assume two or more conditions makes the approach somewhat more difficult. For these reasons, alternative methods are available, as described next.

27.1.2 Finite-Difference Methods

The most common alternatives to the shooting method are *finite-difference approaches*. In these techniques, finite divided differences are substituted for the derivatives in the original equation. Thus, a linear differential equation is transformed into a set of simultaneous algebraic equations that can be solved using the methods from Part Three.

For the case of Fig. 27.2, the finite-divided-difference approximation for the second derivative is (recall Fig. 23.3)

$$\frac{d^2T}{dx^2} = \frac{T_{i+1} - 2T_i + T_{i-1}}{\Delta x^2}$$

This approximation can be substituted into Eq. (27.1) to give

$$\frac{T_{i+1} - 2T_i + T_{i-1}}{\Delta x^2} - h'(T_i - T_a) = 0$$

Collecting terms gives

$$-T_{i-1} + (2 + h'\Delta x^2)T_i - T_{i+1} = h'\Delta x^2 T_a \qquad (27.3)$$

This equation applies for each of the interior nodes of the rod. The first and last interior nodes, T_{i-1} and T_{i+1}, respectively, are specified by the boundary conditions. Therefore, the resulting set of linear algebraic equations will be tridiagonal. As such, it can be solved with the efficient algorithms that are available for such systems (Sec. 11.1).

EXAMPLE 27.3

Finite-Difference Approximation of Boundary-Value Problems

Problem Statement. Use the finite-difference approach to solve the same problem as in Example 27.1.

Solution. Employing the parameters in Example 27.1, we can write Eq. (27.3) for the rod from Fig. 27.2. Using four interior nodes with a segment length of $\Delta x = 2$ m results in the following equations:

$$\begin{bmatrix} 2.04 & -1 & 0 & 0 \\ -1 & 2.04 & -1 & 0 \\ 0 & -1 & 2.04 & -1 \\ 0 & 0 & -1 & 2.04 \end{bmatrix} \begin{Bmatrix} T_1 \\ T_2 \\ T_3 \\ T_4 \end{Bmatrix} = \begin{Bmatrix} 40.8 \\ 0.8 \\ 0.8 \\ 200.8 \end{Bmatrix}$$

which can be solved for

$$\{T\}^T = \lfloor 65.9698 \quad 93.7785 \quad 124.5382 \quad 159.4795 \rfloor$$

Table 27.1 provides a comparison between the analytical solution [Eq. (27.2)] and the numerical solutions obtained in Examples 27.1 and 27.3. Note that there are some discrepancies among the approaches. For both numerical methods, these errors can be mitigated by decreasing their respective step sizes. Although both techniques perform well for the present case, the finite-difference approach is preferred because of the ease with which it can be extended to more complex cases.

The fixed (or *Dirichlet*) boundary condition used in the previous example is but one of several types that are commonly employed in engineering and science. A common alternative, called the *Neumann boundary condition*, is the case where the derivative is given.

TABLE 27.1 Comparison of the exact analytical solution with the shooting and finite-difference methods.

x	True	Shooting Method	Finite-Difference Method
0	40	40	40
2	65.9518	65.9520	65.9698
4	93.7478	93.7481	93.7785
6	124.5036	124.5039	124.5382
8	159.4534	159.4538	159.4795
10	200	200	200

We can use the heated rod model to demonstrate how a derivative boundary condition can be incorporated into the finite-difference approach,

$$0 = \frac{d^2T}{dx^2} + h'(T_\infty - T)$$

However, in contrast to our previous discussions, we will prescribe a derivative boundary condition at one end of the rod,

$$\frac{dT}{dx}(0) = T_a'$$

$$T(L) = T_b$$

Thus, we have a derivative boundary condition at one end of the solution domain and a fixed boundary condition at the other.

As was done in Example 27.3, the rod is divided into a series of nodes and a finite-difference version of the differential equation (Eq. 27.3) is applied to each interior node. However, because its temperature is not specified, the node at the left end must also be included. Writing Eq. (27.3) for this node gives

$$-T_{-1} + (2 + h'\Delta x^2)T_0 - T_1 = h'\Delta x^2 T_\infty \qquad (27.3a)$$

Notice that an imaginary node (-1) lying to the left of the rod's end is required for this equation. Although this exterior point might seem to represent a difficulty, it actually serves as the vehicle for incorporating the derivative boundary condition into the problem. This is done by representing the first derivative in the x dimension at the zero end of the rod by the centered difference

$$\frac{dT}{dx} = \frac{T_1 - T_{-1}}{2\Delta x}$$

which can be solved for

$$T_{-1} = T_1 - 2\Delta x \frac{dT}{dx}$$

Now we have a formula for T_{-1} that actually reflects the impact of the derivative. It can be substituted into Eq. (27.3a) to give

$$(2 + h'\Delta x^2)T_0 - 2T_1 = h'\Delta x^2 T_\infty - 2\Delta x \frac{dT}{dx} \qquad (27.3b)$$

Consequently, we have incorporated the derivative into the heat balance.

A common example of a derivative boundary condition is the situation where the end of the rod is insulated. In this case, the derivative is set to zero. This conclusion follows directly from Fourier's law, which states that the heat flux is directly proportional to the temperature gradient. Thus, insulating a boundary means that the heat flux (and consequently the gradient) must be zero.

Aside from the shooting and finite-difference methods, there are other techniques available for solving boundary-value problems. Some of these will be described in Part Eight. These include steady-state (Chap. 29) and transient (Chap. 30) solution of two-dimensional boundary-value problems using finite differences and steady-state solutions of the one-dimensional problem with the finite-element approach (Chap. 31).

27.2 EIGENVALUE PROBLEMS

Eigenvalue, or characteristic-value, problems are a special class of boundary-value problems that are common in engineering problem contexts involving vibrations, elasticity, and other oscillating systems. In addition, they are used in a wide variety of engineering contexts beyond boundary-value problems. Before describing numerical methods for solving these problems, we will present some general background information. This includes discussion of both the mathematics and the engineering significance of eigenvalues.

27.2.1 Mathematical Background

Part Three dealt with methods for solving sets of linear algebraic equations of the general form

$$[A]\{X\} = \{B\}$$

Such systems are called *nonhomogeneous* because of the presence of the vector $\{B\}$ on the right-hand side of the equality. If the equations comprising such a system are linearly independent (that is, have a nonzero determinant), they will have a unique solution. In other words, there is one set of x values that will make the equations balance.

In contrast, a homogeneous linear algebraic system has the general form

$$[A]\{X\} = 0$$

Although nontrivial solutions (that is, solutions other than all x's $= 0$) of such systems are possible, they are generally not unique. Rather, the simultaneous equations establish relationships among the x's that can be satisfied by various combinations of values.

Eigenvalue problems associated with engineering are typically of the general form

$$\begin{aligned}
(a_{11} - \lambda)x_1 + \quad a_{12}x_2 + \cdots + \quad a_{1n}x_n &= 0 \\
a_{21}x_1 + (a_{22} - \lambda)x_2 + \cdots + \quad a_{2n}x_n &= 0 \\
\vdots \qquad \qquad \vdots \qquad \qquad \vdots \qquad \qquad \\
a_{n1}x_1 + \quad a_{n2}x_2 + \cdots + (a_{nn} - \lambda)x_n &= 0
\end{aligned}$$

where λ is an unknown parameter called the *eigenvalue*, or *characteristic value*. A solution $\{X\}$ for such a system is referred to as an *eigenvector*. The above set of equations may also be expressed concisely as

$$[[A] - \lambda[I]]\{X\} = 0 \tag{27.4}$$

The solution of Eq. (27.4) hinges on determining λ. One way to accomplish this is based on the fact that the determinant of the matrix $[[A] - \lambda[I]]$ must equal zero for nontrivial solutions to be possible. Expanding the determinant yields a polynomial in λ. The roots of this polynomial are the solutions for the eigenvalues. An example of this approach will be provided in the next section.

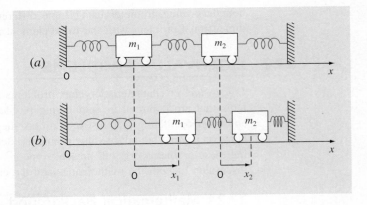

FIGURE 27.5
Positioning the masses away from equilibrium creates forces in the springs that upon release lead to oscillations of the masses. The positions of the masses can be referenced to local coordinates with origins at their respective equilibrium positions.

27.2.2 Physical Background

The mass-spring system in Fig. 27.5a is a simple context to illustrate how eigenvalues occur in physical problem settings. It also will help to illustrate some of the mathematical concepts introduced in the previous section.

To simplify the analysis, assume that each mass has no external or damping forces acting on it. In addition, assume that each spring has the same natural length l and the same spring constant k. Finally, assume that the displacement of each spring is measured relative to its own local coordinate system with an origin at the spring's equilibrium position (Fig. 27.5a). Under these assumptions, Newton's second law can be employed to develop a force balance for each mass (recall Sec. 12.4),

$$m_1 \frac{d^2 x_1}{dt^2} = -kx_1 + k(x_2 - x_1)$$

and

$$m_2 \frac{d^2 x_2}{dt^2} = -k(x_2 - x_1) - kx_2$$

where x_i is the displacement of mass i away from its equilibrium position (Fig. 27.5b). These equations can be expressed as

$$m_1 \frac{d^2 x_1}{dt^2} - k(-2x_1 + x_2) = 0 \tag{27.5a}$$

$$m_2 \frac{d^2 x_2}{dt^2} - k(x_1 - 2x_2) = 0 \tag{27.5b}$$

From vibration theory, it is known that solutions to Eq. (27.5) can take the form

$$x_i = A_i \sin(\omega t) \tag{27.6}$$

where A_i = the amplitude of the vibration of mass i and ω = the frequency of the vibration, which is equal to

$$\omega = \frac{2\pi}{T_p} \tag{27.7}$$

where T_p is the period. From Eq. (27.6) it follows that

$$x_i'' = -A_i\omega^2 \sin(\omega t) \tag{27.8}$$

Equations (27.6) and (27.8) can be substituted into Eq. (27.5), which, after collection of terms, can be expressed as

$$\left(\frac{2k}{m_1} - \omega^2\right)A_1 - \frac{k}{m_1}A_2 = 0 \tag{27.9a}$$

$$-\frac{k}{m_2}A_1 + \left(\frac{2k}{m_2} - \omega^2\right)A_2 = 0 \tag{27.9b}$$

Comparison of Eq. (27.9) with Eq. (27.4) indicates that at this point, the solution has been reduced to an eigenvalue problem.

EXAMPLE 27.4

Eigenvalues and Eigenvectors for a Mass-Spring System

Problem Statement. Evaluate the eigenvalues and the eigenvectors of Eq. (27.9) for the case where $m_1 = m_2 = 40$ kg and $k = 200$ N/m.

Solution. Substituting the parameter values into Eq. (27.9) yields

$$(10 - \omega^2)A_1 - 5A_2 = 0$$
$$-5A_1 + (10 - \omega^2)A_2 = 0$$

The determinant of this system is [recall Eq. (9.3)]

$$(\omega^2)^2 - 20\omega^2 + 75 = 0$$

which can be solved by the quadratic formula for $\omega^2 = 15$ and 5 s^{-2}. Therefore, the frequencies for the vibrations of the masses are $\omega = 3.873$ s^{-1} and 2.236 s^{-1}, respectively. These values can be used to determine the periods for the vibrations with Eq. (27.7). For the first mode, $T_p = 1.62$ s, and for the second, $T_p = 2.81$ s.

As stated in Sec. 27.2.1, a unique set of values cannot be obtained for the unknowns. However, their ratios can be specified by substituting the eigenvalues back into the equations. For example, for the first mode ($\omega^2 = 15$ s^{-2}), $A_1 = -A_2$. For the second mode ($\omega^2 = 5$ s^{-2}), $A_1 = A_2$.

This example provides valuable information regarding the behavior of the system in Fig. 27.5. Aside from its period, we know that if the system is vibrating in the first mode, the amplitude of the second mass will be equal but of opposite sign to the amplitude of the first. As in Fig. 27.6a, this vibration moves the masses apart and then together indefinitely.

In the second mode, the two masses have equal amplitudes at all times. Thus, as in Fig. 27.6b, they vibrate back and forth in unison. It should be noted that the configuration of the amplitudes provides guidance on how to set their initial values to attain pure

FIGURE 27.6
The principal modes of vibration of two equal masses connected by three identical springs between fixed walls.

motion in either of the two modes. Any other configuration will lead to superposition of the modes (recall Chap. 19).

27.2.3 A Boundary-Value Problem

Now that you have been introduced to eigenvalues, we turn to the type of problem that is the subject of the present chapter: boundary-value problems for ordinary differential equations. Figure 27.7 shows a physical system that can serve as a context for examining this type of problem.

The curvature of a slender column subject to an axial load P can be modeled by

$$\frac{d^2y}{dx^2} = \frac{M}{EI} \tag{27.10}$$

where d^2y/dx^2 specifies the curvature, M = the bending moment, E = the modulus of elasticity, and I = the moment of inertia of the cross section about its neutral axis. Considering the free body in Fig. 27.7b, it is clear that the bending moment at x is $M = -Py$. Substituting this value into Eq. (27.10) gives

$$\frac{d^2y}{dx^2} + p^2y = 0 \tag{27.11}$$

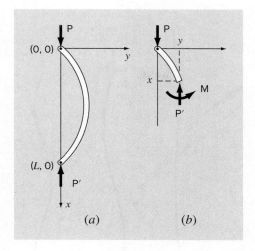

FIGURE 27.7
(a) A slender rod. (b) A free-body diagram of a rod.

where

$$p^2 = \frac{P}{EI} \qquad (27.12)$$

For the system in Fig. 27.7, subject to the boundary conditions

$$y(0) = 0 \qquad (27.13a)$$
$$y(L) = 0 \qquad (27.13b)$$

the general solution for Eq. (27.11) is

$$y = A \sin(px) + B \cos(px) \qquad (27.14)$$

where A and B are arbitrary constants that are to be evaluated via the boundary conditions. According to the first condition [Eq. (27.13a)],

$$0 = A \sin(0) + B \cos(0)$$

Therefore, we conclude that $B = 0$.

According to the second condition [Eq. (27.13b)],

$$0 = A \sin(pL) + B \cos(pL)$$

But, since $B = 0$, $A \sin(pL) = 0$. Because $A = 0$ represents a trivial solution, we conclude that $\sin(pL) = 0$. For this equality to hold,

$$pL = n\pi \qquad \text{for } n = 1, 2, 3, \ldots \qquad (27.15)$$

Thus, there are an infinite number of values that meet the boundary condition. Equation (27.15) can be solved for

$$p = \frac{n\pi}{L} \qquad \text{for } n = 1, 2, 3, \ldots \qquad (27.16)$$

which are the eigenvalues for the column.

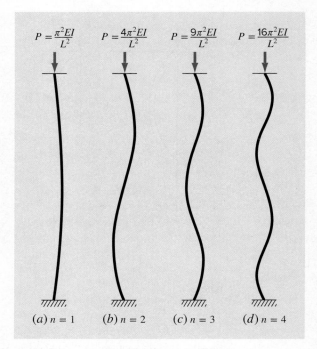

FIGURE 27.8
The first four eigenvalues for the slender rod from Fig. 27.7.

Figure 27.8, which shows the solutions for the first four eigenvalues, can provide insight into the physical significance of the results. Each eigenvalue corresponds to a way in which the column buckles. Combining Eqs. (27.12) and (27.16) gives

$$P = \frac{n^2\pi^2 EI}{L^2} \qquad \text{for } n = 1, 2, 3, \dots \tag{27.17}$$

These can be thought of as *buckling loads* because they represent the levels at which the column moves into each succeeding buckling configuration. In a practical sense, it is usually the first value that is of interest because failure will usually occur when the column first buckles. Thus, a critical load can be defined as

$$P = \frac{\pi^2 EI}{L^2}$$

which is formally known as *Euler's formula.*

EXAMPLE 27.5   **Eigenvalue Analysis of an Axially Loaded Column**

Problem Statement.   An axially loaded wooden column has the following characteristics: $E = 10 \times 10^9$ Pa, $I = 1.25 \times 10^{-5}$ m^4, and $L = 3$ m. Determine the first eight eigenvalues and the corresponding buckling loads.

Solution. Equations (27.16) and (27.17) can be used to compute

n	p, m^{-2}	P, kN
1	1.0472	137.078
2	2.0944	548.311
3	3.1416	1233.701
4	4.1888	2193.245
5	5.2360	3426.946
6	6.2832	4934.802
7	7.3304	6716.814
8	8.3776	8772.982

The critical buckling load is, therefore, 137.078 kN.

Although analytical solutions of the sort obtained above are useful, they are often difficult or impossible to obtain. This is usually true when dealing with complicated systems or those with heterogeneous properties. In such cases, numerical methods of the sort described next are the only practical alternative.

27.2.4 The Polynomial Method

Equation (27.11) can be solved numerically by substituting a centered finite-divided-difference approximation (Fig. 23.3) for the second derivative to give

$$\frac{y_{i+1} - 2y_i + y_{i-1}}{h^2} + p^2 y_i = 0$$

which can be expressed as

$$y_{i-1} - (2 - h^2 p^2)y_i + y_{i+1} = 0 \qquad (27.18)$$

Writing this equation for a series of nodes along the axis of the column yields a homogeneous system of equations. For example, if the column is divided into five segments (that is, four interior nodes), the result is

$$\begin{bmatrix} (2 - h^2 p^2) & -1 & 0 & 0 \\ -1 & (2 - h^2 p^2) & -1 & 0 \\ 0 & -1 & (2 - h^2 p^2) & -1 \\ 0 & 0 & -1 & (2 - h^2 p^2) \end{bmatrix} \begin{Bmatrix} y_1 \\ y_2 \\ y_3 \\ y_4 \end{Bmatrix} = 0 \qquad (27.19)$$

Expansion of the determinant of the system yields a polynomial, the roots of which are the eigenvalues. This approach, called the *polynomial method,* is performed in the following example.

EXAMPLE 27.6   ## The Polynomial Method

Problem Statement.   Employ the polynomial method to determine the eigenvalues for the axially loaded column from Example 27.5 using (a) one, (b) two, (c) three, and (d) four interior nodes.

Solution.

(a) Writing Eq. (27.18) for one interior node yields ($h = 3/2$)

$$-(2 - 2.25p^2)y_1 = 0$$

Thus, for this simple case, the eigenvalue is analyzed by setting the determinant equal to zero

$$2 - 2.25p^2 = 0$$

and solving for $p = \pm 0.9428$, which is about 10% less than the exact value of 1.0472 obtained in Example 27.5.

(b) For two interior nodes ($h = 3/3$), Eq. (27.18) is written as

$$\begin{bmatrix} (2 - p^2) & -1 \\ -1 & (2 - p^2) \end{bmatrix} \begin{Bmatrix} y_1 \\ y_2 \end{Bmatrix} = 0$$

Expansion of the determinant gives

$$(2 - p^2)^2 - 1 = 0$$

which can be solved for $p = \pm 1$ and ± 1.73205. Thus, the first eigenvalue is now about 4.5% low, and a second eigenvalue is obtained that is about 17% low.

(c) For three interior points ($h = 3/4$), Eq. (27.18) yields

$$\begin{bmatrix} 2 - 0.5625p^2 & -1 & 0 \\ -1 & 2 - 0.5625p^2 & -1 \\ 0 & -1 & 2 - 0.5625p^2 \end{bmatrix} \begin{Bmatrix} y_1 \\ y_2 \\ y_3 \end{Bmatrix} = 0 \qquad \text{(E27.6.1)}$$

The determinant can be set equal to zero and expanded to give

$$(2 - 0.5625p^2)^3 - 2(2 - 0.5625p^2) = 0$$

For this equation to hold, $2 - 0.5625p^2 = 0$ and $2 - 0.5625p^2 = \sqrt{2}$. Therefore, the first three eigenvalues can be determined as

$$p = \pm 1.0205 \qquad |\varepsilon_t| = 2.5\%$$
$$p = \pm 1.8856 \qquad |\varepsilon_t| = 10\%$$
$$p = \pm 2.4637 \qquad |\varepsilon_t| = 22\%$$

(d) For four interior points ($h = 3/5$), the result is Eq. (27.19) with $2 - 0.36p^2$ on the diagonal. Setting the determinant equal to zero and expanding it gives

$$(2 - 0.36p^2)^4 - 3(2 - 0.36p^2)^2 + 1 = 0$$

which can be solved for the first four eigenvalues

$$p = \pm 1.0301 \qquad |\varepsilon_t| = 1.6\%$$
$$p = \pm 1.9593 \qquad |\varepsilon_t| = 6.5\%$$
$$p = \pm 2.6967 \qquad |\varepsilon_t| = 14\%$$
$$p = \pm 3.1702 \qquad |\varepsilon_t| = 24\%$$

TABLE 27.2 The results of applying the polynomial method to an axially loaded column. The numbers in parentheses represent the absolute value of the true percent relative error.

Eigenvalue	True	Polynomial Method			
		h = 3/2	h = 3/3	h = 3/4	h = 3/5
1	1.0472	0.9428 (10%)	1.0000 (4.5%)	1.0205 (2.5%)	1.0301 (1.6%)
2	2.0944		1.7321 (21%)	1.8856 (10%)	1.9593 (65%)
3	3.1416			2.4637 (22%)	2.6967 (14%)
4	4.1888				3.1702 (24%)

Table 27.2, which summarizes the results of this example, illustrates some fundamental aspects of the polynomial method. As the segmentation is made more refined, additional eigenvalues are determined and the previously determined values become progressively more accurate. Thus, the approach is best suited for cases where the lower eigenvalues are required.

27.2.5 The Power Method

The *power method* is an iterative approach that can be employed to determine the largest eigenvalue. With slight modification, it can also be employed to determine the smallest and the intermediate values. It has the additional benefit that the corresponding eigenvector is obtained as a by-product of the method.

Determination of the Largest Eigenvalue. To implement the power method, the system being analyzed must be expressed in the form

$$[A]\{X\} = \lambda\{X\} \tag{27.20}$$

As illustrated by the following example, Eq. (27.20) forms the basis for an iterative solution technique that eventually yields the highest eigenvalue and its associated eigenvector.

EXAMPLE 27.7 | **Power Method for Highest Eigenvalue**

Problem Statement.   Employ the power method to determine the highest eigenvalue for part **(c)** of Example 27.6.

Solution.   The system is first written in the form of Eq. (27.20),

$$
\begin{aligned}
3.556x_1 - 1.778x_2 \qquad\qquad &= \lambda x_1 \\
-1.778x_1 + 3.556x_2 - 1.778x_3 &= \lambda x_2 \\
-1.778x_2 + 3.556x_3 &= \lambda x_3
\end{aligned}
$$

Then, assuming the x's on the left-hand side of the equation are equal to 1,

$$
\begin{aligned}
3.556(1) - 1.778(1) &= 1.778 \\
-1.778(1) + 3.556(1) - 1.778(1) &= 0 \\
-1.778(1) + 3.556(1) &= 1.778
\end{aligned}
$$

Next, the right-hand side is normalized by 1.778 to make the largest element equal to

$$
\left\{ \begin{array}{c} 1.778 \\ 0 \\ 1.778 \end{array} \right\} = 1.778 \left\{ \begin{array}{c} 1 \\ 0 \\ 1 \end{array} \right\}
$$

Thus, the first estimate of the eigenvalue is 1.778. This iteration can be expressed concisely in matrix form as

$$
\begin{bmatrix} 3.556 & -1.778 & 0 \\ -1.778 & 3.556 & -1.778 \\ 0 & -1.778 & 3.556 \end{bmatrix} \left\{ \begin{array}{c} 1 \\ 1 \\ 1 \end{array} \right\} = \left\{ \begin{array}{c} 1.778 \\ 0 \\ 1.778 \end{array} \right\} = 1.778 \left\{ \begin{array}{c} 1 \\ 0 \\ 1 \end{array} \right\}
$$

The next iteration consists of multiplying $[A]$ by $\lfloor 1 \quad 0 \quad 1 \rfloor^T$ to give

$$
\begin{bmatrix} 3.556 & -1.778 & 0 \\ -1.778 & 3.556 & -1.778 \\ 0 & -1.778 & 3.556 \end{bmatrix} \left\{ \begin{array}{c} 1 \\ 0 \\ 1 \end{array} \right\} = \left\{ \begin{array}{c} 3.556 \\ -3.556 \\ 3.556 \end{array} \right\} = 3.556 \left\{ \begin{array}{c} 1 \\ -1 \\ 1 \end{array} \right\}
$$

Therefore, the eigenvalue estimate for the second iteration is 3.556, which can be employed to determine the error estimate

$$
|\varepsilon_a| = \left| \frac{3.556 - 1.778}{3.556} \right| 100\% = 50\%
$$

The process can then be repeated.
Third iteration:

$$
\begin{bmatrix} 3.556 & -1.778 & 0 \\ -1.778 & 3.556 & -1.778 \\ 0 & -1.778 & 3.556 \end{bmatrix} \left\{ \begin{array}{c} 1 \\ -1 \\ 1 \end{array} \right\} = \left\{ \begin{array}{c} 5.334 \\ -7.112 \\ 5.334 \end{array} \right\} = -7.112 \left\{ \begin{array}{c} -0.75 \\ 1 \\ -0.75 \end{array} \right\}
$$

where $|\varepsilon_a| = 150\%$ (which is high because of the sign change).
Fourth iteration:

$$
\begin{bmatrix} 3.556 & -1.778 & 0 \\ -1.778 & 3.556 & -1.778 \\ 0 & -1.778 & 3.556 \end{bmatrix} \left\{ \begin{array}{c} -0.75 \\ 1 \\ -0.75 \end{array} \right\} = \left\{ \begin{array}{c} -4.445 \\ 6.223 \\ -4.445 \end{array} \right\} = 6.223 \left\{ \begin{array}{c} -0.714 \\ 1 \\ -0.714 \end{array} \right\}
$$

where $|\varepsilon_a| = 214\%$ (again inflated because of sign change).
Fifth iteration:

$$
\begin{bmatrix} 3.556 & -1.778 & 0 \\ -1.778 & 3.556 & -1.778 \\ 0 & -1.778 & 3.556 \end{bmatrix} \left\{ \begin{array}{c} -0.714 \\ 1 \\ -0.714 \end{array} \right\} = \left\{ \begin{array}{c} -4.317 \\ 6.095 \\ -4.317 \end{array} \right\} = 6.095 \left\{ \begin{array}{c} -0.708 \\ 1 \\ -0.708 \end{array} \right\}
$$

Thus, the normalizing factor is converging on the value of 6.070 ($= 2.4637^2$) obtained in part (c) of Example 27.6.

Note that there are some instances where the power method will converge to the second-largest eigenvalue instead of to the largest. James, Smith, and Wolford (1985) provide an illustration of such a case. Other special cases are discussed in Fadeev and Fadeeva (1963).

Determination of the Smallest Eigenvalue.   There are often cases in engineering where we are interested in determining the smallest eigenvalue. Such was the case for the rod in Fig. 27.7, where the smallest eigenvalue could be used to identify a critical buckling load. This can be done by applying the power method to the matrix inverse of $[A]$. For this case, the power method will converge on the largest value of $1/\lambda$—in other words, the smallest value of λ.

EXAMPLE 27.8   Power Method for Lowest Eigenvalue

Problem Statement.   Employ the power method to determine the lowest eigenvalue for part (c) of Example 27.6.

Solution.   After dividing Eq. E27.6.1 by h^2 ($= 0.5625$), its matrix inverse can be evaluated as

$$[A]^{-1} = \begin{bmatrix} 0.422 & 0.281 & 0.141 \\ 0.281 & 0.562 & 0.281 \\ 0.141 & 0.281 & 0.422 \end{bmatrix}$$

Using the same format as in Example 27.7, the power method can be applied to this matrix.
First iteration:

$$\begin{bmatrix} 0.422 & 0.281 & 0.141 \\ 0.281 & 0.562 & 0.281 \\ 0.141 & 0.281 & 0.422 \end{bmatrix} \begin{Bmatrix} 1 \\ 1 \\ 1 \end{Bmatrix} = \begin{Bmatrix} 0.884 \\ 1.124 \\ 0.884 \end{Bmatrix} = 1.124 \begin{Bmatrix} 0.751 \\ 1 \\ 0.751 \end{Bmatrix}$$

Second iteration:

$$\begin{bmatrix} 0.422 & 0.281 & 0.141 \\ 0.281 & 0.562 & 0.281 \\ 0.141 & 0.281 & 0.422 \end{bmatrix} \begin{Bmatrix} 0.751 \\ 1 \\ 0.751 \end{Bmatrix} = \begin{Bmatrix} 0.704 \\ 0.984 \\ 0.704 \end{Bmatrix} = 0.984 \begin{Bmatrix} 0.715 \\ 1 \\ 0.715 \end{Bmatrix}$$

where $|\varepsilon_a| = 14.6\%$.
Third iteration:

$$\begin{bmatrix} 0.422 & 0.281 & 0.141 \\ 0.281 & 0.562 & 0.281 \\ 0.141 & 0.281 & 0.422 \end{bmatrix} \begin{Bmatrix} 0.715 \\ 1 \\ 0.715 \end{Bmatrix} = \begin{Bmatrix} 0.684 \\ 0.964 \\ 0.684 \end{Bmatrix} = 0.964 \begin{Bmatrix} 0.709 \\ 1 \\ 0.709 \end{Bmatrix}$$

where $|\varepsilon_a| = 4\%$.
Thus, after only three iterations, the result is converging on the value of 0.9602, which is the reciprocal of the smallest eigenvalue, 1.0205 ($= \sqrt{1/0.9602}$), obtained in Example 27.6c.

Determination of Intermediate Eigenvalues. After finding the largest eigenvalue, it is possible to determine the next highest by replacing the original matrix by one that includes only the remaining eigenvalues. The process of removing the largest known eigenvalue is called *deflation*. The technique outlined here, *Hotelling's method,* is designed for symmetric matrices. This is because it exploits the orthogonality of the eigenvectors of such matrices, which can be expressed as

$$\{X\}_i^T \{X\}_j = \begin{cases} 0 & \text{for } i \neq j \\ 1 & \text{for } i = j \end{cases} \tag{27.21}$$

where the components of the eigenvector $\{X\}$ have been normalized so that $\{X\}^T\{X\} = 1$, that is, so that the sum of the squares of the components equals 1. This can be accomplished by dividing each of the elements by the normalizing factor

$$\sqrt{\sum_{k=1}^{n} x_k^2}$$

Now, a new matrix $[A]_2$ can be computed as

$$[A]_2 = [A]_1 - \lambda_1 \{X\}_1 \{X\}_1^T \tag{27.22}$$

where $[A]_1$ = the original matrix and λ_1 = the largest eigenvalue. If the power method is applied to this matrix, the iteration process will converge to the second largest eigenvalue, λ_2. To show this, first postmultiply Eq. (27.22) by $\{X\}_1$,

$$[A]_2\{X\}_1 = [A]_1\{X\}_1 - \lambda_1 \{X\}_1 \{X\}_1^T \{X\}_1$$

Invoking the orthogonality principle converts this equation to

$$[A]_2\{X\}_1 = [A]_1\{X\}_1 - \lambda_1 \{X\}_1$$

where the right-hand side is equal to zero according to Eq. (27.20). Thus, $[A]_2\{X\}_1 = 0$. Consequently, $\lambda = 0$ and $\{X\} = \{X\}_1$ is a solution to $[A]_2\{X\} = \lambda\{X\}$. In other words, matrix $[A]_2$ has eigenvalues of 0, $\lambda_2, \lambda_3, \ldots, \lambda_n$. The largest eigenvalue, λ_1, has been replaced by a 0 and, therefore, the power method will converge on the next biggest, λ_2.

The above process can be repeated by generating a new matrix $[A]_3$, etc. Although in theory this process could be continued to determine the remaining eigenvalues, it is limited by the fact that errors in the eigenvectors are passed along at each step. Thus, it is only of value in determining several of the highest eigenvalues. Although this is somewhat of a shortcoming, such information is precisely what is required in many engineering problems.

27.2.6 Other Methods

A wide variety of additional methods are available for solving eigenvalue problems. Most are based on a two-step process. The first step involves transforming the original matrix to a simpler form (for example, tridiagonal) that retains all the original eigenvalues. Then, iterative methods are used to determine these eigenvalues.

Many of these approaches are designed for special types of matrices. In particular, a variety of techniques are devoted to symmetric systems. For example, *Jacobi's method* transforms a symmetric matrix to a diagonal matrix by eliminating off-diagonal terms in a systematic fashion. Unfortunately, the method requires an infinite number of operations because the removal of each nonzero element often creates a new nonzero value at a previous zero element. Although an infinite time is required to create all nonzero off-diagonal elements, the matrix will eventually tend toward a diagonal form. Thus, the approach is iterative in that it is repeated until the off-diagonal terms are "sufficiently" small.

Given's method also involves transforming a symmetric matrix into a simpler form. However, in contrast to the Jacobi method, the simpler form is tridiagonal. In addition, it differs in that the zeros that are created in off-diagonal positions are retained. Consequently, it is finite and, thus, more efficient than Jacobi's method.

Householder's method also transforms a symmetric matrix into a tridiagonal form. It is a finite method and is more efficient than Given's approach in that it reduces whole rows and columns of off-diagonal elements to zero.

Once a tridiagonal system is obtained from Given's or Householder's method, the remaining step involves finding the eigenvalues. A direct way to do this is to expand the determinant. The result is a sequence of polynomials that can be evaluated iteratively for the eigenvalues.

Aside from symmetric matrices, there are also techniques that are available when all eigenvalues of a general matrix are required. These include the *LR method* of Rutishauser and the *QR method* of Francis. Although the QR method is less efficient, it is usually the preferred approach because it is more stable. As such, it is considered to be the best general-purpose solution method.

Finally, it should be mentioned that the aforementioned techniques are often used in tandem to capitalize on their respective strengths. For example, Given's and Householder's methods can also be applied to nonsymmetric systems. The result will not be tridiagonal but rather a special type of matrix called the *Hessenberg form*. One approach is to exploit the speed of Householder's approach by employing it to transform the matrix to this form and then use the stable QR algorithm to find the eigenvalues. Additional information on these and other issues related to eigenvalues can be found in Ralston and Rabinowitz (1978), Wilkinson (1965), Fadeev and Fadeeva (1963), and Householder (1953, 1964). Computer codes can be found in a number of sources including Press et al. (2007). Rice (1983) discusses available software packages.

27.3 ODES AND EIGENVALUES WITH SOFTWARE PACKAGES

Software packages have great capabilities for solving ODEs and determining eigenvalues. This section outlines some of the ways in which they can be applied for this purpose.

27.3.1 Excel

Excel's direct capabilities for solving eigenvalue problems and ODEs are limited. However, if some programming is done (for example, macros), they can be combined with Excel's visualization and optimization tools to implement some interesting applications. Section 28.1 provides an example of how the Excel Solver can be used for parameter estimation of an ODE.

27.3.2 MATLAB

As might be expected, the standard MATLAB software package has excellent capabilities for determining eigenvalues and eigenvectors. However, it also has built-in functions for solving ODEs. The standard ODE solvers include two functions to implement the adaptive step-size Runge-Kutta Fehlberg method (recall Sec. 25.5.2). These are ode23, which uses second- and third-order formulas to attain medium accuracy, and ode45, which uses fourth- and fifth-order formulas to attain higher accuracy. The following example illustrates how they can be used to solve a system of ODEs.

EXAMPLE 27.9   **Using MATLAB for Eigenvalues and ODEs**

Problem Statement.   Explore how MATLAB can be used to solve the following set of nonlinear ODEs from $t = 0$ to 20:

$$\frac{dx}{dt} = 1.2x - 0.6xy \qquad \frac{dy}{dt} = -0.8y + 0.3xy$$

where $x = 2$ and $y = 1$ at $t = 0$. As we will see in the next chapter (Sec. 28.2), such equations are referred to as *predator-prey equations*.

Solution.   Before obtaining a solution with MATLAB, you must use a text processor to create an M-file containing the right-hand side of the ODEs. This M-file will then be accessed by the ODE solver [where $x = y(1)$ and $y = y(2)$]:

```
function yp = predprey(t,y)
yp = [1.2*y(1) -0.6*y(1)*y(2); -0.8*y(2)+0.3*y(1)*y(2)];
```

We store this M-file under the name: predprey.m.

Next, start up MATLAB, and enter the following commands to specify the integration range and the initial conditions:

```
>> tspan = [0,20];
>> y0 = [2,1];
```

The solver can then be invoked by

```
>> [t,y]=ode23('predprey',tspan,y0);
```

This command will then solve the differential equations in predprey.m over the range defined by tspan using the initial conditions found in y0. The results can be displayed by simply typing

```
>> plot(t,y)
```

which yields Fig. 27.9.

FIGURE 27.9
Solution of predator-prey model with MATLAB.

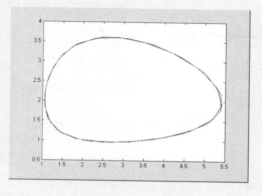

FIGURE 27.10
State-space plot of predator-prey model with MATLAB.

In addition, it is also instructive to generate a state-space plot, that is, a plot of the dependent variables versus each other, by typing

```
>> plot(y(:,1),y(:,2))
```

which yields Fig. 27.10.

MATLAB also has a range of functions designed for stiff systems. These include `ode15s` and `ode23s`. As in the following example, they succeed where the standard functions fail.

EXAMPLE 27.10 **MATLAB for Stiff ODEs**

Problem Statement. *Van der Pol's equation* can be written as

$$\frac{dy_1}{dt} = y_2$$

$$\frac{dy_2}{dt} = \mu(1 - y_1^2)y_2 - y_1$$

As the parameter μ gets large, the system becomes progressively stiffer. Given the initial conditions $y_1(0) = y_2(0) = 1$, use MATLAB to solve the following two cases:

(a) For $\mu = 1$, use `ode45` to solve from $t = 0$ to 20.
(b) For $\mu = 1000$, use `ode23s` to solve from $t = 0$ to 3000.

Solution.

(a) An M-file can be created to hold the differential equations,

```
function yp = vanderpol(t,y)
yp=[y(2);1*(1-y(1)^2)*y(2)-y(1)];
```

Then, as in Example 27.9, `ode45` can be invoked and the results plotted (Fig. 27.11),

```
>> tspan=[0,20];
>> y0=[1,1];
>> [t,y]=ode45('vanderpol',tspan,y0);
>> plot(t,y(:,1))
```

(b) If a standard solver like `ode45` is used for the stiff case ($\mu = 1000$), it will fail miserably (try it, if you like). However, `ode23s` does an efficient job. After revising the M-file to reflect the new value of μ, the solution can be obtained and graphed (Fig. 27.12),

```
>> tspan=[0,3000];
>> y0=[1,1];
```

FIGURE 27.11
Nonstiff form of Van der Pol's equation solved with MATLAB's ode45 function.

FIGURE 27.12
Stiff form of Van der Pol's equation solved with MATLAB's ode23s function.

```
>> [t,y]=ode23S('vanderpol',tspan,y0);
>> plot(t,y(:,1))
```

Notice how this solution has much sharper edges than for case **(a)**. This is a visual manifestation of the "stiffness" of the solution.

For eigenvalues, the capabilities are also very easy to apply. Recall that, in our discussion of stiff systems in Chap. 26, we presented the stiff system defined by Eq. (26.6). Such linear ODEs can be written as an eigenvalue problem of the form

$$\begin{bmatrix} 5 - \lambda & -3 \\ -100 & 301 - \lambda \end{bmatrix} \begin{Bmatrix} e_1 \\ e_2 \end{Bmatrix} = \{0\}$$

where λ and $\{e\}$ = the eigenvalue and the eigenvector, respectively.

MATLAB can then be employed to evaluate both the eigenvalues (d) and eigenvectors (v) with the following simple commands:

```
>> a=[5 -3;-100 301];
>> [v,d]=eig(a)

v =
    -0.9477      0.0101
    -0.3191     -0.9999

d =
     3.9899            0
          0     302.0101
```

Thus, we see that the eigenvalues are of quite different magnitudes, which is typical of a stiff system.

The eigenvalues can be interpreted by recognizing that the general solution for a system of ODEs can be represented as the sum of exponentials. For example, the solution for the present case would be of the form

$$y_1 = c_{11}e^{-3.9899t} + c_{12}e^{-302.0101t}$$
$$y_2 = c_{21}e^{-3.9899t} + c_{22}e^{-302.0101t}$$

where c_{ij} = the part of the initial condition for y_i that is associated with the jth eigenvalue. It should be noted that the c's can be evaluated from the initial conditions and the eigenvectors. Any good book on differential equations, for example, Boyce and DiPrima (1992), will provide an explanation of how this can be done.

Because, for the present case, all the eigenvalues are positive (and hence negative in the exponential function), the solution consists of a series of decaying exponentials. The one with the largest eigenvalue (in this case, 302.0101) would dictate the step size if an explicit solution technique were used.

27.3.3 Mathcad

Mathcad has a number of different functions that solve differential equations and determine eigenvalues and eigenvectors. The most basic technique employed by Mathcad to solve systems of first-order differential equations is a fixed step-size fourth-order Runge-Kutta algorithm. This is provided by the **rkfixed** function. Although this is a good all-purpose integrator, it is not always efficient. Therefore, Mathcad supplies **Rkadapt**, which is a variable step-sized version of **rkfixed**. It is well suited for functions that change rapidly in some regions and slowly in others. Similarly, if you know your solution is a smooth function, then you may find that the Mathcad **Bulstoer** function works well. This function employs the Bulirsch-Stoer method and is often both efficient and highly accurate for smooth functions.

Stiff differential equations are at the opposite end of the spectrum. Under these conditions the **rkfixed** function may be very inefficient or unstable. Therefore, Mathcad provides two special methods specifically designed to handle stiff systems. These functions are called **Stiffb** and **Stiffr** and are based on a modified Bulirsch-Stoer method for stiff systems and the Rosenbrock method.

As an example, let's use Mathcad to solve the following nonlinear ODEs,

$$\frac{dy_1}{dt} = 1.2y_1 - 0.6y_1y_2$$

$$\frac{dy_2}{dt} = -0.8y_2 + 0.3y_1y_2$$

with the initial conditions, $y_1 = 2$ and $y_2 = 1$. This system, called *Lotka-Volterra equations*, is used by environmental engineers and ecologists to evaluate the interactions of predators (y_2) and prey (y_1).

As in Fig. 27.13, the definition symbol is first used to define the vector D(u,y) holding the right-hand sides of the ODEs for input to **rkfixed**. Note that y_1 and y_2 in the ODEs are changed to y_0 and y_1 to comply with Mathcad requirements. In addition,

FIGURE 27.13
Mathcad screen to solve a system of ODEs.

we define the initial conditions (y_0), the integration limit (tf), and the number of values we want to generate (npts). The solutions for **rkfixed** with 200 steps between $t = 0$ and tf are stored in the ysol matrix. The solution is displayed graphically in the plot in Fig. 27.13.

Next, we can illustrate how Mathcad evaluates eigenvalues and eigenvectors. The function **eigenvals(M)** returns the eigenvalues of the square matrix **M**. The function **eigenvecs(M)** returns a matrix containing normalized eigenvectors corresponding to the eigenvectors of **M** whereas **eigenvec(M,e)** returns the eigenvector corresponding to the eigenvalue **e**. We can illustrate these functions for the system given by [recall Eq. (26.6)]

$$\frac{dy_1}{dt} = -5y_1 + 3y_2$$

$$\frac{dy_2}{dt} = 100y_1 - 301y_2$$

The results are shown in Fig. 27.14. Because the eigenvalues (aa) are of different magnitudes, the system is stiff. Note that bb holds the specific eigenvector associated with the smaller eigenvalue. The result cc is a matrix containing both eigenvectors as its columns.

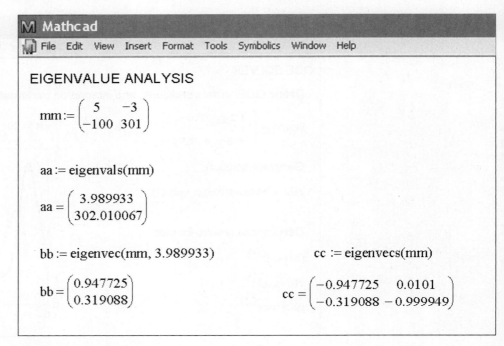

FIGURE 27.14
Mathcad screen to solve for the eigenvalues of a system of ODEs.

PROBLEMS

27.1 A steady-state heat balance for a rod can be represented as

$$\frac{d^2T}{dx^2} - 0.15T = 0$$

Obtain an analytical solution for a 10-m rod with $T(0) = 240$ and $T(10) = 150$.

27.2 Use the shooting method to solve Prob. 27.1.

27.3 Use the finite-difference approach with $\Delta x = 1$ to solve Prob. 27.1.

27.4 Use the shooting method to solve

$$7\frac{d^2y}{dx^2} - 2\frac{dy}{dx} - y + x = 0$$

with the boundary conditions $y(0) = 5$ and $y(20) = 8$.

27.5 Solve Prob. 27.4 with the finite-difference approach using $\Delta x = 2$.

27.6 Use the shooting method to solve

$$\frac{d^2T}{dx^2} - 1 \times 10^{-7}(T + 273)^4 + 4(150 - T) = 0 \qquad (P27.6.1)$$

Obtain a solution for these boundary conditions: $T(0) = 200$ and $T(0.5) = 100$.

27.7 Differential equations like the one solved in Prob. 27.6 can often be simplified by linearizing their nonlinear terms. For example, a first-order Taylor series expansion can be used to linearize the quartic term in Eq. (P27.6.1) as

$$1 \times 10^{-7}(T + 273)^4 = 1 \times 10^{-7}(T_b + 273)^4 + \\ 4 \times 10^{-7}(T_b + 273)^3(T - T_b)$$

where T_b is a base temperature about which the term is linearized. Substitute this relationship into Eq. (P27.6.1), and then solve the resulting linear equation with the finite-difference approach. Employ $T_b = 150$ and $\Delta x = 0.01$ to obtain your solution.

27.8 Repeat Example 27.4 but for three masses. Produce a plot like Fig. 27.6 to identify the principal modes of vibration. Change all the k's to 240.

27.9 Repeat Example 27.6, but for five interior points ($h = 3/6$).

27.10 Use minors to expand the determinant of

$$\begin{bmatrix} 2-\lambda & 8 & 10 \\ 8 & 4-\lambda & 5 \\ 10 & 5 & 7-\lambda \end{bmatrix}$$

27.11 Use the power method to determine the highest eigenvalue and corresponding eigenvector for Prob. 27.10.

27.12 Use the power method to determine the lowest eigenvalue and corresponding eigenvector for Prob. 27.10.

27.13 Develop a user-friendly computer program to implement the shooting method for a linear second-order ODE. Test the program by duplicating Example 27.1.

27.14 Use the program developed in Prob. 27.13 to solve Probs. 27.2 and 27.4.

27.15 Develop a user-friendly computer program to implement the finite-difference approach for solving a linear second-order ODE. Test it by duplicating Example 27.3.

27.16 Use the program developed in Prob. 27.15 to solve Probs. 27.3 and 27.5.

27.17 Develop a user-friendly program to solve for the largest eigenvalue with the power method. Test it by duplicating Example 27.7.

27.18 Develop a user-friendly program to solve for the smallest eigenvalue with the power method. Test it by duplicating Example 27.8.

27.19 Use the Excel Solver to directly solve (that is, without linearization) Prob. 27.6 using the finite-difference approach. Employ $\Delta x = 0.1$ to obtain your solution.

27.20 Use MATLAB to integrate the following pair of ODEs from $t = 0$ to 100:

$$\frac{dy_1}{dt} = 0.4y_1 - 1.8y_1y_2 \qquad \frac{dy_2}{dt} = 0.05y_1y_2 - 0.15y_2$$

where $y_1 = 1$ and $y_2 = 0.05$ at $t = 0$. Develop a state-space plot (y_1 versus y_2) of your results.

27.21 The following differential equation can be used to analyze the vibrations of an automobile shock absorber:

$$1.3 \times 10^6 \frac{d^2x}{dt^2} + 1 \times 10^7 \frac{dx}{dt} + 1.75 \times 10^9 x = 0$$

Transform this equation into a pair of ODEs. **(a)** Use MATLAB to solve these equations from $t = 0$ to 0.4 for the case where $x = 0.5$ and $dx/dt = 0$ at $t = 0$. **(b)** Use MATLAB to determine the eigenvalues and eigenvectors for the system.

27.22 Use MATLAB or Mathcad to integrate

$$\frac{dx}{dt} = -\sigma x + \sigma y$$

$$\frac{dy}{dt} = rx - y - xz$$

$$\frac{dz}{dt} = -bz + xy$$

where $\sigma = 10$, $b = 2.666667$, and $r = 28$. Employ initial conditions of $x = y = z = 5$ and integrate from $t = 0$ to 20.

27.23 Use finite differences to solve the ordinary differential equation

$$\frac{d^2u}{dx^2} + 7\frac{du}{dx} - u = 2$$

given boundary conditions $u(0) = 10$ and $u(2) = 1$. Plot the results of u versus x. Use $\Delta x = 0.1$.

27.24 Solve the nondimensionalized ODE using finite-difference methods that describe the temperature distribution in a circular rod with internal heat source S,

$$\frac{d^2T}{dr^2} + \frac{1}{r}\frac{dT}{dr} + S = 0$$

over the range $0 \le r \le 1$, with the boundary conditions

$$T(r = 1) = 1 \qquad \frac{dT}{dr}\bigg|_{r=0} = 0$$

for $S = 1$, 10, and 20 K/m². Plot the temperature versus radius.

27.25 Derive the set of differential equations for a three mass–four spring system (Fig. P27.25) that describes their time motion. Write the three differential equations in matrix form,

[Acceleration vector] + [k/m matrix][displacement vector x] = 0

Note that each equation has been divided by the mass. Solve for the eigenvalues and natural frequencies for the following values of mass and spring constants: $k_1 = k_4 = 15$ N/m, $k_2 = k_3 = 35$ N/m, and $m_1 = m_2 = m_3 = 1.5$ kg.

FIGURE P27.25

27.26 Consider the mass-spring system in Fig. P27.26. The frequencies for the mass vibrations can be determined by solving for the eigenvalues and by applying $M\ddot{x} + kx = 0$, which yields

$$\begin{bmatrix} m_1 & 0 & 0 \\ 0 & m_2 & 0 \\ 0 & 0 & m_3 \end{bmatrix} \begin{Bmatrix} \ddot{x}_1 \\ \ddot{x}_2 \\ \ddot{x}_3 \end{Bmatrix} + \begin{bmatrix} 2k & -k & -k \\ -k & 2k & -k \\ -k & -k & 2k \end{bmatrix} \begin{Bmatrix} x_1 \\ x_2 \\ x_3 \end{Bmatrix} = \begin{Bmatrix} 0 \\ 0 \\ 0 \end{Bmatrix}$$

Applying the guess $x = x_0 e^{i\omega t}$ as a solution, we get the following matrix:

$$\begin{bmatrix} 2k - m_1\omega^2 & -k & -k \\ -k & 2k - m_2\omega^2 & -k \\ -k & -k & 2k - m_3\omega^2 \end{bmatrix} \begin{Bmatrix} x_{01} \\ x_{02} \\ x_{03} \end{Bmatrix} e^{i\omega t} = \begin{Bmatrix} 0 \\ 0 \\ 0 \end{Bmatrix}$$

Use MATLAB's `eig` command to solve for the eigenvalues of the $k - m\omega^2$ matrix above. Then use these eigenvalues to solve for the frequencies (ω). Let $m_1 = m_2 = m_3 = 1$ kg and $k = 2$ N/m.

FIGURE P27.26

27.27 The following nonlinear, parasitic ODE was suggested by Hornbeck (1975):

$$\frac{dy_1}{dt} = 5(y_1 - t^2)$$

If the initial condition is $y_1(0) = 0.08$, obtain a solution from $t = 0$ to 5:
(a) Analytically.
(b) Using the fourth-order RK method with a constant step size of 0.03125.
(c) Using the MATLAB function `ode45`.
(d) Using the MATLAB function `ode23s`.
(e) Using the MATLAB function `ode23tb`.
Present your results in graphical form.

27.28 A heated rod with a uniform heat source can be modeled with the *Poisson equation*,

$$\frac{d^2 T}{dx^2} = -f(x)$$

Given a heat source $f(x) = 25$ and the boundary conditions $T(x = 0) = 40$ and $T(x = 10) = 200$, solve for the temperature distribution with **(a)** the shooting method and **(b)** the finite-difference method ($\Delta x = 2$).

27.29 Repeat Prob. 27.28, but for the following heat source: $f(x) = 0.12x^3 - 2.4x^2 + 12x$.

27.30 Suppose that the position of a falling object is governed by the following differential equation,

$$\frac{d^2 x}{dt^2} + \frac{c}{m}\frac{dx}{dt} - g = 0$$

where c = a first-order drag coefficient = 12.5 kg/s, m = mass = 70 kg, and g = gravitational acceleration = 9.81 m/s². Use the

shooting method to solve this equation for position and velocity given the boundary conditions $x(0) = 0$ and $x(12) = 500$.

27.31 Repeat Example 27.3, but insulate the left end of the rod. That is, change the boundary condition at the left end of the rod to $T'(0) = 0$.

27.32 Suppose that three individuals are connected by bungee cords. Figure P27.32a shows them being held in place vertically so that each cord is fully extended but unstretched. We can define three distances, x_1, x_2, and x_3, as measured downward from each of the unstretched positions. Assuming that drag is minimal and the cords act like undamped linear springs (i.e., they follow Hooke's law), after the jumpers are released, gravity takes hold and they will oscillate up and down as in Fig. P27.32b. Develop ODEs based on force balance for each jumper (recall Example 9.11). Determine the resulting eigenvalues and eigenvectors that would eventually characterize the jumpers' oscillating motions and relative positions after they were released from their starting positions. Use the following parameters for your analysis:

Jumper	Mass (kg)	Spring Constant (N/m)	Unstretched Cord Length (m)
Top (1)	60	50	20
Middle (2)	70	100	20
Bottom (3)	80	50	20

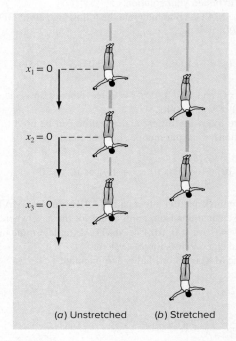

FIGURE P27.32
Three individuals connected by bungee cords.

CHAPTER

28

Case Studies: Ordinary Differential Equations

The purpose of this chapter is to solve some ordinary differential equations using the numerical methods presented in the preceding chapters of Part Seven. The equations originate from practical engineering applications. Many of these applications result in nonlinear differential equations that cannot be solved using analytic techniques. Therefore, numerical methods are usually required. Thus, the techniques for the numerical solution of ordinary differential equations are fundamental capabilities that characterize good engineering practice. The problems in this chapter illustrate some of the trade-offs associated with various methods developed in Part Seven.

Section 28.1 derives from a chemical engineering problem context. It demonstrates how the transient behavior of chemical reactors can be simulated. It also illustrates how optimization can be used to estimate parameters for ODEs.

Sections 28.2 and 28.3, which are taken from civil and electrical engineering, respectively, deal with the solution of systems of equations. In both cases, high accuracy is demanded, and as a consequence, a fourth-order RK scheme is used. In addition, the electrical engineering application also deals with determining eigenvalues.

Section 28.4 employs a variety of different approaches to investigate the behavior of a swinging pendulum. This problem also utilizes two simultaneous equations. An important aspect of this example is that it illustrates how numerical methods allow nonlinear effects to be incorporated easily into an engineering analysis.

28.1 USING ODEs TO ANALYZE THE TRANSIENT RESPONSE OF A REACTOR (CHEMICAL/BIO ENGINEERING)

Background. In Sec. 12.1, we analyzed the steady state of a series of reactors. In addition to steady-state computations, we might also be interested in the transient response of a completely mixed reactor. To do this, we have to develop a mathematical expression for the accumulation term in Eq. (12.1).

Accumulation represents the change in mass in the reactor per change in time. For a constant-volume system, it can be simply formulated as

$$\text{Accumulation} = V \frac{dc}{dt} \tag{28.1}$$

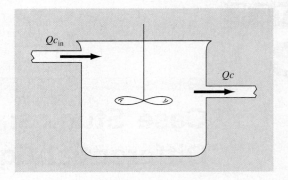

FIGURE 28.1
A single, completely mixed reactor with an inflow and an outflow.

where V = volume and c = concentration. Thus, a mathematical formulation for accumulation is volume times the derivative of c with respect to t.

In this application we will incorporate the accumulation term into the general mass-balance framework we developed in Sec. 12.1. We will then use that model to simulate the dynamics of a single reactor and a system of reactors. In the latter case, we will show how the system's eigenvalues can be determined and provide insight into its dynamics. Finally, we will illustrate how optimization can be used to estimate the parameters of mass-balance models.

Solution.   Equations (28.1) and (12.1) can be used to represent the mass balance for a single reactor such as the one shown in Fig. 28.1:

$$V\frac{dc}{dt} = Qc_{in} - Qc \qquad (28.2)$$

Accumulation = inputs − outputs

Equation (28.2) can be used to determine transient, or time-variable, solutions for the reactor. For example, if $c = c_0$ at $t = 0$, calculus can be employed to analytically solve Eq. (28.2) for

$$c = c_{in}(1 - e^{-(Q/V)t}) + c_0 e^{-(Q/V)t}$$

If $c_{in} = 50$ mg/m^3, $Q = 5$ m^3/min, $V = 100$ m^3, and $c_0 = 10$ mg/m^3, the equation is

$$c = 50(1 - e^{-0.05t}) + 10e^{-0.05t}$$

Figure 28.2 shows this exact, analytical solution.

Euler's method provides an alternative approach for solving Eq. (28.2). Figure 28.2 includes two solutions with different step sizes. As the step size is decreased, the numerical solution converges on the analytical solution. Thus, for this case, the numerical method can be used to check the analytical result.

Besides checking the results of an analytical solution, numerical solutions have added value in those situations where analytical solutions are impossible or so difficult that they are impractical. For example, aside from a single reactor, numerical methods have utility when simulating the dynamics of systems of reactors. For example, ODEs

FIGURE 28.2
Plot of analytical and numerical solutions of Eq. (28.2). The numerical solutions are obtained with Euler's method using different step sizes.

can be written for the five coupled reactors in Fig. 12.3. The mass balance for the first reactor can be written as

$$V_1 \frac{dc_1}{dt} = Q_{01}c_{01} + Q_{31}c_3 - Q_{12}c_1 - Q_{15}c_1$$

or, substituting parameters (note that $Q_{01}c_{01} = 50$ mg/min, $Q_{03}c_{03} = 160$ mg/min, $V_1 = 50$ m^3, $V_2 = 20$ m^3, $V_3 = 40$ m^3, $V_4 = 80$ m^3, and $V_5 = 100$ m^3),

$$\frac{dc_1}{dt} = -0.12c_1 + 0.02c_3 + 1$$

Similarly, balances can be developed for the other reactors as

$$\frac{dc_2}{dt} = 0.15c_1 - 0.15c_2$$

$$\frac{dc_3}{dt} = 0.025c_2 - 0.225c_3 + 4$$

$$\frac{dc_4}{dt} = 0.0125c_2 + 0.1c_3 - 0.1375c_4 + 0.025c_5$$

$$\frac{dc_5}{dt} = 0.03c_1 + 0.01c_2 - 0.04c_5$$

Suppose that at $t = 0$ all the concentrations in the reactors are at zero. Compute how the concentrations will increase over the next hour.

The equations can be integrated with the fourth-order RK method for systems of equations, and the results are depicted in Fig. 28.3. Notice that each of the reactors shows a different transient response to the introduction of chemical. These responses can be parameterized by a 90% response time, t_{90}, which measures the time required for each reactor to reach 90% of its ultimate steady-state level. The times range from about

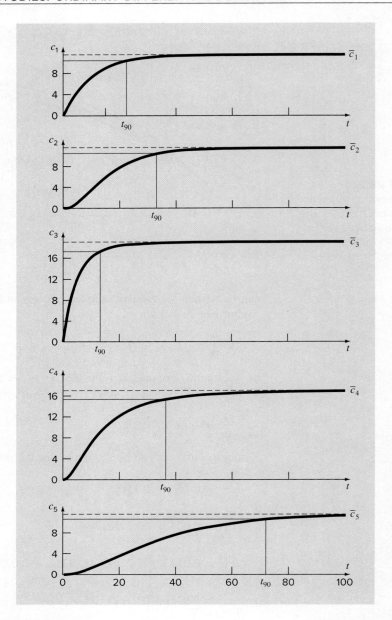

FIGURE 28.3
Plots of transient, or dynamic, response of the network of reactors from Fig. 12.3. Note that all the reactors eventually approach their steady-state concentrations previously computed in Sec. 12.1. In addition, the time to steady state is parameterized by the 90% response time t_{90}.

13 min for reactor 3 to about 72 min for reactor 5. The response times of reactors 4 and 5 are of particular concern because the two outflow streams for the system exit these tanks. Thus, a chemical engineer designing the system might change the flows or volumes of the reactors to speed up the response of these tanks while still maintaining the desired outputs. Numerical methods of the sort described in this part of the book can prove useful in these design calculations.

Further insight into the system's response characteristics can be developed by computing its eigenvalues. First, the system of ODEs can be written as an eigenvalue problem:

$$
\begin{bmatrix}
0.12 - \lambda & 0 & -0.02 & 0 & 0 \\
-0.15 & 0.15 - \lambda & 0 & 0 & 0 \\
0 & -0.025 & 0.225 - \lambda & 0 & 0 \\
0 & -0.0125 & -0.1 & 0.1375 - \lambda & -0.025 \\
-0.03 & -0.01 & 0 & 0 & 0.04 - \lambda
\end{bmatrix}
\begin{Bmatrix}
e_1 \\ e_2 \\ e_3 \\ e_4 \\ e_5
\end{Bmatrix}
= \{0\}
$$

where λ and $\{e\}$ = the eigenvalue and the eigenvector, respectively.

A package like MATLAB software can be used to very conveniently generate the eigenvalues and eigenvectors,

```
a=[0.12 0.0 -0.02 0.0 0.0;-.15 0.15 0.0 0.0 0.0;0.0
-0.025 0.225 0.0 0.0; 0.0 -0.0125 -.1 0.1375 -0.025;-0.03
-0.01 0.0 0.0 0.04];

[evect, lambda] =eig(a)
evect =
            0          0     -0.2490    -0.1228    -0.1059
            0          0     -0.8444     0.2983     0.5784
            0          0     -0.1771     0.5637     0.3041
       1.0000     0.2484     -0.3675    -0.7604    -0.7493
            0     0.9687      0.2419     0.0041    -0.0190


lambda =
       0.1375          0          0          0          0
            0     0.0400          0          0          0
            0          0     0.1058          0          0
            0          0          0     0.2118          0
            0          0          0          0     0.1775
```

The eigenvalues can be interpreted by recognizing that the general solution for a system of ODEs can be represented as the sum of exponentials. For example, for reactor 1, the general solution would be of the form

$$
c_1 = c_{11}e^{-\lambda_1 t} + c_{12}e^{-\lambda_2 t} + c_{13}e^{-\lambda_3 t} + c_{14}e^{-\lambda_4 t} + c_{15}e^{-\lambda_5 t}
$$

where c_{ij} = the part of the initial condition for reactor i that is associated with the jth eigenvalue. Thus, because, for the present case, all the eigenvalues are positive (and hence negative in the exponential function), the solution consists of a series of decaying exponentials. The one with the smallest eigenvalue (in our case, 0.04) will be the slowest.

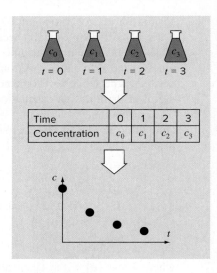

FIGURE 28.4
A simple experiment to collect rate data for a chemical compound that decays with time (reprinted from Chapra 1997).

In some cases, the engineer performing this analysis could be able to relate this eigen-value back to the system parameters. For example, the ratio of the outflow from reactor 5 to its volume is $(Q_{55} + Q_{54})/V_5 = 4/100 = 0.04$. Such information can then be used to modify the system's dynamic performance.

The final topic we would like to review within the present context is *parameter estimation.* One area where this occurs frequently is in *reaction kinetics,* that is, the quantification of chemical reaction rates.

A simple example is depicted in Fig. 28.4. A series of beakers are set up containing a chemical compound that decays over time. At time intervals, the concentration in one of the beakers is measured and recorded. Thus, the result is a table of times and concentrations.

One model that is commonly used to describe such data is

$$\frac{dc}{dt} = -kc^n \qquad (28.3)$$

where k = a reaction rate and n = the order of the reaction. Chemical engineers use concentration-time data of the sort depicted in Fig. 28.4 to estimate k and n. One way to do this is to guess values of the parameters and then solve Eq. (28.3) numerically. The predicted values of concentration can be compared with the measured concentrations and an assessment of the fit made. If the fit is deemed inadequate (for example, by examining a plot or a statistical measure like the sum of the squares of the residuals), the guesses are adjusted and the procedure repeated until a decent fit is attained.

The following data can be fit in this fashion:

t, d	0	1	3	5	10	15	20
c, mg/L	12	10.7	9	7.1	4.6	2.5	1.8

	A	B	C	D	E	F	G	H
1	Fitting of reaction rate							
2	data with the integral/least-squares approach							
3	k	0.091528						
4	n	1.044425						
5	dt	1						
6	t	k1	k2	k3	k4	cp	cm	(cp-cm)^2
7	0	−1.22653	−1.16114	−1.16462	−1.10248	12	12	0
8	1	−1.10261	−1.04409	−1.04719	−0.99157	10.83658	10.7	0.018653
9	2	−0.99169	−0.93929	−0.94206	−0.89225	9.790448		
10	3	−0.89235	−0.84541	−0.84788	−0.80325	8.849344	9	0.022697
11	4	−0.80334	−0.76127	−0.76347	−0.72346	8.002317		
12	5	−0.72354	−0.68582	−0.68779	−0.65191	7.239604	7.1	0.019489
13	6	−0.65198	−0.61814	−0.61989	−0.5877	6.552494		
14	7	−0.58776	−0.55739	−0.55895	−0.53005	5.933207		
15	8	−0.53011	−0.50283	−0.50424	−0.47828	5.374791		
16	9	−0.47833	−0.45383	−0.45508	−0.43175	4.871037		
17	10	−0.4318	−0.40978	−0.4109	−0.38993	4.416389	4.6	0.033713
18	11	−0.38997	−0.37016	−0.37117	−0.35231	4.005877		
19	12	−0.35234	−0.33453	−0.33543	−0.31846	3.635053		
20	13	−0.31849	−0.30246	−0.30326	−0.28798	3.299934		
21	14	−0.28801	−0.27357	−0.2743	−0.26054	2.996949		
22	15	−0.26056	−0.24756	−0.24821	−0.23581	2.7229	2.5	0.049684
23	16	−0.23583	−0.22411	−0.22469	−0.21352	2.474917		
24	17	−0.21354	−0.20297	−0.20349	−0.19341	2.250426		
25	18	−0.19343	−0.18389	−0.18436	−0.17527	2.047117		
26	19	−0.17529	−0.16668	−0.16711	−0.1589	1.862914		
27	20	−0.15891	−0.15115	−0.15153	−0.14412	1.695953	1.8	0.010826
28								
29							SSR =	0.155062

FIGURE 28.5
The application of a spreadsheet and numerical methods to determine the order and rate coefficient of reaction data. This application was performed with an Excel spreadsheet.

The solution to this problem is shown in Fig. 28.5. An Excel spreadsheet was used to perform the computation.

Initial guesses for the reaction rate and order are entered into cells B3 and B4, respectively, and the time step for the numerical calculation is typed into cell B5. For this case, a column of calculation times is entered into column A starting at 0 (cell A7) and ending at 20 (cell A27). The k_1 through k_4 coefficients of the fourth-order RK method are then calculated in the block B7..E27. These are then used to determine the predicted concentrations (the c_p values) in column F. The measured values (c_m) are entered in column G adjacent to the corresponding predicted values. These are then used in conjunction with the predicted values to compute the squared residual in column H. These values are then summed in cell H29.

At this point, the Excel Solver can be used to determine the best parameter values. Once you have accessed the Solver, you are prompted for a target, or solution, cell (H29), queried whether you want to maximize or minimize the target cell (minimize), and

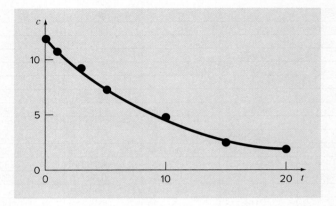

FIGURE 28.6
Plot of fit generated with the integral/least-squares approach.

prompted for the cells that are to be varied (B3..B4). You then activate the algorithm [s(olve)], and the results are as in Fig. 28.5. As shown, the values in cells B3..B4 ($k = 0.0915$ and $n = 1.044$) minimize the sum of the squares of the residuals (SSR = 0.155) between the predicted and measured data. A plot of the fit along with the data is shown in Fig. 28.6.

28.2 PREDATOR-PREY MODELS AND CHAOS (CIVIL/ENVIRONMENTAL ENGINEERING)

Background. Environmental engineers deal with a variety of problems involving systems of nonlinear ordinary differential equations. In this section, we will focus on two of these applications. The first relates to the so-called predator-prey models that are used to study the cycling of nutrient and toxic pollutants in aquatic food chains and biological treatment systems. The second are equations derived from fluid dynamics that are used to simulate the atmosphere. Aside from their obvious application to weather prediction, such equations have also been used to study air pollution and global climate change.

Predator-prey models were developed independently in the early part of the twentieth century by the Italian mathematician Vito Volterra and the American biologist Alfred J. Lotka. These equations are commonly called *Lotka-Volterra equations*. The simplest example is the following pair of ODEs:

$$\frac{dx}{dt} = ax - bxy \tag{28.4}$$

$$\frac{dy}{dt} = -cy + dxy \tag{28.5}$$

where x and y = the number of prey and of predators, respectively, a = the prey growth rate, c = the predator death rate, and b and d = the rate characterizing the effect of the predator-prey interaction on prey death and predator growth, respectively. The multiplicative terms (that is, those involving xy) are what make such equations nonlinear.

An example of a simple model based on atmospheric fluid dynamics is the *Lorenz equations* developed by the American meteorologist Edward Lorenz,

$$\frac{dx}{dt} = -\sigma x + \sigma y \tag{28.6}$$

$$\frac{dy}{dt} = rx - y - xz \tag{28.7}$$

$$\frac{dz}{dt} = -bz + xy \tag{28.8}$$

Lorenz developed these equations to relate the intensity of atmospheric fluid motion, x, to temperature variations y and z in the horizontal and vertical directions, respectively. As with the predator-prey model, we see that the nonlinearity is localized in simple multiplicative terms (xz and xy).

Use numerical methods to obtain solutions for these equations. Plot the results to visualize how the dependent variables change temporally. In addition, plot the dependent variables versus each other to see whether any interesting patterns emerge.

Solution.   Use the following parameter values for the predator-prey simulation: $a = 1.2$, $b = 0.6$, $c = 0.8$, and $d = 0.3$. Employ initial conditions of $x = 2$ and $y = 1$ and integrate from $t = 0$ to 30. We will use the fourth-order RK method with double precision to obtain solutions.

The results using a step size of 0.1 are shown in Fig. 28.7. Note that a cyclical pattern emerges. Thus, because predator population is initially small, the prey grows exponentially. At a certain point, the prey become so numerous, that the predator population begins to grow. Eventually, the increased predators cause the prey to decline. This decrease, in turn, leads to a decrease of the predators. Eventually, the process repeats. Notice that, as expected, the predator peak lags the prey. Also, observe that the process has a fixed period; that is, it repeats in a set time.

Now, if the parameters used to simulate Fig. 28.7 were changed, although the general pattern would remain the same, the magnitudes of the peaks, lags, and period would change. Thus, there are an infinite number of cycles that could occur.

A phase-plane representation is useful in discerning the underlying structure of the model. Rather than plotting x and y versus t, we can plot x versus y. This plot illustrates

FIGURE 28.7

Time-domain representation of numbers of prey and predators for the Lotka-Volterra model.

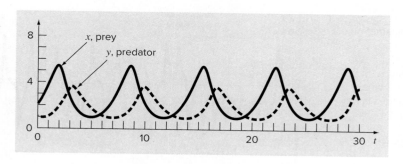

the way that the state variables (x and y) interact, and is referred to as a *phase-plane representation*.

Figure 28.8 shows the phase-plane representation for the case we are studying. Thus, the interaction between the predator and the prey defines a closed counterclockwise orbit. Notice that there is a critical or rest point at the center of the orbit. The exact location of this point can be determined by setting Eqs. (28.4) and (28.5) to steady state ($dy/dt = dx/dt = 0$) and solving for $(x, y) = (0, 0)$ and $(c/d, a/b)$. The former is the trivial result that if we start with neither predators nor prey, nothing will happen. The latter is the more interesting outcome that if the initial conditions are set at $x = c/d$ and $y = a/b$, the derivative will be zero and the populations will remain constant.

Now, let us use the same approach to investigate the trajectories of the Lorenz equations with the following parameter values: $\sigma = 10$, $b = 2.666667$, and $r = 28$. Employ initial conditions of $x = y = z = 5$ and integrate from $t = 0$ to 20. Again, we will use the fourth-order RK method with double precision to obtain solutions.

The results shown in Fig. 28.9 are quite different from the behavior of the Lotka-Volterra equations. The variable x seems to be undergoing an almost random pattern of oscillations, bouncing around from negative values to positive values. However, even

FIGURE 28.8

Phase-plane representation for the Lotka-Volterra model.

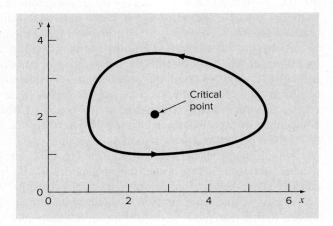

FIGURE 28.9

Time-domain representation of x versus t for the Lorenz equations. The solid time series is for the initial conditions (5, 5, 5). The dashed line is where the initial condition for x is perturbed slightly (5.001, 5, 5).

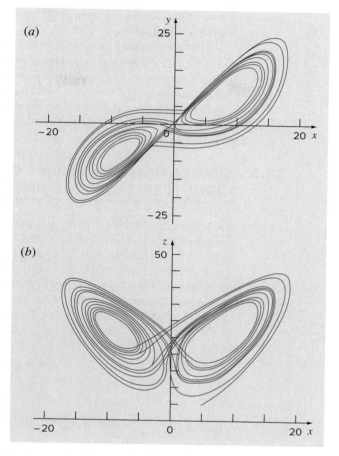

FIGURE 28.10
Phase-plane representation for the Lorenz equations. (a) *xy* projection and (b) *xz* projection.

though the patterns seem random, the frequency of the oscillation and the amplitudes seem fairly consistent.

Another interesting feature can be illustrated by changing the initial condition for x slightly (from 5 to 5.001). The results are superimposed as the dashed line in Fig. 28.9. Although the solutions track on each other for a time, after about $t = 12.5$ they diverge significantly. Thus, we can see that the Lorenz equations are quite sensitive to their initial conditions. In his original study, this led Lorenz to the conclusion that long-range weather forecasts might be impossible!

Finally, let us examine the phase-plane plots. Because we are dealing with three independent variables, we are limited to projections. Figure 28.10 shows projections in the xy and the xz planes. Notice how a structure is manifest when perceived from the phase-plane perspective. The solution forms orbits around what appear to be critical points.

These points are called *strange attractors* in the jargon of mathematicians who study such nonlinear systems.

Solutions such as the type we have explored for the Lorenz equations are referred to as *chaotic* solutions. The study of chaos and nonlinear systems presently represents an exciting area of analysis that has implications to mathematics as well as to science and engineering.

From a numerical perspective, the primary point is the sensitivity of such solutions to initial conditions. Thus, different numerical algorithms, computer precision, and integration time steps can all have an impact on the resulting numerical solution.

28.3 SIMULATING TRANSIENT CURRENT FOR AN ELECTRIC CIRCUIT (ELECTRICAL ENGINEERING)

Background.   Electric circuits where the current is time-variable rather than constant are common. A transient current is established in the right-hand loop of the circuit shown in Fig. 28.11 when the switch is suddenly closed.

Equations that describe the transient behavior of the circuit in Fig. 28.11 are based on Kirchhoff's law, which states that the algebraic sum of the voltage drops around a closed loop is zero (recall Sec. 8.3). Thus,

$$L\frac{di}{dt} + Ri + \frac{q}{C} - E(t) = 0 \tag{28.9}$$

where $L(di/dt)$ = voltage drop across the inductor, L = inductance (H), R = resistance (Ω), q = charge on the capacitor (C), C = capacitance (F), $E(t)$ = time-variable voltage source (V), and

$$i = \frac{dq}{dt} \tag{28.10}$$

Equations (28.9) and (28.10) are a pair of first-order linear differential equations that can be solved analytically. For example, if $E(t) = E_0 \sin \omega t$ and $R = 0$,

$$q(t) = \frac{-E_0}{L(p^2 - \omega^2)} \frac{\omega}{p} \sin pt + \frac{E_0}{L(p^2 - \omega^2)} \sin \omega t \tag{28.11}$$

FIGURE 28.11
An electric circuit where the current varies with time.

where $p = 1/\sqrt{LC}$. The values of q and dq/dt are zero for $t = 0$. Use a numerical approach to solve Eqs. (28.9) and (28.10) and compare the results with Eq. (28.11).

Solution.   This problem involves a rather long integration range and demands the use of a highly accurate scheme to solve the differential equation if good results are expected. Let us assume that $L = 1$ H, $E_0 = 1$ V, $C = 0.25$ C, and $\omega^2 = 3.5$ s^2. This gives $p = 2$, and Eq. (28.11) becomes

$$q(t) = -1.8708 \sin(2t) + 2 \sin(1.8708t)$$

for the analytical solution. This function is plotted in Fig. 28.12. The rapidly changing nature of the function places a severe requirement on any numerical procedure to find $q(t)$. Furthermore, because the function exhibits a slowly varying periodic nature as well as a rapidly varying component, long integration ranges are necessary to portray the solution. Thus, we expect that a high-order method is preferred for this problem.

However, we can try both Euler and fourth-order RK methods and compare the results. Using a step size of 0.1 s gives a value for q at $t = 10$ s of -6.638 with Euler's method and a value of -1.9897 with the fourth-order RK method. These results compare to an exact solution of -1.996 C.

Figure 28.13 shows the results of Euler integration every 1.0 s compared to the exact solution. Note that only every tenth output point is plotted. It is seen that the global error increases as t increases. This divergent behavior intensifies as t approaches infinity.

In addition to directly simulating a network's transient response, numerical methods can also be used to determine its eigenvalues. For example, Fig. 28.14 shows an LC network for which Kirchhoff's voltage law can be employed to develop the following

FIGURE 28.12

Computer screen showing the plot of the function represented by Eq. (28.11).

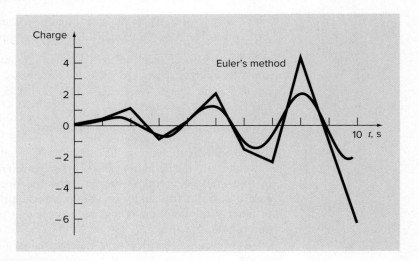

FIGURE 28.13
Results of Euler integration versus exact solution. Note that only every tenth output point is plotted.

FIGURE 28.14
An *LC* network.

system of ODEs:

$$-L_1 \frac{di_1}{dt} - \frac{1}{C_1} \int_{-\infty}^{t} (i_1 - i_2)\, dt = 0$$

$$-L_2 \frac{di_2}{dt} - \frac{1}{C_2} \int_{-\infty}^{t} (i_2 - i_3)\, dt + \frac{1}{C_1} \int_{-\infty}^{t} (i_1 - i_2)\, dt = 0$$

$$-L_3 \frac{di_3}{dt} - \frac{1}{C_3} \int_{-\infty}^{t} i_3 dt + \frac{1}{C_2} \int_{-\infty}^{t} (i_2 - i_3)\, dt = 0$$

Notice that we have represented the voltage drop across the capacitor as

$$V_C = \frac{1}{C} \int_{-\infty}^{t} i\, dt$$

This is an alternative and equivalent expression to the relationship used in Eq. (28.9) and introduced in Sec. 8.3.

The system of ODEs can be differentiated and rearranged to yield

$$L_1 \frac{d^2 i_1}{dt^2} + \frac{1}{C_1}(i_1 - i_2) = 0$$

$$L_2 \frac{d^2 i_2}{dt^2} + \frac{1}{C_2}(i_2 - i_3) - \frac{1}{C_1}(i_1 - i_2) = 0$$

$$L_3 \frac{d^2 i_3}{dt^2} + \frac{1}{C_3}i_3 - \frac{1}{C_2}(i_2 - i_3) = 0$$

Comparison of this system with the one in Eq. (27.5) indicates an analogy between a spring-mass system and an *LC* circuit. As was done with Eq. (27.5), the solution can be assumed to be of the form

$$i_j = A_j \sin(\omega t)$$

This solution along with its second derivative can be substituted into the simultaneous ODEs. After simplification, the result is

$$\left(\frac{1}{C_1} - L_1\omega^2\right)A_1 \qquad -\frac{1}{C_2}A_2 \qquad\qquad\qquad = 0$$

$$-\frac{1}{C_1}A_1 \quad + \left(\frac{1}{C_1} + \frac{1}{C_2} - L_2\omega^2\right)A_2 \qquad -\frac{1}{C_2}A_3 \qquad = 0$$

$$-\frac{1}{C_2}A_2 \qquad + \left(\frac{1}{C_2} + \frac{1}{C_3} - L_3\omega^2\right)A_3 = 0$$

Thus, we have formulated an eigenvalue problem. Further simplification results for the special case where the *C*'s and *L*'s are constant. For this situation, the system can be expressed in matrix form as

$$\begin{bmatrix} 1 - \lambda & -1 & 0 \\ -1 & 2 - \lambda & -1 \\ 0 & -1 & 2 - \lambda \end{bmatrix} \begin{Bmatrix} A_1 \\ A_2 \\ A_3 \end{Bmatrix} = \{0\} \tag{28.12}$$

where

$$\lambda = LC\omega^2 \tag{28.13}$$

Numerical methods can be employed to determine values for the eigenvalues and eigenvectors. MATLAB is particularly convenient in this regard. The following MATLAB session has been developed to do this:

```
>>a=[1 -1 0; -1 2 -1; 0 -1 2]

a =

     1   -1    0
    -1    2   -1
     0   -1    2
```

```
>>[v,d]=eig(a)

v =

        0.7370        0.5910        0.3280
        0.5910       -0.3280       -0.7370
        0.3280       -0.7370        0.5910


d =

        0.1981             0             0
             0        1.5550             0
             0             0        3.2470
```

The matrix v consists of the system's three eigenvectors (arranged as columns), and d is a matrix with the corresponding eigenvalues on the diagonal. Thus, the package computes that the eigenvalues are $\lambda = 0.1981$, 1.555, and 3.247. These values in turn can be substituted into Eq. (28.13) to solve for the natural circular frequencies of the system

$$
\omega = \begin{Bmatrix} \dfrac{0.4450}{\sqrt{LC}} \\[2mm] \dfrac{1.2470}{\sqrt{LC}} \\[2mm] \dfrac{1.8019}{\sqrt{LC}} \end{Bmatrix}
$$

Aside from providing the natural frequencies, the eigenvalues can be substituted into Eq. (28.12) to gain further insight into the circuit's physical behavior. For example, substituting $\lambda = 0.1981$ yields

$$
\begin{bmatrix} 0.8019 & -1 & 0 \\ -1 & 1.8019 & -1 \\ 0 & -1 & 1.8019 \end{bmatrix} \begin{Bmatrix} i_1 \\ i_2 \\ i_3 \end{Bmatrix} = \{0\}
$$

Although this system does not have a unique solution, it will be satisfied if the currents are in fixed ratios, as in

$$0.8019i_1 = i_2 = 1.8019i_3 \tag{28.14}$$

Thus, as depicted in Fig. 28.15a, they oscillate in the same direction with different magnitudes. Observe that if we assume that $i_1 = 0.737$, we can use Eq. (28.14) to compute the other currents, with the result

$$
\{i\} = \begin{Bmatrix} 0.737 \\ 0.591 \\ 0.328 \end{Bmatrix}
$$

which is the first column of the v matrix calculated with MATLAB.

$$(a)\ \omega = \frac{0.4451}{\sqrt{LC}} \qquad (b)\ \omega = \frac{1.2470}{\sqrt{LC}} \qquad (c)\ \omega = \frac{1.8019}{\sqrt{LC}}$$

FIGURE 28.15
A visual representation of the natural modes of oscillation of the *LC* network of Fig. 28.14. Note that the diameters of the circular arrows are proportional to the magnitudes of the currents for the three loops.

In a similar fashion, the second eigenvalue of $\lambda = 1.555$ can be substituted and the result evaluated to yield

$$-1.8018i_1 = i_2 = 2.247i_3$$

As depicted in Fig. 28.15b, the first loop oscillates in the opposite direction from the second and third. Finally, the third mode can be determined as

$$-0.445i_1 = i_2 = -0.8718i_3$$

Consequently, as in Fig. 28.15c, the first and third loops oscillate in the opposite direction from the second.

28.4   THE SWINGING PENDULUM (MECHANICAL/AEROSPACE ENGINEERING)

Background.   Mechanical engineers (as well as all other engineers) are frequently faced with problems concerning the periodic motion of free bodies. The engineering approach to such problems ultimately requires that the position and velocity of the body be known as a function of time. These functions of time invariably are the solutions of ordinary differential equations. The differential equations are usually based on Newton's laws of motion.

As an example, consider the simple pendulum shown previously in Fig. PT7.1. The particle of weight W is suspended on a weightless rod of length l. The only forces acting on the particle are its weight and the tension R in the rod. The position of the particle at any time is completely specified in terms of the angle θ and l.

The free-body diagram in Fig. 28.16 shows the forces on the particle and the acceleration. It is convenient to apply Newton's laws of motion in the x direction tangent to the path of the particle:

$$\Sigma F = -W \sin \theta = \frac{W}{g} a$$

where g = the gravitational constant (32.2 ft/s^2) and a = the acceleration in the x direction. The angular acceleration of the particle (α) becomes

$$\alpha = \frac{a}{l}$$

FIGURE 28.16
A free-body diagram of the swinging pendulum showing the forces on the particle and the acceleration.

Therefore, in polar coordinates ($\alpha = d^2\theta/dt^2$),

$$-W \sin \theta = \frac{Wl}{g}\alpha = \frac{Wl}{g}\frac{d^2\theta}{dt^2}$$

or

$$\frac{d^2\theta}{dt^2} + \frac{g}{l}\sin\theta = 0 \tag{28.15}$$

This apparently simple equation is a second-order nonlinear differential equation. In general, such equations are difficult or impossible to solve analytically. You have two choices regarding further progress. First, the differential equation might be reduced to a form that can be solved analytically (recall Sec. PT7.1.1), or second, a numerical approximation technique can be used to solve the differential equation directly. We will examine both of these alternatives in this example.

Solution.   Proceeding with the first approach, we note that the series expansion for $\sin\theta$ is given by

$$\sin\theta = \theta - \frac{\theta^3}{3!} + \frac{\theta^5}{5!} - \frac{\theta^7}{7!} + \cdots \tag{28.16}$$

For small angular displacements, $\sin\theta$ is approximately equal to θ when expressed in radians. Therefore, for small displacements, Eq. (28.15) becomes

$$\frac{d^2\theta}{dt^2} + \frac{g}{l}\theta = 0 \tag{28.17}$$

which is a second-order linear differential equation. This approximation is very important because Eq. (28.17) is easy to solve analytically. The solution, based on the theory of differential equations, is given by

$$\theta(t) = \theta_0 \cos\sqrt{\frac{g}{l}}t \tag{28.18}$$

where θ_0 = the displacement at $t = 0$ and where it is assumed that the velocity ($v = d\theta/dt$) is zero at $t = 0$. The time required for the pendulum to complete one cycle of oscillation is called the period and is given by

$$T = 2\pi\sqrt{\frac{l}{g}} \tag{28.19}$$

Figure 28.17 shows a plot of the displacement θ and velocity $d\theta/dt$ as a function of time, as calculated from Eq. (28.18) with $\theta_0 = \pi/4$ and $l = 2$ ft. The period, as calculated from Eq. (28.19), is 1.5659 s.

The above calculations essentially are a complete solution of the motion of the pendulum. However, you must also consider the accuracy of the results because of the

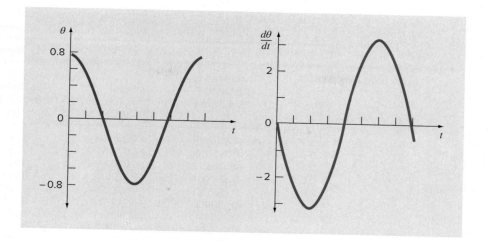

FIGURE 28.17
Plot of displacement θ and velocity $d\theta/dt$ as a function of time t, as calculated from Eq. (28.18). θ_0 is $\pi/4$ and the length is 2 ft.

assumptions inherent in Eq. (28.17). To evaluate the accuracy, it is necessary to obtain a numerical solution for Eq. (28.15), which is a more complete physical representation of the motion. Any of the methods discussed in Chaps. 25 and 26 could be used for this purpose—for example, the Euler and fourth-order RK methods. Equation (28.15) must be transformed into two first-order equations to be compatible with the above methods. This is accomplished as follows. The velocity v is defined by

$$\frac{d\theta}{dt} = v \tag{28.20}$$

and, therefore, Eq. (28.15) can be expressed as

$$\frac{dv}{dt} = -\frac{g}{l}\sin\theta \tag{28.21}$$

Equations (28.20) and (28.21) comprise a coupled system of two ordinary differential equations. The numerical solutions by the Euler method and the fourth-order RK method give the results shown in Table 28.1, which compares the analytic solution for the linear equation of motion [Eq. (28.18)] in column (*a*) with the numerical solutions in columns (*b*), (*c*), and (*d*).

The Euler and fourth-order RK methods yield different results and both disagree with the analytic solution, although the fourth-order RK method for the nonlinear case is closer to the analytic solution than is the Euler method. To properly evaluate the difference between the linear and nonlinear models, it is important to determine the accuracy of the numerical results. This is accomplished in three ways. First, the Euler numerical solution is easily recognized as inadequate because it overshoots the initial condition at $t = 0.8$ s. This clearly violates conservation of energy. Second, columns (*c*) and (*d*) in Table 28.1 show the solutions of the fourth-order RK method

TABLE 28.1 Comparison of a linear analytical solution of the swinging pendulum problem with three nonlinear numerical solutions.

Time, s	Linear Analytical Solution (a)	Nonlinear Numerical Solutions		
		Euler (h = 0.05) (b)	4th-Order RK (h = 0.05) (c)	4th-Order RK (h = 0.01) (d)
0.0	0.785398	0.785398	0.785398	0.785398
0.2	0.545784	0.615453	0.566582	0.566579
0.4	−0.026852	0.050228	0.021895	0.021882
0.6	−0.583104	−0.639652	−0.535802	−0.535820
0.8	−0.783562	−1.050679	−0.784236	−0.784242
1.0	−0.505912	−0.940622	−0.595598	−0.595583
1.2	0.080431	−0.299819	−0.065611	−0.065575
1.4	0.617698	0.621700	0.503352	0.503392
1.6	0.778062	1.316795	0.780762	0.780777

TABLE 28.2 Comparison of the period of an oscillating body calculated from linear and nonlinear models.

Initial Displacement, θ_0	Period, s	
	Linear Model ($T = 2\pi\sqrt{l/g}$)	Nonlinear Model [Numerical Solution of Eq. (28.15)]
$\pi/16$	1.5659	1.57
$\pi/4$	1.5659	1.63
$\pi/2$	1.5659	1.85

for step sizes of 0.05 and 0.01. Because these vary in the fourth decimal place, it is reasonable to assume that the solution with a step size of 0.01 is also accurate with this degree of certainty. Third, for the 0.01-s step-size case, θ obtains a local maximum value of 0.785385 at $t = 1.63$ s (not shown in Table 28.1). This indicates that the pendulum returns to its original position with four-place accuracy with a period of 1.63 s. These considerations allow you to safely assume that the difference between columns (a) and (d) in Table 28.1 truly represents the difference between the linear and nonlinear models.

Another way to characterize the difference between the linear and nonlinear models is on the basis of period. Table 28.2 shows the period of oscillation as calculated by the linear model and the nonlinear model for three different initial displacements. It is seen that the calculated periods agree closely when θ is small because θ is a good approximation for $\sin \theta$ in Eq. (28.16). This approximation deteriorates when θ becomes large.

These analyses are typical of cases you will routinely encounter as an engineer. The utility of the numerical techniques becomes particularly significant in nonlinear problems, and in many cases real-world problems are nonlinear.

PROBLEMS

Chemical/Bio Engineering

28.1 Perform the first computation in Sec. 28.1, but for the case where $h = 10$. Use the Heun method (without iteration) and the fourth-order RK method to obtain solutions.

28.2 Perform the second computation in Sec. 28.1, but for the system described in Prob. 12.4.

28.3 A mass balance for a chemical in a completely mixed reactor can be written as

$$V\frac{dc}{dt} = F - Qc - kVc^2$$

where V = volume (12 m³), c = concentration (g/m³), F = feed rate (175 g/min), Q = flow rate (1 m³/min), and k = a second-order reaction rate (0.15 m³/g/min). If $c(0) = 0$, solve the ODE until the concentration reaches a stable level. Use the midpoint method ($h = 0.5$) and plot your results.

Challenge question: If one ignores the fact that concentrations must be positive, find a range of initial conditions such that you obtain a very different trajectory than was obtained with $c(0) = 0$. Relate your results to the steady-state solutions.

28.4 If $c_{in} = c_b(1 - e^{-0.12t})$, calculate the outflow concentration of a conservative substance (no reaction) for a single, completely mixed reactor as a function of time. Use Heun's method (without iteration) to perform the computation. Employ values of $c_b = 40$ mg/m³, $Q = 6$ m³/min, $V = 100$ m³, and $c_0 = 20$ mg/m³. Perform the computation from $t = 0$ to 100 min using $h = 2$. Plot your results along with the inflow concentration versus time.

28.5 Brackish water with a salt concentration of 8000 g/m³ is pumped into a well-mixed tank at a rate of 0.6 m³/hr. Because of faulty design work, water is evaporating from the tank at a rate of 0.035 m³/hr. The salt solution leaves the tank at a rate of 0.6 m³/hr.
(a) If the tank originally contains 1 m³ of the inlet solution, how long after the outlet pump is turned on will the tank run dry?
(b) Use numerical methods to determine the salt concentration in the tank as a function of time.

28.6 A spherical ice cube (an "ice sphere") that is 6 cm in diameter is removed from a 0°C freezer and placed on a mesh screen at room temperature, $T_a = 20$°C. What will be the diameter of the ice cube as a function of time out of the freezer (assuming that all the water that has melted immediately drips through the screen)? The heat transfer coefficient h for a sphere in a still room is about 3 W/(m² K). The heat flux from the ice sphere to the air is given by

$$\text{Flux} = \frac{q}{A} = h(T_a - T)$$

where q = heat and A = surface area of the sphere. Use a numerical method to make your calculation. Note that the latent heat of fusion is 333 kJ/kg and the density of ice is approximately 0.917 kg/m³.

28.7 The following equations define the concentrations of three reactants:

$$\frac{dc_a}{dt} = -10c_ac_c + c_b$$

$$\frac{dc_b}{dt} = 10c_ac_c - c_b$$

$$\frac{dc_c}{dt} = -10c_ac_c + c_b - 2c_c$$

If the initial conditions are $c_a = 50$, $c_b = 0$, and $c_c = 40$, find the concentrations for the times from 0 to 3 s.

28.8 Compound A diffuses through a 4-cm-long tube and reacts as it diffuses. The equation governing diffusion with reaction is

$$D\frac{d^2A}{dx^2} - kA = 0$$

At one end of the tube, there is a large source of A at a concentration of 0.1 M. At the other end of the tube there is an adsorbent material that quickly absorbs any A, making the concentration 0 M. If $D = 1.5 \times 10^{-6}$ cm²/s and $k = 5 \times 10^{-6}$ s⁻¹, what is the concentration of A as a function of distance in the tube?

28.9 In the investigation of a homicide or accidental death, it is often important to estimate the time of death. From the experimental observations, it is known that the surface temperature of an object changes at a rate proportional to the difference between the temperature of the object and that of the surrounding environment or ambient temperature. This is known as Newton's law of cooling. Thus, if $T(t)$ is the temperature of the object at time t, and T_a is the constant ambient temperature:

$$\frac{dT}{dt} = -K(T - T_a)$$

where $K > 0$ is a constant of proportionality. Suppose that at time $t = 0$ a corpse is discovered and its temperature is measured to be T_o. We assume that at the time of death, the body temperature, T_d, was at the normal value of 37°C. Suppose that the temperature of the corpse when it was discovered was 29.5°C, and that 2 hours later, it is 23.5°C. The ambient temperature is 20°C.
(a) Determine K and the time of death.
(b) Solve the ODE numerically and plot the results.

28.10 The reaction $A \rightarrow B$ takes place in two reactors in series. The reactors are well mixed but are not at steady state. The unsteady-state mass balance for each reactor is shown below:

$$\frac{dCA_1}{dt} = \frac{1}{\tau}(CA_0 - CA_1) - kCA_1$$

$$\frac{dCB_1}{dt} = -\frac{1}{\tau}CB_1 + kCA_1$$

$$\frac{dCA_2}{dt} = \frac{1}{\tau}(CA_1 - CA_2) - kCA_2$$

$$\frac{dCB_2}{dt} = \frac{1}{\tau}(CB_1 - CB_2) + kCA_2$$

where CA_0 = concentration of A at the inlet of the first reactor, CA_1 = concentration of A at the outlet of the first reactor (and inlet of the second), CA_2 = concentration of A at the outlet of the second reactor, CB_1 = concentration of B at the outlet of the first reactor (and inlet of the second), CB_2 = concentration of B in the second reactor, τ = residence time for each reactor, and k = the rate constant for reaction of A to produce B. If CA_0 is equal to 20, find the concentrations of A and B in both reactors during their first 10 minutes of operation. Use $k = 0.12$/min and $\tau = 5$ min and assume that the initial conditions of all the dependent variables are zero.

28.11 A nonisothermal batch reactor can be described by the following equations:

$$\frac{dC}{dt} = -e^{(-10/(T+273))}C$$

$$\frac{dT}{dt} = 1000e^{(-10/(T+273))}C - 10(T - 20)$$

where C is the concentration of the reactant and T is the temperature of the reactor. Initially the reactor is at 15°C and has a concentration of reactant C of 1.0 gmol/L. Find the concentration and temperature of the reactor as a function of time.

28.12 The following system is a classic example of stiff ODEs that can occur in the solution of chemical reaction kinetics:

$$\frac{dc_1}{dt} = -0.013c_1 - 1000c_1c_3$$

$$\frac{dc_2}{dt} = -2500c_2c_3$$

$$\frac{dc_3}{dt} = -0.013c_1 - 1000c_1c_3 - 2500c_2c_3$$

Solve these equations from $t = 0$ to 50 with initial conditions $c_1(0) = c_2(0) = 1$ and $c_3(0) = 0$. If you have access to MATLAB software, use both standard (for example, ode45) and stiff (for example, ode23s) functions to obtain your solutions.

28.13 A *biofilm* with a thickness L_f (cm) grows on the surface of a solid (Fig. P28.13). After traversing a diffusion layer of thickness

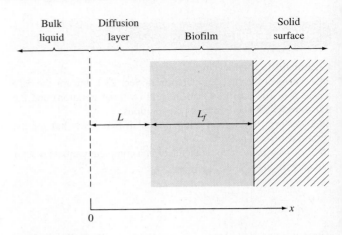

FIGURE P28.13
A biofilm growing on a solid surface.

L (cm), a chemical compound A diffuses into the biofilm, where it is subject to an irreversible first-order reaction that converts it to a product, B.

Steady-state mass balances can be used to derive the following ordinary differential equations for compound A:

$$D\frac{d^2c_a}{dx^2} = 0 \qquad 0 \le x < L$$

$$D_f\frac{d^2c_a}{dx^2} - kc_a = 0 \qquad L \le x < L + L_f$$

where D = the diffusion coefficient in the diffusion layer = 0.8 cm^2/d, D_f = the diffusion coefficient in the biofilm = 0.64 cm^2/d, and k = the first-order rate for the conversion of A to B = 0.1/d. The following boundary conditions hold:

$$c_a = c_{a0} \qquad \text{at } x = 0$$

$$\frac{dc_a}{dx} = 0 \qquad \text{at } x = L + L_f$$

where c_{a0} = the concentration of A in the bulk liquid = 100 mol/L. Use the finite-difference method to compute the steady-state distribution of A from $x = 0$ to $L + L_f$, where $L = 0.008$ cm and $L_f = 0.004$ cm. Employ centered finite differences with $\Delta x = 0.001$ cm.

28.14 The following differential equation describes the steady-state concentration of a substance that reacts with first-order kinetics in an axially dispersed plug-flow reactor (Fig. P28.14),

$$D\frac{d^2c}{dx^2} - U\frac{dc}{dx} - kc = 0$$

FIGURE P28.14

An axially dispersed plug-flow reactor.

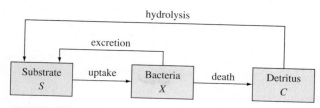

FIGURE P28.16

where D = the dispersion coefficient (m²/hr), c = concentration (mol/L), x = distance (m), U = the velocity (m/hr), and k = the reaction rate (/hr). The boundary conditions can be formulated as

$$Uc_{in} = Uc(x = 0) - D\frac{dc}{dx}(x = 0)$$

$$\frac{dc}{dx}(x = L) = 0$$

where c_{in} = the concentration in the inflow (mol/L) and L = the length of the reactor (m). These are called *Danckwerts boundary conditions*. Use the finite-difference approach to solve for concentration as a function of distance given the following parameters: $D = 5000$ m²/hr, $U = 100$ m/hr, $k = 2$/hr, $L = 100$ m, and $c_{in} = 100$ mol/L. Employ centered finite-difference approximations with $\Delta x = 10$ m to obtain your solutions. Compare your numerical results with the analytical solution,

$$c = \frac{Uc_{in}}{(U - D\lambda_1)\lambda_2 e^{\lambda_2 L} - (U - D\lambda_2)\lambda_1 e^{\lambda_1 L}}$$
$$\times (\lambda_2 e^{\lambda_2 L} e^{\lambda_1 x} - \lambda_1 e^{\lambda_1 L} e^{\lambda_2 x})$$

where

$$\frac{\lambda_1}{\lambda_2} = \frac{U}{2D}\left(1 \pm \sqrt{1 + \frac{4kD}{U^2}}\right)$$

28.15 A series of first-order, liquid-phase reactions create a desirable product (B) and an undesirable by-product (C)

$$A \xrightarrow{k_1} B \xrightarrow{k_2} C$$

If the reactions take place in an axially dispersed plug-flow reactor (Fig. P28.14), steady-state mass balances can be used to develop the following second-order ODEs,

$$D\frac{d^2c_a}{dx^2} - U\frac{dc_a}{dx} - k_1c_a = 0$$

$$D\frac{d^2c_b}{dx^2} - U\frac{dc_b}{dx} + k_1c_a - k_2c_b = 0$$

$$D\frac{d^2c_c}{dx^2} - U\frac{dc_c}{dx} + k_2c_b = 0$$

Use the finite-difference approach to solve for the concentration of each reactant as a function of distance given: $D = 0.1$ m²/min, $U = 1$ m/min, $k_1 = 3$/min, $k_2 = 1$/min, $L = 0.5$ m, $c_{a,in} = 10$ mol/L. Employ centered finite-difference approximations with $\Delta x = 0.05$ m to obtain your solutions and assume Danckwerts boundary conditions, as described in Prob. 28.14. Also, compute the sum of the reactants as a function of distance. Do your results make sense?

28.16 Bacteria growing in a batch reactor utilize a soluble food source (substrate) as depicted in Fig. P28.16. The uptake of the substrate is represented by a logistic model with Michaelis-Menten limitation. Death of the bacteria produces detritus which is subsequently converted to the substrate by hydrolysis. In addition, the bacteria also excrete some substrate directly. Death, hydrolysis, and excretion are all simulated as first-order reactions.

Mass balances can be written as

$$\frac{dX}{dt} = \mu_{max}\left(1 - \frac{X}{K}\right)\left(\frac{S}{K_s + S}\right)X - k_dX - k_eX$$

$$\frac{dC}{dt} = k_dX - k_hC$$

$$\frac{dS}{dt} = k_eX + k_hC - \mu_{max}\left(1 - \frac{X}{K}\right)\left(\frac{S}{K_s + S}\right)X$$

where X, C, and S = the concentrations (mg/L) of bacteria, detritus, and substrate, respectively; μ_{max} = maximum growth rate (d⁻¹); K = the logistic carrying capacity (mg/L); K_s = the Michaelis-Menten half-saturation constant (mg/L), k_d = death rate (d⁻¹); k_e = excretion rate (d⁻¹); and k_h = hydrolysis rate (d⁻¹). Simulate the concentrations from $t = 0$ to 100 d given the initial conditions $X(0) = 1$ mg/L, $S(0) = 100$ mg/L, and $C(0) = 0$ mg/L. Employ the following parameters in your calculation: $\mu_{max} = 10$/d, $K = 10$ mg/L, $K_s = 10$ mg/L, $k_d = 0.1$/d, $k_e = 0.1$/d, and $k_h = 0.1$/d.

28.17 A pharmacokinetic model can be developed to predict the time evolution of alcohol levels in human subjects. As depicted in Fig. P28.17, according to the model, after ingestion into the stomach, alcohol passes through the small intestine followed by

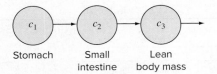

FIGURE P28.17
A three-compartment pharmacokinetic model of the metabolism of alcohol in a human subject.

TABLE P28.17 Typical model parameter values for a reference man and woman.

Parameter	Units	Man	Woman
V_m	g/(L h)	0.47	0.48
K_m	g/L	0.38	0.41
k	hr^{-1}	5.55	4.96
k_a	hr^{-1}	7.05	4.96
a	g^2/h^2	0.42	0.75

distribution across the lean body mass. Mass balances around the three compartments results in the following system of ODEs,

$$\frac{dc_1}{dt} = \frac{k}{1 + ac_1^2}c_1 \qquad \frac{dc_2}{dt} = \frac{k}{1 + ac_1^2}c_1 - k_ac_2$$

$$\frac{dc_3}{dt} = k_ac_2 - \frac{V_m}{K_m + c_3}c_3$$

where t = time, k = the rate constant for gastric emptying if no alcohol has been administered, a = a constant parameterizing the feedback control accounting for stomach alcohol content, k_a = the rate constant for intestinal emptying, V_m = maximum velocity of alcohol elimination from the body mass, and K_m = a half-saturation constant. Using the parameter values in Table P28.17, simulate and graph the alcohol levels in each of the compartments over a 6-hour period given initial stomach levels of $c_1(0) = 0.455$ and 0.703 g/L for the reference man and woman, respectively.

28.18 In bioengineering and biochemistry, the *Michaelis-Menten model* is among the best-known representation of enzyme kinetics. As depicted in Fig. P28.18, it involves an enzyme, E, binding to a substrate, S, to form a complex, ES, via a reversible reaction. The enzyme in turn releases a product, P, and regenerates the original enzyme in an irreversible catalytic reaction. The k's are rate constants, where the subscripts denote the forward (f), reverse (r), and catalytic (c) rates.

Based on the *law of mass action* and mass conservation, the following system of nonlinear ODEs, can be written to describe the rate of change of the reactant and product concentrations with time:

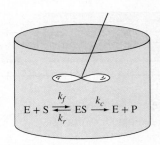

FIGURE P28.18
The reaction of an enzyme (E) and a substrate (S) to produce a product (P) in a completely mixed batch reactor following Michaelis-Menten kinetics.

$$\frac{d[S]}{dt} = -k_f[E][S] + k_r([E]_0 - [E])$$

$$\frac{d[E]}{dt} = -k_f[E][S] + k_r([E]_0 - [E]) + k_c([E]_0 - [E])$$

$$\frac{d[P]}{dt} = k_c([E]_0 - [E])$$

where the brackets designate molar concentrations and $[E]_0$ = the initial enzyme concentration. Note that the total amount of enzyme is conserved,

$$[E]_0 = [E] + [ES]$$

Thus, [ES] can be computed at each time as $[E]_0 - [E]$. Compute and plot the curves for all four variables versus time for the initial conditions $[S]_0 = [E]_0 = 1$ and $[P]_0 = 0$, for two cases: **(a)** $k_f = k_r = k_c = [E]_0 = 1$ and **(b)** $k_f = k_r = [E]_0 = 1$ and $k_c = 10$.

28.19 The following differential equations define the concentrations of three reactants in a closed system (Fig. P28.19),

$$\frac{dc_1}{dt} = -k_{12}c_1 + k_{21}c_2 + k_{31}c_3$$

$$\frac{dc_2}{dt} = k_{12}c_1 - k_{21}c_2 - k_{32}c_2$$

$$\frac{dc_3}{dt} = k_{32}c_2 - k_{31}c_3$$

FIGURE P28.19

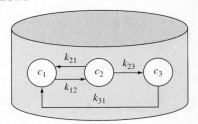

An experiment with initial conditions of $c_1(0) = 100$ and $c_2(0) = c_3(0) = 0$ yields the following data:

t	1	2	3	4	5	6	8	9	10	12	15
c_1	85.3	66.6	60.6	56.1	49.1	45.3	41.9	37.8	33.7	34.4	35.1
c_2	16.9	18.7	24.1	20.9	18.9	19.9	20.6	13.9	19.1	14.5	15.4
c_3	4.7	7.9	20.1	22.8	32.5	37.7	42.4	47	50.5	52.3	51.3

Integrate the equations and use optimization to estimate the values of the k's that minimize the sum of the squares of the discrepancies between the model predictions and the data. Employ initial guesses of 0.15 for all the k's.

Civil/Environmental Engineering

28.20 Perform the same computation for the Lotka-Volterra model in Sec. 28.2, but use **(a)** Euler's method, **(b)** Heun's method (without iterating the corrector), **(c)** the fourth-order RK method, and **(d)** the MATLAB ode45 function. In all cases, use single-precision variables and a step size of 0.1, and simulate from $t = 0$ to 20. Develop phase-plane plots for all cases.

28.21 Perform the same computation for the Lorenz equations in Sec. 28.2, but use **(a)** Euler's method, **(b)** Heun's method (without iterating the corrector), **(c)** the fourth-order RK method, and **(d)** the MATLAB ode45 function. In all cases, use single-precision variables and a step size of 0.1 and simulate from $t = 0$ to 20. Develop phase-plane plots for all cases.

28.22 The following equation can be used to model the deflection of a sailboat mast subject to a wind force:

$$\frac{d^2y}{dz^2} = \frac{f}{2EI}(L - z)^2$$

where f = wind force, E = modulus of elasticity, L = mast length, and I = moment of inertia. Calculate the deflection if $y = 0$ and $dy/dz = 0$ at $z = 0$. Use parameter values of $f = 60$, $L = 30$, $E = 1.25 \times 10^8$, and $I = 0.05$ for your computation.

28.23 Perform the same computation as in Prob. 28.22, but rather than using a constant wind force, employ a force that varies with height according to (recall Sec. 24.2)

$$f(z) = \frac{200z}{5 + z}e^{-2z/30}$$

28.24 An environmental engineer is interested in estimating the mixing that occurs between a stratified lake and an adjacent embayment (Fig. P28.24). A conservative tracer is instantaneously mixed with the bay water, and then the tracer concentration is monitored over the ensuing period in all three segments. The values are

t	0	2	4	6	8	12	16	20
c_1	0	15	11	7	6	3	2	1
c_2	0	3	5	7	7	6	4	2
c_3	100	48	26	16	10	4	3	2

Using mass balances, the system can be modeled as the following simultaneous ODEs:

$$V_1\frac{dc_1}{dt} = -Qc_1 + E_{12}(c_2 - c_1) + E_{13}(c_3 - c_1)$$

$$V_2\frac{dc_2}{dt} = E_{12}(c_1 - c_2)$$

$$V_3\frac{dc_3}{dt} = E_{13}(c_1 - c_3)$$

where V_i = volume of segment i, Q = flow, and E_{ij} = diffusive mixing rate between segments i and j. Use the data and the differential equations to estimate the E's if $V_1 = 1 \times 10^7$, $V_2 = 8 \times 10^6$, $V_3 = 5 \times 10^6$, and $Q = 4 \times 10^6$. Employ Euler's method with a step size of 0.1 for your analysis.

28.25 Population-growth dynamics are important in a variety of planning studies for areas such as transportation and water-resource engineering. One of the simplest models of such growth incorporates the assumption that the rate of change of the population p is proportional to the existing population at any time t:

$$\frac{dp}{dt} = Gp \tag{P28.25.1}$$

FIGURE P28.24

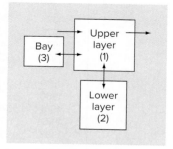

where G = a growth rate (per year). This model makes intuitive sense because the greater the population, the greater the number of potential parents. At time $t = 0$, an island has a population of 6000 people. If $G = 0.075$ per year, employ Heun's method (without iteration) to predict the population at $t = 20$ years, using a step size of 0.5 year. Plot p versus t on standard and semilog graph paper. Determine the slope of the line on the semilog plot. Discuss your results.

28.26 Although the model in Prob. 28.25 works adequately when population growth is unlimited, it breaks down when factors such as food shortages, pollution, and lack of space inhibit growth. In such cases, the growth rate itself can be thought of as being inversely proportional to population. One model of this relationship is

$$G = G'(p_{max} - p) \qquad (P28.26.1)$$

where G' = a population-dependent growth rate (per people-year) and p_{max} = the maximum sustainable population. Thus, when population is small ($p \ll p_{max}$), the growth rate will be at a high constant rate of $G'p_{max}$. For such cases, growth is unlimited and Eq. (P28.26.1) is essentially identical to Eq. (P28.25.1). However, as population grows (that is, p approaches p_{max}), G decreases until at $p = p_{max}$ it is zero. Thus, the model predicts that, when the population reaches the maximum sustainable level, growth is nonexistent, and the system is at a steady state. Substituting Eq. (P28.26.1) into Eq. (P28.25.1) yields

$$\frac{dp}{dt} = G'(p_{max} - p)p$$

For the same island as in Prob. 28.25, employ Heun's method (without iteration) to predict the population at $t = 20$ years, using a step size of 0.5 year. Employ values of $G' = 10^{-5}$ per people-year and $p_{max} = 20,000$ people. At time $t = 0$, the island has a population of 6000 people. Plot p versus t and interpret the shape of the curve.

28.27 Isle Royale National Park is a 210-square-mile archipelago composed of a single large island and many small islands in Lake Superior. Moose arrived around 1900 and by 1930, their population approached 3000, ravaging vegetation. In 1949, wolves crossed an ice bridge from Ontario. Since the late 1950s, the numbers of the moose and wolves have been tracked. (Dash indicates no data.)

Year	Moose	Wolves	Year	Moose	Wolves
1960	700	22	1972	836	23
1961	—	22	1973	802	24
1962	—	23	1974	815	30
1963	—	20	1975	778	41
1964	—	25	1976	641	43
1965	—	28	1977	507	33
1966	881	24	1978	543	40
1967	—	22	1979	675	42
1968	1000	22	1980	577	50
1969	1150	17	1981	570	30
1970	966	18	1982	590	13
1971	674	20	1983	811	23

(a) Integrate the Lotka-Volterra equations from 1960 through 2020. Determine the coefficient values that yield an optimal fit. Compare your simulation with these data using a time-series approach, and comment on the results.

(b) Plot the simulation of (a), but use a phase-plane approach.

(c) After 1993, suppose that the wildlife managers trap one wolf per year and transport it off the island. Predict how the populations of both the wolves and moose would evolve to the year 2020. Present your results as both time-series and phase-plane plots. For this case, as well as for (d), use the following coefficients: $a = 0.3$, $b = 0.01111$, $c = 0.2106$, and $d = 0.0002632$.

(d) Suppose that in 1993, some poachers snuck onto the island and killed 50% of the moose. Predict how the populations of both the wolves and moose would evolve to the year 2020. Present your results as both time-series and phase-plane plots.

28.28 A cable is hanging from two supports at A and B (Fig. P28.28). The cable is loaded with a distributed load whose magnitude varies with x as

$$w = w_o\left[1 + \sin\left(\frac{\pi x}{2l_A}\right)\right]$$

where $w_o = 1000$ lb/ft. The slope of the cable (dy/dx) = 0 at $x = 0$, which is the lowest point for the cable. It is also the point where the tension in the cable is a minimum of T_o. The differential equation that governs the cable is

$$\frac{d^2y}{dx^2} = \frac{w_o}{T_o}\left[1 + \sin\left(\frac{\pi x}{2l_A}\right)\right]$$

Solve this equation using a numerical method and plot the shape of the cable (y versus x). For the numerical solution, the value of T_o is unknown, so the solution must use an iterative technique, similar to the shooting method, to converge on a correct value of h_A for various values of T_o.

FIGURE P28.28

FIGURE P28.29

28.29 The basic differential equation of the elastic curve for a cantilever beam (Fig. P28.29) is given as

$$EI\frac{d^2y}{dx^2} = -P(L - x)$$

where E = the modulus of elasticity and I = the moment of inertia. Solve for the deflection of the beam using a numerical method. The following parameter values apply: E = 30,000 ksi, I = 800 in⁴, P = 1 kip, L = 10 ft. Compare your numerical results to the analytical solution,

$$y = -\frac{PLx^2}{2EI} + \frac{Px^3}{6EI}$$

28.30 The basic differential equation of the elastic curve for a uniformly loaded beam (Fig. P28.30) is given as

$$EI\frac{d^2y}{dx^2} = \frac{wLx}{2} - \frac{wx^2}{2}$$

where E = the modulus of elasticity and I = the moment of inertia. Solve for the deflection of the beam using **(a)** the finite-difference approach (Δx = 2 ft) and **(b)** the shooting method. The following parameter values apply: E = 30,000 ksi, I = 800 in⁴, w = 1 kip/ft, L = 10 ft. Compare your numerical results to the analytical solution,

$$y = \frac{wLx^3}{12EI} - \frac{wx^4}{24EI} - \frac{wL^3x}{24EI}$$

28.31 In Prob. 28.30, the basic differential equation of the elastic curve for a uniformly loaded beam was formulated as

$$EI\frac{d^2y}{dx^2} = \frac{wLx}{2} - \frac{wx^2}{2}$$

Note that the right-hand side represents the moment as a function of x. An equivalent approach can be formulated in terms of the fourth derivative of deflection as

$$EI\frac{d^4y}{dx^4} = -w$$

For this formulation, four boundary conditions are required. For the supports shown in Fig. P28.30, the conditions are that the end displacements are zero, $y(0) = y(L) = 0$, and that the end moments are zero, $y''(0) = y''(L) = 0$. Solve for the deflection of the beam using the finite-difference approach (Δx = 0.6 m). The following parameter values apply: E = 200 GPa, I = 30,000 cm⁴, w = 15 kN/m, and L = 3 m. Compare your numerical results with the analytical solution given in Prob. 28.30.

28.32 A pond drains through a pipe, as shown in Fig. P28.32. Under a number of simplifying assumptions, the following differential equation describes how depth changes with time:

$$\frac{dh}{dt} = -\frac{\pi d^2}{4A(h)}\sqrt{2g(h + e)}$$

where h = depth (m), t = time (s), d = pipe diameter (m), $A(h)$ = pond surface area as a function of depth (m²), g = gravitational constant (= 9.81 m/s²), and e = depth of pipe outlet below the pond bottom (m). Based on the following area-depth table, solve this differential equation to determine how long it takes for the pond to empty given that $h(0)$ = 6 m, d = 0.25 m, and e = 1 m.

h, m	6	5	4	3	2	1	0
$A(h)$, 10⁴ m²	1.17	0.97	0.67	0.45	0.32	0.18	0

28.33 Engineers and scientists use mass-spring models to gain insight into the dynamics of structures under the influence of

FIGURE P28.30

FIGURE P28.32

FIGURE P28.33

FIGURE P28.36
An unconfined, or "phreatic," aquifer.

disturbances such as earthquakes. Figure P28.33 shows such a representation for a three-story building. For this case, the analysis is limited to horizontal motion of the structure. Force balances can be developed for this system as

$$\left(\frac{k_1 + k_2}{m_1} - \omega^2\right)X_1 \qquad -\frac{k_2}{m_1} \qquad X_2 \qquad\qquad = 0$$

$$-\frac{k_2}{m_2} \qquad X_1 + \left(\frac{k_2 + k_3}{m_2} - \omega^2\right)X_2 - \frac{k_3}{m_2} \qquad X_3 = 0$$

$$-\frac{k_3}{m_3} \qquad X_2 + \left(\frac{k_3}{m_3} - \omega^2\right)X_3 = 0$$

Determine the eigenvalues and eigenvectors and graphically represent the modes of vibration for the structure by displaying the amplitudes versus height for each of the eigenvectors. Normalize the amplitudes so that the displacement of the third floor is 1.

28.34 Repeat the computations in Prob. 23.33 but remove the third floor.

28.35 Repeat the computations in Prob. 23.33 but add a fourth floor with a mass of $m_4 = 6000$ kg connected with the third floor by a spring with $k_4 = 1200$ kN/m.

28.36 Under a number of simplifying assumptions, the steady-state height of the water table in a one-dimensional, unconfined groundwater aquifer (Fig. P28.36) can be modeled with the following second-order ODE,

$$K\bar{h}\frac{d^2h}{dx^2} + N = 0$$

where x = distance (m), K = hydraulic conductivity (m/d), h = height of the water table (m), $\bar{h}$ = the average height of the water table (m), and N = infiltration rate (m/d).

Solve for the height of the water table for $x = 0$ to 1000 m where $h(0) = 10$ m and $h(1000) = 5$ m. Use the following parameters for the calculation: $K = 1$ m/d and $N = 0.0001$ m/d. Set the average

height of the water table as the average of the boundary conditions. Obtain your solution with **(a)** the shooting method and **(b)** the finite-difference method ($\Delta x = 100$ m).

28.37 In Prob. 28.36, a linearized groundwater model was used to simulate the height of the water table for an unconfined aquifer. A more realistic result can be obtained by using the following nonlinear ODE:

$$\frac{d}{dx}\left(Kh\frac{dh}{dx}\right) + N = 0$$

where x = distance (m), K = hydraulic conductivity (m/d), h = height of the water table (m), and N = infiltration rate (m/d). Solve for the height of the water table for the same case as in Prob. 28.36. That is, solve from $x = 0$ to 1000 m with $h(0) = 10$ m, $h(1000) = 5$ m, $K = 1$ m/d, and $N = 0.0001$ m/d. Obtain your solution with **(a)** the shooting method and **(b)** the finite-difference method ($\Delta x = 100$ m).

28.38 The Lotka-Volterra equations described in Sec. 28.2 have been refined to include additional factors that impact predator-prey dynamics. For example, over and above predation, prey population can be limited by other factors such as space. Space limitation can be incorporated into the model as a carrying capacity (recall the logistic model described in Prob. 28.16) as in

$$\frac{dx}{dt} = a\left(1 - \frac{x}{K}\right)x - bxy$$

$$\frac{dy}{dt} = -cy + dxy$$

where K = the carrying capacity. Use the same parameter values and initial conditions as in Sec. 28.2 to integrate these equations from $t = 0$ to 100 using ode45.

(a) Employ the very large value of $K = 10^8$ to validate that you obtain the same results as in Sec. 28.2.

(b) Compare **(a)** with the more realistic carrying capacity of $K = 200$. Discuss your results.

28.39 The growth of floating, unicellular algae below a sewage treatment plant discharge can be modeled with the following simultaneous ODEs:

$$\frac{da}{dt} = [k_g(n, p) - k_d - k_s]a$$

$$\frac{dn}{dt} = r_{nc}k_hc - r_{na}k_g(n, p)a$$

$$\frac{dp}{dt} = r_{pc}k_hc - r_{pa}k_g(n, p)a$$

$$\frac{dc}{dt} = r_{ca}k_da - k_hc$$

where t = travel time (d), a = algal chlorophyll concentration (μgA/L), n = inorganic nitrogen concentration (μgN/L), p = inorganic phosphorus concentration (μgP/L), c = detritus concentration (μgC/L), k_d = algal death rate (d^{-1}), k_s = algal settling rate (d^{-1}), k_h = detrital hydrolysis rate (d^{-1}), r_{nc} = nitrogen-to-carbon ratio (μgN/μgC), r_{pc} = phosphorus-to-carbon ratio (μgP/μgC), r_{na} = nitrogen-to-chlorophyll ratio (μgN/μgA), r_{pa} = phosphorus-to-chlorophyll ratio (μgP/μgA), and $k_g(n, p)$ = algal growth rate (d^{-1}), which can be computed with

$$k_g(n, p) = k_g \min\left\{\frac{p}{k_{sp} + p}, \frac{n}{k_{sn} + n}\right\}$$

where k_g = the algal growth rate at excess nutrient levels (d^{-1}), k_{sp} = the phosphorus half-saturation constant (μgP/L), and k_{sn} = the nitrogen half-saturation constant (μgN/L). Use the ode45 and ode15s functions to solve these equations from $t = 0$ to 50 d given the initial conditions $a = 1$, $n = 4000$, $p = 800$, and $c = 0$. Note that the parameters are $k_d = 0.1$, $k_s = 0.15$, $k_h = 0.025$, $r_{nc} = 0.18$, $r_{pc} = 0.025$, $r_{na} = 7.2$, $r_{pa} = 1$, $r_{ca} = 40$, $k_g = 0.5$, $k_{sp} = 2$, and $k_{sn} = 15$. Develop plots of both solutions and interpret the results.

28.40 The following ODEs have been proposed as a model of an epidemic:

$$\frac{dS}{dt} = -aSI$$

$$\frac{dI}{dt} = aSI - rI$$

$$\frac{dR}{dt} = rI$$

where S = the susceptible individuals, I = the infected, R = the recovered, a = the infection rate, and r = the recovery rate. A city has 10,000 people, all of whom are susceptible.

(a) If a single infectious individual enters the city at $t = 0$, compute the progression of the epidemic until the number of infected

individuals falls below 10. Use the following parameters: $a = 0.002/$(person·week) and $r = 0.15/$d. Develop time-series plots of all the state variables. Also generate a phase-plane plot of S versus I versus R.

(b) Suppose that after recovery, there is a loss of immunity that causes recovered individuals to become susceptible. This reinfection mechanism can be computed as ρR, where ρ = the reinfection rate. Modify the model to include this mechanism and repeat the computations in **(a)** using $\rho = 0.015/$d.

Electrical Engineering

28.41 Perform the same computation as in the first part of Sec. 28.3, but with $R = 0.025$ Ω.

28.42 Solve the ODE in the first part of Sec. 8.3 from $t = 0$ to 0.5 using numerical techniques if $q = 0.1$ and $i = -3.281515$ at $t = 0$. Use $R = 50$ along with the other parameters from Sec. 8.3.

28.43 For a simple RL circuit, Kirchhoff's voltage law requires that (if Ohm's law holds)

$$L\frac{di}{dt} + Ri = 0$$

where i = current, L = inductance, and R = resistance. Solve for i, if $L = 1$, $R = 1.5$, and $i(0) = 0.5$. Solve this problem analytically and with a numerical method. Present your results graphically.

28.44 In contrast to the case in Prob. 28.43, real resistors may not always obey Ohm's law. For example, the voltage drop may be nonlinear and the circuit dynamics described by a relationship such as

$$L\frac{di}{dt} + R\left[\frac{i}{I} - \left(\frac{i}{I}\right)^3\right] = 0$$

where all other parameters are as defined in Prob. 28.43 and I is a known reference current equal to 1. Solve for i as a function of time under the same conditions as specified in Prob. 28.43.

28.45 Develop an eigenvalue problem for an LC network similar to the one in Fig. 28.14, but with only two loops. That is, omit the i_3 loop. Draw the network, illustrating how the currents oscillate in their primary modes.

28.46 Just as Fourier's law and the heat balance can be employed to characterize temperature distribution, analogous relationships are available to model field problems in other areas of engineering. For example, electrical engineers use a similar approach when modeling electrostatic fields. Under a number of simplifying assumptions, an analog of Fourier's law can be represented in one-dimensional form as

$$D = -\varepsilon\frac{dV}{dx}$$

where D is called the electric flux density vector, ε = permittivity of the material, and V = electrostatic potential. Similarly, a Poisson equation for electrostatic fields can be represented in one dimension as

$$\frac{d^2V}{dx^2} = -\frac{\rho_v}{\varepsilon}$$

where ρ_v = charge density. Use the finite-difference technique with $\Delta x = 2$ to determine V for a wire where $V(0) = 1000$, $V(20) = 0$, $\varepsilon = 2$, $L = 20$, and $\rho_v = 30$.

Mechanical/Aerospace Engineering

28.47 Perform the same computation as in Sec. 28.4 but for a 1-m-long pendulum.

28.48 The rate of cooling of a body can be expressed as

$$\frac{dT}{dt} = -k(T - T_a)$$

where T = temperature of the body (°C), T_a = temperature of the surrounding medium (°C), and k = the proportionality constant (min⁻¹). Thus, this equation specifies that the rate of cooling is proportional to the difference in temperature between the body and the surrounding medium. If a metal ball heated to 90°C is dropped into water that is held at a constant value of $T_a = 20°C$, use a numerical method to compute how long it takes the ball to cool to 40°C if $k = 0.25$ min⁻¹.

28.49 The rate of heat flow (conduction) between two points on a cylinder heated at one end is given by

$$\frac{dQ}{dt} = \lambda A \frac{dT}{dx}$$

where λ = a constant, A = the cylinder's cross-sectional area, Q = heat flow, T = temperature, t = time, and x = distance from the heated end. Because the equation involves two derivatives, we will simplify this equation by letting

$$\frac{dT}{dx} = \frac{100(L - x)(20 - t)}{100 - xt}$$

where L is the length of the rod. Combine the two equations and compute the heat flow for $t = 0$ to 25 s. The initial condition is $Q(0) = 0$ and the parameters are $\lambda = 0.5$ cal cm/s, $A = 12$ cm², $L = 20$ cm, and $x = 2.5$ cm. Plot your results.

28.50 Repeat the falling parachutist problem (Example 1.2), but with the upward force due to drag as a second-order rate:

$$F_u = -cv^2$$

where $c = 0.225$ kg/m. Solve for $t = 0$ to 30, plot your results, and compare with those of Example 1.2.

28.51 Suppose that, after falling for 13 s, the parachutist from Examples 1.1 and 1.2 pulls the rip cord. At this point, assume that the drag coefficient is instantaneously increased to a constant value of 55 kg/s. Compute the parachutist's velocity from $t = 0$ to

30 s with Heun's method (without iteration of the corrector) using a step size of 2 s. Plot v versus t for $t = 0$ to 30 s.

28.52 The following ordinary differential equation describes the motion of a damped spring-mass system (Fig. P28.52):

$$m\frac{d^2x}{dt^2} + a\left|\frac{dx}{dt}\right|\frac{dx}{dt} + bx^3 = 0$$

where x = displacement from the equilibrium position, t = time, $m = 1$ kg mass, and $a = 5$ N/(m/s)². The damping term is nonlinear and represents air damping.

The spring is a cubic spring and is also nonlinear with $b = 5$ N/m³. The initial conditions are

Initial velocity $\qquad \dfrac{dx}{dt} = 0.5$ m/s

Initial displacement $\qquad x = 1$ m

Solve this equation using a numerical method over the time period $0 \le t \le 8$ s. Plot the displacement and velocity versus time and plot the phase-plane representation (velocity versus displacement) for all the following cases:

(a) A similar linear equation,

$$m\frac{d^2x}{dt^2} + 2\frac{dx}{dt} + 5x = 0$$

(b) The nonlinear equation with only a nonlinear spring term,

$$\frac{d^2x}{dt^2} + 2\frac{dx}{dt} + bx^3 = 0$$

(c) The nonlinear equation with only a nonlinear damping term,

$$m\frac{d^2x}{dt^2} + a\left|\frac{dx}{dt}\right|\frac{dx}{dt} + 5x = 0$$

(d) The full nonlinear equation where both the damping and spring terms are nonlinear,

$$m\frac{d^2x}{dt^2} + a\left|\frac{dx}{dt}\right|\frac{dx}{dt} + bx^3 = 0$$

FIGURE P28.52

FIGURE P28.53

28.53 A forced damped spring-mass system (Fig. P28.53) has the following ordinary differential equation of motion:

$$m\frac{d^2x}{dt^2} + a\left|\frac{dx}{dt}\right|\frac{dx}{dt} + kx = F_o \sin(\omega t)$$

where x = displacement from the equilibrium position, t = time, m = 2 kg mass, a = 5 N/(m/s)^2, and k = 6 N/m. The damping term is nonlinear and represents air damping. The forcing function $F_o \sin(\omega t)$ has values of F_o = 2.5 N and ω = 0.5 rad/sec. The initial conditions are

Initial velocity	$\dfrac{dx}{dt} = 0$ m/s
Initial displacement	$x = 1$ m

Solve this equation using a numerical method over the time period $0 \le t \le 15$ s. Plot the displacement and velocity versus time, and plot the forcing function on the same curve. Also, develop a separate plot of velocity versus displacement.

28.54 The temperature distribution in a tapered conical cooling fin (Fig. P28.54) is described by the following differential equation, which has been nondimensionalized,

$$\frac{d^2u}{dx^2} + \left(\frac{2}{x}\right)\left(\frac{du}{dx} - pu\right) = 0$$

where u = temperature ($0 \le u \le 1$), x = axial distance ($0 \le x \le 1$), and p is a nondimensional parameter that describes the heat transfer and geometry,

$$p = \frac{hL}{k}\sqrt{1 + \frac{4}{2m^2}}$$

where h = a heat transfer coefficient, k = thermal conductivity, L = the length or height of the cone, and m = the slope of the cone wall. The equation has the boundary conditions

$$u(x = 0) = 0 \qquad u(x = 1) = 1$$

Solve this equation for the temperature distribution using finite-difference methods. Use second-order accurate finite-difference

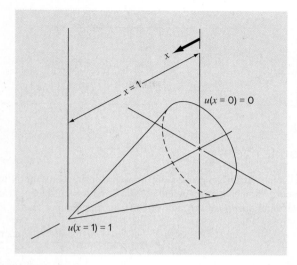

FIGURE P28.54

analogues for the derivatives. Write a computer program to obtain the solution and plot temperature versus axial distance for various values of p = 10, 20, 50, and 100.

28.55 The dynamics of a forced spring-mass-damper system can be represented by the following second-order ODE:

$$m\frac{d^2x}{dt^2} + c\frac{dx}{dt} + k_1 x + k_3 x^3 = P\cos(\omega t)$$

where m = 1 kg, c = 0.4 N · s/m, P = 0.5 N, and ω = 0.5/s. Use a numerical method to solve for displacement (x) and velocity ($v = dx/dt$) as a function of time with the initial conditions $x = v = 0$. Express your results graphically as time-series plots (x and v versus t) and a phase-plane plot (v versus x). Perform simulations for both **(a)** linear ($k_1 = 1$; $k_3 = 0$) and **(b)** nonlinear ($k_1 = 1$; $k_3 = 0.5$) springs.

28.56 The differential equation for the velocity of a bungee jumper is different depending on whether the jumper has fallen to a distance where the cord is fully extended and begins to stretch. Thus, if the distance fallen is less than the cord length, the jumper is only subject to gravitational and drag forces. Once the cord begins to stretch, the spring and dampening forces of the cord must also be included. These two conditions can be expressed by the following equations:

$$\frac{dv}{dt} = g - \text{sign}(v)\frac{c_d}{m}v^2 \qquad\qquad x \le L$$

$$\frac{dv}{dt} = g - \text{sign}(v)\frac{c_d}{m}v^2 - \frac{k}{m}(x - L) - \frac{\gamma}{m}v \quad x > L$$

where v = velocity (m/s), t = time (s), g = gravitational constant (= 9.81 m/s^2), sign(v) = function that returns −1, 0, and 1 for negative, zero, and positive x, respectively, c_d = second-order drag

FIGURE P28.57

coefficient (kg/m), m = mass (kg), k = cord spring constant (N/m), γ = cord dampening coefficient (N s/m), and L = cord length (m). Determine the position and velocity of the jumper given the

following parameters: $L = 30$ m, $m = 68.1$ kg, $c_d = 0.25$ kg/m, $k = 40$ N/m, and $\gamma = 8$ kg/s. Perform the computation from $t = 0$ to 50 s and assume that the initial conditions are $x(0) = v(0) = 0$.

28.57 Two masses are attached to a wall by linear springs (Fig. P28.57). Force balances based on Newton's second law can be written as

$$\frac{d^2x_1}{dt^2} = -\frac{k_1}{m_1}(x_1 - L_1) + \frac{k_2}{m_1}(x_2 - x_1 - w_1 - L_2)$$

$$\frac{d^2x_2}{dt^2} = -\frac{k_2}{m_2}(x_2 - x_1 - w_1 - L_2)$$

where k = the spring constants, m = mass, L = the length of the unstretched spring, and w = the width of the mass. Compute the positions of the masses as a function of time using the following parameter values: $k_1 = k_2 = 5$, $m_1 = m_2 = 2$, $w_1 = w_2 = 5$, and $L_1 = L_2 = 2$. Set the initial conditions as $x_1 = L_1$ and $x_2 = L_1 + w_1 + L_2 + 6$. Perform the simulation from $t = 0$ to 20. Construct time-series plots of both the displacements and the velocities. In addition, produce a phase-plane plot of x_1 versus x_2.

28.58 Suppose that a parachutist with linear drag ($m = 70$ kg, $c_d = 12.5$ kg/s) jumps from an airplane flying at an altitude of 200 m with a horizontal velocity of 180 m/s relative to the ground.

(a) Write a system of four differential equations for x, y, dx/dt, and dy/dt.

(b) If the initial horizontal position is defined as $x = 0$, use Excel, MATLAB, Mathcad, or some other computing tool to implement Euler's method with $\Delta t = 1$ s to compute the jumper's position over the first 10 seconds.

(c) Develop plots of y versus t and y versus x. Use the plots to graphically estimate when and where the jumper would hit the ground if the chute failed to open.

(d) At what angle would the parachutist be traveling in the last whole second before impact?

28.59 As depicted in Fig. P28.59, a double pendulum consists of a pendulum attached to another pendulum. We indicate the upper and lower pendulums by subscripts 1 and 2, respectively, and we place the origin at the pivot point of the upper pendulum with y increasing upwards. We further assume that the system oscillates in a vertical plane subject to gravity, that the pendulum rods are massless and rigid, and that the pendulum masses are considered to be point masses. Under these assumptions, force balances can be used to derive the following equations of motion:

$$\frac{d^2\theta_1}{dt^2} = \frac{-g(2m_1 + m_2)\sin\theta_1 - m_2 g \sin(\theta_1 - 2\theta_2) - 2\sin(\theta_1 - \theta_2)m_2((d\theta_2/dt)^2 L_2 + (d\theta_1/dt)^2 L_1 \cos(\theta_1 - \theta_2))}{L_1[2m_1 + m_2 - m_2 \cos(2\theta_1 - 2\theta_2)]}$$

$$\frac{d^2\theta_2}{dt^2} = \frac{2\sin(\theta_1 - \theta_2)((d\theta_1/dt)^2 L_1(m_1 + m_2) + g(m_1 + m_2)\cos(\theta_1) + (d\theta_2/dt)^2 L_2 m_2 \cos(\theta_1 - \theta_2))}{L_2[2m_1 + m_2 - m_2 \cos(2\theta_1 - 2\theta_2)]}$$

where the subscripts 1 and 2 designate the top and bottom pendulum, respectively, θ = angle (radians) with 0 = vertical downwards and counterclockwise positive, t = time (s), g = gravitational acceleration (= 9.81 m/s²), m = mass (kg), and L = length (m). Note that the x and y coordinates of the masses are functions of the angles, as in

$$x_1 = L_1 \sin\theta_1 \qquad y_1 = -L_1 \cos\theta_1$$
$$x_2 = x_1 + L_2 \sin\theta_2 \qquad y_2 = y_1 - L_2 \cos\theta_2$$

Numerically solve for the angles and angular velocities of the masses as a function of time from $t = 0$ to 25 s. Create a time-series plot of your results. Test your code for the following:

Case 1 (small displacement): $L_1 = L_2 = 1$ m, $m_1 = m_2 = 0.25$ kg, with initial conditions $\theta_1 = 0.5$ m and $\theta_2 = d\theta_1/dt = d\theta_1/dt = 0$.

Case 2 (large displacement): $L_1 = L_2 = 1$ m, $m_1 = 0.5$ kg, $m_2 = 0.25$ kg, with initial conditions $\theta_1 = 1$ m and $\theta_2 = d\theta_1/dt = d\theta_1/dt = 0$.

FIGURE P28.59
A double pendulum.

EPILOGUE: PART SEVEN

PT7.4 TRADE-OFFS

Table PT7.3 contains trade-offs associated with numerical methods for the solution of initial-value ordinary differential equations. The factors in this table must be evaluated by the engineer when selecting a method for each particular applied problem.

Simple self-starting techniques such as Euler's method can be used if the problem requirements involve a short range of integration. In this case, adequate accuracy may be obtained using small step sizes to avoid large truncation errors, and the round-off errors may be acceptable. Euler's method may also be appropriate for cases where the mathematical model has an inherently high level of uncertainty or has coefficients or forcing functions with significant errors, as might arise during a measurement process. In this case, the accuracy of the model itself simply does not justify the effort involved to employ a more complicated numerical method. Finally, the simpler techniques may be best when the problem or simulation need only be performed a few times. In these applications, it is probably best to use a simple method that is easy to program and

TABLE PT7.3 Comparison of the characteristics of alternative methods for the numerical solution of ODEs. The comparisons are based on general experience and do not account for the behavior of special functions.

Method	Starting Values	Iterations Required	Global Error	Ease of Changing Step Size	Programming Effort	Comments
One-step						
Euler's	1	No	$O(h)$	Easy	Easy	Good for quick estimates
Heun's	1	Yes	$O(h^2)$	Easy	Moderate	—
Midpoint	1	No	$O(h^2)$	Easy	Moderate	—
Second-order Ralston	1	No	$O(h^2)$	Easy	Moderate	The second-order RK method that minimizes truncation error
Fourth-order RK	1	No	$O(h^4)$	Easy	Moderate	Widely used
Adaptive fourth-order RK or RK-Fehlberg	1	No	$O(h^5)^*$	Easy	Moderate to difficult	Error estimate allows step-size adjustment
Multistep						
Non-self-starting Heun	2	Yes	$O(h^3)^*$	Difficult	Moderate to difficult[†]	Simple multistep method
Milne's	4	Yes	$O(h^5)^*$	Difficult	Moderate to difficult[†]	Sometimes unstable
Fourth-order Adams	4	Yes	$O(h^5)^*$	Difficult	Moderate to difficult[†]	

*Provided the error estimate is used to modify the solution.
[†]With variable step size.

understand, despite the fact that the method may be computationally inefficient and relatively time-consuming to run on the computer.

If the range of integration of the problem is long enough to involve a large number of steps, then it may be necessary and appropriate to use a more accurate technique than Euler's method. The fourth-order RK method is popular and reliable for many engineering problems. In these cases, it may also be advisable to estimate the truncation error for each step as a guide to selecting the best step size. This can be accomplished with the adaptive RK or fourth-order Adams approaches. If the truncation errors are extremely small, it may be wise to increase the step size to save computer time. On the other hand, if the truncation error is large, the step size should be decreased to avoid accumulation of error. Milne's method should be avoided if significant stability problems are expected. The Runge-Kutta method is simple to program and convenient to use but may be less efficient than the multistep methods. However, the Runge-Kutta method is usually employed in any event to obtain starting values for the multistep methods.

A large number of engineering problems may fall into an intermediate range of integration interval and accuracy requirement. In these cases, the second-order RK and the non-self-starting Heun methods are simple to use and are relatively efficient and accurate.

Stiff systems involve equations with slowly and rapidly varying components. Special techniques are usually required for the adequate solution of stiff equations. For example, implicit approaches are often used. You can consult Enright et al. (1975), Gear (1971), and Shampine and Gear (1979) for additional information regarding these techniques.

A variety of techniques are available for solving eigenvalue problems. For small systems or where only a few of the smallest or largest eigenvalues are required, simple approaches such as the polynomial and the power methods are available. For symmetric systems, Jacobi's, Given's, or Householder's method can be employed. Finally, the QR method represents a general approach for finding all the eigenvalues of symmetric and nonsymmetric matrices.

PT7.5 IMPORTANT RELATIONSHIPS AND FORMULAS

Table PT7.4 summarizes important information that was presented in Part Seven. This table can be consulted to quickly access important relationships and formulas.

PT7.6 ADVANCED METHODS AND ADDITIONAL REFERENCES

Although we have reviewed a number of techniques for solving ordinary differential equations, there is additional information that is important in engineering practice. The question of *stability* was introduced in Sec. 26.2.4. This topic has general relevance to all methods for solving ODEs. Further discussion of the topic can be pursued in Carnahan, Luther, and Wilkes (1969), Gear (1971), and Hildebrand (1974).

In Chap. 27, we introduced methods for solving *boundary-value* problems. Isaacson and Keller (1966), Keller (1968), Na (1979), and Scott and Watts (1976) can be consulted for additional information on standard boundary-value problems. Additional material on eigenvalues can be found in Ralston and Rabinowitz (1978), Wilkinson (1965), Fadeev and Fadeeva (1963), and Householder (1953, 1964).

In summary, the foregoing is intended to provide you with avenues for deeper exploration of the subject. Additionally, all the above references provide descriptions of the basic techniques covered in Part Seven. We urge you to consult these alternative sources to broaden your understanding of numerical methods for the solution of differential equations.

TABLE PT7.4 Summary of important information presented in Part Seven.

Method	Formulation	Graphic Interpretation	Errors
Euler (first-order RK)	$y_{i+1} = y_i + hk_1$ $k_1 = f(x_i, y_i)$		Local error $\cong O(h^2)$ Global error $\cong O(h)$
Ralston's second-order RK	$y_{i+1} = y_i + h(\frac{1}{3}k_1 + \frac{2}{3}k_2)$ $k_1 = f(x_i, y_i)$ $k_2 = f(x_i + \frac{3}{4}h, y_i + \frac{3}{4}hk_1)$		Local error $\cong O(h^3)$ Global error $\cong O(h^2)$
Classic fourth-order RK	$y_{i+1} = y_i + h(\frac{1}{6}k_1 + \frac{1}{3}k_2 + \frac{1}{3}k_3 + \frac{1}{6}k_4)$ $k_1 = f(x_i, y_i)$ $k_2 = f(x_i + \frac{1}{2}h, y_i + \frac{1}{2}hk_1)$ $k_3 = f(x_i + \frac{1}{2}h, y_i + \frac{1}{2}hk_2)$ $k_4 = f(x_i + h, y_i + hk_3)$		Local error $\cong O(h^5)$ Global error $\cong O(h^4)$
Non-self-starting Heun	Predictor: (midpoint method) $y_{i+1}^0 = y_{i-1}^m + 2hf(x_i, y_i^m)$		Predictor modifier: $E_p \cong \frac{4}{5}(y_{i,u}^m + y_{i,u}^0)$
	Corrector: (trapezoidal rule) $y_{i+1}^j = y_i^m + h\dfrac{f(x_i, y_i^m) + f(x_{i+1}, y_{i+1}^{j-1})}{2}$		Corrector modifier: $E_c \cong -\dfrac{y_{i+1,u}^m - y_{i+1,u}^0}{5}$
Fourth-order Adams	Predictor: (fourth Adams-Bashforth) $y_{i+1}^0 = y_i^m + h(\frac{55}{24}f_i^m - \frac{59}{24}f_{i-1}^m + \frac{37}{24}f_{i-2}^m - \frac{9}{24}f_{i-3}^m)$		Predictor modifier: $E_p \cong \frac{251}{270}(y_{i,u}^m - y_{i,u}^0)$
	Corrector: (fourth Adams-Moulton) $y_{i+1}^j = y_i^m + h(\frac{9}{24}f_{i+1}^{j-1} + \frac{19}{24}f_i^m - \frac{5}{24}f_{i-1}^m + \frac{1}{24}f_{i-2}^m)$		Corrector modifier: $E_c \cong -\frac{19}{270}(y_{i+1,u}^m - y_{i+1,u}^0)$

PART EIGHT

PARTIAL DIFFERENTIAL EQUATIONS

PT8.1 MOTIVATION

Given a function u that depends on both x and y, the partial derivative of u with respect to x at an arbitrary point (x, y) is defined as

$$\frac{\partial u}{\partial x} = \lim_{\Delta x \to 0} \frac{u(x + \Delta x, y) - u(x, y)}{\Delta x} \tag{PT8.1}$$

Similarly, the partial derivative with respect to y is defined as

$$\frac{\partial u}{\partial y} = \lim_{\Delta y \to 0} \frac{u(x, y + \Delta y) - u(x, y)}{\Delta y} \tag{PT8.2}$$

An equation involving partial derivatives of an unknown function of two or more independent variables is called a *partial differential equation*, or *PDE*. For example,

$$\frac{\partial^2 u}{\partial x^2} + 2xy \frac{\partial^2 u}{\partial y^2} + u = 1 \tag{PT8.3}$$

$$\frac{\partial^3 u}{\partial x^2 \partial y} + x \frac{\partial^2 u}{\partial y^2} + 8u = 5y \tag{PT8.4}$$

$$\left(\frac{\partial^2 u}{\partial x^2}\right)^3 + 6 \frac{\partial^3 u}{\partial x \partial y^2} = x \tag{PT8.5}$$

$$\frac{\partial^2 u}{\partial x^2} + xu \frac{\partial u}{\partial y} = x \tag{PT8.6}$$

The *order* of a PDE is that of the highest-order partial derivative appearing in the equation. For example, Eqs. (PT8.3) and (PT8.4) are second- and third-order, respectively.

A partial differential equation is said to be *linear* if it is linear in the unknown function and all its derivatives, with coefficients depending only on the independent variables. For example, Eqs. (PT8.3) and (PT8.4) are linear, whereas Eqs. (PT8.5) and (PT8.6) are not.

TABLE PT8.1 Categories into which linear, second-order partial differential equations in two variables can be classified.

$B^2 - 4AC$	Category	Example
< 0	Elliptic	Laplace equation (steady state with two spatial dimensions), $$\frac{\partial^2 T}{\partial x^2} + \frac{\partial^2 T}{\partial y^2} = 0$$
= 0	Parabolic	Heat-conduction equation (time variable with one spatial dimension), $$\frac{\partial T}{\partial t} = k' \frac{\partial^2 T}{\partial x^2}$$
> 0	Hyperbolic	Wave equation (time variable with one spatial dimension), $$\frac{\partial^2 y}{\partial x^2} = \frac{1}{c^2} \frac{\partial^2 y}{\partial t^2}$$

Because of their widespread application in engineering, our treatment of PDEs will focus on linear, second-order equations. For two independent variables, such equations can be expressed in the following general form:

$$A\frac{\partial^2 u}{\partial x^2} + B\frac{\partial^2 u}{\partial x \partial y} + C\frac{\partial^2 u}{\partial y^2} + D = 0 \tag{PT8.7}$$

where A, B, and C are functions of x and y and D is a function of x, y, u, $\partial u/\partial x$, and $\partial u/\partial y$. Depending on the values of the coefficients of the second-derivative terms—A, B, C—Eq. (PT8.7) can be classified into one of three categories (Table PT8.1). This classification, which is based on the method of characteristics, for example, see Vichnevetsky (1981) or Lapidus and Pinder (1981), is useful because each category relates to specific and distinct engineering problem contexts that demand special solution techniques. It should be noted that for cases where A, B, and C depend on x and y, the equation may actually fall into a different category, depending on the location in the domain for which the equation holds. For simplicity, we will limit the present discussion to PDEs that remain exclusively in one of the categories.

PT8.1.1 PDEs and Engineering Practice

Each of the categories of partial differential equations in Table PT8.1 conforms to specific kinds of engineering problems. The initial sections of the following chapters will be devoted to deriving each type of equation for a particular engineering problem context. For the time being, we will discuss the general properties and applications of each type and show how they can be employed in different physical contexts.

Elliptic equations are typically used to characterize *steady-state* systems. As in the *Laplace equation* in Table PT8.1, this is indicated by the absence of a time derivative. Thus, these equations are typically employed to determine the steady-state distribution of an unknown in two spatial dimensions.

FIGURE PT8.1
Three steady-state distribution problems that can be characterized by elliptic PDEs. (*a*) Temperature distribution on a heated plate, (*b*) seepage of water under a dam, and (*c*) the electric field near the point of a conductor.

A simple example is the heated plate in Fig. PT8.1*a*. For this case, the boundaries of the plate are held at different temperatures. Because heat flows from regions of high to low temperature, the boundary conditions set up a potential that leads to heat flow from the hot to the cool boundaries. If sufficient time elapses, such a system will eventually reach the stable, or steady-state, distribution of temperature depicted in Fig. PT8.1*a*. The Laplace equation, along with appropriate boundary conditions, provides a means to determine this distribution. By analogy, the same approach can be employed to tackle other problems involving potentials, such as seepage of water under a dam (Fig. PT8.1*b*) or the distribution of an electric field (Fig. PT8.1*c*).

In contrast to the elliptic category, *parabolic equations* determine how an unknown varies in both space and time. This is manifested by the presence of both spatial and temporal derivatives in the *heat-conduction equation* from Table PT8.1. Such cases are referred to as *propagation problems* because the solution "propagates," or changes, in time.

A simple example is a long, thin rod that is insulated everywhere except at its end (Fig. PT8.2*a*). The insulation is employed to avoid complications due to heat loss along the rod's length. As was the case for the heated plate in Fig. PT8.1*a*, the ends of the rod are set at fixed temperatures. However, in contrast to Fig. PT8.1*a*, the rod's thinness allows us to assume that heat is distributed evenly over its cross section—that is, laterally. Consequently, lateral heat flow is not an issue, and the problem reduces to studying the conduction of heat along the rod's longitudinal axis. Rather than focusing on the steady-state distribution in two spatial dimensions, the problem shifts to determining how the one-dimensional spatial distribution changes as a function of time (Fig. PT8.2*b*). Thus, the solution consists of a series of spatial distributions corresponding to the state

FIGURE PT8.2
(*a*) A long, thin rod that is insulated everywhere but at its end. The dynamics of the one-dimensional distribution of temperature along the rod's length can be described by a parabolic PDE. (*b*) The solution, consisting of distributions corresponding to the state of the rod at various times.

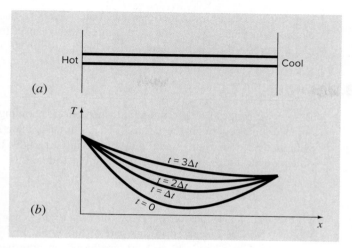

FIGURE PT8.3
A taut string vibrating at a low amplitude is a simple physical system that can be characterized by a hyperbolic PDE.

of the rod at various times. Using an analogy from photography, the elliptic case yields a portrait of a system's stable state, whereas the parabolic case provides a motion picture of how it changes from one state to another. As with the other types of PDEs described herein, parabolic equations can be used to characterize a wide variety of other engineering problem contexts by analogy.

The final class of PDEs, the *hyperbolic* category, also deals with *propagation problems*. However, an important distinction manifested by the wave equation in Table PT8.1 is that the unknown is characterized by a second derivative with respect to time. As a consequence, the solution oscillates.

The vibrating string in Fig. PT8.3 is a simple physical model that can be described with the wave equation. The solution consists of a number of characteristic states with which the string oscillates. A variety of engineering systems such as vibrations of rods and beams, motion of fluid waves, and transmission of sound and electrical signals can be characterized by this model.

PT8.1.2 Precomputer Methods for Solving PDEs

Prior to the advent of digital computers, engineers relied on analytical or exact solutions of partial differential equations. Aside from the simplest cases, these solutions often required a great deal of effort and mathematical sophistication. In addition, many physical systems could not be solved directly but had to be simplified using linearizations, simple geometric representations, and other idealizations. Although these solutions are elegant and yield insight, they are limited with respect to how faithfully they represent real systems—especially those that are highly nonlinear and irregularly shaped.

PT8.2 ORIENTATION

Before we proceed to the numerical methods for solving partial differential equations, some orientation might be helpful. The following material is intended to provide you with an overview of the material discussed in Part Eight. In addition, we have formulated objectives to focus your studies in the subject area.

PT8.2.1 Scope and Preview

Figure PT8.4 provides an overview of Part Eight. Two broad categories of numerical methods will be discussed in this part of this book. Finite-difference approaches, which are covered in Chaps. 29 and 30, are based on approximating the solution at a finite number of points. In contrast, the finite-element method, covered in Chap. 31, approximates the solution in pieces, or "elements." Various parameters are adjusted until these approximations conform to the underlying differential equation in an optimal sense.

Chapter 29 is devoted to *finite-difference* solutions of *elliptic equations.* Before launching into the methods, we derive the Laplace equation for the physical problem context of the temperature distribution for a heated plate. Then, a standard solution approach, the *Liebmann method,* is described. We will illustrate how this approach is used to compute the distribution of the primary scalar variable, temperature, as well as a secondary vector variable, heat flux. The third section of the chapter deals with *boundary conditions*. This material includes procedures to handle different types of conditions as well as irregular boundaries.

In *Chap. 30,* we turn to *finite-difference* solutions of *parabolic equations.* As with the discussion of elliptic equations, we first provide an introduction to a physical problem context, the heat-conduction equation for a one-dimensional rod. Then we introduce both explicit and implicit algorithms for solving this equation. This is followed by an efficient and reliable implicit method—the *Crank-Nicolson technique.* Finally, we describe a particularly effective approach for solving two-dimensional parabolic equations—the *alternating-direction implicit,* or *ADI, method.*

Note that, because they are somewhat beyond the scope of this book, we have chosen to omit hyperbolic equations. The epilogue of this part of the book contains references related to this type of PDE.

In *Chap. 31,* we turn to the other major approach for solving PDEs—the *finite-element method.* Because it is so fundamentally different from the finite-difference approach, we devote the initial section of the chapter to a general overview. Then we show how the finite-element method is used to compute the steady-state temperature distribution of a heated rod. Finally, we provide an introduction to some of the issues involved in extending such an analysis to two-dimensional problem contexts.

Chapter 32 is devoted to applications from all fields of engineering. Finally, a short review section is included at the end of Part Eight. This epilogue summarizes important information related to PDEs. This material includes a discussion of trade-offs that are

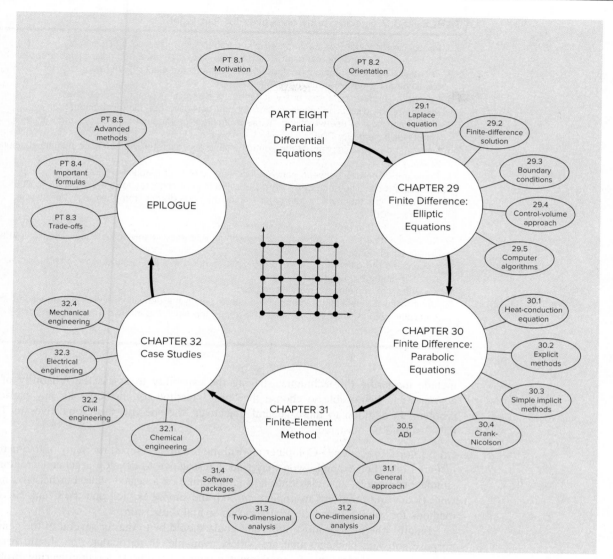

FIGURE PT8.4
Schematic representation of the organization of material in Part Eight: Partial Differential Equations.

relevant to their implementation in engineering practice. The epilogue also includes references for advanced topics.

PT8.2.2 Goals and Objectives

Study Objectives.   After completing Part Eight, you should have greatly enhanced your capability to confront and solve partial differential equations. General study goals should

TABLE PT8.2 Specific study objectives for Part Eight.

1. Recognize the difference between elliptic, parabolic, and hyperbolic PDEs.
2. Understand the fundamental difference between finite-difference and finite-element approaches.
3. Recognize that the Liebmann method is equivalent to the Gauss-Seidel approach for solving simultaneous linear algebraic equations.
4. Know how to determine secondary variables for two-dimensional field problems.
5. Recognize the distinction between Dirichlet and derivative boundary conditions.
6. Understand how to use weighting factors to incorporate irregular boundaries into a finite-difference scheme for PDEs.
7. Understand how to implement the control-volume approach for implementing numerical solutions of PDEs.
8. Know the difference between convergence and stability of parabolic PDEs.
9. Understand the difference between explicit and implicit schemes for solving parabolic PDEs.
10. Recognize how the stability criteria for explicit methods detract from their utility for solving parabolic PDEs.
11. Know how to interpret computational molecules.
12. Recognize how the ADI approach achieves high efficiency in solving parabolic equations in two spatial dimensions.
13. Understand the difference between the direct method and the method of weighted residuals for deriving element equations.
14. Know how to implement Galerkin's method.
15. Understand the benefits of integration by parts during the derivation of element equations; in particular, recognize the implications of lowering the highest derivative from a second to a first derivative.

include mastering the techniques, having the capability to assess the reliability of the answers, and being able to choose the "best" method (or methods) for any particular problem. In addition to these general objectives, the specific study objectives in Table PT8.2 should be mastered.

Computer Objectives.   Computer algorithms can be developed for many of the methods in Part Eight. For example, you may find it instructive to develop a general program to simulate the steady-state distribution of temperature on a heated plate. Further, you might want to develop programs to implement both the simple explicit and the Crank-Nicolson methods for solving parabolic PDEs in one spatial dimension.

Finally, one of your most important goals should be to master several of the general-purpose software packages that are widely available. In particular, you should become adept at using these tools to implement numerical methods for engineering problem solving.

CHAPTER

29

Finite Difference: Elliptic Equations

Elliptic equations in engineering are typically used to characterize steady-state, boundary-value problems. Before demonstrating how they can be solved, we will illustrate how a simple case—the Laplace equation—is derived from a physical problem context.

29.1 THE LAPLACE EQUATION

As mentioned in the introduction to this part of the book, the Laplace equation can be used to model a variety of problems involving the potential of an unknown variable. Because of its simplicity and general relevance to most areas of engineering, we will use a heated plate as our fundamental context for deriving and solving this elliptic PDE. Homework problems and engineering applications (Chap. 32) will be employed to illustrate the applicability of the model to other engineering problem contexts.

Figure 29.1 shows an element on the face of a thin rectangular plate of thickness Δz. The plate is insulated everywhere but at its edges, where the temperature can be set at a prescribed level. The insulation and the thinness of the plate mean that heat transfer is limited to the x and y dimensions. At steady state, the flow of heat into the element over a unit time period Δt must equal the flow out, as in

$$q(x)\Delta y\, \Delta z\, \Delta t + q(y)\, \Delta x\, \Delta z\, \Delta t = q(x + \Delta x)\Delta y\, \Delta z\, \Delta t \\ + q(y + \Delta y)\Delta x\, \Delta z\, \Delta t \tag{29.1}$$

where $q(x)$ and $q(y)$ = the heat fluxes at x and y, respectively [cal/(cm^2 s)]. Dividing by Δz and Δt and collecting terms yields

$$[q(x) - q(x + \Delta x)]\Delta y + [q(y) - q(y + \Delta y)]\Delta x = 0$$

Multiplying the first term by $\Delta x/\Delta x$ and the second by $\Delta y/\Delta y$ gives

$$\frac{q(x) - q(x + \Delta x)}{\Delta x}\Delta x\, \Delta y + \frac{q(y) - q(y + \Delta y)}{\Delta y}\Delta y\, \Delta x = 0 \tag{29.2}$$

Dividing by $\Delta x\, \Delta y$ and taking the limit results in

$$-\frac{\partial q}{\partial x} - \frac{\partial q}{\partial y} = 0 \tag{29.3}$$

where the partial derivatives result from the definitions in Eqs. (PT8.1) and (PT8.2).

FIGURE 29.1
A thin plate of thickness Δz. An element is shown about which a heat balance is taken.

Equation (29.3) is a partial differential equation that is an expression of the conservation of energy for the plate. However, unless heat fluxes are specified at the plate's edges, it cannot be solved. Because temperature boundary conditions are given, Eq. (29.3) must be reformulated in terms of temperature. The link between flux and temperature is provided by *Fourier's law of heat conduction,* which can be represented as

$$q_i = -k\rho C \frac{\partial T}{\partial i} \tag{29.4}$$

where q_i = heat flux in the direction of the i dimension [cal/(cm^2 s)], k = coefficient of *thermal diffusivity* (cm^2/s), ρ = density of the material (g/cm^3), C = heat capacity of the material [cal/(g °C)], and T = temperature (°C), which is defined as

$$T = \frac{H}{\rho C V}$$

where H = heat (cal) and V = volume (cm^3). Sometimes the term in front of the differential in Eq. (29.4) is treated as a single term,

$$k' = k\rho C \tag{29.5}$$

where k' is referred to as the *coefficient of thermal conductivity* [cal/(s cm °C)]. In either case, both k and k' are parameters that reflect how well the material conducts heat.

Fourier's law is sometimes referred to as a *constitutive equation*. It is given this label because it provides a mechanism that defines the system's internal interactions. Inspection of Eq. (29.4) indicates that Fourier's law specifies that heat flux perpendicular to

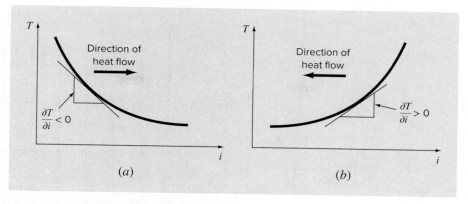

FIGURE 29.2

Graphical depiction of a temperature gradient. Because heat moves "downhill" from high to low temperature, the flow in (a) is from left to right in the positive i direction. However, due to the orientation of Cartesian coordinates, the slope is negative for this case. Thus, a negative gradient leads to a positive flow. This is the origin of the minus sign in Fourier's law of heat conduction. The reverse case is depicted in (b), where the positive gradient leads to a negative heat flow from right to left.

the i axis is proportional to the gradient or slope of temperature in the i direction. The negative sign ensures that a positive flux in the direction of i results from a negative slope from high to low temperature (Fig. 29.2). Substituting Eq. (29.4) into Eq. (29.3) results in

$$\frac{\partial^2 T}{\partial x^2} + \frac{\partial^2 T}{\partial y^2} = 0 \tag{29.6}$$

which is the *Laplace equation*. Note that for the case where there are sources or sinks of heat within the two-dimensional domain, the equation can be represented as

$$\frac{\partial^2 T}{\partial x^2} + \frac{\partial^2 T}{\partial y^2} = f(x, y) \tag{29.7}$$

where $f(x, y)$ is a function describing the sources or sinks of heat. Equation (29.7) is referred to as the *Poisson equation*.

29.2 SOLUTION TECHNIQUE

The numerical solution of elliptic PDEs such as the Laplace equation proceeds in the reverse manner of the derivation of Eq. (29.6) from the preceding section. Recall that the derivation of Eq. (29.6) employed a balance around a discrete element to yield an algebraic difference equation characterizing heat flux for a plate. Taking the limit turned this difference equation into a differential equation [Eq. (29.3)].

For the numerical solution, finite-difference representations based on treating the plate as a grid of discrete points (Fig. 29.3) are substituted for the partial derivatives in Eq. (29.6). As described next, the PDE is transformed into an algebraic difference equation.

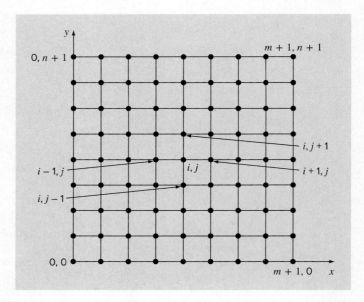

FIGURE 29.3
A grid used for the finite-difference solution of elliptic PDEs in two independent variables, such as the Laplace equation.

29.2.1 The Laplacian Difference Equation

Central differences based on the grid scheme from Fig. 29.3 are (recall Fig. 23.3)

$$\frac{\partial^2 T}{\partial x^2} = \frac{T_{i+1,j} - 2T_{i,j} + T_{i-1,j}}{\Delta x^2}$$

and

$$\frac{\partial^2 T}{\partial y^2} = \frac{T_{i,j+1} - 2T_{i,j} + T_{i,j-1}}{\Delta y^2}$$

which have errors of $O[\Delta(x)^2]$ and $O[\Delta(y)^2]$, respectively. Substituting these expressions into Eq. (29.6) gives

$$\frac{T_{i+1,j} - 2T_{i,j} + T_{i-1,j}}{\Delta x^2} + \frac{T_{i,j+1} - 2T_{i,j} + T_{i,j-1}}{\Delta y^2} = 0$$

For the square grid in Fig. 29.3, $\Delta x = \Delta y$, and by collection of terms, the equation becomes

$$T_{i+1,j} + T_{i-1,j} + T_{i,j+1} + T_{i,j-1} - 4T_{i,j} = 0 \tag{29.8}$$

This relationship, which holds for all interior points on the plate, is referred to as the *Laplacian difference equation*.

In addition, boundary conditions along the edges of the plate must be specified to obtain a unique solution. The simplest case is where the temperature at the boundary is set at a fixed value. This is called a *Dirichlet boundary condition*. Such is the case for Fig. 29.4, where

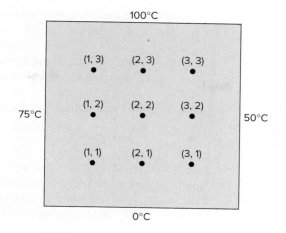

FIGURE 29.4
A heated plate where boundary temperatures are held at constant levels. This case is called a Dirichlet boundary condition.

the edges are held at constant temperatures. For the case illustrated in Fig. 29.4, a balance for node (1, 1) is, according to Eq. (29.8),

$$T_{21} + T_{01} + T_{12} + T_{10} - 4T_{11} = 0 \tag{29.9}$$

However, $T_{01} = 75$ and $T_{10} = 0$, and therefore, Eq. (29.9) can be expressed as

$$-4T_{11} + T_{12} + T_{21} = -75$$

Similar equations can be developed for the other interior points. The result is the following set of nine simultaneous equations with nine unknowns:

$$
\begin{array}{rrrrrrrrrr}
4T_{11} & -T_{21} & & -T_{12} & & & & & & = 75 \\
-T_{11} & +4T_{21} & -T_{31} & & -T_{22} & & & & & = 0 \\
& -T_{21} & +4T_{31} & & & -T_{32} & & & & = 50 \\
-T_{11} & & & +4T_{12} & -T_{22} & & -T_{13} & & & = 75 \\
& -T_{21} & & -T_{12} & +4T_{22} & -T_{32} & & -T_{23} & & = 0 \\
& & -T_{31} & & -T_{22} & +4T_{32} & & & -T_{33} & = 50 \\
& & & -T_{12} & & & +4T_{13} & -T_{23} & & = 175 \\
& & & & -T_{22} & & -T_{13} & +4T_{23} & -T_{33} & = 100 \\
& & & & & -T_{32} & & -T_{23} & +4T_{33} & = 150 \\
\end{array}
\tag{29.10}
$$

29.2.2 The Liebmann Method

Most numerical solutions of the Laplace equation involve systems that are much larger than Eq. (29.10). For example, a 10-by-10 grid involves 100 linear algebraic equations. Solution techniques for these types of equations were discussed in Part Three.

Notice that there is a maximum of five unknown terms per line in Eq. (29.10). For larger-sized grids, this means that a significant number of the terms will be zero. When applied to such sparse systems, full-matrix elimination methods waste great amounts of computer memory storing these zeros. For this reason, approximate methods provide a viable approach for obtaining solutions for elliptical equations. The most commonly employed approach is *Gauss-Seidel*, which when applied to PDEs is also referred to as *Liebmann's method*. In this technique, Eq. (29.8) is expressed as

$$T_{i,j} = \frac{T_{i+1,j} + T_{i-1,j} + T_{i,j+1} + T_{i,j-1}}{4} \tag{29.11}$$

and solved iteratively for $j = 1$ to n and $i = 1$ to m. Because Eq. (29.8) is diagonally dominant, this procedure will eventually converge on a stable solution (recall Sec. 11.2.1). Overrelaxation is sometimes employed to accelerate the rate of convergence by applying the following formula after each iteration:

$$T_{i,j}^{new} = \lambda T_{i,j}^{new} + (1 - \lambda)T_{i,j}^{old} \tag{29.12}$$

where $T_{i,j}^{new}$ and $T_{i,j}^{old}$ are the values of $T_{i,j}$ from the present and the previous iteration, respectively, and λ is a weighting factor that is set between 1 and 2.

As with the conventional Gauss-Seidel method, the iterations are repeated until the absolute values of all the percent relative errors $(\varepsilon_a)_{i,j}$ fall below a prespecified stopping criterion, ε_s. These percent relative errors are estimated by

$$|(\varepsilon_a)_{i,j}| = \left| \frac{T_{i,j}^{new} - T_{i,j}^{old}}{T_{i,j}^{new}} \right| 100\% \tag{29.13}$$

EXAMPLE 29.1　　**Temperature of a Heated Plate with Fixed Boundary Conditions**

Problem Statement.　Use Liebmann's method (Gauss-Seidel) to solve for the temperature of the heated plate in Fig. 29.4. Employ overrelaxation with a value of 1.5 for the weighting factor and iterate to $\varepsilon_s = 1\%$.

Solution.　Equation (29.11) at $i = 1, j = 1$ is

$$T_{11} = \frac{0 + 75 + 0 + 0}{4} = 18.75$$

and applying overrelaxation yields

$$T_{11} = 1.5(18.75) + (1 - 1.5)0 = 28.125$$

For $i = 2, j = 1$,

$$T_{21} = \frac{0 + 28.125 + 0 + 0}{4} = 7.03125$$

$$T_{21} = 1.5(7.03125) + (1 - 1.5)0 = 10.54688$$

For $i = 3, j = 1$,

$$T_{31} = \frac{50 + 10.54688 + 0 + 0}{4} = 15.13672$$

$$T_{31} = 1.5(15.13672) + (1 - 1.5)0 = 22.70508$$

The computation is repeated for the other rows to give

$T_{12} = 38.67188$	$T_{22} = 18.45703$	$T_{32} = 34.18579$
$T_{13} = 80.12696$	$T_{23} = 74.46900$	$T_{33} = 96.99554$

Because all the $T_{i,j}$'s are initially zero, all ε_a's for the first iteration will be 100%. For the second iteration the results are

$T_{11} = 32.51953$	$T_{21} = 22.35718$	$T_{31} = 28.60108$
$T_{12} = 57.95288$	$T_{22} = 61.63333$	$T_{32} = 71.86833$
$T_{13} = 75.21973$	$T_{23} = 87.95872$	$T_{32} = 67.68736$

The error for $T_{1,1}$ can be estimated as [Eq. (29.13)]

$$|(\varepsilon_a)_{1,1}| = \left| \frac{32.51953 - 28.12500}{32.51953} \right| 100\% = 13.5\%$$

Because this value is above the stopping criterion of 1%, the computation is continued. The ninth iteration gives the result

$T_{11} = 43.00061$	$T_{21} = 33.29755$	$T_{31} = 33.88506$
$T_{12} = 63.21152$	$T_{22} = 56.11238$	$T_{32} = 52.33999$
$T_{13} = 78.58718$	$T_{23} = 76.06402$	$T_{33} = 69.71050$

where the maximum error is 0.71%.

Figure 29.5 shows the results. As expected, a gradient is established as heat flows from high to low temperatures.

FIGURE 29.5

Temperature distribution for a heated plate subject to fixed boundary conditions.

29.2.3 Secondary Variables

Because its distribution is described by the Laplace equation, temperature is considered to be the primary variable in the heated-plate problem. For this case, as well as for other problems involving PDEs, secondary variables may also be of interest. As a matter of fact, in certain engineering contexts, the secondary variable may actually be more important.

For the heated plate, a secondary variable is the rate of heat flux across the plate's surface. This quantity can be computed from Fourier's law. Centered finite-difference approximations for the first derivatives (recall Fig. 23.3) can be substituted into Eq. (29.4) to give the following values for heat flux in the x and y dimensions:

$$q_x = -k' \frac{T_{i+1,j} - T_{i-1,j}}{2\,\Delta x} \tag{29.14}$$

and

$$q_y = -k' \frac{T_{i,j+1} - T_{i,j-1}}{2\,\Delta y} \tag{29.15}$$

The resultant heat flux can be computed from these two quantities using

$$q_n = \sqrt{q_x^2 + q_y^2} \tag{29.16}$$

where the direction of q_n is given by

$$\theta = \tan^{-1}\left(\frac{q_y}{q_x}\right) \tag{29.17}$$

for $q_x > 0$ and

$$\theta = \tan^{-1}\left(\frac{q_y}{q_x}\right) + \pi \tag{29.18}$$

for $q_x < 0$. Recall that the angle can be expressed in degrees by multiplying it by $180°/\pi$. If $q_x = 0$, θ is $\pi/2$ (90°) or $3\pi/2$ (270°), depending on whether q_y is positive or negative, respectively.

EXAMPLE 29.2 **Flux Distribution for a Heated Plate**

Problem Statement.   Employ the results of Example 29.1 to determine the distribution of heat flux for the heated plate from Fig. 29.4. Assume that the plate is 40×40 cm and is made out of aluminum [$k' = 0.49$ cal/(s cm °C)].

Solution.   For $i = j = 1$, Eq. (29.14) can be used to compute

$$q_x = \left(-0.49\,\frac{\text{cal}}{\text{s cm °C}}\right) \frac{(33.29755 - 75)°C}{2(10\ \text{cm})} = 1.022\ \text{cal/(cm}^2\,\text{s)}$$

and [Eq. (29.15)]

$$q_y = \left(-0.49\,\frac{\text{cal}}{\text{s cm °C}}\right) \frac{(63.21152 - 0)°C}{2(10\ \text{cm})} = -1.549\ \text{cal/(cm}^2\,\text{s)}$$

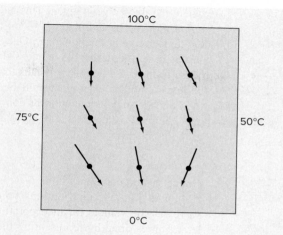

FIGURE 29.6
Heat flux for a plate subject to fixed boundary temperatures. Note that the lengths of the arrows are proportional to the magnitude of the flux.

The resultant flux can be computed with Eq. (29.16):

$$q_n = \sqrt{(1.022)^2 + (-1.549)^2} = 1.856 \text{ cal/(cm}^2\text{ s)}$$

and the angle of its trajectory by Eq. (29.17)

$$\theta = \tan^{-1}\left(\frac{-1.549}{1.022}\right) = -0.98758 \times \frac{180°}{\pi} = -56.584°$$

Thus, at this point, the heat flux is directed down and to the right. Values at the other grid points can be computed; the results are displayed in Fig. 29.6.

29.3 BOUNDARY CONDITIONS

Because it is free of complicating factors, the rectangular plate with fixed boundary conditions has been an ideal context for showing how elliptic PDEs can be solved numerically. We will now elaborate on other issues that will expand our capabilities to address more realistic problems. These involve boundaries at which the derivative is specified and boundaries that are irregularly shaped.

29.3.1 Derivative Boundary Conditions

The fixed, or Dirichlet, boundary condition discussed to this point is but one of several types that are used with partial differential equations. A common alternative is the case

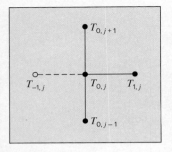

FIGURE 29.7
A boundary node $(0, j)$ on the left edge of a heated plate. To approximate the derivative normal to the edge (that is, the x derivative), an imaginary point $(-1, j)$ is located a distance Δx beyond the edge.

where the derivative is given. This is commonly referred to as a *Neumann boundary condition*. For the heated-plate problem, this amounts to specifying the heat flux rather than the temperature at the boundary. One example is the situation where the edge is insulated. In this case, the derivative is zero. This conclusion is drawn directly from Eq. (29.4) because insulating a boundary means that the heat flux (and consequently the gradient) must be zero. Another example would be where heat is lost across the edge by predictable mechanisms such as radiation or convection.

Figure 29.7 depicts a node $(0, j)$ at the left edge of a heated plate. Applying Eq. (29.8) at the point gives

$$T_{1,j} + T_{-1,j} + T_{0,j+1} + T_{0,j-1} - 4T_{0,j} = 0 \tag{29.19}$$

Notice that an imaginary point $(-1, j)$ lying outside the plate is required for this equation. Although this exterior fictitious point might seem to represent a problem, it actually serves as the vehicle for incorporating the derivative boundary condition into the problem. This is done by representing the first derivative in the x dimension at $(0, j)$ by the finite divided difference

$$\frac{\partial T}{\partial x} \cong \frac{T_{1,j} - T_{-1,j}}{2\,\Delta x}$$

which can be solved for

$$T_{-1,j} = T_{1,j} - 2\,\Delta x \frac{\partial T}{\partial x}$$

Now we have a relationship for $T_{-1,j}$ that actually includes the derivative. It can be substituted into Eq. (29.19) to give

$$2T_{1,j} - 2\,\Delta x \frac{\partial T}{\partial x} + T_{0,j+1} + T_{0,j-1} - 4T_{0,j} = 0 \tag{29.20}$$

Thus, we have incorporated the derivative into the balance.

Similar relationships can be developed for derivative boundary conditions at the other edges. The following example shows how this is done for the heated plate.

EXAMPLE 29.3   **Heated Plate with an Insulated Edge**

Problem Statement.   Repeat the same problem as in Example 29.1, but with the lower edge insulated.

Solution.   The general equation to characterize a derivative at the lower edge (that is, at $j = 0$) of a heated plate is

$$T_{i+1,0} + T_{i-1,0} + 2T_{i,1} - 2\,\Delta y \frac{\partial T}{\partial y} - 4T_{i,0} = 0$$

For an insulated edge, the derivative is zero and the equation becomes

$$T_{i+1,0} + T_{i-1,0} + 2T_{i,1} - 4T_{i,0} = 0$$

The simultaneous equations for temperature distribution on the plate in Fig. 29.4 with an insulated lower edge can be written in matrix form as

$$
\begin{bmatrix}
4 & -1 & & -2 & & & & & & & & \\
-1 & 4 & -1 & & -2 & & & & & & & \\
& -1 & 4 & & & -2 & & & & & & \\
-1 & & & 4 & -1 & & -1 & & & & & \\
& -1 & & -1 & 4 & -1 & & -1 & & & & \\
& & -1 & & -1 & 4 & -1 & & -1 & & & \\
& & & -1 & & & 4 & -1 & & -1 & & \\
& & & & -1 & & -1 & 4 & -1 & & -1 & \\
& & & & & -1 & & -1 & 4 & & & -1 \\
& & & & & & -1 & & & 4 & -1 & \\
& & & & & & & -1 & & -1 & 4 & -1 \\
& & & & & & & & -1 & & -1 & 4
\end{bmatrix}
\begin{Bmatrix}
T_{10} \\ T_{20} \\ T_{30} \\ T_{11} \\ T_{21} \\ T_{31} \\ T_{12} \\ T_{22} \\ T_{32} \\ T_{13} \\ T_{23} \\ T_{33}
\end{Bmatrix}
=
\begin{Bmatrix}
75 \\ 0 \\ 50 \\ 75 \\ 0 \\ 50 \\ 75 \\ 0 \\ 50 \\ 175 \\ 100 \\ 150
\end{Bmatrix}
$$

Note that because of the derivative boundary condition, the matrix is increased to 12×12, in contrast to the 9×9 system in Eq. (29.10), to account for the three unknown temperatures along the plate's lower edge. These equations can be solved for

$T_{10} = 71.91$	$T_{20} = 67.01$	$T_{30} = 59.54$
$T_{11} = 72.81$	$T_{21} = 68.31$	$T_{31} = 60.57$
$T_{12} = 76.01$	$T_{22} = 72.84$	$T_{32} = 64.42$
$T_{13} = 83.41$	$T_{23} = 82.63$	$T_{33} = 74.26$

FIGURE 29.8

Temperature and flux distribution for a heated plate subject to fixed boundary conditions except for an insulated lower edge.

These results and computed fluxes (for the same parameters as in Example 29.2) are displayed in Fig. 29.8. Note that, because the lower edge is insulated, the plate's temperature is higher than for Fig. 29.5, where the bottom edge temperature is fixed at zero. In addition, the heat flow (in contrast to Fig. 29.6) is now deflected to the right and moves parallel to the insulated wall.

29.3.2 Irregular Boundaries

Although the rectangular plate from Fig. 29.4 has served well to illustrate the fundamental aspects of solving elliptic PDEs, many engineering problems do not exhibit such an idealized geometry. For example, a great many systems have irregular boundaries (Fig. 29.9).

Figure 29.9 is a system that can serve to illustrate how nonrectangular boundaries can be handled. As depicted, the plate's lower left boundary is circular. Notice that we have affixed parameters—α_1, α_2, β_1, β_2—to each of the lengths surrounding the node. Of course, for the plate depicted in Fig. 29.9, $\alpha_2 = \beta_2 = 1$. However, we will retain these parameters throughout the following derivation so that the resulting equation is generally applicable to any irregular boundary—not just one on the lower left-hand corner of a heated plate. The first derivatives in the x dimension can be approximated as

$$\left(\frac{\partial T}{\partial x}\right)_{i-1,i} \cong \frac{T_{i,j} - T_{i-1,j}}{\alpha_1 \, \Delta x} \tag{29.21}$$

and

$$\left(\frac{\partial T}{\partial x}\right)_{i,i+1} \cong \frac{T_{i+1,j} - T_{i,j}}{\alpha_2 \, \Delta x} \tag{29.22}$$

FIGURE 29.9
A grid for a heated plate with an irregularly shaped boundary. Note how weighting coefficients are used to account for the nonuniform spacing in the vicinity of the nonrectangular boundary.

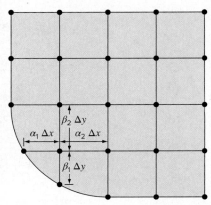

The second derivatives can be developed from these first derivatives. For the x dimension, the second derivative is

$$\frac{\partial^2 T}{\partial x^2} = \frac{\partial}{\partial x}\left(\frac{\partial T}{\partial x}\right) = \frac{\left(\dfrac{\partial T}{\partial x}\right)_{i,i+1} - \left(\dfrac{\partial T}{\partial x}\right)_{i-1,i}}{\dfrac{\alpha_1\,\Delta x + \alpha_2\,\Delta x}{2}} \tag{29.23}$$

Substituting Eqs. (29.21) and (29.22) into (29.23) gives

$$\frac{\partial^2 T}{\partial x^2} = 2\,\frac{\dfrac{T_{i-1,j} - T_{i,j}}{\alpha_1\,\Delta x} - \dfrac{T_{i+1,j} - T_{i,j}}{\alpha_2\,\Delta x}}{\alpha_1\,\Delta x + \alpha_2\,\Delta x}$$

Collecting terms yields

$$\frac{\partial^2 T}{\partial x^2} = \frac{2}{\Delta x^2}\left[\frac{T_{i-1,j} - T_{i,j}}{\alpha_1(\alpha_1 + \alpha_2)} + \frac{T_{i+1,j} - T_{i,j}}{\alpha_2(\alpha_1 + \alpha_2)}\right]$$

A similar equation can be developed in the y dimension:

$$\frac{\partial^2 T}{\partial y^2} = \frac{2}{\Delta y^2}\left[\frac{T_{i,j-1} - T_{i,j}}{\beta_1(\beta_1 + \beta_2)} + \frac{T_{i,j+1} - T_{i,j}}{\beta_2(\beta_1 + \beta_2)}\right]$$

Substituting these equations in Eq. (29.6) yields

$$\frac{2}{\Delta x^2}\left[\frac{T_{i-1,j} - T_{i,j}}{\alpha_1(\alpha_1 + \alpha_2)} + \frac{T_{i+1,j} - T_{i,j}}{\alpha_2(\alpha_1 + \alpha_2)}\right]$$

$$+ \frac{2}{\Delta y^2}\left[\frac{T_{i,j-1} - T_{i,j}}{\beta_1(\beta_1 + \beta_2)} + \frac{T_{i,j+1} - T_{i,j}}{\beta_2(\beta_1 + \beta_2)}\right] = 0 \tag{29.24}$$

As illustrated in the following example, Eq. (29.24) can be applied to any node that lies adjacent to an irregular, Dirichlet-type boundary.

EXAMPLE 29.4   **Heated Plate with an Irregular Boundary**

Problem Statement.   Repeat the same problem as in Example 29.1, but with the lower edge as depicted in Fig. 29.9.

Solution.   For the case in Fig. 29.9, $\Delta x = \Delta y$, $\alpha_1 = \beta_1 = 0.732$, and $\alpha_2 = \beta_2 = 1$. Substituting these values into Eq. (29.24) yields the following balance for node (1, 1):

$$0.788675(T_{01} - T_{11}) + 0.57735(T_{21} - T_{11})$$

$$+ 0.788675(T_{10} - T_{11}) + 0.57735(T_{12} - T_{11}) = 0$$

Collecting terms, we can express this equation as

$$-4T_{11} + 0.8453T_{21} + 0.8453T_{12} = -1.1547T_{01} - 1.1547T_{10}$$

The simultaneous equations for temperature distribution on the plate in Fig. 29.9 with a lower-edge boundary temperature of 75 can be written in matrix form as

$$
\begin{bmatrix}
4 & -0.845 & & -0.845 & & & & & \\
-1 & 4 & -1 & & -1 & & & & \\
& -1 & 4 & & & -1 & & & \\
-1 & & & 4 & -1 & & -1 & & \\
& -1 & & -1 & 4 & -1 & & -1 & \\
& & -1 & & -1 & 4 & & & -1 \\
& & & -1 & & & 4 & -1 & \\
& & & & -1 & & -1 & 4 & -1 \\
& & & & & -1 & & -1 & -4 \\
\end{bmatrix}
\begin{Bmatrix}
T_{11} \\ T_{21} \\ T_{31} \\ T_{12} \\ T_{22} \\ T_{32} \\ T_{13} \\ T_{23} \\ T_{33}
\end{Bmatrix}
=
\begin{Bmatrix}
173.2 \\ 75 \\ 125 \\ 75 \\ 0 \\ 50 \\ 175 \\ 100 \\ 150
\end{Bmatrix}
$$

These equations can be solved for

$$
\begin{aligned}
T_{11} &= 74.98 & T_{21} &= 72.76 & T_{31} &= 66.07 \\
T_{12} &= 74.23 & T_{22} &= 75.00 & T_{32} &= 66.52 \\
T_{13} &= 83.93 & T_{23} &= 83.48 & T_{33} &= 75.00
\end{aligned}
$$

These results along with the computed fluxes are displayed in Fig. 29.10. Note that the fluxes are computed in the same fashion as in Sec. 29.2.3, with the exception that $(\alpha_1 + \alpha_2)$ and $(\beta_1 + \beta_2)$ are substituted for the 2's in the denominators of Eqs. (29.14) and (29.15), respectively. Section 32.3 illustrates how this is done.

FIGURE 29.10
Temperature and flux distribution for a heated plate with a circular boundary.

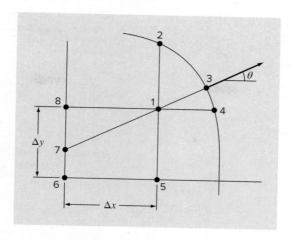

FIGURE 29.11
A curved boundary where the normal gradient is specified.

Derivative conditions for irregularly shaped boundaries are more difficult to formulate. Figure 29.11 shows a point near an irregular boundary where the normal derivative is specified.

The normal derivative at node 3 can be approximated by the gradient between nodes 1 and 7,

$$\frac{\partial T}{\partial \eta}\bigg|_3 = \frac{T_1 - T_7}{L_{17}}$$ (29.25)

When θ is less than 45° as shown, the distance from node 7 to 8 is $\Delta x \tan \theta$, and linear interpolation can be used to estimate

$$T_7 = T_8 + (T_6 - T_8)\frac{\Delta x \tan \theta}{\Delta y}$$

The length L_{17} is equal to $\Delta x/\cos \theta$. This length, along with the approximation for T_7, can be substituted into Eq. (29.25) to give

$$T_1 = \left(\frac{\Delta x}{\cos \theta}\right)\frac{\partial T}{\partial \eta}\bigg|_3 + T_6\frac{\Delta x \tan \theta}{\Delta y} + T_8\left(1 - \frac{\Delta x \tan \theta}{\Delta y}\right)$$ (29.26)

Such an equation provides a means for incorporating the normal gradient into the finite-difference approach. For cases where θ is greater than 45°, a different equation would be used. The determination of this formula will be left as a homework exercise.

29.4 THE CONTROL-VOLUME APPROACH

To summarize, the finite-difference, or Taylor series, approach divides the continuum into nodes (Fig. 29.12a). The underlying partial differential equation is written for each of these nodes. Finite-difference approximations are then substituted for the derivatives to convert the equations to an algebraic form.

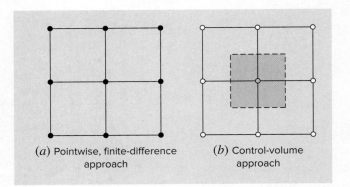

FIGURE 29.12
Two different perspectives for developing approximate solutions of PDEs: (*a*) finite difference and (*b*) control volume.

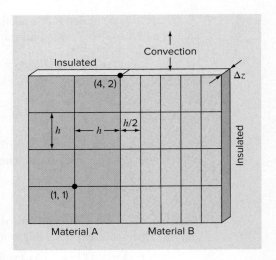

FIGURE 29.13
A heated plate with unequal grid spacing, two materials, and mixed boundary conditions.

Such an approach is quite simple and straightforward for orthogonal (that is, rectangular) grids and constant coefficients. However, the approach becomes a more difficult endeavor for derivative conditions on irregularly shaped boundaries.

Figure 29.13 is an example of a system where additional difficulties arise. This plate is made of two different materials and has unequal grid spacing. In addition, half of its top edge is subject to convective heat transfer, whereas half is insulated. Developing equations for node (4, 2) would require some additional derivation beyond the approaches developed to this point.

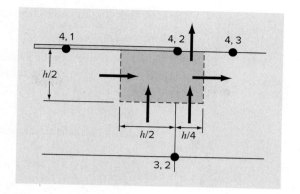

FIGURE 29.14
A control volume for node (4, 2) with arrows indicating heat transfer through the boundaries.

The *control-volume approach* (also called the *volume-integral approach*) offers an alternative way to numerically approximate PDEs that is especially useful for cases such as Fig. 29.13. As in Fig. 29.12b, the approach resembles the pointwise approach in that points are determined across the domain. However, rather than approximating the PDE at a point, the approximation is applied to a volume surrounding the point. For an orthogonal grid, the volume is formed by the perpendicular lines through the midpoint of each line joining adjacent nodes. A heat balance can then be developed for each volume in a fashion similar to Eq. (29.1).

As an example, we will apply the control-volume approach to node (4, 2). First, the volume is defined by bisecting the lines joining the nodes. As in Fig. 29.14, the volume has conductive heat transfer through its left, right, and lower boundaries and convective heat transfer through half of its upper boundary. Notice that the transfer through the lower boundary involves both materials.

A steady-state heat balance for the volume can be written in qualitative terms as

$$0 = \begin{pmatrix} \text{left-side} \\ \text{conduction} \end{pmatrix} - \begin{pmatrix} \text{right-side} \\ \text{conduction} \end{pmatrix} + \begin{pmatrix} \text{lower conduction} \\ \text{material "a"} \end{pmatrix}$$
$$+ \begin{pmatrix} \text{lower conduction} \\ \text{material "b"} \end{pmatrix} - \begin{pmatrix} \text{upper} \\ \text{convection} \end{pmatrix} \tag{29.27}$$

Now the conduction flux rate can be represented by the finite-difference version of Fourier's law. For example, for the left-side conduction gain, it would be

$$q = -k'_a \frac{T_{42} - T_{41}}{h}$$

where q has units of cal/cm^2/s. This flux rate must be then multiplied by the area across which it enters ($\Delta z \times h/2$) to give the rate of heat entering the volume per unit time,

$$Q = -k'_a \frac{T_{42} - T_{41}}{h} \frac{h}{2} \Delta z$$

where Q has units of cal/s.

The heat flux due to convection can be formulated as

$$q = h_c(T_a - T_{42})$$

where h_c = a heat convection coefficient [cal/(s cm^2 °C)] and T_a = the air temperature (°C). Again, multiplication by the proper area yields the rate of heat flow per time,

$$Q = h_c(T_a - T_{42})\frac{h}{4}\Delta z$$

The other transfers can be developed in a similar fashion and substituted into Eq. (29.27) to yield

$$0 = -k_a'\frac{T_{42} - T_{41}}{h}\frac{h}{2}\Delta z + k_b'\frac{T_{43} - T_{42}}{h/2}\frac{h}{2}\Delta z$$

(left-side conduction)(right-side conduction)

$$-k_a'\frac{T_{42} - T_{32}}{h}\frac{h}{2}\Delta z - k_b'\frac{T_{42} - T_{32}}{h}\frac{h}{4}\Delta z + h_c(T_a - T_{42})\frac{h}{4}\Delta z$$

$$\left(\begin{matrix}\text{lower conduction}\\ \text{material "a"}\end{matrix}\right)\left(\begin{matrix}\text{lower conduction}\\ \text{material "b"}\end{matrix}\right)\text{(upper convection)}$$

Parameter values can then be substituted to yield the final heat balance equation. For example, if $\Delta z = 0.5$ cm, $h = 10$ cm, $k_a' = 0.3$ cal/(s cm °C), $k_b' = 0.5$ cal/(s cm °C), and $h_c = 0.1$ cal/(s cm^2 °C), the equation becomes

$$0.5875T_{42} - 0.075T_{41} - 0.25T_{43} - 0.1375T_{32} = 2.5$$

To make this equation comparable to the standard Laplacian, it can be multiplied by $4/0.5875$ so that the coefficient of the base node has a coefficient of 4,

$$4T_{42} - 0.510638T_{41} - 1.702128T_{43} - 0.93617T_{32} = 17.02128$$

For the standard cases covered to this point, the control-volume and pointwise finite-difference approaches yield identical results. For example, for node (1, 1) in Fig. 29.13, the heat balance will be

$$0 = -k_a'\frac{T_{11} - T_{01}}{h}h\,\Delta z + k_a'\frac{T_{21} - T_{11}}{h}h\,\Delta z - k_a'\frac{T_{11} - T_{10}}{h}h\,\Delta z + k_a'\frac{T_{12} - T_{11}}{h}h\,\Delta z$$

which simplifies to the standard Laplacian,

$$0 = 4T_{11} - T_{01} - T_{21} - T_{12} - T_{10}$$

We will look at other standard cases (for example, the derivative boundary condition) and explore the control-volume approach in additional detail in the problems at the end of this chapter.

29.5   SOFTWARE TO SOLVE ELLIPTIC EQUATIONS

Modifying a computer program to include derivative boundary conditions for rectangular systems is a relatively straightforward task. It merely involves ensuring that additional equations are generated to characterize the boundary nodes at which the derivatives are specified. In addition, the code must be modified so that these equations incorporate the derivative, as seen in Eq. (29.20).

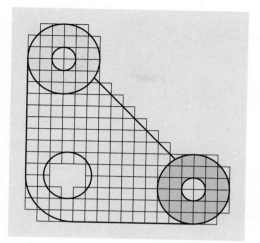

FIGURE 29.15
A finite-difference grid superimposed on an irregularly shaped gasket.

Developing general software to characterize systems with irregular boundaries is a much more difficult proposition. For example, a fairly involved algorithm would be required to model the simple gasket depicted in Fig. 29.15. This would involve two major modifications. First, a scheme would have to be developed to conveniently input the configuration of the nodes and to identify which were at the boundary. Second, an algorithm would be required to generate the proper simultaneous equations on the basis of the input information. The net result is that general software for solving elliptic (and for that matter, all) PDEs is relatively complicated.

One method used to simplify such efforts is to require a very fine grid. For such cases, it is often assumed that the closest node serves as the boundary point. In this way, the analysis does not have to consider the weighting parameters from Sec. 29.3.2. Although this introduces some error, the use of a sufficiently fine mesh can make the resulting discrepancy negligible. However, this involves a trade-off due to the computational burden introduced by the increased number of simultaneous equations.

As a consequence of these considerations, numerical analysts have developed alternative approaches that differ radically from finite-difference methods. Although these finite-element methods are more conceptually difficult, they can much more easily accommodate irregular boundaries. We will turn to these methods in Chap. 31. Before doing this, however, we will first describe finite-difference approaches for another category of PDEs—parabolic equations.

PROBLEMS

29.1 Use Liebmann's method to solve for the temperature of the square heated plate in Fig. 29.4, but with the upper boundary condition increased to 175°C and the left boundary insulated. Use a relaxation factor of 1.2 and iterate to $\varepsilon_s = 1\%$.

29.2 Use Liebmann's method to solve for the temperature of the square heated plate in Fig. 29.4, but with the upper boundary condition increased to 150°C and the left boundary decreased to 50°C. Use a relaxation factor of 1.2 and iterate to $\varepsilon_s = 1\%$.

29.3 Compute the fluxes for Prob. 29.2 using the parameters from Example 29.3.

29.4 Repeat Example 29.1, but use 49 interior nodes (that is, $\Delta x = \Delta y = 5$ cm).

FIGURE P29.8

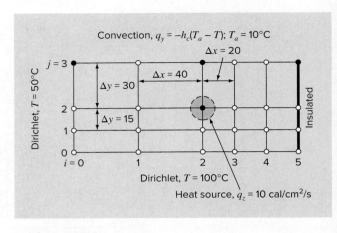

FIGURE P29.9

29.5 Repeat Prob. 29.4, but for the case where the lower edge is insulated.

29.6 Repeat Examples 29.1 and 29.3, but for the case where the flux at the lower edge is directed downward with a value of 2 cal/cm² s.

29.7 Repeat Example 29.4 for the case where both the lower left and the upper right corners are rounded in the same fashion as the lower left corner of Fig. 29.9. Note that all boundary temperatures on the upper and right sides are fixed at 100°C and all on the lower and left sides are fixed at 50°C.

29.8 With the exception of the boundary conditions, the plate in Fig. P29.8 has the exact same characteristics as the plate used in Examples 29.1 through 29.3. Simulate both the temperatures and fluxes for the plate.

29.9 Write equations for the darkened nodes in the grid in Fig. P29.9. Note that all units are cgs. The coefficient of thermal conductivity for the plate is 0.75 cal/(s cm °C), the convection coefficient is $h_c = 0.015$ cal/(cm² °C s), and the thickness of the plate is 0.5 cm.

29.10 Write equations for the darkened nodes in the grid in Fig. P29.10. Note that all units are cgs. The convection coefficient is $h_c = 0.01$ cal/(cm² °C s) and the thickness of the plate is 2 cm.

29.11 Apply the control-volume approach to develop the equation for node $(0, j)$ in Fig. 29.7.

29.12 Derive an equation like Eq. (29.26) for the case where θ is greater than 45° in Fig. 29.11.

29.13 Develop a user-friendly computer program to implement Liebmann's method for a rectangular plate with Dirichlet boundary conditions. Design the program so that it can compute both

FIGURE P29.10

temperature and flux. Test the program by duplicating the results of Examples 29.1 and 29.2.

29.14 Employ the program from Prob. 29.13 to solve Probs. 29.2 and 29.3.

29.15 Employ the program from Prob 29.13 to solve Prob. 29.4.

29.16 Use the control-volume approach and derive the node equation for node $(2, 2)$ in Fig. 29.13 and include a heat source at this point. Use the following values for the constants: $\Delta z = 0.25$ cm, $h = 10$ cm, $k_A = 0.25$ W/cm °C, and $k_B = 0.45$ W/cm °C. The heat source comes only from material A at the rate of $= 6$ W/cm³.

29.17 Calculate heat flux (W/cm²) for node $(2, 2)$ in Fig. 29.13 using finite-difference approximations for the temperature gradients

FIGURE P29.18

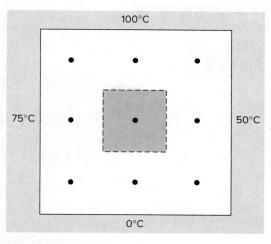

FIGURE P29.20

at this node. Calculate the flux in the horizontal direction in materials A and B, and determine if these two fluxes should be equal. Also, calculate the vertical flux in materials A and B. Should these two fluxes be equal? Use the following values for the constants: $\Delta z = 0.5$ cm, $h = 10$ cm, $k_A = 0.25$ W/cm °C, $k_B = 0.45$ W/cm °C, and nodal temperatures: $T_{22} = 51.6$°C, $T_{21} = 74.2$°C, $T_{23} = 45.3$°C, $T_{32} = 38.6$°C, and $T_{12} = 87.4$°C.

29.18 Compute the temperature distribution for the L-shaped plate in Fig. P29.18.

29.19 The Poisson equation can be written in three dimensions as

$$\frac{\partial^2 T}{\partial x^2} + \frac{\partial^2 T}{\partial y^2} + \frac{\partial^2 T}{\partial z^2} = f(x, y, z)$$

Solve for the distribution of temperature within a unit (1×1) cube with zero boundary conditions and $f = -10$. Employ $\Delta x = \Delta y = \Delta z = 1/6$.

29.20 Determine the temperature distribution and fluxes for the plate depicted in Fig. P29.20. The plate is $60 \times 60 \times 1$ cm, is made out of aluminum [$k' = 0.49$ cal/(s cm °C)], with an input of 10 cal/s into the middle node.

29.21 Repeat Prob. 29.20, but with the bottom edge insulated.

Finite Difference: Parabolic Equations

Chapter 29 dealt with steady-state PDEs. We now turn to the parabolic equations that are employed to characterize time-variable problems. In the latter part of this chapter, we will illustrate how this is done in two spatial dimensions for the heated plate. Before doing this, we will first show how the simpler one-dimensional case is approached.

30.1 THE HEAT-CONDUCTION EQUATION

In a fashion similar to the derivation of the Laplace equation [Eq. (29.6)], conservation of heat can be used to develop a heat balance for the differential element in the long, thin insulated rod shown in Fig. 30.1. However, rather than examine the steady-state case, the present balance also considers the amount of heat stored in the element over a unit time period Δt. Thus, the balance is in the form: inputs − outputs = storage, or

$$q(x)\, \Delta y\, \Delta z\, \Delta t - q(x + \Delta x)\, \Delta y\, \Delta z\, \Delta t = \Delta x\, \Delta y\, \Delta z \rho C\, \Delta T$$

Dividing by the volume of the element (= $\Delta x\, \Delta y\, \Delta z$) and Δt gives

$$\frac{q(x) - q(x + \Delta x)}{\Delta x} = \rho C \frac{\Delta T}{\Delta t}$$

Taking the limit yields

$$-\frac{\partial q}{\partial x} = \rho C \frac{\partial T}{\partial t}$$

FIGURE 30.1
A thin rod, insulated at all points except at its ends.

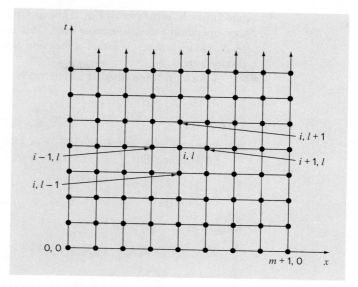

FIGURE 30.2
A grid used for the finite-difference solution of parabolic PDEs in two independent variables such as the heat-conduction equation. Note how, in contrast to Fig. 29.3, this grid is open-ended in the temporal dimension.

Substituting Fourier's law of heat conduction [Eq. (29.4)] results in

$$k\frac{\partial^2 T}{\partial x^2} = \frac{\partial T}{\partial t}$$

(30.1)

which is the *heat-conduction equation.*

Just as with elliptic PDEs, parabolic equations can be solved by substituting finite divided differences for the partial derivatives. However, in contrast to elliptic PDEs, we must now consider changes in time as well as in space. Whereas elliptic equations were bounded in all relevant dimensions, parabolic PDEs are temporally open-ended (Fig. 30.2). Because of their time-variable nature, solutions to these equations involve a number of new issues, notably stability. This, as well as other aspects of parabolic PDEs, will be examined in the following sections as we present two fundamental solution approaches—explicit and implicit schemes.

30.2 EXPLICIT METHODS

The heat-conduction equation requires approximations for the second derivative in space and the first derivative in time. The former is represented in the same fashion as for the Laplace equation by a centered finite-divided difference:

$$\frac{\partial^2 T}{\partial x^2} = \frac{T_{i+1}^l - 2T_i^l + T_{i-1}^l}{\Delta x^2}$$

(30.2)

which has an error (recall Fig. 23.3) of $O[(\Delta x)^2]$. Notice the slight change in notation of the superscripts used to denote time. This is done so that a second subscript can be used to designate a second spatial dimension when the approach is expanded to two spatial dimensions.

A forward finite-divided difference is used to approximate the time derivative

$$\frac{\partial T}{\partial t} = \frac{T_i^{l+1} - T_i^l}{\Delta t} \tag{30.3}$$

which has an error (recall Fig. 23.1) of $O(\Delta t)$.

Substituting Eqs. (30.2) and (30.3) into Eq. (30.1) yields

$$k\frac{T_{i+1}^l - 2T_i^l + T_{i-1}^l}{(\Delta x)^2} = \frac{T_i^{l+1} - T_i^l}{\Delta t} \tag{30.4}$$

which can be solved for

$$T_i^{l+1} = T_i^l + \lambda(T_{i+1}^l - 2T_i^l + T_{i-1}^l) \tag{30.5}$$

where $\lambda = k\,\Delta t/(\Delta x)^2$.

This equation can be written for all the interior nodes on the rod. It then provides an explicit means to compute values at each node for a future time based on the present values at the node and its neighbors. Notice that this approach is actually a manifestation of Euler's method for solving systems of ODEs. That is, if we know the temperature distribution as a function of position at an initial time, we can compute the distribution at a future time based on Eq. (30.5).

A computational molecule for the explicit method is depicted in Fig. 30.3, showing the nodes that constitute the spatial and temporal approximations. This molecule can be contrasted with others in this chapter to illustrate the differences between approaches.

FIGURE 30.3
A computational molecule for the explicit form.

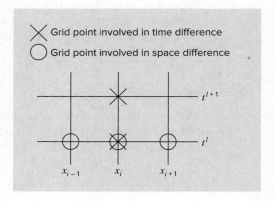

EXAMPLE 30.1

Explicit Solution of the One-Dimensional Heat-Conduction Equation

Problem Statement. Use the explicit method to solve for the temperature distribution of a long, thin rod with a length of 10 cm and the following values: $k' = 0.49$ cal/(s cm °C), $\Delta x = 2$ cm, and $\Delta t = 0.1$ s. At $t = 0$, the temperature of the rod is zero and the boundary conditions are fixed for all times at $T(0) = 100$°C and $T(10) = 50$°C. Note that the rod is aluminum with $C = 0.2174$ cal/(g °C) and $\rho = 2.7$ g/cm^3. Therefore, $k = 0.49/(2.7 \cdot 0.2174) = 0.835$ cm^2/s and $\lambda = 0.835(0.1)/(2)^2 = 0.020875$.

Solution. Applying Eq. (30.5) gives the following value at $t = 0.1$ s for the node at $x = 2$ cm:

$$T_1^1 = 0 + 0.020875[0 - 2(0) + 100] = 2.0875$$

At the other interior points, $x = 4$, 6, and 8 cm, the results are

$$T_2^1 = 0 + 0.020875[0 - 2(0) + 0] = 0$$
$$T_3^1 = 0 + 0.020875[0 - 2(0) + 0] = 0$$
$$T_4^1 = 0 + 0.020875[50 - 2(0) + 0] = 1.0438$$

At $t = 0.2$ s, the values at the four interior nodes are computed as

$$T_1^2 = 2.0875 + 0.020875[0 - 2(2.0875) + 100] = 4.0878$$
$$T_2^2 = 0 + 0.020875[0 - 2(0) + 2.0875] = 0.043577$$
$$T_3^2 = 0 + 0.020875[1.0438 - 2(0) + 0] = 0.021788$$
$$T_4^2 = 1.0438 + 0.020875[50 - 2(1.0438) + 0] = 2.0439$$

The computation is continued, and the results at 3-s intervals are depicted in Fig. 30.4. The general rise in temperature with time indicates that the computation captures the conduction of heat from the boundaries into the bar.

FIGURE 30.4
Temperature distribution in a long, thin rod as computed with the explicit method described in Sec. 30.2.

30.2.1 Convergence and Stability

Convergence means that as Δx and Δt approach zero, the results of the finite-difference technique approach the true solution. *Stability* means that errors at any stage of the computation are not amplified but are attenuated as the computation progresses. It can be shown (Carnahan et al., 1969) that the explicit method is both convergent and stable if $\lambda \leq 1/2$, or

$$\Delta t \leq \frac{1}{2} \frac{\Delta x^2}{k} \tag{30.6}$$

In addition, it should be noted that setting $\lambda \leq 1/2$ could result in a solution in which errors do not grow, but oscillate. Setting $\lambda \leq 1/4$ ensures that the solution will not oscillate. It is also known that setting $\lambda = 1/6$ tends to minimize truncation error (Carnahan et al., 1969).

Figure 30.5 is an example of instability caused by violating Eq. (30.6). This plot is for the same case as in Example 30.1 but with $\lambda = 0.735$, which is considerably greater than 0.5. As in Fig. 30.5, the solution undergoes progressively increasing oscillations. This situation will continue to deteriorate as the computation continues.

Although satisfaction of Eq. (30.6) will alleviate the instabilities of the sort manifested in Fig. 30.5, it also places a strong limitation on the explicit method. For example, suppose that Δx is halved to improve the approximation of the spatial second derivative. According to Eq. (30.6), the time step must be quartered to maintain convergence and stability. Thus, to perform comparable computations, the time steps must be increased by a factor of 4. Furthermore, the computation for each of these time steps will take twice as long because halving Δx doubles the total number of nodes for which equations must be written. Consequently, for the one-dimensional case, halving Δx results in an eightfold increase in the number of calculations. Thus, the computational burden may be large to attain acceptable accuracy. As will be described shortly, other techniques are available that do not suffer from such severe limitations.

30.2.2 Derivative Boundary Conditions

As was the case for elliptic PDEs (recall Sec. 29.3.1), derivative boundary conditions can be readily incorporated into parabolic equations. For a one-dimensional rod, this necessitates adding two equations to characterize the heat balance at the end nodes. For example, the node at the left end ($i = 0$) would be represented by

$$T_0^{l+1} = T_0^l + \lambda(T_1^l - 2T_0^l + T_{-1}^l)$$

Thus, an imaginary point is introduced at $i = -1$ (recall Fig. 29.7). However, as with the elliptic case, this point provides a vehicle for incorporating the derivative boundary condition into the analysis. Problem 30.2 at the end of the chapter deals with this exercise.

30.2.3 Higher-Order Temporal Approximations

The general idea of re-expressing the PDE as a system of ODEs is sometimes called the *method of lines*. Obviously, one way to improve on the Euler approach used above would be to employ a more accurate integration scheme for solving the ODEs. For example, the Heun method can be employed to obtain second-order temporal accuracy. An example of

FIGURE 30.5
An illustration of instability. Solution of Example 30.1 but with $\lambda = 0.735$.

this approach is called *MacCormack's method*. This and other improved explicit methods are discussed elsewhere (for example, Hoffman 1992).

30.3   A SIMPLE IMPLICIT METHOD

As noted previously, explicit finite-difference formulations have problems related to stability. In addition, as depicted in Fig. 30.6, they exclude information that has a bearing on the solution. Implicit methods overcome both these difficulties at the expense of somewhat more complicated algorithms.

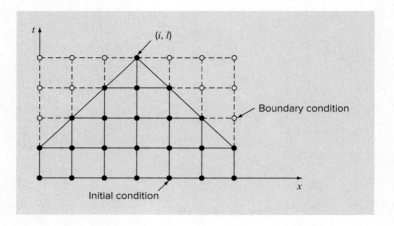

FIGURE 30.6
Representation of the effect of other nodes on the finite-difference approximation at node (i, l) using an explicit finite-difference scheme. The shaded nodes have an influence on (i, l), whereas the unshaded nodes, which in reality affect (i, l), are excluded.

FIGURE 30.7
Computational molecules demonstrating the fundamental differences between (a) explicit and (b) implicit methods.

The fundamental difference between explicit and implicit approximations is depicted in Fig. 30.7. For the explicit method, we approximate the spatial derivative at time level l (Fig. 30.7a). Recall that when we substituted this approximation into the partial differential equation, we obtained a difference equation (30.4) with a single unknown T_i^{l+1}. Thus, we can solve "explicitly" for this unknown as in Eq. (30.5).

In implicit methods, the spatial derivative is approximated at an advanced time level $l + 1$. For example, the second derivative would be approximated by (Fig. 30.7b)

$$\frac{\partial^2 T}{\partial x^2} \cong \frac{T_{i+1}^{l+1} - 2T_i^{l+1} + T_{i-1}^{l+1}}{(\Delta x)^2} \tag{30.7}$$

which is second-order accurate. When this relationship is substituted into the original PDE, the resulting difference equation contains several unknowns. Thus, it cannot be solved explicitly by simple algebraic rearrangement as was done in going from Eq. (30.4) to (30.5). Instead, the entire system of equations must be solved simultaneously. This is possible because, along with the boundary conditions, the implicit formulations result in a set of linear algebraic equations with the same number of unknowns. Thus, the method reduces to the solution of a set of simultaneous equations at each point in time.

To illustrate how this is done, substitute Eqs. (30.3) and (30.7) into Eq. (30.1) to give

$$k\frac{T_{i+1}^{l+1} - 2T_i^{l+1} + T_{i-1}^{l+1}}{(\Delta x)^2} = \frac{T_i^{l+1} - T_i^l}{\Delta t}$$

which can be expressed as

$$-\lambda T_{i-1}^{l+1} + (1 + 2\lambda)T_i^{l+1} - \lambda T_{i+1}^{l+1} = T_i^l \tag{30.8}$$

where $\lambda = k\,\Delta t/(\Delta x)^2$. This equation applies to all but the first and the last interior nodes, which must be modified to reflect the boundary conditions. For the case where the temperature levels at the ends of the rod are given, the boundary condition at the left end of the rod ($i = 0$) can be expressed as

$$T_0^{l+1} = f_0(t^{l+1}) \tag{30.9}$$

where $f_0(t^{l+1})$ = a function describing how the boundary temperature changes with time. Substituting Eq. (30.9) into Eq. (30.8) gives the difference equation for the first interior node ($i = 1$):

$$(1 + 2\lambda)T_1^{l+1} - \lambda T_2^{l+1} = T_1^l + \lambda f_0(t^{l+1}) \tag{30.10}$$

Similarly, for the last interior node ($i = m$),

$$-\lambda T_{m-1}^{l+1} + (1 + 2\lambda)T_m^{l+1} = T_m^l + \lambda f_{m+1}(t^{l+1}) \tag{30.11}$$

where $f_{m+1}(t^{l+1})$ describes the specified temperature changes at the right end of the rod ($i = m + 1$).

When Eqs. (30.8), (30.10), and (30.11) are written for all the interior nodes, the resulting set of m linear algebraic equations has m unknowns. In addition, the method has the added bonus that the system is tridiagonal. Thus, we can utilize the extremely efficient solution algorithms (recall Sec. 11.1.1) that are available for tridiagonal systems.

EXAMPLE 30.2   **Simple Implicit Solution of the Heat-Conduction Equation**

Problem Statement.   Use the simple implicit finite-difference approximation to solve the same problem as in Example 30.1.

Solution.   For the rod from Example 30.1, $\lambda = 0.020875$. Therefore, at $t = 0$, Eq. (30.10) can be written for the first interior node as

$$1.04175T_1^1 - 0.020875T_2^1 = 0 + 0.020875(100)$$

or

$$1.04175T_1^1 - 0.020875T_2^1 = 2.0875$$

In a similar fashion, Eqs. (30.8) and (30.11) can be applied to the other interior nodes. This leads to the following set of simultaneous equations:

$$
\begin{bmatrix}
1.04175 & -0.020875 & & \\
-0.020875 & 1.04175 & -0.020875 & \\
& -0.020875 & 1.04175 & -0.020875 \\
& & -0.020875 & 1.04175
\end{bmatrix}
\begin{Bmatrix}
T_1^1 \\
T_2^1 \\
T_3^1 \\
T_4^1
\end{Bmatrix}
=
\begin{Bmatrix}
2.0875 \\
0 \\
0 \\
1.04375
\end{Bmatrix}
$$

which can be solved for the temperatures at $t = 0.1$ s:

$T_1^1 = 2.0047$
$T_2^1 = 0.0406$
$T_3^1 = 0.0209$
$T_4^1 = 1.0023$

Notice how in contrast to Example 30.1, all the points have changed from the initial condition during the first time step.

To solve for the temperatures at $t = 0.2$, the right-hand-side vector must be modified to account for the results of the first step, as in

$$
\begin{Bmatrix}
4.09215 \\
0.04059 \\
0.02090 \\
2.04069
\end{Bmatrix}
$$

The simultaneous equations can then be solved for the temperatures at $t = 0.2$ s:

$T_1^2 = 3.9305$
$T_2^2 = 0.1190$
$T_3^2 = 0.0618$
$T_4^2 = 1.9653$

Whereas the implicit method described is stable and convergent, it has the defect that the temporal difference approximation is first-order accurate, whereas the spatial difference approximation is second-order accurate (Fig. 30.8). In the next section we present an alternative implicit method that remedies the situation.

Before proceeding, it should be mentioned that, although the simple implicit method is unconditionally stable, there is an accuracy limit to the use of large time steps. Consequently, it is not that much more efficient than the explicit approaches for most time-variable problems.

Where it does shine is for steady-state problems. Recall from Chap. 29 that a form of Gauss-Seidel (Liebmann's method) can be used to obtain steady-state solutions for elliptic equations. An alternative approach would be to run a time-variable solution until it reached a steady state. In such cases, because inaccurate intermediate results are not an issue, implicit methods allow you to employ larger time steps, and hence, can generate steady-state results in an efficient manner.

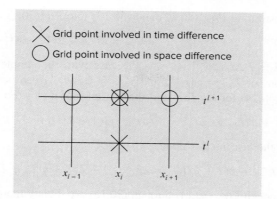

FIGURE 30.8
A computational molecule for
the simple implicit method.

30.4   THE CRANK-NICOLSON METHOD

The *Crank-Nicolson method* provides an alternative implicit scheme that is second-order accurate in both space and time. To provide this accuracy, difference approximations are developed at the midpoint of the time increment (Fig. 30.9). To do this, the temporal first derivative can be approximated at $t^{l+1/2}$ by

$$\frac{\partial T}{\partial t} \cong \frac{T_i^{l+1} - T_i^l}{\Delta t} \tag{30.12}$$

The second derivative in space can be determined at the midpoint by averaging the difference approximations at the beginning (t^l) and at the end (t^{l+1}) of the time increment:

$$\frac{\partial^2 T}{\partial x^2} \cong \frac{1}{2}\left[\frac{T_{i+1}^l - 2T_i^l + T_{i+1}^l}{(\Delta x)^2} + \frac{T_{i+1}^{l+1} - 2T_i^{l+1} + T_{i-1}^{l+1}}{(\Delta x)^2}\right] \tag{30.13}$$

FIGURE 30.9
A computational molecule for
the Crank-Nicolson method.

Substituting Eqs. (30.12) and (30.13) into Eq. (30.1) and collecting terms gives

$$-\lambda T_{i-1}^{l+1} + 2(1 + \lambda)T_i^{l+1} - \lambda T_{i+1}^{l+1} = \lambda T_{i-1}^l + 2(1 - \lambda)T_i^l + \lambda T_{i+1}^l \qquad (30.14)$$

where $\lambda = k\,\Delta t/(\Delta x)^2$. As was the case with the simple implicit approach, boundary conditions of $T_0^{l+1} = f_0(t^{l+1})$ and $T_{m+1}^{l+1} = f_{m+1}(t^{l+1})$ can be prescribed to derive versions of Eq. (30.14) for the first and the last interior nodes. For the first interior node,

$$2(1 + \lambda)T_1^{l+1} - \lambda T_2^{l+1} = \lambda f_0(t^l) + 2(1 - \lambda)T_1^l + \lambda T_2^l + \lambda f_0(t^{l+1}) \qquad (30.15)$$

and for the last interior node,

$$-\lambda T_{m-1}^{l+1} + 2(1 + \lambda)T_m^{l+1} = \lambda f_{m+1}(t^l) + 2(1 - \lambda)T_m^l + \lambda T_{m-1}^l + \lambda f_{m+1}(t^{l+1}) \qquad (30.16)$$

Although Eqs. (30.14) through (30.16) are slightly more complicated than Eqs. (30.8), (30.10), and (30.11), they are also tridiagonal and, therefore, efficient to solve.

EXAMPLE 30.3   Crank-Nicolson Solution to the Heat-Conduction Equation

Problem Statement. Use the Crank-Nicolson method to solve the same problem as in Examples 30.1 and 30.2.

Solution. Equations (30.14) through (30.16) can be employed to generate the following tridiagonal set of equations:

$$
\begin{bmatrix}
2.04175 & -0.020875 & & \\
-0.020875 & 2.04175 & -0.020875 & \\
& -0.020875 & 2.04175 & -0.020875 \\
& & -0.020875 & 2.04175
\end{bmatrix}
\begin{Bmatrix}
T_1^1 \\ T_2^1 \\ T_3^1 \\ T_4^1
\end{Bmatrix}
=
\begin{Bmatrix}
4.175 \\ 0 \\ 0 \\ 2.0875
\end{Bmatrix}
$$

which can be solved for the temperatures at $t = 0.1$ s:

$T_1^1 = 2.0450$
$T_2^1 = 0.0210$
$T_3^1 = 0.0107$
$T_4^1 = 1.0225$

To solve for the temperatures at $t = 0.2$ s, the right-hand-side vector must be changed to

$$
\begin{Bmatrix}
8.1801 \\ 0.0841 \\ 0.0427 \\ 4.0901
\end{Bmatrix}
$$

The simultaneous equations can then be solved for

$T_1^2 = 4.0073$
$T_2^2 = 0.0826$
$T_3^2 = 0.0422$
$T_4^2 = 2.0036$

30.4.1 Comparison of One-Dimensional Methods

Equation (30.1) can be solved analytically. For example, a solution is available for the case where the rod's temperature is initially at zero. At $t = 0$, the boundary condition at $x = L$ is instantaneously increased to a constant level of T while $T(0)$ is held at zero. For this case, the temperature can be computed with

$$T = \bar{T} \left[\frac{x}{L} + \sum_{n=1}^{\infty} \frac{2}{n\pi} (-1)^n \sin\left(\frac{n \times x}{L}\right) \exp\left(\frac{-n^2\pi^2 kt}{L^2}\right) \right] \tag{30.17}$$

where L = total length of the rod. This equation can be employed to compute the evolution of the temperature distribution for each boundary condition. Then, the total solution can be determined by superposition.

EXAMPLE 30.4

Comparison of Analytical and Numerical Solutions

Problem Statement.   Compare the analytical solution from Eq. (30.17) with numerical results obtained with the explicit, simple implicit, and Crank-Nicolson techniques. Perform the comparison for the rod employed in Examples 30.1, 30.2, and 30.3.

Solution.   Recall from the previous examples that $k = 0.835$ cm^2/s, $L = 10$ cm, and $\Delta x = 2$ cm. For this case, Eq. (30.17) can be used to predict that the temperature at $x = 2$ cm, and $t = 10$ s would equal 64.8018. Table 30.1 presents numerical predictions of $T(2, 10)$. Notice that a range of time steps are employed. These results indicate a number of properties of the numerical methods. First, it can be seen that the explicit method is unstable for high values of λ. This instability is not manifested by either implicit approach. Second, the Crank-Nicolson method converges more rapidly as λ is decreased and provides moderately accurate results even when λ is relatively high. These outcomes are as expected because Crank-Nicolson is second-order accurate with respect to both independent variables. Finally, notice that as λ decreases, the methods seem to be converging on a value of 64.73 that is different than the analytical result of 64.80. This should not be surprising because a fixed value of $\Delta x = 2$ is used to characterize the x dimension. If both Δx and Δt were decreased as λ was decreased (that is, more spatial segments were used), the numerical solution would more closely approach the analytical result.

TABLE 30.1 Comparison of three methods of solving a parabolic PDE: the heated rod. The results shown are for temperature at $t = 10$ s at $x = 2$ cm for the rod from Examples 30.1 through 30.3. Note that the analytical solution is $T(2, 10) = 64.8018$.

Δt	λ	Explicit	Implicit	Crank-Nicolson
10	2.0875	208.75	53.01	79.77
5	1.04375	−9.13	58.49	64.79
2	0.4175	67.12	62.22	64.87
1	0.20875	65.91	63.49	64.77
0.5	0.104375	65.33	64.12	64.74
0.2	0.04175	64.97	64.49	64.73

The Crank-Nicolson method is often used for solving linear parabolic PDEs in one spatial dimension. Its advantages become even more pronounced for more complicated applications such as those involving unequally spaced meshes. Such nonuniform spacing is often advantageous where we have foreknowledge that the solution varies rapidly in local portions of the system. Further discussion of such applications and the Crank-Nicolson method in general can be found elsewhere (Ferziger 1981; Lapidus and Pinder 1981; Hoffman 1992).

30.5 PARABOLIC EQUATIONS IN TWO SPATIAL DIMENSIONS

The heat-conduction equation can be applied to more than one spatial dimension. For two dimensions, its form is

$$\frac{\partial T}{\partial t} = k\left(\frac{\partial^2 T}{\partial x^2} + \frac{\partial^2 T}{\partial y^2}\right) \tag{30.18}$$

One application of this equation is to model the temperature distribution on the face of a heated plate. However, rather than characterizing its steady-state distribution, as was done in Chap. 29, Eq. (30.18) provides a means to compute the plate's temperature distribution as it changes in time.

30.5.1 Standard Explicit and Implicit Schemes

An explicit solution can be obtained by substituting finite-difference approximations of the form of Eqs. (30.2) and (30.3) into Eq. (30.18). However, as with the one-dimensional case, this approach is limited by a stringent stability criterion. For the two-dimensional case, the criterion is

$$\Delta t \le \frac{1}{8}\frac{(\Delta x)^2 + (\Delta y)^2}{k}$$

Thus, for a uniform grid ($\Delta x = \Delta y$), $\lambda = k\Delta t/(\Delta x)^2$ must be less than or equal to $1/4$. Consequently, halving the step size results in a fourfold increase in the number of nodes and a 16-fold increase in computational effort.

As was the case with one-dimensional systems, implicit techniques offer alternatives that guarantee stability. However, the direct application of implicit methods such as the Crank-Nicolson technique leads to the solution of $m \times n$ simultaneous equations. Additionally, when written for two or three spatial dimensions, these equations lose the valuable property of being tridiagonal. Thus, matrix storage and computation time can become exorbitantly large. The method described in the next section offers one way around this dilemma.

30.5.2 The ADI Scheme

The alternating-direction implicit, or ADI, scheme provides a means for solving parabolic equations in two spatial dimensions using tridiagonal matrices. To do this, each time

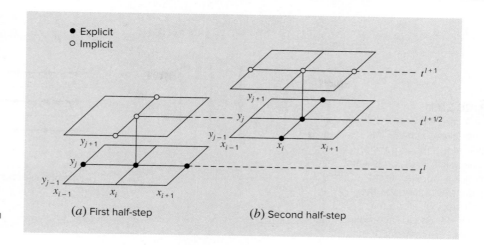

FIGURE 30.10
The two half-steps used in implementing the alternating-direction implicit scheme for solving parabolic equations in two spatial dimensions.

increment is executed in two steps (Fig. 30.10). For the first step, Eq. (30.18) is approximated by

$$\frac{T_{i,j}^{l+1/2} - T_{i,j}^{l}}{\Delta t/2} = k\left[\frac{T_{i+1,j}^{l} - 2T_{i,j}^{l} + T_{i-1,j}^{l}}{(\Delta x)^2} + \frac{T_{i,j+1}^{l+1/2} - 2T_{i,j}^{l+1/2} + T_{i,j-1}^{l+1/2}}{(\Delta y)^2}\right] \tag{30.19}$$

Thus, the approximation of $\partial^2 T/\partial x^2$ is written explicitly—that is, at the base point t^l where values of temperature are known. Consequently, only the three temperature terms in the approximation of $\partial^2 T/\partial y^2$ are unknown. For the case of a square grid ($\Delta y = \Delta x$), this equation can be expressed as

$$-\lambda T_{i,j-1}^{l+1/2} + 2(1+\lambda)T_{i,j}^{l+1/2} - \lambda T_{i,j+1}^{l+1/2} = \lambda T_{i-1,j}^{l} + 2(1-\lambda)T_{i,j}^{l} + \lambda T_{i+1,j}^{l} \tag{30.20}$$

which, when written for the system, results in a tridiagonal set of simultaneous equations. For the second step from $t^{l+1/2}$ to t^{l+1}, Eq. (30.18) is approximated by

$$\frac{T_{i,j}^{l+1} - T_{i,j}^{l+1/2}}{\Delta t/2} = k\left[\frac{T_{i+1,j}^{l+1} - 2T_{i,j}^{l+1} + T_{i-1,j}^{l+1}}{(\Delta x)^2} + \frac{T_{i,j+1}^{l+1/2} - 2T_{i,j}^{l+1/2} + T_{i,j-1}^{l+1/2}}{(\Delta y)^2}\right] \tag{30.21}$$

In contrast to Eq. (30.19), the approximation of $\partial^2 T/\partial x^2$ is now implicit. Thus, the bias introduced by Eq. (30.19) will be partially corrected. For a square grid, Eq. (30.21) can be written as

$$-\lambda T_{i-1,j}^{l+1} + 2(1+\lambda)T_{i,j}^{l+1} - \lambda T_{i+1,j}^{l+1} = \lambda T_{i,j-1}^{l+1/2} + 2(1-\lambda)T_{i,j}^{l+1/2} + \lambda T_{i,j+1}^{l+1/2} \tag{30.22}$$

Again, when written for a two-dimensional grid, the equation results in a tridiagonal system (Fig. 30.11). As in the following example, this leads to an efficient numerical solution.

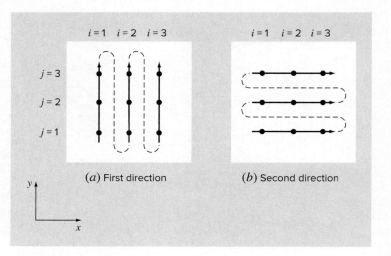

FIGURE 30.11
The ADI method only results in tridiagonal equations if it is applied along the dimension that is implicit. Thus, on the first step (*a*), it is applied along the *y* dimension and, on the second step (*b*), along the *x* dimension. These "alternating directions" are the root of the method's name.

EXAMPLE 30.5　　　**ADI Method**

Problem Statement.　Use the ADI method to solve for the temperature of the plate in Examples 29.1 and 29.2. At $t = 0$, assume that the temperature of the plate is zero and the boundary temperatures are instantaneously brought to the levels shown in Fig. 29.4. Employ a time step of 10 s. Recall from Example 30.1 that the coefficient of thermal diffusivity for aluminum is $k = 0.835$ cm^2/s.

Solution.　A value of $\Delta x = 10$ cm was employed to characterize the 40×40-cm plate from Examples 29.1 and 29.2. Therefore, $\lambda = 0.835(10)/(10)^2 = 0.0835$. For the first step to $t = 5$ (Fig. 30.11*a*), Eq. (30.20) is applied to nodes (1, 1), (1, 2), and (1, 3) to yield the following tridiagonal equations:

$$\begin{bmatrix} 2.167 & -0.0835 & \\ -0.0835 & 2.167 & -0.0835 \\ & -0.0835 & 2.167 \end{bmatrix} \begin{Bmatrix} T_{1,1} \\ T_{1,2} \\ T_{1,3} \end{Bmatrix} = \begin{Bmatrix} 6.2625 \\ 6.2625 \\ 14.6125 \end{Bmatrix}$$

which can be solved for

$$T_{1,1} = 3.01597 \qquad T_{1,2} = 3.2708 \qquad T_{1,3} = 6.8692$$

In a similar fashion, tridiagonal equations can be developed and solved for

$$T_{2,1} = 0.1274 \qquad T_{2,2} = 0.2900 \qquad T_{2,3} = 4.1291$$

and

$$T_{3,1} = 2.0181 \qquad T_{3,2} = 2.2477 \qquad T_{3,3} = 6.0256$$

For the second step to $t = 10$ (Fig. 30.11*b*), Eq. (30.22) is applied to nodes (1, 1), (2, 1), and (3, 1) to yield

$$\begin{bmatrix} 2.167 & -0.0835 & \\ -0.0835 & 2.167 & -0.0835 \\ & -0.0835 & 2.167 \end{bmatrix} \begin{Bmatrix} T_{1,1} \\ T_{2,1} \\ T_{3,1} \end{Bmatrix} = \begin{Bmatrix} 12.0639 \\ 0.2577 \\ 8.0619 \end{Bmatrix}$$

60.76 52.57 53.02	72.82 68.17 64.12	76.54 73.29 67.68
41.09 27.20 31.94	55.26 45.32 44.86	60.30 52.25 49.67
28.56 14.57 20.73	37.40 25.72 28.69	40.82 30.43 31.96
$(a)\, t = 100$ s	$(b)\, t = 200$ s	$(c)\, t = 300$ s

FIGURE 30.12
Solution for the heated plate of Example 30.5 at $(a)\, t = 100$ s, $(b)\, t = 200$ s, and $(c)\, t = 300$ s.

which can be solved for

$$T_{1,1} = 5.5855 \qquad T_{2,1} = 0.4782 \qquad T_{3,1} = 3.7388$$

Tridiagonal equations for the other rows can be developed and solved for

$$T_{1,2} = 6.1683 \qquad T_{2,2} = 0.8238 \qquad T_{3,2} = 4.2359$$

and

$$T_{1,3} = 13.1120 \qquad T_{2,3} = 8.3207 \qquad T_{3,3} = 11.3606$$

The computation can be repeated, and the results for $t = 100, 200,$ and 300 s are depicted in Fig. 30.12a through c, respectively. As expected, the temperature of the plate rises. After a sufficient time elapses, the temperature will approach the steady-state distribution of Fig. 29.5.

The ADI method is but one of a group of techniques called splitting methods. Some of these represent efforts to circumvent shortcomings of ADI. Discussion of other splitting methods as well as more information on ADI can be found elsewhere (Ferziger 1981; Lapidus and Pinder 1981).

PROBLEMS

30.1 Repeat Example 30.1, but use the midpoint method to generate your solution.

30.2 Repeat Example 30.1, but for the case where the rod is initially at 50°C and the derivative at $x = 0$ is equal to 1 and at $x = 10$ is equal to 0. Interpret your results.

30.3 (a) Repeat Example 30.1, but for a time step of $\Delta t = 0.05$ s. Compute results to $t = 0.2$. (b) In addition, perform the same computation with the Heun method (without iteration of the corrector) with a much smaller step size of $\Delta t = 0.001$ s. Assuming that the results of (b) are a valid approximation of the true solution, determine percent relative errors for the results obtained in Example 30.1 as well as for part (a).

30.4 Repeat Example 30.2, but for the case where the derivative at $x = 10$ is equal to zero.

30.5 Repeat Example 30.3, but for $\Delta x = 1$ cm.

30.6 Repeat Example 30.5, but for the plate described in Prob. 29.2.

30.7 The advection-diffusion equation is used to compute the distribution of concentration along the length of a rectangular chemical reactor (see Sec. 32.1),

$$\frac{\partial c}{\partial t} = D\frac{\partial^2 c}{\partial x^2} - U\frac{\partial c}{\partial x} - kc$$

where c = concentration (mg/m³), t = time (min), D = a diffusion coefficient (m²/min), x = distance along the tank's longitudinal axis (m) where $x = 0$ at the tank's inlet, U = velocity in the x direction (m/min), and k = a reaction rate (min⁻¹) whereby the chemical decays to another form. Develop an explicit scheme to solve this equation numerically. Test it for $k = 0.2$, $D = 75$, and $U = 2.5$ for a tank of length 20 m. Use $\Delta x = 2$ m and a step size of $\Delta t = 0.005$. Assume that the inflow concentration is 100 and that the initial concentration in the tank is zero. Perform the simulation from $t = 0$ to 100 and plot the final resulting concentrations versus x.

30.8 Develop a user-friendly computer program for the simple explicit method from Sec. 30.2. Test it by duplicating Example 30.1.

30.9 Modify the program in Prob. 30.8 so that it employs either Dirichlet or derivative boundary conditions. Test it by solving Prob. 30.2.

30.10 Develop a user-friendly computer program to implement the simple implicit scheme from Sec. 30.3. Test it by duplicating Example 30.2.

30.11 Develop a user-friendly computer program to implement the Crank-Nicolson method from Sec. 30.4. Test it by duplicating Example 30.3.

30.12 Develop a user-friendly computer program for the ADI method described in Sec. 30.5. Test it by duplicating Example 30.5.

30.13 The nondimensional form for the transient heat conduction in an insulated rod (Eq. 30.1) can be written as

$$\frac{\partial^2 u}{\partial \bar{x}^2} = \frac{\partial u}{\partial \bar{t}}$$

where nondimensional space, time, and temperature are defined as

$$\bar{x} = \frac{x}{L} \qquad \bar{t} = \frac{T}{(\rho C L^2/k)} \qquad u = \frac{T - T_o}{T_L - T_o}$$

where L = the rod length, k = thermal conductivity of the rod material, ρ = density, C = specific heat, T_o = temperature at $x = 0$, and T_L = temperature at $x = L$. This yields the following boundary and initial conditions:

Boundary conditions	$u(0, \bar{t}) = 0$	$u(1, \bar{t}) = 0$
Initial conditions	$u(\bar{x}, 0) = 0$	$0 \le \bar{x} \le 1$

Solve this nondimensional equation for the temperature distribution using finite-difference methods and a second-order accurate Crank-Nicolson formulation to integrate in time. Write a computer program to obtain the solution. Increase the value of $\Delta\bar{t}$ by 10% for each time step to more quickly obtain the steady-state solution, and select values of $\Delta\bar{x}$ and $\Delta\bar{t}$ for good accuracy. Plot the nondimensional temperature versus nondimensional length for various values of nondimensional times.

30.14 The problem of transient radial heat flow in a circular rod in nondimensional form is described by

$$\frac{\partial^2 u}{\partial \bar{r}^2} + \frac{1}{\bar{r}}\frac{\partial u}{\partial \bar{r}} = \frac{\partial u}{\bar{t}}$$

Boundary conditions	$u(1, \bar{t}) = 1$	$\frac{\partial u}{\partial \bar{t}}(0, \bar{t}) = 0$
Initial conditions	$u(\bar{x}, 0) = 0$	$0 \le \bar{x} \le 1$

Solve the nondimensional transient radial heat-conduction equation in a circular rod for the temperature distribution at various times as the rod temperature approaches steady state. Use second-order accurate finite-difference analogues for the derivatives, with a Crank-Nicolson formulation. Write a computer program for the solution. Select values of $\Delta\bar{r}$ and $\Delta\bar{t}$ for good accuracy. Plot the temperature u versus radius $\bar{r}$ for various times $\bar{t}$.

30.15 Solve the following PDE:

$$\frac{\partial^2 u}{\partial x^2} + b\frac{\partial u}{\partial x} = \frac{\partial u}{\partial t}$$

Boundary conditions	$u(0, t) = 0$	$u(1, t) = 0$
Initial conditions	$u(x, 0) = 0$	$0 \le x \le 1$

Use second-order accurate finite-difference analogues for the derivatives with a Crank-Nicolson formulation to integrate in time. Write a computer program for the solution. Increase the value of Δt by 10% for each time step to more quickly obtain the steady-state solution, and select values of Δx and Δt for good accuracy. Plot u versus x for various values of t. Solve for these values of b: 4, 2, 0, −2, −4.

30.16 Determine the temperatures along a 1-m horizontal rod described by the heat-conduction equation (Eq. 30.1). Assume that the right boundary is insulated and that the left boundary ($x = 0$) is represented by

$$-k'\frac{\partial T}{\partial x}\bigg|_{x=0} = h(T_a - T_0)$$

where k' = coefficient of thermal conductivity (W/m °C), h = convective heat transfer coefficient (W/m² °C), T_a = ambient temperature (°C), and T_0 = temperature of the rod at $x = 0$ (°C). Solve for temperature as a function of time using a spatial step of $\Delta x = 1$ cm and the following parameter values: $k = 2 \times 10^{-5}$ m²/s, $k' = 10$ W/m °C, $h = 25$ W/m² °C, and $T_a = 50$ °C. Assume that the initial temperature of the rod is zero.

CHAPTER

31

Finite-Element Method

To this juncture, we have employed *finite-difference* methods to solve partial differential equations. In these methods, the solution domain is divided into a grid of discrete points or nodes (Fig. 31.1*b*). The PDE is then written for each node and its derivatives replaced by finite-divided differences. Although such "pointwise" approximation is conceptually easy to understand, it has a number of shortcomings. In particular, it becomes harder to apply for systems with irregular geometry, unusual boundary conditions, or heterogenous composition.

The *finite-element* method provides an alternative that is better suited for such systems. In contrast to finite-difference techniques, the finite-element method divides the solution domain into simply shaped regions, or "elements" (Fig. 31.1*c*). An approximate solution for

FIGURE 31.1

(*a*) A gasket with irregular geometry and nonhomogeneous composition. (*b*) Such a system is very difficult to model with a finite-difference approach. This is due to the fact that complicated approximations are required at the boundaries of the system and at the boundaries between regions of differing composition. (*c*) A finite-element discretization is much better suited for such systems.

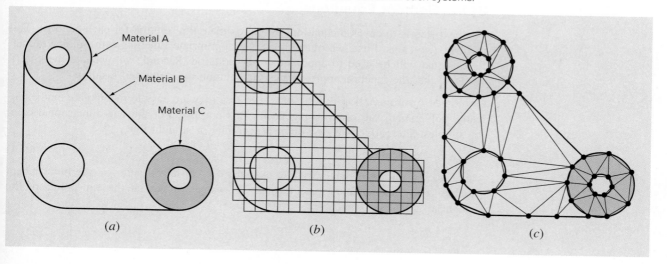

(*a*)   (*b*)   (*c*)

the PDE can be developed for each of these elements. The total solution is then generated by linking together, or "assembling," the individual solutions, taking care to ensure continuity at the interelement boundaries. Thus, the PDE is satisfied in a piecewise fashion.

As in Fig. 31.1c, the use of elements, rather than a rectangular grid, provides a much better approximation for irregularly shaped systems. Further, values of the unknown can be generated continuously across the entire solution domain rather than at isolated points.

Because a comprehensive description is beyond the scope of this book, this chapter provides a general introduction to the finite-element method. Our primary objective is to make you comfortable with the approach and cognizant of its capabilities. In this spirit, the following section is devoted to a general overview of the steps involved in a typical finite-element solution of a problem. This is followed by a simple example: a steady-state, one-dimensional heated rod. Although this example does not involve PDEs, it allows us to develop and demonstrate major aspects of the finite-element approach unencumbered by complicating factors. We can then discuss some issues involved in employing the finite-element method to solve PDEs.

31.1   THE GENERAL APPROACH

Although the particulars will vary, the implementation of the finite-element approach usually follows a standard step-by-step procedure. The following provides a brief overview of each of these steps. The application of these steps to engineering problem contexts will be developed in subsequent sections.

31.1.1 Discretization

This step involves dividing the solution domain into finite elements. Figure 31.2 provides examples of elements employed in one, two, and three dimensions. The points of intersection of the lines that make up the sides of the elements are referred to as nodes and the sides themselves are called *nodal lines* or *planes*.

31.1.2 Element Equations

The next step is to develop equations to approximate the solution for each element. This involves two steps. First, we must choose an appropriate function with unknown coefficients that will be used to approximate the solution. Second, we evaluate the coefficients so that the function approximates the solution in an optimal fashion.

Choice of Approximation Functions.   Because they are easy to manipulate mathematically, polynomials are often employed for this purpose. For the one-dimensional case, the simplest alternative is a first-order polynomial or straight line,

$$u(x) = a_0 + a_1 x \tag{31.1}$$

where $u(x)$ = the dependent variable, a_0 and a_1 = constants, and x = the independent variable. This function must pass through the values of $u(x)$ at the end points of the element at x_1 and x_2. Therefore,

$$u_1 = a_0 + a_1 x_1$$
$$u_2 = a_0 + a_1 x_2$$

FIGURE 31.2
Examples of elements employed in (a) one, (b) two, and (c) three dimensions.

where $u_1 = u(x_1)$ and $u_2 = u(x_2)$. These equations can be solved using Cramer's rule for

$$a_0 = \frac{u_1 x_2 - u_2 x_1}{x_2 - x_1} \qquad a_1 = \frac{u_2 - u_1}{x_2 - x_1}$$

These results can then be substituted into Eq. (31.1), which, after collection of terms, can be written as

$$u = N_1 u_1 + N_2 u_2 \tag{31.2}$$

where

$$N_1 = \frac{x_2 - x}{x_2 - x_1} \tag{31.3}$$

and

$$N_2 = \frac{x - x_1}{x_2 - x_1} \tag{31.4}$$

Equation (31.2) is called an *approximation,* or *shape, function,* and N_1 and N_2 are called *interpolation functions.* Close inspection reveals that Eq. (31.2) is, in fact, the Lagrange

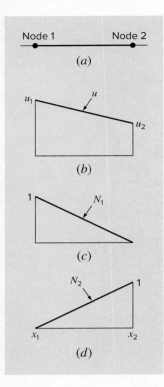

FIGURE 31.3
(b) A linear approximation or shape function for (a) a line element. The corresponding interpolation functions are shown in (c) and (d).

first-order interpolating polynomial. It provides a means to predict intermediate values (that is, to interpolate) between given values u_1 and u_2 at the nodes.

Figure 31.3 shows the shape function along with the corresponding interpolation functions. Notice that the sum of the interpolation functions is equal to 1.

In addition, the fact that we are dealing with linear equations facilitates operations such as differentiation and integration. Such manipulations will be important in later sections. The derivative of Eq. (31.2) is

$$\frac{du}{dx} = \frac{dN_1}{dx}u_1 + \frac{dN_2}{dx}u_2 \tag{31.5}$$

According to Eqs. (31.3) and (31.4), the derivatives of the N's can be calculated as

$$\frac{dN_1}{dx} = -\frac{1}{x_2 - x_1} \qquad \frac{dN_2}{dx} = \frac{1}{x_2 - x_1} \tag{31.6}$$

and, therefore, the derivative of u is

$$\frac{du}{dx} = \frac{1}{x_2 - x_1}(-u_1 + u_2) \tag{31.7}$$

In other words, it is a divided difference representing the slope of the straight line connecting the nodes.

The integral can be expressed as

$$\int_{x_1}^{x_2} u \, dx = \int_{x_1}^{x_2} (N_1 u_1 + N_2 u_2) \, dx$$

Each term on the right-hand side is merely the integral of a right triangle with base $x_2 - x_1$ and height u. That is,

$$\int_{x_1}^{x_2} Nu \, dx = \frac{1}{2}(x_2 - x_1)u$$

Thus, the entire integral is

$$\int_{x_1}^{x_2} u \, dx = \frac{u_1 + u_2}{2}(x_2 - x_1) \tag{31.8}$$

In other words, it is simply the trapezoidal rule.

Obtaining an Optimal Fit of the Function to the Solution. Once the interpolation function is chosen, the equation governing the behavior of the element must be developed. This equation represents a fit of the function to the solution of the underlying differential equation. Several methods are available for this purpose. Among the most common are the direct approach, the method of weighted residuals, and the variational approach. The outcome of all of these methods is analogous to curve fitting. However, instead of fitting functions to data, these methods specify relationships between the unknowns in Eq. (31.2) that satisfy the underlying PDE in an optimal fashion.

Mathematically, the resulting element equations will often consist of a set of linear algebraic equations that can be expressed in matrix form,

$$[k]\{u\} = \{F\} \tag{31.9}$$

where $[k]$ = an *element property*, or *stiffness, matrix*, $\{u\}$ = a column vector of unknowns at the nodes, and $\{F\}$ = a column vector reflecting the effect of any external influences applied at the nodes. Note that, in some cases, the equations can be nonlinear. However, for the elementary examples described herein, and for many practical problems, the systems are linear.

31.1.3 Assembly

After the individual element equations are derived, they must be linked together or assembled to characterize the unified behavior of the entire system. The assembly process is governed by the concept of continuity. That is, the solutions for contiguous elements are matched so that the unknown values (and sometimes the derivatives) at their common nodes are equivalent. Thus, the total solution will be continuous.

When all the individual versions of Eq. (31.9) are finally assembled, the entire system is expressed in matrix form as

$$[k]\{u'\} = \{F'\} \tag{31.10}$$

where $[K]$ = the *assemblage property matrix* and $\{u'\}$ and $\{F'\}$ column vectors for unknowns and external forces that are marked with primes to denote that they are an assemblage of the vectors $\{u\}$ and $\{F\}$ from the individual elements.

31.1.4 Boundary Conditions

Before Eq. (31.10) can be solved, it must be modified to account for the system's boundary conditions. These adjustments result in

$$[\bar{k}]\{u'\} = \{\bar{F}'\} \tag{31.11}$$

where the overbars signify that the boundary conditions have been incorporated.

31.1.5 Solution

Solutions of Eq. (31.11) can be obtained with techniques described previously in Part Three, such as *LU* decomposition. In many cases, the elements can be configured so that the resulting equations are banded. Thus, the highly efficient solution schemes available for such systems can be employed.

31.1.6 Postprocessing

Upon obtaining a solution, it can be displayed in tabular form or graphically. In addition, secondary variables can be determined and displayed.

Although the preceding steps are very general, they are common to most implementations of the finite-element approach. In the following section, we illustrate how they can be applied to obtain numerical results for a simple physical system—a heated rod.

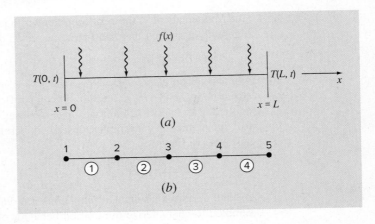

FIGURE 31.4
(a) A long, thin rod subject to fixed boundary conditions and a continuous heat source along its axis. (b) The finite-element representation consisting of four equal-length elements and five nodes.

31.2   FINITE-ELEMENT APPLICATION IN ONE DIMENSION

Figure 31.4 shows a system that can be modeled by a one-dimensional form of Poisson's equation

$$\frac{d^2 T}{dx^2} = -f(x) \tag{31.12}$$

where $f(x)$ = a function defining a heat source along the rod and where the ends of the rod are held at fixed temperatures,

$$T(0, t) = T_1$$

and

$$T(L, t) = T_2$$

Notice that this is not a partial differential equation but rather is a boundary-value ODE. This simple model is used because it will allow us to introduce the finite-element approach without some of the complications involved with, for example, a two-dimensional PDE.

EXAMPLE 31.1   **Analytical Solution for a Heated Rod**

Problem Statement.   Solve Eq. (31.12) for a 10-cm rod with boundary conditions of $T(0, t) = 40$ and $T(10, t) = 200$ and a uniform heat source of $f(x) = 10$.

Solution.   The equation to be solved is

$$\frac{d^2 T}{dx^2} = -10$$

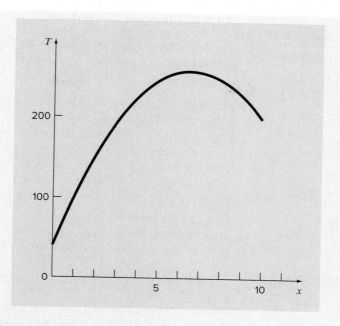

FIGURE 31.5
The temperature distribution along a heated rod subject to a uniform heat source and held at fixed end temperatures.

Assume a solution of the form

$$T = ax^2 + bx + c$$

which can be differentiated twice to give $T'' = 2a$. Substituting this result into the differential equation gives $a = -5$. The boundary conditions can be used to evaluate the remaining coefficients. For the first condition at $x = 0$,

$$40 = -5(0)^2 + b(0) + c$$

or $c = 40$. Similarly, for the second condition,

$$200 = -5(10)^2 + b(10) + 40$$

which can be solved for $b = 66$. Therefore, the final solution is

$$T = -5x^2 + 66x + 40$$

The results are plotted in Fig. 31.5.

31.2.1 Discretization

A simple configuration to model the system is a series of equal-length elements (Fig. 31.4b). Thus, the system is treated as four equal-length elements and five nodes.

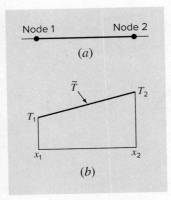

FIGURE 31.6
(a) An individual element.
(b) The approximation function
used to characterize the
temperature distribution along
the element.

31.2.2 Element Equations

An individual element is shown in Fig. 31.6a. The distribution of temperature for the element can be represented by the approximation function

$$\widetilde{T} = N_1 T_1 + N_2 T_2 \tag{31.13}$$

where N_1 and N_2 = linear interpolation functions specified by Eqs. (31.3) and (31.4), respectively. Thus, as depicted in Fig. 31.6b, the approximation function amounts to a linear interpolation between the two nodal temperatures.

As noted in Sec. 31.1, there are a variety of approaches for developing the element equation. In this section, we employ two of these. First, a *direct approach* will be used for the simple case where $f(x) = 0$. Then, because of its general applicability in engineering, we will devote most of the section to the *method of weighted residuals*.

The Direct Approach. For the case where $f(x) = 0$, a direct method can be employed to generate the element equations. The relationship between heat flux and temperature gradient can be represented by Fourier's law:

$$q = -k' \frac{dT}{dx}$$

where q = flux [cal/(cm^2 s)] and k' = the coefficient of thermal conductivity [cal/(s cm °C)]. If a linear approximation function is used to characterize the element's temperature, the heat flow into the element through node 1 can be represented by

$$q_1 = k' \frac{T_1 - T_2}{x_2 - x_1}$$

where q_1 is heat flux at node 1. Similarly, for node 2,

$$q_2 = k' \frac{T_2 - T_1}{x_2 - x_1}$$

These two equations express the relationship of the element's internal temperature distribution (as reflected by the nodal temperatures) to the heat flux at its ends. As such, they constitute our desired element equations. They can be simplified further by recognizing that Fourier's law can be used to couch the end fluxes themselves in terms of the temperature gradients at the boundaries. That is,

$$q_1 = -k' \frac{dT(x_1)}{dx} \qquad q_2 = k' \frac{dT(x_2)}{dx}$$

which can be substituted into the element equations to give

$$\frac{1}{x_2 - x_1} \begin{bmatrix} 1 & -1 \\ -1 & 1 \end{bmatrix} \begin{Bmatrix} T_1 \\ T_2 \end{Bmatrix} = \begin{Bmatrix} -\dfrac{dT(x_1)}{dx} \\ \dfrac{dT(x_2)}{dx} \end{Bmatrix} \tag{31.14}$$

Notice that Eq. (31.14) has been cast in the format of Eq. (31.9). Thus, we have succeeded in generating a matrix equation that describes the behavior of a typical element in our system.

The direct approach has great intuitive appeal. Additionally, in areas such as mechanics, it can be employed to solve meaningful problems. However, in other contexts, it is often difficult or impossible to derive finite-element equations directly. Consequently, as described next, more general mathematical techniques are required.

The Method of Weighted Residuals. The differential equation [Eq. (31.12)] can be re-expressed as

$$0 = \frac{d^2 T}{dx^2} + f(x)$$

The approximate solution [Eq. (31.13)] can be substituted into this equation. Because Eq. (31.13) is not the exact solution, the left side of the resulting equation will not be zero but will equal a residual,

$$R = \frac{d^2 \widetilde{T}}{dx^2} + f(x) \tag{31.15}$$

The *method of weighted residuals* (*MWR*) consists of finding a minimum for the residual according to the general formula

$$\int_D R W_i \, dD = 0 \qquad i = 1, 2, \ldots, m \tag{31.16}$$

where D = the solution domain and W_i = linearly independent weighting functions.

At this point, there are a variety of choices that could be made for the weighting function (Box 31.1). The most common approach for the finite-element method is to employ the interpolation functions N_i as the weighting functions. When these are substituted into Eq. (31.16), the result is referred to as Galerkin's method,

$$\int_D R N_i \, dD = 0 \qquad i = 1, 2, \ldots, m$$

For our one-dimensional rod, Eq. (31.15) can be substituted into this formulation to give

$$\int_{x_1}^{x_2} \left[\frac{d^2 \widetilde{T}}{dx^2} + f(x) \right] N_i \, dx \qquad i = 1, 2$$

which can be re-expressed as

$$\int_{x_1}^{x_2} \frac{d^2 \widetilde{T}}{dx^2} N_i(x) \, dx = - \int_{x_1}^{x_2} f(x) N_i(x) \, dx \qquad i = 1, 2 \tag{31.17}$$

At this point, a number of mathematical manipulations will be applied to simplify and evaluate Eq. (31.17). Among the most important is the simplification of the left-hand side using integration by parts. Recall from calculus that this operation can be expressed generally as

$$\int_a^b u \, dv = uv \Big|_a^b - \int_a^b v \, du$$

Box 31.1 Alternative Residual Schemes for the MWR

Several choices can be made for the weighting functions of Eq. (31.16). Each represents an alternative approach for the MWR.

In the *collocation approach,* we choose as many locations as there are unknown coefficients. Then, the coefficients are adjusted until the residual vanishes at each of these locations. Consequently, the approximating function will yield perfect results at the chosen locations but will have a nonzero residual elsewhere. Thus, it is akin to the interpolation methods in Chap. 18. Note that collocation amounts to using the weighting function

$$W = \delta(x - x_i) \qquad \text{for } i = 1, 2, \ldots, n$$

where n = the number of unknown coefficients and $\delta(x - x_i)$ = the *Dirac delta function* that vanishes everywhere but at $x = x_i$, where it equals 1.

In the *subdomain method,* the interval is divided into as many segments, or "subdomains," as there are unknown coefficients. Then, the coefficients are adjusted until the average value of the residual is zero in each subdomain. Thus, for each subdomain, the weighting function is equal to 1 and Eq. (31.16) is

$$\int_{x_{i-1}}^{x_i} R \, dx = 0 \qquad \text{for } i = 1, 2, \ldots, n$$

where x_{i-1} and x_i are the bounds of the subdomain.

For the *least-squares approach,* the coefficients are adjusted so as to minimize the integral of the square of the residual. Thus, the weighting functions are

$$W_i = \frac{\partial R}{\partial a_i}$$

which can be substituted into Eq. (31.16) to give

$$\int_D R \frac{\partial R}{\partial a_i} \, dD = 0 \qquad i = 1, 2, \ldots, n$$

or

$$\frac{\partial}{\partial a_i} \int_D R^2 \, dD = 0 \qquad i = 1, 2, \ldots, n$$

Comparison of the formulation with those of Chap. 17 shows that this is the continuous form of regression.

Galerkin's method employs the interpolation functions N_i as weighting functions. Recall that these functions always sum to 1 at any position in an element. For many problem contexts, Galerkin's method yields the same results as are obtained by variational methods. Consequently, it is the most commonly employed version of MWR used in finite-element analysis.

If u and v are chosen properly, the new integral on the right-hand side will be easier to evaluate than the original one on the left-hand side. This can be done for the term on the left-hand side of Eq. (31.17) by choosing $N_i(x)$ as u and $(d^2 T/dx^2)$ dx as dv to yield

$$\int_{x_1}^{x_2} N_i(x) \frac{d^2 \tilde{T}}{dx^2} dx = N_i(x) \frac{d\tilde{T}}{dx}\bigg|_{x_1}^{x_2} - \int_{x_1}^{x_2} \frac{d\tilde{T}}{dx} \frac{dN_i}{dx} dx \qquad i = 1, 2 \tag{31.18}$$

Thus, we have taken the significant step of lowering the highest-order term in the formulation from a second to a first derivative.

Next, we can evaluate the individual terms that we have created in Eq. (31.18). For $i = 1$, the first term on the right-hand side of Eq. (31.18) can be evaluated as

$$N_1(x) \frac{d\tilde{T}}{dx}\bigg|_{x_1}^{x_2} = N_1(x_2) \frac{d\tilde{T}(x_2)}{dx} - N_1(x_1) \frac{d\tilde{T}(x_1)}{dx}$$

However, recall from Fig. 31.3 that $N_1(x_2) = 0$ and $N_1(x_1) = 1$, and therefore,

$$N_1(x) \frac{d\tilde{T}}{dx}\bigg|_{x_1}^{x_2} = -\frac{d\tilde{T}(x_1)}{dx} \tag{31.19}$$

Similarly, for $i = 2$,

$$N_2(x) \frac{d\tilde{T}}{dx}\bigg|_{x_1}^{x_2} = \frac{d\tilde{T}(x_2)}{dx} \tag{31.20}$$

Thus, the first term on the right-hand side of Eq. (31.18) represents the natural boundary conditions at the ends of the elements.

Now, before proceeding let us regroup by substituting our results back into the original equation. Substituting Eqs. (31.18) through (31.20) into Eq. (31.17) and rearranging gives for $i = 1$,

$$\int_{x_1}^{x_2} \frac{d\tilde{T}}{dx} \frac{dN_1}{dx} \, dx = -\frac{d\tilde{T}(x_1)}{dx} + \int_{x_1}^{x_2} f(x) N_1(x) \, dx \tag{31.21}$$

and for $i = 2$,

$$\int_{x_1}^{x_2} \frac{d\tilde{T}}{dx} \frac{dN_2}{dx} \, dx = \frac{d\tilde{T}(x_2)}{dx} + \int_{x_1}^{x_2} f(x) N_2(x) \, dx \tag{31.22}$$

Notice that the integration by parts has led to two important outcomes. First, it has incorporated the boundary conditions directly into the element equations. Second, it has lowered the highest-order evaluation from a second to a first derivative. This latter outcome yields the significant result that the approximation functions need to preserve continuity of value but not slope at the nodes.

Also notice that we can now begin to ascribe some physical significance to the individual terms we have derived. On the right-hand side of each equation, the first term represents one of the element's boundary conditions and the second is the effect of the system's forcing function—in the present case, the heat source $f(x)$. As will now become evident, the left-hand side embodies the internal mechanisms that govern the element's temperature distribution. That is, in terms of the finite-element method, the left-hand side will become the element property matrix.

To see this, let us concentrate on the terms on the left-hand side. For $i = 1$, the term is

$$\int_{x_1}^{x_2} \frac{d\tilde{T}}{dx} \frac{dN_1}{dx} \, dx \tag{31.23}$$

Recall from Sec. 31.1.2 that the linear nature of the shape function makes differentiation and integration simple. Substituting Eqs. (31.6) and (31.7) into Eq. (31.23) gives

$$\int_{x_1}^{x_2} \frac{T_1 - T_2}{(x_2 - x_1)^2} \, dx = \frac{1}{x_2 - x_1}(T_1 - T_2) \tag{31.24}$$

Similar substitutions for $i = 2$ [Eq. (31.22)] yield

$$\int_{x_1}^{x_2} \frac{-T_1 + T_2}{(x_2 - x_1)^2} \, dx = \frac{1}{x_2 - x_1}(-T_1 + T_2) \tag{31.25}$$

Comparison with Eq. (31.14) shows that these are similar to the relationships that were developed with the direct method using Fourier's law. This can be made even clearer by re-expressing Eqs. (31.24) and (31.25) in matrix form as

$$\frac{1}{x_2 - x_1}\begin{bmatrix} 1 & -1 \\ -1 & 1 \end{bmatrix}\begin{Bmatrix} T_1 \\ T_2 \end{Bmatrix}$$

Substituting this result into Eqs. (31.21) and (31.22) and expressing the result in matrix form gives the final version of the element equations

$$\underbrace{\frac{1}{x_2 - x_1}\begin{bmatrix} 1 & -1 \\ -1 & 1 \end{bmatrix}\{T\}}_{\text{Element stiffness matrix}} = \underbrace{\begin{Bmatrix} -\dfrac{dT(x_1)}{dx} \\[2ex] \dfrac{dT(x_2)}{ds} \end{Bmatrix}}_{\substack{\text{Boundary} \\ \text{condition}}} + \underbrace{\begin{Bmatrix} \displaystyle\int_{x_1}^{x_2} f(x)N_1(x)\,dx \\[2ex] \displaystyle\int_{x_1}^{x_2} f(x)N_2(x)\,dx \end{Bmatrix}}_{\text{External effects}} \qquad (31.26)$$

Note that aside from the direct and the weighted residual methods, the element equations can also be derived using variational calculus (for example, see Allaire 1985). For the present case, this approach yields equations that are identical to those derived above.

EXAMPLE 31.2   Element Equation for a Heated Rod

Problem Statement.   Employ Eq. (31.26) to develop the element equations for a 10-cm rod with boundary conditions of $T(0, t) = 40$ and $T(10, t) = 200$ and a uniform heat source of $f(x) = 10$. Employ four equal-size elements of length $= 2.5$ cm.

Solution.   The heat source term in the first row of Eq. (31.26) can be evaluated by substituting Eq. (31.3) and integrating to give

$$\int_0^{2.5} 10\frac{2.5 - x}{2.5}\,dx = 12.5$$

Similarly, Eq. (31.4) can be substituted into the heat source term of the second row of Eq. (31.26), which can also be integrated to yield

$$\int_0^{2.5} 10\frac{x - 0}{2.5}\,dx = 12.5$$

These results along with the other parameter values can be substituted into Eq. (31.26) to give

$$0.4T_1 - 0.4T_2 = -\frac{dT}{dx}(x_1) + 12.5$$

and

$$-0.4T_1 + 0.4T_2 = \frac{dT}{dx}(x_2) + 12.5$$

31.2.3 Assembly

Before the element equations are assembled, a global numbering scheme must be established to specify the system's topology, or spatial layout. As in Table 31.1, this defines the

TABLE 31.1 The system topology for the finite-element segmentation scheme from Fig. 31.4b.

Element	Node Numbers	
	Local	Global
1	1	1
	2	2
2	1	2
	2	3
3	1	3
	2	4
4	1	4
	2	5

FIGURE 31.7
The assembly of the equations for the total system.

$$(a) \quad \begin{bmatrix} 0.4 & -0.4 & 0 & 0 & 0 \\ -0.4 & 0.4 & 0 & 0 & 0 \\ 0 & 0 & 0 & 0 & 0 \\ 0 & 0 & 0 & 0 & 0 \\ 0 & 0 & 0 & 0 & 0 \end{bmatrix} \begin{Bmatrix} T_1 \\ T_2 \\ 0 \\ 0 \\ 0 \end{Bmatrix} = \begin{Bmatrix} -dT(x_1)/dx + 12.5 \\ dT(x_2)/dx + 12.5 \\ 0 \\ 0 \\ 0 \end{Bmatrix}$$

$$(b) \quad \begin{bmatrix} 0.4 & -0.4 & 0 & 0 & 0 \\ -0.4 & 0.4 & +0.4 & -0.4 & 0 \\ 0 & -0.4 & 0.4 & 0 & 0 \\ 0 & 0 & 0 & 0 & 0 \\ 0 & 0 & 0 & 0 & 0 \end{bmatrix} \begin{Bmatrix} T_1 \\ T_2 \\ T_3 \\ 0 \\ 0 \end{Bmatrix} = \begin{Bmatrix} -dT(x_1)/dx + 12.5 \\ 12.5 + 12.5 \\ dT(x_3)/dx + 12.5 \\ 0 \\ 0 \end{Bmatrix}$$

$$(c) \quad \begin{bmatrix} 0.4 & -0.4 & 0 & 0 & 0 \\ -0.4 & 0.8 & -0.4 & 0 & 0 \\ 0 & -0.4 & 0.4 & +0.4 & -0.4 \\ 0 & 0 & -0.4 & 0.4 & 0 \\ 0 & 0 & 0 & 0 & 0 \end{bmatrix} \begin{Bmatrix} T_1 \\ T_2 \\ T_3 \\ T_4 \\ 0 \end{Bmatrix} = \begin{Bmatrix} -dT(x_1)/dx + 12.5 \\ 25 \\ 12.5 + 12.5 \\ dT(x_4)/dx + 12.5 \\ 0 \end{Bmatrix}$$

$$(d) \quad \begin{bmatrix} 0.4 & -0.4 & 0 & 0 & 0 \\ -0.4 & 0.8 & -0.4 & 0 & 0 \\ 0 & -0.4 & 0.8 & -0.4 & 0 \\ 0 & 0 & -0.4 & 0.4 & +0.4 & -0.4 \\ 0 & 0 & 0 & -0.4 & 0.4 \end{bmatrix} \begin{Bmatrix} T_1 \\ T_2 \\ T_3 \\ T_4 \\ T_5 \end{Bmatrix} = \begin{Bmatrix} -dT(x_1)/dx + 12.5 \\ 25 \\ 25 \\ 12.5 + 12.5 \\ dT(x_5)/dx + 12.5 \end{Bmatrix}$$

$$(e) \quad \begin{bmatrix} 0.4 & -0.4 & 0 & 0 & 0 \\ -0.4 & 0.8 & -0.4 & 0 & 0 \\ 0 & -0.4 & 0.8 & -0.4 & 0 \\ 0 & 0 & -0.4 & 0.8 & -0.4 \\ 0 & 0 & 0 & -0.4 & 0.4 \end{bmatrix} \begin{Bmatrix} T_1 \\ T_2 \\ T_3 \\ T_4 \\ T_5 \end{Bmatrix} = \begin{Bmatrix} -dT(x_1)/dx + 12.5 \\ 25 \\ 25 \\ 25 \\ dT(x_5)/dx + 12.5 \end{Bmatrix}$$

connectivity of the element mesh. Because the present case is one-dimensional, the numbering scheme might seem so predictable that it is trivial. However, for two- and three-dimensional problems it offers the only means to specify which nodes belong to which elements.

Once the topology is specified, the element equation (31.26) can be written for each element using the global coordinates. Then they can be added one at a time to assemble the total system matrix (note that this process is explored further in Sec. 32.4). The process is depicted in Fig. 31.7.

31.2.4 Boundary Conditions

Notice that, as the equations are assembled, the internal boundary conditions cancel. Thus, the final result for $\{F\}$ in Fig. 31.7e has boundary conditions for only the first and the last nodes. Because T_1 and T_5 are given, these natural boundary conditions at the ends of the bar, $dT(x_1)/dx$ and $dT(x_5)/dx$, represent unknowns. Therefore, the equations can be re-expressed as

$$
\begin{aligned}
\frac{dT}{dx}(x_1) \quad -0.4T_2 & & & & &= -3.5 \\
0.8T_2 & -0.4T_3 & & & &= 41 \\
-0.4T_2 & +0.8T_3 & -0.4T_4 & & &= 25 \\
& -0.4T_3 & +0.8T_4 & & &= 105 \\
& & -0.4T_4 & -\frac{dT}{dx}(x_5) &= -67.5
\end{aligned}
\tag{31.27}
$$

FIGURE 31.8
Results of applying the finite-element approach to a heated bar. The exact solution is also shown.

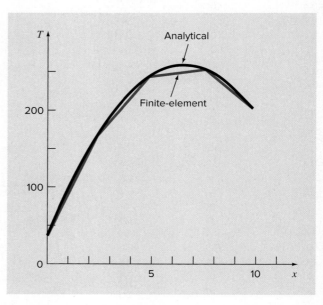

31.2.5 Solution

Equation (31.27) can be solved for

$$\frac{dT}{dx}(x_1) = 66 \qquad T_2 = 173.75 \qquad T_3 = 245$$

$$T_4 = 253.75 \qquad \frac{dT}{dx}(x_5) = -34$$

31.2.6 Postprocessing

The results can be displayed graphically. Figure 31.8 shows the finite-element results along with the exact solution. Notice that the finite-element calculation captures the overall trend of the exact solution and, in fact, provides an exact match at the nodes. However, a discrepancy exists in the interior of each element due to the linear nature of the shape functions.

31.3 TWO-DIMENSIONAL PROBLEMS

Although the mathematical "bookkeeping" increases markedly, the extension of the finite-element approach to two dimensions is conceptually similar to the one-dimensional applications discussed to this point. It thus follows the same steps as were outlined in Sec. 31.1.

31.3.1 Discretization

A variety of simple elements such as triangles or quadrilaterals are usually employed for the finite-element mesh in two dimensions. In the present discussion, we will limit ourselves to triangular elements of the type depicted in Fig. 31.9.

31.3.2 Element Equations

Just as for the one-dimensional case, the next step is to develop an equation to approximate the solution for the element. For a triangular element, the simplest approach is the linear polynomial [compare with Eq. (31.1)]

$$u(x, y) = a_0 + a_{1,1}x + a_{1,2}y \tag{31.28}$$

FIGURE 31.9
A triangular element.

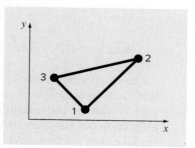

where $u(x, y)$ = the dependent variable, the a's = coefficients, and x and y = independent variables. This function must pass through the values of $u(x, y)$ at the triangle's nodes (x_1, y_1), (x_2, y_2), and (x_3, y_3). Therefore,

$$u_1(x, y) = a_0 + a_{1,1}x_1 + a_{1,2}y_1$$
$$u_2(x, y) = a_0 + a_{1,1}x_2 + a_{1,2}y_2$$
$$u_3(x, y) = a_0 + a_{1,1}x_3 + a_{1,2}y_3$$

or in matrix form,

$$\begin{bmatrix} 1 & x_1 & y_1 \\ 1 & x_2 & y_2 \\ 1 & x_3 & y_3 \end{bmatrix} \begin{Bmatrix} a_0 \\ a_{1,1} \\ a_{1,2} \end{Bmatrix} = \begin{Bmatrix} u_1 \\ u_2 \\ u_3 \end{Bmatrix}$$

which can be solved for

$$a_0 = \frac{1}{2A_e}[u_1(x_2y_3 - x_3y_2) + u_2(x_3y_1 - x_1y_3) + u_3(x_1y_2 - x_2y_1)] \qquad (31.29)$$

$$a_{1,1} = \frac{1}{2A_e}[u_1(y_2 - y_3) + u_2(y_3 - y_1) + u_3(y_1 - y_2)] \qquad (31.30)$$

$$a_{1,2} = \frac{1}{2A_e}[u_1(x_3 - x_2) + u_2(x_1 - x_3) + u_3(x_2 - x_1)] \qquad (31.31)$$

where A_e is the area of the triangular element,

$$A_e = \frac{1}{2}[(x_2y_3 - x_3y_2) + (x_3y_1 - x_1y_3) + (x_1y_2 - x_2y_1)]$$

Equations (31.29) through (31.31) can be substituted into Eq. (31.28). After collection of terms, the result can be expressed as

$$u = N_1u_1 + N_2u_2 + N_3u_3 \qquad (31.32)$$

where

$$N_1 = \frac{1}{2A_e}[(x_2y_3 - x_3y_2) + (y_2 - y_3)x + (x_3 - x_2)y]$$

$$N_2 = \frac{1}{2A_e}[(x_3y_1 - x_1y_3) + (y_3 - y_1)x + (x_1 - x_3)y]$$

$$N_3 = \frac{1}{2A_e}[(x_1y_2 - x_2y_1) + (y_1 - y_2)x + (x_2 - x_1)y]$$

Equation (31.32) provides a means to predict intermediate values for the element on the basis of the values at its nodes. Figure 31.10 shows the shape function along with the corresponding interpolation functions. Notice that the sum of the interpolation functions is always equal to 1.

As with the one-dimensional case, various methods are available for developing element equations based on the underlying PDE and the approximating functions. The resulting equations are considerably more complicated than Eq. (31.26). However, because the

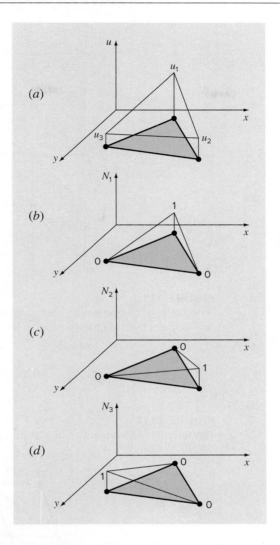

FIGURE 31.10
(a) A linear approximation function for a triangular element. The corresponding interpolation functions are shown in (b) through (d).

approximating functions are usually lower-order polynomials like Eq. (31.28), the terms of the final element matrix will consist of lower-order polynomials and constants.

31.3.3 Boundary Conditions and Assembly

The incorporation of boundary conditions and the assembly of the system matrix also become more complicated when the finite-element technique is applied to two- and three-dimensional problems. However, as with the derivation of the element matrix, the

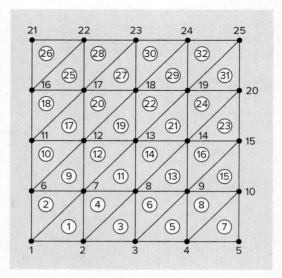

FIGURE 31.11
A numbering scheme for the nodes and elements of a finite-element approximation of the heated plate that was previously characterized by finite differences in Chap. 29.

FIGURE 31.12
The temperature distribution of a heated plate as calculated with a finite-element method.

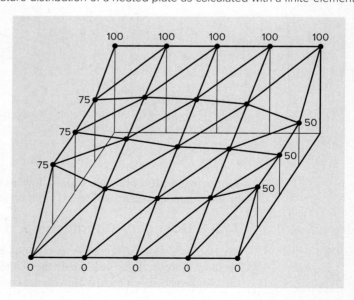

difficulty relates to the mechanics of the process rather than to conceptual complexity. For example, the establishment of the system topology, which was trivial for the one-dimensional case, becomes a matter of great importance in two and three dimensions. In particular, the choice of a numbering scheme will dictate the bandedness of the resulting system matrix and hence the efficiency with which it can be solved. Figure 31.11 shows a scheme that was developed for the heated plate formerly solved by finite-difference methods in Chap. 29.

31.3.4 Solution and Postprocessing

Although the mechanics are complicated, the system matrix is merely a set of n simultaneous equations that can be used to solve for the values of the dependent variable at the n nodes. Figure 31.12 shows a solution that corresponds to the finite-difference solution from Fig. 29.5.

31.4   SOLVING PDEs WITH SOFTWARE PACKAGES

Software packages have some capabilities for directly solving PDEs. However, as described in the following sections, many of the solutions are limited to simple problems. This is particularly true of two- and three-dimensional cases. For these situations, generic packages (that is, ones not expressly developed to solve PDEs such as finite-element packages) are often limited to simple rectangular domains.

Although this might seem limiting, simple applications can be of great utility in a pedagogical sense. This is particularly true when the packages' visualization tools are used to display calculation results.

31.4.1 Excel

Although Excel does not have the direct capability to solve PDEs, it is a nice environment to develop simple solutions of elliptic PDEs. For example, the orthogonal layout of the spreadsheet cells (Fig. 31.13b) is directly analogous to the grid used in Chap. 29 to model the heated plate (Fig. 31.13a).

FIGURE 31.13

The analogy between (a) a rectangular grid and (b) the cells of a spreadsheet.

	A	B	C	D	E
1	87.5	100	100	100	75
2	75	78.57	76.12	69.64	50
3	75	63.17	56.25	52.46	50
4	75	42.86	33.26	33.93	50
5	37.5	0	0	0	25

$$T_{13} = \frac{100 + T_{23} + T_{12} + 75}{4}$$

$$B2 = \frac{B1 + C2 + B3 + A2}{4}$$

(a) Grid

(b) Spreadsheet

As in Fig. 31.13*b*, the Dirichlet boundary conditions can first be entered along the periphery of the cell block. The formula for the Liebmann method can be implemented by entering Eq. (29.11) in one of the cells in the interior (like cell B2 in Fig. 31.13*b*). Thus, the value for the cell can be computed as a function of its adjacent cells. Then the cell can be copied to the other interior cells. Because of the relative nature of the Excel copy command, all the other cells will properly be dependent on their adjacent cells.

Once you have copied the formula, you will probably get an error message: **Cannot resolve circular references.** You can rectify this by selecting **File, Options** and clicking on the **Formulas** category. Then, go to the **Calculation options** section and enable the **Iterative calculation** check box. This will allow the spreadsheet to recalculate (the default is 100 iterations) and solve Liebmann's method iteratively. After this occurs, strike the F9 key to manually recalculate the sheet until the answers do not vary. This means that the solution has converged.

Once the problem has been solved, Excel's graphics tools can be used to visualize the results. An example is shown in Fig. 31.14*a*. For this case, we have

- Used a finer grid.
- Made the lower boundary insulated.
- Added a heat source of 150 to the middle of the plate (cell E5).

FIGURE 31.14
(*a*) Excel solution of the Poisson equation for a plate with an insulated lower edge and a heat source. (*b*) A "topographic map" and (*c*) a 3-D display of the temperatures.

	A	B	C	D	E	F	G	H	I
1	87.5	100.0	100.0	100.0	100.0	100.0	100.0	100.0	75.0
2	75.0	89.2	95.8	99.1	99.7	96.6	89.9	77.6	50.0
3	75.0	86.2	94.7	100.9	103.1	96.7	85.5	70.3	50.0
4	75.0	85.7	96.1	106.7	115.3	101.4	85.2	68.2	50.0
5	75.0	85.5	97.4	114.3	150.0	108.6	85.6	67.3	50.0
6	75.0	84.0	93.4	103.4	111.6	97.4	81.3	65.6	50.0
7	75.0	82.2	88.9	94.2	95.6	88.1	76.6	63.6	50.0
8	75.0	80.9	85.9	88.9	88.4	82.8	73.5	62.2	50.0
9	75.0	80.4	84.9	87.3	86.3	81.1	72.4	61.7	50.0

(*a*)

(*b*) (*c*)

The numerical results from Fig. 31.14a can then be displayed with Excel's Chart Wizard. Figure 31.14b and c show 3-D surface plots. The y orientation of these is normally the reverse of the spreadsheet. Thus, the top high-temperature edge (100) would normally be displayed at the bottom of the plot. We reversed the y values on our sheet prior to plotting so that the graphs would be consistent with the spreadsheet.

Notice how the graphs help you visualize what is going on. Heat flows down from the source toward the boundaries, forming a mountainlike shape. Heat also flows from the high-temperature boundary down to the two side edges. Notice how the heat flows preferentially toward the lower-temperature edge (50). Finally, notice how the temperature gradient in the y dimension goes to zero at the lower insulated edge ($\partial T/\partial y \rightarrow 0$).

31.4.2 MATLAB

Although the standard MATLAB software package does not presently have great capabilities for solving PDEs, M-files and functions can certainly be developed for this purpose. In addition, its display capabilities are very nice, particularly for visualization of 2-D spatial problems.

To illustrate this capability, we first set up the Excel spreadsheet in Fig. 31.14a. These results can be saved as a text (Tab delimited) file with a name like `plate.txt`. This file can then be moved to the MATLAB directory.

Once in MATLAB, the file can be loaded by typing

```
>> load plate.txt
```

Next, the gradients can be simply calculated as

```
>> [px,py]=gradient(plate);
```

Note that this is the simplest method to compute gradients using default values of dx = dy = 1. Therefore, the directions and relative magnitudes will be correct.

Finally, a series of commands can be used to develop the plot. The command `contour` develops a contour plot of the data. The command `clabel` adds contour labels to the plot. Finally, `quiver` takes the gradient data and adds it to the plot as arrows,

```
>> cs=contour(plate);clabel(cs);hold on
>> quiver(-px,-py);hold off
```

Note that the minus signs are added because of the minus sign in Fourier's law [Eq. (29.4)]. As seen in Fig. 31.15, the resulting plot provides an excellent representation of the solution.

Note that any file in the proper format can be entered into MATLAB and displayed in this way. This sharing of files between tools is becoming commonplace. In addition, files can be created in one location on one tool, transmitted over the Internet to another location, where the file might be displayed with another tool. This is one of the exciting aspects of modern numerical applications.

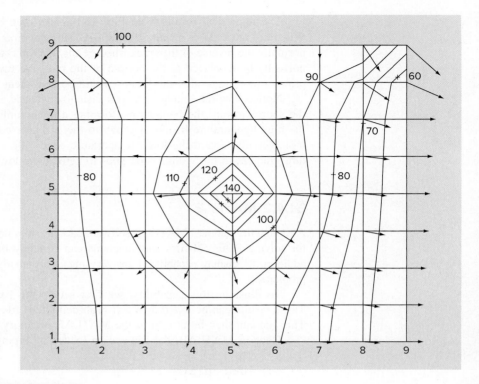

FIGURE 31.15
MATLAB-generated contour plots for the heated plate calculated with Excel (Fig. 31.14).

31.4.3 Mathcad

Mathcad has two functions that can solve Poisson's equation. You can use the **relax** function when you know the value of the unknown on all four sides of a square region. This function solves a system of linear algebraic equations using Gauss-Seidel iteration with overrelaxation to speed the rate of convergence. For the special case where there are internal sources or sinks, and the unknown function is zero on all four sides of the square, you can use the **multigrid** function, which is usually faster than **relax**. Both of these functions return a square matrix where the location of the element in the matrix corresponds to its location within the square region. The value of the element approximates the value of the solution of Poisson's equation at this point.

Figure 31.16 shows an example where a square plate contains heat sources while the boundary is maintained at zero. The first step is to establish dimensions for the temperature grid and the heat source matrix. The temperature grid has dimensions $(R + 1) \times (R + 1)$ while the heat source matrix is $R \times R$. For example, a 3×3 temperature grid has 4 (2×2) possible heat sources. In this case, we establish a 33×33 temperature grid and a 32×32 heat source matrix. The Mathcad command $M_{RR} := 0$ (with $R = 32$) establishes the dimensions of the source matrix and sets all the elements to zero. Next, the location and strength of two heat sources are established. Finally, S is the resulting temperature distribution

FIGURE 31.16
Mathcad screen to determine the solution of an elliptic PDE.

as calculated by the **multigrid** function. The second argument of **multigrid** is a parameter that controls the numerical accuracy. As suggested by Mathcad Help, a value of 2 generally gives a good approximation of the solution.

The temperature distribution can be displayed with surface, contour, or vector-field plots. These plots can be placed anywhere on the worksheet by clicking to the desired location. This places a red crosshair at that location. Then, use the Insert/Graph pull-down menu to place an empty plot on the worksheet with placeholders for the expressions to be graphed and for the ranges of variables. Simply type S in the placeholder on the z axis. Mathcad does the rest to produce the graphs shown in Fig. 31.16. Once the graph has been created, you can use the Format/Surface Plot and Format/Contour Plot pull-down menus to change the color or add titles, labels, and other features.

PROBLEMS

31.1 Repeat Example 31.1, but for $T(0, t) = 75$ and $T(10, t) = 150$ and a uniform heat source of 15.

31.2 Repeat Example 31.2, but for boundary conditions of $T(0, t) = 75$ and $T(10, t) = 150$ and a heat source of 15.

31.3 Apply the results of Prob. 31.2 to compute the temperature distribution for the entire rod using the finite-element approach.

31.4 Use Galerkin's method to develop an element equation for a steady-state version of the advection-diffusion equation described

in Prob. 30.7. Express the final result in the format of Eq. (31.26) so that each term has a physical interpretation.

31.5 A version of the Poisson equation that occurs in mechanics is the following model for the vertical deflection of a bar with a distributed load $P(x)$:

$$A_c E \frac{\partial^2 u}{\partial x^2} = P(x)$$

where A_c = cross-sectional area, E = Young's modulus, u = deflection, and x = distance measured along the bar's length. If the bar is rigidly fixed ($u = 0$) at both ends, use the finite-element method to model its deflections for $A_c = 0.1$ m², $E = 200 \times 10^9$ N/m², $L = 10$ m, and $P(x) = 1000$ N/m. Employ a value of $\Delta x = 2$ m.

31.6 Develop a user-friendly program to model the steady-state distribution of temperature in a rod with a constant heat source using the finite-element method. Set up the program so that unequally spaced nodes may be used.

31.7 Use Excel to perform the same computation as in Fig. 31.14, but insulate the right-hand edge and add a heat sink of -150 at cell C7.

31.8 Use MATLAB or Mathcad to develop a contour plot with flux arrows for the Excel solution from Prob. 31.7.

31.9 Use Excel to model the temperature distribution of the slab shown in Fig. P31.9. The slab is 0.02 m thick and has a thermal conductivity of 3 W/(m °C).

FIGURE P31.9

31.10 Use MATLAB or Mathcad to develop a contour plot with flux arrows for the Excel solution from Prob. 31.9.

31.11 Find the temperature distribution in a rod (Fig. P31.11) with internal heat generation using the finite-element method. Derive the element nodal equations using Fourier heat conduction and heat conservation relationships:

$$q_k = -kA \frac{dT}{\partial x}$$

and

$$\sum [q_k + f(x)] = 0$$

where q_k = heat flow (W), k = thermal conductivity [W/(m °C)], A = cross-sectional area (m²), and $f(x)$ = heat source (W/cm). The rod has a value of $kA = 100$ W m/°C. The rod is 50 cm long, the x-coordinate is zero at the left end and positive to the right. Divide the rod into five elements (six nodes, each 10 cm long). The

FIGURE P31.11

left end of the rod has a fixed temperature gradient and the temperature is a variable. The right end has a fixed temperature and the gradient is a variable. The heat source $f(x)$ has a constant value. Thus, the conditions are

$$\left.\frac{dT}{dx}\right|_{x=0} = 0.25°\text{C/m} \quad T|_{x=50} = 100°\text{C} \quad f(x) = 30 \text{ W/cm}$$

Develop the nodal equations that must be solved for the temperatures and temperature gradients at each of the six nodes. Assemble the equations, insert the boundary conditions, and solve the resulting set for the unknowns.

31.12 Find the temperature distribution in a rod (Fig. P31.12) with internal heat generation using the finite-element method. Derive the element nodal equations using Fourier heat conduction and heat conservation relationships:

$$q_k = -kA \frac{dT}{\partial x}$$

and

$$\sum [q_k + f(x)] = 0$$

where q_k = heat flow (W), k = thermal conductivity [W/(m °C)], A = cross-sectional area (m²), and $f(x)$ = heat source (W/cm). The

FIGURE P31.12

rod is 50 cm long, the x-coordinate is zero at the left end and positive to the right. The rod is also linearly tapered with a value of $kA = 100$ and 50 W m/°C at $x = 0$ and at $x = 50$, respectively. Divide the rod into five elements (six nodes, each 10 cm long). Both ends of the rod have fixed temperatures. The heat source $f(x)$ has a constant value. Thus, the conditions are

$$T\,|_{x=0} = 100°C \quad T\,|_{x=50} = 50°C \quad f(x) = 30 \text{ W/cm}$$

The tapered areas must be treated as if they were constant over the length of an element. Therefore, average the kA values at each end of the node and take that average as a constant over the node. Develop the nodal equations that must be solved for the temperatures and temperature gradients at each of the six nodes. Assemble the equations, insert the boundary conditions, and solve the resulting set for the unknowns.

31.13 Use a software package to solve for the temperature distribution of the L-shaped plate in Fig. P29.18. Display your results as a contour plot with flux arrows.

31.14 Use a software package like Excel or Matlab to determine the temperature distribution and fluxes for the plate depicted in Fig. P31.14. The plate is $60 \times 60 \times 1$ cm, is made out of aluminum $[k' = 0.49 \text{ cal/(s cm °C)}]$, and has an input of 10 cal/s into the middle node.

31.15 Repeat Prob. 31.14, but with the bottom edge insulated.

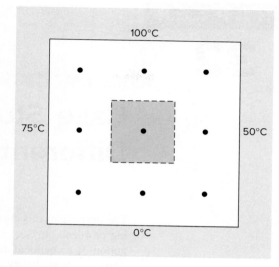

FIGURE P31.14

CHAPTER

32

Case Studies: Partial Differential Equations

The purpose of this chapter is to apply the methods from Part Eight to practical engineering problems. In Sec. 32.1, a parabolic PDE is used to compute the time-variable distribution of a chemical along the longitudinal axes of a rectangular reactor. This example illustrates how the instability of a solution can be due to the nature of the PDE rather than to properties of the numerical method.

Sections 32.2 and 32.3 involve applications of the Poisson and Laplace equations to civil and electrical engineering problems, respectively. Among other things, this will allow you to see similarities as well as differences between field problems in these areas of engineering. In addition, they can be contrasted with the heated-plate problem that has served as our prototype system in this part of the book. Section 32.2 deals with the deflection of a square plate, whereas Sec. 32.3 is devoted to computing the voltage distribution and charge flux for a two-dimensional surface with a curved edge.

Section 32.4 presents a finite-element analysis as applied to a series of springs. This application is closer in spirit to finite-element applications in mechanics and structures than was the temperature field problem used to illustrate the approach in Chap. 31.

32.1 ONE-DIMENSIONAL MASS BALANCE OF A REACTOR (CHEMICAL/BIO ENGINEERING)

Background.   Chemical engineers make extensive use of idealized reactors in their design work. In Secs. 12.1 and 28.1, we focused on single or coupled well-mixed reactors. These are examples of *lumped-parameter systems* (recall Sec. PT3.1.2).

FIGURE 32.1
An elongated reactor with a single entry and exit point. A mass balance is developed around a finite segment along the tank's longitudinal axis in order to derive a differential equation for the concentration.

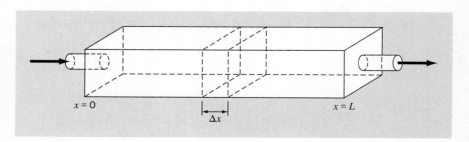

$x = 0$ Δx $x = L$

Figure 32.1 depicts an elongated reactor with a single entry and exit point. This reactor can be characterized as a *distributed-parameter system*. If it is assumed that the chemical being modeled is subject to first-order decay[1] and the tank is well-mixed vertically and laterally, a mass balance can be written for a finite segment of length Δx, as in

$$V\frac{\Delta c}{\Delta t} = \underbrace{Qc(x)}_{\text{Flow in}} - \underbrace{Q\left[c(x) + \frac{\partial c(x)}{\partial x}\Delta x\right]}_{\text{Flow out}} - \underbrace{DA_c\frac{\partial c(x)}{\partial x}}_{\text{Dispersion in}}$$

$$+ \underbrace{DA_c\left[\frac{\partial c(x)}{\partial x} + \frac{\partial}{\partial x}\frac{\partial c(x)}{\partial x}\Delta x\right]}_{\text{Dispersion out}} - \underbrace{kVc}_{\text{Decay reaction}} \qquad (32.1)$$

where V = volume (m³), Q = flow rate (m³/h), c is concentration (moles/m³), D is a dispersion coefficient (m²/h), A_c is the tank's cross-sectional area (m²), and k is the first-order decay coefficient (h⁻¹). Note that the dispersion terms are based on *Fick's first law*,

$$\text{Flux} = -D\frac{\partial c}{\partial x} \qquad (32.2)$$

which is directly analogous to Fourier's law for heat conduction [recall Eq. (29.4)]. It specifies that turbulent mixing tends to move mass from regions of high to low concentration. The parameter D, therefore, reflects the magnitude of turbulent mixing.

If Δx and Δt are allowed to approach zero, Eq. (32.1) becomes

$$\frac{\partial c}{\partial t} = D\frac{\partial^2 c}{\partial x^2} - U\frac{\partial c}{\partial x} - kc \qquad (32.3)$$

where $U = Q/A_c$ is the velocity of the water flowing through the tank. The mass balance for Fig. 32.1 is, therefore, now expressed as a parabolic partial differential equation. Equation (32.3) is sometimes referred to as the *advection-diffusion equation* with first-order reaction. At steady state, it is reduced to a second-order ODE,

$$0 = D\frac{d^2 c}{dx^2} - U\frac{d^2 c}{dx} - kc \qquad (32.4)$$

Prior to $t = 0$, the tank is filled with water that is devoid of the chemical. At $t = 0$, the chemical is injected into the reactor's inflow at a constant level of c_{in}. Thus, the following boundary conditions hold:

$$Qc_{\text{in}} = Qc_0 - DA_c\frac{\partial c_0}{\partial x}$$

and

$$c'(L, t) = 0$$

The second condition specifies that the chemical leaves the reactor purely as a function of flow through the outlet pipe. That is, it is assumed that dispersion in the reactor does

---

[1]That is, the chemical decays at a rate that is linearly proportional to how much chemical is present.

not affect the exit rate. Under these conditions, use numerical methods to solve Eq. (32.4) for the steady-state levels in the reactor. Note that this is an ODE boundary-value problem. Then solve Eq. (32.3) to characterize the transient response—that is, how the levels change in time as the system approaches the steady state. This application involves a PDE.

Solution. A steady-state solution can be developed by substituting centered finite differences for the first and the second derivatives in Eq. (32.4) to give

$$0 = D\frac{c_{i+1} - 2c_i + c_{i-1}}{\Delta x^2} - U\frac{c_{i+1} - c_{i-1}}{2\Delta x} - kc_i$$

Collecting terms gives

$$-\left(\frac{D}{U\Delta x} + \frac{1}{2}\right)c_{i-1} + \left(\frac{2D}{U\Delta x} + \frac{k\Delta x}{U}\right)c_0 - \left(\frac{D}{U\Delta x} - \frac{1}{2}\right)c_{i+1} = 0 \qquad (32.5)$$

This equation can be written for each of the system's nodes. At the reactor's ends, this process introduces nodes that lie outside the system. For example, at the inlet node ($i = 0$),

$$-\left(\frac{D}{U\Delta x} + \frac{1}{2}\right)c_{-1} + \left(\frac{2D}{U\Delta x} + \frac{k\Delta x}{U}\right)c_0 - \left(\frac{D}{U\Delta x} - \frac{1}{2}\right)c_1 = 0 \qquad (32.6)$$

The c_{-1} can be removed by invoking the first boundary condition. At the inlet, the following mass balance must hold:

$$Qc_{\text{in}} = Qc_0 - DA_c\frac{\partial c_0}{\partial x}$$

where c_0 = concentration at $x = 0$. Thus, this boundary condition specifies that the amount of chemical carried into the tank by advection through the pipe must be equal to the amount carried away from the inlet by both advection and turbulent dispersion in the tank. A finite divided difference can be substituted for the derivative

$$Qc_{\text{in}} = Qc_0 - DA_c\frac{c_1 - c_{-1}}{2\Delta x}$$

which can be solved for

$$c_{-1} = c_1 + \frac{2\Delta xU}{D}c_{\text{in}} - \frac{2\Delta xU}{D}c_0$$

which can be substituted into Eq. (32.6) to give

$$\left(\frac{2D}{U\Delta x} + \frac{k\Delta x}{U} + 2 + \frac{\Delta xU}{D}\right)c_0 - \left(\frac{2D}{U\Delta x}\right)c_1 = \left(2 + \frac{\Delta xU}{D}\right)c_{\text{in}} \qquad (32.7)$$

A similar derivation can be performed for the outlet, where the original difference equation is

$$-\left(\frac{D}{U\Delta x} + \frac{1}{2}\right)c_{n-1} + \left(\frac{2D}{U\Delta x} + \frac{k\Delta x}{U}\right)c_n - \left(\frac{D}{U\Delta x} - \frac{1}{2}\right)c_{n+1} = 0 \qquad (32.8)$$

The boundary condition at the outlet is

$$Qc_n - DA_c\frac{dc_n}{dx} = Qc_n$$

As with the inlet, a divided difference can be used to approximate the derivative:

$$Qc_n - DA_c\frac{c_{n+1} - c_{n-1}}{2\,\Delta x} = Qc_n \tag{32.9}$$

Inspection of this equation leads us to conclude that $c_{n+1} = c_{n-1}$. In other words, the slope at the outlet must be zero for Eq. (32.9) to hold. Substituting this result into Eq. (32.8) and simplifying gives

$$-\left(\frac{2D}{U\Delta x}\right)c_{n-1} + \left(\frac{2D}{U\Delta x} + \frac{k\,\Delta x}{U}\right)c_n = 0 \tag{32.10}$$

Equations (32.5), (32.7), and (32.10) now form a system of n tridiagonal equations with n unknowns. For example, if $D = 2$, $U = 1$, $\Delta x = 2.5$, $k = 0.2$, and $c_{in} = 100$, the system is

$$\begin{bmatrix} 5.35 & -1.6 & & & \\ -1.3 & 2.1 & -0.3 & & \\ & -1.3 & 2.1 & -0.3 & \\ & & -1.3 & 2.1 & -0.3 \\ & & & -1.6 & 2.1 \end{bmatrix} \begin{Bmatrix} c_0 \\ c_1 \\ c_2 \\ c_3 \\ c_4 \end{Bmatrix} = \begin{Bmatrix} 325 \\ 0 \\ 0 \\ 0 \\ 0 \end{Bmatrix}$$

which can be solved for

$$c_0 = 76.44 \qquad c_1 = 52.47 \qquad c_2 = 36.06$$
$$c_3 = 25.05 \qquad c_4 = 19.09$$

These results are plotted in Fig. 32.2. As expected, the concentration decreases due to the decay reaction as the chemical flows through the tank. In addition to the above computation, Fig. 32.2 shows another case with $D = 4$. Notice how increasing the turbulent mixing tends to flatten the curve.

FIGURE 32.2
Concentration versus distance along the longitudinal axis of a rectangular reactor for a chemical that decays with first-order kinetics.

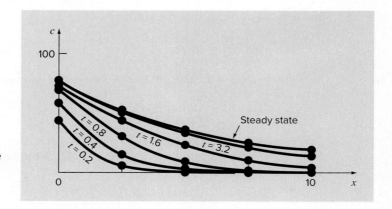

FIGURE 32.3
Concentration versus distance at different times during the buildup of chemical in a reactor.

In contrast, if dispersion is decreased, the curve will become steeper as mixing becomes less important relative to advection and decay. It should be noted that if dispersion is decreased too much, the computation will become subject to numerical errors. This type of error is referred to as *static instability* to contrast it with the *dynamic instability* due to too large a time step during a dynamic computation. The criterion to avoid this static instability is

$$\Delta x \le \frac{2D}{U}$$

Thus, the criterion becomes more stringent (lower Δx) for cases where advection dominates over dispersion.

Aside from steady-state computations, numerical methods can be used to generate time-variable solutions of Eq. (32.3). Figure 32.3 shows results for $D = 2$, $U = 1$, $\Delta x = 2.5$, $k = 0.2$, and $c_{in} = 100$, where the concentration in the tank is 0 at time zero. As expected, the immediate impact is near the inlet. With time, the solution eventually approaches the steady-state level.

It should be noted that in such dynamic calculations, the time step is constrained by a stability criterion expressed as (Chapra 1997)

$$\Delta t \le \frac{(\Delta x)^2}{2D + k(\Delta x)^2}$$

Thus, the reaction term acts to make the time step smaller.

32.2 DEFLECTIONS OF A PLATE (CIVIL/ENVIRONMENTAL ENGINEERING)

Background.   A square plate with simply supported edges is subject to an areal load q (Fig. 32.4). The deflection in the z dimension can be determined by solving the elliptic PDE (see Carnahan, Luther, and Wilkes 1969)

$$\frac{\partial^4 z}{\partial x^4} + 2\frac{\partial^4 z}{\partial x^2 \partial y^2} + \frac{\partial^4 z}{\partial y^4} = \frac{q}{D} \tag{32.11}$$

FIGURE 32.4
A simply supported square plate subject to an areal load.

subject to the boundary conditions that, at the edges, the deflection and slope normal to the boundary are zero. The parameter D is the flexural rigidity,

$$D = \frac{E \, \Delta z^3}{12(1 - \sigma^2)} \tag{32.12}$$

where E = the modulus of elasticity, Δz = the plate's thickness, and σ = Poisson's ratio. If a new variable is defined as

$$u = \frac{\partial^2 z}{\partial x^2} + \frac{\partial^2 z}{\partial y^2}$$

Eq. (32.11) can be re-expressed as

$$\frac{\partial^2 u}{\partial x^2} + \frac{\partial^2 u}{\partial y^2} = \frac{q}{D} \tag{32.13}$$

Therefore, the problem reduces to successively solving two Poisson equations. First, Eq. (32.13) can be solved for u subject to the boundary condition that $u = 0$ at the edges. Then, the results can be employed in conjunction with

$$\frac{\partial^2 z}{\partial x^2} + \frac{\partial^2 z}{\partial y^2} = u \tag{32.14}$$

to solve for z subject to the condition that $z = 0$ at the edges.

Develop a computer program to determine the deflections for a square plate subject to a constant areal load. Test the program for a plate with 2-m-long edges, $q = 33.6$ kN/m^2, $\sigma = 0.3$, $\Delta z = 10^{-2}$ m, and $E = 2 \times 10^{11}$ Pa. Employ $\Delta x = \Delta y = 0.5$ m for your test run.

Solution. Finite-divided differences can be substituted into Eq. (32.13) to give

$$\frac{u_{i+1,j} - 2u_{i,j} + u_{i-1,j}}{\Delta x^2} + \frac{u_{i,j+1} - 2u_{i,j} + u_{i,j-1}}{\Delta y^2} = \frac{q}{D} \tag{32.15}$$

Equation (32.12) can be used to compute $D = 1.832 \times 10^4$ N/m. This result, along with the other system parameters, can be substituted into Eq. (32.15) to give

$$u_{i+1,j} + u_{i-1,j} + u_{i,j+1} + u_{i,j-1} - 4u_{i,j} = 0.458$$

This equation can be written for all the nodes with the boundaries set at $u = 0$. The resulting equations are

$$
\begin{bmatrix}
-4 & 1 & & 1 & & & & & \\
1 & -4 & 1 & & 1 & & & & \\
& 1 & -4 & & & 1 & & & \\
1 & & & -4 & 1 & & 1 & & \\
& 1 & & 1 & -4 & 1 & & 1 & \\
& & 1 & & 1 & -4 & & & 1 \\
& & & 1 & & & -4 & 1 & \\
& & & & 1 & & 1 & -4 & 1 \\
& & & & & 1 & & 1 & -4
\end{bmatrix}
\begin{Bmatrix}
u_{1,1} \\ u_{2,1} \\ u_{3,1} \\ u_{1,2} \\ u_{2,2} \\ u_{3,2} \\ u_{1,3} \\ u_{2,3} \\ u_{3,3}
\end{Bmatrix}
=
\begin{Bmatrix}
0.458 \\ 0.458 \\ 0.458 \\ 0.458 \\ 0.458 \\ 0.458 \\ 0.458 \\ 0.458 \\ 0.458
\end{Bmatrix}
$$

which can be solved for

$$
\begin{array}{lll}
u_{1,1} = -0.315 & u_{1,2} = -0.401 & u_{1,3} = -0.315 \\
u_{2,1} = -0.401 & u_{2,2} = -0.515 & u_{2,3} = -0.401 \\
u_{3,1} = -0.315 & u_{3,2} = -0.401 & u_{3,3} = -0.315
\end{array}
$$

These results in turn can be substituted into Eq. (32.14), which can be written in finite-difference form and solved for

$$
\begin{array}{lll}
z_{1,1} = 0.063 & z_{1,2} = 0.086 & z_{1,3} = 0.063 \\
z_{2,1} = 0.086 & z_{2,2} = 0.118 & z_{2,3} = 0.086 \\
z_{3,1} = 0.063 & z_{3,2} = 0.086 & z_{3,3} = 0.063
\end{array}
$$

32.3 TWO-DIMENSIONAL ELECTROSTATIC FIELD PROBLEMS (ELECTRICAL ENGINEERING)

Background.   Just as Fourier's law and the heat balance can be employed to characterize temperature distribution, analogous relationships are available to model field problems in other areas of engineering. For example, electrical engineers use a similar approach when modeling electrostatic fields.

Under a number of simplifying assumptions, an analog of Fourier's law can be represented in one-dimensional form as

$$
D = -\varepsilon \frac{dV}{dx}
$$

where D is called the electric flux density vector, ε = permittivity of the material, and V = electrostatic potential.

Similarly, a Poisson equation for electrostatic fields can be represented in two dimensions as

$$
\frac{\partial^2 V}{\partial x^2} + \frac{\partial^2 V}{\partial y^2} = -\frac{\rho_v}{\varepsilon} \tag{32.16}
$$

where ρ_v = volumetric charge density.

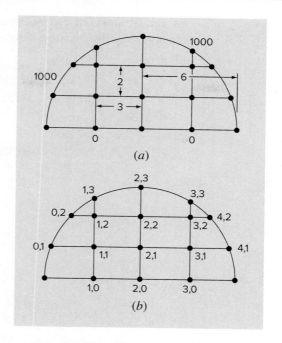

FIGURE 32.5

(*a*) A two-dimensional system with a voltage of 1000 along the circular boundary and a voltage of 0 along the base. (*b*) The nodal numbering scheme.

Finally, for regions containing no free charge (that is $\rho_v = 0$), Eq. (32.16) reduces to a Laplace equation,

$$\frac{\partial^2 V}{\partial x^2} + \frac{\partial^2 V}{\partial y^2} = 0 \tag{32.17}$$

Employ numerical methods to solve Eq. (32.17) for the situation depicted in Fig. 32.5. Compute the values for both V and for D if $\varepsilon = 2$.

Solution.   Using the approach outlined in Sec. 29.3.2, Eq. (29.24) can be written for node (1, 1) as

$$\frac{2}{\Delta x^2}\left[\frac{V_{1,1} - V_{0,1}}{\alpha_1(\alpha_1 + \alpha_2)} + \frac{V_{1,1} - V_{2,1}}{\alpha_2(\alpha_1 + \alpha_2)}\right] + \frac{2}{\Delta y^2}\left[\frac{V_{1,1} - V_{0,1}}{\beta_1(\beta_1 + \beta_2)} + \frac{V_{1,1} - V_{2,1}}{\beta_2(\beta_1 + \beta_2)}\right] = 0$$

According to the geometry depicted in Fig. 32.5, $\Delta x = 3$, $\Delta y = 2$, $\beta_1 = \beta_2 = \alpha_2 = 1$, and $\alpha_1 = 0.94281$. Substituting these values yields

$$0.12132\,V_{1,1} - 121.32 + 0.11438\,V_{1,1} - 0.11438\,V_{2,1} + 0.25\,V_{1,1}$$
$$+ 0.25\,V_{1,1} - 0.25\,V_{1,2} = 0$$

Collecting terms gives

$$0.73570V_{1,1} - 0.11438V_{2,1} - 0.25V_{1,2} = 121.32$$

A similar approach can be applied to the remaining interior nodes. The resulting simultaneous equations can be expressed in matrix form as

$$
\begin{bmatrix}
0.73570 & -0.11438 & & -0.25000 & & \\
-0.11111 & 0.72222 & -0.11111 & & -0.25000 & \\
& -0.11438 & 0.73570 & & & -0.25000 \\
-0.31288 & & & 1.28888 & -0.14907 & \\
& -0.25000 & & -0.11111 & 0.72222 & -0.11111 \\
& & -0.31288 & & -0.14907 & 1.28888
\end{bmatrix}
$$

$$
\times
\begin{Bmatrix}
V_{1,1} \\
V_{2,1} \\
V_{3,1} \\
V_{1,2} \\
V_{2,2} \\
V_{3,2}
\end{Bmatrix}
=
\begin{Bmatrix}
121.32 \\
0 \\
121.32 \\
826.92 \\
250 \\
826.92
\end{Bmatrix}
$$

which can be solved for

$$V_{1,1} = 521.19 \qquad V_{2,1} = 421.85 \qquad V_{3,1} = 521.19$$
$$V_{1,2} = 855.47 \qquad V_{2,2} = 755.40 \qquad V_{3,2} = 855.47$$

These results are depicted in Fig. 32.6a.

To compute the flux (recall Sec. 29.2.3), Eqs. (29.14) and (29.15) must be modified to account for the irregular boundaries. For the present example, the modifications

FIGURE 32.6

The results of solving the Laplace equation with correction factors for the irregular boundaries. (a) Potential and (b) flux.

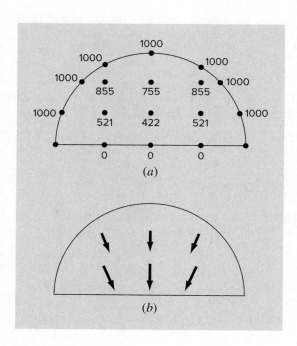

result in

$$D_x = -\varepsilon \frac{V_{i+1,j} - V_{i-1,j}}{(\alpha_1 + \alpha_2)\Delta x}$$

and

$$D_y = -\varepsilon \frac{V_{i,j+1} - V_{i,j-1}}{(\beta_1 + \beta_2)\Delta y}$$

For node (1, 1), these formulas can be used to compute the x and y components of the flux

$$D_x = -2\frac{421.85 - 1000}{(0.94281 + 1)3} = 198.4$$

and

$$D_y = -2\frac{855.47 - 0}{(1 + 1)2} = -427.7$$

which in turn can be used to calculate the electric flux density vector,

$$D = \sqrt{198.4^2 + (-427.7)^2} = 471.5$$

which has the direction

$$\theta = \tan^{-1}\left(\frac{-427.7}{198.4}\right) = -65.1°$$

The results for the other nodes are

Node	D_x	D_y	D	θ
2, 1	0.0	−377.7	377.7	−90
3, 1	−198.4	−427.7	471.5	245.1
1, 2	109.4	−299.6	281.9	−69.1
2, 2	0.0	−289.1	289.1	−90.1
3, 2	−109.4	−299.6	318.6	249.9

The fluxes are displayed in Fig. 32.6b.

32.4 FINITE-ELEMENT SOLUTION OF A SERIES OF SPRINGS (MECHANICAL/AEROSPACE ENGINEERING)

Background. Figure 32.7 shows a series of interconnected springs. One end is fixed to a wall, whereas the other is subject to a constant force F. Using the step-by-step procedure outlined in Chap. 31, a finite-element approach can be employed to determine the displacements of the springs.

Solution.

Discretization. The way to partition this system is obviously to treat each spring as an element. Thus, the system consists of four elements and five nodes (Fig. 32.7b).

(a)

FIGURE 32.7
(a) A series of interconnected springs. One end is fixed to a wall, whereas the other is subject to a constant force F.
(b) The finite-element representation. Each spring represents an element. Therefore, the system consists of four elements and five nodes.

(b)

FIGURE 32.8
A free-body diagram of an element of the spring system.

Element equations. Because this system is so simple, its element equations can be written directly without recourse to mathematical approximations. This is an example of the direct approach for deriving elements.

Figure 32.8 shows an individual element. The relationship between force F and displacement x can be represented mathematically by Hooke's law:

$$F = kx$$

where k = the spring constant, which can be interpreted as the force required to cause a unit displacement. If a force F_1 is applied at node 1, the following force balance must hold:

$$F = k(x_1 - x_2)$$

where x_1 = displacement of node 1 from its equilibrium position and x_2 = displacement of node 2 from its equilibrium position. Thus, $x_2 - x_1$ represents how much the spring is elongated or compressed relative to equilibrium (Fig. 32.8).

This equation can also be written as

$$F_1 = kx_1 - kx_2$$

For a stationary system, a force balance also necessitates that $F_1 = -F_2$ and, therefore,

$$F_2 = -kx_1 + kx_2$$

These two simultaneous equations specify the behavior of the element in response to prescribed forces. They can be written in matrix form as

$$\begin{bmatrix} k & -k \\ -k & k \end{bmatrix} \begin{Bmatrix} x_1 \\ x_2 \end{Bmatrix} = \begin{Bmatrix} F_1 \\ F_2 \end{Bmatrix}$$

or

$$[k]\{x\} = \{F\} \tag{32.18}$$

where the matrix $[k]$ is the element property matrix. For this case, it is also referred to as the *element stiffness matrix*. Notice that Eq. (32.18) has been cast in the format of Eq. (31.9). Thus, we have succeeded in generating a matrix equation that describes the behavior of a typical element in our system.

Before proceeding to the next step—the assembly of the total solution—we will introduce some notation. The elements of $[k]$ and $\{F\}$ are conventionally superscripted and subscripted, as in

$$\begin{bmatrix} k_{11}^{(e)} & -k_{12}^{(e)} \\ -k_{21}^{(e)} & k_{22}^{(e)} \end{bmatrix} \begin{Bmatrix} x_1 \\ x_2 \end{Bmatrix} = \begin{Bmatrix} F_1^{(e)} \\ F_2^{(e)} \end{Bmatrix}$$

where the superscript (e) designates that these are the element equations. The k's are also subscripted as k_{ij} to denote their location in the ith row and jth column of the matrix. For the present case, they can also be physically interpreted as representing the force required at node i to induce a unit displacement at node j.

Assembly. Before the element equations are assembled, all the elements and nodes must be numbered. This global numbering scheme specifies a system configuration, or topology (note that the present case uses a scheme identical to that in Table 31.1). That is, it documents which nodes belong to which element. Once the topology is specified, the equations for each element can be written with reference to the global coordinates.

The element equations can then be added one at a time to assemble the total system. The final result can be expressed in matrix form as [recall Eq. (31.10)]

$$[k]\{x'\} = \{F'\}$$

where

$$[k] = \begin{bmatrix} k_{11}^{(1)} & -k_{12}^{(1)} & & & \\ -k_{21}^{(1)} & k_{22}^{(1)} + k_{11}^{(2)} & -k_{12}^{(2)} & & \\ & -k_{21}^{(2)} & k_{22}^{(2)} + k_{11}^{(3)} & -k_{12}^{(3)} & \\ & & -k_{21}^{(3)} & k_{22}^{(3)} + k_{11}^{(4)} & -k_{12}^{(4)} \\ & & & -k_{21}^{(4)} & k_{22}^{(4)} \end{bmatrix} \tag{32.19}$$

and

$$\{F'\} = \begin{Bmatrix} F_1^{(1)} \\ 0 \\ 0 \\ 0 \\ F_2^{(4)} \end{Bmatrix}$$

and $\{x'\}$ and $\{F'\}$ are the expanded displacement and force vectors, respectively. Notice that, as the equations were assembled, the internal forces cancel. Thus, the final result for $\{F'\}$ has zeros for all but the first and last nodes.

Before proceeding to the next step, we must comment on the structure of the assemblage property matrix [Eq. (32.19)]. Notice that the matrix is tridiagonal. This is a direct result of the particular global numbering scheme that was chosen (Table 31.1) prior to assemblage. Although it is not very important in the present context, the attainment of such a banded, sparse system can be a decided advantage for more complicated problem settings. This is due to the efficient schemes that are available for solving such systems.

Boundary Conditions. The present system is subject to a single boundary condition, $x_1 = 0$. Introduction of this condition and applying the global renumbering scheme reduces the system to (k's = 1)

$$\begin{bmatrix} 2 & -1 & & \\ -1 & 2 & -1 & \\ & -1 & 2 & -1 \\ & & -1 & 1 \end{bmatrix} \begin{Bmatrix} x_2 \\ x_3 \\ x_4 \\ x_5 \end{Bmatrix} = \begin{Bmatrix} 0 \\ 0 \\ 0 \\ F \end{Bmatrix}$$

The system is now in the form of Eq. (31.11) and is ready to be solved.

Although reduction of the equations is certainly a valid approach for incorporating boundary conditions, it is usually preferable to leave the number of equations intact when performing the solution on the computer. Whatever the method, once the boundary conditions are incorporated, we can proceed to the next step—the solution.

Generating Solution. Using one of the approaches from Part Three, such as the efficient tridiagonal solution technique delineated in Chap. 11, the system can be solved for (with all k's = 1 and $F = 1$)

$$x_2 = 1 \qquad x_3 = 2 \qquad x_4 = 3 \qquad x_5 = 4$$

Postprocessing. The results can now be displayed graphically. As in Fig. 32.9, the results are as expected. Each spring is elongated a unit displacement.

FIGURE 32.9

(*a*) The original spring system. (*b*) The system after the application of a constant force. The displacements are indicated in the space between the two systems.

PROBLEMS

Chemical/Bio Engineering

32.1 Perform the same computation as in Sec. 32.1, but use $\Delta x = 1.25$.

32.2 Develop a finite-element solution for the steady-state system of Sec. 32.1.

32.3 Compute mass fluxes for the steady-state solution of Sec. 32.1 using Fick's first law.

32.4 Compute the steady-state distribution of concentration for the tank shown in Fig. P32.4. The PDE governing this system is

$$D\left(\frac{\partial^2 c}{\partial x^2} + \frac{\partial^2 c}{\partial y^2}\right) - kc = 0$$

and the boundary conditions are as shown. Employ a value of 0.6 for D and 0.1 for k.

32.5 Two plates are 10 cm apart, as shown in Fig. P32.5. Initially, both plates and the fluids are still. At $t = 0$, the top plate is moved at a constant velocity of 8 cm/s. The equations governing the motions of the fluids are

$$\frac{\partial v_{oil}}{\partial t} = \mu_{oil}\frac{\partial^2 v_{oil}}{\partial x^2} \quad \text{and} \quad \frac{\partial v_{water}}{\partial t} = \mu_{water}\frac{\partial^2 v_{water}}{\partial x^2}$$

and the following relationships hold true at the oil-water interface:

$$v_{oil} = v_{water} \quad \text{and} \quad \mu_{oil}\frac{\partial v_{oil}}{\partial x} = \mu_{water}\frac{\partial v_{water}}{\partial x}$$

What is the velocity of the two fluid layers at $t = 0.5$, 1, and 1.5 s at distances $x = 2$, 4, 6, and 8 cm from the bottom plate? Note that μ_{water} and $\mu_{oil} = 1$ and 3 cP, respectively.

32.6 The displacement of a uniform membrane subject to a tension and a uniform pressure can be described by the Poisson equation

$$\frac{\partial^2 z}{\partial x^2} + \frac{\partial^2 z}{\partial y^2} = -\frac{P}{T}$$

Solve for the displacement of a 1-cm-square membrane that has $P/T = 0.7$/cm and is fastened so that it has zero displacement along its four boundaries. Employ $\Delta x = \Delta y = 0.1$ cm. Display your results as a contour plot.

Civil/Environmental Engineering

32.7 Perform the same computation as in Sec. 32.2, but use $\Delta x = \Delta y = 0.4$ m.

32.8 The flow through porous media can be described by the Laplace equation

$$\frac{\partial^2 h}{\partial x^2} + \frac{\partial^2 h}{\partial y^2} = 0$$

where h is head. Use numerical methods to determine the distribution of head for the system shown in Fig. P32.8.

FIGURE P32.4

FIGURE P32.5

FIGURE P32.8

32.9 The velocity of water flow through the porous media can be related to head by D'Arcy's law

$$q_n = -K\frac{dh}{dn}$$

where K is the hydraulic conductivity and q_n is discharge velocity in the n direction. If $K = 5 \times 10^{-4}$ cm/s, compute the water velocities for Prob. 32.8.

Electrical Engineering
32.10 Perform the same computation as in Sec. 32.3 but for the system depicted in Fig. P32.10.
32.11 Perform the same computation as in Sec. 32.3 but for the system depicted in Fig. P32.11.

FIGURE P32.10

FIGURE P32.11

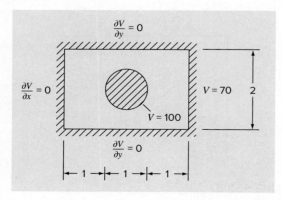

32.12 Use Poisson's equation to compute the electric potential over a unit square (1×1) plate with zero voltage at the edges and point charge sources of $\rho_v/\varepsilon\,(0.7, 0.7) = 1.5$ and $\rho_v/\varepsilon\,(0.3, 0.3) = -2$. Employ $\Delta x = \Delta y = 0.1$, and display your results as a contour plot.

Mechanical/Aerospace Engineering
32.13 Perform the same computation as in Sec. 32.4, but change the force to 1.5 and the spring constants to

Spring	1	2	3	4
k	0.75	1.5	0.5	2

32.14 Perform the same computation as in Sec. 32.4, but use a force of 2 and five springs with the following spring constants:

Spring	1	2	3	4	5
k	0.25	0.5	1.5	1	2

32.15 An insulated composite rod is formed of two parts arranged end to end, and both halves are of equal length. Part a has thermal conductivity k_a, for $0 \le x \le 1/2$, and part b has thermal conductivity k_b, for $1/2 \le x \le 1$. The nondimensional transient heat conduction equations that describe the temperature u over the length x of the composite rod are

$$\frac{\partial^2 u}{\partial x^2} = \frac{\partial u}{\partial t} \qquad 0 \le x \le 1/2$$

$$r\frac{\partial^2 u}{\partial x^2} = \frac{\partial u}{\partial t} \qquad 1/2 \le x \le 1$$

where u = temperature, x = axial coordinate, t = time, and $r = k_a/k_b$. The boundary and initial conditions are

Boundary conditions	$u(0, t) = 1$	$u(1, t) = 1$
	$\left(\dfrac{\partial u}{\partial x}\right)_a = \left(\dfrac{\partial u}{\partial x}\right)_b$	$x = 1/2$
Initial conditions	$u(x, 0) = 0$	$0 < x < 1$

Solve this set of equations for the temperature distribution as a function of time. Use second-order accurate finite-difference analogues for the derivatives with a Crank-Nicolson formulation to integrate in time. Write a computer program for the solution, and select values of Δx and Δt for good accuracy. Plot the temperature u versus length x for various values of time t. Generate a separate curve for the following values of the parameter $r = 1, 0.1, 0.01, 0.001$, and 0.
32.16 Solve the nondimensional transient heat conduction equation in two dimensions, which represents the transient temperature distribution in an insulated plate. The governing equation is

$$\frac{\partial^2 u}{\partial x^2} + \frac{\partial^2 u}{\partial y^2} = \frac{\partial u}{\partial t}$$

where u = temperature, x and y are spatial coordinates, and t = time. The boundary and initial conditions are

Boundary conditions	$u(x, 0, t) = 0$	$u(x, 1, t) = 1$
	$u(0, y, t) = 0$	$u(1, y, t) = 1$
Initial condition	$u(x, y, 0) = 0$	$0 \le x < 1 \qquad 0 \le y < 1$

Solve using the alternating-direction implicit (ADI) technique (Sec. 30.5.2). Write a computer program to implement the solution. Plot the results using a three-dimensional plotting routine where the horizontal plane contains the x and y axes and the z axis is the dependent variable u. Construct several plots at various times, including the following: (a) the initial conditions; (b) one intermediate time, approximately halfway to steady state; and (c) the steady-state condition.

EPILOGUE: PART EIGHT

PT8.3 TRADE-OFFS

The primary trade-offs associated with the use of numerical methods for the solution of partial differential equations involve choosing between *finite-difference* and *finite-element* approaches. The finite-difference methods are conceptually easier to understand. In addition, they are easy to program for systems that can be approximated with uniform grids. However, they are difficult to apply to systems with complicated geometries.

Finite-difference approaches can be divided into categories depending on the type of PDE that is being solved. *Elliptic PDEs* can be approximated by a set of linear algebraic equations. Consequently, the *Liebmann method* (which, in fact, is Gauss-Seidel) can be employed to obtain a solution iteratively.

One-dimensional parabolic PDEs can be solved in two fundamentally different ways: explicit or implicit approaches. The *explicit method* steps out in time in a fashion that is similar to Euler's technique for solving ODEs. It has the advantage that it is simple to program but has the shortcoming of a very stringent stability criterion. In contrast, stable implicit methods are available. These typically involve solving simultaneous tridiagonal algebraic equations at each time step. One of these approaches, the *Crank-Nicolson method,* is both accurate and stable and, therefore, is widely used for one-dimensional linear parabolic problems.

Two-dimensional parabolic PDEs can also be modeled explicitly. However, their stability constraints are even more severe than for the one-dimensional case. Special implicit approaches, which are generally referred to as splitting methods, have been developed to circumvent this shortcoming. These approaches are both efficient and stable. One of the most common is the *ADI,* or *alternating-direction implicit, method.*

All the above *finite-difference* approaches become unwieldy when applied to systems involving nonuniform shapes and heterogeneous conditions. Finite-element methods are available that handle such systems in a superior fashion.

Although the *finite-element method* is based on some fairly straightforward ideas, the mechanics of generating a good finite-element code for two- and three-dimensional problems is not a trivial exercise. In addition, this method can be computationally expensive for large problems. However, it is vastly superior to finite-difference approaches for systems involving complicated shapes. Consequently, its expense and conceptual "overhead" are often justified because of the detail of the final solution.

PT8.4 IMPORTANT RELATIONSHIPS AND FORMULAS

Table PT8.3 summarizes important information that was presented regarding the finite-difference methods in Part Eight. This table can be consulted to quickly access important relationships and formulas.

TABLE PT8.3 Summary of finite-difference methods.

	Computational Molecule	Equation
Elliptic PDEs Liebmann's method		$T_{i,j} = \dfrac{T_{i+1,j} + T_{i-1,j} + T_{i,j+1} + T_{i,j-1}}{4}$
Parabolic PDEs (one-dimensional) Explicit method		$T_i^{l+1} = T_i^l + \lambda(T_{i+1}^l - 2T_i^l + T_{i-1}^l)$
Implicit method		$-\lambda T_{i-1}^{l+1} + (1 + 2\lambda) T_i^{l+1} - \lambda T_{i+1}^{l+1} = T_i^l$
Crank-Nicolson method		$-\lambda T_{i-1}^{l+1} + 2(1 + \lambda) T_i^{l+1} - \lambda T_{i+1}^{l+1}$ $= \lambda T_{i-1}^l + 2(1 - \lambda) T_i^l + \lambda T_{i+1}^l$

PT8.5   ADVANCED METHODS AND ADDITIONAL REFERENCES

Carnahan, Luther, and Wilkes (1969); Rice (1983); Ferziger (1981); and Lapidus and Pinder (1981) provide useful surveys of methods and software for solving PDEs. You can also consult Ames (1977), Gladwell and Wait (1979), Vichnevetsky (1981, 1982), and Zienkiewicz (1971) for more in-depth treatments. Additional information on the finite-element method can be found in Allaire (1985), Huebner and Thornton (1982), Stasa (1985), and Baker (1983). In addition to elliptic and parabolic PDEs, numerical methods are also available to solve hyperbolic equations. Nice introductions and summaries of some of these methods can be found in Lapidus and Pinder (1981), Ferziger (1981), Forsythe and Wasow (1960), and Hoffman (1992).

The Fourier Series

The Fourier series can be expressed in a number of different formats. Two equivalent trigonometric expressions are

$$f(t) = a_0 + \sum_{k=1}^{\infty} [a_k \cos (k\omega_0 t) + b_k \sin (k\omega_0 t)]$$

or

$$f(t) = a_0 + \sum_{k=1}^{\infty} [c_k \cos (k\omega_0 t + \theta_k)]$$

where the coefficients are related by (see Fig. A.1)

$$c_k = \sqrt{a_k^2 + b_k^2}$$

and

$$\theta_k = -\tan^{-1} \left(\frac{b_k}{a_k} \right)$$

In addition to the trigonometric formats, the series can also be expressed in terms of the exponential function as

$$f(t) = \tilde{c}_0 + \sum_{k=1}^{\infty} [\tilde{c}_k e^{ik\omega_0 t} + \tilde{c}_{-k} e^{-ik\omega_0 t}] \qquad (A.1)$$

where (see Fig. A.2)

$$\tilde{c}_0 = a_0$$

$$\tilde{c}_k = \frac{1}{2} (a_k - ib_k) = |\tilde{c}_k| e^{i\phi_k}$$

$$\tilde{c}_{-k} = \frac{1}{2} (a_k + ib_k) = |\tilde{c}_k| e^{-i\phi_k}$$

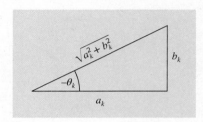

FIGURE A.1
Relationships between rectangular and polar forms of the Fourier series coefficients.

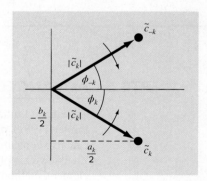

FIGURE A.2
Relationships between complex exponential and real coefficients of the Fourier series.

where $|\tilde{c}_0| = a_0$ and

$$|\tilde{c}_k| = \frac{1}{2}\sqrt{a_k^2 + b_k^2} = \frac{c_k}{2}$$

and

$$\phi_k = \tan^{-1}\left(\frac{-b_k}{a_k}\right)$$

Note that the tilde signifies that the coefficient is a complex number.

Each term in Eq. (A.1) can be visualized as a rotating phasor (the arrows in Fig. A.2). Terms with a positive subscript rotate in a counterclockwise direction, whereas those with a negative subscript rotate clockwise. The coefficients $\tilde{c}_k$ and $\tilde{c}_{-k}$ specify the position of the phasor at $t = 0$. The infinite summation of the spinning phasors, which are allowed to rotate at $t = 0$, is then equal to $f(t)$.

Getting Started with MATLAB

MATLAB software is a computer program that provides the user with a convenient environment for many types of calculations—in particular, those that are related to matrix manipulations. MATLAB operates interactively, executing the user's commands one-by-one as they are entered. A series of commands may be saved as a script and run like an interpretive program. MATLAB has a large number of built-in functions; however, it is possible for users to build their own functions made up of MATLAB commands and functions. The primary features of MATLAB are built-in vector and matrix computations including:

- Vector-matrix arithmetic.
- Matrix inversion and eigenvalue/vector analysis.
- Complex arithmetic and polynomial operations.
- Statistical calculations.
- Graphical displays.
- Control system design.
- Fitting process models from test data.

MATLAB has a number of optional toolboxes that provide specialized functions. These include: signal processing, control systems, system identification, optimization, and statistics.

MATLAB is available in versions that run on PCs, Macs, and workstations. The modern version that runs on PCs does so in the Windows environment. The seven exercises that follow are meant to give you the flavor of computing with MATLAB; they do not constitute a comprehensive tutorial. There are additional tutorial materials in the MATLAB manuals. A number of textbooks now feature MATLAB exercises. Also, on-line information is available for any command or function by typing: help *name*, where *name* identifies the command. Do not just look through these exercises; try them all and try variations that occur to you. Check the answers that MATLAB gives and make sure you understand them and they are correct. That is the effective way to learn MATLAB.

1. Assignment of Values to Variable Names

Assignment of values to scalar variables is similar to what is done in other computer languages. Try typing

```
a = 4
```

and

```
A = 6
```

Note how the assignment echoes to confirm what you have done. This is a characteristic of MATLAB. The echo can be suppressed by terminating the command line with the semicolon (;) character. Try typing

```
b = -3;
```

MATLAB treats names in a case-sensitive manner, that is, the name a is not the same as the name A. To illustrate this, enter

```
a
```

and

```
A
```

See how their values are distinct. They are distinct names.

Variable names in MATLAB generally represent matrix quantities. A row vector can be assigned as follows:

```
a = [1 2 3 4 5]
```

The echo confirms the assignment again. Notice how the new assignment of a has taken over. A column vector can be entered in several ways. Try them.

```
b = [ 1 ; 2 ; 3 ; 4 ; 5 ]
```

or

```
b = [ 1;
      2;
      3;
      4;
      5 ]
```

or, by transposing a row vector with the ' operator,

```
b = [ 1 2 3 4 5 ]'
```

A two-dimensional matrix of values can be assigned as follows:

```
A = [ 1 2 3 ; 4 5 6 ; 7 8 8 ]
```

or

```
A = [ 1 2 3 ;
      4 5 6 ;
      7 8 8  ]
```

The values stored by a variable can be examined at any time by typing the name alone, for example,

```
b
```

or

```
A
```

Also, a list of all current variables can be obtained by entering the command

```
who
```

or, with more detail, enter

```
whos
```

There are several predefined variables, for example, pi.

It is also possible to assign complex values to variables, since MATLAB handles complex arithmetic automatically. To do this, it is convenient to assign a variable name, usually either i or j, to the square root of -1.

```
i = sqrt(-1)
```

Then, a complex value can be assigned, like

```
x = 2 + i*4
```

2. Mathematical Operations

Operations with scalar quantities are handled in a straightforward manner, similar to how they are handled in computer languages. The common operators, in order of priority, are

```
^     Exponentiation
* /   Multiplication and division
\     Left division (applies to matrices)
+ -   Addition and subtraction
```

These operators will work in calculator fashion. Try

```
2 * pi
```

Also, scalar real variables can be included:

```
y = pi / 4
y ^ 2.45
```

Results of calculations can be assigned to a variable, as in the next-to-last example, or simply displayed, as in the last example.

Calculations can also involve complex quantities. Using the x defined above, try

```
3 * x
1 / x
x ^ 2
x + y
```

The real power of MATLAB is illustrated in its ability to carry out matrix calculations. The inner product of two vectors (dot product) can be calculated using the * operator,

```
a * b
```

and likewise, the outer product

```
b * a
```

To illustrate vector-matrix multiplication, first redefine a and b,

```
a = [1 2 3]
```

and

```
b = [4 5 6]'
```

Now, try

```
a * A
```

or

```
A * b
```

What happens when the dimensions are not those required by the operations? Try

```
A * a
```

Matrix-matrix multiplication is carried out in likewise fashion:

```
A * A
```

Mixed operations with scalars are also possible:

```
A / pi
```

It is important to always remember that MATLAB will apply the simple arithmetic operators in vector-matrix fashion if possible. At times, you will want to carry out calculations item-by-item in a matrix or vector. MATLAB provides for that too. For example,

```
A ^ 2
```

results in matrix multiplication of A with itself. What if you want to square each element of A? That can be done with

```
A .^ 2
```

The . preceding the ^ operator signifies that the operation is to be carried out item-by-item. The MATLAB manual calls these *array operations*.

When the division operator (/) is used with matrices, the use of a matrix inverse is implied. Therefore, if A is a square, nonsingular matrix, then b/A corresponds to the right multiplication of b by the inverse of A. A longer way to do this used the *inv* function, that is, b*inv(A); however, using the division operator is more efficient since X = b/A is actually solved as the set of equations X*A=b using a decomposition/elimination scheme.

The "left division" operator (\ , the backslash character) is used in matrix operations also. As above, A\b corresponds to the left multiplication of b by the inverse of A. This is actually solved as the set of equations A*X=b, a common engineering calculation.

For example, if c is a column vector with values 0.1, 1.0, and 10, the solution of A * x=c, where A has been set above, can be obtained by typing

```
c = [0.1 1.0 10]'
x = A\c
```

Try that.

3. Use of Built-In Functions

MATLAB and its toolboxes have a rich collection of built-in functions. You can use online help to find out more about them. One of their important properties is that they will operate directly on vector and matrix quantities. For example, try

```
log(A)
```

and you will see that the natural logarithm function is applied in array style, element by element, to the matrix A. Most functions, like *sqrt, abs, sin, acos, tanh,* and *exp,* operate in array fashion. Certain functions, like the exponential and square root functions, have matrix definitions also. MATLAB will evaluate the matrix version when the letter m is appended to the function name. Try

```
sqrtm(A)
```

A common use of functions is to evaluate a formula for a series of arguments. Create a column vector t that contains values from 0 to 100 in steps of 5,

```
t = [100]'
```

Check the number of items in the t array with the *length* function,

```
length(t)
```

Now, say that you want to evaluate a formula y = f(t), where the formula is computed for each value of the t array, and the result is assigned to a corresponding position in the y array. For example,

```
y = t .^ 0.34 - log10(t) + 1 ./ t
```

Done! [Note the array operators include adjacent decimal points.] This is similar to creating a column of the t values on a spreadsheet and copying a formula down an adjacent column to evaluate y values.

4. Graphics

MATLAB's graphics capabilities have similarities to those of a spreadsheet program. Graphs can be created quickly and conveniently; however, there is not much flexibility to customize them.

For example, to create a graph of the t,y arrays from the data above, enter

```
plot(t, y)
```

That's it! You can customize the graph a bit with commands like the following:

```
title('Plot of y versus t')
xlabel('Values of t')
ylabel('Values of y')
grid
```

The graph appears in a separate window and can be printed or transferred via the clipboard (in PCs with Windows or Macs) to other programs.

There are other features of graphics that are useful, for example, plotting objects instead of lines, families of curves plots, plotting on the complex plane, multiple graphs windows, log-log or semilog plots, three-dimensional mesh plots, and contour plots.

5. Polynomials

There are many MATLAB functions that allow you to operate on arrays as if their entries were coefficients or roots of polynomial equations. For example, enter

```
c = [1 1 1 1]
```

and then

```
r = roots(c)
```

and the roots of the polynomial $x^3 + x^2 + x + 1 = 0$ should be printed and are also stored in the r array. The polynomial coefficients can be computed from the roots with the *poly* function,

```
poly(r)
```

and a polynomial can be evaluated for a given value of x. For example,

```
polyval(c, 1.32)
```

If another polynomial, $2x^2 - 0.4x - 1$, is represented by the array d,

```
d = [2 -0.4 -1]
```

the two polynomials can be multiplied symbolically with the convolution function, *conv,* to yield the coefficients of the product polynomial,

```
cd = conv(c,d)
```

The deconvolution function, *deconv,* can be used to divide one polynomial into another, for example,

```
[q,r] = deconv(c,d)
```

The q result is the quotient, and the r result is the remainder.

There are other polynomial functions that may become useful to you, such as the *residue* function for partial fraction expansion.

6. Statistical Analysis

The Statistics Toolbox contains many features for statistical analysis; however, common statistical calculations can be performed with MATLAB's basic function set. You can generate a series of (pseudo) random numbers with the *rand* function. Either a uniform (*rand*) or normal (*randn*) distribution is available:

```
n = 0:5:1000;
```

(Did you forget the ; ?)

```
num = randn(size(n));
```

You probably understand why using the semicolon at the end of the commands above is important, especially if you neglected to do so.

If you would like to see a plot of noise, try

```
plot(num)
```

These are supposed to be normally distributed numbers with a mean of zero and variance (and standard deviation) of one. Check by

```
mean(num)
```

and

```
std(num)
```

No one is perfect! You can find minimum and maximum values,

```
min(num)
max(num)
```

There is a convenient function for plotting a histogram of the data:

```
hist(num,20)
```

where 20 is the number of bins.

If you would like to fit a polynomial to some data by least squares, you can use the *polyfit* function. Try the following example:

```
t = 05
y = [−0.45 0.56 2.34 5.6 9.45 24.59]
coef = polyfit(t, y, 3)
```

The values in `coef` are the fitted polynomial coefficients. To generate the computed value of y, use

```
yc = polyval(coef,t)
```

and to plot the data versus the fitted curve, enter

```
plot(t,yc,t,y,'o')
```

The plot of the continuous curve is piecewise linear; therefore, it does not look very smooth. Improve the smoothness as follows:

```
t1 = [0 : 0.05 : 5];
yc = polyval(coef, t1);
plot(t1, yc, t, y, 'o')
```

7. This and That

There are many, many other features to MATLAB. Some of these you will find useful; perhaps others you will never use. We encourage you to explore and experiment.

To save a copy of your session, MATLAB has a useful capability called *diary*. You issue the command

```
diary problem1
```

and MATLAB opens a disk file in which it stores all the subsequent commands and results (not graphs) of your session. You can turn the diary feature off:

```
diary off
```

and back on with the same file:

```
diary on
```

After you leave MATLAB, the diary file is available to you. It is common to use an editor or word processor to clean up the diary file (getting rid of all those errors you made before anyone can see them!) and then print the file to obtain a hard copy of the important parts of your work session, for example, key numerical results.

Exit MATLAB with the `quit` or `exit` command. It is possible to save the current state of your work with the `save` command. It is also possible to reload that state with the `load` command.

Getting Started with Mathcad

Mathcad has a unique way to handle equations, numbers, text, and graphs. It works like a scratch pad and pencil. The screen interface is a blank worksheet on which you can enter equations, graph data or functions, and annotate operations with text. It uses standard mathematical symbols to represent operators when possible. Therefore, you may find that the Mathcad interface is quite natural and familiar.

Mathcad can solve a wide range of mathematical problems either numerically or symbolically. The symbolic capabilities of Mathcad have relatively little application in this text, although they may be used to check our numerical results. Therefore, they will not be covered in detail in this overview. Mathcad has a comprehensive set of operators and functions that allow you to perform many of the numerical methods covered in this text. It also has a programming language that allows you to write your own multiline procedures and subprograms. The following discussion provides a brief description of the features of Mathcad you will find most useful for this text.

THE BASICS OF MATHCAD

Applications in this text will require that you be able to create your own worksheets. To facilitate your efforts, let's go over the main features of the Mathcad application window.

The Main Menu

This is your gateway to Mathcad. It also provides commands that handle the details of editing and managing your worksheets. For example, click on the **File** and **Tools** menus to see some of the functionality available to you.

The Standard Toolbar

Several toolbars should be automatically displayed just below the **Main** menu. As the name implies, the **Standard** toolbar provides shortcuts for many common tasks, from worksheet opening and file saving to bringing up lists of built-in functions and units. Depending on what you are doing in your worksheet, one or more of the buttons on this toolbar may appear grayed out. If you let your mouse hover over each of the buttons on the palette, you will see a description of the button's function.

The Math Palette

The **Math Palette** may automatically be displayed at the top of the screen. If not, just select **View, Toolbars, Math** and it will appear. The buttons and their functions are listed below:

🖩	Calculator		Boolean
	Graph		Programming
	Vector and Matrix	$\alpha\beta$	Greek Symbol
$x=$	Evaluation		Symbolic Keyword
	Calculus		

Click on one of these buttons to bring up the full palette. You can use the palettes to insert math symbols and operations directly into your Mathcad worksheet.

ENTERING TEXT AND MATHEMATICAL OPERATIONS

Entering Text

To create a text region, click in a blank area of the screen to position the red crosshair cursor and type a double quote ["]. Now you can type whatever you like, just as in a word processor. As the region grows, a black box appears around the text. The box has resizing "handles" on the right and bottom edges of the rectangle. Once you are done, click outside the text region to go back to inputting math operations. The black selection box disappears when you are no longer working in the text region.

Mathematical Operations

Type	See on Screen
1+	$1 + \blacksquare$

Click somewhere in the upper-left-hand corner of the worksheet, and the red crosshair should move to where you click. After you type the number 1 and the + sign you will see a little black box delimited by blue editing lines. In Mathcad this black box is called a placeholder. If you continue typing, whatever you type next will appear in the placeholder. For example, type 2 in the placeholder, then press the equals key (=) to see the result.

$$1 + 2 = 3$$

The basic arithmetic operators are listed below, along with their keystrokes and **Calculator Palette** button equivalents.

Operation	Keystroke	Palette	Example
Addition	+	+	$2 + 2 = 4$
Subtraction	−	−	$2 - 2 = 0$
Multiplication	*	×	$2 \cdot 2 = 4$
Division	/	÷	$\frac{2}{2} = 1$
Exponentiation	^	x^Y	$2^2 = 4$

Notice that operations in a Mathcad worksheet appear in familiar notation—multiplication as a dot, division with a fraction bar, exponents in a raised position, and so on. Calculations are computed internally to 15 places, but you can show fewer places in the answer. To change the default display of numerical and symbolic results in a worksheet, click in a blank area of the worksheet. Then select **Result** from the **Format** menu to display the **Result Format** dialog box, and choose your default settings. Make sure that the box labeled "Set as default" is checked, and click OK. If you just want to change the display of a particular result, click on the equation, and follow the same steps.

Here are a few more examples that demonstrate Mathcad features.

$$\sqrt{\frac{1.837 \cdot 10^3}{100 + 3^5}} = 2.3142353232$$

Most standard engineering and mathematical functions are built in.

$$\log(1347.2) \cdot \sin\left(\frac{3}{5} \cdot \pi\right) = 2.976$$

Mathcad's functions and operators easily handle complex numbers.

$$(2.3 + 4.7i)^3 + e^{3-2i} = -148.613 - 47.498i$$

MATHEMATICAL FUNCTIONS AND VARIABLES

The definition symbol := is used to define a function or variable in Mathcad. For example, click an empty worksheet to position the red crosshair in a blank area and type:

Type	See on Screen
f(x):x^2	$f(x) := x^2$

The definition symbol is also located on the **Evaluation** selection of the **Math Palette.** When you change a definition function or variable, Mathcad immediately recalculates any new values that depend on it. Once you've defined a function like f(x), you can use it in a number of ways, for example:

$$f(x) := x^2$$

Now you can insert a numerical value as the argument of f(x),

$$f(10) = 100$$

or define a variable and insert it as the argument of f(x),

$$x := 3$$

$$f(x) = 9$$

You can even define another function in terms of f(x):

$$g(y) := f(y) + 6$$

$$g(x) = 15$$

Note that you can define a function using expressions you build up from the keyboard or from the palettes of math operators. You can also include any of Mathcad's hundreds of built-in functions. To see a list of built-in functions along with brief descriptions, select **Function** from the **Insert** menu, or click on the *f(x)* button. You can also type the name of any built-in function directly from the keyboard. The following are just a few examples that use some of Mathcad's built-in functions.

Trig and Logs

$$\ln(26) = 3.258 \qquad \csc(45 \cdot .deg) = 1.414$$

Matrix Functions

$$\text{identity}(3) = \begin{bmatrix} 1 & 0 & 0 \\ 0 & 1 & 0 \\ 0 & 0 & 1 \end{bmatrix}$$

Probability Distributions

$$\text{pnorm}(2,0,1) = 0.977$$

Range Variables

In Mathcad you will find yourself wanting to work with a range of values for many applications—for example, to define a series of values to plot. Mathcad therefore provides the range operator (..), which can be entered by typing a semicolon (;) on the keyboard.

The first and last numbers establish the endpoints of the range variable, and the second number sets the increment. For example,

Type	See on Screen
z:0,0.5;2	z:=0,0.5..2
z =	z =

$$\begin{array}{|c|} \hline 0 \\ \hline 0.5 \\ \hline 1 \\ \hline 1.5 \\ \hline 2 \\ \hline \end{array}$$

Matrix Computations and Operations

To enter a matrix, click on the 3×3 matrix icon in the **Matrix Palette** (or choose **Matrix** from the **Insert** menu), choose the number of rows and columns, then fill in the placeholders. For example,

$$A := \begin{bmatrix} 4 & 5 & 1 \\ 5 & 0 & -12 \\ -7 & 2 & 8 \end{bmatrix}$$

To compute the inverse,

Type	See on Screen
A^–1=	$A^{-1} = \begin{bmatrix} 0.074 & -0.117 & -0.184 \\ 0.135 & 0.12 & 0.163 \\ 0.031 & -0.132 & -0.077 \end{bmatrix}$

Mathcad has a comprehensive set of commands to perform various matrix operations. For example, to find the determinant, type a vertical bar (|) or use the button on the **Matrix Palette.**

$$|A| = 326$$

Units

Mathcad can also handle units. To see the built-in units, choose **Unit** from the **Insert** menu, or click on the appropriate toolbar button. Let's start with a simple example. Open a new worksheet in Mathcad and type Mass:75kg[*Enter*]. You should see

$$\text{Mass} := 75 \text{ kg}$$

You could also have typed `Mass:75*kg[`*Enter*`]`, multiplying the quantity times the unit. Now, enter `g:9.8m/s^2` and you should have

$$g := 9.8 \frac{m}{s^2}$$

To see how Mathcad manages the units with calculations, enter `Mass*g=` and the result should be displayed with combined units as

$$Mass \cdot g = 735 \ N$$

Mathcad uses the SI unit system by default, but you can change that from **Tools, Worksheet Options, Unit System.** Alternate systems include CGS, MKS, and US. Instead of typing in the unit, you can also insert it from a list. Try the following. Type `Temp:273.16*` then click on **Insert, Unit.** Select **Temperature** from the upper box and **Kelvin (K)** from the lower box and click OK.

NUMERICAL METHODS FUNCTION

Mathcad has a number of special built-in functions that perform a variety of numerical operations of particular interest to readers of this book. Examples of the development and application of these functions are described in detail in the text. Here we will provide a brief list of some of the more important functions just to give you an overview of the capabilities. We illustrate their use in the relevant sections of this book.

Function Name	Use
root	Solves f(x) = 0
polyroots	Finds all roots of a polynomial
find	Solves a system of nonlinear algebraic equations
minerr	Returns a minimum error solution of a system of equations
lsolve	Solves a system of linear algebraic equations
linterp	Linear interpolation
cspline	Cubic spline interpolation
regress	Polynomial regression
genfit	General nonlinear regression
fft	Fourier transform
ifft	Inverse Fourier transform
rkfixed	Solves a system of differential equations using a fixed step-size fourth-order Runge-Kutta method
rkadapt	Solves a system of differential equations using a variable step-size fourth-order Runge-Kutta method
sbval	Solves a two-point boundary-value problem
eigenvals	Finds eigenvalues
eigenvecs	Finds eigenvectors
relax	Solves Poisson's equation for a square domain

MULTILINE PROCEDURES AND SUBPROGRAMS

The **Programming Palette** in Mathcad provides the capability for multiline procedures or subprograms with standard control structures such as FOR and WHILE loops, branching, recursion, and more. Subprograms can be integrated with Mathcad's worksheets and can operate on scalars, vectors, arrays, and even arrays of arrays.

CREATING GRAPHS

Mathcad's graphics capabilities are particularly important to engineering work. The first type of graph to know about is the *QuickPlot*. Start with a new worksheet and enter the following formula. Leave the formula selected.

$$e^{\frac{-x}{4}} \cdot (2 - x) - 1|$$

From the menu, select **Insert, Graph, X-Y Plot.** Click away from the formula and you should see an automatic QuickPlot.

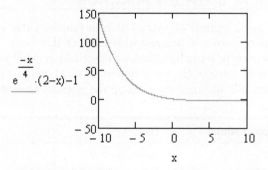

This plot can now be adjusted as desired. For example, click in the plot to select it and change the *x* limits to 0 to 1. This should appear as

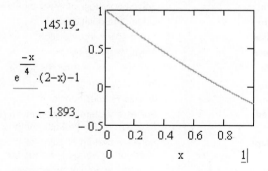

Then, click away from the plot, and the *y* axis will rescale automatically. Double-click on the plot, and the **Formatting Currently Selected X-Y Plot** dialog box should appear. Check the boxes for X and Y gridlines and click OK. Click away from the plot, and you should now have

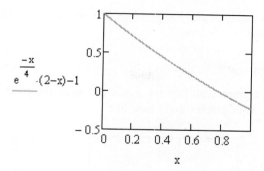

$e^{\frac{-x}{4}} \cdot (2-x) - 1$

Mathcad graphs one point for each value of the range variable x. This variable was created automatically in this case. The x-y points are joined by short straight-line segments. You can create your own range variables for the x axis.

Instead of using a formula for the y axis, you can use a function. To illustrate this, enter the following function definition *above* your plot:

$$f(x) := e^{\frac{-x}{4}} \cdot (2-x) - 1$$

and change the y axis from the formula to **f(x)**. You should have the same plot, but now it is in terms of the function **f(x)** instead of the direct formula.

A function can also be used for the x axis. Try another example below your current graph. Make the following definitions:

$$N := 100 \qquad \theta := 0, \frac{2 \cdot x}{N} .. 2 \cdot x$$

$$x(\theta) := \cos(\theta) \qquad y(\theta) := \sin(\theta)$$

Insert a blank plot by pressing the @ key [*Shift-2*]. Enter $y(\theta)$ in the y-axis placeholder and $x(\theta)$ in the x-axis placeholder and click away from the graph. This should yield the plot of a circle:

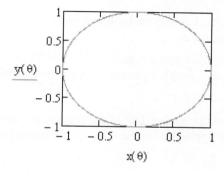

It is also possible to plot the elements of a vector. Create the following vector of binomial probabilities:

$$i := 0 .. 10$$
$$p_i := 0.4^i \cdot 0.6^{(10-i)}$$

and insert a graph with **i** on the x axis and $\mathbf{p_i}$ on the y axis (you can use the [key for the subscript). Your graph should look like

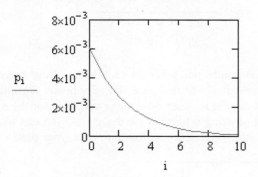

Of course, you can plot one vector against another too, as long as they have the same number of elements. The vectors could contain data instead of a mathematical formula.

There are many axis settings that can be adjusted by double-clicking on the graph. These are self-explanatory to a great extent, and you can become familiar with them through practice.

There are many other styles of plots that can be generated by Mathcad. These include polar, surface, contour, 3D bar, scatter, vector-field plots, and graphical animations.

SYMBOLIC MATHEMATICS

An intriguing and valuable feature of Mathcad is its capability to carry out symbolic math manipulations. The symbolic capabilities include

- Algebraic manipulations.
- Calculus: differentiation and integration.
- Solving algebraic equations and systems of such equations.

and more advanced features:

- Symbolic Fourier, Laplace, and z transforms.
- Symbolic optimization.

We will review the first group here.

Let's start with a simple example of symbolic algebra. Enter the expression

$$(x + 2) \cdot (x - 1) \cdot (x + 4)$$

Expand the horizontal editing line to encompass the entire expression and then select **Expand** from the **Symbolics** menu. You should see below:

$$x^3 + 5 \cdot x^2 + 2 \cdot x - 8$$

Now, enter the polynomial $x^3 + 3x^2 + 3x + 1$ and follow the same procedure, except pick **Factor** from the **Symbolics** menu. You should have

$$x^3 + 3 \cdot x^2 + 3 \cdot x + 1$$
$$(x + 1)^3$$

Another way to carry out symbolic commands is with keystrokes. Enter the following expression:

$$\boxed{\frac{x^2 - 3 \cdot x - 4}{x - 4} + 2 \cdot x - 5|}$$

Then press the *Ctrl-Shift-.* key combination and type `simplify` into the placeholder that appears. You should get

$$\frac{x^2 - 3 \cdot x - 4}{x - 4} + 2 \cdot x - 5 \text{ simplify} \rightarrow 3 \cdot x - 4$$

This is a different style of symbolic evaluation with the keyword retained and the result appearing out to the right. You've seen the use of three important symbolic operators:

- expand   Expand all powers and products of sums.
- factor   Factor into a product of simpler functions.
- simplify   Simplify by performing arithmetic, canceling common factors, using identities, and simplifying powers.

Additional algebraic features include expansion to a series, partial fraction expansion, and extracting coefficients of a polynomial into a vector.

Now, let's experiment with simple differentiation. Enter the expression

$$\boxed{e^{\frac{-E}{R \cdot T}|}}$$

and leave the T selected (or click on it to select it). Then select **Variable** and **Differentiate** from the **Symbolics** menu. You should have the result

$$e^{\frac{-E}{R \cdot T}}$$

$$\frac{E \cdot e^{-\frac{E}{R \cdot T}}}{R \cdot T^2}$$

Another way to do this is to use the differentiation operator from the **Calculus** toolbar. Enter the following expression:

$$\boxed{\frac{d^2}{dx^2}\left(x^2 \cdot \sin(x)\right)}$$

and then click on the [→] button on the **Evaluation** toolbar (you can also press *Ctrl-.*). The result should be

$$\frac{d^2}{dx^2}\left(x^2 \cdot \sin(x)\right) \rightarrow 2 \cdot \sin(x) - x^2 \cdot \sin(x) + 4 \cdot x \cdot \cos(x)$$

Symbolic integrals can be determined either in indefinite or definite form. For an indefinite integral, start by pressing *Ctrl-i* or click on the [∫] button on the **Calculus** toolbar. Then enter the desired function and differential followed by clicking on the [→] button to produce the following result:

$$\int e^{-x} dx \rightarrow -e^{-x}$$

Mathcad can also compute limits symbolically. The appropriate buttons are on the **Calculus** toolbar.

LEARNING MORE ABOUT MATHCAD

In this brief introduction we have covered only the Mathcad basics. Further help is available in the Mathcad software package in a variety of forms.

ToolTips

Let your mouse pointer hover over a palette or toolbar button for a few seconds. You will see an explanatory tooltip displayed near the button. Look also on the message line at the bottom of the Mathcad application window for helpful tips and shortcuts.

Resource Center and QuickSheets

To help you get going fast and keep you learning, Mathcad comes complete with Quick-Sheets. These provide mathematical shortcuts for frequently used analyses—from graphing a function to solving simultaneous equations to the analysis of variance. There are numerous QuickSheets. To open the QuickSheets section, choose **QuickSheets** from the Help option on the **Main** menu.

Online Help

Online Help provides detailed, step-by-step instructions for using all of Mathcad's features. Help is available at any time by simply going to the **Help** button on the **Main** menu. There you will find several links including the Mathcad website and Mathcad training.

BIBLIOGRAPHY

Al-Khafaji, A. W., and J. R. Tooley, *Numerical Methods in Engineering Practice,* Holt, Rinehart and Winston, New York, 1986.

Allaire, P. E., *Basics of the Finite Element Method,* William C. Brown, Dubuque, IA, 1985.

Ames, W. F., *Numerical Methods for Partial Differential Equations,* Academic Press, New York, 1977.

Ang, A. H.-S., and W. H. Tang, *Probability Concepts in Engineering Planning and Design, Vol. 1: Basic Principles,* Wiley, New York, 1975.

APHA (American Public Health Association). 1992. *Standard Methods for the Examination of Water and Wastewater,* 18th ed., Washington, DC.

Atkinson, K. E., *An Introduction to Numerical Analysis,* Wiley, New York, 1978.

Atkinson, L. V., and P. J. Harley, *An Introduction to Numerical Methods with Pascal,* Addison-Wesley, Reading, MA, 1983.

Baker, A. J., *Finite Element Computational Fluid Mechanics,* McGraw-Hill, New York, 1983.

Bathe, K.-J., and E. L. Wilson, *Numerical Methods in Finite Element Analysis,* Prentice-Hall, Englewood Cliffs, NJ, 1976.

Booth, G. W., and T. L. Peterson, "Nonlinear Estimation," I.B.M. Share Program Pa. No. 687 WLNL1, 1958.

Boyce, W. E., and R. C. DiPrima, *Elementary Differential Equations and Boundary Value Problems,* 5th ed. Wiley, New York, 1992.

Branscomb, L. M., "Electronics and Computers: An Overview," *Science,* 215:755, 1982.

Brent, R. P., *Algorithms for Minimization Without Derivatives,* Prentice-Hall, Englewood Cliffs, NJ, 1973.

Brigham, E. O., *The Fast Fourier Transform,* Prentice-Hall, Englewood Cliffs, NJ, 1974.

Burden, R. L., and J. D. Faires, *Numerical Analysis,* 8th ed., PWS Publishing, Boston, 2005.

Butcher, J. C., "On Runge-Kutta Processes of Higher Order," *J. Austral. Math. Soc.,* 4:179, 1964.

Carnahan, B., H. A. Luther, and J. O. Wilkes, *Applied Numerical Methods,* Wiley, New York, 1969.

Cash, J. R., and A. H. Karp, *ACM Transactions on Mathematical Software,* 16:201–222, 1990.

Chapra, S. C., *Surface Water-Quality Modeling,* McGraw-Hill, New York, 1997.

Chapra, S. C., *Applied Numerical Methods with MATLAB,* 2d ed., McGraw-Hill, New York, 2007.

Cheney, W., and D. Kincaid, *Numerical Mathematics and Computing,* 6th ed., Brooks/Cole, Monterey, CA, 2008.

Chirlian, P. M., *Basic Network Theory,* McGraw-Hill, New York, 1969.

Cooley, J. W., P. A. W. Lewis, and P. D. Welch, "Historical Notes on the Fast Fourier Transform," *IEEE Trans. Audio Electroacoust.,* AU-15(2):76–79, 1977.

Dantzig, G. B., *Linear Programming and Extensions,* Princeton University Press, Princeton, NJ, 1963.

Davis, H. T., *Introduction to Nonlinear Differential and Integral Equations,* Dover, New York, 1962.

Davis, L., *Handbook of Genetic Algorithms,* Van Nostrand Reinhold, New York, 1991.

Davis, P. J., and P. Rabinowitz, *Methods of Numerical Integration,* Academic Press, New York, 1975.

Dennis, J. E., and R. B. Schnabel, *Numerical Methods for Unconstrained Optimization and Nonlinear Equations,* Society for Industrial and Applied Mathematics (SIAM), Philadelphia, PA, 1996.

Dekker, T. J., "Finding a Zero by Means of Successive Linear Interpolation," in *Constructive Aspects of the Fundamental Theorem of Algebra,* B. Dejon and P. Henrici (editors), Wiley-Interscience, New York, 1969, pp. 37–48.

Dijkstra, E. W., "Go To Statement Considered Harmful," *Commun. ACM,* 11(3):147–148, 1968.

Draper, N. R., and H. Smith, *Applied Regression Analysis,* 2d ed., Wiley, New York, 1981.

Enright, W. H., T. E. Hull, and B. Lindberg, "Comparing Numerical Methods for Stiff Systems of ODE's," *BIT,* 15:10, 1975.

Fadeev, D. K., and V. N. Fadeeva, *Computational Methods of Linear Algebra,* Freeman, San Francisco, 1963.

Ferziger, J. H., *Numerical Methods for Engineering Application,* Wiley, New York, 1981.

Fletcher, R., *Practical Methods of Optimization: 1: Unconstrained Optimization,* Wiley, Chichester, UK, 1980.

Fletcher, R., *Practical Methods of Optimization: 2: Constrained Optimization,* Wiley, Chichester, 1981.

Forsythe, G. E., and W. R. Wasow, *Finite-Difference Methods for Partial Differential Equations,* Wiley, New York, 1960.

Forsythe, G. E., M. A. Malcolm, and C. B. Moler, *Computer Methods for Mathematical Computation,* Prentice-Hall, Englewood Cliffs, NJ, 1977.

Fylstra, D., L. S. Lasdon, J. Watson, and A. Waren, "Design and Use of the Microsoft Excel Solver," *Interfaces,* 28(5):29–55, 1998.

Gabel, R. A., and R. A. Roberts, *Signals and Linear Systems,* Wiley, New York, 1987.

Gear, C. W., *Numerical Initial-Value Problems in Ordinary Differential Equations,* Prentice-Hall, Englewood Cliffs, NJ, 1971.

Gerald, C. F., and P. O. Wheatley, *Applied Numerical Analysis,* 7th ed., Addison-Wesley, Reading, MA, 2004.

Gill, P. E., W. Murray, and M. H. Wright, *Practical Optimization,* Academic Press, London, 1981.

Gladwell, J., and R. Wait, *A Survey of Numerical Methods of Partial Differential Equations,* Oxford University Press, New York, 1979.

Goldberg, D. E., *Genetic Algorithms in Search, Optimization and Machine Learning,* Addison-Wesley, Reading, MA, 1989.

Guest, P. G., *Numerical Methods of Curve Fitting,* Cambridge University Press, New York, 1961.

Hammersley, J. M., and D. C. Handscomb, *Monte Carlo Methods*, Methuen, London, 1964.

Hamming, R. W., *Numerical Methods for Scientists and Engineers,* 2d ed., McGraw-Hill, New York, 1973.

Hartley, H. O., "The Modified Gauss-Newton Method for Fitting Non-linear Regression Functions by Least Squares," *Technometrics,* 3:269–280, 1961.

Hayt, W. H., and J. E. Kemmerly, *Engineering Circuit Analysis,* McGraw-Hill, New York, 1986.

Heideman, M. T., D. H. Johnson, and C. S. Burrus, "Gauss and the History of the Fast Fourier Transform," *IEEE ASSP Mag.,*1(4):14–21, 1984.

Henrici, P. H., *Elements of Numerical Analysis,* Wiley, New York, 1964.

Hildebrand, F. B., *Introduction to Numerical Analysis,* 2d ed., McGraw-Hill, New York, 1974.

Hillier, F. S., and G. J. Lieberman, *Introduction to Operations Research*, 8th ed., McGraw-Hill, New York, 2005.

Hoffman, J., *Numerical Methods for Engineers and Scientists,* McGraw-Hill, New York, 1992.

Holland, J. H., *Adaptation in Natural and Artificial Systems,* University of Michigan Press, Ann Arbor, MI, 1975.

Hornbeck, R. W., *Numerical Methods,* Quantum, New York, 1975.

Householder, A. S., *Principles of Numerical Analysis,* McGraw-Hill, New York, 1953.

Householder, A. S., *The Theory of Matrices in Numerical Analysis,* Blaisdell, New York, 1964.

Householder, A. S., *The Numerical Treatment of a Single Nonlinear Equation,* McGraw-Hill, New York, 1970.

Huebner, K. H., and E. A. Thornton, *The Finite Element Method for Engineers,* Wiley, New York, 1982.

Hull, T. E., and A. L. Creemer, "The Efficiency of Predictor-Corrector Procedures," *J. Assoc. Comput. Mach.,* 10:291, 1963.

Isaacson, E., and H. B. Keller, *Analysis of Numerical Methods,* Wiley, New York, 1966.

Jacobs, D. (ed.), *The State of the Art in Numerical Analysis,* Academic Press, London, 1977.

James, M. L., G. M. Smith, and J. C. Wolford, *Applied Numerical Methods for Digital Computations with FORTRAN and CSMP,* 3d ed., Harper & Row, New York, 1985.

Keller, H. B., *Numerical Methods for Two-Point Boundary-Value Problems,* Wiley, New York, 1968.

Lapidus, L., and G. F. Pinder, *Numerical Solution of Partial Differential Equations in Science and Engineering,* Wiley, New York, 1981.

Lapidus, L., and J. H. Seinfield, *Numerical Solution of Ordinary Differential Equations,* Academic Press, New York, 1971.

Lapin, L. L., *Probability and Statistics for Modern Engineering,* Brooks/Cole, Monterey, CA, 1983.

Lasdon, L. S., and S. Smith, "Solving Large Nonlinear Programs Using GRG," *ORSA Journal on Computing,* 4(1):2–15, 1992.

Lasdon, L. S., A. Waren, A. Jain, and M. Ratner, "Design and Testing of a Generalized Reduced Gradient Code for Nonlinear Programming," *ACM Transactions on Mathematical Software,* 4(1):34–50, 1978.

Lawson, C. L., and R. J. Hanson, *Solving Least Squares Problems,* Prentice-Hall, Englewood Cliffs, NJ, 1974.

Luenberger, D. G., *Introduction to Linear and Nonlinear Programming,* Addison-Wesley, Reading, MA, 1984.

Lyness, J. M., "Notes on the Adaptive Simpson Quadrature Routine," *J. Assoc. Comput. Mach.,* 16:483, 1969.

Malcolm, M. A., and R. B. Simpson, "Local Versus Global Strategies for Adaptive Quadrature," *ACM Trans. Math. Software,* 1:129, 1975.

Maron, M. J., *Numerical Analysis, A Practical Approach,* Macmillan, New York, 1982.

Milton, J. S., and J. C. Arnold, *Introduction to Probability and Statistics: Principles and Applications for Engineering and the Computing Sciences,* 4th ed., McGraw-Hill, New York, 2002.

Moler, C. B., *Numerical Computing with MATLAB,* SIAM, Philadelphia, 2004.

Muller, D. E., "A Method for Solving Algebraic Equations Using a Digital Computer," *Math. Tables Aids Comput.,* 10:205, 1956.

Na, T. Y., *Computational Methods in Engineering Boundary Value Problems,* Academic Press, New York, 1979.

Newman, M. E. J., and G. T. Barkema, *Monte Carlo Methods in Statistical Physics*, Clarendon Press, Oxford, UK, 1999.

Noyce, R. N., "Microelectronics," *Sci. Am.,* 237:62, 1977.

Oppenheim, A. V., and R. Schafer, *Digital Signal Processing,* Prentice-Hall, Englewood Cliffs, NJ, 1975.

Ortega, J., and W. Rheinboldt, *Iterative Solution of Nonlinear Equations in Several Variables,* Academic Press, New York, 1970.

Ortega, J. M., *Numerical Analysis—A Second Course,* Academic Press, New York, 1972.

Prenter, P. M., *Splices and Variational Methods,* Wiley, New York, 1975.

Press, W. H., B. P. Flanner, S. A. Teukolsky, and W. T. Vetterling, *Numerical Recipes: The Art of Scientific Computing,* 3d ed. Cambridge University Press, Cambridge, 2007.

Rabinowitz, P., "Applications of Linear Programming to Numerical Analysis," *SIAM Rev.,* 10:121–159, 1968.

Ralston, A., "Runge-Kutta Methods with Minimum Error Bounds," *Match. Comp.,* 16:431, 1962.

Ralston, A., and P. Rabinowitz, *A First Course in Numerical Analysis,* 2d ed., McGraw-Hill, New York, 1978.

Ramirez, R. W., *The FFT, Fundamentals and Concepts,* Prentice-Hall, Englewood Cliffs, NJ, 1985.

Rao, S. S., *Engineering Optimization: Theory and Practice,* 3d ed., Wiley-Interscience, New York, 1996.

Revelle, C. S., E. E. Whitlach, and J. R. Wright, *Civil and Environmental Systems Engineering,* Prentice-Hall, Englewood Cliffs, NJ, 1997.

Rice, J. R., *Numerical Methods, Software and Analysis,* McGraw-Hill, New York, 1983.

Robert, C. P., and G. Casella, *Monte Carlo Statistical Methods*, 2d ed., Springer, New York, 2004.

Ruckdeschel, F. R., *BASIC Scientific Subroutine,* Vol. 2, Byte/McGraw-Hill, Peterborough, NH, 1981.

Scarborough, J. B., *Numerical Mathematical Analysis,* 6th ed., Johns Hopkins Press, Baltimore, MD, 1966.

Scott, M. R., and H. A. Watts, "A Systematized Collection of Codes for Solving Two-Point Boundary-Value Problems," in *Numerical Methods for Differential Equations,* L. Lapidus and W. E. Schiesser (eds.), Academic Press, New York, 1976.

Shampine, L. F., and R. C. Allen, Jr., *Numerical Computing: An Introduction,* Saunders, Philadelphia, 1973.

Shampine, L. F., and C. W. Gear, "A User's View of Solving Stiff Ordinary Differential Equations," *SIAM Review,* 21:1, 1979.

Simmons, E. F., *Calculus with Analytical Geometry,* McGraw-Hill, New York, 1985.

Stark, P. A., *Introduction to Numerical Methods,* Macmillan, New York, 1970.

Stasa, F. L., *Applied Finite Element Analysis for Engineers,* Holt, Rinehart and Winston, New York, 1985.

Stewart, G. W., *Introduction to Matrix Computations,* Academic Press, New York, 1973.

Swokowski, E. W., *Calculus with Analytical Geometry,* 2d ed., Prindle, Weber and Schmidt, Boston, 1979.

Taylor, J. R., *An Introduction to Error Analysis,* University Science Books, Mill Valley, CA, 1982.

Tewarson, R. P., *Sparse Matrices,* Academic Press, New York, 1973.

Thomas, G. B., Jr., and R. L. Finney, *Calculus and Analytical Geometry,* 5th ed., Addison-Wesley, Reading, MA, 1979.

Van Valkenburg, M. E., *Network Analysis,* Prentice-Hall, Englewood Cliffs, NJ, 1974.

Varga, R., *Matrix Iterative Analysis,* Prentice-Hall, Englewood Cliffs, NJ, 1962.

Vichnevetsky, R., *Computer Methods for Partial Differential Equations, Vol. 1: Elliptical Equations and the Finite Element Method,* Prentice-Hall, Englewood Cliffs, NJ, 1981.

Vichnevetsky, R., *Computer Methods for Partial Differential Equations, Vol. 2: Initial Value Problems,* Prentice-Hall, Englewood Cliffs, NJ, 1982.

Wilkinson, J. H., *The Algebraic Eigenvalue Problem,* Oxford University Press, Fair Lawn, NJ, 1965.

Wilkinson, J. H., and C. Reinsch, *Linear Algebra: Handbook for Automatic Computation,* Vol. 11, Springer-Verlag, Berlin, 1971.

Wold, S., "Spline Functions in Data Analysis," *Technometrics,* 16(1):1–11, 1974.

Yakowitz, S., and F. Szidarovsky, *An Introduction to Numerical Computation,* Macmillan, New York, 1986.

Young, D. M., *Iterative Solution of Large Linear Systems,* Academic Press, New York, 1971.

Zienkiewicz, O. C., *The Finite Element Method in Engineering Science,* McGraw-Hill, London, 1971.

INDEX